# Theoretical and Mathematical Physics

The series founded in 1975 and formerly (until 2005) entitled *Texts and Monographs in Physics* (TMP) publishes high-level monographs in theoretical and mathematical physics. The change of title to *Theoretical and Mathematical Physics* (TMP) signals that the series is a suitable publication platform for both the mathematical and the theoretical physicist. The wider scope of the series is reflected by the composition of the editorial board, comprising both physicists and mathematicians.

The books, written in a didactic style and containing a certain amount of elementary background material, bridge the gap between advanced textbooks and research monographs. They can thus serve as basis for advanced studies, not only for lectures and seminars at graduate level, but also for scientists entering a field of research.

T0182055

Jouni Suhonen

# From Nucleons
# to Nucleus

## Concepts of Microscopic Nuclear Theory

With 103 Figures

 Springer

Professor Jouni Suhonen

Department of Physics
University of Jyväskylä
P.O. Box 35
40014 Jyväskylä
Finland
E-mail: Jouni.Suhonen@phys.jyu.fi

ISSN 0172-5998

ISBN   978-3-642-08025-8          e-ISBN   978-3-540-48861-3

Springer is a part of Springer Science+Business Media
springer.com
© Springer-Verlag Berlin Heidelberg 2007
Softcover reprint of the hardcover 1st edition 2007

Cover design: WMXDesign GmbH, Heidelberg

To my dear family Tiina, Olli and Hannu

# Preface

This book was born out of the need to gather and develop material for a two-term course on microscopic nuclear physics. As I started my Nuclear Physics II and III at the dawn of the present millenium, I realized that available material consisted either of qualitative introductory textbooks or of handbooks aimed for the professional practitioner. Neither of these two categories matched my idea of a guided pedagogical, hands-on approach to a quantitative description of the structure and decay of atomic nuclei.

The goal was to create a book that would contain an introduction to theory, worked-out examples and end-of-the-chapter exercises. At the same time the book should serve as a reference work for up-to-date applications of nuclear structure methods. With this vision in mind I set out to first produce hand-written lecture notes. On the next round of the two lecture courses the notes were transformed into typed hand-outs, which finally grew into this textbook.

This book builds on the premise that the reader has taken lecture courses on introductory nuclear physics and basic quantum mechanics. A good number of postgraduate and advanced undergraduate students, both theorists and experimentalists, have taken these courses. The style and contents of the book have been greatly influenced by their comments and criticism.

In each chapter I first derive the basic theoretical framework, apply it through worked-out examples and, in the end, discuss the physical implications and limitations of the theory. The formal derivations help understand the approximations and limitations behind the nuclear models that are introduced. However, the details of the derivations are not compulsory reading for someone who wants to go directly to the applications. In fact, in my lectures I skipped details of derivations and used the results as cookbook recipes, an approach particularly suitable for experimentally oriented students.

Even though the nuclear models introduced are generally valid, the examples and exercises are restricted to light and medium-heavy nuclei, up to the $0f$-$1p$-$0g_{9/2}$ major shell. The reason for this is pedagogical: the small model space makes numerical problems tractable; a pocket calculator suffices in most cases. The applications are greatly simplified by numerical tables of two-body interaction matrix elements and of single-particle matrix elements of various electromagnetic and beta-decay operators. Even the most ambitious numerical applications amount to diagonalizing small symmetric matrices. In this way the calculational bulk does not obscure the physical insight to be gained from the exercises.

Only spherical nuclei are considered. However, nuclear structure, discussed in terms of particles, holes and quasiparticles, is boundlessly rich even under this constraint. Detailed single-particle and collective features emerge in electromagnetic and beta-decay transitions. The worked-out examples show in a tangible way how the quality of the computed nuclear wave functions can be probed by applying them to electromagnetic and beta-decay rates and comparing the results with experimental data.

As a textbook this work is self-contained. It introduces in the first four chapters the mathematical machinery needed for the theoretical derivations and their applications. The main theoretical tools are angular momentum coupling, tensor algebra, calculation of matrix elements of spherical tensor operators, the notion of the nuclear mean field and the subtleties of occupation number representation.

The book is divided into two parts. Part I is comprised of the first eleven chapters, and it deals with particles and holes. Chapters 5–7 discuss the simple single-particle shell model. At the same time details of electromagnetic and beta-decay transitions are introduced. Chapters 8–11 treat different configuration mixing schemes for particle, hole and particle–hole excitations in nuclei lying at the borders of closed major shells.

Chapters 12–19 form Part II of the book. They introduce Bogoliubov–Valatin quasiparticles and proceed to treat the quasiparticle mean field and configuration mixing of quasiparticle excitations. In all configuration mixing calculations I use the same two-body interaction, namely the surface delta interaction, SDI. The SDI is simple to derive and use, but at the same time it is realistic enough to do service in serious computing.

I wish to acknowledge the great help of my assistants Jussi Toivanen, Markus Kortelainen and Eero Holmlund, all PhDs by now, for their great pedagogical work in guiding the students through a good part of the exercises of this book. I owe my sincere thanks to Eero Holmlund also for producing the 100-odd figures of the book. I also want to extend my thanks to Dr. Matias Aunola, Mr. Heikki Heiskanen and Mr. Mika Mustonen for their contributions and corrections to the book.

A highly indispensable contribution to this book comes from Professor Emeritus Pertti Lipas, who initially guided my way to nuclear physics with his excellent lectures on the subject. As regards this book, he checked all the theory and worked-out numerical examples, corrected errors and introduced pedagogical improvements, revised the English language and style, and produced the final LATEXtypescript including the demanding layout of equations and tables. Without doubt, Pertti Lipas is the person whom I owe the greatest debt for bringing this project to conclusion.

Finally, a project of this scale has demanded a great deal of extra time outside the office hours, time that I would normally have dedicated to my family. My deepest and loving thanks go to my wife Tiina and sons Olli and Hannu for their patience during the writing of my lecture notes and the final manuscript of this book.

Jyväskylä, Finland                                              *Jouni Suhonen*
July 2006

# Contents

# Part I

## Particles and Holes

# 1

# Angular Momentum Coupling

## Prologue

In nuclear physics, as also in atomic and molecular physics, the entities to be described consist of sub-entities with some orbital angular momentum and spin. The angular momentum of the entity is built, then, of the angular momenta of the sub-entities. This building process leads to quantum-mechanical angular momentum coupling. This chapter presents the basic machinery for treating angular momentum and its coupling. Clebsch–Gordan coefficients and $3j$ symbols are introduced. It is shown that Clebsch–Gordan coefficients and $3j$ symbols relate to the coupling of two angular momenta. Increasing the number of angular momenta to be coupled leads to more complicated coupling patterns. Transformations between different coupling schemes are mediated by $6j$ symbols and $9j$ symbols, relating to the coupling of three and four angular momenta respectively. A host of special relations and special cases are listed to make the manipulations of angular momentum algebra easier and more straightforward in the physical applications of the chapters to come.

## 1.1 Clebsch–Gordan Coefficients and $3j$ Symbols

We start by discussing a system of two angular momenta. These can be the angular momenta of two different particles or two different angular momenta (e.g. the orbital angular momentum and spin) of one single particle. The two angular momenta can be coupled to produce a total angular momentum $J$. The notion of angular momentum can be related to abstract rotations of state vectors in an abstract Hilbert space. This property naturally relates angular momentum to the rotation group, the components of angular momentum being the generators of the corresponding group algebra. For our purposes these connections to the underlying group structure and group algebra are not essential, but they make the following, rather abstract definition of an angular momentum vector more justifiable.

The angular momentum is, as we have learned in the basic courses of classical physics, a three-component vector. To transfer the classical Hamiltonian function to the quantum-mechanical Hamiltonian operator, one has to postulate certain commutation relations between the key observables of a physical system, e.g. the commutation relations of the positions and momenta of particles. To carry the notion of classical angular momentum over to quantum mechanics goes along the same lines by postulating commutation rules of some key observables. In the case of angular momentum these observables are the three components of the angular momentum operator. Related to this, we postulate that the quantity $\boldsymbol{J}$ is an *angular momentum* (vector) if its components satisfy the requirements

$$J_k^\dagger = J_k \ , \quad k = 1, 2, 3 \ , \quad [J_i, J_j] = i\hbar \sum_k \epsilon_{ijk} J_k \ , \tag{1.1}$$

where $\epsilon_{ijk}$ is the antisymmetric three-dimensional Levi-Civita permutation symbol. It has the following properties: $\epsilon_{ijk} = 0$ if two of the indices are the same; if all the indices are different, we have $\epsilon_{123} = \epsilon_{231} = \epsilon_{312} = 1$, in cyclic permutation; the value of the remaining non-zero permutation symbols is $-1$. It is to be noted that the components 1–3 are a running numbering of the usual $x$, $y$ and $z$ components in a Cartesian coordinate system. Thus we have for these indices $1 \equiv x$, $2 \equiv y$ and $3 \equiv z$.

Since the angular momentum vector, defined above, is an operator, one can find its eigenstates. These eigenstates of $\boldsymbol{J}$ satisfy the eigenvalue equations

$$\boldsymbol{J}^2 |j\, m\rangle = j(j+1)\hbar^2 |j\, m\rangle \ , \tag{1.2}$$

$$J_z |j\, m\rangle = m\hbar |j\, m\rangle \ , \tag{1.3}$$

where orthogonality and normalization (making together *orthonormality*) are assumed:

$$\langle j\, m | j'\, m'\rangle = \delta_{jj'}\delta_{mm'} \ , \quad m = -j, -j+1, \ldots, j-1, j \ . \tag{1.4}$$

The quantum numbers $j$ and $m$ correspond to the operators $\boldsymbol{J}^2$ and $J_z$ respectively. Hence $j$ corresponds to the length of the angular momentum vector and $m$ to its projection on the $z$-axis of a Cartesian coordinate system. As is known from textbooks on quantum mechanics, the eigenstates of $\boldsymbol{J}^2$ and $J_z$ can be derived elegantly by using the  ladder technique, i.e. by starting from the simple and obvious, so-called 'streched states' and then applying successively the raising and lowering ladder operators to produce the rest of the states belonging to a multiplet with a definite $j$ quantum number. These raising and lowering operators of the angular momentum are defined as

$$J_\pm = J_x \pm iJ_y \ . \tag{1.5}$$

Their commutation relations are

$$[J_+, J_-] = 2\hbar J_z \, , \quad [J_\pm, J_z] = \mp \hbar J_\pm \, , \tag{1.6}$$

as can be easily verified by using the basic commutation relations (1.1) for the components of the angular momentum. One can use the commutation rules of the ladder operators to derive their raising and lowering properties

$$J_\pm |j\, m\rangle = \hbar \sqrt{j(j+1) - m(m \pm 1)} \, |j\; m \pm 1\rangle$$
$$= \hbar \sqrt{(j \pm m + 1)(j \mp m)} \, |j\; m \pm 1\rangle \, . \tag{1.7}$$

The coordinate representation of the abstract angular momentum eigenstates $|j\, m\rangle$ for a single particle are the usual *spherical harmonics*, defined in e.g. [1–7].

Having defined the basic properties of an angular momentum operator, we can start combining angular momenta. Let $\boldsymbol{J}_1$ and $\boldsymbol{J}_2$ be two commuting angular momentum vectors, e.g. angular momenta of two different particles or, say, the orbital angular momentum and spin angular momentum of a single particle. For these angular momenta we thus have

$$[\boldsymbol{J}_1, \boldsymbol{J}_2] = 0 \, , \quad \text{or} \quad [J_{1k}, J_{2l}] = 0 \quad \text{for all } k, l = 1, 2, 3 \, , \tag{1.8}$$

where, for convenience, the running numbering is used for the Cartesian components of angular momentum. Each of the angular momenta has its eigenstates, which satisfy

$$\boldsymbol{J}_k^2 |j_k\, m_k\rangle = j_k(j_k + 1)\hbar^2 |j_k\, m_k\rangle \, , \tag{1.9}$$
$$J_{kz} |j_k\, m_k\rangle = m_k \hbar |j_k\, m_k\rangle \, , \quad k = 1, 2 \, . \tag{1.10}$$

The sum $\boldsymbol{J} = \boldsymbol{J}_1 + \boldsymbol{J}_2$ of the two commuting angular momentum vectors is also an angular momentum vector since it satisfies the basic properties (1.1) of angular momentum (verifying this is left as an exercise for the reader). From basic quantum mechanics we know that when an operator (typically the Hamiltonian) is the sum of two or more commuting operators, then the eigenvectors are products of eigenvectors of each individual term of the sum. So for the total angular momentum we can write directly the product eigenstates as

$$|j_1\, m_1\, j_2\, m_2\rangle = |j_1\, m_1\rangle |j_2\, m_2\rangle \, . \tag{1.11}$$

These states are eigenstates of the operator set

$$\{ \boldsymbol{J}_1^2, \, J_{1z}, \, \boldsymbol{J}_2^2, \, J_{2z} \} \, , \tag{1.12}$$

which constitutes a complete set of commuting operators. The explicit eigenvalue equations are

$$\boldsymbol{J}_k^2 |j_1\, m_1\, j_2\, m_2\rangle = j_k(j_k + 1)\hbar^2 |j_1\, m_1\, j_2\, m_2\rangle \, , \tag{1.13}$$
$$J_{kz} |j_1\, m_1\, j_2\, m_2\rangle = m_k \hbar |j_1\, m_1\, j_2\, m_2\rangle \, , \quad k = 1, 2 \, . \tag{1.14}$$

The product states (1.11) form an orthonormal set if the members of the product form separately orthonormal sets in their respective subspaces. It follows that

$$\langle j_1 \, m_1 \, j_2 \, m_2 | j_1' \, m_1' \, j_2' \, m_2' \rangle = \langle j_1 \, m_1 | j_1' \, m_1' \rangle \langle j_2 \, m_2 | j_2' \, m_2' \rangle$$
$$= \delta_{j_1 j_1'} \delta_{m_1 m_1'} \delta_{j_2 j_2'} \delta_{m_2 m_2'} \, . \tag{1.15}$$

Since the angular momenta in (1.11) are not coupled to form a well-defined total angular momentum, we say:

> The complete set of states $\{|j_1 \, m_1 \, j_2 \, m_2\rangle\}$ is called the *uncoupled basis*.

Now it is time to start coupling the two angular momenta. The coupled angular momentum is a member of the following set of mutually commuting angular momenta:

$$\{\boldsymbol{J}_1^2, \, \boldsymbol{J}_2^2, \, \boldsymbol{J}^2, \, J_z\} \, . \tag{1.16}$$

The four operators have a complete set of common eigenvectors. The explicit eigenvalue equations are

$$\boldsymbol{J}_k^2 |j_1 \, j_2 \, j \, m\rangle = j_k(j_k + 1)\hbar^2 |j_1 \, j_2 \, j \, m\rangle \, , \quad k = 1, 2 \, , \tag{1.17}$$
$$\boldsymbol{J}^2 |j_1 \, j_2 \, j \, m\rangle = j(j + 1)\hbar^2 |j_1 \, j_2 \, j \, m\rangle \, , \tag{1.18}$$
$$J_z |j_1 \, j_2 \, j \, m\rangle = m\hbar |j_1 \, j_2 \, j \, m\rangle \, . \tag{1.19}$$

The orthonormality condition reads

$$\langle j_1 \, j_2 \, j \, m | j_1 \, j_2 \, j' \, m' \rangle = \delta_{jj'} \delta_{mm'} \, . \tag{1.20}$$

Due to the angular momentum coupling accomplished, we state:

> The complete set of states $\{|j_1 \, j_2 \, j \, m\rangle\}$ is called the *coupled basis*.

It is worth noting that the set (1.12) contains two angular momentum vectors and their projections on the $z$-axis. On the other hand, the set (1.16) contains three angular momentum vectors and only one projection on the $z$-axis, namely the projection of the coupled angular momentum vector. This means that coupling two angular momenta leaves only their 'lengths' as sharp, measurable quantities, while their $z$ projections become fuzzy, without any sharp quantum number related to them.

Each of the basis sets, the uncoupled and the coupled one, forms a complete set of states since the operators are all Hermitian. This means that an identity operator can be formed by using the states of either set. Using the identity operator constructed from the uncoupled basis, we can write

$$|j_1 \, j_2 \, j \, m\rangle = \sum_{m_1 m_2} |j_1 \, m_1 \, j_2 \, m_2\rangle \langle j_1 \, m_1 \, j_2 \, m_2 | j_1 \, j_2 \, j \, m\rangle$$
$$\equiv \sum_{m_1 m_2} (j_1 \, m_1 \, j_2 \, m_2 | j \, m) |j_1 \, m_1 \, j_2 \, m_2\rangle \, , \tag{1.21}$$

where the quantity $(j_1 \, m_1 \, j_2 \, m_2 | j \, m)$ is called the *Clebsch–Gordan coefficient*. The same notation is used in e.g. [8,9]. Other notations can be found in books on quantum mechanics, such as [1,2,4–7]. Yet different types of notation occur in nuclear physics textbooks [10–19]. From (1.21) it is clear that a Clebsch–Gordan coefficient is the overlap between an uncoupled and a coupled state. In this way the coefficients define a linear transformation from the uncoupled basis to the coupled one. Because the norm of the states is preserved, this transformation is also unitary.

The basic properties of angular momentum coupling and Clebsch–Gordan coefficients are listed below.

- The projection quantum numbers have to fulfil the addition law

$$(j_1 \, m_1 \, j_2 \, m_2 | j \, m) = 0 \quad \text{unless} \quad m_1 + m_2 = m \; . \tag{1.22}$$

- The coupled angular momenta have to fulfil the so-called *triangular condition* defined as

$$|j_1 - j_2| \le j \le j_1 + j_2 \quad \text{(triangular condition)} \tag{1.23}$$

and denoted by $\Delta(j_1 j_2 j)$.

- The quantum numbers of the angular momenta to be summed can separately be integers or half integers. However, for the sum applies the restriction $j_1 + j_2 + j = $ integer, which leads to the following allowed values of the total angular momentum:

$$j = |j_1 - j_2|, \; |j_1 - j_2| + 1, \; \ldots, \; j_1 + j_2 - 1, \; j_1 + j_2 \; . \tag{1.24}$$

- The Clebsch–Gordan coefficients are chosen to be *real*, and so that

$$(j_1 \, j_1 \, j_2 \, j_2 | j_1 + j_2 \, j_1 + j_2) = +1 \; , \quad (j_1 \, m_1 \, j_2 \, -j_2 | j \, m) \ge 0 \; . \tag{1.25}$$

These conditions fix the phases of all Clebsch–Gordan coefficients. This phase convention needs to be taken into account when calculating numerical values of the coefficients, by using either the ladder operations (1.7) or certain recursion relations (see Exercise 1.101).

## 1.2 More on Clebsch–Gordan Coefficients; $3j$ Symbols

In the following we list the most important properties of Clebsch–Gordan coefficients. Proofs of these properties are straightforward and are left as an exercise for the reader. Later in the section we introduce $3j$ symbols as a more symmetric alternative to the Clebsch–Gordan coefficients.

### 1.2.1 Clebsch–Gordan Coefficients

By starting from the overlap of two states of the coupled basis and expanding them in the uncoupled basis, one can prove the following, so-called *orthogonality* property:

$$\sum_{m_1 m_2} (j_1 \, m_1 \, j_2 \, m_2 | j \, m)(j_1 \, m_1 \, j_2 \, m_2 | j' \, m') = \delta_{jj'} \delta_{mm'} \quad \text{(orthogonality)} .$$

$$(1.26)$$

Starting from the completeness of the coupled basis

$$\sum_{jm} |j_1 \, j_2 \, j \, m\rangle \langle j_1 \, j_2 \, j \, m| = 1 , \qquad (1.27)$$

one can immediately derive the so-called *completeness* property of the Clebsch–Gordan coefficients:

$$\sum_{jm} (j_1 \, m_1 \, j_2 \, m_2 | j \, m)(j_1 \, m_1' \, j_2 \, m_2' | j \, m) = \delta_{m_1 m_1'} \delta_{m_2 m_2'} \quad \text{(completeness)} .$$

$$(1.28)$$

It follows from this relation or directly from (1.27) that

$$|j_1 \, m_1 \, j_2 \, m_2\rangle = \sum_{jm} (j_1 \, m_1 \, j_2 \, m_2 | j \, m) |j_1 \, j_2 \, j \, m\rangle , \qquad (1.29)$$

which is the inverse of the basic Clebsch–Gordan relation (1.21).

There exist useful symmetry relations for Clebsch–Gordan coefficients. The exchange of the coupling order of the two angular momenta introduces a phase factor. This means that the coupling coefficients for $J_2$ coupled with $J_1$ to produce $J$ are but a phase factor times the coupling coefficients for $J_1$ coupled with $J_2$ to produce $J$, i.e.

$$(j_2 \, m_2 \, j_1 \, m_1 | j \, m) = (-1)^{j_1 + j_2 - j} (j_1 \, m_1 \, j_2 \, m_2 | j \, m) . \qquad (1.30)$$

The sign of all the projection quantum numbers in a Clebsch–Gordan coefficient can be changed and the original Clebsch–Gordan coefficient, to within a phase, is recovered:

$$(j_1 \, -m_1 \, j_2 \, -m_2 | j \, -m) = (-1)^{j_1 + j_2 - j} (j_1 \, m_1 \, j_2 \, m_2 | j \, m) . \qquad (1.31)$$

For zero values of the projection quantum numbers we have the symmetry restriction

$$(j_1 \, 0 \, j_2 \, 0 | j \, 0) = 0 \quad \text{unless} \quad j_1 + j_2 + j = \text{even} , \qquad (1.32)$$

which comes from the fact that for zero projections the angular momentum quantum numbers all have to be integers, and the symmetry (1.30) then requires that $j_1 + j_2 - j = \text{even}$, which is equivalent to (1.32) for integer values of the $j$ quantum numbers.

These symmetries come in handy in actual calculations when deriving compact expressions for physical quantities by using angular momentum algebra. In the later chapters, where physical applications of angular momentum algebra are discussed, this will be much in evidence.

There are further special properties of the Clebsch–Gordan coefficients which are very useful in applications. A particularly simple coefficient is

$$(j\, m\, 0\, 0 | j\, m) = 1 \,, \tag{1.33}$$

which just records that nothing happens when a zero angular momentum is coupled with a non-zero one. When two angular momenta are coupled to zero angular momentum, the two angular momenta have to be necessarily the same due to the addition property (1.24). Furthermore, the net result of the coupling is given by the simple Clebsch–Gordan coefficient

$$(j\, m\, j\, m' | 0\, 0) = (-1)^{j-m}\, \widehat{j}^{\,-1} \delta_{m,-m'} \,, \tag{1.34}$$

which contains the usual abbreviation, the 'hat factor'

$$\widehat{j} \equiv \sqrt{2j+1} \,. \tag{1.35}$$

Sometimes it is necessary to change the coupling order of the angular momenta so that the total angular momentum is to be coupled with either one of the original angular momenta to produce the remaining one. This change in the coupling order produces a phase factor and a ratio of two hat factors in the following two ways:

$$(j_1\, m_1\, j_2\, m_2 | j\, m) = (-1)^{j_2+m_2} \frac{\widehat{j}}{\widehat{j_1}} (j_2\, -m_2\, j\, m | j_1\, m_1) \,, \tag{1.36}$$

$$(j_1\, m_1\, j_2\, m_2 | j\, m) = (-1)^{j_1-m_1} \frac{\widehat{j}}{\widehat{j_2}} (j\, m\, j_1\, -m_1 | j_2\, m_2) \,. \tag{1.37}$$

### 1.2.2 More Symmetry: $3j$ Symbols

The symmetry properties (1.30) and (1.31) of the Clebsch–Gordan coefficients carry a phase factor which itself is not the most symmetric possible. A more symmetric phase factor can be produced by redefinig the Clebsch–Gordan coefficients in a suitable way. This leads us to the so-called $3j$ symbols. They are obtained from the Clebsch–Gordan coefficients by the following definition:

$$\begin{pmatrix} j_1 & j_2 & j_3 \\ m_1 & m_2 & m_3 \end{pmatrix} \equiv (-1)^{j_1-j_2-m_3}\, \widehat{j_3}^{\,-1} (j_1\, m_1\, j_2\, m_2 | j_3\, -m_3) \,. \tag{1.38}$$

Since these coefficients look more symmetric and behave more symmetrically than the Clebsch–Gordan coefficients, they can be easily tabulated. A very

convenient tabulation is given in the tables of Rotenberg et al. [20]. These
tables are easy to use for extracting values of specific $3j$ symbols.

The basic symmetry properties of the $3j$ symbols are the following. Exchange of the coupling order of the two angular momenta simply produces the
symmetric phase factor

$$\begin{pmatrix} j_2 & j_1 & j_3 \\ m_2 & m_1 & m_3 \end{pmatrix} = (-1)^{j_1+j_2+j_3} \begin{pmatrix} j_1 & j_2 & j_3 \\ m_1 & m_2 & m_3 \end{pmatrix} , \qquad (1.39)$$

and similarly for the change of sign of the projection quantum numbers,

$$\begin{pmatrix} j_1 & j_2 & j_3 \\ -m_1 & -m_2 & -m_3 \end{pmatrix} = (-1)^{j_1+j_2+j_3} \begin{pmatrix} j_1 & j_2 & j_3 \\ m_1 & m_2 & m_3 \end{pmatrix} . \qquad (1.40)$$

Furthermore, the triangular condition for the lengths of the angular momentum vectors has to be satisfied, as does the additivity of the projection quantum numbers:

$$\begin{pmatrix} j_1 & j_2 & j_3 \\ m_1 & m_2 & m_3 \end{pmatrix} = 0 \quad \text{unless} \quad \begin{cases} \Delta(j_1 j_2 j_3) , \\ m_1 + m_2 + m_3 = 0 . \end{cases} \qquad (1.41)$$

The following simple properties can be derived directly from the corresponding properties (1.32) and (1.34) of the Clebsch–Gordan coefficients:

$$\begin{pmatrix} j_1 & j_2 & 0 \\ m_1 & m_2 & 0 \end{pmatrix} = (-1)^{j_1-m_1} \hat{j}_1^{-1} \delta_{j_1 j_2} \delta_{m_1,-m_2} , \qquad (1.42)$$

$$\begin{pmatrix} j_1 & j_2 & j_3 \\ 0 & 0 & 0 \end{pmatrix} = 0 \quad \text{unless} \quad j_1 + j_2 + j_3 = \text{even} . \qquad (1.43)$$

It is useful to invert the relation (1.38) to enable simplification of some
expressions occurring in angular momentum algebra. This inverted relation
reads

$$(j_1\, m_1\, j_2\, m_2 | j_3\, m_3) = (-1)^{j_2-j_1-m_3} \hat{j}_3 \begin{pmatrix} j_1 & j_2 & j_3 \\ m_1 & m_2 & -m_3 \end{pmatrix} . \qquad (1.44)$$

### 1.2.3 Relations for $3j$ Symbols

Further relations for the $3j$ symbols, useful in the simplification of complicated
results of angular momentum algebra or in obtaining numerical values of $3j$
symbols, can be obtained from [2]:

$$\begin{pmatrix} j+\frac{1}{2} & j & \frac{1}{2} \\ m & -m-\frac{1}{2} & \frac{1}{2} \end{pmatrix} = (-1)^{j-m-\frac{1}{2}} \sqrt{\frac{j-m+\frac{1}{2}}{(2j+2)(2j+1)}} , \qquad (1.45)$$

$$\begin{pmatrix} j+1 & j & 1 \\ m & -m-1 & 1 \end{pmatrix} = (-1)^{j-m-1} \sqrt{\frac{(j-m)(j-m+1)}{(2j+3)(2j+2)(2j+1)}} , \qquad (1.46)$$

$$\begin{pmatrix} j+1 & j & 1 \\ m & -m & 0 \end{pmatrix} = (-1)^{j-m-1} \sqrt{\frac{2(j+m+1)(j-m+1)}{(2j+3)(2j+2)(2j+1)}} \,, \qquad (1.47)$$

$$\begin{pmatrix} j & j & 1 \\ m & -m-1 & 1 \end{pmatrix} = (-1)^{j-m} \sqrt{\frac{2(j-m)(j+m+1)}{2j(2j+1)(2j+2)}} \,, \qquad (1.48)$$

$$\begin{pmatrix} j & j & 1 \\ m & -m & 0 \end{pmatrix} = (-1)^{j-m} \frac{m}{\sqrt{j(j+1)(2j+1)}} \,, \qquad (1.49)$$

$$\begin{pmatrix} j & j & 2 \\ m & -m & 0 \end{pmatrix} = (-1)^{j-m} \frac{3m^2 - j(j+1)}{\sqrt{j(j+1)(2j-1)(2j+1)(2j+3)}} \,. \qquad (1.50)$$

## 1.3 The 6$j$ Symbol

The coupling of two angular momenta discussed above can be extended to a discussion of the coupling of three and four angular momenta. The related coupling coefficients bear the generic name *vector coupling coefficients* or *Wigner nj symbols*. In the case of more than two angular momenta the situation is complicated by the fact that the angular momenta can be coupled in various different ways, depending on the order of coupling. Let us first see which complications this brings in the case of three angular momenta.[1]

Let $J_1$, $J_2$ and $J_3$ be three commuting angular momentum vectors, so that $J = J_1 + J_2 + J_3$ is the total angular momentum. We have three possible ways to couple the three angular momenta to yield $J$, namely

$$J = J_{12} + J_3 \,, \quad J_{12} \equiv J_1 + J_2 \,, \qquad (1.51)$$

$$J = J_1 + J_{23} \,, \quad J_{23} \equiv J_2 + J_3 \,, \qquad (1.52)$$

$$J = J_{13} + J_2 \,, \quad J_{13} \equiv J_1 + J_3 \,. \qquad (1.53)$$

It can be shown that

- the values of the quantum number $j$ (corresponding to $J$) do not depend on the coupling order, and that

- the states corresponding to different coupling schemes are not the same, i.e.

$$|j_1 j_2 (j_{12}) j_3 ; j\,m\rangle \neq |j_1, j_2 j_3 (j_{23}) ; j\,m\rangle \neq |j_1 j_3 (j_{13}) j_2 ; j\,m\rangle \,. \qquad (1.54)$$

---

[1] A complication of another nature arises when coupling angular momenta of identical particles. The wave function of a fermion system must be antisymmetric. This eliminates geometrically possible values of total angular momentum. For the two-particle case see Chap. 5; for examples of three and four particles see Subsect. 12.4.4. There is an algorithm, the so-called $m$ table, to find which total angular momenta are possible for a system of identical particles. (See e.g. [21].)

In all of the above cases (1.51)–(1.53), the number of linearly independent states is the same and each of the three sets forms a complete set of states, with the completeness relations

$$1 = \sum_{j_{12}} |j_1 j_2 (j_{12}) j_3 ; j m\rangle\langle j_1 j_2 (j_{12}) j_3 ; j m| , \tag{1.55}$$

$$1 = \sum_{j_{23}} |j_1, j_2 j_3 (j_{23}) ; j m\rangle\langle j_1, j_2 j_3 (j_{23}) ; j m| , \tag{1.56}$$

$$1 = \sum_{j_{13}} |j_1 j_3 (j_{13}) j_2 ; j m\rangle\langle j_1 j_3 (j_{13}) j_2 ; j m| . \tag{1.57}$$

A change from the basis (1.51) to the basis (1.52) can be accomplished in the following way:

$$
\begin{aligned}
|j_1, j_2 j_3 (j_{23}) ; j m\rangle &= \sum_{j_{12}} |j_1 j_2 (j_{12}) j_3 ; j m\rangle \\
&\quad \times \langle j_1 j_2 (j_{12}) j_3 ; j m|j_1, j_2 j_3 (j_{23}) ; j m\rangle \\
&\equiv \sum_{j_{12}} (-1)^{j_1+j_2+j_3+j}\, \widehat{j_{12}}\widehat{j_{23}} \begin{Bmatrix} j_1 & j_2 & j_{12} \\ j_3 & j & j_{23} \end{Bmatrix} |j_1 j_2 (j_{12}) j_3 ; j m\rangle ,
\end{aligned}
\tag{1.58}
$$

where the array with braces is called the $6j$ *symbol*. Its numerical values have been tabulated in e.g. [20]. As can be seen, a $6j$ symbol is proportional to the overlap of two state vectors related to two different coupling schemes of three angular momenta. This makes it part of a unitary transformation, which preserves the norms of the states. Furthermore, in distinction from the $3j$ symbol, there are no projection quantum numbers involved. The reason is that the $6j$ symbols are related to a transformation between two basis sets where all the states have a good total angular momentum, whereas the $3j$ symbol relates to a transformation between a basis of states with a good total angular momentum and a basis of states with no total angular momentum defined.

Starting from the defining Eq. (1.58), one can derive the following explicit expression for the $6j$ symbol:

$$
\begin{aligned}
\begin{Bmatrix} j_1 & j_2 & j_{12} \\ j_3 & j & j_{23} \end{Bmatrix} &= \sum_{\substack{m_1 m_2 m_3 \\ m_{12} m_{23} m}} (-1)^{j_3+j+j_{23}-m_3-m-m_{23}} \begin{pmatrix} j_1 & j_2 & j_{12} \\ m_1 & m_2 & m_{12} \end{pmatrix} \\
&\quad \times \begin{pmatrix} j_1 & j & j_{23} \\ m_1 & -m & m_{23} \end{pmatrix} \begin{pmatrix} j_3 & j_2 & j_{23} \\ m_3 & m_2 & -m_{23} \end{pmatrix} \begin{pmatrix} j_3 & j & j_{12} \\ -m_3 & m & m_{12} \end{pmatrix} .
\end{aligned}
\tag{1.59}
$$

This explicit form of the $6j$ symbol becomes handy when angular momentum algebra produces summations over all the projection quantum numbers of a set of four $3j$ symbols. This happens for some reduced matrix elements (see Sect. 2.2), which are not dependent on any projection quantum numbers. Then a summation over all of the projection quantum numbers should reduce to a quantity containing only quantum numbers associated with the lengths of the angular momentum vectors involved. In the following, we discuss symmetry properties of the $6j$ symbols.

### 1.3.1 Symmetry Properties of the $6j$ Symbol

The angular momenta involved in a $6j$ symbol have to satisfy certain triangular conditions. From the bracket on the second line of (1.58) one can extract all the triangular conditions needed. The left-hand side gives the conditions $\Delta(j_1 j_2 j_{12})$ and $\Delta(j_{12} j_3 j)$, and the right-hand side gives $\Delta(j_2 j_3 j_{23})$ and $\Delta(j_1 j_{23} j)$. These conditions can be summarized in the following compact way:

$$\begin{Bmatrix} j_1 & j_2 & j_3 \\ l_1 & l_2 & l_3 \end{Bmatrix} = 0 \quad \text{unless} \quad \begin{cases} \Delta(j_1 j_2 j_3), \ \Delta(l_1 l_2 j_3), \\ \Delta(l_1 j_2 l_3), \ \Delta(j_1 l_2 l_3). \end{cases} \tag{1.60}$$

There are further symmetries on permutations of the entries in a $6j$ symbol. Exchange of any two columns leave the value of the symbol unchanged:

$$\begin{Bmatrix} j_1 & j_2 & j_3 \\ l_1 & l_2 & l_3 \end{Bmatrix} = \begin{Bmatrix} j_2 & j_1 & j_3 \\ l_2 & l_1 & l_3 \end{Bmatrix} = \begin{Bmatrix} j_3 & j_1 & j_2 \\ l_3 & l_1 & l_2 \end{Bmatrix} = \cdots. \tag{1.61}$$

Furthermore, the value of a $6j$ symbol is unaltered by the exchange of the upper and lower arguments in each of any two columns:

$$\begin{Bmatrix} j_1 & j_2 & j_3 \\ l_1 & l_2 & l_3 \end{Bmatrix} = \begin{Bmatrix} l_1 & l_2 & j_3 \\ j_1 & j_2 & l_3 \end{Bmatrix} = \begin{Bmatrix} j_1 & l_2 & l_3 \\ l_1 & j_2 & j_3 \end{Bmatrix} = \cdots. \tag{1.62}$$

Altogether these exchanges create 24 symmetry transformations of the symmetry group of the regular tetrahedron (see e.g. [2]). Each of them corresponds to a rotation, reflection or both of a tetrahedron with its sides labelled by the quantum numbers of the corresponding $6j$ symbol.

The many years of development of angular momentum algebra have produced various symbols related to the $6j$ symbol. One of the most commonly used, besides the $6j$ symbol, is the so-called *Racah symbol* or Racah $W$ coefficient. It can be defined as

$$W(j_1 j_2 l_2 l_1; j_3 l_3) = (-1)^{j_1+j_2+l_1+l_2} \begin{Bmatrix} j_1 & j_2 & j_3 \\ l_1 & l_2 & l_3 \end{Bmatrix}. \tag{1.63}$$

A useful special case of the $6j$ symbol occurs when one of the angular momenta is zero. This makes one of the coupling schemes trivial, and the transformation to the remaining coupling scheme is simplified. This simplification appears in the $6j$ symbol as

$$\boxed{\begin{Bmatrix} j_1 & j_2 & j_3 \\ 0 & j_3' & j_2' \end{Bmatrix} = \frac{(-1)^{j_1+j_2+j_3}}{\widehat{j_2}\widehat{j_3}} \, \delta_{j_2 j_2'} \delta_{j_3 j_3'} \Delta(j_1 j_2 j_3).} \tag{1.64}$$

## 1.3.2 Relations for 6j Symbols

Below are listed miscellaneous results on $6j$ symbols (for more relations, see e.g. [2,22]). These relations turn out to be quite convenient when one tries to simplify complicated expressions arising from angular momentum algebra in physical applications. In the present work most of these relations have been used in working out final results on nuclear structure and transitions. These useful relations are

$$\sum_j (-1)^{j+j'+j''} \hat{j}^2 \begin{Bmatrix} j_1 & j_2 & j' \\ j_3 & j_4 & j \end{Bmatrix} \begin{Bmatrix} j_1 & j_3 & j'' \\ j_2 & j_4 & j \end{Bmatrix} = \begin{Bmatrix} j_1 & j_2 & j' \\ j_4 & j_3 & j'' \end{Bmatrix} , \qquad (1.65)$$

$$\sum_j \hat{j}^2 \begin{Bmatrix} j_1 & j_2 & j' \\ j_3 & j_4 & j \end{Bmatrix} \begin{Bmatrix} j_1 & j_2 & j'' \\ j_3 & j_4 & j \end{Bmatrix} = \frac{\delta_{j'j''}}{\hat{j}'^2} \Delta(j_1 j_2 j') \Delta(j_3 j_4 j') \quad \text{(unitarity)} , \qquad (1.66)$$

$$\sum_j (-1)^{j_1+j_2+j} \hat{j}^2 \begin{Bmatrix} j_1 & j_1 & j' \\ j_2 & j_2 & j \end{Bmatrix} = \hat{j}_1 \hat{j}_2 \delta_{j'0} , \qquad (1.67)$$

$$\sum_j \hat{j}^2 \begin{Bmatrix} j_1 & j_2 & j' \\ j_1 & j_2 & j \end{Bmatrix} = (-1)^{2j_1+2j_2} \Delta(j_1 j_2 j') , \qquad (1.68)$$

$$\hat{j}_1 \hat{j}_2 \begin{pmatrix} j_1 & j_2 & j_3 \\ 0 & 0 & 0 \end{pmatrix} \begin{Bmatrix} j_1 & j_2 & j_3 \\ l_1 & l_2 & \frac{1}{2} \end{Bmatrix} = -\frac{1+(-1)^{j_1+j_2+j_3}}{2} \begin{pmatrix} l_2 & l_1 & j_3 \\ \frac{1}{2} & -\frac{1}{2} & 0 \end{pmatrix}$$
$$\times \Delta(j_1 \tfrac{1}{2} l_2) \Delta(j_2 \tfrac{1}{2} l_1) , \quad (1.69)$$

$$\hat{j}_1 \begin{pmatrix} j_1 & j_2 & j_3 \\ 0 & 0 & 0 \end{pmatrix} \begin{Bmatrix} j_1 & j_2 & j_3 \\ j & j_1 & 1 \end{Bmatrix} = -(-1)^j \frac{1+(-1)^{j_1+j_2+j_3}}{\sqrt{2}} \begin{pmatrix} j & 1 & j_2 \\ 1 & -1 & 0 \end{pmatrix}$$
$$\times \begin{pmatrix} j & j_1 & j_3 \\ 1 & -1 & 0 \end{pmatrix} \Delta(j_1 j_2 j_3) , \quad (1.70)$$

$$\hat{j}_1 \hat{j}_2 \begin{pmatrix} j_1 & j_2 & j_3 \\ 0 & 0 & 0 \end{pmatrix} \begin{Bmatrix} j_1 & j_2 & j_3 \\ j_2 & j_1 & 1 \end{Bmatrix} = \frac{1+(-1)^{j_1+j_2+j_3}}{2} \begin{pmatrix} j_1 & j_2 & j_3 \\ 1 & -1 & 0 \end{pmatrix}$$
$$\times \Delta(j_1 \tfrac{1}{2} l_2) \Delta(j_2 \tfrac{1}{2} l_1) . \quad (1.71)$$

## 1.3.3 Explicit Expressions for 6j Symbols

There are reasonably simple explicit expressions for some $6j$ symbols. They are often useful for simplifying final results to the level where numerical values can be extracted by use of a pocket calculator. Examples of the use of these expressions are given in several instances in the present work. We note additionally that numerical values can be obtained by using recursion relations of the $6j$ symbols, listed in e.g. [2,7]. We list the following explicit expressions:

$$\left\{ \begin{matrix} j_1 & j_2 & j_3 \\ 1 & j_3 & j_2 \end{matrix} \right\} = (-1)^{j_1+j_2+j_3} \frac{j_1(j_1+1) - j_2(j_2+1) - j_3(j_3+1)}{\sqrt{4j_2(j_2+1)(2j_2+1)j_3(j_3+1)(2j_3+1)}},$$
$$(1.72)$$

$$\left\{ \begin{matrix} j_1 & j_2 & j_3 \\ \frac{1}{2} & j_3 - \frac{1}{2} & j_2 + \frac{1}{2} \end{matrix} \right\} = (-1)^s \sqrt{\frac{(j_1+j_3-j_2)(j_1+j_2-j_3+1)}{2(2j_2+1)(2j_2+2)j_3(2j_3+1)}}, \quad (1.73)$$

$$\left\{ \begin{matrix} j_1 & j_2 & j_3 \\ \frac{1}{2} & j_3 - \frac{1}{2} & j_2 - \frac{1}{2} \end{matrix} \right\} = (-1)^s \sqrt{\frac{(s+1)(j_2+j_3-j_1)}{4j_2(2j_2+1)j_3(2j_3+1)}}, \quad (1.74)$$

$$\left\{ \begin{matrix} j_1 & j_2 & j_3 \\ 1 & j_3 - 1 & j_2 - 1 \end{matrix} \right\} = (-1)^s \sqrt{\frac{s(s+1)(s-2j_1-1)(s-2j_1)}{4j_2(2j_2-1)(2j_2+1)j_3(2j_3-1)(2j_3+1)}},$$
$$(1.75)$$

$$\left\{ \begin{matrix} j_1 & j_2 & j_3 \\ 1 & j_3 - 1 & j_2 \end{matrix} \right\} = (-1)^s \sqrt{\frac{2(s+1)(s-2j_1)(s-2j_2)(s-2j_3+1)}{4j_2(2j_2+1)(2j_2+2)j_3(2j_3-1)(2j_3+1)}}, \quad (1.76)$$

$$\left\{ \begin{matrix} j_1 & j_2 & j_3 \\ 1 & j_3 - 1 & j_2 + 1 \end{matrix} \right\}$$
$$= (-1)^s \sqrt{\frac{(s-2j_2-1)(s-2j_2)(s-2j_3+1)(s-2j_3+2)}{2(2j_2+1)(2j_2+2)(2j_2+3)j_3(2j_3-1)(2j_3+1)}}, \quad (1.77)$$

where the sum of the angular momenta is written as

$$s = j_1 + j_2 + j_3 . \tag{1.78}$$

## 1.4 The 9$j$ Symbol

We now investigate the problem of coupling four angular momentum vectors to a given total angular momentum. As in the case of three angular momenta, there are several different ways to do the coupling. To begin with, let $\boldsymbol{J}_1$, $\boldsymbol{J}_2$, $\boldsymbol{J}_3$ and $\boldsymbol{J}_4$ be four commuting angular momentum vectors, with $\boldsymbol{J} = \boldsymbol{J}_1 + \boldsymbol{J}_2 + \boldsymbol{J}_3 + \boldsymbol{J}_4$ their vector sum.

We have now several possible ways to couple the four angular momenta to the total angular momentum $\boldsymbol{J}$. Let us look more closely at the following two ways:

$$\boldsymbol{J} = \boldsymbol{J}_{12} + \boldsymbol{J}_{34}, \quad \boldsymbol{J}_{12} \equiv \boldsymbol{J}_1 + \boldsymbol{J}_2, \quad \boldsymbol{J}_{34} \equiv \boldsymbol{J}_3 + \boldsymbol{J}_4, \tag{1.79}$$
$$\boldsymbol{J} = \boldsymbol{J}_{13} + \boldsymbol{J}_{24}, \quad \boldsymbol{J}_{13} \equiv \boldsymbol{J}_1 + \boldsymbol{J}_3, \quad \boldsymbol{J}_{24} \equiv \boldsymbol{J}_2 + \boldsymbol{J}_4. \tag{1.80}$$

As in the case of the $6j$ symbols, after making the correspondence $j_i \leftrightarrow \boldsymbol{J}_i$, $j \leftrightarrow \boldsymbol{J}$ between the angular momenta and their quantum numbers, it can be shown that

- the values of the quantum number $j$ do not depend on the coupling order, and that
- the states corresponding to different coupling schemes are not the same, i.e.

$$|j_1 j_2 (j_{12}) j_3 j_4 (j_{34}) ; j\, m\rangle \neq |j_1 j_3 (j_{13}) j_2 j_4 (j_{24}) ; j\, m\rangle . \qquad (1.81)$$

In both of the cases (1.79) and (1.80) the number of linearly independent states is the same and the two sets both form complete sets of states. Their completeness relations are

$$1 = \sum_{j_{12} j_{34}} |j_1 j_2 (j_{12}) j_3 j_4 (j_{34}) ; j\, m\rangle \langle j_1 j_2 (j_{12}) j_3 j_4 (j_{34}) ; j\, m| , \qquad (1.82)$$

$$1 = \sum_{j_{13} j_{24}} |j_1 j_3 (j_{13}) j_2 j_4 (j_{24}) ; j\, m\rangle \langle j_1 j_3 (j_{13}) j_2 j_4 (j_{24}) ; j\, m| . \qquad (1.83)$$

A change from the basis (1.79) to the basis (1.80) can be accomplished, as in the case of the $6j$ symbols, by a linear unitary transformation, which is written as

$$
\begin{aligned}
|j_1 j_3 (j_{13}) &j_2 j_4 (j_{24}) ; j\, m\rangle = \sum_{j_{12} j_{34}} |j_1 j_2 (j_{12}) j_3 j_4 (j_{34}) ; j\, m\rangle \\
&\times \langle j_1 j_2 (j_{12}) j_3 j_4 (j_{34}) ; j\, m | j_1 j_3 (j_{13}) j_2 j_4 (j_{24}) ; j\, m\rangle \\
&\equiv \sum_{j_{12} j_{34}} \widehat{j_{12}} \widehat{j_{34}} \widehat{j_{13}} \widehat{j_{24}} \left\{ \begin{array}{ccc} j_1 & j_2 & j_{12} \\ j_3 & j_4 & j_{34} \\ j_{13} & j_{24} & j \end{array} \right\} |j_1 j_2 (j_{12}) j_3 j_4 (j_{34}) ; j\, m\rangle ,
\end{aligned}
\qquad (1.84)
$$

where the array with braces is called the $9j$ *symbol*. Its numerical values are tabulated in e.g. [23]. Numerical values of $9j$ symbols can be derived from the values of $6j$ symbols because, as we see below, a $9j$ symbol can be written as a sum over products of three $6j$ symbols. In addition, several special cases of $9j$ symbols reduce to simple expressions involving one or two $6j$ or $3j$ symbols or both.

Starting from (1.84), one can derive the following explicit expression for the $9j$ symbol:

$$
\begin{aligned}
\left\{ \begin{array}{ccc} j_1 & j_2 & j_{12} \\ j_3 & j_4 & j_{34} \\ j_{13} & j_{24} & j \end{array} \right\} &= \sum_{\substack{m_1 m_2 m_3 m_4 \\ m_{12} m_{34} \\ m_{13} m_{24} m}} \begin{pmatrix} j_1 & j_2 & j_{12} \\ m_1 & m_2 & m_{12} \end{pmatrix} \begin{pmatrix} j_3 & j_4 & j_{34} \\ m_3 & m_4 & m_{34} \end{pmatrix} \\
&\times \begin{pmatrix} j_{13} & j_{24} & j \\ m_{13} & m_{24} & m \end{pmatrix} \begin{pmatrix} j_1 & j_3 & j_{13} \\ m_1 & m_3 & m_{13} \end{pmatrix} \begin{pmatrix} j_2 & j_4 & j_{24} \\ m_2 & m_4 & m_{24} \end{pmatrix} \begin{pmatrix} j_{12} & j_{34} & j \\ m_{12} & m_{34} & m \end{pmatrix} .
\end{aligned}
\qquad (1.85)
$$

This can be used for simplification purposes as an intermediate step in applications of angular momentum algebra to physical problems. As a special case of (1.84) we can derive the following relation between the $9j$ and $6j$ symbols:

$$\begin{Bmatrix} j_1 & j_2 & j_{12} \\ j_3 & j_4 & j_{34} \\ j_{13} & j_{24} & 0 \end{Bmatrix} = \delta_{j_{12}j_{34}}\delta_{j_{13}j_{24}}\widehat{j_{12}}^{-1}\widehat{j_{13}}^{-1}(-1)^{j_{12}+j_{13}+j_2+j_3}\begin{Bmatrix} j_1 & j_2 & j_{12} \\ j_4 & j_3 & j_{13} \end{Bmatrix}.$$

$$(1.86)$$

### 1.4.1 Symmetry Properties of the 9j Symbol

As can be seen directly from the $3j$ coefficients in (1.85), there are six triangular conditions to be satisfied by the nine angular momentum quantum numbers inside the $9j$ symbol. These coupling rules are very symmetrically arranged inside the $9j$ symbol. The angular momenta $j_1$ and $j_2$ are coupled to $j_{12}$, and $j_3$ and $j_4$ to $j_{34}$, in our first set of basis states. These couplings are symbolized in the first two rows of the $9j$ symbol. In our second set of basis states, the angular momenta $j_1$ and $j_3$ are coupled to $j_{13}$, and $j_2$ and $j_4$ to $j_{24}$. These couplings are symbolized in the first two columns of the $9j$ symbol. The third column and third row of the $9j$ symbol relate to the final couplings of the intermediate angular momenta $j_{12}$ and $j_{34}$, and $j_{13}$ and $j_{24}$, to the total angular momentum $j$. This means that the value of a $9j$ symbol is zero unless the triangular rule is valid for *all* of its columns and rows, i.e.

$$\begin{Bmatrix} j_1 & j_2 & j_{12} \\ j_3 & j_4 & j_{34} \\ j_{13} & j_{24} & j \end{Bmatrix} = 0 \quad \text{unless} \quad \begin{cases} \Delta(j_1j_2j_{12}),\ \Delta(j_3j_4j_{34}),\ \Delta(j_{13}j_{24}j), \\ \Delta(j_1j_3j_{13}),\ \Delta(j_2j_4j_{24}),\ \Delta(j_{12}j_{34}j). \end{cases}$$

$$(1.87)$$

It follows from the definition of the $9j$ symbol that its value does not change when its columns and rows are interchanged, i.e.

$$\begin{Bmatrix} j_1 & j_2 & j_{12} \\ j_3 & j_4 & j_{34} \\ j_{13} & j_{24} & j \end{Bmatrix} = \begin{Bmatrix} j_1 & j_3 & j_{13} \\ j_2 & j_4 & j_{24} \\ j_{12} & j_{34} & j \end{Bmatrix}. \qquad (1.88)$$

Furthermore, the exchange of any two columns or any two rows changes the symbol only by a phase factor,

$$\begin{Bmatrix} j_2 & j_1 & j_{12} \\ j_4 & j_3 & j_{34} \\ j_{24} & j_{13} & j \end{Bmatrix} = (-1)^{\Sigma}\begin{Bmatrix} j_1 & j_2 & j_{12} \\ j_3 & j_4 & j_{34} \\ j_{13} & j_{24} & j \end{Bmatrix} = (-1)^{\Sigma}\begin{Bmatrix} j_4 & j_3 & j_{34} \\ j_2 & j_1 & j_{12} \\ j_{24} & j_{13} & j \end{Bmatrix} = \cdots,$$

$$(1.89)$$

with

$$\Sigma \equiv j_1 + j_2 + j_{12} + j_3 + j_4 + j_{34} + j_{13} + j_{24} + j \qquad (1.90)$$

being the sum of all the angular momentum quantum numbers inside the $9j$ symbol.

### 1.4.2 Relations for $9j$ Symbols

Below are listed some useful results on $9j$ symbols. For more relations, see e.g. [2,7,22]. These relations are useful in simplifying results of angular momentum algebra and in obtaining numerical values for $9j$ symbols. Numerical values can also be found by using recursion relations of the $9j$ symbols, extensively listed in [7]. Our list is as follows:

$$\sum_j \hat{j}^2 \begin{Bmatrix} j_1 & j_2 & J \\ j_3 & j_4 & J' \\ J & J' & j \end{Bmatrix} = \frac{\delta_{j_2 j_3}}{2j_2 + 1} , \qquad (1.91)$$

$$\sum_{j_{13} j_{24}} \widehat{j_{13}}^2 \, \widehat{j_{24}}^2 \begin{Bmatrix} j_1 & j_2 & j_{12} \\ j_3 & j_4 & j_{34} \\ j_{13} & j_{24} & j \end{Bmatrix} \begin{Bmatrix} j_1 & j_2 & j'_{12} \\ j_3 & j_4 & j'_{34} \\ j_{13} & j_{24} & j \end{Bmatrix} = \frac{\delta_{j_{12} j'_{12}} \delta_{j_{34} j'_{34}}}{\widehat{j_{12}}^2 \, \widehat{j_{34}}^2} , \qquad (1.92)$$

$$\sum_{j_{13} j_{24}} (-1)^{j_2 + j_4 + j_{24}} \, \widehat{j_{13}}^2 \, \widehat{j_{24}}^2 \begin{Bmatrix} j_1 & j_2 & j_{12} \\ j_3 & j_4 & j_{34} \\ j_{13} & j_{24} & j \end{Bmatrix} \begin{Bmatrix} j_1 & j_4 & j_{14} \\ j_3 & j_2 & j_{23} \\ j_{13} & j_{24} & j \end{Bmatrix}$$

$$= (-1)^{j_2 - j_4 - j_{34} + j_{23}} \begin{Bmatrix} j_1 & j_4 & j_{14} \\ j_2 & j_3 & j_{23} \\ j_{12} & j_{34} & j \end{Bmatrix} , \qquad (1.93)$$

$$\begin{Bmatrix} j_1 & j_2 & j_{12} \\ j_3 & j_4 & j_{34} \\ j_{13} & j_{24} & j \end{Bmatrix} = (-1)^{2j_1 + j} \sum_{j'} (2j' + 1) \begin{Bmatrix} j_1 & j_2 & j_{12} \\ j_{34} & j & j' \end{Bmatrix}$$

$$\times \begin{Bmatrix} j_3 & j_4 & j_{34} \\ j_2 & j' & j_{24} \end{Bmatrix} \begin{Bmatrix} j_{13} & j_{24} & j \\ j' & j_1 & j_3 \end{Bmatrix} , \qquad (1.94)$$

$$\begin{Bmatrix} j_1 & j_2 & j \\ j_3 & j_4 & j \\ j' & j' & 1 \end{Bmatrix} = \frac{(-1)^{j_2 + j_3 + j + j'}}{\hat{j} \hat{j}'}$$

$$\times \frac{j_1(j_1 + 1) - j_2(j_2 + 1) - j_3(j_3 + 1) + j_4(j_4 + 1)}{2\sqrt{j(j + 1)j'(j' + 1)}} \begin{Bmatrix} j_1 & j_2 & j \\ j_4 & j_3 & j' \end{Bmatrix} , \qquad (1.95)$$

$$\begin{Bmatrix} j_1 & j_2 & j - \frac{1}{2} \\ j_3 & j_4 & j + \frac{1}{2} \\ \frac{1}{2} & \frac{1}{2} & 1 \end{Bmatrix} = -\frac{\hat{j}}{\sqrt{3}} \begin{Bmatrix} j_1 & j & j_4 \\ \frac{1}{2} & j_2 & j - \frac{1}{2} \end{Bmatrix} \begin{Bmatrix} j_4 & j & j_1 \\ \frac{1}{2} & j_3 & j + \frac{1}{2} \end{Bmatrix} , \qquad (1.96)$$

$$\hat{l} \hat{l}' \begin{pmatrix} l & l' & L \\ 0 & 0 & 0 \end{pmatrix} \begin{Bmatrix} l & l' & L \\ j & j' & L \\ \frac{1}{2} & \frac{1}{2} & J \end{Bmatrix} = \frac{1 + (-1)^{l+l'+L}}{2} \frac{(\pm 1)^{l'+j'+\frac{1}{2}}}{\sqrt{2\hat{L}\hat{J}}}$$

$$\times \begin{pmatrix} j & j' & L \\ -\frac{1}{2} & \frac{1}{2} & -J \end{pmatrix} \qquad \text{for } J = \tfrac{1}{2} \pm \tfrac{1}{2} \quad (1.97)$$

$$\hat{l}\,\hat{l}'\begin{pmatrix} l & l' & L \\ 0 & 0 & 0 \end{pmatrix}\begin{Bmatrix} l & l' & L \\ j & j' & L' \\ \tfrac{1}{2} & \tfrac{1}{2} & 1 \end{Bmatrix} = \frac{(-1)^L}{\sqrt{6}}\,\frac{1+(-1)^{l+l'+L}}{2}\begin{pmatrix} j & j' & L' \\ \tfrac{1}{2} & -\tfrac{1}{2} & 0 \end{pmatrix}$$

$$\times\left[\frac{\hat{j}^2+(-1)^{j+j'+L'}\,\hat{j}'^2}{\sqrt{2L'(L'+1)}}\begin{pmatrix} L' & 1 & L \\ 1 & -1 & 0 \end{pmatrix}+(-1)^{l+j+\tfrac{1}{2}}\begin{pmatrix} L' & 1 & L \\ 0 & 0 & 0 \end{pmatrix}\right].\quad (1.98)$$

## Epilogue

With this selection of useful formulas, compiled for the convenience of the reader, we can end this chapter. All the tools needed for angular momentum coupling have been exposed and discussed for future applications. In the next chapter we take up the important machinery of spherical tensors and their reduced matrix elements. The formalism makes use of many of the formulas presented and in this way justifies the lengthy listings of relations with complicated appearance.

## Exercises

**1.1.** Show that the vector sum $\boldsymbol{J} = \boldsymbol{J}_1 + \boldsymbol{J}_2$ of two angular momentum vectors, $\boldsymbol{J}_1$ and $\boldsymbol{J}_2$, is also an angular momentum vector.

**1.2.** The structure of the nucleus $^6$Li is roughly determined by two nucleons, one proton and one neutron, in the $0p_{3/2}$ orbital.

(a) What are the possible parities and angular momenta of the states of $^6$Li on this simple scheme?
(b) The spin and parity of the ground state of $^6$Li are measured to be $1^+$. Write down explicitly the ground-state wave function for all three possible magnetic substates by using numerical values of the Clebsch–Gordan coefficients.

**1.3.** Take $j = 1$ and write down the matrix representation of the operators $\boldsymbol{J}^2$, $J_\pm$, $J_z$, $J_x$ and $J_y$.

**1.4.** The Hamiltonian of a *rigid rotor* can be written in a principal-axis coordinate system as

$$H = \frac{J_1^2}{2\mathcal{I}_1} + \frac{J_2^2}{2\mathcal{I}_2} + \frac{J_3^2}{2\mathcal{I}_3}, \qquad (1.99)$$

where $\mathcal{I}_1$, $\mathcal{I}_2$ and $\mathcal{I}_3$ are the moments of inertia with respect to the principal axes 1, 2 and 3. Determine the eigenenergies of this Hamiltonian in the case of $j = 1$ by using the matrix representation of Exercise 1.3.

**1.5.** In the case that $\mathcal{I}_1 = \mathcal{I}_2$ the eigenenergies of the Hamiltonian (1.99) are easy to determine without use of a matrix representation. Determine these eigenenergies for arbitrary $j$ and compare with the result of Exercise 1.4 in the case $j = 1$.

**1.6.** Show that the coupled and uncoupled bases have the same dimension.

**1.7.** Two angular momenta, $\boldsymbol{J}_1$ ($j_1 = 1$) and $\boldsymbol{J}_2$ ($j_2 = \frac{1}{2}$), are coupled to total angular momentum $\boldsymbol{J}$. By starting from the 'stretched' state

$$|j = \tfrac{3}{2}, m = \tfrac{3}{2}\rangle = |j_1 = 1, m_1 = 1\rangle |j_2 = \tfrac{1}{2}, m_2 = \tfrac{1}{2}\rangle \qquad (1.100)$$

find the Clebsch–Gordan coefficients for the state $|j = \tfrac{3}{2}, m = \tfrac{1}{2}\rangle$. Check your result by consulting the tables of Rotenberg et al. [20] or the website `http://www.sct.gu.edu.au/research/laserP/java/ClebschGordan/`.

**1.8.** Derive the orthogonality and completeness relations (1.26) and (1.28).

**1.9.** Derive the completeness and orthogonality properties of $3j$ symbols by starting from the corresponding ones for the Clebsch–Gordan coefficients.

**1.10.** Derive the relations (1.36) and (1.37) by exploiting the relation between the Clebsch–Gordan coefficient and the $3j$ symbol.

**1.11.** By operating with the ladder operators (1.7) on the definition (1.21) of a Clebsch–Gordan coefficient, derive the following recursion relation

$$\sqrt{j(j+1) - m(m \pm 1)}(j_1\, m_1\, j_2\, m_2 | j\, m \pm 1)$$
$$= \sqrt{j_1(j_1 + 1) - m_1(m_1 \mp 1)}(j_1\, m_1 \mp 1\, j_2\, m_2 | j\, m)$$
$$+ \sqrt{j_2(j_2 + 1) - m_2(m_2 \mp 1)}(j_1\, m_1\, j_2\, m_2 \mp 1 | j\, m) . \qquad (1.101)$$

**1.12.** Derive (1.34) by starting from (1.33).

**1.13.** Verify that the state

$$|\Psi\rangle = \sum_{m_1 m_2 m_3} \begin{pmatrix} j_1 & j_2 & j_3 \\ m_1 & m_2 & m_3 \end{pmatrix} |j_1\, m_1\rangle |j_2\, m_2\rangle |j_3\, m_3\rangle \qquad (1.102)$$

has total angular momentum zero.

**1.14.** Derive the following property of the $3j$ symbols:

$$\sum_{m_1 m_2 m_3} \begin{pmatrix} j_1 & j_2 & j_3 \\ m_1 & m_2 & m_3 \end{pmatrix}^2 = \Delta(j_1 j_2 j_3) . \qquad (1.103)$$

**1.15.** Perform the coupling of the angular momenta $\boldsymbol{J}_1$ ($j_1 = 1$), $\boldsymbol{J}_2$ ($j_2 = \frac{1}{2}$) and $\boldsymbol{J}_3$ ($j_3 = \frac{1}{2}$) in the following two ways:
(a) $\boldsymbol{J}_{12} = \boldsymbol{J}_1 + \boldsymbol{J}_2$ and $\boldsymbol{J} = \boldsymbol{J}_{12} + \boldsymbol{J}_3$,
(b) $\boldsymbol{J}_{23} = \boldsymbol{J}_2 + \boldsymbol{J}_3$ and $\boldsymbol{J} = \boldsymbol{J}_1 + \boldsymbol{J}_{23}$.
Write down the corresponding state vectors for $j = 1, m = 1$, the quantum number $j$ corresponding to the vector $\boldsymbol{J}$. Compare the two sets of state vectors and comment on the result.

**1.16.** Exploit the wave functions derived in Exercise 1.15 and determine the values of the $6j$ symbols

$$\begin{Bmatrix} 1 & \frac{1}{2} & \frac{1}{2} \\ \frac{1}{2} & 1 & 1 \end{Bmatrix} \quad \text{and} \quad \begin{Bmatrix} 1 & \frac{1}{2} & \frac{3}{2} \\ \frac{1}{2} & 1 & 1 \end{Bmatrix}$$

by using the definition (1.58) of the $6j$ symbol.

**1.17.** Derive (1.59) by starting from the definition (1.58) of the $6j$ symbol.

**1.18.** Derive (1.64) by starting from (1.59).

**1.19.** Derive (1.85) by starting from the definition (1.84) of the $9j$ symbol.

**1.20.** Show that the two sets $\{\boldsymbol{J}_1^2, J_{1z}, \boldsymbol{J}_2^2, J_{2z}\}$ and $\{\boldsymbol{J}_1^2, \boldsymbol{J}_2^2, \boldsymbol{J}^2, J_z\}$ both are sets of commuting operators.

**1.21.** The low-energy structure of $^{28}$F can be thought of as resulting from a valence proton occupying the $0d_{5/2}$ or $1s_{2/2}$ orbital, and a valence neutron hole occupying the $0d_{3/2}$ orbital. What are the possible angular momenta and parities emerging for the states of $^{28}$F on this scheme? Write down the corresponding wave functions.

**1.22.** Consider two nucleons in a p orbital (orbital angular momentum $l = 1$). The coupling to a given total angular momentum can be done by first coupling the spins and then the orbital angular momenta to form $ss$ and $ll$ couplings to total spin $S$ and total orbital angular momentum $L$. As a final step, $L$ and $S$ can be coupled to a total angular momentum $J$. This procedure is called the $LS$ coupling scheme. On the other hand, one can perform for each of the nucleons the coupling of the spin and orbital angular momenta, $ls$, to yield a total single-particle angular momentum $j$. The two single-particle angular momenta are then coupled to total angular momentum $J$. This way of coupling is called the $jj$ coupling scheme. The transformation mediating the connection between the states of these two coupling schemes is carried by $9j$ symbols. One mediator of this transformation for two nucleons in the p orbital and coupled to total angular momentum $J = 1$ is

$$\begin{Bmatrix} 1 & \frac{1}{2} & \frac{3}{2} \\ 1 & \frac{1}{2} & \frac{3}{2} \\ 2 & 1 & 1 \end{Bmatrix}.$$

Find the value of this $9j$ symbol by applying the appropriate relations listed at the end of Sect. 1.2.

# Tensor Operators and the Wigner–Eckart Theorem

## Prologue

In this chapter we pave the way to the use of the coupling methods of Chap. 1 for manipulating operators and their matrix elements. To enable smooth application of the angular momentum methods, we introduce so-called spherical tensor operators. Spherical tensors can be related to Cartesian tensors. A Cartesian tensor of a given Cartesian rank can be reduced to spherical tensors of several spherical ranks. There is a very convenient procedure, the so-called Wigner–Eckart theorem, to separate the part containing the projection quantum numbers from the rest of the matrix element of a spherical tensor operator. The remaining piece, called the reduced matrix element, is rotationally invariant and contains the physics of the matrix element.

## 2.1 Spherical Tensor Operators

Spherical tensor operators are introduced in this section. In preparation to defining them we give a short review of rotations in coordinate space. These rotations, represented by a rotation operator $R$ or its matrix R, induce unitary transformations (complex rotations) in the Hilbert space of angular momentum eigenstates. These eigenstates can be used to write down the matrix representation, consisting of so-called Wigner $D$ functions, of the unitary rotations. The $D$ functions can, in turn, be used to define the spherical tensor operators.

### 2.1.1 Rotations of the Coordinate Axes

The treatment of coordinate rotations and their representation in an abstract Hilbert space is far from exhaustive in the present work. Also, extensive listings of the properties of $D$ functions have been omitted. Details can be found in the standard literature on quantum mechanics and angular momentum algebra, e.g. [1–7]. Explicit properties of $D$ functions are not needed in this book.

We begin by considering a rotation of the coordinate axes in three-dimensional space. The position vector of a point P can be expressed in a Cartesian coordinate system as

$$\boldsymbol{r} = \sum_{i=1}^{3} x^i \hat{\boldsymbol{e}}_i \, , \tag{2.1}$$

where the $x^i$ are the *contravariant*[1] coordinates of the point ($x^1 \equiv x$, $x^2 \equiv y$, $x^3 \equiv z$) with an orthogonal system of *covariant* unit basis vectors $\{\hat{\boldsymbol{e}}_i\} \equiv \{\hat{\boldsymbol{i}}, \hat{\boldsymbol{j}}, \hat{\boldsymbol{k}}\}$. A change of basis $\{\hat{\boldsymbol{e}}_i\} \rightarrow \{\hat{\boldsymbol{e}}'_i\}$ yields new coordinates $x'^i$ of the point P such that

$$\boldsymbol{r}' = \sum_{i=1}^{3} x'^i \hat{\boldsymbol{e}}'_i \, , \tag{2.2}$$

where $\boldsymbol{r}'$ describes the new *components* of the position vector. Regarding the whole vector we have $\boldsymbol{r} = \boldsymbol{r}'$; a vector as a whole is independent of any coordinate system used to describe its components. The notation may be misleading, but it is common in the literature. In this change of the coordinate system the new coordinates become functions of the old coordinates,

$$x'^i = x'^i(\{x^j\}) \, . \tag{2.3}$$

By definition, the components of any *vector* $\boldsymbol{V}$ transform in the same way as the components of the position vector $\boldsymbol{r}$ under a rotation of the coordinate axes,

$$\boldsymbol{V} = \sum_{i=1}^{3} V^i \hat{\boldsymbol{e}}_i = \sum_{i=1}^{3} V'^i \hat{\boldsymbol{e}}'_i = \boldsymbol{V}' \, . \tag{2.4}$$

Thus, on the basis of (2.3), we have

$$V'^i = \sum_{j=1}^{3} \frac{\partial x'^i}{\partial x^j} V^j \equiv \sum_{j=1}^{3} R^i_{\ j} V^j \, . \tag{2.5}$$

The transformation coefficients $R^i_{\ j}$ are the elements of a matrix R, the *rotation matrix*.

In a similar way we have for the nine contravariant components $T^{ij}$ of a *second-rank tensor* $\boldsymbol{T}$ the transformation

---

[1] These are the usual components of a vector in a Cartesian coordinate system. The reader need not be frightened by the distinction made between contravariant and covariant components. In any case, the Cartesian components are discussed only as a means of introducing the spherical ones.

$$T'^{ij} = \sum_{k,l=1}^{3} \frac{\partial x'^i}{\partial x^k} \frac{\partial x'^j}{\partial x^l} T^{kl} = \sum_{k,l=1}^{3} R^i_k R^j_l T^{kl} . \qquad (2.6)$$

Written as a matrix equation this becomes $T' = \mathsf{R}T\mathsf{R}^{-1}$, a form known as a *similarity transformation*.

Using the rotation matrix $\mathsf{R}$ we state (2.5) as $V' = \mathsf{R}V$. For the position vector $r$ we then have $r' = \mathsf{R}r$, or in detail

$$\begin{pmatrix} x'^1 \\ x'^2 \\ x'^3 \end{pmatrix} = \begin{pmatrix} R^1_1 & R^1_2 & R^1_3 \\ R^2_1 & R^2_2 & R^2_3 \\ R^3_1 & R^3_2 & R^3_3 \end{pmatrix} \begin{pmatrix} x^1 \\ x^2 \\ x^3 \end{pmatrix} . \qquad (2.7)$$

The rotation matrix $\mathsf{R}$ is an *orthogonal* matrix, reflecting the fact that a vector's length is preserved in rotations. The matrix elements $R^i_j$ are commonly parametrized by the *Euler angles* $(\alpha, \beta, \gamma)$,

$$R^i_j = R^i_j(\alpha, \beta, \gamma) . \qquad (2.8)$$

### 2.1.2 Wigner $D$ Functions and Spherical Tensors

As mentioned at the beginning of this section, a rotation $\mathsf{R}$ in a three-dimensional Cartesian space induces a *unitary transformation* $\mathsf{U}$ of angular momentum eigenstates $|j\, m\rangle$ in an abstract Hilbert space $\mathcal{H}$ of dimension $2j + 1$. Hence there exists a correspondence $\mathsf{R} \leftrightarrow \mathsf{U}$ (for a detailed discussion see e.g. [6] The transformation $\mathsf{U}$ is a complex rotation in $\mathcal{H}$. The matrix elements of the $(2j + 1) \times (2j + 1)$ dimensional matrix representation of $\mathsf{U}$ are the so-called *Wigner $D$ functions* $D^j_{m'm}$:

$$\langle j\, m' | U\big(R(\alpha, \beta, \gamma)\big) | j\, m\rangle \equiv D^j_{m'm}(\alpha, \beta, \gamma) , \qquad (2.9)$$

where $U$ is the unitary rotation operator and $R$ with its detailed parameters $\alpha, \beta, \gamma$ signifies the rotation of the coordinate axes. These functions were introduced by Wigner in [24]. As their special case one obtains for $m = 0$ the spherical harmonic $Y^*_{jm'}(\beta, \alpha)$ multiplied by a constant. A further special case is $D^l_{00}(\alpha, \beta, \gamma) = P_l(\cos \beta)$, where $P_l$ is the usual Legendre polynomial. The rotational nature of the $D$ functions is apparent in the transformation formula

$$\boxed{U(R)|j\, m\rangle = \sum_{m'=-j}^{j} D^j_{m'm}(R)|j\, m'\rangle ,} \qquad (2.10)$$

which is equivalent to (2.9) and shows how an angular momentum eigenstate $|j\, m\rangle$ transforms under the complex rotation generated by the operator $U$.

In quantum mechanics, transformations of state vectors also generate compatible transformations of the operators (observables). If $U$ transforms the state vectors, i.e. $|\Psi'\rangle = U|\Psi\rangle$, then the operators $O$ are transformed as $O' = UOU^\dagger$. A special class of these operators consists of *spherical tensor operators* $\boldsymbol{T}_L$ with $2L + 1$ components (projections on the $z$-axis) $T_{L,-L}, T_{L,-L+1}, \ldots, T_{L,L-1}, T_{LL}$ which transform as

$$U(R)T_{LM}U^\dagger(R) = \sum_{M'=-L}^{L} T_{LM'} D_{M'M}^L(R) \,. \qquad (2.11)$$

Here $L$ is said to be the *rank* of the spherical tensor $\boldsymbol{T}_L$.

We note the following properties of spherical tensor operators.

- The spherical tensor operators of the few lowest ranks are named such that the names reflect their physical correspondence and meaning. The most frequently encountered cases are

$$L = 0 : scalar \ (monopole) \ \text{operator}$$
$$L = 1 : vector \ (dipole) \ \text{operator}$$
$$L = 2 : quadrupole \ (moment) \ \text{operator}$$
$$L = 3 : octupole \ (moment) \ \text{operator}$$

The various 'poles' can refer e.g. to the multipole expansion of the electromagnetic field, the nuclear two-body interaction or the nuclear shape.

- It can be shown that the definition (2.11) of a spherical tensor is equivalent to the commutation relations

$$[J_z, T_{JM}] = M\hbar T_{JM} \,,$$
$$[J_\pm, T_{JM}] = \hbar\sqrt{(J \pm M + 1)(J \mp M)}\, T_{J,M\pm 1} \,. \qquad (2.12)$$

Here $J_\pm = J_x \pm \mathrm{i}J_y$ are the ladder operators of the angular momentum $\boldsymbol{J}$. Their commutation relations are

$$[J_+, J_-] = 2\hbar J_z \,, \quad [J_z, J_\pm] = \pm\hbar J_\pm \,. \qquad (2.13)$$

The commutation relations (2.12) can be used to test whether a set of objects form a spherical tensor. An example of such use is provided by the creation and annihilation operators in Chap. 4. Other examples are given as exercises at the end of this chapter.

- For the maximal and minimal values of the projection quantum number we have the natural rule $[J_\pm, T_{J,\pm J}] = 0$.

### 2.1.3 Contravariant and Covariant Components of Spherical Tensors

Similarly to Cartesian vectors and tensors, one can define the contravariant and covariant components of a spherical tensor. Again, one need not worry about this terminology, since the angular momentum algebra of the rest of this book, and most of the literature, is based on the covariant components, but without explicit reference to their covariant nature. However, for completeness, we introduce in this subsection both types of component and discuss their basic properties. A detailed exposition of the topic is given in [7].

We define the *contravariant spherical components* of a vector $\boldsymbol{V}$ as

$$V^{\pm 1} = \mp \frac{1}{\sqrt{2}} (V_x \mp iV_y) , \quad V^0 = V_z , \tag{2.14}$$

and the *covariant spherical components* as

$$V_{\pm 1} = \mp \frac{1}{\sqrt{2}} (V_x \pm iV_y) , \quad V_0 = V_z , \tag{2.15}$$

where $V_x$, $V_y$ and $V_z$ are the usual (contravariant) Cartesian components of the vector. The covariant components (2.15) are the ones which are commonly defined as spherical components in the literature. These are the spherical components used also in this work. In particular, for the position vector $\boldsymbol{r}$ we have

$$r_{\pm 1} = \mp \frac{1}{\sqrt{2}} (x \pm iy) , \quad r_0 = z , \tag{2.16}$$

where $x$, $y$ and $z$ are the usual Cartesian coordinates. In fact, one can easily relate $r_\mu$ to the spherical harmonic $Y_{1\mu}$ as

$$r_\mu = \sqrt{\frac{4\pi}{3}} \, rY_{1\mu}(\theta, \varphi) , \quad \mu = 0, \pm 1 , \quad r = |\boldsymbol{r}| . \tag{2.17}$$

This shows that the spherical harmonic $\boldsymbol{Y}_1$ acts like a vector. More generally, the spherical harmonic $\boldsymbol{Y}_l$ is a spherical tensor of rank $l$, with $2l + 1$ spherical components (see the exercises at the end of this chapter). The following note contains a summary.

- The spherical components $J_\mu$ ($\mu = 0, \pm 1$) of an angular momentum vector $\boldsymbol{J}$ form a spherical tensor of rank 1.

- The spherical components $r_\mu$ ($\mu = 0, \pm 1$) of the position vector $\boldsymbol{r}$ form a spherical tensor of rank 1.

- The spherical components $Y_{\lambda\mu}$ ($\mu = -\lambda, -\lambda+1, \ldots, \lambda-1, \lambda$) of a spherical harmonic $\boldsymbol{Y}_\lambda$ form a spherical tensor of rank $\lambda$.

As stated in the prologue to this chapter, a Cartesian tensor of a given Cartesian rank can generally be reduced to spherical tensors of several spherical ranks. A scalar and a vector have the same rank (0 and 1 respectively) as a spherical or a Cartesian tensor, but a Cartesian tensor of the second rank can be decomposed into a scalar (the trace of the Cartesian tensor), a vector (the antisymmetric part of the Cartesian tensor) and a quadrupole spherical tensor (the symmetric part of the Cartesian tensor). One can express this difference by stating that a Cartesian tensor is *reducible* to spherical tensors of several ranks, the spherical tensors being *irreducible*.

We complete our discussion of contravariant and covariant spherical components by looking at unit vectors and products of vectors.

The contravariant components of the *unit basis vectors* are defined as

$$\hat{e}^{\pm 1} = \mp \frac{1}{\sqrt{2}}(\hat{e}_x \mp i\hat{e}_y), \quad \hat{e}^0 = \hat{e}_z, \tag{2.18}$$

and the corresponding covariant components as

$$\boxed{\hat{e}_{\pm 1} = \mp \frac{1}{\sqrt{2}}(\hat{e}_x \pm i\hat{e}_y), \quad \hat{e}_0 = \hat{e}_z,} \tag{2.19}$$

where $\hat{e}_x = \hat{i}$, $\hat{e}_y = \hat{j}$ and $\hat{e}_z = \hat{k}$ are the usual unit basis vectors of a Cartesian coordinate system. Equations (2.18) and (2.19) give the orthogonality relations

$$\hat{e}_\mu \cdot \hat{e}^\nu = \hat{e}_\mu \cdot \hat{e}_\nu^* = \delta_{\mu\nu}. \tag{2.20}$$

Some further relations are

$$\hat{e}^\mu = \hat{e}_\mu^* = (-1)^\mu \hat{e}_{-\mu}, \quad \hat{e}_\mu = \hat{e}^{\mu *} = (-1)^\mu \hat{e}^{-\mu}. \tag{2.21}$$

A vector $\boldsymbol{V}$ can be expressed in these two bases as

$$\boldsymbol{V} = \sum_\mu V^\mu \hat{e}_\mu = \sum_\mu V_\mu \hat{e}^\mu, \quad V_\mu = \boldsymbol{V} \cdot \hat{e}_\mu, \quad V^\mu = \boldsymbol{V} \cdot \hat{e}^\mu. \tag{2.22}$$

The components of $\boldsymbol{V}$ satisfy the relations

$$V_\mu = (-1)^\mu V^{-\mu}, \quad V^\mu = (-1)^\mu V_{-\mu}, \quad V_\mu^* = (\boldsymbol{V}^*)^\mu, \quad V^{\mu *} = (\boldsymbol{V}^*)_\mu, \tag{2.23}$$

where the possibility of $\boldsymbol{V}$ being complex has been allowed.

The scalar, or dot, product of two vectors $\boldsymbol{A}$ and $\boldsymbol{B}$ can be written as

$$\boldsymbol{A} \cdot \boldsymbol{B} = \sum_\mu A^\mu B_\mu = \sum_\mu A_\mu B^\mu = \sum_\mu (-1)^\mu A_\mu B_{-\mu} = \sum_\mu (-1)^\mu A^\mu B^{-\mu}. \tag{2.24}$$

For the vector, or cross, product of these vectors we have

$$[\boldsymbol{A} \times \boldsymbol{B}]_\mu = -i\sqrt{2} \sum_{\nu\nu'} (1\,\nu\,1\,\nu'|1\,\mu) A_\nu B_{\nu'}, \tag{2.25}$$

$$[\boldsymbol{A} \times \boldsymbol{B}]^\mu = i\sqrt{2} \sum_{\nu\nu'} (1\,\nu\,1\,\nu'|1\,\mu) A^\nu B^{\nu'}. \tag{2.26}$$

## 2.2 The Wigner–Eckart Theorem

In this section we address the very important issue of simplifying the calculation of matrix elements of operators related to different observables of a nucleus. This simplification is achieved by expanding the relevant observables in terms of spherical tensors and then exploiting their special transformation properties. The notion of a spherical tensor is the key to developing an elegant and systematic way of treating complicated matrix elements of observables.

All this boils down to the possibility of defining a reduced matrix element that is independent of all projection quantum numbers, namely those of the initial and final nuclear states and that of the operator. This reduced matrix element contains all the physical information carried by the initial and final nuclear wave functions, as well as the physical content of the observable, represented by the operator sandwiched between these wave functions. In the following, we take the existence of such a reduced matrix element as a fact without presenting a proof of it. For proof see e.g. [2] or [10].

Consider a matrix element $\langle \xi' \, j' \, m' | T_{LM} | \xi \, j \, m \rangle$, where $T_{LM}$ is a spherical tensor of rank $L$, and the $|\xi \, j \, m\rangle$ are quantum states with good angular momentum $j$ and its $z$ projection $m$. The quantity $\xi$ contains all other quantum numbers needed to completely specify the quantum state. In the present context the states could be the initial and final nuclear states of a process, e.g. electromagnetic decay, governed by the observable $T_{LM}$. As asserted above, one can write the matrix element as a product of a geometric factor containing the projection quantum numbers $m'$, $m$ and $M$, and a matrix element which does not contain them,

$$
\begin{aligned}
\langle \xi' \, j' \, m' | T_{LM} | \xi \, j \, m \rangle &= \widehat{j'}^{\,-1} (j \, m \, L \, M | j' \, m')(\xi' \, j' \| \boldsymbol{T}_L \| \xi \, j) \\
&= (-1)^{j'-m'} \begin{pmatrix} j' & L & j \\ -m' & M & m \end{pmatrix} (\xi' \, j' \| \boldsymbol{T}_L \| \xi \, j) \,,
\end{aligned} \tag{2.27}
$$

where $(\xi' \, j' \| \boldsymbol{T}_L \| \xi \, j)$ is called the *reduced matrix element*. This is the *Wigner–Eckart theorem*, introduced by Wigner [24] and Eckart [25] in 1930–1931.

The Wigner–Eckart theorem is basically an existence theorem. It states that it is possible to extract a factor containing all $j$ quantum numbers but no $m$ quantum numbers. The precise definition of the reduced matrix element can vary, and different definitions and notations occur in the literature. However, the definition contained in (2.27) is the one of [2] and coincides with the original definition of Racah [26].

### 2.2.1 Immediate Consequences of the Wigner–Eckart Theorem

We can see directly from the Wigner–Eckart theorem (2.27) that

$$
\langle \xi' \, j' \, m' | T_{LM} | \xi \, j \, m \rangle = \frac{(j \, m \, L \, M | j' \, m')}{(j \, m_0 \, L \, M_0 | j' \, m'_0)} \langle \xi' \, j' \, m'_0 | T_{LM_0} | \xi \, j \, m_0 \rangle \,. \tag{2.28}
$$

Once we know the matrix element for one set of quantum numbers $m = m_0$, $M = M_0$ and $m' = m'_0$, this equation gives us the matrix element for any other set $m$, $M$ and $m'$. To evaluate the general matrix element on the left-hand side, we try to find and evaluate the simplest possible matrix element on the right-hand side. This is accomplished by choosing $M_0 = 0$ and $m'_0 = m_0$, and furthermore $m_0 = 0$ if the Clebsch–Gordan coefficient does not then vanish.

From the Clebsch–Gordan coefficient in (2.27) and the conditions (1.22) and (1.23) for its non-vanishing, we obtain the selection rules

$$\langle \xi' j' m' | T_{LM} | \xi j m \rangle = 0 \quad \text{unless} \quad \begin{cases} \Delta(jLj') , \\ m + M = m' . \end{cases} \tag{2.29}$$

Thus, for a non-zero matrix element of the observable $T_{LM}$, one should be able to couple the angular momentum $j$ of the initial state and the angular momentum $L$ of the observable to the angular momentum $j'$ of the final state. Furthermore, the projection quantum number $m$ of the initial state and $M$ of the observable should sum up to the projection quantum number $m'$ of the final state. So we see that angular momentum conservation is guaranteed by the Wigner–Eckart theorem.

We derive next a symmetry relation for the reduced matrix element $(\xi' j' \| T_L \| \xi j)$. Applying a general property of any matrix element, we can write the left-hand side of (2.27) as

$$\langle \xi' j' m' | T_{LM} | \xi j m \rangle = \langle \xi j m | (T_{LM})^\dagger | \xi' j' m' \rangle^* . \tag{2.30}$$

Now assume that the tensor operator $T_L$ has the property

$$(T_{LM})^\dagger = (-1)^M T_{L,-M} . \tag{2.31}$$

Such operators are called *Hermitian tensor operators*[2]; their $M = 0$ component satisfies the normal Hermiticity criterion: $(T_{L0})^\dagger = T_{L0}$. Applying the Wigner–Eckart theorem to both sides of (2.30) leads to the symmetry relation

$$(\xi j \| T_L \| \xi' j') = (-1)^{j-j'} (\xi' j' \| T_L \| \xi j)^* . \tag{2.32}$$

Below we list explicit expressions for the most frequently occurring reduced matrix elements. Verification of these values is left as an exercise for the reader (see the exercises at the end of this chapter). The operators in these basic matrix elements are the identity operator, which we denote simply by 1, the angular momentum operator $J$, the spherical harmonic $Y_L$ (viewed now as a spherical tensor operator) and, as a special case of angular momentum, the spin-$\frac{1}{2}$ operator $S$. The results are

---

[2] A Hermitian tensor operator is consistent with the Condon–Shortley (CS) phase convention introduced in Sect. 3.3. See also Exercise 11.50.

$$(j'\|\mathbf{1}\|j) = \delta_{jj'}\hat{j} \,, \tag{2.33}$$

$$(j'\|\mathbf{J}\|j) = \delta_{jj'}\hbar\sqrt{j(j+1)(2j+1)} \,, \tag{2.34}$$

$$(l'\|\mathbf{Y}_L\|l) = (-1)^{l'}\frac{\hat{l'}\,\hat{L}\,\hat{l}}{\sqrt{4\pi}}\begin{pmatrix} l' & L & l \\ 0 & 0 & 0 \end{pmatrix} \,, \tag{2.35}$$

$$(\tfrac{1}{2}\|\mathbf{S}\|\tfrac{1}{2}) = \sqrt{\tfrac{3}{2}}\,\hbar \,. \tag{2.36}$$

The result (2.35) is very important for the multipole decomposition of a physical field. It is used extensively in this work for the electromagnetic field in Chap. 6 and for the nuclear two-body interaction in Chap. 8. We make use of this formula also for the transition matrix elements of forbidden beta decays in Chap. 7.

### 2.2.2 Pauli Spin Matrices

Although we have not paid attention to explicit matrix representation of the angular momentum operators, we now note the matrices for spin $\frac{1}{2}$. These matrices are nothing but the matrices of general angular momentum $\mathbf{J}$ for the case $j = \frac{1}{2}$, derivable from (1.3), (1.5) and (1.7). However, as a matter of convenience and historical convention, these matrices are expressed as matrices of the operator $\boldsymbol{\sigma}$ such that the factor $\frac{1}{2}\hbar$ is removed:

$$\boxed{\mathbf{J} = \mathbf{S} = \tfrac{1}{2}\hbar\boldsymbol{\sigma} \,.} \tag{2.37}$$

The elements of the resulting 2-by-2 matrices are $\langle\frac{1}{2}m|\boldsymbol{\sigma}|\frac{1}{2}m'\rangle$, where $m, m' = \pm\frac{1}{2}$. Another historical convention is that the matrices are denoted by the same symbols as their operators. So the matrices of the operators $\sigma_x$, $\sigma_y$ and $\sigma_z$ are

$$\boxed{\sigma_x = \begin{pmatrix} 0 & 1 \\ 1 & 0 \end{pmatrix} \,, \quad \sigma_y = \begin{pmatrix} 0 & -\mathrm{i} \\ \mathrm{i} & 0 \end{pmatrix} \,, \quad \sigma_z = \begin{pmatrix} 1 & 0 \\ 0 & -1 \end{pmatrix} \,.} \tag{2.38}$$

These matrices are known as *Pauli spin matrices*. They were introduced by Pauli [27] in 1927. Equations (2.36) and (2.37) yield the reduced matrix element

$$\boxed{(\tfrac{1}{2}\|\boldsymbol{\sigma}\|\tfrac{1}{2}) = \sqrt{6} \,.} \tag{2.39}$$

The Pauli spin matrices are Hermitian, $\sigma_i^\dagger = \sigma_i$ $(i = x, y, z)$. Their commutation relations are

$$[\sigma_i, \sigma_j] = 2\mathrm{i}\epsilon_{ijk}\sigma_k \,. \tag{2.40}$$

Further relations are

$$\{\sigma_i, \sigma_j\} = 0 \text{ for all } i \neq j \,, \quad \sigma_i^2 = 1 \text{ for all } i \,, \tag{2.41}$$

where $\{A, B\} \equiv AB + BA$ is the *anticommutator* of objects $A$ and $B$.

The following properties of the Pauli spin matrices are sometimes useful:

$$\sigma(\sigma \cdot A) = A 1_2 - i\sigma \times A , \tag{2.42}$$

$$(\sigma \cdot A)(\sigma \cdot B) = (A \cdot B) 1_2 + i\sigma \cdot (A \times B) , \tag{2.43}$$

$$(\sigma \times A) \cdot (\sigma \times B) = 2(A \cdot B) 1_2 + i\sigma \cdot (A \times B) , \tag{2.44}$$

$$\sigma \times \sigma \cdot A = 2i\sigma \cdot A , \tag{2.45}$$

$$\sigma \times (\sigma \times A) = -2A 1_2 + i\sigma \times A , \tag{2.46}$$

where $1_2$ is the $2 \times 2$ unit matrix and $A$ and $B$ are arbitrary vectors.

One can also form the (covariant) spherical components of the Pauli matrices by using the definition (2.15). The commutation and anticommutation rules, (2.40) and (2.41), of the Cartesian Pauli matrices are then carried over to

$$[\sigma_\mu, \sigma_\nu] = -2\sqrt{2}(1\,\mu\,1\,\nu|1\,M)\sigma_M , \tag{2.47}$$

$$\{\sigma_\mu, \sigma_\nu\} = 2(-1)^\mu \delta_{\mu,-\nu} 1_2 . \tag{2.48}$$

Finally, the product of two Pauli matrices can be cast into the form

$$\sigma_\mu \sigma_\nu = (-1)^\mu \delta_{\mu,-\nu} 1_2 - 2\sqrt{2}(1\,\mu\,1\,\nu|1\,M)\sigma_M . \tag{2.49}$$

Equations (2.47)–(2.49), valid for the spherical components, are extremely useful in the simplification of complicated commutators and products of the Pauli matrices.


## 2.3 Matrix Elements of Coupled Tensor Operators

We now present useful theorems for products of two spherical tensors. In particular we state expressions for the reduced matrix elements of these products. Proofs are omitted here but can be worked out as exercises. The theorems are applied extensively in later chapters. We start by defining the tensor product of two spherical tensors.

Let $T_{L_1}$ and $T_{L_2}$ be two spherical tensor operators of rank $L_1$ and $L_2$ respectively. The spherical tensor $T_L$ is said to be their *tensor product* of rank $L$ if

$$T_{LM} = \sum_{M_1 M_2} (L_1\,M_1\,L_2\,M_2|L\,M)\,T_{L_1 M_1} T_{L_2 M_2} \equiv [T_{L_1} T_{L_2}]_{LM} . \tag{2.50}$$

There are two important special cases of this. The first is the *scalar* or *dot product* of two spherical tensors of the same rank,

$$T_L \cdot S_L \equiv (-1)^L \hat{L}\,[T_L S_L]_{00} = \sum_M (-1)^M T_{LM} S_{L,-M} . \tag{2.51}$$

For vectors, $L = 1$, this coincides with the familiar dot product, $\boldsymbol{T}_1 \cdot \boldsymbol{S}_1 = \boldsymbol{T} \cdot \boldsymbol{S}$. The second special case is the usual *vector* or *cross product*

$$(\boldsymbol{T} \times \boldsymbol{S})_M = -\mathrm{i}\sqrt{2}\,[\boldsymbol{T}_1 \boldsymbol{S}_1]_{1M} \; . \tag{2.52}$$

Having defined the tensor product of two tensor operators, (2.50), we are ready to state and discuss two important theorems on the matrix elements of coupled spherical tensors.

## 2.3.1 Theorem I

Let $\boldsymbol{J}_1$ and $\boldsymbol{J}_2$ be two *commuting* angular momentum operators and the set $\{|j_1 j_2 j m\rangle\}$ the corresponding coupled basis. Let $\boldsymbol{T}_{L_1}$ and $\boldsymbol{T}_{L_2}$ be two *commuting* spherical tensors such that $\boldsymbol{T}_{L_1}$ operates in the Hilbert space spanned by the basis states $\{|j_1 m_1\rangle\}$ and $\boldsymbol{T}_{L_2}$ operates in the Hilbert space spanned by the basis states $\{|j_2 m_2\rangle\}$. Let

$$T_{LM} = [\boldsymbol{T}_{L_1} \boldsymbol{T}_{L_2}]_{LM} \tag{2.53}$$

be the tensor product to be considered. Then the reduced matrix element of $\boldsymbol{T}_L$ is

$$(j_1 j_2 j \|\boldsymbol{T}_L\| j_1' j_2' j') = \widehat{j}\,\widehat{L}\,\widehat{j}' \left\{ \begin{array}{ccc} j_1 & j_2 & j \\ j_1' & j_2' & j' \\ L_1 & L_2 & L \end{array} \right\} (j_1 \|\boldsymbol{T}_{L_1}\| j_1')(j_2 \|\boldsymbol{T}_{L_2}\| j_2') \; . \tag{2.54}$$

Equation (2.54) makes it possible to calculate the reduced matrix element of a tensor product of two commuting operators from the reduced matrix elements of each member of the tensor product, evaluated in its own subspace. The two separate reduced matrix elements are combined by a $9j$ symbol to produce the desired final matrix element. In this way the original, usually rather complicated reduced matrix element can be expressed as a product of two, usually much simpler, matrix elements. The two commuting operators can refer e.g. to two different particles or to the spatial and spin variables of one particle.

For the scalar product (2.51) of two commuting spherical tensors we have the following special case of (2.54):

$$(j_1 j_2 j \|\boldsymbol{T}_L \cdot \boldsymbol{S}_L\| j_1' j_2' j') = \delta_{jj'}(-1)^{j_2+j+j_1'}\,\widehat{j} \left\{ \begin{array}{ccc} j_1 & j_2 & j \\ j_2' & j_1' & L \end{array} \right\}$$
$$\times (j_1 \|\boldsymbol{T}_L\| j_1')(j_2 \|\boldsymbol{S}_L\| j_2') \; . \tag{2.55}$$

Important basic results are obtained when Theorem I, (2.54), is applied to a single particle with spin. The spin–orbit-coupled wave functions are of the form $|n\,l\,\tfrac{1}{2}\,j\,m\rangle$. With the radial matrix element factored out, the quantum numbers appearing in the reduced matrix element for these states are $l$, $\tfrac{1}{2}$ and $j$.

The key to the following relations is to write the spin or spatial operator as a tensor product of that operator and the identity operator 1, of spherical rank $L = 0$, operating in the other space. Thus we write the Pauli spin tensor operator $\boldsymbol{\sigma}$, defined in (2.37) and of spherical tensor rank $L = 1$, as $\boldsymbol{\sigma} = [\mathbf{1}\boldsymbol{\sigma}]_1$. So we have a tensor product of two commuting spherical tensors, the identity 1 operating in coordinate space and $\boldsymbol{\sigma}$ operating in spin space. The resulting reduced matrix element is

$$(l\,\tfrac{1}{2}\,j\|\boldsymbol{\sigma}\|l'\,\tfrac{1}{2}\,j') = \sqrt{6}\,\delta_{ll'}\,\widehat{j}\,\widehat{j'}(-1)^{l+j+\frac{3}{2}}\left\{\begin{matrix}\tfrac{1}{2} & \tfrac{1}{2} & 1 \\ j' & j & l\end{matrix}\right\}, \qquad (2.56)$$

where the basic reduced matrix elements (2.33) and (2.39) have been inserted. The matrix element (2.56) is directly the one used as the single-particle matrix element of the Gamow–Teller beta-decay operator in Sect. 7.2.

For the spherical tensor $\boldsymbol{Y}_\lambda = [\boldsymbol{Y}_\lambda \mathbf{1}]_\lambda$ we find similarly

$$(l\,\tfrac{1}{2}\,j\|\boldsymbol{Y}_\lambda\|l'\,\tfrac{1}{2}\,j') = \frac{1}{\sqrt{4\pi}}(-1)^{j'-\frac{1}{2}+\lambda}\,\frac{1+(-1)^{l+l'+\lambda}}{2}\,\widehat{j}\,\widehat{j'}\,\widehat{\lambda}\begin{pmatrix}j & j' & \lambda \\ \tfrac{1}{2} & -\tfrac{1}{2} & 0\end{pmatrix}, \qquad (2.57)$$

where (2.33) and (2.35) have been inserted. This particular matrix element appears in the single-particle matrix elements of the electromagnetic decay operators in Sect. 6.1 and in the interaction matrix elements of a separable two-body nucleon–nucleon force in Sect. 8.2.

Along the same lines, one obtains for the *orbital* angular momentum operator $\boldsymbol{L}$ the result

$$(l\,\tfrac{1}{2}\,j\|\boldsymbol{L}\|l'\,\tfrac{1}{2}\,j') = \delta_{ll'}\,\widehat{j}\,\widehat{j'}\,\sqrt{l(l+1)(2l+1)}(-1)^{l+j'+\frac{3}{2}}\left\{\begin{matrix}l & l & 1 \\ j' & j & \tfrac{1}{2}\end{matrix}\right\}\hbar. \qquad (2.58)$$

As special cases of (2.56) and (2.58) we have the diagonal matrix elements

$$(l\,\tfrac{1}{2}\,j\|\boldsymbol{\sigma}\|l\,\tfrac{1}{2}\,j) = \sqrt{\frac{2j+1}{j(j+1)}}\left[j(j+1) - l(l+1) + \tfrac{3}{4}\right], \qquad (2.59)$$

$$(l\,\tfrac{1}{2}\,j\|\boldsymbol{L}\|l\,\tfrac{1}{2}\,j) = \sqrt{\frac{2j+1}{4j(j+1)}}\left[j(j+1) + l(l+1) - \tfrac{3}{4}\right]\hbar, \qquad (2.60)$$

These are used in the computation of the reduced matrix element of the magnetic dipole operator in Sect. 6.2.

## 2.3.2 Theorem II

Let $\boldsymbol{T}_{L_1}$ and $\boldsymbol{T}_{L_2}$ be two spherical tensors that both operate in a Hilbert space spanned by the basis states $\{|\alpha\,j\,m\rangle\}$, where $\alpha$ is a set of additional quantum numbers needed to fully specify the basis states. Let

$$T_{LM} = [T_{L_1}T_{L_2}]_{LM} \tag{2.61}$$

be the tensor product to be considered. Then the reduced matrix element of $T_L$ is

$$
\boxed{
\begin{aligned}
(\alpha\, j\|T_L\|\alpha'\, j') &= (-1)^{j+L+j'}\widehat{L} \sum_{\alpha''j''} \begin{Bmatrix} L_1 & L_2 & L \\ j' & j & j'' \end{Bmatrix} \\
&\times (\alpha\, j\|T_{L_1}\|\alpha''\, j'')(\alpha''\, j''\|T_{L_2}\|\alpha'\, j') \, .
\end{aligned}
}
\tag{2.62}
$$

In Theorem II, (2.62), the separation of the two parts of the reduced matrix element is not as 'clean' as in Theorem I, (2.54). This is due to the fact that the completeness relation $1 = \sum_{\alpha j}|\alpha\, j\rangle\langle\alpha\, j|$ has to be used to achieve the separation present in Theorem II.

As a special case of (2.62) we have for the scalar product of two spherical tensors operating in the same Hilbert space the expression

$$(\alpha\, j\|T_L \cdot S_L\|\alpha'\, j') = \delta_{jj'}\widehat{j}^{-1} \sum_{\alpha''j''} (-1)^{j''-j}(\alpha\, j\|T_L\|\alpha''\, j'')(\alpha''\, j''\|S_L\|\alpha'\, j') \, . \tag{2.63}$$

A straightforward application of this gives

$$(l\|L^2\|l') = \delta_{ll'}l(l+1)\widehat{l}\hbar^2 \, . \tag{2.64}$$

## Epilogue

This chapter was closed by introducing two important theorems on reduced matrix elements of coupled spherical tensor operators. The two key quantities of this chapter, namely the spherical tensors and their reduced matrix elements, defined by the Wigner–Eckart theorem, play an important role in subsequent physical applications. These tensor methods play a dominant role also in nuclear shell theory, as is extensively discussed in [10]. The life of a nuclear physicist would be much harder without spherical tensors and reduced matrix elements.

## Exercises

**2.1.** Starting from the explicit expression of the spherical harmonic $Y_{1m}(\theta,\varphi)$ (see, e.g. [6]), derive (2.17).

**2.2.** By using the commutation relations (2.12), show that the spherical components $J_\mu$, $\mu = 0, \pm 1$, of the angular momentum operator $J$ form a spherical tensor of rank 1.

**2.3.** By using the commutation relations (2.12), show that the spherical components $Y_{\lambda\mu}$, $\mu = -\lambda, -\lambda+1, \ldots, \lambda-1, \lambda$, of the spherical harmonic $\mathbf{Y}_\lambda$ form a spherical tensor of rank $\lambda$.

**2.4.** Verify the orthogonality relations (2.20).

**2.5.** Verify that the spherical components of the vector product of two vectors, $\mathbf{A}$ and $\mathbf{B}$, satisfy the relation (2.25).

**2.6.** By using the Wigner–Eckart theorem derive the *Landé formula*

$$\langle \xi\, j\, m | \mathbf{A} | \xi\, j\, m' \rangle = \frac{\langle \xi\, j\, m | \mathbf{A} \cdot \mathbf{J} | \xi\, j\, m \rangle}{j(j+1)} \langle \xi\, j\, m | \mathbf{J} | \xi\, j\, m' \rangle \,, \qquad (2.65)$$

where $\mathbf{A}$ is an arbitrary vector operator and $\mathbf{J}$ is the angular momentum operator.

**2.7.** Derive the symmetry relation (2.32).

**2.8.** Derive the reduced matrix elements (2.33) and (2.34) by using the Wigner–Eckart theorem.

**2.9.** Derive the reduced matrix element (2.35) by using the Wigner–Eckart theorem and the *Gaunt formula*

$$\int_0^{2\pi} d\varphi \int_0^\pi \sin\theta \, d\theta \, Y_{l_1 m_1}^*(\theta,\varphi) \, Y_{l_2 m_2}(\theta,\varphi) \, Y_{l_3 m_3}(\theta,\varphi)$$

$$= (-1)^{m_1} \frac{\widehat{l_1}\, \widehat{l_2}\, \widehat{l_3}}{\sqrt{4\pi}} \begin{pmatrix} l_1 & l_2 & l_3 \\ 0 & 0 & 0 \end{pmatrix} \begin{pmatrix} l_1 & l_2 & l_3 \\ -m_1 & m_2 & m_3 \end{pmatrix} \,. \qquad (2.66)$$

**2.10.** Evaluate the reduced matrix element $(n'\, l' \| \mathbf{r} \| n\, l)$.

**2.11.** Derive the Pauli spin matrices (2.38) by using general properties of angular momentum from Sect. 1.1.

**2.12.** Verify the commutation relations (2.40) and the anticommutation relations (2.41) of the Pauli spin matrices.

**2.13.** Verify as many as you can of the relations (2.42)–(2.46).

**2.14.** Verify the commutation relations (2.47) and the anticommutation relations (2.48) of the spherical components of the Pauli spin matrices.

**2.15.** Show that the scalar product of two vectors coincides with the scalar product of two rank-1 spherical tensors, as stated in (2.51).

**2.16.** Establish the relation (2.52) between the vector product of two vectors and the rank-1 product of two rank-1 spherical tensors.

**2.17.** Derive the formula (2.54).

**2.18.** Derive (2.55) from (2.54).

**2.19.** Derive the reduced matrix elements (2.56) and (2.58) from (2.54) and the reduced matrix elements (2.33)–(2.36).

**2.20.** Derive the expression (2.57). Hint: Feel free to use the relations for $6j$ and $9j$ symbols listed in Sects. 1.3 and 1.4.

**2.21.** Verify the formulas (2.59) and (2.60) by starting from (2.56) and (2.58) and using explicit expressions given for the $6j$ symbol in Sect. 1.3.

**2.22.** The nuclear magnetic dipole moment $\mu$ is defined as

$$\mu(\alpha, J) \equiv \langle \alpha\, J,\, M = J | \mu_z | \alpha\, J,\, M = J \rangle , \qquad (2.67)$$

where $J$ is the total angular momentum and $M$ its $z$ projection, and $\alpha$ denotes all the other quantum numbers identifying the state. The details of the vector operator $\boldsymbol{\mu}$ depend on the description of the nuclear system. Express $\mu(\alpha, J)$ as a product of an algebraic factor and the reduced matrix element of $\boldsymbol{\mu}$. What is the reason for the fact that $\mu$ can differ from zero only for $J \geq \frac{1}{2}$?

**2.23.** Evaluate the reduced matrix element $(n'\, l'\, \frac{1}{2}\, j' \| \boldsymbol{r} \| n\, l\, \frac{1}{2}\, j)$. The electric dipole operator can be written as $-e\boldsymbol{r}$. Can you find a reason why no single-particle state can have a non-zero electric dipole moment, defined analogously to (2.67)?

**2.24.** The quadrupole moment $Q$ of a nucleus is defined as

$$Q(\alpha, J) \equiv \sqrt{\frac{16\pi}{5}} \langle \alpha\, J,\, M = J | Q_{20} | \alpha\, J,\, M = J \rangle , \qquad (2.68)$$

where $J$ is the total angular momentum and $M$ its $z$ projection, and $\alpha$ denotes all the other quantum numbers identifying the state. Here $\boldsymbol{Q}_2$ is the electric quadrupole operator, a tensor operator of rank 2. Express $Q(\alpha, J)$ as a product of an algebraic factor and the reduced matrix element of $\boldsymbol{Q}_2$. Deduce the angular momentum values for which $Q$ is non-vanishing.

**2.25.** The excited states of atoms and nuclei usually de-excite via spontaneous photon emission (multipole radiation). The magnetic substates of the angular momentum are not measured and thus a convenient observable turns out to be the *reduced transition probability*

$$B(\sigma\lambda;\, J \to J') \equiv \sum_{\mu M'} \left| \langle \alpha'\, J'\, M' | \mathcal{M}_{\sigma\lambda\mu} | \alpha\, J\, M \rangle \right|^2 , \qquad (2.69)$$

where $J$ is the total angular momentum and $M$ its $z$ projection, and $\alpha$ denotes all the other quantum numbers needed to specify the state. The electromagnetic transition operator $\mathcal{M}_{\sigma\lambda}$ is a rank-$\lambda$ spherical tensor of electric or magnetic origin, as indicated by $\sigma = \mathrm{E}$ or $\sigma = \mathrm{M}$ (see Chap. 6). Show that one can write

$$B(\sigma\lambda;\ J \rightarrow J') = \frac{1}{2J + 1}\left|(\alpha'\ J'\|\boldsymbol{M}_{\sigma\lambda}\|\alpha\ J)\right|^2 . \tag{2.70}$$

**2.26.** By applying the appropriate reduced matrix elements of Sect. 2.3, find the single-particle value $(l\,\frac{1}{2}\,j\|\boldsymbol{\mu}_{\mathrm{sp}}\|l\,\frac{1}{2}\,j)$ of the magnetic dipole moment for the cases $j = l + \frac{1}{2}$ and $j = l - \frac{1}{2}$. The single-particle magnetic-moment operator is

$$\boldsymbol{\mu}_{\mathrm{sp}} = \frac{\mu_{\mathrm{N}}}{\hbar}\left[g_l\boldsymbol{J} + (g_s - g_l)\boldsymbol{S}\right] , \tag{2.71}$$

where $\mu_{\mathrm{N}}$ is the nuclear magneton and $g_l$ and $g_s$ are the orbital and spin gyromagnetic ratios respectively (see Chap. 6).

**2.27.** Find the single-particle value

$$Q_{\mathrm{sp}}(nlj) = -e\,\frac{2j - 1}{2j + 2}\langle r^2\rangle_{nl} \tag{2.72}$$

of the electric quadrupole moment by starting from (2.68) and using the single-particle quadrupole operator

$$Q_{20} = er^2 Y_{20}(\theta, \varphi) . \tag{2.73}$$

Note: The quantity $\langle r^2\rangle_{nl}$ is used to designate the integral over the radial wave functions of the single-particle states.

**2.28.** Prove Theorem II stated in (2.62).

**2.29.** Derive the special case (2.63) from (2.62).

**2.30.** Derive the reduced matrix element (2.64) by starting from (2.63).

# 3

# The Nuclear Mean Field and Many-Nucleon Configurations

## Prologue

After the two preceding chapters, throughout impregnated with messy-looking, though necessary mathematics, we are finally entering the realm of basic concepts of nuclear structure physics. While the preceding chapters may have been a shock to the reader not familiar with the fine details of angular momentum coupling, the present chapter should offer a soothing soft landing to the basic philosophy behind the nuclear shell model, namely the nuclear mean field.

The mean field is discussed both phenomenologically and as an expression of self-consistency. The many-body theory of the subsequent chapters is based on the notions of a single-particle basis and an antisymmetric many-particle state, called the Slater determinant, formed of them. It turns out that a major part of the nucleon–nucleon interactions can be included in the single-particle energies of the mean field. Thus the Slater determinants represent a long leap towards a proper many-body wave function for strongly interacting nucleons.

## 3.1 The Nuclear Mean Field

A nucleus of mass number $A$, neutron number $N$ and proton number (atomic number) $Z$, consists of $A$ strongly interacting nucleons, $N$ neutrons and $Z$ protons. In addition to the strong nuclear force, the protons feel also the Coulomb force. In the present work the nucleons are considered to be point particles without any internal structure. This is an excellent approximation when the purpose is to study nuclear structure at low energies. Along the same lines, the nuclear forces are described without attention to the basic mechanisms underlying them, i.e. the mesonic or quark degrees of freedom. The two-nucleon interaction is described by two-body interaction matrix elements, without a detailed account of the methods used to obtain them.

The $A$-nucleon Schrödinger equation cannot be solved exactly, at least for $A > 10$. Therefore one has to look for reasonable approximate methods to solve this many-body problem of strongly interacting particles. An elegant approximation is to convert the strongly interacting system of particles into a system of weakly interacting *quasiparticles* quasiparticles are often referred to as particles approximation the system of quasiparticles can be treated as a set of $A$ non-interacting quasiparticles. The remaining interactions, called *residual interactions*, can be treated in perturbation theory. The transformation from particles to quasiparticles is not easy, and its success depends on the nuclear system under discussion.

In this section we discuss the *mean-field* (or Hartree–Fock) quasiparticles. They can be introduced in the following way. Let $H$ be the nuclear many-body Hamiltonian consisting of kinetic energy $T$ and potential energy $V$, i.e.

$$H = T + V = \sum_{i=1}^{A} t(\boldsymbol{r}_i) + \sum_{\substack{i,j=1 \\ i<j}}^{A} v(\boldsymbol{r}_i, \boldsymbol{r}_j) = \sum_{i=1}^{A} \frac{-\hbar^2}{2m_{\mathrm{N}}} \nabla_i^2 + \sum_{\substack{i,j=1 \\ i<j}}^{A} v(\boldsymbol{r}_i, \boldsymbol{r}_j) \,, \quad (3.1)$$

where $m_{\mathrm{N}}$ is the mass of a nucleon (here we assume that the masses of a proton and a neutron are the same, i.e. $m_{\mathrm{N}}c^2 \approx 940\,\mathrm{MeV}$), and $\boldsymbol{r}_i$ denotes the coordinates of nucleon $i$. A summed single-particle potential energy, so far undefined, can be added and subtracted,

$$H = \left[ T + \sum_{i=1}^{A} v(\boldsymbol{r}_i) \right] + \left[ V - \sum_{i=1}^{A} v(\boldsymbol{r}_i) \right] \equiv H_{\mathrm{MF}} + V_{\mathrm{RES}} \,, \quad (3.2)$$

where

$$H_{\mathrm{MF}} = T + \sum_{i=1}^{A} v(\boldsymbol{r}_i) \equiv T + V_{\mathrm{MF}} = \sum_{i=1}^{A} [t(\boldsymbol{r}_i) + v(\boldsymbol{r}_i)] \equiv \sum_{i=1}^{A} h(\boldsymbol{r}_i) \quad (3.3)$$

is the nuclear *mean-field Hamiltonian* and

$$V_{\mathrm{RES}} = V - \sum_{i=1}^{A} v(\boldsymbol{r}_i) = \sum_{\substack{i,j=1 \\ i<j}}^{A} v(\boldsymbol{r}_i, \boldsymbol{r}_j) - \sum_{i=1}^{A} v(\boldsymbol{r}_i) \quad (3.4)$$

is the *residual interaction*. It is presumed that the residual interaction is much reduced in strength from the original interaction $V$.

### 3.1.1 The Mean-Field Approximation

In the *mean-field approximation* each nucleon can be viewed as moving in an external field created by the remaining $A-1$ nucleons. This external potential $V_{\mathrm{MF}}$ can be thought of as a time average, during a suitably defined short time interval $\Delta T$, of the interactions between the nucleon and its $A-1$ neighbours:

$$V_{\mathrm{MF}} = \sum_{i=1}^{A} v(\boldsymbol{r}_i) , \quad v(\boldsymbol{r}_i) = \frac{1}{\Delta T} \int_{T}^{T+\Delta T} \mathrm{d}t \sum_{\substack{j=1 \\ j \neq i}}^{A} v\big(\boldsymbol{r}_i(t), \boldsymbol{r}_j(t)\big) . \qquad (3.5)$$

It must be understood that the time-average idea can only provide intuitive guidance and cannot be implemented in practice.

In the mean-field approximation a strongly interacting many-fermion system becomes a system of $A$ *non-interacting* fermions (quasiparticles) in an *external* potential $v(\boldsymbol{r})$. For such an external potential it is easy to find the stationary one-particle states; it is the usual potential well problem of elementary quantum mechanics. From these one-particle states it is again easy to construct the $A$-particle wave function, as demonstrated below.

The mean-field Hamiltonian $H_{\mathrm{MF}}$ is easy to treat since the corresponding $A$-nucleon Schrödinger equation

$$H_{\mathrm{MF}} \Psi_0(\boldsymbol{r}_1, \boldsymbol{r}_2, \ldots, \boldsymbol{r}_A) = E \Psi_0(\boldsymbol{r}_1, \boldsymbol{r}_2, \ldots, \boldsymbol{r}_A) \qquad (3.6)$$

can be separated by using the ansatz wave function

$$\Psi_0(\boldsymbol{r}_1, \boldsymbol{r}_2, \ldots, \boldsymbol{r}_A) = \phi_{\alpha_1}(\boldsymbol{r}_1)\phi_{\alpha_2}(\boldsymbol{r}_2) \cdots \phi_{\alpha_A}(\boldsymbol{r}_A) . \qquad (3.7)$$

Substituting this ansatz into the Schrödinger equation (3.6) yields $A$ identical one-nucleon Schrödinger equations

$$h(\boldsymbol{r})\phi_\alpha(\boldsymbol{r}) = \varepsilon_\alpha \phi_\alpha(\boldsymbol{r}) , \quad h(\boldsymbol{r}) = t(\boldsymbol{r}) + v(\boldsymbol{r}) = \frac{-\hbar^2}{2m_{\mathrm{N}}}\nabla^2 + v(\boldsymbol{r}) , \qquad (3.8)$$

with the separation constants $\varepsilon_{\alpha_i}$ satisfying the condition

$$E = \sum_{i=1}^{A} \varepsilon_{\alpha_i} . \qquad (3.9)$$

The solution of the many-nucleon Schrödinger equation is thus a product of single-particle wave functions obtained by solving a one-nucleon Schrödinger equation for an external potential well. In this way the mean-field concept has turned the complicated many-nucleon problem into a simple one-nucleon one.

The problem remains how to determine the mean field, in particular an *optimal* mean field that minimizes the residual interaction between the quasiparticles. To solve the problem we seek an optimal set $\{\phi_\alpha(\boldsymbol{r})\}$ of one-quasiparticle states. This is a Rayleigh–Ritz variational problem where the variations $\phi_\alpha(\boldsymbol{r}) \to \phi_\alpha(\boldsymbol{r}) + \delta\phi_\alpha(\boldsymbol{r})$ of the single-particle orbitals are determined by minimizing the ground-state energy of the nucleus

$$E_{\mathrm{gs}} = \langle \Psi_0 | H | \Psi_0 \rangle , \quad H = T + V_{\mathrm{MF}} + V_{\mathrm{RES}} . \qquad (3.10)$$

As the starting point of the variational calculation it is customary to use either a product ansatz

$$\Psi_0(\boldsymbol{r}_1, \boldsymbol{r}_2, \ldots, \boldsymbol{r}_A) = \prod_{i=1}^{A} \phi_{\alpha_i}(\boldsymbol{r}_i) \,, \qquad (3.11)$$

constituting the *Hartree method*, or an antisymmetrized product ansatz

$$\Psi_0(\boldsymbol{r}_1, \boldsymbol{r}_2, \ldots, \boldsymbol{r}_A) = \mathcal{A} \left[ \prod_{i=1}^{A} \phi_{\alpha_i}(\boldsymbol{r}_i) \right] \,, \qquad (3.12)$$

constituting the *Hartree–Fock method*. The antisymmetrized ansatz wave function is called the *Slater determinant* of the given single-particle states. Here $\mathcal{A}$ is an antisymmetrization operator that performs the sign-accompanied permutations of the single-particle orbitals in the product wave function; $\mathcal{A}$ also carries a normalization factor. For example, for three particles in single-particle states labelled $1, 2, 3$ the normalized antisymmetric state, or Slater determinant, is

$$\Psi_0(\boldsymbol{r}_1, \boldsymbol{r}_2, \boldsymbol{r}_3) = \frac{1}{\sqrt{6}} \begin{vmatrix} \phi_1(\boldsymbol{r}_1) & \phi_1(\boldsymbol{r}_2) & \phi_1(\boldsymbol{r}_3) \\ \phi_2(\boldsymbol{r}_1) & \phi_2(\boldsymbol{r}_2) & \phi_2(\boldsymbol{r}_3) \\ \phi_3(\boldsymbol{r}_1) & \phi_3(\boldsymbol{r}_2) & \phi_3(\boldsymbol{r}_3) \end{vmatrix} \,. \qquad (3.13)$$

The energy (3.10) has to be varied under the constraint that the normalization of $\Psi_0$ is preserved, $\langle \Psi_0 | \Psi_0 \rangle = 1$. This leads to the constrained variational problem

$$\delta \left( \frac{\langle \Psi_0 | H | \Psi_0 \rangle}{\langle \Psi_0 | \Psi_0 \rangle} \right) = 0 \,, \qquad (3.14)$$

which can be transformed into an unconstrained one by using the method of Lagrange undetermined multipliers. After performing the variation, it turns out that the undetermined multipliers are nothing but the single-particle energies $\varepsilon_\alpha$. One ends up with the following *Hartree(–Fock) equation*:

$$\boxed{\begin{array}{c} \dfrac{-\hbar^2}{2m_{\mathrm{N}}} \nabla^2 \phi_\alpha(\boldsymbol{r}) + V_{\mathrm{H(F)}}(\{\phi_i(\boldsymbol{r})\}) \phi_\alpha(\boldsymbol{r}) = \varepsilon_\alpha \phi_\alpha(\boldsymbol{r}) \,, \\[2mm] i = 1, 2, \ldots, A \,, \qquad \alpha = 1, 2, \ldots, \infty \,. \end{array}} \qquad (3.15)$$

This equation is like the Schrödinger equation except that the simple potential term $V(\boldsymbol{r})$ is replaced with a functional of the unknown wave functions,

$$V(\boldsymbol{r}) \to V_{\mathrm{H(F)}}(\{\phi_i(\boldsymbol{r})\}) \,. \qquad (3.16)$$

Here $V_{\mathrm{H}}$ refers to the Hartree mean field and $V_{\mathrm{HF}}$ to the Hartree–Fock mean field; the two alternatives are carried in parallel.

Equation (3.15) is nonlinear and therefore much more difficult to solve than the Schrödinger equation. The solution can only be carried out by *iteration*. This means that we start from a set of guessed single-particle wave functions $\{\phi_i^{(0)}(\boldsymbol{r})\}_{i=1}^{A}$ and use them to calculate the initial potential term $V_{\mathrm{H(F)}}^{(0)}$. As the following step we solve the equation for a complete set of new wave functions $\{\phi_\alpha^{(1)}(r)\}_{\alpha=1}^\infty$ and eigenenergies $\varepsilon_\alpha^{(1)}$. With this new set of eigen-functions we generate the next potential $V_{\text{H(F)}}^{(1)}$ and solve (3.15) for the next set of eigenfunctions and eigenenergies. In a schematic way, this process can be depicted as

$$\phi_i^{(0)} \longrightarrow V_{\text{H(F)}}^{(0)} \longrightarrow \phi_\alpha^{(1)}, \varepsilon_\alpha^{(1)} \xrightarrow{\phi_i^{(1)}} V_{\text{H(F)}}^{(1)} \longrightarrow \cdots \longrightarrow \phi_\alpha^{(n)}, \varepsilon_\alpha^{(n)} . \tag{3.17}$$

This process is repeated until *self-consistency* is achieved. This means that the wave functions (or eigenenergies) do not differ more than a preset limit in two consecutive iterations, i.e.

$$\|\phi_\alpha^{(n-1)} - \phi_\alpha^{(n)}\| < \text{preset limit} , \tag{3.18}$$

where $\|\cdots\|$ denotes the norm of a wave function.

When the iteration has converged and self-consistency been achieved, one has generated a *self-consistent mean field* $v_{\text{H(F)}}(r)$ and the associated eigenstates $\phi_\alpha(r)$ and eigenenergies $\varepsilon_\alpha$, all *simultaneously*. The generated set of eigenenergies can be schematically drawn inside the generated self-consistent mean-field potential well, as shown in Fig. 3.1. We may also note that for a finite potential well there will be, in addition to the bound states shown, an infinite number of unbound states.

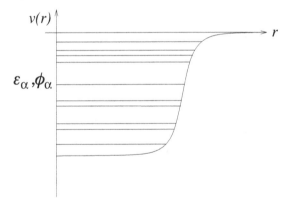

**Fig. 3.1.** Schematic view of a central mean-field potential and its single-particle eigenenergies

In our example the generated mean-field potential is central, i.e. only a function of $r$. Central mean-field potentials describe *spherical* nuclei, and they are the ones we address in this work. Deviations from centrality show up as a dependence of the potential on the polar and azimuthal angles $\theta, \varphi$ of the spherical coordinates; such potentials are used to describe *deformed* nuclei. The Hartree–Fock equation (3.15) is derived and further discussed in Sect. 4.5.

### 3.1.2 Phenomenological Potentials

Often one just selects a particular type of mean-field potential without going through the steps leading to self-consistency. The use of such phenomenological potentials is a practical shortcut, taken at the expense of theoretical precisenes. The simplest frequently used potential is the three-dimensional *harmonic oscillator* potential

$$v_{\mathrm{HO}}(r) = -V_1 + kr^2 = -V_1 + \tfrac{1}{2}m_{\mathrm{N}}\omega^2 r^2 \,, \tag{3.19}$$

where $V_1$ and $k$ are parameters to be fitted for best result. A common, more realistic choice is the *Woods–Saxon* potential

$$v_{\mathrm{WS}}(r) = \frac{-V_0}{1 + \mathrm{e}^{(r-R)/a}} \,. \tag{3.20}$$

Its usual parametrization is

$$R = r_0 A^{1/3} = 1.27 A^{1/3} \,\mathrm{fm} \quad \text{(nuclear radius)} \,, \tag{3.21}$$

$$a = 0.67 \,\mathrm{fm} \quad \text{(surface diffuseness)} \,, \tag{3.22}$$

$$V_0 = \left(51 \pm 33\frac{N-Z}{A}\right) \mathrm{MeV} \,, \tag{3.23}$$

the + sign being for a proton and the − sign for a neutron. When not making a distinction between the nucleons one may use $V_0 = 57\,\mathrm{MeV}$ as a suitable average value.

It is possible to select the oscillator energy spacing $\hbar\omega$ and the depth $V_1$ so that the resulting potential $v_{\mathrm{HO}}$ is roughly equivalent to a given Woods–Saxon potential $v_{\mathrm{WS}}$; see Fig. 3.2. To find such an oscillator potential, we choose $\hbar\omega$ to simulate the Woods–Saxon major spacing. Then we fix $V_1$ so that the 0s state occurs at the same energy in both potentials.

The equivalent harmonic oscillator potential reproduces quite nicely the wave functions of the Woods–Saxon potential near the bottom of the wells, but when approaching zero energy, the differences grow (see Subsect. 3.2.2). Near zero energy the Woods–Saxon wave functions have a long tail extending far beyond the nuclear radius $R$, whereas the harmonic oscillator wave functions decrease sharply beyond the potential wall.

### 3.1.3 The Spin–Orbit Interaction

The central potential alone does not reproduce the experimentally observed qualitative behaviour of the single-particle energies in the mean field. The observed energies bunch into groups, or shells, similarly to the atomic case. These groups of states and the energy gaps between them are depicted in Fig. 3.3.

The nucleon numbers at which energy gaps occur are called *magic numbers*. Their experimentally known values are 2, 8, 20, 28, 50, 82, 126; further values

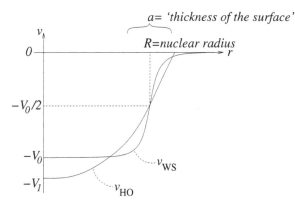

**Fig. 3.2.** Two phenomenological mean-field potentials: the Woods–Saxon potential $v_{WS}$ and the approximately equivalent harmonic oscillator potential $v_{HO}$

are uncertain. To reproduce these numbers theoretically, one needs to add to the mean-field potential a term resulting from the *spin–orbit interaction*. This spin–orbit interaction, or coupling or force, splits the states of the same orbital angular momentum quantum number $l$ into two, with total single-particle angular momenta $j = l + \frac{1}{2}$ and $j = l - \frac{1}{2}$.

The nuclear spin–orbit interaction does not have the same origin as the spin–orbit interaction in atoms. The atomic spin–orbit force has a well-known electromagnetic origin and causes part of the atomic fine structure, on the millielectronvolts scale, while the energy differences are on the electronvolts scale. The atomic shells are not significantly affected by spin–orbit effects. In nuclei the spin–orbit splitting is on the million electronvolts scale, as are the single-particle energy differences. Yet another difference is that the splitting order is opposite in nuclei to that in atoms. In fact, the origin of the nuclear spin–orbit force is not well understood, and phenomenological descriptions must be relied upon.

## 3.2 Woods–Saxon Wave Functions

We now give a numerical solution of the Schrödinger equation with the Woods–Saxon potential (3.20) together with the Coulomb potential and spin-orbit coupling. The solution is constructed in a basis of harmonic oscillator wave functions. We therefore state their analytic properties in considerable detail. A busy reader can skip this part without losing track of the rest of the book. Only a few basic equations of this section are needed later, and references to them are given in the appropriate places.

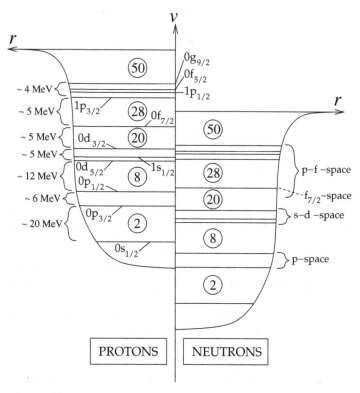

**Fig. 3.3.** Schematic view of mean-field potentials for protons and neutrons. The magic numbers and shells of states are indicated with rough energies

We solve the Schrödinger equation with the complete Hamiltonian

$$h = \frac{-\hbar^2}{2m_{\mathrm{N}}}\nabla^2 + v(r) + v_{\mathrm{LS}}(r)\boldsymbol{L}\cdot\boldsymbol{S}$$

$$= \frac{-\hbar^2}{2m_{\mathrm{N}}}\left(\nabla_r^2 - \frac{\boldsymbol{L}^2/\hbar^2}{r^2}\right) + v_{\mathrm{WS}}(r) + v_{\mathrm{C}}(r) + v_{\mathrm{LS}}(r)\boldsymbol{L}\cdot\boldsymbol{S}\,, \qquad (3.24)$$

where the radial derivative has the usual form (see e.g. [6])

$$\nabla_r^2 \equiv \frac{1}{r^2}\frac{\mathrm{d}}{\mathrm{d}r}\left(r^2\frac{\mathrm{d}}{\mathrm{d}r}\right)\,. \qquad (3.25)$$

The Woods–Saxon term $v_{\mathrm{WS}}(r)$ is given in (3.20), and we use the parameter values (3.21)–(3.23). The Coulombic part of the potential is

$$v_{\mathrm{C}}(r) = \frac{Ze^2}{4\pi\epsilon_0}\begin{cases}\dfrac{3-(r/R)^2}{2R}\,, & r \le R\,, \\[2ex] \dfrac{1}{r}\,, & r > R\,,\end{cases} \qquad (3.26)$$

which is the static Coulomb potential of a uniformly charged sphere of radius $R$, to be taken from (3.21). The term $v_C(r)$ applies only to protons; for neutrons it is zero.

For the spin–orbit term we use [12]

$$v_{\mathrm{LS}}(r) = v_{\mathrm{LS}}^{(0)} \left(\frac{r_0}{\hbar}\right)^2 \frac{1}{r} \left[\frac{\mathrm{d}}{\mathrm{d}r} \frac{1}{1 + e^{(r-R)/a}}\right] . \qquad (3.27)$$

The second pair of parentheses serves to indicate that the derivative does not operate on the wave function. The $r$ dependence is phenomenological. The derivative makes the spin–orbit effect peak in the nuclear surface region. However, the radial dependence is often neglected so that $v_{\mathrm{LS}}(r)$ is replaced by a constant; see Exercise 3.11. In (3.27) we use the Woods–Saxon parameters (3.21)–(3.23) and take the strength to be

$$v_{\mathrm{LS}}^{(0)} = 0.44 V_0 . \qquad (3.28)$$

Let us examine the angular momentum dependence of the wave functions generated by the Hamiltonian (3.24), which contains the terms with $L^2$ and $L \cdot S$. We assert that the angular momentum eigenstates of these operators are the states $|l \frac{1}{2} j m\rangle$ of the coupled basis defined in (1.16–1.20) with $j_1 = l$, $j_2 = \frac{1}{2}$. The eigenvalue Eqs. (1.17)–(1.19) become

$$L^2 |l \tfrac{1}{2} j m\rangle = l(l+1)\hbar^2 |l \tfrac{1}{2} j m\rangle , \qquad (3.29)$$

$$S^2 |l \tfrac{1}{2} j m\rangle = \tfrac{3}{4}\hbar^2 |l \tfrac{1}{2} j m\rangle , \qquad (3.30)$$

$$J^2 |l \tfrac{1}{2} j m\rangle = j(j+1)\hbar^2 |l \tfrac{1}{2} j m\rangle , \qquad (3.31)$$

$$J_z |l \tfrac{1}{2} j m\rangle = m\hbar |l \tfrac{1}{2} j m\rangle . \qquad (3.32)$$

When we write $J^2 = (L + S)^2 = L^2 + 2L \cdot S + S^2$ and use the preceding equations, we see that

$$L \cdot S |l \tfrac{1}{2} j m\rangle = \tfrac{1}{2} \left[j(j+1) - l(l+1) - \tfrac{3}{4}\right] \hbar^2 |l \tfrac{1}{2} j m\rangle . \qquad (3.33)$$

Equations (3.29 and 3.33) prove our assertion and give the eigenvalues of the relevant operators $L^2$ and $L \cdot S$.

This leads us to seek eigenfunctions of the Hamiltonian (3.24) in the form $|n l \frac{1}{2} j m\rangle$, where $n$ completes the labelling. The Schrödinger equation for this ansatz is

$$
\begin{aligned}
h|n l \tfrac{1}{2} j m\rangle &= \left\{ \frac{-\hbar^2}{2m_{\mathrm{N}}} \left[\nabla_r^2 - \frac{l(l+1)}{r^2}\right] + v_{\mathrm{WS}}(r) + v_C(r) \right. \\
&\quad \left. + \tfrac{1}{2} \left[j(j+1) - l(l+1) - \tfrac{3}{4}\right] \hbar^2 v_{\mathrm{LS}}(r) \right\} |n l \tfrac{1}{2} j m\rangle \\
&\equiv h_{lj}(r)|n l \tfrac{1}{2} j m\rangle \\
&= \varepsilon_{nlj}|n l \tfrac{1}{2} j m\rangle , \qquad (3.34)
\end{aligned}
$$

where the notation $h_{lj}(r)$ indicates that the angular momentum quantum numbers $l$ and $j$ are parameters of the radial Hamiltonian. It should be intuitively clear that the energy does not depend on the projection quantum number $m$ because all directions are equivalent for a spherically symmetric Hamiltonian. This can also be formally proved by means of the Wigner–Eckart theorem (2.27). Equation (3.34) is in fact the radial Schrödinger equation for our Woods–Saxon Hamiltonian, and we can here replace the complete wave function[1] $|n\, l\, \frac{1}{2}\, j\, m\rangle = f_{nlj}(r)\, |l\, \frac{1}{2}\, j\, m\rangle$ by the radial function $f_{nlj}(r)$:

$$h_{lj}(r)f_{nlj}(r) = \varepsilon_{nlj} f_{nlj}(r) \,. \tag{3.35}$$

This is the differential equation to be solved for the eigenvalues $\varepsilon_{nlj}$ and eigenfunctions $f_{nlj}$. The eigenfunctions are orthogonal with respect to $n$; orthogonality with respect to $l, j$ is carried by the angular momentum states $|l\, \frac{1}{2}\, j\, m\rangle$. We select the phases to be real and normalize, so that we have

$$\int_0^\infty r^2 \mathrm{d}r\, f_{nlj}(r)f_{n'lj}(r) = \delta_{nn'} \,. \tag{3.36}$$

### 3.2.1 Harmonic Oscillator Wave Functions

The differential Eq. (3.35) could be solved by direct numerical methods. However, we choose to solve it by seeking the wave functions as a linear combinations of harmonic oscillator wave functions $g_{nl}(r)$:

$$f_{nlj}(r) = \sum_\nu A_\nu^{(nlj)} g_{\nu l}(r) \,, \quad \sum_\nu \left[A_\nu^{(nlj)}\right]^2 = 1 \,, \tag{3.37}$$

where we have included the normalization condition. The solution is found by forming the Hamiltonian matrix in the harmonic oscillator basis and diagonalizing it (see e.g. [6]). The angular momentum part is the same for the Woods–Saxon and oscillator wave functions, so we may ignore it in the diagonalization calculation. We are thus concerned with the matrix elements

$$\int_0^\infty r^2 \mathrm{d}r\, g_{\nu'l}(r)h_{lj}(r)g_{\nu l}(r) \equiv \langle \nu'|h_{lj}(r)|\nu\rangle \,. \tag{3.38}$$

This method of solution provides a direct comparison between Woods–Saxon and harmonic oscillator wave functions through the expansion coefficients $A_\nu^{(nlj)}$ that emerge from the process.

Let us review the oscillator functions $g_{nl}(r)$. For the potential (3.19) they are solutions of the radial Schrödinger equation

---

[1] As is customary, we are here ignoring the difference between a state vector and its coordinate representation, the wave function. The difference is made clear in Sect. 3.3.

$$\frac{-\hbar^2}{2m_N}\left[\nabla_r^2 - \frac{l(l+1)}{r^2}\right]g_{nl}(r) - V_1 + \tfrac{1}{2}m_N\omega^2 r^2 g_{nl}(r) = \varepsilon_{nl}g_{nl}(r) \,, \quad (3.39)$$

where

$$2n + l = N = 0, 1, 2, 3, \dots \,, \quad (3.40)$$

$$\varepsilon_{nl} = -V_1 + (N + \tfrac{3}{2})\hbar\omega = -V_1 + (2n + l + \tfrac{3}{2})\hbar\omega \quad (3.41)$$

are the radial quantum numbers and energy eigenvalues. Here we use the convention where the *principal quantum number* $n = 0, 1, 2, 3, \dots$ indicates the number of nodes of the wave function, i.e. the number of places where the wave function crosses zero. This convention is used also by some other authors, for example [17]. Many others, e.g. [9, 10, 12, 16] use the convention $\tilde{n} = n + 1 = 1, 2, 3, \dots$, where also the zero of the wave function at the origin is counted as a node.

The function can be written explicitly as [28]

$$g_{nl}(r) = \sqrt{\frac{2n!}{b^3\Gamma(n+l+\tfrac{3}{2})}}\left(\frac{r}{b}\right)^l e^{-r^2/2b^2} L_n^{(l+\frac{1}{2})}(r^2/b^2) \,, \quad (3.42)$$

where $L_n^{(l+\frac{1}{2})}(x)$ is the *associated Laguerre polynomial* [29]. The parameter $b$ is called the *oscillator length*. It characterizes the width of the oscillator potential and is given by

$$\boxed{b \equiv \sqrt{\frac{\hbar}{m_N\omega}} = \frac{\hbar c}{\sqrt{(m_N c^2)(\hbar\omega)}} \approx \frac{197.33}{\sqrt{940 \times \hbar\omega\,[\mathrm{MeV}]}}\,\mathrm{fm}\,.} \quad (3.43)$$

The value of $\hbar\omega$ can be obtained from a simple argument [16] resulting in the formulas

$$\hbar\omega = 41 A^{-1/3}\,\mathrm{MeV}\,, \quad b = 1.005 A^{1/6}\,\mathrm{fm}\,, \quad (3.44)$$

or from the more refined Blomqvist–Molinari formula [30], which gives satisfactory agreement with observed charge radii,

$$\boxed{\hbar\omega = \left(45 A^{-1/3} - 25 A^{-2/3}\right)\mathrm{MeV}\,.} \quad (3.45)$$

In the following, we adopt this more refined value of $\hbar\omega$ to evaluate $b$.

The orthonormality relation of the oscillator functions is

$$\int_0^\infty r^2\mathrm{d}r\, g_{nl}(r)g_{n'l}(r) = \delta_{nn'}\,. \quad (3.46)$$

In the present work their phase has been chosen to satisfy the requirements [17, 28]

$$g_{nl}(r) \xrightarrow{r\to\infty} (-1)^n \times \text{positive function}\,, \quad (3.47)$$

$$g_{nl}(r) \xrightarrow{r\to 0} r^l > 0\,. \quad (3.48)$$

Some other authors use the convention $G_{nl}(r) = (-1)^n g_{nl}(r)$, which leads to positive functions at large distances and functions of varying sign, $(-1)^n$, near the origin.

The first three associated Laguerre polynomials are

$$L_0^{(l+\frac{1}{2})}(x) = 1 , \tag{3.49}$$

$$L_1^{(l+\frac{1}{2})}(x) = l - x + \tfrac{3}{2} , \tag{3.50}$$

$$L_2^{(l+\frac{1}{2})}(x) = \tfrac{1}{2} \left[ (l + \tfrac{3}{2})(l + \tfrac{5}{2}) - 2(l + \tfrac{5}{2})x + x^2 \right] , \tag{3.51}$$

and further polynomials can be derived e.g. by using the recursion relation [29]

$$L_n^{(l+\frac{1}{2})}(x) = L_n^{(l+\frac{3}{2})}(x) - L_{n-1}^{(l+\frac{3}{2})}(x) . \tag{3.52}$$

Numerical values for the $g_{nl}(r)$ can be obtained conveniently by exploiting the auxiliary functions $v_{nl}(r)$ [10], defined through

$$g_{nl}(r) = \sqrt{\frac{2^{l+2-n}(2n + 2l + 1)!!}{b^3 \sqrt{\pi} n! [(2l + 1)!!]^2}} \left(\frac{r}{b}\right)^l e^{-r^2/2b^2} v_{nl}(r^2/b^2) , \tag{3.53}$$

and using the recursion relations

$$v_{n,l-1}(x) = v_{n-1,l-1}(x) - 2x v_{n-1,l}(x)/(2l + 1) , \tag{3.54}$$

$$v_{nl}(x) = \left[ (2l + 1)v_{n,l-1}(x) + 2n v_{n-1,l}(x) \right] /(2n + 2l + 1) . \tag{3.55}$$

By using (3.43) and (3.53)–(3.55), one can generate the radial wave functions $g_{nl}(r)$.

### 3.2.2 Diagonalization of the Woods–Saxon Hamiltonian

The matrix elements (3.38) are computed by numerical integration. To avoid prior numerical differentiation by the term $\nabla_r^2$ in the Woods–Saxon radial Hamiltonian $h_{lj}(r)$, we take $\nabla_r^2 g_{ln}(r)$ from the harmonic oscillator equation (3.39). This leaves us with

$$\langle \nu' | h_{lj}(r) | \nu \rangle = \int_0^\infty r^2 dr\, g_{\nu'l}(r) g_{\nu l}(r) \left[ \frac{\hbar^2}{2m_N} \left( \frac{4n + 2l + 3}{b^2} - \frac{r^2}{b^4} \right) \right.$$

$$\left. + v_{\text{WS}}(r) + v_{\text{C}}(r) + \tfrac{1}{2} \left[ j(j + 1) - l(l + 1) - \tfrac{3}{4} \right] \hbar^2 v_{\text{LS}}(r) \right] , \tag{3.56}$$

where the explicit angular momentum dependence of the spin–orbit term is taken from (3.34).

In principle the dimension of the matrix of the Woods–Saxon Hamiltonian is infinite. The practical dimensions are determined by the convergence of

**Table 3.1.** Woods–Saxon energies $\varepsilon_{nlj}$ and oscillator amplitudes $A_\nu^{(nlj)}$, defined in (3.35)–(3.37), for neutron single-particle states in $^{16}$O

| $nlj$ | $\varepsilon_{nlj}$ (MeV) | $\nu = 0$ | $\nu = 1$ | $\nu = 2$ | $\nu = 3$ | $\nu = 4$ | $\nu = 5$ | $\nu = 6$ |
|---|---|---|---|---|---|---|---|---|
| $0s_{1/2}$ | $-31.091$ | 0.999 | 0.011 | 0.004 | $-0.015$ | $-0.003$ | $-0.002$ | 0.000 |
| $0p_{3/2}$ | $-18.612$ | 0.999 | 0.004 | 0.035 | $-0.024$ | $-0.001$ | $-0.006$ | 0.001 |
| $0p_{1/2}$ | $-13.466$ | 0.997 | $-0.010$ | 0.065 | $-0.026$ | 0.006 | $-0.009$ | 0.002 |
| $0d_{5/2}$ | $-6.359$ | 0.992 | $-0.057$ | 0.098 | $-0.054$ | 0.020 | $-0.021$ | 0.001 |
| $1s_{1/2}$ | $-3.970$ | $-0.007$ | 0.943 | $-0.214$ | 0.196 | $-0.126$ | 0.074 | $-0.056$ |
| $0d_{3/2}$ | 1.098 | 0.902 | $-0.242$ | 0.250 | $-0.169$ | 0.129 | $-0.102$ | 0.073 |

**Table 3.2.** Woods–Saxon energies $\varepsilon_{nlj}$ and oscillator amplitudes $A_\nu^{(nlj)}$, defined in (3.35)–(3.37), for proton single-particle states in $^{16}$O

| $nlj$ | $\varepsilon_{nlj}$ (MeV) | $\nu = 0$ | $\nu = 1$ | $\nu = 2$ | $\nu = 3$ | $\nu = 4$ | $\nu = 5$ |
|---|---|---|---|---|---|---|---|
| $0s_{1/2}$ | $-26.445$ | 0.999 | $-0.011$ | 0.005 | $-0.016$ | $-0.002$ | $-0.002$ |
| $0p_{3/2}$ | $-14.451$ | 0.998 | $-0.026$ | 0.039 | $-0.029$ | 0.002 | $-0.007$ |
| $0p_{1/2}$ | $-9.328$ | 0.995 | $-0.045$ | 0.074 | $-0.035$ | 0.011 | $-0.012$ |
| $0d_{5/2}$ | $-2.731$ | 0.985 | $-0.100$ | 0.114 | $-0.071$ | 0.033 | $-0.029$ |
| $1s_{1/2}$ | $-0.709$ | 0.015 | 0.905 | $-0.275$ | 0.235 | $-0.167$ | 0.108 |
| $0d_{3/2}$ | 4.088 | 0.807 | $-0.306$ | 0.309 | $-0.244$ | 0.203 | $-0.168$ |

the summation over $\nu$. Tables 3.1–3.4 give numerical examples of the amplitudes $A_\nu^{(nlj)}$ for the neutrons and protons in the nuclei $^{16}_8$O$_8$ and $^{40}_{20}$Ca$_{20}$. The parameters used are those given in (3.21)–(3.23) and (3.28).

The tables show that the main oscillator component for each $n$ is $\nu = n$. The wave functions of the lowest-lying Woods–Saxon states are seen to match the corresponding oscillator states almost exactly. With increasing energy more than one oscillator wave function acquires a non-negligible amplitude in the expansion, indicating increasing deviation from the near match.

Note that some of the highest-lying states have positive energies. These states are discrete states; they do not belong to the continuum of free states. They are due to the $l(l+1)/r^2$ term of the radial Schrödinger equation. For $l > 0$ the term acts as a 'centrifugal barrier' for positive energies. It gives rise to quasi-stationary, long-lived single-particle states localized within the barrier. In our examples it is easy to understand that these states are indeed quasi-stationary when we look at the barrier heights. The height of the centrifugal barrier is

$$v_{\mathrm{cf}}(R) = \frac{\hbar^2}{2m_{\mathrm{N}}} \frac{l(l+1)}{R^2} \approx 13.2 A^{-2/3} l(l+1) \, \mathrm{MeV} \,. \tag{3.57}$$

For the $0d_{3/2}$ proton state at 4.088 MeV in $^{16}$O (Table 3.1) this gives $v_{\mathrm{cf}}(R) = 12.5$ MeV. Additionally we have for protons the Coulomb barrier

$$v_{\mathrm{C}}(R) = \frac{Ze^2}{4\pi\epsilon_0} \frac{1}{R} \approx 1.15 Z A^{-1/3} \, \mathrm{MeV} \,, \tag{3.58}$$

**Table 3.3.** Woods–Saxon energies $\varepsilon_{nlj}$ and oscillator amplitudes $A_\nu^{(nlj)}$, defined in (3.35)–(3.37), for neutron single-particle states in $^{40}$Ca

| $nlj$ | $\varepsilon_{nlj}$ (MeV) | $\nu = 0$ | $\nu = 1$ | $\nu = 2$ | $\nu = 3$ | $\nu = 4$ | $\nu = 5$ |
|---|---|---|---|---|---|---|---|
| $0s_{1/2}$ | $-38.842$ | 0.998 | $-0.051$ | $-0.035$ | $-0.014$ | 0.001 | 0.002 |
| $0p_{3/2}$ | $-29.541$ | 0.999 | $-0.003$ | $-0.027$ | $-0.025$ | $-0.004$ | 0.001 |
| $0p_{1/2}$ | $-26.942$ | 0.999 | 0.039 | $-0.009$ | $-0.024$ | $-0.006$ | $-0.001$ |
| $0d_{5/2}$ | $-19.614$ | 0.999 | 0.035 | $-0.003$ | $-0.032$ | $-0.007$ | $-0.002$ |
| $1s_{1/2}$ | $-15.684$ | 0.051 | 0.997 | 0.005 | 0.022 | $-0.044$ | $-0.004$ |
| $0d_{3/2}$ | $-14.310$ | 0.996 | 0.070 | 0.038 | $-0.027$ | $-0.005$ | $-0.007$ |
| $0f_{7/2}$ | $-9.323$ | 0.997 | 0.045 | 0.039 | $-0.042$ | $-0.004$ | $-0.010$ |
| $1p_{3/2}$ | $-5.673$ | 0.003 | 0.985 | $-0.088$ | 0.109 | $-0.091$ | 0.027 |
| $1p_{1/2}$ | $-3.320$ | $-0.035$ | 0.966 | $-0.131$ | 0.167 | $-0.116$ | 0.054 |
| $0f_{5/2}$ | $-1.346$ | 0.988 | $-0.009$ | 0.132 | $-0.064$ | 0.031 | $-0.033$ |
| $0g_{9/2}$ | 0.985 | 0.988 | $-0.010$ | 0.120 | $-0.081$ | 0.031 | $-0.039$ |

**Table 3.4.** Woods–Saxon energies $\varepsilon_{nlj}$ and oscillator amplitudes $A_\nu^{(nlj)}$, defined in (3.35)–(3.37), for proton single-particle states in $^{40}$Ca

| $nlj$ | $\varepsilon_{nlj}$ (MeV) | $\nu = 0$ | $\nu = 1$ | $\nu = 2$ | $\nu = 3$ | $\nu = 4$ | $\nu = 5$ |
|---|---|---|---|---|---|---|---|
| $0s_{1/2}$ | $-29.982$ | 0.995 | $-0.087$ | $-0.034$ | $-0.014$ | 0.002 | 0.003 |
| $0p_{3/2}$ | $-21.255$ | 0.998 | $-0.044$ | $-0.026$ | $-0.026$ | $-0.001$ | 0.001 |
| $0p_{1/2}$ | $-18.573$ | 0.999 | $-0.002$ | $-0.009$ | $-0.026$ | $-0.004$ | $-0.001$ |
| $0d_{5/2}$ | $-11.871$ | 0.999 | $-0.013$ | 0.000 | $-0.037$ | $-0.003$ | $-0.003$ |
| $1s_{1/2}$ | $-7.900$ | 0.085 | 0.992 | $-0.068$ | 0.034 | $-0.057$ | 0.006 |
| $0d_{3/2}$ | $-6.489$ | 0.998 | 0.015 | 0.044 | $-0.037$ | 0.000 | $-0.010$ |
| $0f_{7/2}$ | $-2.143$ | 0.997 | $-0.016$ | 0.050 | $-0.056$ | 0.006 | $-0.015$ |
| $1p_{3/2}$ | 1.092 | 0.041 | 0.946 | $-0.205$ | 0.166 | $-0.148$ | 0.076 |
| $0f_{5/2}$ | 5.542 | 0.945 | $-0.128$ | 0.202 | $-0.146$ | 0.105 | $-0.095$ |

which for $^{16}$O gives $v_C(R) = 3.65$ MeV. The combined barrier height is thus four times the single-particle energy.

Figure 3.4 shows some of the Woods–Saxon wave functions tabulated in Tables 3.1–3.4. The examples indicate that a typical Woods–Saxon wave function strongly resembles its main oscillator component. The examples with three components included, $\nu = 0, 1, 2$, show good convergence. Usually full convergence is obtained by including oscillator functions up to $\nu = 8$ in the expansion (3.37). It is in fact often a good approximation in nuclear structure calculations to use pure harmonic oscillator wave functions instead of Woods–Saxon ones. However, oscillator functions are inadequate for bound neutron s states close to zero energy and for $l > 0$ states with positive energy.

Even though we have obtained accurate numerical solutions for the single-particle states in the mean field, it is instructive to examine qualitatively the evolution of the spectrum with the development of the Hamiltonian. Figure 3.5 shows this evolution from the simple harmonic oscillator spectrun to a

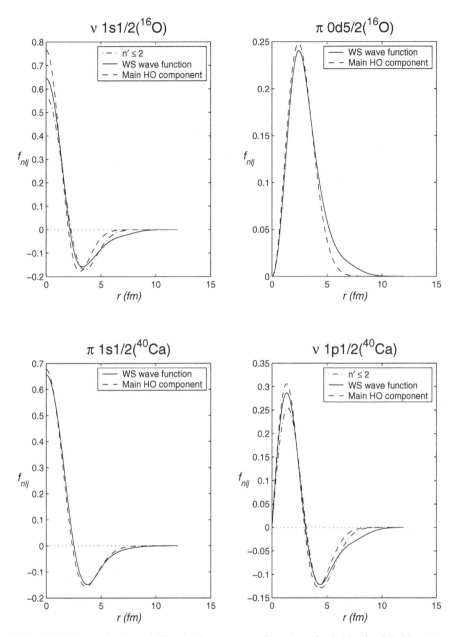

**Fig. 3.4.** Plots of selected Woods–Saxon wave functions (*solid line*) of Tables 3.1–3.4. The *dashed line* gives the leading harmonic oscillator component. The two upper panels are for the neutron $1s_{1/2}$ and proton $0d_{5/2}$ states in $^{16}O$, the two lower panels for the proton $1s_{1/2}$ and neutron $1p_{1/2}$ states in $^{40}Ca$. In the upper left and lower right panels the dash-dotted line gives the result of an expansion including only $\nu \leq 2$

Woods–Saxon spectrum with spin–orbit interaction. As seen from (3.40) and (3.41), in the oscillator potential the single-particle orbitals $nlj$ for $2n+l = N$ are degenerate within one major oscillator shell $N$. All the wave functions within this major shell have the same parity $P = (-1)^l = \pm$ given by the spherical harmonic $Y_{lm}(\theta, \varphi)$ of the angular part of the wave function.

Adding up the number of protons or neutrons that can, according to the Pauli principle, fill the shells in the harmonic oscillator potential, we get the magic numbers 2, 8, 20, 40, 70, 112, .... The Woods–Saxon potential splits the different $l$ values within the same $N$ but leaves the magic numbers unchanged. Finally, switching on the spin–orbit interaction produces the experimentally observed magic numbers.

We can get a clear picture of the energy separation $\Delta_{nl}^{\mathrm{LS}}$ between the *spin–orbit partners* $j_+ = l + \frac{1}{2}$ and $j_- = l - \frac{1}{2}$ by using (3.33):

$$\Delta_{nl}^{\mathrm{LS}} = \varepsilon_{nlj_-} - \varepsilon_{nlj_+} \approx -(l + \tfrac{1}{2})\hbar^2 C_{nl} \tag{3.59}$$

$$C_{nl} = \frac{C_{nlj_+} + C_{nlj_-}}{2} , \qquad C_{nlj} = \int_0^\infty r^2 \mathrm{d}r \, f_{nlj}^2(r) v_{\mathrm{LS}}(r) . \tag{3.60}$$

The function $v_{\mathrm{LS}}(r)$, given in (3.27), is negative because of the derivative, so $C_{nl}$ is negative and the spin–orbit splitting $\Delta_{nl}^{\mathrm{LS}}$ is positive as shown in Fig. 3.5. If $v_{\mathrm{LS}}$ is replaced by a (negative) constant, a possibility mentioned after (3.27), $C_{nlj}$ becomes just that constant because of the normalization (3.36). Then the splitting becomes exactly that given by (3.59), i.e. proportional to $l + \frac{1}{2}$. The spin–orbit splitting in Fig. 3.5 indeed increases visibly linearly with $l$, although the figure is based on an exact Woods–Saxon calculation including the $v_{\mathrm{LS}}(r)$ of (3.27).

## 3.3 Many-Nucleon Configurations

Consider a nucleon moving in a central mean field $v(r)$ with a spin–orbit potential $v_{\mathrm{LS}}(r)\boldsymbol{L} \cdot \boldsymbol{S}$. The central mean field can emerge from a Hartree–Fock calculation or it can be phenomenological, e.g. of the Woods–Saxon type. As an abbreviation we adopt the symbol $\boldsymbol{x}$ to describe both the spatial coordinates $\boldsymbol{r}$ and the spin degree of freedom, represented by a spin-$\frac{1}{2}$ spinor; we call this 'coordinate representation'. The nucleon is described by the Schrödinger equation

$$h(\boldsymbol{x})\phi_\alpha(\boldsymbol{x}) = \varepsilon_\alpha \phi_\alpha(\boldsymbol{x}) , \qquad h(\boldsymbol{x}) = t(\boldsymbol{r}) + v(\boldsymbol{r}) + v_{\mathrm{LS}}(r)\boldsymbol{L} \cdot \boldsymbol{S} , \tag{3.61}$$

where $\alpha$ stands for a complete set of quantum numbers necessary for describing the state.

We assume that (3.61) has been solved, with resulting single-particle states $\phi_\alpha(\boldsymbol{x})$ and single-particle energies $\varepsilon_\alpha$. Due to the presence of the spin–orbit term in the Hamiltonian, the orbital angular momentum and spin cannot be

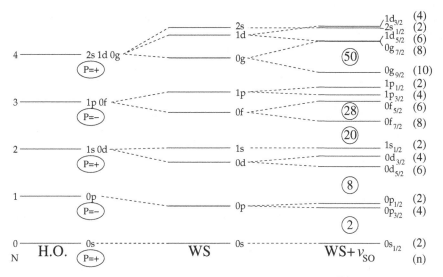

**Fig. 3.5.** Evolution of the neutron single-particle spectrum of $^{136}$Xe from the harmonic oscillator spectrum to the Woods–Saxon spectrum without and with the spin–orbit term in (3.24). The Woods–Saxon parameters are from (3.21)–(3.23) and the additional spin–orbit parameter from (3.28). The equivalent oscillator potential (see Fig. 3.2) has the parameters $\hbar\omega = 6.53\,\text{MeV}$ and $V_1 = 48.6\,\text{MeV}$. The parity $P$ of each oscillator major shell is indicated

treated separately. As can be seen from (3.33), the coupled angular momentum states $|l\,\frac{1}{2}\,j\,m\rangle$ are eigenstates of $\boldsymbol{L} \cdot \boldsymbol{S}$, i.e. the spin–orbit term is diagonal in the coupled basis. Therefore the eigenstates $\phi_\alpha(\boldsymbol{x})$ are necessarily states of the coupled basis.

Single-particle states and single-particle energies are the foundation of the nuclear shell model. They are needed throughout this book, so a convenient notation is desirable. Following Baranger [31] we adopt the notation

$$\boxed{|\phi_\alpha\rangle \equiv |\alpha\rangle \equiv |a\,m_\alpha\rangle\,, \quad a = n_a l_a j_a\,,} \qquad (3.62)$$

Here $l_a$ and $j_a$ have their usual meanings as the quantum numbers for orbital and total angular momenta of the orbital $a$, while $n_a$ is the additional, energy-related quantum number often called principal quantum number; see (3.40). The quantity $m_\alpha$ is the $z$ projection of $j_a$. The coordinate representation of the state vector $|\alpha\rangle$ is the wave function $\langle\boldsymbol{x}|\alpha\rangle = \phi_\alpha(\boldsymbol{x})$. In detail we have the single-particle wave function

$$\boxed{\langle\boldsymbol{x}|\alpha\rangle = \phi_\alpha(\boldsymbol{x}) = \eta_a g_{n_a l_a}(r)\Big[Y_{l_a}(\varOmega)\chi_{\frac{1}{2}}\Big]_{j_a m_\alpha}\,, \quad \varOmega = (\theta,\varphi)\,.} \qquad (3.63)$$

We have chosen here the radial function as the harmonic oscillator function $g_{n_a l_a}(r)$ discussed in Subsect. 3.2.1. This is the usual choice in microscopic nuclear structure calculations. The wave-function coupling of the orbital angular momentum and the spin is expressed in the notation (2.50) defined for tensor products; thus we have explicitly

$$\left[Y_{l_a}(\Omega)\chi_{\frac{1}{2}}\right]_{j_a m_\alpha} = \sum_{mm'} (l_a \, m \, \tfrac{1}{2} \, m' | j_a \, m_\alpha) Y_{l_a m}(\Omega) \chi_{\frac{1}{2}m'} \,, \qquad (3.64)$$

where $Y_{l_a m}(\Omega)$ is an eigenfunction of orbital angular momentum (spherical harmonic) and $\chi_{\frac{1}{2}m}$ is a spin-$\frac{1}{2}$ eigenspinor. In matrix representation the spinors are

$$\chi_{\frac{1}{2}\frac{1}{2}} \equiv \chi_+ \equiv \chi_\uparrow = \begin{pmatrix} 1 \\ 0 \end{pmatrix} \,, \qquad (3.65)$$

$$\chi_{\frac{1}{2},-\frac{1}{2}} \equiv \chi_- \equiv \chi_\downarrow = \begin{pmatrix} 0 \\ 1 \end{pmatrix} \,. \qquad (3.66)$$

The *phase factor* $\eta_a$ in (3.63) depends on the adopted phase convention. There are two widely used phase conventions, namely

$$\eta_a = \begin{cases} 1 & \text{Condon–Shortley phase convention} \,, \\ i^{l_a} & \text{Biedenharn–Rose phase convention} \,. \end{cases} \qquad (3.67)$$

The CS phase convention [32] is common in the literature on the nuclear shell model, whereas the BR convention [33] is used when dealing with BCS (Bardeen–Cooper–Schrieffer) quasiparticles. With the BR phase convention the BCS occupation amplitudes $u_a(\text{BR})$ are all positive, whereas within the CS phase convention the sign is given by $u_a(\text{CS}) = (-1)^{l_a} u_a(\text{BR})$. Here we adopt the Condon–Shortley phase convention, but keep the discussion so general that it is easy to switch between the two conventions.

The single-particle wave functions $\phi_\alpha(\boldsymbol{x})$ form an orthonormal and complete set:

$$\langle \alpha | \beta \rangle = \int \phi_\alpha^\dagger(\boldsymbol{x}) \phi_\beta(\boldsymbol{x}) \mathrm{d}^3 \boldsymbol{r} = \delta_{\alpha\beta} \quad \text{(orthonormality)} \,, \qquad (3.68)$$

$$\sum_\alpha \phi_\alpha(\boldsymbol{x}) \phi_\alpha^\dagger(\boldsymbol{x}') = \delta(\boldsymbol{r} - \boldsymbol{r}') \quad \text{(completeness)} \,, \qquad (3.69)$$

where the bra part contains Hermitian conjugation, rather than mere complex conjugation, because of the presence of the spinor $\chi_{\frac{1}{2}m_s}$. The orthonormality and completeness relations for the spin-$\frac{1}{2}$ spinors are

$$\chi_{\frac{1}{2}m}^\dagger \chi_{\frac{1}{2}m'} = \delta_{mm'} \,, \qquad (3.70)$$

$$\sum_m \chi_{\frac{1}{2}m} \chi_{\frac{1}{2}m}^\dagger = 1_2 \,, \qquad (3.71)$$

where $1_2$ is the unit two-by-two matrix.

The delta symbol in (3.68) contains all the quantum numbers of (3.62), i.e.

$$\delta_{\alpha\beta} = \delta_{n_a n_b} \delta_{l_a l_b} \delta_{j_a j_b} \delta_{m_\alpha m_\beta} \ . \tag{3.72}$$

The $A$-nucleon wave function has to be *antisymmetric* with respect to the exchange of any two protons or any two neutrons, due to the quantum statistics of identical fermions. We write the $A$-nucleon wave function as a product of a proton factor and a neutron factor,

$$\Psi_A(\boldsymbol{x}_1, \boldsymbol{x}_2, \ldots, \boldsymbol{x}_A) = \Psi_Z(\boldsymbol{x}_1, \boldsymbol{x}_2, \ldots, \boldsymbol{x}_Z) \Psi_N(\boldsymbol{y}_1, \boldsymbol{y}_2, \ldots, \boldsymbol{y}_N) \ . \tag{3.73}$$

Applied to the protons, the antisymmetry requirement is

$$\begin{aligned} P_{ij} \Psi_Z(\boldsymbol{x}_1, \ldots, \boldsymbol{x}_i, \ldots, \boldsymbol{x}_j, \ldots, \boldsymbol{x}_Z) &= \Psi_Z(\boldsymbol{x}_1, \ldots, \boldsymbol{x}_j, \ldots, \boldsymbol{x}_i, \ldots, \boldsymbol{x}_Z) \\ &= -\Psi_Z(\boldsymbol{x}_1, \ldots, \boldsymbol{x}_i, \ldots, \boldsymbol{x}_j, \ldots, \boldsymbol{x}_Z) \end{aligned} \tag{3.74}$$

for any exchange $(i,j)$. Antisymmetry is similarly required of $\Psi_N$. As noted already in Subsect. 3.1.1, an antisymmetric wave function can be written as a Slater determinant. The $Z$-proton wave function thus becomes

$$\begin{aligned} \Psi_Z = \Psi_{\pi_1 \pi_2 \cdots \pi_Z}(\boldsymbol{x}_1, \boldsymbol{x}_2, \ldots, \boldsymbol{x}_Z) &= \mathcal{A}\left[\prod_{i=1}^{Z} \phi_{\pi_i}(\boldsymbol{x}_i)\right] \\ &= \frac{1}{\sqrt{Z!}} \begin{vmatrix} \phi_{\pi_1}(\boldsymbol{x}_1) & \phi_{\pi_1}(\boldsymbol{x}_2) & \cdots & \phi_{\pi_1}(\boldsymbol{x}_Z) \\ \phi_{\pi_2}(\boldsymbol{x}_1) & \phi_{\pi_2}(\boldsymbol{x}_2) & \cdots & \phi_{\pi_2}(\boldsymbol{x}_Z) \\ \vdots & \vdots & \ddots & \vdots \\ \phi_{\pi_Z}(\boldsymbol{x}_1) & \phi_{\pi_Z}(\boldsymbol{x}_2) & \cdots & \phi_{\pi_Z}(\boldsymbol{x}_Z) \end{vmatrix} , \end{aligned} \tag{3.75}$$

where the label $\pi_i$ contains all the quantum numbers of a proton orbital.

We note in passing that the antisymmetrizer can be expressed formally as

$$\mathcal{A} = \frac{1}{\sqrt{Z!}} \sum_{P \in S_Z} \text{sign}(P) \prod_{ij} P_{ij} \ , \tag{3.76}$$

where $P_{ij}$ exchanges the pair $(i,j)$, $\text{sign}(P) = -1$ for an odd number of pair exchanges, and $S_Z$ is the permutation group of $Z$ particles that includes all possible $Z$-particle permutations.

Combining the similar proton and neutron parts we can write the total $A$-nucleon wave function as

$$\begin{aligned} &\Psi_{\pi_1 \pi_2 \cdots \pi_Z \nu_1 \nu_2 \cdots \nu_N}(\boldsymbol{x}_1, \boldsymbol{x}_2, \ldots, \boldsymbol{x}_Z, \boldsymbol{y}_1, \boldsymbol{y}_2, \ldots, \boldsymbol{y}_N) \\ &\qquad = \Psi_{\pi_1 \pi_2 \cdots \pi_Z}(\boldsymbol{x}_1, \boldsymbol{x}_2, \ldots, \boldsymbol{x}_Z) \Psi_{\nu_1 \nu_2 \cdots \nu_N}(\boldsymbol{y}_1, \boldsymbol{y}_2, \ldots, \boldsymbol{y}_N) \ , \end{aligned} \tag{3.77}$$

where the labels $\nu_i$ contain the quantum numbers of the neutron single-particle orbitals.

All properties of the antisymmetric $Z$-proton ($N$-neutron) wave functions follow from the properties of the determinant (3.75). Taking $\pi_i = \pi_j$ for a given pair $i, j$ means that the $i$th and $j$th rows of the determinant are equal. Then the determinant vanishes, and so the $Z$-proton wave function vanishes. This is nothing but the *Pauli exclusion principle*: two identical nucleons may not occupy the same quantum state $\phi_\pi$.

Let us discuss now the possible mean-field proton *configurations* given by sets of the quantum numbers $(\pi_1, \pi_2, \ldots, \pi_Z)$. Due to the spherical symmetry of the Hamiltonian, all the $2j_p + 1$ proton single-particle states $|p\,m_\pi\rangle$, $m_\pi = -j_p, -j_p + 1, \ldots, j_p - 1, j_p$ are *degenerate*, i.e. they have the same energy. As an example we can take the neon (Ne) nucleus with $Z = 10$. The proton ground-state configuration can be schematically written as[2]

$$\Psi_0^{(Z=10)}(\boldsymbol{x}_1, \boldsymbol{x}_2, \ldots, \boldsymbol{x}_{10})$$
$$= \mathcal{A}\big[\phi_{0\mathrm{s}_{1/2}}(\boldsymbol{x}_1)\phi_{0\mathrm{s}_{1/2}}(\boldsymbol{x}_2)\phi_{0\mathrm{p}_{3/2}}(\boldsymbol{x}_3)\cdots\phi_{0\mathrm{d}_{5/2}}(\boldsymbol{x}_{10})\big] \ . \quad (3.78)$$

This state is constructed on the principle that the $0\mathrm{s}_{1/2}$ orbital can accommodate two protons, the $0\mathrm{p}_{3/2}$ orbital four protons, the $0\mathrm{p}_{1/2}$ orbital two protons and, finally, the $0\mathrm{d}_{5/2}$ orbital six protons. For simplicity of notation, the proton label $\pi$ has been suppressed since only proton orbitals are involved. Proton configurations for $Z = 10$ are depicted in Fig. 3.6.

When the protons move independently in the mean field, their contribution to the ground-state energy is given by

$$E_0^{(Z=10)} = 2\varepsilon_{0\mathrm{s}_{1/2}} + 4\varepsilon_{0\mathrm{p}_{3/2}} + 2\varepsilon_{0\mathrm{p}_{1/2}} + 2\varepsilon_{0\mathrm{d}_{5/2}} \ . \quad (3.79)$$

Treating the first excited proton state similarly to the ground state we have its wave function

$$\Psi_1^{(Z=10)}(\boldsymbol{x}_1, \boldsymbol{x}_2, \ldots, \boldsymbol{x}_{10})$$
$$= \mathcal{A}\big[\phi_{0\mathrm{s}_{1/2}}(\boldsymbol{x}_1)\phi_{0\mathrm{s}_{1/2}}(\boldsymbol{x}_2)\cdots\phi_{0\mathrm{d}_{5/2}}(\boldsymbol{x}_9)\phi_{1\mathrm{s}_{1/2}}(\boldsymbol{x}_{10})\big] \ , \quad (3.80)$$

with the associated energy

$$E_1^{(Z=10)} = 2\varepsilon_{0\mathrm{s}_{1/2}} + 4\varepsilon_{0\mathrm{p}_{3/2}} + 2\varepsilon_{0\mathrm{p}_{1/2}} + \varepsilon_{0\mathrm{d}_{5/2}} + \varepsilon_{1\mathrm{s}_{1/2}} \ . \quad (3.81)$$

This and some other excited-state configurations are illustrated in Fig. 3.6. So far we have looked only at the protons of the neon nucleus. However, to obtain the states of a given isotope $^A_{10}\mathrm{Ne}_N$, say $^{20}_{10}\mathrm{Ne}_{10}$, one has to take into account also the neutron configurations.

Below we note some further aspects of the preceding discussion.

---

[2] The various possible intermediate and final angular momentum couplings are disregarded.

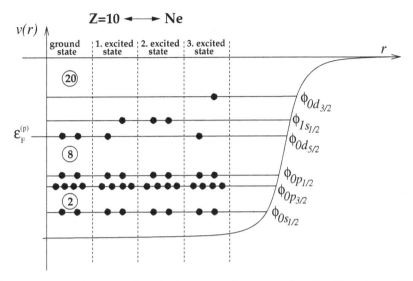

**Fig. 3.6.** Ground-state and excited-state configurations for the 10 protons of the neon nucleus. The symbol $\varepsilon_F^{(\pi)}$ denotes the position of the proton Fermi level, which is located at the last occupied single-particle level in the ground-state configuration

- In our example of proton configurations in Ne the *excitation energy* from the ground-state configuration to the first excited configuration is

$$E_1^{(Z=10)} - E_0^{(Z=10)} = \varepsilon_{1s_{1/2}} - \varepsilon_{0d_{5/2}} \,, \tag{3.82}$$

corresponding to a jump of one proton from the $d_{5/2}$ orbital to the $s_{1/2}$ orbital. The second and third excited states can be interpreted as similar jumps of one or two protons from the ground-state configuration, as shown in Fig. 3.6.

- Each $Z$-proton ($N$-neutron) Slater determinant corresponds to one $Z$-proton ($N$-neutron) configuration.

- The energy of the highest occupied single-particle state in the ground-state configuration is called the *Fermi energy*. In a nucleus, we have both a proton and a neutron Fermi energy. The *Fermi level*, or Fermi surface, is located at the Fermi energy.

Before closing this chapter, let us generalize the orthonormality and completeness relations (3.68) and (3.69) to many-nucleon configurations. The $Z$-proton Slater determinants form an *orthonormal* and *complete* set of wave functions. For orthonormality we have

$$\int d^3r_1 d^3r_2 \cdots d^3r_Z \Psi^\dagger_{\pi_1 \pi_2 \cdots \pi_Z}(\boldsymbol{x}_1, \boldsymbol{x}_2, \ldots, \boldsymbol{x}_Z) \Psi_{\pi'_1 \pi'_2 \cdots \pi'_Z}(\boldsymbol{x}_1, \boldsymbol{x}_2, \ldots, \boldsymbol{x}_Z)$$
$$= \delta_{\pi_1 \pi'_1} \delta_{\pi_2 \pi'_2} \cdots \delta_{\pi_Z \pi'_Z} \quad \text{(orthonormality)} \, .$$

$$(3.83)$$

The completeness relation is

$$\sum_{\pi_1 \cdots \pi_Z} \Psi_{\pi_1 \pi_2 \cdots \pi_Z}(\boldsymbol{x}_1, \boldsymbol{x}_2, \ldots, \boldsymbol{x}_Z) \Psi^\dagger_{\pi_1 \pi_2 \cdots \pi_Z}(\boldsymbol{x}'_1, \boldsymbol{x}'_2, \ldots, \boldsymbol{x}'_Z)$$
$$= \delta(\boldsymbol{r}_1 - \boldsymbol{r}'_1) \delta(\boldsymbol{r}_2 - \boldsymbol{r}'_2) \cdots \delta(\boldsymbol{r}_Z - \boldsymbol{r}'_Z) \quad \text{(completeness)} \, ,$$

$$(3.84)$$

where the sum runs over all possible $Z$-proton configurations. Similar expressions are valid for the neutrons.

As noted in (3.63), the relation between a one-proton state vector $|\pi\rangle$ and its coordinate representation, the wave function $\phi_\pi(\boldsymbol{x})$, is $\phi_\pi(\boldsymbol{x}) = \langle \boldsymbol{x}|\pi\rangle$, where $|\boldsymbol{x}\rangle$ is the state vector describing a particle in the infinitesimal vicinity of $\boldsymbol{x}$ with a definite spin (see e.g. [6]). We now generalize this notion to the case of $Z$ protons:

$$\Psi_{\pi_1 \pi_2 \cdots \pi_Z}(\boldsymbol{x}_1, \boldsymbol{x}_2, \ldots, \boldsymbol{x}_Z) = \langle \boldsymbol{x}_1, \boldsymbol{x}_2, \ldots, \boldsymbol{x}_Z | \pi_1 \, \pi_2 \, \cdots \, \pi_Z \rangle \, , \qquad (3.85)$$

where $|\pi_1 \, \pi_2 \, \cdots \, \pi_Z\rangle$ is an antisymmetrized state vector with the single-particle states $|\pi_1\rangle, \ldots, |\pi_Z\rangle$ occupied. A corresponding expression applies to neutrons.

## Epilogue

In this chapter we introduced and discussed the very basic concept of the nuclear shell model, the nuclear mean field. It was conjectured that a large part of the nucleon–nucleon interaction can be hidden in Slater determinants that describe non-interacting nucleons moving in the mean-field orbitals. These wave functions can be considered as zeroth-order approximations to the true many-body wave functions of $A$ interacting nucleons. In the remaining chapters we present and investigate various approximation schemes for the true many-nucleon wave functions. In all of these more sophisticated schemes the notion of the mean-field single-particle space plays a crucial role.

## Exercises

**3.1.** Using the Wigner–Eckart theorem prove that the eigenvalues of a scalar Hamiltonian like (3.24) are independent of the projection quantum number $m$.

**3.2.** Derive (3.56).

**3.3.** By using (3.21), (3.43) and (3.44) show that

$$\frac{R}{b} \approx 1.26 A^{1/6} \ . \tag{3.86}$$

**3.4.** Derive an expression for the Laguerre polynomial $L_3^{(l+\frac{1}{2})}(x)$ by using the relations (3.53)–(3.55).

**3.5.** Verify the numerical values and $A$ dependence in (3.57) and (3.58).

**3.6.** In (3.60) one can substitute the harmonic oscillator radial wave functions for the Woods–Saxon ones to produce

$$C_{nl} \approx \int_0^\infty r^2 dr \, g_{nl}^2(r) v_{\mathrm{LS}}(r) \ . \tag{3.87}$$

Show that to a fair approximation

$$C_{nl} \approx v_{\mathrm{LS}}^{(0)} \left(\frac{r_0}{\hbar}\right)^2 R g_{nl}^2(R) \ , \tag{3.88}$$

where $R$ is the nuclear radius and the constant $v_{\mathrm{LS}}^{(0)}$ is given in (3.28).

**3.7.** By using (3.88) show that one can write

$$\Delta_{0l}^{\mathrm{LS}} \approx 22.44 A^{-1/6} \exp\left(-1.60 A^{1/3}\right) \frac{2^{l+1}}{\sqrt{\pi}(2l-1)!!} \left(1.60 A^{1/3}\right)^l \ \mathrm{MeV} \ . \tag{3.89}$$

**3.8.** Verify the relations (3.70) and (3.71).

**3.9.** Verify the validity of the orthonormality relation (3.68) for the harmonic oscillator wave functions (3.63).

**3.10.** Verify the validity of the completeness relation (3.69) by assuming the completeness

$$\sum_{n_a l_a m_\alpha} F_{n_a l_a m_\alpha}^*(\boldsymbol{r}) F_{n_a l_a m_\alpha}(\boldsymbol{r}') = \delta(\boldsymbol{r} - \boldsymbol{r}') \ ,$$

$$F_{n_a l_a m_\alpha}(\boldsymbol{r}) \equiv g_{n_a l_a}(r) Y_{l_a m_\alpha}(\Omega) \ , \tag{3.90}$$

of the harmonic oscillator wave functions.

**3.11.** Calculate the values of the overlap intergrals

$$\int_0^\infty g_{0\mathrm{s}}(r) g_{0\mathrm{p}}(r) r^2 dr \ , \quad \int_0^\infty g_{0\mathrm{s}}(r) g_{0\mathrm{d}}(r) r^2 dr \ ,$$

where the $g_{nl}(r)$ are harmonic oscillator wave functions (3.42).

**3.12.** Calculate the spin–orbit splitting of the 0p orbital in the nucleus $^{12}$C. Assume that the spin–orbit interaction is

$$v_{LS} = v_0 \boldsymbol{L} \cdot \boldsymbol{S}/\hbar^2 , \quad v_0 = -20A^{-2/3} \, \text{MeV} , \qquad (3.91)$$

where $v_0$ has been estimated from experimental data.

**3.13.** Construct a harmonic oscillator potential for the neutrons of $^{16}$O such that it is equivalent to the Woods–Saxon potential used to produce the data given in Table 3.1. To achieve this, select $\hbar\omega$ to equal the Woods–Saxon $0p_{3/2}$ excitation energy. Then choose $V_1$ so that the $0s_{1/2}$ energy is the same for both potentials. Sketch a figure like Fig. 3.2.

**3.14.** Consider the protons of the nucleus $^{40}$Ca. Write down the radial Woods–Saxon wave function for each of the orbitals $0s_{1/2}$, $0p_{3/2}$ and $0p_{1/2}$ in terms of harmonic oscillator wave functions. Use the data of Table 3.4 and include the first three terms. Plot the functions.

# 4

## Occupation Number Representation

## Prologue

The first two chapters of this book presented angular momentum algebra as the basic tool of nuclear theory. That includes angular momentum coupling coefficients, spherical tensor operators and reduced matrix elements. In the preceding chapter we introduced the mean-field concept, along with associated many-nucleon wave functions, Slater determinants, describing configurations of non-interacting particles in mean-field single-particle orbitals.

In this chapter we introduce an alternative to the traditional wave function concept, namely the occupation number representation. This new concept can be viewed as a clever book-keeping method for manipulating many-nucleon wave functions and their matrix elements. The method incorporates the notion of a Slater determinant in such a way that mathematical manipulations in actual calculations are much more elegant and tractable than when operating directly with the many-nucleon wave function.

In occupation number representation, state vectors of different particle numbers can be discussed within the same formalism by introducing the so-called Fock space with its particle creation and annihilation operators. The fermionic character of nucleons is taken into account by anticommutation rules for these operators. Towards the end of this chapter we present a powerful method, known as Wick's theorem, for calculating nuclear matrix elements and apply it to derive the Hartree–Fock equation.

## 4.1 Occupation Number Representation of Many-Nucleon States

Manipulation of many-nucleon wave functions, the Slater determinants of Sect. 3.2, can be tedious even if the number of nucleons is small. To make things easier and more systematic, an efficient book-keeping method is needed. The introduction of such a method is the subject of the present chapter.

We start by discussing the basic principles of many-body quantum mechanics from the point of view of particle creation and annihilation. Conventionally this type of approach to the quantum many-body problem is called *second quantization*. A more appropriate name, however, is *occupation number representation*.

### 4.1.1 Fock Space: Particle Creation and Annihilation

States in occupation number representation exist in so-called Fock space. This space is constructed as follows. Consider a given set of single-particle states $\{|\alpha_i\rangle\}$. Build from them the complete set of $N$-particle basis states, Slater determinants:

$$\{|\alpha_1 \alpha_2 \ldots \alpha_N\rangle\}_{\alpha_1 \alpha_2 \cdots \alpha_N} \,.$$

The notation means that all basis vectors with different sets of $N$ single-particle quantum numbers, i.e. all $N$-particle configurations, are included. These vectors span the Hilbert space for $N$ identical fermions in the given single-particle states. The totality of these Hilbert spaces with different values of $N$ is the *Fock space* built on the single-particle states.

The basis vectors of a Fock space, which are *Fock vectors*, can be specified by listing the occupation numbers of the single-particle states, i.e.

$$|\alpha_1 \alpha_2 \ldots \alpha_N\rangle = |n_1 \, n_2 \, n_3 \, \ldots\rangle \,, \tag{4.1}$$

where

$$n_i = \begin{cases} 1 \,, & \text{if } i \in \{\alpha_1, \ldots, \alpha_N\} \\ 0 \,, & \text{if } i \notin \{\alpha_1, \ldots, \alpha_N\} \end{cases} \,, \quad \sum_{i=1}^{\infty} n_i = N \,. \tag{4.2}$$

So, instead of having a set of $N$ single-particle quantum numbers, in occupation number representation we have $N$ occupation numbers $n_i = 1$ for the occupied single-particle orbitals $i$.

Fock space is convenient for handling operators that change particle number. We denote such operators by $c_\alpha^\dagger$ and its Hermitian conjugate $c_\alpha$, and they have the following properties:

$c_\alpha^\dagger$ gives birth to a nucleon in state $|\alpha\rangle$ (*creation operator*),

$c_\alpha$ deletes a nucleon from state $|\alpha\rangle$ (*annihilation operator*).

Their action on a Fock vector $|n_1 \, n_2 \, n_3 \ldots\rangle$ is defined by

$$c_\alpha^\dagger |n_1 \ldots n_\alpha \ldots\rangle = \begin{cases} \eta_\alpha |n_1 \ldots n_\alpha + 1 \ldots\rangle \,, & \text{if } n_\alpha = 0 \,, \\ 0 \,, & \text{if } n_\alpha = 1 \,, \end{cases} \tag{4.3}$$

$$c_\alpha |n_1 \ldots n_\alpha \ldots\rangle = \begin{cases} \eta_\alpha |n_1 \ldots n_\alpha - 1 \ldots\rangle \,, & \text{if } n_\alpha = 1 \,, \\ 0 \,, & \text{if } n_\alpha = 0 \,, \end{cases} \tag{4.4}$$

where

$$\eta_\alpha \equiv (-1)^{\sum_{\beta < \alpha} n_\beta} = \pm 1 . \tag{4.5}$$

The phase factor $\eta_\alpha$ can be justified by using the following construction. Let us write the $N$-particle state as

$$|\alpha_1 \, \alpha_2 \dots \alpha_N\rangle \equiv c^\dagger_{\alpha_1} c^\dagger_{\alpha_2} \cdots c^\dagger_{\alpha_N} |0\rangle = \prod_\alpha \left( c^\dagger_\alpha \right)^{n_\alpha} |0\rangle , \tag{4.6}$$

where the *particle vacuum* $|0\rangle$ is defined as

$$|0\rangle \equiv |0\,0\,0\ldots\rangle . \tag{4.7}$$

Note the convention we adopt for the order of writing creation operators; this is our *standard order*. For the vacuum we have from (4.4)

$$c_\alpha |0\rangle = 0 \quad \text{for all } \alpha . \tag{4.8}$$

The phase factor $\eta_\alpha$ relates to changes in the ordering of creation operators in (4.6). Since the creation operators create, and the annihilation operators annihilate, *fermions*, they satisfy the *anticommutation relations*

$$\begin{aligned} \left\{ c_\alpha, c_\beta \right\} &= 0 , \quad \left\{ c^\dagger_\alpha, c^\dagger_\beta \right\} = 0 \quad \text{for all } \alpha, \beta, \\ \left\{ c_\alpha, c^\dagger_\beta \right\} &= \delta_{\alpha\beta} , \end{aligned} \tag{4.9}$$

where $\{\cdots\}$ is an *anticommutator*:

$$\{A, B\} \equiv AB + BA . \tag{4.10}$$

The phase factor $\eta_\alpha$ depends on the standard order of creation operators.[1] One has to *anticommute* a creation or an annihilation operator through $\sum_{\beta < \alpha} n_\beta$ creation operators when acting on a state vector $|n_1 \, n_2 \, n_3 \ldots\rangle$ of $N$ particles to produce a state vector of $N+1$ or $N-1$ particles with its creation operators in standard order.

Equation (4.6) defines the order in which an ordered set, labelled $1, 2, \ldots$, of creation operators are applied to the vacuum state. However, given a set of creation operators, in practice one has to label them first in a definite order. In this work we label according to increasing single-particle energy. Thus our labelling order is $0s_{1/2}$, $0p_{3/2}$, $0p_{1/2}$, $0d_{5/2}$, $1s_{1/2}$, $0d_{3/2}$, $0f_{7/2}$, .... The degenerate angular momentum substates we order in a sequence of increasing $m$ quantum number.

---

[1] The standard order opposite to ours is also known in the literature, e.g. [17], so that $|\alpha_1 \, \alpha_2 \dots \alpha_N\rangle \equiv c^\dagger_{\alpha_N} c^\dagger_{\alpha_{N-1}} \cdots c^\dagger_{\alpha_1} |0\rangle$.

## 4.1.2 Further Properties of Creation and Annihilation Operators

We list here some specific properties of the creation and annihilation operators. As noted above in passing, creation and annihilation operators are related by Hermitian conjugation,

$$\left(c_\alpha^\dagger\right)^\dagger = c_\alpha . \tag{4.11}$$

The coordinate representation, i.e. the wave function, of a single-particle state is obtained as

$$\phi_\alpha(\boldsymbol{x}) = \langle \boldsymbol{x}|\alpha\rangle = \langle \boldsymbol{x}|c_\alpha^\dagger|0\rangle . \tag{4.12}$$

Equation (3.63) shows that the two different phase conventions (3.67) for single-particle wave functions imply corresponding phase conventions for creation operators. The Condon–Shortley creation operator $c_\alpha^\dagger(\mathrm{CS})$ and the Biedenharn–Rose one $c_\alpha^\dagger(\mathrm{BR})$ are related through

$$c_\alpha^\dagger(\mathrm{BR}) = \mathrm{i}^{l_a} c_\alpha^\dagger(\mathrm{CS}) , \tag{4.13}$$

leading to

$$c_\alpha(\mathrm{BR}) = (-1)^{l_a} \mathrm{i}^{l_a} c_\alpha(\mathrm{CS}) . \tag{4.14}$$

The commutation relations of creation and annihilation operators are, however, the *same* for the CS and BR phase conventions, as given by (4.9).

So far we have considered only one given species of identical fermions, either protons or neutrons. The wave function of $N$ protons or $N$ neutrons is of the form (4.6); in coordinate representation it is a Slater determinant like (3.75).[2] Such a wave function is antisymmetric under the exchange of any two particles. A wave function containing both protons and neutrons is of the form (3.77), with no antisymmetry requirement between proton and neutron labels. However, in occupation number representation it is convenient to treat protons and neutrons on an equal footing, so that their mutual creation and annihilation operators anticommute. We thus extend the anticommutation relations (4.9) to include

$$\left\{c_\pi, c_\nu\right\} = 0 , \quad \left\{c_\pi^\dagger, c_\nu^\dagger\right\} = 0 , \quad \left\{c_\pi, c_\nu^\dagger\right\} = 0 . \tag{4.15}$$

This extension is essentially notational and has no physical consequences. In the isospin formalism, introduced in Sect. 5.5, protons and neutrons are treated as members of the same species, the nucleon. Then the anticommutation relations (4.9) apply so that $\alpha$ and $\beta$ include the isospin labels.

One operator that frequently appears in computations of nuclear properties is the *particle number operator* $\hat{n}$, defined as

$$\hat{n} = \hat{n}_\mathrm{p} + \hat{n}_\mathrm{n} = \sum_\pi \hat{n}_\pi + \sum_\nu \hat{n}_\nu = \sum_\pi c_\pi^\dagger c_\pi + \sum_\nu c_\nu^\dagger c_\nu , \tag{4.16}$$

---

[2] States of the form (4.6) are frequently called Slater determinants even when not using coordinate representation.

where $\hat{n}_\mathrm{p}$ and $\hat{n}_\mathrm{n}$ are the particle number operators for protons and neutrons respectively. The sums in these operators run over all proton (labels $\pi$) or neutron (labels $\nu$) single-particle states. The eigenvalue equations for the number operators are

$$\hat{n}_\mathrm{p}|\pi_1\,\pi_2\ldots\pi_Z\,\nu_1\,\nu_2\ldots\nu_N\rangle = Z|\pi_1\,\pi_2\ldots\pi_Z\,\nu_1\,\nu_2\ldots\nu_N\rangle\,, \tag{4.17}$$

$$\hat{n}_\mathrm{n}|\pi_1\,\pi_2\ldots\pi_Z\,\nu_1\,\nu_2\ldots\nu_N\rangle = N|\pi_1\,\pi_2\ldots\pi_Z\,\nu_1\,\nu_2\ldots\nu_N\rangle\,. \tag{4.18}$$

## 4.2 Operators and Their Matrix Elements

In this section we discuss the occupation number representation of one- and two-body operators. In addition, the matrix elements of spherical tensors are discussed in considerable detail.

### 4.2.1 Occupation Number Representation of One-Body Operators

Let $t$ be a *one-body operator*, e.g. the kinetic energy of a particle. A one-body operator is an operator which acts on the coordinates, including spin, of only one particle at a time. If this is particle number $i$, the one-body operator is $t(i)$. In coordinate representation it is $t(\boldsymbol{x}_i)$, where $\boldsymbol{x}_i$ comprises both the space and spin variables. The total effect of an operator $T$ acting on a nucleus is obtained by summing the actions on individual nucleons, i.e. $T = \sum_{i=1}^{A} t(\boldsymbol{x}_i)$. A familiar example of a one-body operator is the kinetic-energy operator which depends only on the space variables $\boldsymbol{r}_i$.

It can be proved that a general one-body operator can be expressed in occupation number representation as

$$\boxed{\begin{aligned} T &= \sum_{i=1}^{A} t(\boldsymbol{x}_i) = \sum_{\alpha\beta} t_{\alpha\beta}c_\alpha^\dagger c_\beta\,, \\ t_{\alpha\beta} &\equiv \langle\alpha|T|\beta\rangle = \int \phi_\alpha^\dagger(\boldsymbol{x})t(\boldsymbol{x})\phi_\beta(\boldsymbol{x})\mathrm{d}^3\boldsymbol{r}\,. \end{aligned}} \tag{4.19}$$

**Example 4.1**
The angular momentum operators $J_z$ and $J_\pm$ can be expressed in occupation number representation as

$$J_z = \hbar\sum_\alpha m_\alpha c_\alpha^\dagger c_\alpha\,, \quad J_\pm = \hbar\sum_\alpha m_\alpha^\mp c_\alpha^\dagger c_{\alpha\mp1}\,, \tag{4.20}$$

where

$$m_\alpha^\pm \equiv \sqrt{(j_a \pm m_\alpha + 1)(j_a \mp m_\alpha)}\,, \quad c_{\alpha\pm1} \equiv c_{a,m_\alpha\pm1}\,. \tag{4.21}$$

This representation of the angular momentum operators can be conveniently used in the commutation relations (2.12) to test whether a set of operators form a spherical tensor. In this way one can show that the operators $c_\alpha^\dagger$ and $\tilde{c}_\alpha$ are spherical tensors of rank $j_a$. Note that this expands the concept of spherical tensor to odd-half-integer ranks.

For a one-body spherical tensor operator $T_{\lambda\mu}$ one can derive the following, very useful formula:

$$T_{\lambda\mu} = \sum_{\alpha\beta} \langle\alpha|T_{\lambda\mu}|\beta\rangle c_\alpha^\dagger c_\beta = \hat{\lambda}^{-1} \sum_{ab} (a\|\boldsymbol{T}_\lambda\|b) \left[ c_a^\dagger \tilde{c}_b \right]_{\lambda\mu} , \qquad (4.22)$$

where the tilde operator

$$\tilde{c}_\alpha \equiv (-1)^{j_a+m_\alpha} c_{-\alpha} , \quad c_{-\alpha} = c_{a,-m_\alpha} \qquad (4.23)$$

is an annihilation operator with the proper behaviour of a spherical tensor of rank $j_a$. The matrix element $\langle\alpha|T_{\lambda\mu}|\beta\rangle$ is the *single-particle matrix element*, and the matrix element $(a\|\boldsymbol{T}_\lambda\|b)$ is the *reduced single-particle matrix element*. The matrix elements carry the information about the properties of the one-body operator involved.

The single-particle matrix elements completely characterize the operator; they have nothing to do with the many-body aspects of nuclear structure. The many-nucleon aspects are probed by the latter part of $T_{\lambda\mu}$, namely the part containing the particle creation and annihilation operators. One can imagine that the one-body operator probes the nucleus by scattering particles from one single-particle orbital to another. To each scattering it attaches an amplitude, the single-particle matrix element, characterizing the scattering properties of the operator itself.

### 4.2.2 Matrix Elements of One-Body Operators

When evaluating electromagnetic or beta decay rates, one needs to compute matrix elements of the one-body operators in (4.22), of the type

$$\langle \xi_f J_f M_f | T_{\lambda\mu} | \xi_i J_i M_i \rangle = \sum_{\alpha\beta} \langle\alpha|T_{\lambda\mu}|\beta\rangle \langle \xi_f J_f M_f | c_\alpha^\dagger c_\beta | \xi_i J_i M_i \rangle , \qquad (4.24)$$

where $\xi$ represents one or more additional quantum numbers needed to completely specify each state of the system. Applying the Wigner–Eckart theorem to this yields the very important expression

$$(\xi_f J_f \|\boldsymbol{T}_\lambda\| \xi_i J_i) = \hat{\lambda}^{-1} \sum_{ab} (a\|\boldsymbol{T}_\lambda\|b)(\xi_f J_f \| \left[ c_a^\dagger \tilde{c}_b \right]_\lambda \| \xi_i J_i) . \qquad (4.25)$$

The reduced matrix element (4.25) is called the *transition amplitude*; when referring to a decay process it can also be called the *decay amplitude*. The matrix element $\langle \xi_f J_f M_f | c_\alpha^\dagger c_\beta | \xi_i J_i M_i \rangle$ is called the *one-body transition density*, and the reduced matrix element $(\xi_f J_f \| \left[ c_a^\dagger \tilde{c}_b \right]_\lambda \| \xi_i J_i)$ is called the *reduced one-body transition density.* The transition densities characterize the many-nucleon properties of the initial $(i)$ and final $(f)$ nuclear states. They do not contain any information about the transition operator beyond its one-body character.

### 4.2.3 Occupation Number Representation of Two-Body Operators

A *two-body operator* $v(i,j)$, e.g. the potential energy of interaction, acts simultaneously on the observables of two particles $i$ and $j$. One can sum all the pairwise actions of this operator to produce the total action $V$. In coordinate representation its expression is

$$V = \sum_{i<j} v(\boldsymbol{x}_i, \boldsymbol{x}_j) = \tfrac{1}{2} \sum_{i \neq j} v(\boldsymbol{x}_i, \boldsymbol{x}_j) \,, \tag{4.26}$$

where we allow also a spin dependence for the two-body operator. It can be shown that in occupation number representation

$$\boxed{\begin{aligned} V &= \tfrac{1}{2} \sum_{\alpha\beta\gamma\delta} v_{\alpha\beta\gamma\delta} c_\alpha^\dagger c_\beta^\dagger c_\delta c_\gamma \,, \\ v_{\alpha\beta\gamma\delta} &= \int \phi_\alpha^\dagger(\boldsymbol{x}_1)\phi_\beta^\dagger(\boldsymbol{x}_2)v(\boldsymbol{x}_1,\boldsymbol{x}_2)\phi_\gamma(\boldsymbol{x}_1)\phi_\delta(\boldsymbol{x}_2)\mathrm{d}^3 r_1 \mathrm{d}^3 r_2 \,. \end{aligned}} \tag{4.27}$$

In addition to spatial coordinates and spin variables the two-body operator $v(i,j)$ can contain isospin dependence. In that case the single-particle wave functions carry also isospin labels. An isospin-dependent two-nucleon interaction is an important practical example.

It is convenient to write the two-nucleon interaction $V$ by using an *antisymmetrized two-nucleon interaction matrix element* $\bar{v}_{\alpha\beta\gamma\delta}$:

$$\boxed{V = \tfrac{1}{4} \sum_{\alpha\beta\gamma\delta} \bar{v}_{\alpha\beta\gamma\delta} c_\alpha^\dagger c_\beta^\dagger c_\delta c_\gamma \,, \quad \bar{v}_{\alpha\beta\gamma\delta} \equiv v_{\alpha\beta\gamma\delta} - v_{\alpha\beta\delta\gamma} \,,} \tag{4.28}$$

with the symmetry properties[3]

$$\bar{v}_{\alpha\beta\gamma\delta} = -\bar{v}_{\beta\alpha\gamma\delta} = -\bar{v}_{\alpha\beta\delta\gamma} = \bar{v}_{\beta\alpha\delta\gamma} = \bar{v}_{\gamma\delta\alpha\beta}^* \,. \tag{4.29}$$

These symmetries are helpful in actual calculations. It is worth noting that the antisymmetrized matrix element (4.29) is the same as the *normalized* and *antisymmetrized* two-body matrix element defined by some authors, e.g. [17]:

$$\bar{v}_{\alpha\beta\gamma\delta} = \text{n.as.}\langle\alpha\beta|V|\gamma\delta\rangle_{\text{n.as.}} \,. \tag{4.30}$$

---

[3] See also (13.127).

## 4.3 Evaluation of Many-Nucleon Matrix Elements

We now discuss the evaluation of many-particle matrix elements of the type

$$\langle \alpha_1\, \alpha_2 \ldots \alpha_N | \mathcal{O} | \beta_1\, \beta_2 \ldots \beta_{N'} \rangle \,,$$

where $\mathcal{O}$ is an operator corresponding to a nuclear observable. Computation of matrix elements of this type is considerably simplified by resorting to a particular formalism that paves the way to Wick's theorem. This formalism consists of several auxiliary operations, called normal ordering, contraction, etc. These operations are discussed in detail in this section.

### 4.3.1 Normal Ordering

The development of our mathematical machinery starts by introducing the normal-ordered product $\mathcal{N}[\cdots]$. Let $\prod(\cdots)$ be a product of creation operators from the set $\{A_k^\dagger\}$ and annihilation operators from the set $\{A_k\}$ (annihilation operators relative to a vacuum $|\Psi_0\rangle$),

$$\prod = \prod\left(\{A_k\}, \{A_l^\dagger\}\right) \,.$$

Then $\mathcal{N}[\prod]$ is the *normal-ordered product* of the operators if

$$\boxed{\mathcal{N}[\prod] = (-1)^P \prod \left(\text{creation} \times \text{annihilation}\right) \,,} \qquad (4.31)$$

where $P$ is the number of transpositions (exchanges of places of two particles) needed to transport all the annihilation operators to the right of the creation operators. For fermion single-particle operators, transpositions of a creation operator and an annihilation operator ignore the $\delta_{\alpha\beta}$ part of the anticommutation rule in (4.9). Thus in the normal-ordering process all fermion creation and annihilation operators are taken to anticommute. This is just a definition of a mathematical procedure, not a violation of the fermion nature of the operators.

The final normal-ordered product is not uniquely defined since one can anticommute among the creation operators and among the annihilation operators according to (4.9). All forms obtained by such anticommutation are considered to be equivalent.

**Example 4.2**
Consider the following normal ordering of a product, where the annihilation operators $c_\alpha$ annihilate the *particle vacuum* $|0\rangle$:

$$\mathcal{N}\left[c_\alpha^\dagger c_\beta c_\gamma^\dagger c_\delta\right] = (-1)^1 c_\alpha^\dagger c_\gamma^\dagger c_\beta c_\delta = (-1)^2 c_\alpha^\dagger c_\gamma^\dagger c_\delta c_\beta \,.$$

The resulting two forms of the normal-ordered product are equivalent. The first phase factor arises from the single transposition exchanging the places of $c_\beta$ and $c_\gamma^\dagger$. This is the relevant transposition. The second transposition is allowed by the anticommutation rule of two annihilation operators and thus produces an equivalent form of normal ordering.

A frequently used alternative notation for the normal ordering is

$$\mathcal{N}[ABC\cdots] \equiv \;: ABC\cdots :\,, \tag{4.32}$$

where $A, B, C, \ldots$ are arbitrary operators and the two colons separate the product to be normal ordered.

It is important to recognize that normal ordering is always relative to a given vacuum $|\Psi_0\rangle$. Usually, but not always, the annihilation operators involved are defined with respect to this particular vacuum. In this case the vacuum expectation value of the normal-ordered product clearly vanishes. It is also clear that the normal ordering is a linear operation, i.e.

$$\mathcal{N}\big[\lambda_1 \textstyle\prod +\lambda_2 \prod{}'\big] = \lambda_1 \mathcal{N}\big[\textstyle\prod\big] + \lambda_2 \mathcal{N}\big[\textstyle\prod{}'\big]\,, \tag{4.33}$$

$$\mathcal{N}\big[\textstyle\prod (\prod{}' + \prod{}'')\big] = \mathcal{N}\big[\textstyle\prod\prod{}'\big] + \mathcal{N}\big[\textstyle\prod\prod{}''\big] \tag{4.34}$$

for any c-numbers $\lambda_1$ and $\lambda_2$ and any products involved.

### 4.3.2 Contractions

Now we are ready to enter the second phase of development of the machinery for matrix element computation. For a given species of fermion, let $A$ be any creation or annihilation operator and $B$ likewise. Then their *contraction* is defined as

$$\boxed{\overset{\sqcap}{AB} \equiv AB - \mathcal{N}[AB]\,.} \tag{4.35}$$

This is a c-number, as is easily seen. In fact, performing a contraction can be viewed as an operation with respect to the vacuum specified by the normal ordering in the definition (4.35). It is also easily seen that

$$\{A, B\} = 0 \quad \text{implies} \quad \overset{\sqcap}{AB} = 0\,, \tag{4.36}$$

i.e. that the contraction of two anticommuting operators vanishes. Furthermore, if $A$ and $B$ are operators associated with the vacuum $|\Psi_0\rangle$, we have $\langle\Psi_0|\mathcal{N}[AB]|\Psi_0\rangle = 0$. It follows that

$$\langle\Psi_0|AB|\Psi_0\rangle = \langle\Psi_0|\overset{\sqcap}{AB}|\Psi_0\rangle = \overset{\sqcap}{AB}\,.$$

So from the fact that a contraction is a c-number we have the important result

$$\boxed{\overset{\sqcap}{AB} = \langle\Psi_0|AB|\Psi_0\rangle\,.} \tag{4.37}$$

Here the contraction, the related normal ordering and the operators are defined with respect to the *same* vacuum $|\Psi_0\rangle$.

**Example 4.3**
Let us apply (4.37) to particle creation and annihilation operators with respect to the particle vacuum $|0\rangle$:

$$\overset{\frown}{c_k c_l^\dagger} = \langle 0|c_k c_l^\dagger|0\rangle = \delta_{kl} \;, \quad \overset{\frown}{c_k^\dagger c_l} = 0 \;, \quad \overset{\frown}{c_k^\dagger c_l^\dagger} = 0 \;, \quad \overset{\frown}{c_k c_l} = 0 \;. \qquad (4.38)$$

In the general case we define

$$\overset{\frown}{AB} \equiv \langle \phi_0|AB|\phi_0\rangle \qquad (4.39)$$

as a contraction with respect to the vacuum $|\phi_0\rangle$, even if $A$ and $B$ were defined with respect to some other vacuum $|\Psi_0\rangle$. Examples of the use of this definition come later, e.g. in the case where the contractions of particle creation and annihilation operators are taken with respect to the particle–hole vacuum.

### 4.3.3 Wick's Theorem

Next we combine normal-ordered products with contractions. A normal-ordered product containing contractions is of the form

$$\mathcal{N}[\overset{\frown}{AB\overset{\frown}{C}DE} \cdots XYZ] \equiv (-1)^P \overset{\frown}{BE}\,\overset{\frown}{DY}\mathcal{N}[AC \cdots XZ] \;, \qquad (4.40)$$

where $P$ is that number of operator transpositions which takes all contracted pairs to the left of the normal-ordered operator product. With the concepts defined, we now state without proof *Wick's theorem* [34]:

$$
\begin{aligned}
&ABCDEF\cdots \\
&= \mathcal{N}[ABCDEF\cdots] \\
&\quad + \mathcal{N}[\overset{\frown}{A}BCDEF\cdots] + \mathcal{N}[A\overset{\frown}{BC}DEF\cdots] + \text{all other 1-contractions} \\
&\quad + \mathcal{N}[\overset{\frown}{A}\overset{\frown}{BC}DEF\cdots] + \text{all other 2-contractions} \\
&\quad + \cdots \\
&\quad + \text{all normal-ordered terms with } n \text{ contractions} \\
&\quad + \cdots \\
&\quad + \text{all terms with all pairs contracted} \;.
\end{aligned}
$$

$$(4.41)$$

Wick's theorem is valid for any vacuum $|\phi_0\rangle$ when the normal ordering is taken with respect to that vacuum. Since the vacuum expectation value of all normal-ordered terms vanishes we have the compact result

$$\langle\phi_0|ABCDEF\cdots|\phi_0\rangle = \sum_{\substack{\text{all contraction}\\\text{combinations}}} (-1)^{\text{no. of contraction line crossings}}$$

$$\times \text{ product with all pairs contracted .}$$

(4.42)

This formula means that one has to form all possible completely contracted products and sum them with appropriate phase factors. The phase factor is $+1$ if the number of intersections of contraction lines is even and $-1$ if it is odd. Next we show some simple examples of the use of this powerful result.

**Example 4.4**
The normalization of the one-particle states $c_\alpha^\dagger|0\rangle$ can be verified as follows:

$$\langle\alpha|\beta\rangle = \langle0|c_\alpha c_\beta^\dagger|0\rangle = \langle0|\overbracket{c_\alpha c_\beta^\dagger}|0\rangle = (-1)^0 \overbracket{c_\alpha c_\beta^\dagger} = \delta_{\alpha\beta} \;.$$

In this trivial example there are no crossings of contraction lines and the value of the contraction is given in (4.38).

**Example 4.5**
The calculation of the matrix element of a one-body operator $T$ for the one-particle states $c_\alpha^\dagger|0\rangle$ proceeds as follows:

$$\langle\alpha|T|\beta\rangle = \sum_{\alpha'\beta'} t_{\alpha'\beta'} \langle0|c_\alpha c_{\alpha'}^\dagger c_{\beta'} c_\beta^\dagger|0\rangle$$

$$= \sum_{\alpha'\beta'} t_{\alpha'\beta'} \langle0|\overbracket{c_\alpha c_{\alpha'}^\dagger}\,\overbracket{c_{\beta'} c_\beta^\dagger} + \overbracket{c_\alpha c_{\alpha'}^\dagger c_{\beta'} c_\beta^\dagger} + \overbracket{c_\alpha c_{\alpha'}^\dagger}\,\overbracket{c_{\beta'} c_\beta^\dagger}|0\rangle$$

$$= \sum_{\alpha'\beta'} t_{\alpha'\beta'} \big[(-1)^0 \overbracket{c_\alpha c_{\alpha'}^\dagger}\,\overbracket{c_{\beta'} c_\beta^\dagger} + (-1)^1 \overbracket{c_\alpha c_{\beta'}}\,\overbracket{c_{\alpha'}^\dagger c_\beta^\dagger} + (-1)^0 \overbracket{c_\alpha c_\beta^\dagger}\,\overbracket{c_{\alpha'}^\dagger c_{\beta'}}\big]$$

$$= \sum_{\alpha'\beta'} t_{\alpha'\beta'} [\delta_{\alpha\alpha'}\delta_{\beta'\beta} - 0 + 0] = t_{\alpha\beta} \;,$$

(4.43)

where we have taken the contractions from (4.38). This result in fact proves (4.19) for the case of one-particle states.

**Example 4.6**
The overlap of the two-particle states $|\alpha\beta\rangle = c_\alpha^\dagger c_\beta^\dagger|0\rangle$, $\alpha \neq \beta$, is calculated as follows:

$$\langle\alpha\beta|\gamma\delta\rangle = \langle0|c_\beta c_\alpha c_\gamma^\dagger c_\delta^\dagger|0\rangle = \langle0|\overbracket{c_\beta c_\alpha c_\gamma^\dagger c_\delta^\dagger}|0\rangle + \langle0|\overbracket{c_\beta c_\alpha c_\gamma^\dagger c_\delta^\dagger}|0\rangle$$

$$= \delta_{\beta\delta}\delta_{\alpha\gamma} - \delta_{\beta\gamma}\delta_{\alpha\delta} \;.$$

(4.44)

This verifies that $\langle\alpha\beta|\alpha\beta\rangle = 1$. For the two-body interaction $V$, as expressed in (4.28), a similar calculation gives

$$\langle\alpha\beta|V|\gamma\delta\rangle = \bar{v}_{\alpha\beta\gamma\delta} \;.$$

(4.45)

This proves (4.27), via (4.28), for the case of two-particle states. It also confirms that the matrix element (4.30) is indeed computed with antisymmetrized and normalized two-particle states.

## 4.4 Particle–Hole Representation

In this section we define the particle operators $c_\alpha^\dagger, c_\alpha$ and hole operators $h_\alpha^\dagger, h_\alpha$, and the associated vacuum which is annihilated by annihilation operators of both types. This vacuum is called the particle–hole vacuum (or the Hartree–Fock vacuum) and is denoted by $|\mathrm{HF}\rangle$. This vacuum contains $A$ non-interacting nucleons, namely $Z$ protons and $N$ neutrons. The nucleus is assumed to be in its ground state, i.e. the energy levels filled up to the proton and neutron Fermi levels. Then the nucleus is said to be in its particle–hole ground state. The situation is illustrated in Fig. 4.1.

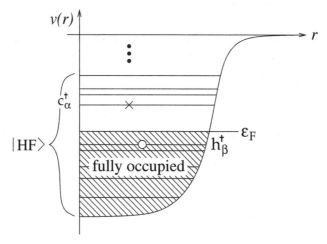

**Fig. 4.1.** Particle–hole vacuum and a particle–hole excitation

This type of a vacuum is reasonable to define for nuclei at closed major shells. These *closed-shell nuclei* or *doubly magic nuclei* display a hierarchy of particle–hole excitations, as is discussed at the end of this section. When departing from a full major shell, this hierarchy is lost. Such *open-shell nuclei* are treated by other methods. One of the most successful of them is the use of BCS quasiparticles, extensively discussed in the latter half of this book.

The operators corresponding to the particle–hole vacuum are defined separately below and above the Fermi surface. The orbitals above the Fermi surface are *particle orbitals* and those below are *hole orbitals*. The notation is

$$
\boxed{
\begin{array}{ll}
c_\alpha^\dagger , \ c_\alpha & \text{when } \varepsilon_\alpha > \varepsilon_F , \\[4pt]
h_\beta^\dagger = \tilde c_\beta , \ h_\beta = \tilde c_\beta^\dagger & \text{when } \varepsilon_\beta \le \varepsilon_F ,
\end{array}
}
\tag{4.46}
$$

where

$$
\tilde c_\beta = (-1)^{j_b + m_\beta} c_{-\beta} , \qquad -\beta = (b, -m_\beta) . \tag{4.47}
$$

The $c_\alpha^\dagger$ create particles above the Fermi surface, whereas the $h_\beta^\dagger$ annihilate particles and thus create holes below the Fermi surface. In this introductory section we take the index $\alpha$ to denote an orbital above the Fermi surface ($\varepsilon_\alpha > \varepsilon_F$) and the index $\beta$ to denote one below ($\varepsilon_\beta \le \varepsilon_F$). In subsequent sections this convention is abandoned and the location of an orbital relative to the Fermi surface, whenever known, is indicated explicitly, e.g. under a summation symbol.

### 4.4.1 Properties of Particle and Hole Operators

Equations (4.46) and (4.47) lead to the following useful relations between the particle and hole operators:

$$
c_\beta = -\tilde h_\beta^\dagger , \ c_\beta^\dagger = -\tilde h_\beta \quad (\varepsilon_\beta \le \varepsilon_F) . \tag{4.48}
$$

The hole operators $h_\beta^\dagger$ and $\tilde h_\beta$ are spherical tensors of rank $j_b$ since they are constructed from the spherical tensors $\tilde c_\beta$ and $c_\beta^\dagger$, respectively, in the way shown in (4.46) and (4.48). The relations (4.46)–(4.48) constitute the *particle–hole representation*. The particles and holes thus defined can be considered to be *Hartree–Fock quasiparticles*. For these quasiparticles we have

$$
c_\alpha|\mathrm{HF}\rangle = 0 , \ h_\beta|\mathrm{HF}\rangle = 0 \quad (\varepsilon_\alpha > \varepsilon_F , \quad \varepsilon_\beta \le \varepsilon_F) , \tag{4.49}
$$

with the *particle–hole vacuum* defined as

$$
|\mathrm{HF}\rangle = c_{\pi1}^\dagger c_{\pi2}^\dagger \cdots c_{\pi Z}^\dagger c_{\nu1}^\dagger c_{\nu2}^\dagger \cdots c_{\nu N}^\dagger |0\rangle . \tag{4.50}
$$

Thus the particle–hole vacuum consists of $Z$ protons occupying the $Z$ lowest proton orbitals and $N$ neutrons occupying the $N$ lowest neutron orbitals. The last occupied orbitals for protons and neutrons are the proton and neutron Fermi levels.

The concept of quasiparticle can be elaborated further by defining a unitary transformation $\mathsf{U}$ from particle operators to Hartree–Fock quasiparticle operators:

$$
\begin{pmatrix} c_\alpha^\dagger \\ \tilde c_\alpha \\ h_\beta^\dagger \\ \tilde h_\beta \end{pmatrix}
=
\begin{pmatrix}
1 & 0 & 0 & 0 \\
0 & 1 & 0 & 0 \\
0 & 0 & 0 & 1 \\
0 & 0 & -1 & 0
\end{pmatrix}
\begin{pmatrix} c_\alpha^\dagger \\ \tilde c_\alpha \\ c_\beta^\dagger \\ \tilde c_\beta \end{pmatrix}
\quad (\varepsilon_\alpha > \varepsilon_F , \quad \varepsilon_\beta \le \varepsilon_F) . \tag{4.51}
$$

The transformation matrix satisfies $U^\dagger U = 1_4$, where $1_4$ is the four-by-four unit matrix. This transformation is simple in that it defines each quasiparticle operator in terms of just one particle operator. More complicated unitary transformations to quasiparticles have been devised to convert a complicated many-body problem to a more tractable one. They are discussed in Part II of this book.

The advantage of the particle and hole creation operators, defined in (4.46), are their simple contraction properties relative to the particle–hole vacuum. These contractions read

$$
\begin{aligned}
\overline{c_\alpha c_{\alpha'}^\dagger} &= \langle \mathrm{HF}|c_\alpha c_{\alpha'}^\dagger|\mathrm{HF}\rangle = \delta_{\alpha\alpha'} \quad (\varepsilon_\alpha, \varepsilon_{\alpha'} > \varepsilon_\mathrm{F})\,, \\
\overline{h_\beta h_{\beta'}^\dagger} &= \langle \mathrm{HF}|h_\beta h_{\beta'}^\dagger|\mathrm{HF}\rangle = \delta_{\beta\beta'} \quad (\varepsilon_\beta, \varepsilon_{\beta'} \le \varepsilon_\mathrm{F})\,, \\
\text{others} &= 0\,.
\end{aligned}
\tag{4.52}
$$

The contractions of the particle operators $c_\alpha^\dagger$ and $c_\alpha$ with respect to the particle–hole vacuum are

$$
\begin{aligned}
\overline{c_\beta^\dagger c_{\beta'}} &= \langle \mathrm{HF}|c_\beta^\dagger c_{\beta'}|\mathrm{HF}\rangle = \langle \mathrm{HF}|\tilde{h}_\beta \tilde{h}_{\beta'}^\dagger|\mathrm{HF}\rangle = \delta_{\beta\beta'} \quad (\varepsilon_\beta \le \varepsilon_\mathrm{F})\,, \\
\overline{c_\alpha c_{\alpha'}^\dagger} &= \langle \mathrm{HF}|c_\alpha c_{\alpha'}^\dagger|\mathrm{HF}\rangle = \delta_{\alpha\alpha'} \quad (\varepsilon_\alpha > \varepsilon_\mathrm{F})\,, \\
\text{others} &= 0\,.
\end{aligned}
\tag{4.53}
$$

## Example 4.7

Contractions with respect to the particle–hole vacuum are evaluated as shown in this example. We calculate the expectation value of the two-body interaction (4.28) in the particle–hole vacuum:

$$
\begin{aligned}
\langle \mathrm{HF}|V|\mathrm{HF}\rangle &= \tfrac{1}{4} \sum_{\alpha\beta\gamma\delta} \bar{v}_{\alpha\beta\gamma\delta} \langle \mathrm{HF}|c_\alpha^\dagger c_\beta^\dagger c_\delta c_\gamma|\mathrm{HF}\rangle \\
&= \tfrac{1}{4} \sum_{\alpha\beta\gamma\delta} \bar{v}_{\alpha\beta\gamma\delta} \langle \mathrm{HF}|\tilde{h}_\alpha \tilde{h}_\beta \tilde{h}_\delta^\dagger \tilde{h}_\gamma^\dagger|\mathrm{HF}\rangle \\
&= \tfrac{1}{4} \sum_{\alpha\beta\gamma\delta} \bar{v}_{\alpha\beta\gamma\delta} \left( \tilde{h}_\alpha \tilde{h}_\beta \tilde{h}_\delta^\dagger \tilde{h}_\gamma^\dagger + \tilde{h}_\alpha \tilde{h}_\beta \tilde{h}_\delta^\dagger \tilde{h}_\gamma^\dagger \right) \\
&= \tfrac{1}{4} \sum_{\alpha\beta\gamma\delta} \bar{v}_{\alpha\beta\gamma\delta} (\delta_{\alpha\gamma}\delta_{\beta\delta} - \delta_{\alpha\delta}\delta_{\beta\gamma}) \\
&= \tfrac{1}{4} \sum_{\substack{\beta\beta' \\ \varepsilon_\beta \le \varepsilon_\mathrm{F} \\ \varepsilon_{\beta'} \le \varepsilon_\mathrm{F}}} (\bar{v}_{\beta\beta'\beta\beta'} - \bar{v}_{\beta\beta'\beta'\beta}) = \tfrac{1}{2} \sum_{\substack{\beta\beta' \\ \varepsilon_\beta \le \varepsilon_\mathrm{F} \\ \varepsilon_{\beta'} \le \varepsilon_\mathrm{F}}} \bar{v}_{\beta\beta'\beta\beta'}\,.
\end{aligned}
\tag{4.54}
$$

### 4.4.2 Particle–Hole Representation of Operators and Excitations

The particle representation of a general one-body operator was given in (4.19). It is a suitable starting point for the *particle–hole representation* of such an operator. The orbitals are divided into particle and hole categories and the transformations (4.48) are applied below the Fermi surface. It can be shown (see Exercises) that in the particle–hole representation a general one-body operator $T$ becomes

$$T = \sum_{\alpha\beta} t_{\alpha\beta} c_\alpha^\dagger c_\beta = \sum_{\substack{\beta \\ \varepsilon_\beta \leq \varepsilon_F}} t_{\beta\beta} + \sum_{\substack{\alpha\alpha' \\ \varepsilon_\alpha > \varepsilon_F \\ \varepsilon_{\alpha'} > \varepsilon_F}} t_{\alpha\alpha'} c_\alpha^\dagger c_{\alpha'} - \sum_{\substack{\alpha\beta \\ \varepsilon_\alpha > \varepsilon_F \\ \varepsilon_\beta \leq \varepsilon_F}} t_{\alpha\beta} c_\alpha^\dagger \tilde{h}_\beta^\dagger$$

$$- \sum_{\substack{\beta\alpha \\ \varepsilon_\beta \leq \varepsilon_F \\ \varepsilon_\alpha > \varepsilon_F}} t_{\beta\alpha} \tilde{h}_\beta c_\alpha - \sum_{\substack{\beta\beta' \\ \varepsilon_\beta \leq \varepsilon_F \\ \varepsilon_{\beta'} \leq \varepsilon_F}} t_{\beta\beta'} \tilde{h}_{\beta'}^\dagger \tilde{h}_\beta . \tag{4.55}$$

The operators in $T$ are already normal-ordered with respect to the particle–hole vacuum $|\mathrm{HF}\rangle$, so that we can write

$$T = \sum_{\substack{\beta \\ \varepsilon_\beta \leq \varepsilon_F}} t_{\beta\beta} + \mathcal{N}[T]_{\mathrm{HF}} , \tag{4.56}$$

where $\mathcal{N}[\cdots]_{\mathrm{HF}}$ denotes normal ordering with respect to $|\mathrm{HF}\rangle$.

Two-body operators $V$ can be represented in the particle–hole language in a similar way. However, this representation is not needed later in the book and is therefore omitted here, but can be found in e.g. [12].

Consider next excitations of the particle–hole vacuum. In Fig. 4.1 the particle–hole vacuum is shown (enclosed by curly bracket) together with the operational regimes of the particle creation operators $c_\alpha^\dagger$ and hole creation operators $h_\beta^\dagger$. The simplest excitation of a closed-shell nucleus is a *particle–hole excitation*

$$|\mathrm{p}_\alpha \mathrm{h}_\beta\rangle = c_\alpha^\dagger h_\beta^\dagger |\mathrm{HF}\rangle . \tag{4.57}$$

This excitation is shown in Fig. 4.1. It can be viewed as the creation of a particle and a hole on the particle–hole vacuum, or as exciting a particle from the particle–hole vacuum into the empty orbitals above the Fermi level. The description of the excited states of a closed-shell nucleus relies on the treatment of this type of excitations, and higher, many-particle–many-hole excitations within different approximation schemes, including their mixing through the residual interactions. These matters will be discussed extensively later.

Additional interesting nuclei near closed major shells are those whose ground states are of one-particle, one-hole, two-particle or two-hole type,

$$c_\alpha^\dagger |\mathrm{HF}\rangle , \quad h_\beta^\dagger |\mathrm{HF}\rangle , \quad c_\alpha^\dagger c_{\alpha'}^\dagger |\mathrm{HF}\rangle , \quad h_\beta^\dagger h_{\beta'}^\dagger |\mathrm{HF}\rangle . \tag{4.58}$$

Such nuclei will be discussed in Chap. 5 and in some later chapters, in the context of configuration mixing of the basic excitations (4.57) and (4.58).

## 4.5 Hartree–Fock Equation from Wick's Theorem

The Hartree–Fock equation can be derived elegantly by applying Wick's theorem in the particle–hole representation. This means that we use normal ordering and contraction with respect to the particle–hole vacuum $|\text{HF}\rangle$. However, instead of using the particle–hole representation of the Hamiltonian, we use its particle representation.[4] Thus we exploit the contractions (4.53).

### 4.5.1 Derivation of the Hartree–Fock Equation

Starting directly from Wick's theorem (4.41), we obtain for the potential part of the Hamiltonian

$$
4V = \sum_{\alpha\beta\gamma\delta} \bar{v}_{\alpha\beta\gamma\delta} c_\alpha^\dagger c_\beta^\dagger c_\delta c_\gamma = \sum_{\alpha\beta\gamma\delta} \bar{v}_{\alpha\beta\gamma\delta} \mathcal{N}\left[c_\alpha^\dagger c_\beta^\dagger c_\delta c_\gamma\right]
$$

$$
- \sum_{\alpha\beta\gamma\delta} \bar{v}_{\alpha\beta\gamma\delta} \langle\text{HF}|c_\alpha^\dagger c_\delta|\text{HF}\rangle \mathcal{N}\left[c_\beta^\dagger c_\gamma\right] + \sum_{\alpha\beta\gamma\delta} \bar{v}_{\alpha\beta\gamma\delta} \langle\text{HF}|c_\alpha^\dagger c_\gamma|\text{HF}\rangle \mathcal{N}\left[c_\beta^\dagger c_\delta\right]
$$

$$
+ \sum_{\alpha\beta\gamma\delta} \bar{v}_{\alpha\beta\gamma\delta} \langle\text{HF}|c_\beta^\dagger c_\delta|\text{HF}\rangle \mathcal{N}\left[c_\alpha^\dagger c_\gamma\right] - \sum_{\alpha\beta\gamma\delta} \bar{v}_{\alpha\beta\gamma\delta} \langle\text{HF}|c_\beta^\dagger c_\gamma|\text{HF}\rangle \mathcal{N}\left[c_\alpha^\dagger c_\delta\right]
$$

$$
- \sum_{\alpha\beta\gamma\delta} \bar{v}_{\alpha\beta\gamma\delta} \langle\text{HF}|c_\alpha^\dagger c_\delta|\text{HF}\rangle \langle\text{HF}|c_\beta^\dagger c_\gamma|\text{HF}\rangle
$$

$$
+ \sum_{\alpha\beta\gamma\delta} \bar{v}_{\alpha\beta\gamma\delta} \langle\text{HF}|c_\alpha^\dagger c_\gamma|\text{HF}\rangle \langle\text{HF}|c_\beta^\dagger c_\delta|\text{HF}\rangle
$$

$$
= \sum_{\alpha\beta\gamma\delta} \bar{v}_{\alpha\beta\gamma\delta} \mathcal{N}\left[c_\alpha^\dagger c_\beta^\dagger c_\delta c_\gamma\right] - \sum_{\substack{\alpha\beta\gamma \\ \varepsilon_\alpha \leq \varepsilon_F}} \bar{v}_{\alpha\beta\gamma\alpha} \mathcal{N}\left[c_\beta^\dagger c_\gamma\right]
$$

$$
+ \sum_{\substack{\alpha\beta\delta \\ \varepsilon_\alpha \leq \varepsilon_F}} \bar{v}_{\alpha\beta\alpha\delta} \mathcal{N}\left[c_\beta^\dagger c_\delta\right] + \sum_{\substack{\alpha\beta\gamma \\ \varepsilon_\beta \leq \varepsilon_F}} \bar{v}_{\alpha\beta\gamma\beta} \mathcal{N}\left[c_\alpha^\dagger c_\gamma\right] - \sum_{\substack{\alpha\beta\delta \\ \varepsilon_\beta \leq \varepsilon_F}} \bar{v}_{\alpha\beta\beta\delta} \mathcal{N}\left[c_\alpha^\dagger c_\delta\right]
$$

$$
- \sum_{\substack{\alpha\beta \\ \varepsilon_\alpha \leq \varepsilon_F \\ \varepsilon_\beta \leq \varepsilon_F}} \bar{v}_{\alpha\beta\beta\alpha} + \sum_{\substack{\alpha\beta \\ \varepsilon_\alpha \leq \varepsilon_F \\ \varepsilon_\beta \leq \varepsilon_F}} \bar{v}_{\alpha\beta\alpha\beta} \ . \tag{4.59}
$$

With renamed summation indices and the symmetry properties (4.29) of the antisymmetrized two-body matrix elements, this is reduced to

$$
4V = \sum_{\alpha\beta\gamma\delta} \bar{v}_{\alpha\beta\gamma\delta} \mathcal{N}\left[c_\alpha^\dagger c_\beta^\dagger c_\delta c_\gamma\right] + 4 \sum_{\substack{\alpha\beta\gamma \\ \varepsilon_\alpha \leq \varepsilon_F}} \bar{v}_{\alpha\beta\alpha\gamma} \mathcal{N}\left[c_\beta^\dagger c_\gamma\right] + 2 \sum_{\substack{\alpha\beta \\ \varepsilon_\alpha \leq \varepsilon_F \\ \varepsilon_\beta \leq \varepsilon_F}} \bar{v}_{\alpha\beta\alpha\beta} \ . \tag{4.60}
$$

Applied to the second normal-ordered product, the definition of contraction (4.35) gives

---

[4] The particle–hole representation of the Hamiltonian could be used, but it is very complicated and would not serve our purposes here.

$$\mathcal{N}[c_\beta^\dagger c_\gamma] = c_\beta^\dagger c_\gamma - \langle \mathrm{HF}|c_\beta^\dagger c_\gamma|\mathrm{HF}\rangle = c_\beta^\dagger c_\gamma - \delta_{\beta\gamma} \quad (\varepsilon_\beta \leq \varepsilon_\mathrm{F}) \,. \tag{4.61}$$

This leads to the following expression for the Hamiltonian:

$$
\begin{aligned}
H = T + V &= \sum_{\alpha\beta} t_{\alpha\beta} c_\alpha^\dagger c_\beta + \tfrac{1}{4} \sum_{\alpha\beta\gamma\delta} \bar{v}_{\alpha\beta\gamma\delta} \mathcal{N}[c_\alpha^\dagger c_\beta^\dagger c_\delta c_\gamma] \\
&+ \sum_{\substack{\alpha\beta\gamma \\ \varepsilon_\alpha \leq \varepsilon_\mathrm{F}}} \bar{v}_{\alpha\beta\alpha\gamma}(c_\beta^\dagger c_\gamma - \delta_{\beta\gamma}) + \tfrac{1}{2} \sum_{\substack{\alpha\beta \\ \varepsilon_\alpha \leq \varepsilon_\mathrm{F} \\ \varepsilon_\beta \leq \varepsilon_\mathrm{F}}} \bar{v}_{\alpha\beta\alpha\beta} \\
&= \tfrac{1}{4} \sum_{\alpha\beta\gamma\delta} \bar{v}_{\alpha\beta\gamma\delta} \mathcal{N}[c_\alpha^\dagger c_\beta^\dagger c_\delta c_\gamma] + \sum_{\alpha\beta} \left( t_{\alpha\beta} + \sum_{\substack{\gamma \\ \varepsilon_\gamma \leq \varepsilon_\mathrm{F}}} \bar{v}_{\gamma\alpha\gamma\beta} \right) c_\alpha^\dagger c_\beta \\
&\quad - \tfrac{1}{2} \sum_{\substack{\alpha\beta \\ \varepsilon_\alpha \leq \varepsilon_\mathrm{F} \\ \varepsilon_\beta \leq \varepsilon_\mathrm{F}}} \bar{v}_{\alpha\beta\alpha\beta} \,. \tag{4.62}
\end{aligned}
$$

The second term is seen to be a one-body term. We abbreviate its matrix element as

$$t_{\alpha\beta} + \sum_{\substack{\gamma \\ \varepsilon_\gamma \leq \varepsilon_\mathrm{F}}} \bar{v}_{\gamma\alpha\gamma\beta} \equiv T_{\alpha\beta} \,. \tag{4.63}$$

To proceed, we change the single-particle basis so that the set of creation operators $\{c_\alpha^\dagger\}$ is transformed to a new set $\{b_\alpha^\dagger\}$. This is accomplished by a unitary transformation $\mathsf{U}$:

$$c_\alpha^\dagger = \sum_{\alpha'} U_{\alpha\alpha'}^* b_{\alpha'}^\dagger \,, \quad c_\alpha = \sum_{\alpha'} U_{\alpha\alpha'} b_{\alpha'} \,. \tag{4.64}$$

Substituting these into the one-body part of (4.62) gives

$$\sum_{\alpha\beta} T_{\alpha\beta} c_\alpha^\dagger c_\beta = \sum_{\alpha'\beta'} \sum_{\alpha\beta} T_{\alpha\beta} U_{\alpha\alpha'}^* U_{\beta\beta'} b_{\alpha'}^\dagger b_{\beta'} \equiv \sum_{\alpha'\beta'} T_{\alpha'\beta'}' b_{\alpha'}^\dagger b_{\beta'} \,, \tag{4.65}$$

where $T'$ is the one-body operator expressed in the new basis. We now require the new basis to be such that $T'$ is diagonal, i.e.

$$T_{\alpha'\beta'}' = \varepsilon_{\alpha'} \delta_{\alpha'\beta'} \,. \tag{4.66}$$

Combining (4.65) and (4.66) in matrix form we have

$$\mathsf{U}^\dagger \mathsf{T} \mathsf{U} = \mathsf{T}' = \mathrm{diag}(\varepsilon_1, \varepsilon_2, \dots) \,, \tag{4.67}$$

where $\mathrm{diag}(\varepsilon_1, \varepsilon_2, \dots)$ is a diagonal matrix with diagonal elements $\varepsilon_1, \varepsilon_2, \dots$. This means that the unitary transformation $\mathsf{U}$ is chosen such that it diagonalizes the one-body part of the Hamiltonian. We can now drop the prime from $T'$ and rename the indices so that (4.66) becomes

$$T_{\alpha\beta} = \varepsilon_\alpha \delta_{\alpha\beta} .$$  (4.68)

Substituting back from (4.63) we finally have

$$\boxed{t_{\alpha\beta} + \sum_{\substack{\gamma \\ \varepsilon_\gamma \leq \varepsilon_F}} \bar{v}_{\gamma\alpha\gamma\beta} = \varepsilon_\alpha \delta_{\alpha\beta} .}$$  (4.69)

This is the *Hartree–Fock equation* [35, 36] for the computation of the eigenenergies $\varepsilon_\alpha$.

### 4.5.2 Residual Interaction; Ground-State Energy

If one manages at the outset to choose that particular set of single-particle orbitals $c_\alpha^\dagger |0\rangle$ which satisfies the Hartree–Fock equation (4.69), then the one-body part $T_{\alpha\beta}$ of the Hamiltonian is immediately diagonal and the operators $c_\alpha^\dagger$ are identified with the operators $b_\alpha^\dagger$ that appear in (4.65). In this case the Hamiltonian can be written concisely as

$$\boxed{H = H_{\mathrm{HF}} + V_{\mathrm{RES}} .}$$  (4.70)

The part $H_{\mathrm{HF}}$ is the total single-particle energy for the Hartree–Fock mean field,

$$\boxed{H_{\mathrm{HF}} = \sum_\alpha \varepsilon_\alpha c_\alpha^\dagger c_\alpha ,}$$  (4.71)

where we have used (4.62), (4.63) and (4.68). The residual interaction is what is left over of the Hamiltonian (4.62), namely

$$\boxed{V_{\mathrm{RES}} = \tfrac{1}{4} \sum_{\alpha\beta\gamma\delta} \bar{v}_{\alpha\beta\gamma\delta} \mathcal{N}\big[c_\alpha^\dagger c_\beta^\dagger c_\delta c_\gamma\big]_{\mathrm{HF}} - \tfrac{1}{2} \sum_{\substack{\alpha\beta \\ \varepsilon_\alpha \leq \varepsilon_F \\ \varepsilon_\beta \leq \varepsilon_F}} \bar{v}_{\alpha\beta\alpha\beta} ,}$$  (4.72)

where $\mathcal{N}[\cdots]_{\mathrm{HF}}$ denotes normal ordering with respect to the particle–hole vacuum $|\mathrm{HF}\rangle$.

The last term of (4.72) is important for the absolute ground-state energies; for the excitation energies it cancels out. It is remarkable that the derivation of the Hartree–Fock equation via Wick's theorem gives as a by-product an explicit expression for the residual interaction $V_{\mathrm{RES}}$. This expression is useful when calculating corrections, e.g. by perturbation theory, to the mean-field approximation of nuclear structure.

In the mean-field approximation the nuclear ground state is the Hartree–Fock vacuum $|\mathrm{HF}\rangle$. We calculate the ground-state energy as the expectation value of the Hamiltonian:

$$E_{HF} = \langle HF|H|HF \rangle = \langle HF|H_{HF}|HF \rangle - \frac{1}{2} \sum_{\substack{\alpha\beta \\ \varepsilon_\alpha \leq \varepsilon_F \\ \varepsilon_\beta \leq \varepsilon_F}} \bar{v}_{\alpha\beta\alpha\beta}$$

$$= \sum_\alpha \varepsilon_\alpha \langle HF|c_\alpha^\dagger c_\alpha|HF \rangle - \frac{1}{2} \sum_{\substack{\alpha\beta \\ \varepsilon_\alpha \leq \varepsilon_F \\ \varepsilon_\beta \leq \varepsilon_F}} \bar{v}_{\alpha\beta\alpha\beta} = \sum_{\substack{\alpha \\ \varepsilon_\alpha \leq \varepsilon_F}} \varepsilon_\alpha - \frac{1}{2} \sum_{\substack{\alpha\beta \\ \varepsilon_\alpha \leq \varepsilon_F \\ \varepsilon_\beta \leq \varepsilon_F}} \bar{v}_{\alpha\beta\alpha\beta} .$$

(4.73)

From the Hartree–Fock equation (4.69) we have

$$\varepsilon_\alpha = t_{\alpha\alpha} + \sum_{\substack{\beta \\ \varepsilon_\beta \leq \varepsilon_F}} \bar{v}_{\beta\alpha\beta\alpha} ,$$ (4.74)

which gives for the Hartree–Fock ground-state energy

$$E_{HF} = \sum_{\substack{\alpha \\ \varepsilon_\alpha \leq \varepsilon_F}} t_{\alpha\alpha} + \frac{1}{2} \sum_{\substack{\alpha\beta \\ \varepsilon_\alpha \leq \varepsilon_F \\ \varepsilon_\beta \leq \varepsilon_F}} \bar{v}_{\alpha\beta\alpha\beta} .$$ (4.75)

The Hartree–Fock equation (4.69) can be cast into coordinate representation (see Exercises). The result has the form

$$\frac{-\hbar^2}{2m_N} \nabla^2 \phi_\alpha(\boldsymbol{x}) + V_{HF}\left(\{\phi_\beta(\boldsymbol{x})\}_{\varepsilon_\beta \leq \varepsilon_F}\right)\phi_\alpha(\boldsymbol{x}) = \varepsilon_\alpha \phi_\alpha(\boldsymbol{x}) .$$ (4.76)

The potential $V_{HF}$ cannot be expressed by itself but only as it acts on $\phi_\alpha$:

$$V_{HF}\left(\{\phi_\beta(\boldsymbol{x})\}_{\varepsilon_\beta \leq \varepsilon_F}\right)\phi_\alpha(\boldsymbol{x}) = v_H(\boldsymbol{x})\phi_\alpha(\boldsymbol{x}) - \int d^3r' v_F(\boldsymbol{x}', \boldsymbol{x})\phi_\alpha(\boldsymbol{x}') ,$$

$$v_H(\boldsymbol{x}) = \sum_{\substack{\beta \\ \varepsilon_\beta \leq \varepsilon_F}} \int d^3r' \phi_\beta^\dagger(\boldsymbol{x}')v(\boldsymbol{x}', \boldsymbol{x})\phi_\beta(\boldsymbol{x}') ,$$

$$v_F(\boldsymbol{x}', \boldsymbol{x}) = \sum_{\substack{\beta \\ \varepsilon_\beta \leq \varepsilon_F}} \phi_\beta^\dagger(\boldsymbol{x}')v(\boldsymbol{x}', \boldsymbol{x})\phi_\beta(\boldsymbol{x}) .$$

(4.77)

The first term with the *local* potential $v_H(\boldsymbol{x})$ is called the Hartree or *direct* term, and the second term with the *non-local* potential $v_F(\boldsymbol{x}', \boldsymbol{x})$ is called the Fock or *exchange term*. The formal solution of this non-linear equation was discussed in Sect. 3.1.

## 4.6 Hartree–Fock Eigenvalue Problem

The Hartree–Fock equation (4.69) can be converted to an eigenvalue problem of a Hermitian matrix. To do so, the wave functions $\phi_\alpha(\boldsymbol{x}) = \langle \boldsymbol{x}|\alpha \rangle$ in (4.76) are expanded in terms of some basis states, usually harmonic oscillator wave functions. Thus we seek solutions of the Hartree–Fock equation in the form

$$|\alpha\rangle = \sum_j C_j^\alpha |j\rangle ,$$  (4.78)

where the basis $\{|j\rangle\}$ is orthonormal and complete. These conditions lead to the following constraints on the expansion coefficients $C_j^\alpha$ of the Hartree–Fock wave function (4.78):

$$\sum_j \left(C_j^\alpha\right)^* C_j^\beta = \delta_{\alpha\beta} \quad \text{(orthonormality)} ,$$  (4.79)

$$\sum_\alpha C_i^\alpha \left(C_j^\alpha\right)^* = \delta_{ij} \quad \text{(completeness)} .$$  (4.80)

The coefficients $C_j^\alpha$ are determined by performing a variation of the Hartree–Fock ground-state energy $E_{\text{HF}}$ as given in (4.75). In this procedure we seek the minimum of $E_{\text{HF}}$ using the $C_j^\alpha$ as variational parameters subject to the normalization condition (4.79). This constrained variational problem is solved by the technique of Lagrange undetermined multipliers. The variational condition yields

$$\frac{\partial}{\partial\left(C_j^\alpha\right)^*}\left[E_{\text{HF}} - \sum_{\alpha'} \varepsilon_{\alpha'} \sum_{j'} \left(C_{j'}^{\alpha'}\right)^* C_{j'}^{\alpha'}\right] = 0 ,$$  (4.81)

where the Lagrange multipliers are denoted by $\varepsilon_{\alpha'}$. For the differentiation the normalization condition (4.79) is relaxed but must be restored at the end. Noting the simple relations (4.43) and (4.45) and substituting the states (4.78) into $E_{\text{HF}}$ we obtain

$$\begin{aligned}
E_{\text{HF}} &= \sum_{\substack{\alpha \\ \varepsilon_\alpha \leq \varepsilon_{\text{F}}}} \langle\alpha|T|\alpha\rangle + \frac{1}{2} \sum_{\substack{\alpha\beta \\ \varepsilon_\alpha \leq \varepsilon_{\text{F}} \\ \varepsilon_\beta \leq \varepsilon_{\text{F}}}} \langle\alpha\beta|V|\alpha\beta\rangle \\
&= \sum_{\substack{\alpha \\ \varepsilon_\alpha \leq \varepsilon_{\text{F}}}} \sum_{jj'} \left(C_j^\alpha\right)^* C_{j'}^\alpha \langle j|T|j'\rangle \\
&\quad + \frac{1}{2} \sum_{\substack{\alpha\beta \\ \varepsilon_\alpha \leq \varepsilon_{\text{F}} \\ \varepsilon_\beta \leq \varepsilon_{\text{F}}}} \sum_{jj'} \sum_{j_1 j_2} \left(C_j^\alpha\right)^* \left(C_{j_1}^\beta\right)^* C_{j'}^\alpha C_{j_2}^\beta \langle j j_1|V|j' j_2\rangle .
\end{aligned}$$  (4.82)

We now carry out the differentiation of both terms of (4.81):

$$\frac{\partial E_{\text{HF}}}{\partial\left(C_j^\alpha\right)^*} = \sum_{j'}\left[\langle j|T|j'\rangle + \sum_{\substack{\beta \\ \varepsilon_\beta \leq \varepsilon_{\text{F}}}} \sum_{j_1 j_2} \left(C_{j_1}^\beta\right)^* \langle j j_1|V|j' j_2\rangle C_{j_2}^\beta\right] C_{j'}^\alpha ,$$  (4.83)

$$\frac{\partial}{\partial\left(C_j^\alpha\right)^*} \sum_{\alpha'} \varepsilon_{\alpha'} \sum_{j'} \left(C_{j'}^{\alpha'}\right)^* C_{j'}^{\alpha'} = \varepsilon_\alpha C_j^\alpha .$$  (4.84)

The final result is

$$
\sum_{j'} H_{jj'}^{(\mathrm{HF})} C_{j'}^{\alpha} = \varepsilon_{\alpha} C_{j}^{\alpha} \,,
$$

$$
H_{jj'}^{(\mathrm{HF})} \equiv t_{jj'} + \sum_{\substack{\beta \\ \varepsilon_{\beta} \leq \varepsilon_{\mathrm{F}}}} \sum_{j_1 j_2} \left( C_{j_1}^{\beta} \right)^{*} \bar{v}_{jj_1 j' j_2} C_{j_2}^{\beta} \,. \qquad (4.85)
$$

In this way we have been able to convert the Hartree–Fock equation into an eigenvalue problem for the matrix $\mathsf{H}^{(\mathrm{HF})}$ with elements $H_{jj'}^{(\mathrm{HF})}$. The column vector $\{C_j^{\alpha}\}$ is the representation of the solution $|\alpha\rangle$ of the Hartree–Fock equation in the basis $\{|j\rangle\}$. The Lagrange multipliers $\varepsilon_{\alpha}$ are identified as the Hartree–Fock single-particle energies to be solved as eigenvalues of the matrix $\mathsf{H}^{(\mathrm{HF})}$. As one would expect on physical grounds, the matrix is Hermitian (see Exercises).

The diagonalization procedure produces Hartree–Fock wave functions $\phi_{\alpha}(\boldsymbol{x})$ as linear combinations (4.78) of known standard basis functions $\psi_j(\boldsymbol{x}) = \langle \boldsymbol{x}|j\rangle$. At closed neutron and proton major shells, the natural choice for $\psi_j$ are the harmonic oscillator (radial) wave functions (3.42). Hartree–Fock wave functions of magic nuclei are then expansions mathematically similar to the Woods–Saxon expansion (3.37).

The non-linearity of the Hartree–Fock equation is present irrespective of the method of solution. In particular, the matrix equation (4.85) must be solved iteratively since the solutions $C_j^{\beta}$ themselves are building blocks of the matrix $\mathsf{H}^{(\mathrm{HF})}$ to be diagonalized.

## Epilogue

In this chapter we have developed the powerful formalism of occupation number representation. It is an elegant and efficient alternative to the traditional manipulation of wave functions and their matrix elements. The clever bookkeeping feature of the method has its culmination in Wick's theorem, which provides for efficient evaluation of nuclear matrix elements. In the chapters to follow, occupation number representation is used to describe many-nucleon states and to calculate important matrix elements both for nuclear transitions and for two-body interaction matrix elements.

## Exercises

**4.1.** By using the actions (4.3) and (4.4) of the creation and annihilation operators on Fock vectors verify the anticommutation relations (4.9).

**4.2.** Consider two sets of single-particle states in coordinate representation,

$$\phi_\alpha(\boldsymbol{x}) = \langle \boldsymbol{x} | c_\alpha^\dagger | 0 \rangle , \quad \psi_\alpha(\boldsymbol{x}) = \langle \boldsymbol{x} | b_\alpha^\dagger | 0 \rangle , \tag{4.86}$$

as given by (4.12). Furthermore, let U be a unitary transformation such that

$$\phi_\alpha(\boldsymbol{x}) = \sum_\beta U_{\alpha\beta} \psi_\beta(\boldsymbol{x}) . \tag{4.87}$$

Show that

$$c_\alpha = \sum_\beta U_{\alpha\beta}^* b_\beta , \quad b_\alpha = \sum_\beta U_{\beta\alpha} c_\beta . \tag{4.88}$$

**4.3.** If the $c^\dagger$ and $c$ operators of Exercise 2 satisfy the fermion anticommutation relations (4.9) show that the operators $b^\dagger$ and $b$ also satisfy them.

**4.4.** If a one-body operator $T$ is written in the form (4.19) in the $c^\dagger, c$ basis of Exercise 2 show that the same form is valid in the $b^\dagger, b$ basis related to the previous basis by a unitary transformation U.

**4.5.** By acting with the proton and neutron number operators of (4.16) on the $A$-nucleon wave function verify the results (4.17) and (4.18).

**4.6.** Verify the validity of the representations of the angular momentum operators $J_z$ and $J_\pm$ as given in (4.20) and (4.21).

**4.7.** Show that the operators $c_\alpha^\dagger$ and $\tilde{c}_\alpha$ are spherical tensors of rank $j_a$.

**4.8.** By using the Wigner–Eckart theorem derive the expression on the right-hand side of (4.22).

**4.9.** Verify the validity of (4.25) by starting from (4.22).

**4.10.** Based on Exercise 2.22 and (4.25) what can you say about the electric dipole moment of a nuclear state?

**4.11.** Show that the contraction (4.35) is a c-number.

**4.12.** Show that the contraction of two anticommuting operators vanishes, i.e. verify (4.36).

**4.13.** Let $A = \sum_\alpha x_\alpha A_\alpha$ and $B = \sum_\beta y_\beta B_\beta$ be arbitrary operators built from operators $A_\alpha$ and $B_\beta$ using c-numbers $x_\alpha$ and $y_\beta$. By using the definition (4.35) of the contraction show that it is a linear operation, i.e.

$$\overline{AB} = \sum_{\alpha\beta} x_\alpha y_\beta \overline{A_\alpha B_\beta} . \tag{4.89}$$

**4.14.** By starting from (4.37), where both the operators and the contraction have been defined with respect to the same vacuum $|\Psi_0\rangle$, show that for the same operators the contraction relative to any vacuum $|\phi_0\rangle$ can be written as in (4.39). *Hint*: The operators $A$ and $B$ of (4.37) can be related by a unitary transformation to operators which have been defined relative to the vacuum $|\phi_0\rangle$. Use of (4.89) also helps.

**4.15.** By using the contraction technique verify (4.45).

**4.16.** Verify the particle–hole representation of a one-body operator (4.55).

**4.17.** Let $|\Psi_i\rangle = c_i^\dagger|0\rangle$ and $|\Psi_f\rangle = c_f^\dagger|0\rangle$ be the initial and final one-particle states. Evaluate the *reduced one-body transition density*

$$\left(\Psi_f\|\left[c_a^\dagger \tilde{c}_b\right]_\lambda\|\Psi_i\right) . \tag{4.90}$$

**4.18.** Evaluate the reduced one-body transition density (4.90) when the one-particle states of the previous exercise are replaced by the one-hole states $|\Psi_i\rangle = h_i^\dagger|\text{HF}\rangle$ and $|\Psi_f\rangle = h_f^\dagger|\text{HF}\rangle$.

**4.19.** Let $|\Psi_i\rangle = \left[c_{a_i}^\dagger h_{b_i}^\dagger\right]_{J_i M_i}|\text{HF}\rangle$ and $|\Psi_f\rangle = |\text{HF}\rangle$. Evaluate the reduced one-body transition density (4.90) for these states.

**4.20.** Complete all the details leading to (4.60).

**4.21.** By using the coordinate representations (4.19) and (4.27) of the kinetic energy and potential energy operators derive the coordinate representation, contained in (4.76) and (4.77), of the Hartree–Fock equation (4.69).

**4.22.** The Hartree–Fock equation (4.76) becomes the Hartree equation when the Fock term is omitted from (4.77). Solve the Hartree equation for $N$ nucleons assuming a two-nucleon interaction potential

$$v(\boldsymbol{r}_j, \boldsymbol{r}_l) = -C\delta(\boldsymbol{r}_j - \boldsymbol{r}_l) , \tag{4.91}$$

where $C$ is a constant. This *delta-function interaction* describes a two-nucleon force of zero range. It can be used to mimick the strong short-range nature of the nuclear two-body interaction. Start the iteration (see Subsect. 3.1.1, the scheme (3.17) in particular) with the *plane waves*

$$\phi_j^{(0)}(\boldsymbol{r}) = \frac{1}{\sqrt{V}}e^{i\boldsymbol{k}_j \cdot \boldsymbol{r}} , \tag{4.92}$$

where $V$ is a normalization volume and $\boldsymbol{k}_j$ a wave vector of a plane-wave state below the Fermi surface; ignore spin. The associated unperturbed energy is

$$\varepsilon_j^{(0)} = \frac{\hbar^2 k_j^2}{2m_{\text{N}}} , \quad k_j \equiv |\boldsymbol{k}_j| , \tag{4.93}$$

where $m_{\text{N}}$ is the nucleon mass. What happens when the Fock term is added to the calculation?

**4.23.** By using the contraction technique show that for the one-hole state

$$|\beta_0^{-1}\rangle \equiv h_{\beta_0}^\dagger|\text{HF}\rangle \tag{4.94}$$

the expectation value of the Hamiltonian $H = T + V$ is

$$\langle\beta_0^{-1}|H|\beta_0^{-1}\rangle = \sum_{\substack{\beta\neq\beta_0 \\ \varepsilon_\beta\leq\varepsilon_F}} t_{\beta\beta} + \frac{1}{2}\sum_{\substack{\beta,\beta'\neq\beta_0 \\ \varepsilon_\beta\leq\varepsilon_F \\ \varepsilon_{\beta'}\leq\varepsilon_F}} \bar{v}_{\beta\beta'\beta\beta'} . \tag{4.95}$$

**4.24.** Show that

$$E_{\text{HF}} - \langle\beta_0^{-1}|H|\beta_0^{-1}\rangle = t_{\beta_0\beta_0} + \sum_{\substack{\beta \\ \varepsilon_\beta\leq\varepsilon_F}} \bar{v}_{\beta_0\beta\beta_0\beta} , \tag{4.96}$$

where the hole state is defined in (4.94) and the Hartree–Fock ground-state energy is given in (4.75).

**4.25.** The *separation energy* of a nucleon from the orbital $\beta_0$ is defined for magic nuclei as

$$S_{\beta_0} \equiv E(\beta_0^{-1},\infty) - E_{\text{HF}} , \tag{4.97}$$

where the Hartree–Fock ground-state energy is given in (4.75) and

$$E(\beta_0^{-1},\infty) \equiv \langle\beta_0^{-1}|H|\beta_0^{-1}\rangle + E_\infty . \tag{4.98}$$

Here $E_\infty \approx 0$ is the energy of the separated nucleon at infinity. Show that the separation energy can be written simply as

$$S_{\beta_0} = -\varepsilon_{\beta_0} . \tag{4.99}$$

This result is known as *Koopmans' theorem*.

**4.26.** By using the contraction technique show that the excitation energy of the particle–hole state

$$|\alpha\beta^{-1}\rangle \equiv c_\alpha^\dagger h_\beta^\dagger|\text{HF}\rangle \tag{4.100}$$

is

$$E(\alpha\beta^{-1}) \equiv \langle\alpha\beta^{-1}|H|\alpha\beta^{-1}\rangle - E_{\text{HF}} = \varepsilon_\alpha - \varepsilon_\beta - \bar{v}_{\alpha\beta\alpha\beta} . \tag{4.101}$$

This is the residual interaction energy of the particle and the hole subtracted from the pure particle–hole mean-field energy.

**4.27.** Verify (4.79) and (4.80) by starting from (4.78) and the assumption that the orbitals $|\alpha\rangle$ form a complete orthonormal set of states.

**4.28.** Derive the expression (4.83) for the derivative of the Hartree–Fock ground-state energy by starting from (4.82).

**4.29.** Show that the matrix $H^{\text{HF}}$ with elements given in (4.85) is Hermitian.

# 5

## The Mean-Field Shell Model

## Prologue

Chapter 3 introduced the notion of a nuclear mean field with associated single-particle orbitals. It was explained how the single-particle energies can be obtained either by using an empirical Woods–Saxon potential or by the self-consistent Hartree–Fock approach, extensively discussed in Chap. 4.

In this chapter the mean-field concept is used to discuss the ground state and few lowest excited states of nuclei with the simplest possible structure. These simple nuclei are light to medium-heavy, with mass numbers ranging from $A = 4$ to $A = 54$. They consist of magic nuclei, single-particle and single-hole nuclei, and two-particle and two-hole nuclei. The valence space for the active particles or holes is built on the nuclear core consisting of the lowest, inactive single-particle orbitals of the mean field.

For magic nuclei, i.e. nuclei with completely filled proton and neutron major shells, the ground state is the particle–hole vacuum and the lowest excited states are particle–hole excitations of it. The one- and two-particle nuclei are defined as having one or two particles outside the nuclear core. The one- and two-hole nuclei, on the other hand, consist of one or two holes in the particle–hole vacuum. Except for a few qualitative comments, no residual interaction or configuration mixing is included in the discussion. The isospin representation of the states of two-particle and two-hole nuclei and particle–hole nuclei is introduced.

## 5.1 Valence Space

In this chapter we discuss concrete applications of the mean-field shell model to nuclei in the 0p, 0d-1s and 0f-1p-0g$_{9/2}$ major shells. In this simple approach the nucleons appear non-interacting and occupy single-particle energy levels of the mean field. We discuss the following types of nuclei.

- *Magic nuclei*, where both the protons and neutrons have completely filled major shells at some magic numbers.

- *Semi-magic nuclei*, where protons (neutrons) have a completely filled major shell, and the neutrons (protons) do not.

- Nearly magic nuclei, with one or two valence nucleons or holes.

The simple approach of this chapter does not take into account any residual interaction $V_{\mathrm{RES}}$. Each state is built from a single configuration, i.e. there is no configuration mixing. A key notion in our discussion is the *valence space* or *model space* which consists of all single-particle orbitals actively involved in the generation of configurations of the many-nucleon system considered.

Figure 5.1 shows a typical valence space for protons or neutrons. In this particular case the Fermi energy lies at the magic number 20. Below the Fermi level we have the active hole orbitals of the 0d-1s major shell (with hole creation operators $h_\beta^\dagger$) and above it the active particle orbitals of the 0f-1p-0g$_{9/2}$ shells (with particle creation operators $c_\alpha^\dagger$). The lowest excitations are particle–hole excitations across the Fermi surface from the 0d-1s shells to the 0f-1p-0g$_{9/2}$ shells. All the orbitals below the 0d-1s shell, i.e. below the magic number 8, are considered to be inert. This inactive lowest range of the mean-field orbitals is called the *core*. The core is the effective particle vacuum, i.e.

$$\boxed{c_\alpha|\mathrm{CORE}\rangle = 0 \quad \text{for all } \alpha \text{ in the valence space.}} \tag{5.1}$$

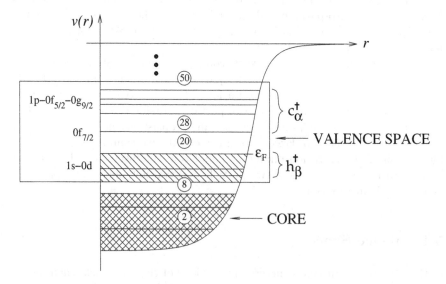

**Fig. 5.1.** Valence space and core. The ground state is the particle–hole vacuum. Excitations are described by creation of holes below, and particles above, the magic shell gap 20.

The 0d-1s major shell forms the filled part of the particle–hole vacuum $|\text{HF}\rangle$ and the 0f-1p-0g$_{9/2}$ orbitals are the empty orbitals of $|\text{HF}\rangle$.

The reason for defining a core is that the computational effort increases very rapidly with an increasing number of single-particle orbitals included in the valence space. The computational burden can be lightened by dropping the lowest-lying orbitals from the valence space and burying them into the inert core. The use of a core is of course an approximation, and its consequences should be examined in each case. For low-lying nuclear states core excitations do not play a role, except in cases where highly collective low-energy states can be built by opening the core. Problems of this kind arise in large shell-model calculations, e.g. when trying to discuss $3^-$ octupole-vibrational states. In general the size of the valence space needed increases with excitation energies to be calculated.

For nuclei with more than 20 protons or neutrons one could simply take as active shells the 0f-1p-0g$_{9/2}$ valence space, pushing the 0d-1s shells into the core. For nuclei with fewer than 20 protons or neutrons the 0d-1s shells could serve as a valence space to study either the particle states beyond the magic core of eight particles or the hole states below the magic number 20.

In the following sections we calculate one- and two-particle and one- and two-hole excitation energies and compare them with experimental data. Throughout this chapter, we assume that the *relative* energies of the single-particle orbitals are the same for protons and neutrons. This is a very good approximation for light nuclei near the valley of beta-decay stability. The experimental data used in the book are taken from the *Table of Isotopes* [37] and the website [38].

## 5.2 One-Particle and One-Hole Nuclei

One-particle and one-hole nuclei allow of the simplest possible theoretical description of their states. This description consists of one particle outside an inert core or one hole in a completely filled valence space. Below we give several examples of such nuclei.

### 5.2.1 Examples of One-Particle Nuclei

The structure of one-particle nuclei within the simple mean-field picture is the following. One-proton states $|\pi\rangle$ and one-neutron states $|\nu\rangle$ are described as

$$|\pi\rangle = c_\pi^\dagger |\text{CORE}\rangle \,, \quad |\nu\rangle = c_\nu^\dagger |\text{CORE}\rangle \,, \tag{5.2}$$

where $|\text{CORE}\rangle$ is the core with its Fermi level at some magic number. The operator $c_\pi^\dagger$ creates a proton and $c_\nu^\dagger$ creates a neutron in one of the available single-particle orbitals above the magic shell gap. These nuclei are always *odd-mass* or *odd-A* nuclei, or simply *odd* nuclei. According to their $Z$ and $N$ they are called *even–odd* or *odd–even* nuclei.

The nuclei $^{17}_{8}O_9$ and $^{17}_{9}F_8$ are examples of one-particle nuclei. In the 0d-1s valence space with the core

$$|\text{CORE}\rangle = |\text{CORE(0s-0p)}\rangle_\pi |\text{CORE(0s-0p)}\rangle_\nu \qquad (5.3)$$

their mean-field structure is given by[1]

$$|^{17}O\,;\,5/2^+_{gs}\rangle = c^\dagger_{\nu 0d_{5/2}}|\text{CORE}\rangle\,, \qquad (5.4)$$

$$|^{17}O\,;\,1/2^+\rangle = c^\dagger_{\nu 1s_{1/2}}|\text{CORE}\rangle\,, \qquad (5.5)$$

$$|^{17}O\,;\,3/2^+\rangle = c^\dagger_{\nu 0d_{3/2}}|\text{CORE}\rangle\,, \qquad (5.6)$$

$$|^{17}F\,;\,5/2^+_{gs}\rangle = c^\dagger_{\pi 0d_{5/2}}|\text{CORE}\rangle\,, \qquad (5.7)$$

$$|^{17}F\,;\,1/2^+\rangle = c^\dagger_{\pi 1s_{1/2}}|\text{CORE}\rangle\,, \qquad (5.8)$$

$$|^{17}F\,;\,3/2^+\rangle = c^\dagger_{\pi 0d_{3/2}}|\text{CORE}\rangle\,. \qquad (5.9)$$

These excitations can be compared to the experimental ones, depicted in Fig. 5.2. The expected ground state $5/2^+$ and the two expected excited states $1/2^+$ and $3/2^+$ are clearly visible in the experimental spectra of the one-proton nucleus $^{17}F$ and the one-neutron nucleus $^{17}O$.

We digress for a moment in anticipation of later chapters. The negative-parity states in the 3–5 MeV range result from configuration mixing of more complicated excitations. They involve opening the core and allowing jumps of nucleons across the $N = 8$ and the $Z = 8$ magic gaps. These states can be produced by coupling either a neutron ($^{17}O$) or a proton ($^{17}F$) to an excitation constructed from many proton and neutron particle–hole excitations across the gap between the 0p and 0d-1s major shells. This kind of multiparticle excitation is interpreted as a vibration, whose quantum is a phonon. The parity of the phonon is negative, so coupling with a particle in the 0d-1s shell of positive parity produces a negative-parity state consisting of two-particle–one-hole components. Because of the parity difference the vibrations do not disturb the one-particle excitations in $^{17}O$ and $^{17}F$, the nuclear Hamiltonian being parity conserving.

Similarly, the $7/2^-$ ground state of $^{41}_{20}Ca_{21}$ and $^{41}_{21}Sc_{20}$ can be described in the simplest possible scheme by taking the $0f_{7/2}$ orbital as the valence space and the orbitals below as the inert core. This choice of valence space produces only the ground state shown in Fig. 5.3. Simple excited states of these nuclei result from leaps of the odd nucleon from the $0f_{7/2}$ orbital to the rest of the 0f-1p-0g$_{9/2}$ space. Because of the large ($\approx 5$ MeV) magic gaps above $N = 28$ and $Z = 28$ these states lie at high excitation energies and are therefore not seen in the figure.

---

[1] Here, and in many places in the sequel, the projection quantum number is omitted as irrelevant.

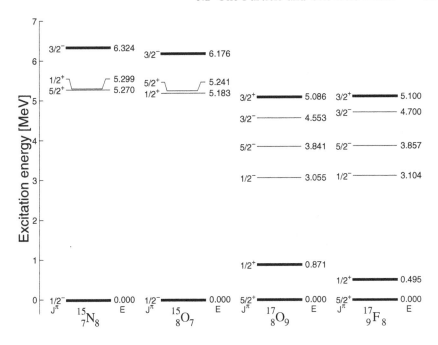

**Fig. 5.2.** Experimental spectra of the one-hole nuclei $^{15}$N and $^{15}$O and the one-particle nuclei $^{17}$O and $^{17}$F. States of the one-particle and one-hole types are shown with a thick line

### 5.2.2 Examples of One-Hole Nuclei

One-hole nuclei are simply described as having one hole in a completely filled valence space. In the following the symbol $|\text{HF}\rangle$ designates the valence space which consists of the filled orbitals of the particle–hole vacuum.

Examples of one-hole nuclei are $^{15}_{7}\text{N}_8$ and $^{15}_{8}\text{O}_7$. The simplest description of them is to use the 0p shell as the valence space, with the $0s_{1/2}$ orbital remaining as the core. The one-hole ground state and excited state of each nucleus can then be written as

$$|^{15}\text{N}\,;\,1/2^-_{\text{gs}}\rangle = h^{\dagger}_{\pi 0p_{1/2}}|\text{HF}\rangle\;,\tag{5.10}$$

$$|^{15}\text{N}\,;\,3/2^-\rangle = h^{\dagger}_{\pi 0p_{3/2}}|\text{HF}\rangle\;,\tag{5.11}$$

$$|^{15}\text{O}\,;\,1/2^-_{\text{gs}}\rangle = h^{\dagger}_{\nu 0p_{1/2}}|\text{HF}\rangle\;,\tag{5.12}$$

$$|^{15}\text{O}\,;\,3/2^-\rangle = h^{\dagger}_{\nu 0p_{3/2}}|\text{HF}\rangle\;,\tag{5.13}$$

where

$$|\text{HF}\rangle = |\text{HF}(0p)\rangle_{\pi}|\text{HF}(0p)\rangle_{\nu}\tag{5.14}$$

is the hole vacuum with its Fermi surface at the $Z = 8$ and $N = 8$ magic numbers. These simple states can be clearly seen in the experimental spectra

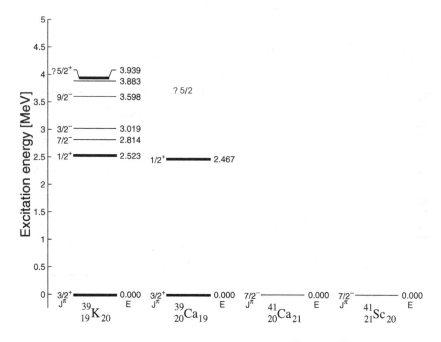

**Fig. 5.3.** Experimental spectra of the one-hole nuclei $^{39}$K and $^{39}$Ca, and the one-particle nuclei $^{41}$Ca and $^{41}$Sc. The one-hole type of states have been drawn with a thick line. A question mark indicates incomplete experimental identification of the spin or energy of the state

of $^{15}$N and $^{15}$O, depicted in Fig. 5.2. The two positive-parity 'intruder' states below the $3/2^-$ state come from exciting particles across the $Z = 8$ or $N = 8$ magic gap to produce a negative-parity phonon that is then coupled to a hole in the negative-parity 0p shell. Due to the parity difference these one-particle–two-hole excitations do not mix with the one-hole states.

In a similar way, the $^{39}_{19}$K$_{20}$ and $^{39}_{20}$Ca$_{19}$ nuclei can be viewed as one-hole nuclei within the simplest possible scheme where the valence space consists of the 0d-1s major shell and the lower-lying 0s and 0p shells form the core. In this case the hole vacuum $|\mathrm{HF}\rangle$ is the 0d-1s shell with its Fermi surface at the $Z = 20$ and $N = 20$ magic numbers. The resulting single-hole states are

$$|^{39}\mathrm{K}\,;\,3/2^+_{\mathrm{gs}}\rangle = h^\dagger_{\pi 0\mathrm{d}_{3/2}}|\mathrm{HF}\rangle\,, \tag{5.15}$$

$$|^{39}\mathrm{K}\,;\,1/2^+\rangle = h^\dagger_{\pi 1\mathrm{s}_{1/2}}|\mathrm{HF}\rangle\,, \tag{5.16}$$

$$|^{39}\mathrm{K}\,;\,5/2^+\rangle = h^\dagger_{\pi 0\mathrm{d}_{5/2}}|\mathrm{HF}\rangle\,, \tag{5.17}$$

$$|^{39}\mathrm{Ca}\,;\,3/2^+_{\mathrm{gs}}\rangle = h^\dagger_{\nu 0\mathrm{d}_{3/2}}|\mathrm{HF}\rangle\,, \tag{5.18}$$

$$|^{39}\mathrm{Ca}\,;\,1/2^+\rangle = h^\dagger_{\nu 1\mathrm{s}_{1/2}}|\mathrm{HF}\rangle\,, \tag{5.19}$$

$$|^{39}\text{Ca}\,;\,5/2^+\rangle = h^\dagger_{\nu 0\text{d}_{5/2}}|\text{HF}\rangle\,. \tag{5.20}$$

These states can be recognized in the experimental spectra of Fig. 5.3. Again, the 'intruding' negative-parity states can be explained as above.

The one-particle and one-hole nuclei can be used to probe experimentally the Hartree–Fock mean field since the excitation energies of their simple states show the energy differences between the mean-field single-particle orbitals. Thus we can read from the experimental spectra of $^{17}\text{O}$ and $^{17}\text{F}$ in Fig. 5.2 that the energy difference between the $0\text{d}_{5/2}$ and $1\text{s}_{1/2}$ single-particle Hartree–Fock orbitals is $\approx 0.7\,\text{MeV}$, and that between the $0\text{d}_{3/2}$ and $0\text{d}_{5/2}$ orbitals is $\approx 5.1\,\text{MeV}$. These experimental Hartree–Fock energies can be used in calculations for nuclei with a few particles occupying the 0d-1s shell. From the hole states of $^{39}\text{K}$ and $^{39}\text{Ca}$ in Fig. 5.3 we obtain similarly that $\varepsilon_{1\text{s}_{1/2}} - \varepsilon_{0\text{d}_{5/2}} \approx 1.4\,\text{MeV}$ and $\varepsilon_{0\text{d}_{3/2}} - \varepsilon_{0\text{d}_{5/2}} \approx 3.9\,\text{MeV}$; these energies are applicable to nuclei with a few holes in the 0d-1s shell. Finally we deduce from the spectra of $^{15}\text{N}$ and $^{15}\text{O}$ in Fig. 5.2 that $\varepsilon_{0\text{p}_{1/2}} - \varepsilon_{0\text{p}_{3/2}} \approx 6.3\,\text{MeV}$, which is applicable to nuclei with a few holes in the 0p shell.

## 5.3 Two-Particle and Two-Hole Nuclei

Two-particle and two-hole nuclei can be described within occupation number representation in a straightforward manner by creating two particles on top of a suitably chosen core (particle vacuum) or by creating two holes into a hole vacuum. These particle and hole vacuums are chosen as was done for the one-particle and one-hole nuclei in Sect. 5.2. Angular momentum coupling is necessary to describe the states of a two-particle or two-hole nucleus. This coupling is discussed in detail below.

### 5.3.1 Examples of Two-Particle Nuclei

In the case of two *like nucleons* outside the core we write the wave function as

$$
\begin{aligned}
|a\,b\,;\,J\,M\rangle &= \mathcal{N}_{ab}(J)\left[c^\dagger_a c^\dagger_b\right]_{JM}|\text{CORE}\rangle \\
&= \mathcal{N}_{ab}(J)\sum_{m_\alpha m_\beta}(j_a\,m_\alpha\,j_b\,m_\beta|J\,M)c^\dagger_\alpha c^\dagger_\beta|\text{CORE}\rangle\,, \\
\mathcal{N}_{ab}(J) &= \frac{\sqrt{1+\delta_{ab}(-1)^J}}{1+\delta_{ab}}\,,
\end{aligned}
\tag{5.21}
$$

where $\mathcal{N}_{ab}(J)$ is a normalization factor and the quantum numbers $\alpha$ and $\beta$ both signify either proton or neutron orbitals. For $a = b$, i.e. two identical nucleons in the same single-particle orbital $a = n_a l_a j_a$, the normalization factor vanishes if $J$ is odd, which means that only states of even $J$ occur. Otherwise we have

$$\mathcal{N}_{ab}(J) = 1 \quad \text{for } a \neq b, \qquad \mathcal{N}_{aa}(J = \text{even}) = \frac{1}{\sqrt{2}}. \qquad (5.22)$$

Calculation of the overlap of two states of the form (5.21) is left as an exercise. Two-particle nuclei described by (5.21) are always *even–even* (or *doubly even*) nuclei.

In the case of one proton and one neutron outside the core the nuclear state is a normalized *proton–neutron* two-particle state of the form

$$|p\,n\,;\,J\,M\rangle = \left[c_p^\dagger c_n^\dagger\right]_{JM}|\text{CORE}\rangle = \sum_{m_\pi m_\nu} (j_p\,m_\pi\,j_n\,m_\nu | J\,M) c_\pi^\dagger c_\nu^\dagger |\text{CORE}\rangle ,$$

$$(5.23)$$

where $\pi \equiv (p, m_\pi)$, $p = n_p l_p j_p$ and $\nu \equiv (n, m_\nu)$, $n = n_n l_n j_n$. Evaluation of the overlap of two states of the form (5.23) is left as an exercise. Two-particle nuclei of this type are always *odd–odd* (or *doubly odd*) nuclei.

The two creation operators in (5.21) and (5.23) always anticommute, as we know from the relations (4.9) and (4.16). Taking into account the symmetry properties of the Clebsch–Gordan coefficients, reviewed in Subsect. 1.2.1, we find for two-particle states the inversion relations

$$|b\,a\,;\,J\,M\rangle = (-1)^{j_a+j_b+J+1}|a\,b\,;\,J\,M\rangle , \qquad (5.24)$$

$$|n\,p\,;\,J\,M\rangle = (-1)^{j_p+j_n+J+1}|p\,n\,;\,J\,M\rangle . \qquad (5.25)$$

Nuclei of the same mass number $A$, known as *isobars*, have many comparable properties. We select isobars with $A = 6$ (Fig. 5.4), $A = 18$ (Fig. 5.5) and $A = 42$ (Fig. 5.6) to examine two-particle nuclei. For the $A = 6$ isobars $^6$He, $^6$Li and $^6$Be the low-energy states can be identified as the two-particle states

$$|^6\text{He}\,;\,0^+,2^+\rangle = \frac{1}{\sqrt{2}}\left[c_{\nu 0p_{3/2}}^\dagger c_{\nu 0p_{3/2}}^\dagger\right]_{0^+,2^+}|\text{CORE}\rangle , \qquad (5.26)$$

$$|^6\text{Li}\,;\,0^+,1^+,2^+,3^+\rangle = \left[c_{\pi 0p_{3/2}}^\dagger c_{\nu 0p_{3/2}}^\dagger\right]_{0^+,1^+,2^+,3^+}|\text{CORE}\rangle , \qquad (5.27)$$

$$|^6\text{Be}\,;\,0^+,2^+\rangle = \frac{1}{\sqrt{2}}\left[c_{\pi 0p_{3/2}}^\dagger c_{\pi 0p_{3/2}}^\dagger\right]_{0^+,2^+}|\text{CORE}\rangle , \qquad (5.28)$$

where the core only contains the $0s_{1/2}$ orbital,

$$|\text{CORE}\rangle = |\text{CORE}(0s)\rangle_\pi |\text{CORE}(0s)\rangle_\nu . \qquad (5.29)$$

As seen in Fig. 5.4, the two-particle states account for the presence of nearly all of the low-energy states in these nuclei. The *Coulomb energy*[2] has been

---

[2] See 17.4.5.

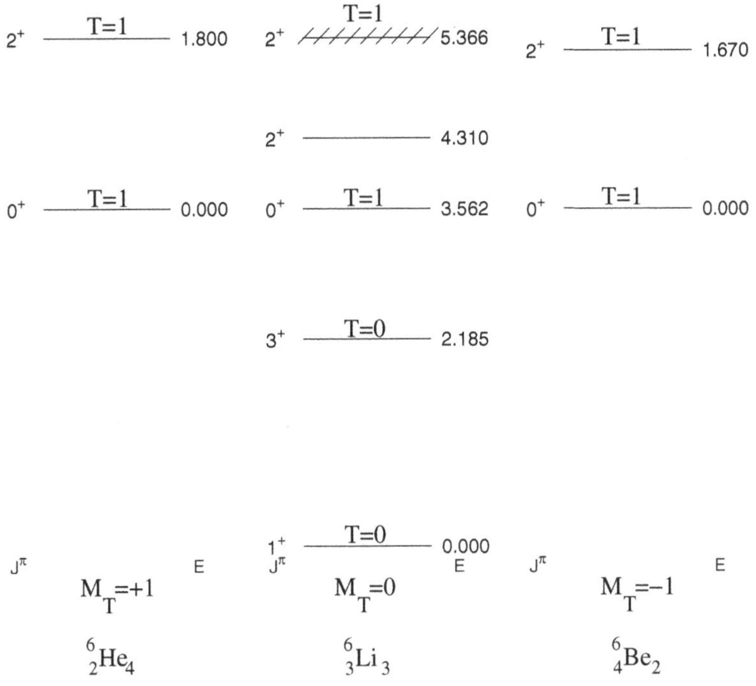

**Fig. 5.4.** Experimental low-energy spectra of the two-particle nuclei $^6$He, $^6$Li and $^6$Be. All experimental energy levels up to the $2^+(T=1)$ level are shown. In $^6$Li the $2^+$ state at 5.366 MeV is unbound. The Coulomb energy has been subtracted and the isospin quantum numbers of the relevant states are displayed

subtracted in the figure, which puts the $0^+$ states at the same level. The isospin quantum numbers, to be discussed in Sect. 5.5, are displayed in the figure.

Our current, highly simplified theory does not give any energy differences between states of different $J$; they are *degenerate*. All the $J^\pi = 0^+, 1^+, 2^+, 3^+$ states in the three isobars are predicted to have the same energy $2\varepsilon_{0p_{3/2}}$. Only the residual interaction, discussed extensively in the later chapters, will lift the degeneracy. However, we note the regularity of the experimental energy differences among comparable states in Fig. 5.4: the $0^+(T=1)$ and $2^+(T=1)$ levels are very nearly equidistant in the three nuclei. A good residual interaction would predict these differences as well as the $1^+$ and $3^+$ energies and the 'extra' $2^+$ state in $^6$Li.

Figure 5.5 shows the low-energy spectra of the isobars $^{18}$O, $^{18}$F and $^{18}$Ne. The counterparts of the following two-particle states are identified among the experimental states:

$$|^{18}\text{O}\,;\,0^+,2^+,4^+\rangle = \frac{1}{\sqrt{2}}\left[c^\dagger_{\nu 0d_{5/2}}c^\dagger_{\nu 0d_{5/2}}\right]_{0^+,2^+,4^+}|\text{CORE}\rangle\,, \quad (5.30)$$

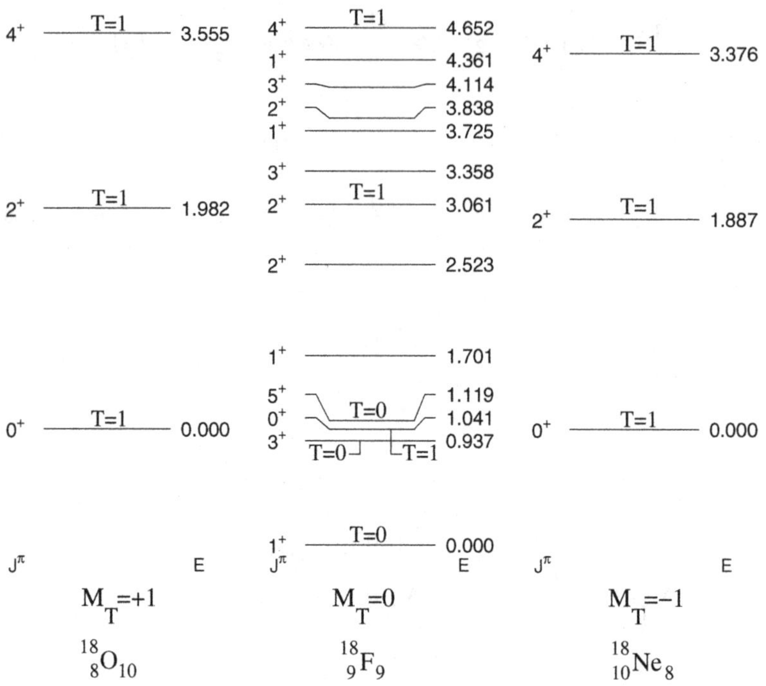

**Fig. 5.5.** Experimental low-energy spectra of the two-particle nuclei $^{18}$O, $^{18}$F and $^{18}$Ne. For $^{18}$F only the positive-parity states are shown. The Coulomb energy has been subtracted and the isospin quantum numbers of the relevant states are displayed

$$|^{18}\mathrm{F}\,;\,0^+,1^+,2^+,3^+,4^+,5^+\rangle = \left[c^\dagger_{\pi 0\mathrm{d}_{5/2}} c^\dagger_{\nu 0\mathrm{d}_{5/2}}\right]_{0^+,1^+,2^+,3^+,4^+,5^+} |\mathrm{CORE}\rangle\,,$$
$$(5.31)$$

$$|^{18}\mathrm{Ne}\,;\,0^+,2^+,4^+\rangle = \frac{1}{\sqrt{2}}\left[c^\dagger_{\pi 0\mathrm{d}_{5/2}} c^\dagger_{\pi 0\mathrm{d}_{5/2}}\right]_{0^+,2^+,4^+} |\mathrm{CORE}\rangle\,,\quad (5.32)$$

where

$$|\mathrm{CORE}\rangle = |\mathrm{CORE}(0\mathrm{s}\text{-}0\mathrm{p})\rangle_\pi |\mathrm{CORE}(0\mathrm{s}\text{-}0\mathrm{p})\rangle_\nu\,. \qquad (5.33)$$

As in the case of the $A = 6$ nuclei, all the $J$ states in (5.30)–(5.32) are degenerate in our simple description. The comparable experimental levels in the three isobars again have nearly the same energies.

Figure 5.6 shows the low-energy spectra of $^{42}$Ca, $^{42}$Sc and $^{42}$Ti. Their $0^+$–$7^+$ states are mainly two-particle states of the form

$$|A = 42\,;\,(0\mathrm{f}_{7/2})^2\,J^+\rangle = \mathcal{N}\left[c^\dagger_{0\mathrm{f}_{7/2}} c^\dagger_{0\mathrm{f}_{7/2}}\right]_{J^+} |\mathrm{CORE}\rangle\,, \qquad (5.34)$$

where the core contains the orbitals within the 0s-0p-0d-1s shells. These nuclei form a special case since the valence space between adjacent magic numbers consists of the $0\mathrm{f}_{7/2}$ orbital only. All nuclei with this valence space are

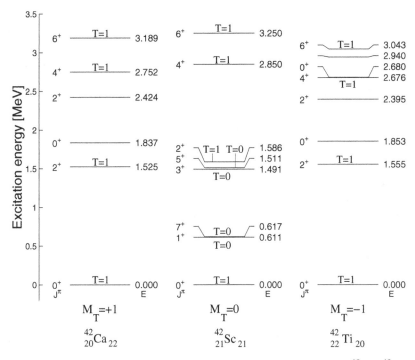

**Fig. 5.6.** Experimental low-energy spectra of the two-particle nuclei $^{42}$Ca, $^{42}$Sc and $^{42}$Ti. For the even–even isobars all experimental levels are shown up to $6^+(T = 1)$, whereas for $^{42}$Sc only the assumed two-particle states are included. The Coulomb energy has been subtracted and the isospin quantum numbers of the relevant states are displayed

traditionally called $f_{7/2}$-shell nuclei. They have been studied by shell-model calculations long before any advanced computing equipment existed.

### 5.3.2 Examples of Two-Hole Nuclei

The wave functions of *two-hole nuclei* are similar to those of two-particle nuclei. For two proton holes or two neutron holes we have, similarly to (5.21),

$$\left|a^{-1}b^{-1}; J M\right\rangle = \mathcal{N}_{ab}(J)\left[h_a^\dagger h_b^\dagger\right]_{JM}|\mathrm{HF}\rangle , \quad \mathcal{N}_{ab}(J) = \frac{\sqrt{1 + \delta_{ab}(-1)^J}}{1 + \delta_{ab}} .$$

$$(5.35)$$

For one proton hole and one neutron hole we have, similarly to (5.23),

$$\left|p^{-1}n^{-1}; J M\right\rangle = \left[h_p^\dagger h_n^\dagger\right]_{JM}|\mathrm{HF}\rangle .$$

$$(5.36)$$

The hole character of the single-particle components is indicated by the superscript $-1$. Two-proton-hole nuclei and two-neutron-hole nuclei are always even–even nuclei, and proton-hole–neutron-hole nuclei are always odd–odd nuclei. Overlaps of wave functions of the types (5.35) and (5.36) are discussed as exercises.

The two hole-creation operators in (5.35) and (5.36) always anticommute because they both are associated with particle annihilation through (4.46) and two particle annihilation operators anticommute according to (4.9). This fact, together with the symmetry properties of the Clebsch–Gordan coefficients, gives the symmetry relations

$$|b^{-1} a^{-1}\,;\, J\,M\rangle = (-1)^{j_a+j_b+J+1}|a^{-1} b^{-1}\,;\, J\,M\rangle\,, \qquad (5.37)$$

$$|n^{-1} p^{-1}\,;\, J\,M\rangle = (-1)^{j_p+j_n+J+1}|p^{-1} n^{-1}\,;\, J\,M\rangle\,. \qquad (5.38)$$

Examples of two-hole nuclei are provided by the mass chains $A = 38$ (Fig. 5.7) and $A = 54$ (Fig. 5.8). The states of $^{38}$Ar, $^{38}$K and $^{38}$Ca are

$$|^{38}\mathrm{Ar}\,;\, 0^+, 2^+\rangle = \frac{1}{\sqrt{2}}\big[h^\dagger_{\pi 0d_{3/2}}\, h^\dagger_{\pi 0d_{3/2}}\big]_{0^+,2^+}|\mathrm{HF}\rangle\,, \qquad (5.39)$$

$$|^{38}\mathrm{K}\,;\, 0^+, 1^+, 2^+, 3^+\rangle = \big[h^\dagger_{\pi 0d_{3/2}}\, h^\dagger_{\nu 0d_{3/2}}\big]_{0^+,1^+,2^+,3^+}|\mathrm{HF}\rangle\,, \qquad (5.40)$$

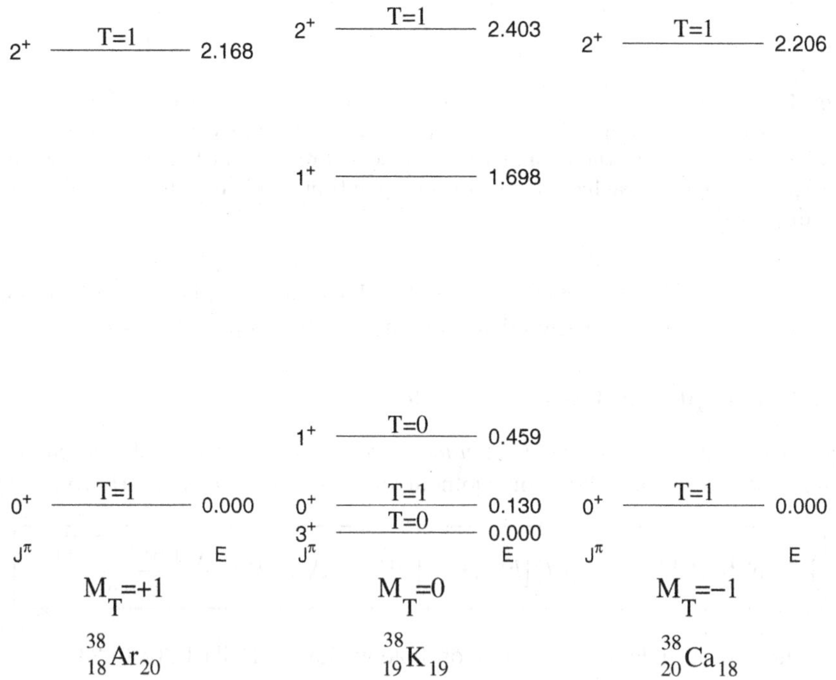

**Fig. 5.7.** Experimental low-energy spectra of the two-hole nuclei $^{38}$Ar, $^{38}$K and $^{38}$Ca. All known energy levels up to the $2^+$ state are shown. The Coulomb energy has been subtracted and the isospin quantum numbers of the relevant states are displayed

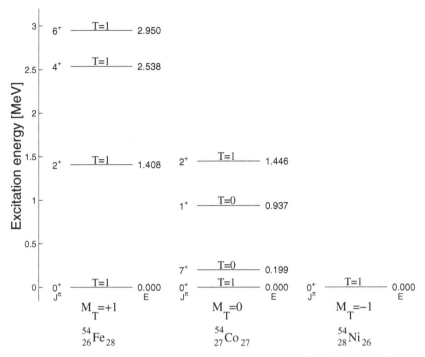

**Fig. 5.8.** Experimental energy levels of the two-hole nuclei $^{54}$Fe, $^{54}$Co and $^{54}$Ni. The Coulomb energy has been subtracted and the isospin quantum numbers displayed

$$|^{38}\mathrm{Ca}\,;\,0^+,2^+\rangle = \frac{1}{\sqrt{2}}\left[h^\dagger_{\nu 0d_{3/2}}h^\dagger_{\nu 0d_{3/2}}\right]_{0^+,2^+}|\mathrm{HF}\rangle\,, \tag{5.41}$$

where the particle–hole vacuum is defined as

$$|\mathrm{HF}\rangle = |\mathrm{HF}(Z=20)\rangle_\pi |\mathrm{HF}(N=20)\rangle_\nu\,. \tag{5.42}$$

As in the case of two-particle nuclei, all these two-hole states are degenerate within our simple approach. Turning on the residual interaction will split the degeneracies in various ways and explain the additional $1^+$ state in $^{38}$K.

Figure 5.8 shows the known experimental energy levels of $^{54}$Fe, $^{54}$Co and $^{54}$Ni. They can be explained as arising from two holes in the $0f_{7/2}$ shell:

$$|A=54\,;\,\left[(0f_{7/2})^{-1}\right]^2 J^+\rangle = \mathcal{N}\left[h^\dagger_{0f_{7/2}}h^\dagger_{0f_{7/2}}\right]_{J^+}|\mathrm{HF}\rangle\,, \tag{5.43}$$

where the particle–hole vacuum corresponds to a state whose Fermi surfaces occur at the magic numbers $Z=28$ and $N=28$.

## 5.4 Particle–Hole Nuclei

The starting point for the description of particle–hole nuclei is the particle–hole vacuum $|\mathrm{HF}\rangle$, which describes the ground state of a doubly magic even–

even nucleus. This even–even nucleus serves as a *reference nucleus* for calculation of its own excited states and states of its neighboring nuclei. All these states are constructed by creating particle–hole excitations on the particle–hole vacuum.

Particle–hole excited states are created by letting one nucleon jump from a state below the Fermi level to a state above it. This can be done in two essentially different ways, which constitute *charge-conserving* and *charge-changing* particle–hole excitations. The former are jumps from a proton (neutron) hole state to a proton (neutron) particle state. They are denoted as $pp^{-1}$ and $nn^{-1}$, and they generate excited states of nuclei with doubly closed major shells. The latter are either proton-particle–neutron-hole ($pn^{-1}$) or neutron-particle–proton-hole ($np^{-1}$) excitations, which describe the ground and excited states of odd–odd nuclei at doubly closed major shells.

The wave functions of the particle–hole nuclei can be written as follows. The excited states of the *even–even closed-shell nuclei* are

$$\left| a\, b^{-1}\,;\, J\,M \right\rangle = \left[ c_a^\dagger h_b^\dagger \right]_{JM} |\mathrm{HF}\rangle = \left[ c_a^\dagger \tilde{c}_b \right]_{JM} |\mathrm{HF}\rangle . \tag{5.44}$$

with the orthogonality and normalization relation

$$\langle a\, b^{-1}\,;\, J\,M | c\, d^{-1}\,;\, J'\,M' \rangle = \delta_{ac}\delta_{bd}\delta_{JJ'}\delta_{MM'} . \tag{5.45}$$

In the particle–hole picture the ground and excited states of *odd–odd nuclei* at closed major shells can be generated by starting from the particle–hole vacuum. The wave functions of the proton-particle–neutron-hole nuclei are

$$\left| p\, n^{-1}\,;\, J\,M \right\rangle = \left[ c_p^\dagger h_n^\dagger \right]_{JM} |\mathrm{HF}\rangle = \left[ c_p^\dagger \tilde{c}_n \right]_{JM} |\mathrm{HF}\rangle \tag{5.46}$$

with the orthonormality condition

$$\langle p\, n^{-1}\,;\, J\,M | p'\, n'^{-1}\,;\, J'\,M' \rangle = \delta_{pp'}\delta_{nn'}\delta_{JJ'}\delta_{MM'} . \tag{5.47}$$

For the neutron-particle–proton-hole we have

$$\left| n\, p^{-1}\,;\, J\,M \right\rangle = \left[ c_n^\dagger h_p^\dagger \right]_{JM} |\mathrm{HF}\rangle = \left[ c_n^\dagger \tilde{c}_p \right]_{JM} |\mathrm{HF}\rangle \tag{5.48}$$

with the orthonormality condition

$$\langle n\, p^{-1}\,;\, J\,M | n'\, p'^{-1}\,;\, J'\,M' \rangle = \delta_{pp'}\delta_{nn'}\delta_{JJ'}\delta_{MM'} . \tag{5.49}$$

The particle-creation and hole-creation operators in (5.44), (5.46) and (5.48) always anticommute because they refer to different single-particle orbitals, one below and one above the Fermi surface. Using the symmetry properties of the Clebsch–Gordan coefficients we then find

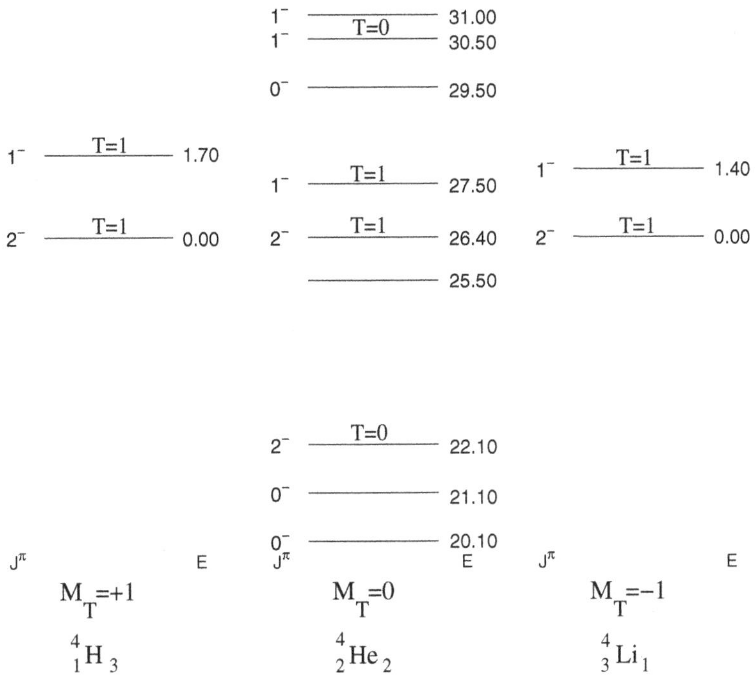

**Fig. 5.9.** All experimentally known energy levels, except the ground state of $^4$He, below the highest $1^-$ states shown in the isobars $^4$H, $^4$He and $^4$Li. The Coulomb energy has been subtracted and the isospin quantum numbers of the relevant states are displayed

$$|b^{-1}\,a\,;\,J\,M\rangle = (-1)^{j_a+j_b+J+1}|a\,b^{-1}\,;\,J\,M\rangle\,, \tag{5.50}$$

$$|n^{-1}\,p\,;\,J\,M\rangle = (-1)^{j_p+j_n+J+1}|p\,n^{-1}\,;\,J\,M\rangle\,, \tag{5.51}$$

$$|p^{-1}\,n\,;\,J\,M\rangle = (-1)^{j_p+j_n+J+1}|n\,p^{-1}\,;\,J\,M\rangle\,. \tag{5.52}$$

As examples of typical particle–hole nuclei we present triplets in the mass chains $A = 4$ (Fig. 5.9), $A = 16$ (Fig. 5.10) and $A = 40$ (Fig. 5.11). For the $A = 4$ nuclei a proton or a neutron can be excited from the 0s shell into the $0\mathrm{p}_{3/2}$ orbital in the 0p major shell. The corresponding charge-changing excitations produce the odd–odd nuclei $^4$H and $^4$Li. Their wave functions are

$$|^4\mathrm{H}\,;\,1^-,2^-\rangle = \left[c^\dagger_{\nu 0\mathrm{p}_{3/2}}h^\dagger_{\pi 0\mathrm{s}_{1/2}}\right]_{1^-,2^-}|\mathrm{HF}\rangle\,, \tag{5.53}$$

$$|^4\mathrm{Li}\,;\,1^-,2^-\rangle = \left[c^\dagger_{\pi 0\mathrm{p}_{3/2}}h^\dagger_{\nu 0\mathrm{s}_{1/2}}\right]_{1^-,2^-}|\mathrm{HF}\rangle\,, \tag{5.54}$$

with the particle–hole vacuum

$$|\mathrm{HF}\rangle = |\mathrm{HF}(Z=2)\rangle_\pi|\mathrm{HF}(N=2)\rangle_\nu\,. \tag{5.55}$$

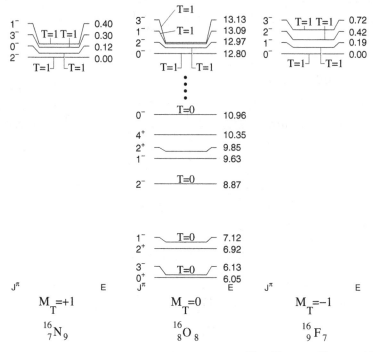

**Fig. 5.10.** Experimental energy levels of the isobars $^{16}$N, $^{16}$O and $^{16}$F. For $^{16}$O the ground state and the states between $0^-(T=0)$ and $0^-(T=1)$ have been omitted. The Coulomb energy has been subtracted and the isospin quantum numbers of the relevant states are displayed

These states are seen in the experimental spectra of the $A = 4$ nuclei in Fig. 5.9. Higher-lying excited $0^-$ and $1^-$ states can be built by exciting the $0s_{1/2}$ nucleon up to the $0p_{1/2}$ orbital, giving the wave functions

$$|^4\mathrm{H}; 0^-, 1^-\rangle = \left[c^\dagger_{\nu 0p_{1/2}} h^\dagger_{\pi 0s_{1/2}}\right]_{0^-,1^-} |\mathrm{HF}\rangle \,, \qquad (5.56)$$

$$|^4\mathrm{Li}; 0^-, 1^-\rangle = \left[c^\dagger_{\pi 0p_{1/2}} h^\dagger_{\nu 0s_{1/2}}\right]_{0^-,1^-} |\mathrm{HF}\rangle \,. \qquad (5.57)$$

For the even–even $^4$He nucleus we have charge-conserving particle–hole excitations, of both the proton–proton and the neutron–neutron type, across the $Z = 2$ and $N = 2$ magic gaps. The basic excitations are

$$\left[c^\dagger_{\pi 0p_{3/2}} h^\dagger_{\pi 0s_{1/2}}\right]_{1^-,2^-} |\mathrm{HF}\rangle \,, \quad \left[c^\dagger_{\nu 0p_{3/2}} h^\dagger_{\nu 0s_{1/2}}\right]_{1^-,2^-} |\mathrm{HF}\rangle \,. \qquad (5.58)$$

One can form two orthogonal and normalized linear combinations of these,

$$\begin{aligned}|^4\mathrm{He}; 1^-, 1^-, 2^-, 2^-\rangle = \frac{1}{\sqrt{2}}\Big( & \left[c^\dagger_{\pi 0p_{3/2}} h^\dagger_{\pi 0s_{1/2}}\right]_{1^-,2^-} |\mathrm{HF}\rangle \\ \pm & \left[c^\dagger_{\nu 0p_{3/2}} h^\dagger_{\nu 0s_{1/2}}\right]_{1^-,2^-} |\mathrm{HF}\rangle \Big) \,, \qquad (5.59)\end{aligned}$$

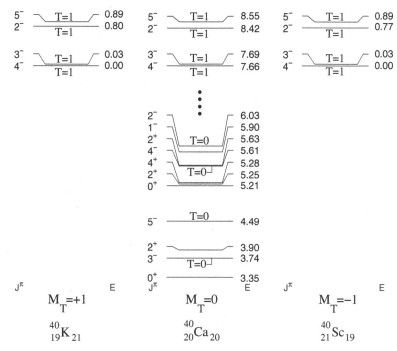

**Fig. 5.11.** Experimental energy levels of the isobars $^{40}$K, $^{40}$Ca and $^{40}$Sc. For $^{40}$Ca the ground state and the states between $2^-(T=0)$ and $4^-(T=1)$ have been omitted. The Coulomb energy has been subtracted and the isospin quantum numbers of the relevant states are displayed

leading to two $1^-$ and two $2^-$ states. These states are seen in the spectrum of $^4$He in Fig. 5.9. In our simple theory the four states are degenerate at a common energy, due to the assumption of the same relative energies for protons and neutrons. The residual interaction will lift the degeneracy, as discussed in Chap. 8. The two linear combinations in (5.59) can be conveniently labelled by the *isospin* quantum number $T$ discussed in the next section.

The discussion of the $A = 16$ isobars, $^{16}$N, $^{16}$O and $^{16}$F, follows the lines of the discussion of the $A = 4$ nuclei. The charge-changing particle–hole excitations yield for the odd–odd nuclei $^{16}_{7}$N$_9$ and $^{16}_{9}$F$_7$ the following low-energy states:

$$|^{16}\mathrm{N}\,;\,2^-,3^-\rangle = \left[c^\dagger_{\nu 0\mathrm{d}_{5/2}} h^\dagger_{\pi 0\mathrm{p}_{1/2}}\right]_{2^-,3^-}|\mathrm{HF}\rangle\,, \tag{5.60}$$

$$|^{16}\mathrm{N}\,;\,0^-,1^-\rangle = \left[c^\dagger_{\nu 1\mathrm{s}_{1/2}} h^\dagger_{\pi 0\mathrm{p}_{1/2}}\right]_{0^-,1^-}|\mathrm{HF}\rangle\,, \tag{5.61}$$

$$|^{16}\mathrm{F}\,;\,2^-,3^-\rangle = \left[c^\dagger_{\pi 0\mathrm{d}_{5/2}} h^\dagger_{\nu 0\mathrm{p}_{1/2}}\right]_{2^-,3^-}|\mathrm{HF}\rangle\,, \tag{5.62}$$

$$|^{16}\mathrm{F}\,;\,0^-,1^-\rangle = \left[c^\dagger_{\pi 1\mathrm{s}_{1/2}} h^\dagger_{\nu 0\mathrm{p}_{1/2}}\right]_{0^-,1^-}|\mathrm{HF}\rangle\,, \tag{5.63}$$

where the particle–hole vacuum is

$$|\mathrm{HF}\rangle = |\mathrm{HF}(Z=8)\rangle_\pi |\mathrm{HF}(N=8)\rangle_\nu \ . \tag{5.64}$$

The $0^-$–$3^-$ states predicted are seen to be present closely bunched in Fig. 5.10. Within our simple scheme, the $0^-$ and $1^-$ levels should lie above the $2^-$ and $3^-$ levels by an energy of $\varepsilon_{1s_{1/2}} - \varepsilon_{0d_{5/2}} \approx 0.7\,\mathrm{MeV}$, as obtained from the analysis of the one-particle spectra in Sect. 5.2. The width of the bunches is of this order but there is no further agreement.

The states of the even–even nucleus $^{16}$O are given by

$$|^{16}\mathrm{O}\,;\,2^-,2^-,3^-,3^-\rangle = \frac{1}{\sqrt{2}}\left(\left[c_{\pi 0d_{5/2}}^\dagger h_{\pi 0p_{1/2}}^\dagger\right]_{2^-,3^-}|\mathrm{HF}\rangle\right.$$
$$\left. \pm \left[c_{\nu 0d_{5/2}}^\dagger h_{\nu 0p_{1/2}}^\dagger\right]_{2^-,3^-}|\mathrm{HF}\rangle\right) , \tag{5.65}$$

$$|^{16}\mathrm{O}\,;\,0^-,0^-,1^-,1^-\rangle = \frac{1}{\sqrt{2}}\left(\left[c_{\pi 1s_{1/2}}^\dagger h_{\pi 0p_{1/2}}^\dagger\right]_{0^-,1^-}|\mathrm{HF}\rangle\right.$$
$$\left. \pm \left[c_{\nu 1s_{1/2}}^\dagger h_{\nu 0p_{1/2}}^\dagger\right]_{0^-,1^-}|\mathrm{HF}\rangle\right) . \tag{5.66}$$

These states are identified in Fig. 5.10. However, the residual interaction is seen to cause wide splittings of the degeneracies, up to 7 MeV for the $3^-$ states.

Finally, for the $A = 40$ nuclei $^{40}$K, $^{40}$Ca and $^{40}$Sc, the experimental situation is depicted in Fig. 5.11. In this case the particle–hole excitations proceed across the $Z = 20$ and $N = 20$ magic gaps, those of lowest energy from the $0d_{3/2}$ orbital below the gap to the $0f_{7/2}$ orbital above. The resulting states are

$$\left[c_{0f_{7/2}}^\dagger h_{0d_{3/2}}^\dagger\right]_{2^-,3^-,4^-,5^-}|\mathrm{HF}\rangle , \tag{5.67}$$

where $|\mathrm{HF}\rangle$ closes at the magic numbers $Z = 20$ and $N = 20$. As can be seen from Fig. 5.11, our simple description is rather good since the relevant states, labelled $T = 1$, are very similarly spaced and within 1 MeV. Contrariwise, the states labelled by $T = 0$ in the spectrum of $^{40}$Ca appear strongly perturbed by the residual interaction.

Further interesting cases of particle–hole nuclei are the $A = 48$ isobars $^{48}_{19}\mathrm{K}_{29}$, $^{48}_{20}\mathrm{Ca}_{28}$ and $^{48}_{21}\mathrm{Sc}_{27}$, with the particle–hole vacuum

$$|\mathrm{HF}\rangle = |\mathrm{HF}(Z=20)\rangle_\pi |\mathrm{HF}(N=28)\rangle_\nu \ . \tag{5.68}$$

Anticipating the following section, we note that in the previous examples we have witnessed good *isospin symmetry*. It shows up in the similarity among the isobars of the group of $T = 1$ states with negative parity and is due to the proton and neutron Fermi surfaces occurring at the same nucleon number. For the $A = 48$ nuclei this parallel is lost because the Fermi levels are at $Z = 20$ and $N = 28$. The active single-particle orbitals are $\pi 0d_{3/2}$, $\pi 0f_{7/2}$, $\nu 0f_{7/2}$ and $\nu 1p_{3/2}$. They give rise to particle–hole states of both parities. Also for the $A = 56$ mass chain, $^{56}_{27}\mathrm{Co}_{29}$, $^{56}_{28}\mathrm{Ni}_{28}$ and $^{56}_{29}\mathrm{Cu}_{27}$, some experimental data are available to enable a discussion in terms of simple particle–hole excitations across the magic gaps.

## 5.5 Isospin Representation of Few-Nucleon Systems

### 5.5.1 General Isospin Formalism

So far we have considered the proton and the neutron as two different particle species, labelled $\pi = (p, m_\pi)$ and $\nu = (n, m_\nu)$. An alternative to this
*proton–neutron formalism (representation)*, without any difference in physical
content, is the *isospin formalism (representation)*. There the $\pi$ and $\nu$ labels
are abandoned, and protons and neutrons are considered to form just two
*isospin states* of a generic nuclear particle, the *nucleon*. The states 'neutron'
and 'proton' are denoted in various ways as follows[3]:

$$
\begin{aligned}
|n\rangle &= |t = \tfrac{1}{2} , m_t = +\tfrac{1}{2}\rangle = \chi^T_{\frac{1}{2},+\frac{1}{2}} = \begin{pmatrix} 1 \\ 0 \end{pmatrix} , \\
|p\rangle &= |t = \tfrac{1}{2} , m_t = -\tfrac{1}{2}\rangle = \chi^T_{\frac{1}{2},-\frac{1}{2}} = \begin{pmatrix} 0 \\ 1 \end{pmatrix} .
\end{aligned}
\tag{5.69}
$$

The matrix representation is equated to the abstract state, consistent with
the convention noted in connection with the Pauli spin matrices (2.38). The
isospin 'length' quantum number is $t$, and $m_t$ is its projection quantum num-
ber. Isospin $t = \tfrac{1}{2}$ is completely analogous to mechanical spin $s = \tfrac{1}{2}\hbar$ except
that isospin is not measured in units of $\hbar$ since it has nothing to do with
angular momentum.

The *isospinors* $\chi^T_{\frac{1}{2}m_t}$ are analogous to the spinors $\chi_{\frac{1}{2}m_s}$ of (3.65) and
(3.66). The isospin vector operator $t$ is defined in analogy to the definition
(1.1) of the angular momentum:

$$
t_k^\dagger = t_k , \quad k = 1, 2, 3, \quad [t_i, t_j] = i \sum_k \epsilon_{ijk} t_k .
\tag{5.70}
$$

The isospin components are often labelled $x, y, z$, but we use $1, 2, 3$ to empha-
size that isospin space (isospace) is distinct from the usual coordinate space.
In the pattern of the general eigenvalue equations (1.2) and (1.3), we have for
isospin $\tfrac{1}{2}$

$$
t^2|n\rangle = \tfrac{3}{4}|n\rangle , \quad t_3|n\rangle = +\tfrac{1}{2}|n\rangle ,
\tag{5.71}
$$
$$
t^2|p\rangle = \tfrac{3}{4}|p\rangle , \quad t_3|p\rangle = -\tfrac{1}{2}|p\rangle .
\tag{5.72}
$$

Exploiting the general analogy to angular momentum, and specifically the
relations (1.5)–(1.7), we have the isospin raising and lowering operators

---

[3] The convention of denoting the neutron by 'isospin up' and the proton by 'isospin
down' is normal in nuclear physics. It leads to a positive total isospin projection
$M_T = \tfrac{1}{2}(N - Z)$ for most nuclei. The opposite convention is used in particle
physics.

$$t_\pm = t_1 \pm \mathrm{i} t_2 \tag{5.73}$$

with the commutation relations

$$[t_+, t_-] = 2t_3 , \quad [t_\pm, t_3] = \mp t_\pm . \tag{5.74}$$

The effect of $t_\pm$ on the proton and neutron is given by

$$t_+ |p\rangle = |n\rangle , \quad t_+ |n\rangle = 0 , \quad t_- |p\rangle = 0 , \quad t_- |n\rangle = |p\rangle . \tag{5.75}$$

Denoting $\boldsymbol{\tau} \equiv 2\boldsymbol{t}$ we have the isospin analogues of the Pauli spin matrices (2.38):

$$\tau_1 = \begin{pmatrix} 0 & 1 \\ 1 & 0 \end{pmatrix} , \quad \tau_2 = \begin{pmatrix} 0 & -\mathrm{i} \\ \mathrm{i} & 0 \end{pmatrix} , \quad \tau_3 = \begin{pmatrix} 1 & 0 \\ 0 & -1 \end{pmatrix} . \tag{5.76}$$

The *total isospin* operator is

$$\boldsymbol{T} = \sum_{i=1}^{A} \boldsymbol{t}(i) \tag{5.77}$$

just like the total angular momentum $\boldsymbol{J} = \sum_i \boldsymbol{s}(i)$ of spin-$\frac{1}{2}$ vectors $\boldsymbol{s}(i)$ for some particles $i$. For the total isospin quantum number $T$ and its projection $M_T$ we have

$$\boldsymbol{T}^2 |T\, M_T\rangle = T(T+1)|T\, M_T\rangle , \quad T_3 |T\, M_T\rangle = M_T |T\, M_T\rangle \tag{5.78}$$

with $M_T = -T, -T+1, \ldots, T-1, T$. The total isospin is integer of half-integer depending on whether $A$ is even or odd. The operators

$$T_\pm = \sum_{i=1}^{A} t_\pm(i) , \quad T_3 = \sum_{i=1}^{A} t_3(i) \tag{5.79}$$

have the properties

$$T_\pm |T\, M_T\rangle = \sqrt{T(T+1) - M_T(M_T \pm 1)}\, |T\, M_T \pm 1\rangle , \tag{5.80}$$

$$T_3 |T\, M_T\rangle = [N \times (+\tfrac{1}{2}) + Z \times (-\tfrac{1}{2})]|T\, M_T\rangle = \tfrac{1}{2}(N - Z)|T\, M_T\rangle , \tag{5.81}$$

so that

$$M_T = \tfrac{1}{2}(N - Z) . \tag{5.82}$$

Before proceeding to further technical details of isospin, let us briefly examine *isospin symmetry* in nuclei. From experimental data it is known that to a good approximation the energy levels of light nuclei are characterized by a unique isospin quantum number $T$, i.e. $T$ is a good quantum number. This is equivalent to the commutation condition

$$[H, \boldsymbol{T}^2] \approx 0 . \tag{5.83}$$

For complete rotation invariance in isospace we would require not only an equality sign in (5.83) but commutation of $H$ with each component of $\boldsymbol{T}$ separately,

$$[H, T_\pm] = 0 , \quad [H, T_3] = 0 . \tag{5.84}$$

These conditions are not satisfied because of the proton–neutron mass difference and the Coulomb interaction between protons. The former leads to an isovector term and the latter to both an isovector and an isotensor term in $H$. Neglecting them leads to complete isospin symmetry, where all the substates $M_T = -T, -T+1, \ldots, T-1, T$ are degenerate. This situation is approximately realized in the examples of this chapter, Figs. 5.4–5.11, where the Coulomb energy has been subtracted.

### 5.5.2 Tensor Operators in Isospin Representation

Single-particle states are written in the Baranger notation (3.62) as $\alpha = (a, m_\alpha)$, where $a = n_a l_a j_a$. In the proton–neutron formalism the distinction between proton and neutron occupation of this orbital was indicated by writing $\pi = (p, m_\pi)$ for protons and $\nu = (n, m_\nu)$ for neutrons. In the isospin formalism we have to add an extra quantum number, namely the isospin projection $m_{t\alpha}$, to fully specify a single-particle state:

$$\alpha = (a, m_\alpha, m_{t\alpha}) . \tag{5.85}$$

It also understood that the quantum number $t = \frac{1}{2}$ is now carried by $a$ just as is $s = \frac{1}{2}$. With the new quantum number $m_t$ present the Kronecker delta of (3.72) is generalized to

$$\delta_{ab} = \delta_{n_a n_b} \delta_{l_a l_b} \delta_{j_a j_b} , \tag{5.86}$$

$$\delta_{\alpha\beta} = \delta_{ab} \delta_{m_\alpha m_\beta} \delta_{m_{t\alpha} m_{t\beta}} . \tag{5.87}$$

Isospins are coupled by Clebsch–Gordan coefficients the same way as angular momenta. The only difference is that the isospin of a single particle has $t = \frac{1}{2}$ as its only value whereas the angular momentum of a single particle can be any of $j = \frac{1}{2}, \frac{3}{2}, \frac{5}{2}, \ldots$ (in units of $\hbar$). Two isospins can thus be coupled to $T = 0$ and $T = 1$ only, according to

$$|ab; J M; T M_T\rangle = \mathcal{N} \sum_{\substack{m_\alpha m_\beta \\ m_{t\alpha} m_{t\beta}}} (j_a m_\alpha j_b m_\beta | J M)(\tfrac{1}{2} m_{t\alpha} \tfrac{1}{2} m_{t\beta} | T M_T)|\alpha\rangle|\beta\rangle . \tag{5.88}$$

Tensor operators exist in isospace as they do in coordinate space. Consider a one-body operator $O_{\lambda\mu}^{TM_T}$. It is the component $M_T$ of a spherical tensor of rank $T$ in isospace. Equation (4.22) gives its expression in occupation number representation as

$$O_{\lambda\mu}^{TM_T} = \widehat{\lambda}^{-1} \sum_{\substack{ab \\ m_{t\alpha} m_{t\beta}}} (a\, m_{t\alpha} \| \boldsymbol{O}_\lambda^{TM_T} \| b\, m_{t\beta}) \big[ c_{am_{t\alpha}}^\dagger \tilde{c}_{bm_{t\beta}} \big]_{\lambda\mu} . \tag{5.89}$$

We wish to bring this into an isospin-reduced form by performing the sums over $m_{t\alpha}$ and $m_{t\beta}$. The double-barred matrix element is reduced in coordinate space only. To reduce it in isospace too, we apply the Wigner–Eckart theorem (2.27) to it:

$$(a\,m_{t\alpha}\|\boldsymbol{O}_\lambda^{TM_T}\|b\,m_{t\beta}) = (-1)^{\frac{1}{2}-m_{t\alpha}} \begin{pmatrix} \frac{1}{2} & T & \frac{1}{2} \\ -m_{t\alpha} & M_T & m_{t\beta} \end{pmatrix} (a\,\|\|\boldsymbol{O}_\lambda^T\|\|\,b)\,. \quad (5.90)$$

The triple-barred matrix element is reduced both in coordinate space and in isospace.

The next step is to address the isospin-tensorial character of the creation and annihilation operators. The creation operator $c_{am_{t\alpha}}^\dagger$ is immediately the component $m_{t\alpha}$ of an isotensor of rank $\frac{1}{2}$. The annihilation operator is defined in (4.23): $\tilde{c}_{am_\alpha} = (-1)^{j_a+m_\alpha} c_{a,-m_\alpha}$. It is a tensor in coordinate space but not in isospace. By analogy, we define the isotensor annihilation operator by

$$\hat{c}_{am_{t\alpha}} \equiv (-1)^{\frac{1}{2}+m_{t\alpha}} \tilde{c}_{a,-m_{t\alpha}}\,. \quad (5.91)$$

The operator $\hat{c}_a$ is a rank-$j_a$ tensor in coordinate space and a rank-$\frac{1}{2}$ tensor in isospace.

Substituting (5.90) and (5.91) into (5.89) and applying relations of Sect. 1.2 finally gives

$$\boxed{O_{\lambda\mu}^{TM_T} = \hat{\lambda}^{-1}\hat{T}^{-1} \sum_{ab}(a\,\|\|\boldsymbol{O}_\lambda^T\|\|\,b)\big[c_a^\dagger \hat{c}_b\big]_{\lambda\mu}^{TM_T}\,,} \quad (5.92)$$

where the superscripts $MT_M$ denote isospin coupling as detailed in (5.88). The only possible values of $T$ are 0 ($M_T = 0$) and 1 ($M_T = 0, \pm 1$), representing *isoscalar* and *isovector* operators respectively. With the values of the $3j$ symbols taken from Sect. 1.2, equation (5.90) gives

$$(a\,\|\|\boldsymbol{O}_\lambda^0\|\|\,b) = \sqrt{2}(a\tfrac{1}{2}\|\boldsymbol{O}_\lambda^{00}\|b\tfrac{1}{2})\,, \quad (5.93)$$

$$(a\,\|\|\boldsymbol{O}_\lambda^1\|\|\,b) = \sqrt{6}(a\tfrac{1}{2}\|\boldsymbol{O}_\lambda^{10}\|b\tfrac{1}{2})\,. \quad (5.94)$$

The doubly reduced nuclear matrix element can now be written as

$$\boxed{\begin{aligned} (\xi_f\,J_f\,T_f\,\|\|\boldsymbol{O}_\lambda^T\|\|\,\xi_i\,J_i\,T_i) &= \hat{\lambda}^{-1}\hat{T}^{-1}\sum_{ab}(a\,\|\|\boldsymbol{O}_\lambda^T\|\|\,b) \\ &\times (\xi_f\,J_f\,T_f\,\|\|\big[c_a^\dagger \hat{c}_b\big]_\lambda^T\|\|\,\xi_i\,J_i\,T_i)\,. \end{aligned}} \quad (5.95)$$

The last factor on the right is the *doubly reduced one-body transition density*. It contains the nuclear structure part of an electromagnetic or beta transition matrix element. Isospin-reduced nuclear matrix elements are used in the shell model literature (see, e.g. [28]). The isospin formalism is often favoured because it reduces computational work as compared with the proton–neutron formalism. The differences in computational effort are discussed in Chap. 8 in connection with the two-body interaction.

### 5.5.3 Isospin Representation of Two-Particle and Two-Hole Nuclei

Wave functions of two-particle nuclei with coupled angular momenta were discussed in Subsect. 5.3.1. We now include isospin coupling and write the wave function in occupation number representation:

$$|ab; JM; T M_T\rangle = \mathcal{N}_{ab}(JT)\left[c_a^\dagger c_b^\dagger\right]_{JM}^{TM_T}|\text{CORE}\rangle ,$$

$$\mathcal{N}_{ab}(JT) = \frac{\sqrt{1 - \delta_{ab}(-1)^{J+T}}}{1 + \delta_{ab}} . \qquad (5.96)$$

The normalization constant $\mathcal{N}_{ab}$ has two possible non-zero values: $\mathcal{N}_{ab} = 1$ for $a \neq b$ and $\mathcal{N}_{aa}(J + T = \text{odd}) = 1/\sqrt{2}$. In the $a = b$ case we have the important result that only certain $(J,T)$ pairs are possible: for $T = 0$ only $J = \text{odd}$, and for $T = 1$ only $J = \text{even}$.

Two-hole wave functions with coupled angular momentum and coupled isospin can be written down immediately following the pattern of (5.96):

$$|a^{-1} b^{-1}; JM; T M_T\rangle = \mathcal{N}_{ab}(JT)\left[h_a^\dagger h_b^\dagger\right]_{JM}^{TM_T}|\text{HF}\rangle . \qquad (5.97)$$

The hole creation operator $h_\alpha^\dagger$ appearing here is like the one defined in (4.46), $h_\alpha^\dagger = \tilde{c}_\alpha$, but now generalized to include isospin. Following (5.91), we thus have

$$h_{am_\alpha m_{t\alpha}}^\dagger = \hat{c}_{am_\alpha m_{t\alpha}} = (-1)^{\frac{1}{2}+m_{t\alpha}} \tilde{c}_{am_\alpha, -m_{t\alpha}}$$

$$= (-1)^{j_a+m_\alpha+\frac{1}{2}+m_{t\alpha}} c_{a,-m_\alpha,-m_{t\alpha}} , \qquad (5.98)$$

where we can use the abbreviation (5.85) and

$$-\alpha \equiv (a, -m_\alpha, -m_{t\alpha}) \qquad (5.99)$$

when it is clear that we are in the isospin formalism. It follows that

$$h_{am_\alpha,\pm\frac{1}{2}}^\dagger = \mp\tilde{c}_{am_\alpha,\mp\frac{1}{2}} . \qquad (5.100)$$

Let us form a one-hole state using a detailed notation as in (5.97):

$$|a^{-1}; Jm_\alpha; \tfrac{1}{2} \pm\tfrac{1}{2}\rangle = h_{am_\alpha,\pm\frac{1}{2}}^\dagger|\text{HF}\rangle = \mp\tilde{c}_{am_\alpha,\mp\frac{1}{2}}|\text{HF}\rangle . \qquad (5.101)$$

With the upper signs we have annihilated a proton ($m_{t\alpha} = -\frac{1}{2}$), with the lower signs a neutron ($m_{t\alpha} = +\frac{1}{2}$). We thus have a one-proton-hole state with $T = \frac{1}{2}$, $M_T = +\frac{1}{2}$ and a one-neutron-hole state with $T = \frac{1}{2}$, $M_T = -\frac{1}{2}$,

$$|a^{-1}; Jm_\alpha; \tfrac{1}{2} +\tfrac{1}{2}\rangle = -|p^{-1}; Jm_\alpha\rangle , \qquad (5.102)$$

$$|a^{-1}; Jm_\alpha; \tfrac{1}{2} -\tfrac{1}{2}\rangle = |n^{-1}; Jm_\alpha\rangle . \qquad (5.103)$$

The relation (5.100) is important in handling states with holes.

The two creation operators in (5.96) and (5.97) always anticommute. Taking into account the symmetry properties of the Clebsch–Gordan coefficients, both for angular momentum and for isospin, we can write the symmetry relations

$$|b\,a\,;\,J\,M\,;\,T\,M_T\rangle = (-1)^{j_a+j_b+J+T}|a\,b\,;\,J\,M\,;\,T\,M_T\rangle\,, \qquad (5.104)$$

$$|b^{-1}\,a^{-1}\,;\,J\,M\,;\,T\,M_T\rangle = (-1)^{j_a+j_b+J+T}|a^{-1}\,b^{-1}\,;\,J\,M\,;\,T\,M_T\rangle\,. \qquad (5.105)$$

**Example 5.1**

Taking $a = b = 0\mathrm{f}_{7/2}$ leads to the following possible two-particle and two-hole states:

$$T = 0\,,\quad J = 1,3,5,7\,,$$
$$T = 1\,,\quad J = 0,2,4,6\,.$$

So we have altogether $4 + 3 \times 4 = 16$ states ($M_T = 0, \pm 1$). In the proton–neutron representation we have

$J = 0,2,4,6$ (proton–proton and neutron–neutron states),

$J = 0,1,2,3,4,5,6,7$ (proton–neutron states),

altogether $2 \times 4 + 8 = 16$ $J$ states. So we have the same number of states, and indeed the same $J$ states, in both representations. This meets the expectations.

All the 16 states are seen in the spectra of the $M_T = +1, 0, -1$ triplet $^{42}$Ca, $^{42}$Sc and $^{42}$Ti in Fig. 5.6, where there are two particles in the $0\mathrm{f}_{7/2}$ shell. A similar, though incomplete, pattern is seen in the spectra of the triplet $^{54}$Fe, $^{54}$Co and $^{54}$Ni in Fig. 5.8, where there are two holes in the $0\mathrm{f}_{7/2}$ shell.

In the $A = 42$ triplet the $M_T = +1, 0, -1$ states for each $J$ are almost degenerate. This means that the Hamiltonian is a very good isospin scalar if its Coulomb term is discarded. Similar observations can be made of the two-particle nuclei $A = 6$ in Fig. 5.4 and $A = 18$ in Fig. 5.5. Also the $A = 38$ triplet of two-hole nuclei in Fig. 5.7 seems to conserve isospin very well.

Let us analyse the isospin structure of two-particle states and deduce its correspondence with the proton–neutron structure. Equation (5.96) and Clebsch–Gordan coefficients from Chap. 1 give

$$|a_1\,a_2\,;\,J\,M\,;\,1 \pm 1\rangle = \mathcal{N}_{a_1 a_2}(J1)\big[c^\dagger_{a_1} c^\dagger_{a_2}\big]^{1,\pm 1}_{JM}|\mathrm{CORE}\rangle$$
$$= \mathcal{N}_{a_1 a_2}(J1)\big[c^\dagger_{a_1,\pm\frac{1}{2}} c^\dagger_{a_2,\pm\frac{1}{2}}\big]_{JM}|\mathrm{CORE}\rangle\,. \qquad (5.106)$$

This means that

$$|a_1\,a_2\,;\,J\,M\,;\,1 +1\rangle = |n_1\,n_2\,;\,J\,M\rangle\,, \qquad (5.107)$$
$$|a_1\,a_2\,;\,J\,M\,;\,1 -1\rangle = |p_1\,p_2\,;\,J\,M\rangle\,. \qquad (5.108)$$

Thus two-neutron nuclei are $M_T = +1$ nuclei and two-proton nuclei are $M_T = -1$ nuclei, as indicated in Figs. 5.4–5.6. Thinking of the neutron as 'isospin up' and of the proton as 'isospin down' gives this result immediately in an elementary way.

Similarly we find for $M_T = 0$

$$|a_1 a_2; JM; T0\rangle = \frac{\mathcal{N}_{a_1 a_2}(JT)}{\sqrt{2}} \Big( \big[ c^\dagger_{a_1, \frac{1}{2}} c^\dagger_{a_2, -\frac{1}{2}} \big]_{JM}$$
$$+ (-1)^{T+1} \big[ c^\dagger_{a_1, -\frac{1}{2}} c^\dagger_{a_2, \frac{1}{2}} \big]_{JM} \Big) |\text{CORE}\rangle . \quad (5.109)$$

In terms of protons and neutrons this means

$$|a_1 a_2; JM; 00\rangle = \frac{\mathcal{N}_{a_1 a_2}(J0)}{\sqrt{2}} \big( |n_1 p_2; JM\rangle - |p_1 n_2; JM\rangle \big) , \quad (5.110)$$

$$|a_1 a_2; JM; 10\rangle = \frac{\mathcal{N}_{a_1 a_2}(J1)}{\sqrt{2}} \big( |n_1 p_2; JM\rangle + |p_1 n_2; JM\rangle \big) . \quad (5.111)$$

These proton–neutron mixtures are the $T = 0$ and $T = 1$ states in an $M_T = 0$ ($N = Z$) two-particle nucleus. We note that in structure they and the other two $T = 1$ states (5.107) and (5.108) are the same as the familiar spin $S = 0$ singlet and $S = 1$ triplet. From Figs. 5.4–5.6 one can see that, evidently as a result of the residual interaction, the $T = 0$ states of the $np - pn$ type tend to lie lower in energy than the $T = 1$ states of the $np + pn$ type in the two-particle proton–neutron $N = Z$ nuclei.

Finally, to complete the two-particle picture, we invert (5.110) and (5.111) to express the proton–neutron states in terms of the isospin states. In the case $a_1 \neq a_2$ all values of $J$ occur for both values of $T$, and we can solve the pair of equations straightforwardly for $|n_1 p_2; JM\rangle$ and $|p_1 n_2; JM\rangle$. In the case $a_1 = a_2$, states of a given $T$ exist only for even $J$ ($T = 1$) or for odd $J$ ($T = 0$). Then the non-vanishing states are

$$|aa; J \text{ even}, M; 10\rangle = |np; JM\rangle = |pn; JM\rangle , \quad (5.112)$$
$$|aa; J \text{ odd}, M; 00\rangle = |np; JM\rangle = -|pn; JM\rangle , \quad (5.113)$$

where we have used the symmetry relation (5.25). We can now write the proton–neutron states in terms of the isospin states for unrestricted $a_1, a_2$ and for all $J$ as

$$|n_1 p_2; JM\rangle = \frac{1}{\sqrt{2}} \Big( \sqrt{1 + \delta_{a_1 a_2}(-1)^J} \, |a_1 a_2; JM; 10\rangle$$
$$+ \sqrt{1 - \delta_{a_1 a_2}(-1)^J} \, |a_1 a_2; JM; 00\rangle \Big) , \quad (5.114)$$

$$|p_1 n_2; JM\rangle = \frac{1}{\sqrt{2}} \Big( \sqrt{1 + \delta_{a_1 a_2}(-1)^J} \, |a_1 a_2; JM; 10\rangle$$
$$- \sqrt{1 - \delta_{a_1 a_2}(-1)^J} \, |a_1 a_2; JM; 00\rangle \Big) . \quad (5.115)$$

It is to be noted that for $a_1 \neq a_2$ these states do not have good isospin and do not occur in the examples of this chapter. However, they will be needed for technical purposes later in the book.

Two-hole states are derived starting from their definition (5.97) and proceeding as above. Additionally we need the relations (5.100). The results are

$$|a_1^{-1} a_2^{-1}; JM; 1+1\rangle = |p_1^{-1} p_2^{-1}; JM\rangle, \qquad (5.116)$$

$$|a_1^{-1} a_2^{-1}; JM; 1-1\rangle = |n_1^{-1} n_2^{-1}; JM\rangle, \qquad (5.117)$$

$$|a_1^{-1} a_2^{-1}; JM; 00\rangle = \frac{\mathcal{N}_{a_1 a_2}(J0)}{\sqrt{2}} \left( |n_1^{-1} p_2^{-1}; JM\rangle - |p_1^{-1} n_2^{-1}; JM\rangle \right), \qquad (5.118)$$

$$|a_1^{-1} a_2^{-1}; JM; 10\rangle = -\frac{\mathcal{N}_{a_1 a_2}(J1)}{\sqrt{2}} \left( |n_1^{-1} p_2^{-1}; JM\rangle + |p_1^{-1} n_2^{-1}; JM\rangle \right). \qquad (5.119)$$

Thus two-proton-hole nuclei are $M_T = +1$ nuclei and two-neutron-hole nuclei are $M_T = -1$ nuclei, as indicated in Figs. 5.7 and 5.8. Proceeding as in the derivation of (5.114) and (5.115), we may invert the last two equations. The result is

$$|p_1^{-1} n_2^{-1}; JM\rangle = -\frac{1}{\sqrt{2}} \left( \sqrt{1 + \delta_{a_1 a_2}(-1)^J} |a_1^{-1} a_2^{-1}; JM; 10\rangle \right.$$
$$\left. + \sqrt{1 - \delta_{a_1 a_2}(-1)^J} |a_1^{-1} a_2^{-1}; JM; 00\rangle \right), \qquad (5.120)$$

$$|n_1^{-1} p_2^{-1}; JM\rangle = -\frac{1}{\sqrt{2}} \left( \sqrt{1 + \delta_{a_1 a_2}(-1)^J} |a_1^{-1} a_2^{-1}; JM; 10\rangle \right.$$
$$\left. - \sqrt{1 - \delta_{a_1 a_2}(-1)^J} |a_1^{-1} a_2^{-1}; JM; 00\rangle \right). \qquad (5.121)$$

### 5.5.4 Isospin Representation of Particle–Hole Nuclei

Angular-momentum-coupled wave functions of particle–hole nuclei were written down in Sect. 5.4. Similarly to the two-particle and two-hole cases we now include isospin in the description of particle–hole nuclei. Using (5.98) we write the general wave function as

$$|a\,b^{-1}; JM; TM_T\rangle = [c_a^\dagger h_b^\dagger]_{JM}^{TM_T} |\text{HF}\rangle = [c_a^\dagger \hat{c}_b]_{JM}^{TM_T} |\text{HF}\rangle. \qquad (5.122)$$

As before, $a$ and $b$ carry the orbital labels but no information about particle species. As noted in Sect. 5.4, we have necessarily $a \neq b$, which makes the normalization simple. The orthonormality condition is

$$\langle a\,b^{-1}\,;\,J\,M\,;\,T\,M_T\,|c\,d^{-1}\,;\,J'\,M'\,;\,T'\,M_T'\rangle = \delta_{ac}\delta_{bd}\delta_{JJ'}\delta_{MM'}\delta_{TT'}\delta_{M_T M_T'}\,.$$
(5.123)

Because $a \neq b$, the operators in (5.120) anticommute. With the Clebsch–Gordan symmetry properties it follows that

$$|b^{-1}\,a\,;\,J\,M\,;\,T\,M_T\rangle = (-1)^{j_a+j_b+J+T}|a\,b^{-1}\,;\,J\,M\,;\,T\,M_T\rangle\,.$$
(5.124)

By expanding the isospin tensor products and inserting the Clebsch–Gordan coefficients we obtain

$$|a_1\,a_2^{-1}\,;\,J\,M\,;\,T\,0\rangle$$
$$= \frac{1}{\sqrt{2}}\left([c_{a_1,\frac{1}{2}}^\dagger h_{a_2,-\frac{1}{2}}^\dagger]_{JM} + (-1)^{T+1}[c_{a_1,-\frac{1}{2}}^\dagger h_{a_2,\frac{1}{2}}^\dagger]_{JM}\right)|\text{HF}\rangle\,.$$
(5.125)

This gives the $M_T = 0$ particle–hole states

$$|a_1\,a_2^{-1}\,;\,J\,M\,;\,0\,0\rangle = \frac{1}{\sqrt{2}}\left(|n_1\,n_2^{-1}\,;\,J\,M\rangle + |p_1\,p_2^{-1}\,;\,J\,M\rangle\right)\,,$$
(5.126)

$$|a_1\,a_2^{-1}\,;\,J\,M\,;\,1\,0\rangle = \frac{1}{\sqrt{2}}\left(|n_1\,n_2^{-1}\,;\,J\,M\rangle - |p_1\,p_2^{-1}\,;\,J\,M\rangle\right)\,,$$
(5.127)

where the sign changes from (5.125) are due to the relation (5.100), as displayed for a single-proton-hole state in (5.102). The $M_T = \pm 1$ particle–hole states are easily found to be

$$|a_1\,a_2^{-1}\,;\,J\,M\,;\,1\,+1\rangle = -|n_1\,p_2^{-1}\,;\,J\,M\rangle\,,$$
(5.128)

$$|a_1\,a_2^{-1}\,;\,J\,M\,;\,1\,-1\rangle = |p_1\,n_2^{-1}\,;\,J\,M\rangle\,.$$
(5.129)

Inverting (5.126) and (5.127) leads to

$$|n_1\,n_2^{-1}\,;\,J\,M\rangle = \frac{1}{\sqrt{2}}\left(|a_1\,a_2^{-1}\,;\,J\,M\,;\,0\,0\rangle + |a_1\,a_2^{-1}\,;\,J\,M\,;\,1\,0\rangle\right)\,,$$
(5.130)

$$|p_1\,p_2^{-1}\,;\,J\,M\rangle = \frac{1}{\sqrt{2}}\left(|a_1\,a_2^{-1}\,;\,J\,M\,;\,0\,0\rangle - |a_1\,a_2^{-1}\,;\,J\,M\,;\,1\,0\rangle\right)\,.$$
(5.131)

These expressions for the neutron particle–hole and proton particle–hole states are important in later work.

Figures 5.9–5.11 illustrate the experimental situation for light particle–hole nuclei. Let us examine the spectra of the $A = 4$ triplet of nuclei in Fig. 5.9. The nucleus ${}^4_1\text{H}_3$ is a neutron-particle–proton-hole nucleus, so it is described by (5.128). This equation gives $T = 1$, $M_T = +1$ as the isospin quantum numbers for the particle–hole states. The nucleus ${}^4_3\text{Li}_1$ is a proton-particle–neutron-hole nucleus. Equation (5.129) gives $T = 1$, $M_T = -1$ for its particle–hole states.

The ground state of ${}^4_2\text{He}_2$, the $M_T = 0$ member of the triplet, is the particle–hole vacuum $|\text{HF}\rangle$. The excited states consist of charge-conserving proton-particle–proton-hole and neutron-particle–neutron-hole excitations, governed by (5.126) and (5.127). The $T = 0$ states are of the $nn^{-1} + pp^{-1}$ type, the $T = 1$ states of the $nn^{-1} - pp^{-1}$ type.

## Epilogue

In this chapter we have discussed many examples of the structure of simple nuclei. These nuclei are either magic or have one or two particles or holes, or a particle–hole pair, in their valence space. The description of these nuclei was kept as simple as possible, without including any configuration mixing. In the next two chapters the simple wave functions are tested against experimental data by using the electromagentic and beta decays as probes. After this, from Chap. 8 on, these simple excitations are allowed to mix among each other through the residual interaction so far neglected.

## Exercises

**5.1.** Write down the wave functions of the low-energy states of the nuclei $^{47}$K and $^{57}$Ni in the mean-field approximation and compare the resulting energy spectra with experimental data. Discuss the successes and failures of the model in these particular cases.

**5.2.** Show that the overlap of two wave functions of the form (5.21) is

$$\langle a\,b;\,J\,M|a'\,b';\,J'\,M'\rangle = \delta_{JJ'}\delta_{MM'}\mathcal{N}_{ab}(J)^2[\delta_{aa'}\delta_{bb'} - (-1)^{j_a+j_b+J}\delta_{ab'}\delta_{ba'}]\,.$$
(5.132)

This overlap leads trivially to the following overlap between proton–neutron two-particle wave functions:

$$\langle p\,n;\,J\,M|p'\,n';\,J'\,M'\rangle = \delta_{JJ'}\delta_{MM'}\delta_{pp'}\delta_{nn'}\,.$$
(5.133)

**5.3.** Derive the normalization constant in (5.21) by exploiting the overlap (5.132).

**5.4.** Derive the symmetry properties (5.24) and (5.25) of two-particle wave functions.

**5.5.** Write down the wave functions of the low-energy states of $^{50}$Ca, $^{50}$Sc and $^{50}$Ti in the mean-field approximation and compare the resulting energy spectra with experimental data. Discuss the quality of the mean-field shell model in the description of these nuclei.

**5.6.** Write down the wave functions of the low-energy states of $^{58}$Ni and $^{58}$Cu in the mean-field approximation. Compare with experimental data and comment.

**5.7.** Show that the overlap between two two-hole wave functions of the form (5.35) is

$$\langle a^{-1}\,b^{-1};\,J\,M|a'^{-1}\,b'^{-1};\,J'\,M'\rangle$$
$$= \delta_{JJ'}\delta_{MM'}\mathcal{N}_{ab}(J)^2[\delta_{aa'}\delta_{bb'} - (-1)^{j_a+j_b+J}\delta_{ab'}\delta_{ba'}]\,. \quad (5.134)$$

This overlap leads trivially to the following overlap between proton–neutron two-hole wave functions:

$$\langle p^{-1} n^{-1} ; J M | p'^{-1} n'^{-1} ; J' M' \rangle = \delta_{JJ'} \delta_{MM'} \delta_{pp'} \delta_{nn'} . \qquad (5.135)$$

**5.8.** Write down the wave functions of the low-energy states of $^{14}$C, $^{14}$N and $^{14}$O in the mean-field approximation. Compare with experimental data and comment.

**5.9.** Write down the wave functions of the low-energy states of $^{46}$K and $^{46}$Ca in the mean-field approximation and compare the resulting energy spectra with experimental data. Discuss the successes and failures of the mean-field description.

**5.10.** Write down the wave functions of the low-energy states of $^{54}$Fe and $^{54}$Co in the mean-field approximation and compare the resulting energy spectra with experimental data. Discuss the successes and failures of the mean-field description.

**5.11.** Derive the overlap (5.45).

**5.12.** Write down the wave functions of the low-energy states of $^{48}$K, $^{48}$Ca and $^{48}$Sc in the mean-field approximation and compare the resulting energy spectra with experimental data. Discuss the quality of the mean-field shell model in the description of these nuclei.

**5.13.** Write down the wave functions of the low-energy states of $^{56}$Co, $^{56}$Ni and $^{56}$Cu in the mean-field approximation. Compare with experimental data and comment.

**5.14.** Derive the commutators (5.74) from the basic commutators (5.70).

**5.15.** Complete the details in the derivation of (5.92).

**5.16.** Show that the overlap between two two-particle wave functions of the form (5.96) is

$$\langle a\, b;\, J M;\, T M_T | a'\, b';\, J' M';\, T' M_T' \rangle = \delta_{JJ'} \delta_{TT'} \delta_{MM'} \delta_{M_T M_T'} \mathcal{N}_{ab}(JT)^2$$
$$\times \left[ \delta_{aa'} \delta_{bb'} + (-1)^{j_a + j_b + J + T} \delta_{ab'} \delta_{ba'} \right] . \qquad (5.136)$$

**5.17.** Derive the normalization constant in (5.96) by exploiting the overlap (5.136).

**5.18.** Show that the overlap of two states of the form (5.97) is

$$\langle a^{-1} b^{-1} ;\, J M;\, T M_T | a'^{-1} b'^{-1} ;\, J' M';\, T' M_T' \rangle$$
$$= \delta_{JJ'} \delta_{TT'} \delta_{MM'} \delta_{M_T M_T'} \mathcal{N}_{ab}(JT)^2 \left[ \delta_{aa'} \delta_{bb'} + (-1)^{j_a + j_b + J + T} \delta_{ab'} \delta_{ba'} \right] .$$
$$\qquad (5.137)$$

**5.19.** Derive the symmetry properties (5.104) and (5.105).

**5.20.** Derive the wave functions (5.116) and (5.117).

**5.21.** Derive the wave functions (5.118) and (5.119).

**5.22.** Based on the material of Subsect. 5.5.3, how would you classify the states of $^{10}$Be, $^{10}$B and $^{10}$C? Form the isospin singlet and triplet states and compare with experimental data.

**5.23.** Derive the overlap (5.123).

**5.24.** Derive the expression (5.125).

**5.25.** Derive the wave functions (5.126) and (5.127).

**5.26.** Derive the wave functions (5.128) and (5.129).

**5.27.** Based on the material of Subsect. 5.5.4, how would you classify the states of $^{12}$B, $^{12}$C and $^{12}$N? Form the isospin singlet and triplet states and compare with experimental data.

**5.28.** Classify the states of $^{4}$H, $^{4}$He and $^{4}$Li according to isospin. Form the isospin singlet and triplet states and compare with experimental data.

# 6

# Electromagnetic Multipole Moments and Transitions

## Prologue

In the preceding chapter we constructed and discussed the simplest possible nuclear wave functions. This construction was done at the mean-field level. No account was taken of configuration mixing caused by the nuclear residual interaction. These simple wave functions produce degeneracies in energy spectra. This is contrary to experimental data, so improved wave functions are called for.

In this chapter we introduce various electromagnetic observables. They probe the structure of the wave functions involved through multipole moments and gamma decays. The elecromagnetic processes are excellent tests of the validity of various assumptions underlying the structure of nuclear states. In particular, the decay properties produced by the simple wave functions of Chap. 5 are good indicators of the degree of validity of the plain mean-field picture.

## 6.1 General Properties of Electromagnetic Observables

The electromagnetic decay processes of nuclei are described as resulting from the interaction of the nucleus with an external electromagnetic field. The complete field consists of the electric field $\boldsymbol{E}$ and magnetic field $\boldsymbol{B}$. The energy density of the electromagnetic field[1] is proportional to $\boldsymbol{E}^2 + c^2\boldsymbol{B}^2$. The interaction between the nucleus and the field is mediated by the four-potential $(\phi, \boldsymbol{A})$. The scalar potential $\phi$ couples to the nuclear charge density $\rho$ and the vector potential $\boldsymbol{A}$ to the nuclear current density $\boldsymbol{j}$. The nuclear current density consists of two parts, namely the orbital part due to the moving charges of protons and the spin part due to the spin magnetism of protons and neutrons.

---

[1] SI units are used throughout the book. Footnotes are inserted to indicate differences when quantities are expressed in Gaussian units. Thus the energy density in Gaussian units is proportional to $\boldsymbol{E}^2 + \boldsymbol{B}^2$.

The electromagnetic radiation field can be expanded in multipoles containing spherical harmonics. The field is quantized in terms of photons. Creation and annihilation of photons is described in occupation number representation. The complete system is nucleus plus field. Its two parts interact weakly, so that the interaction can be treated as a perturbation. Consider the electromagnetic decay of an excited nucleus to its ground state. The unperturbed initial state of the system is the excited nuclear state and the electromagnetic field in its ground state, i.e. no photons. The final state is the nuclear ground state and the electromagnetic field with one photon created into it. The transition from the initial to the final state has been mediated by one of the multipole terms in the expansion of the radiation field. For further discussion and for filling in details see e.g. [16] or [39]. Below we cite the important results without getting involved in the heavy calculations needed to obtain them.

### 6.1.1 Transition Probability and Half-Life

Consider the transition probability per unit time, usually called just *transition probability*, of gamma decay from an initial nuclear state $i$ to a final nuclear state $f$, denoted $T_{fi}$. The *lifetime* of the transition is $1/T_{fi}$ and its *half-life* is

$$t_{1/2} = \frac{\ln 2}{T_{fi}} . \tag{6.1}$$

Gamma transitions proceed by multipole components $\lambda\mu$ of the radiation field. The sources of the field are either of *electric* or *magnetic* type, designated by an index $\sigma$ such that $\sigma = E$ or $\sigma = M$. The $\sigma\lambda\mu$ transition probability, calculated by the 'golden rule' of time-dependent perturbation theory, is[2]

$$T_{fi}^{(\sigma\lambda\mu)} = \frac{2}{\epsilon_0\hbar} \frac{\lambda+1}{\lambda[(2\lambda+1)!!]^2} \left(\frac{E_\gamma}{\hbar c}\right)^{2\lambda+1} \left|\langle \xi_f\, J_f\, m_f | \mathcal{M}_{\sigma\lambda\mu} | \xi_i\, J_i\, m_i \rangle\right|^2 , \tag{6.2}$$

where $E_\gamma$ is the energy of the transition and $\mathcal{M}_{\sigma\lambda\mu}$ is the nuclear operator associated with the multipole radiation field $\sigma\lambda\mu$.

Magnetic substates are normally not observed separately. Averaging (6.2) over the initial substates and summing over the final substates and all $\mu$ yields the transition probability[3]

$$
\begin{aligned}
T_{fi}^{(\sigma\lambda)} &= \frac{1}{2J_i+1} \sum_{m_i\mu m_f} T_{fi}^{(\sigma\lambda\mu)} \\
&= \frac{2}{\epsilon_0\hbar} \frac{\lambda+1}{\lambda[(2\lambda+1)!!]^2} \left(\frac{E_\gamma}{\hbar c}\right)^{2\lambda+1} B(\sigma\lambda\,;\, \xi_i J_i \to \xi_f J_f) ,
\end{aligned}
\tag{6.3}
$$

---

[2] In Gaussian units the constant factor in front is $8\pi/\hbar$.
[3] See Exercise 2.25.

where we have the *reduced transition probability*

$$B(\sigma\lambda\,; \xi_i J_i \to \xi_f J_f) \equiv \frac{1}{2J_i + 1} \left| (\xi_f\, J_f \| \mathcal{M}_{\sigma\lambda} \| \xi_i\, J_i) \right|^2 .$$  (6.4)

The usual notation is

$$\mathcal{M}_{E\lambda} = Q_\lambda\,, \quad \mathcal{M}_{M\lambda} = M_\lambda\,.$$  (6.5)

Written out in detail the components of the electric and magnetic tensor operators are

$$Q_{\lambda\mu} = \zeta^{(E\lambda)} \sum_{j=1}^{A} e(j) r_j^\lambda Y_{\lambda\mu}(\Omega_j)\,,$$  (6.6)

$$M_{\lambda\mu} = \frac{\mu_N}{\hbar c} \zeta^{(M\lambda)} \sum_{j=1}^{A} \left[ \frac{2}{\lambda+1} g_l^{(j)} \boldsymbol{l}(j) + g_s^{(j)} \boldsymbol{s}(j) \right] \cdot \boldsymbol{\nabla}_j [r_j^\lambda Y_{\lambda\mu}(\Omega_j)]\,.$$  (6.7)

Here $e(j)$ is the electric charge, and $\boldsymbol{l}(j)$ and $\boldsymbol{s}(j)$ are the orbital and spin angular momenta, of nucleon $j$. Below we write just $e$ for the charge, with the understanding that it is the fundmental unit of charge for protons and zero for neutrons. However, effective charges can replace these values as discussed in Subsect. 6.1.3.

From (2.17) we can see that the gradient factor in the magnetic dipole operator $M_{1\mu}$ is $\boldsymbol{\nabla}(rY_{1\mu}) \propto \hat{\boldsymbol{e}}_\mu$, so that the operator simply consists of the angular momentum components $l_\mu$ and $s_\mu$. For $\lambda > 1$ the gradient factor is more complicated. However, it commutes with the $\boldsymbol{l}$ and $\boldsymbol{s}$ operators, which simplifies matters.[4] The *g factors*,[5] or *gyromagnetic ratios*, are $g_s^{(j)} = g_p$ for the proton spin and $g_s^{(j)} = g_n$ for the neutron spin. Their values are

$$g_p = 5.586\,, \quad g_n = -3.826\,.$$  (6.8)

The orbital $g$ factors are $g_l^{(j)} = 1$ for protons and $g_l^{(j)} = 0$ for neutrons. These free-nucleon values are used normally in nuclear structure calculations. The *nuclear magneton* is

$$\mu_N = \frac{e\hbar}{2m_p} = 0.10515\ ce\,\mathrm{fm}\,,$$  (6.9)

where $m_p = 938.27\,\mathrm{MeV}/c^2$ is the proton mass.[6]

The phase factors $\zeta^{(E\lambda)}$ and $\zeta^{(M\lambda)}$ are different for the CS and BR phase conventions, introduced in (3.67):

---

[4] For the $\boldsymbol{l}$ term this follows from the fact that $(\boldsymbol{r} \times \boldsymbol{\nabla}) \cdot \boldsymbol{\nabla} f = 0$ for a general function $f(\boldsymbol{r})$.

[5] In this work the $g$ factors are taken to be bare numbers, contrary to e.g. [17].

[6] In Gaussian units the nuclear magneton is $\mu_N = e\hbar/2m_p c$. Then there appears no $c$ in (6.7) or on the right-hand side of (6.9).

$$\zeta^{(\mathrm{E}\lambda)} = \begin{cases} 1 & \text{Condon–Shortley phase convention ,} \\ \mathrm{i}^\lambda & \text{Biedenharn–Rose phase convention ,} \end{cases} \tag{6.10}$$

$$\zeta^{(\mathrm{M}\lambda)} = \begin{cases} 1 & \text{Condon–Shortley phase convention ,} \\ \mathrm{i}^{\lambda-1} & \text{Biedenharn–Rose phase convention .} \end{cases} \tag{6.11}$$

The Biedenharn–Rose complex phases are needed to obtain real single-particle matrix elements for the electromagnetic transition amplitudes, i.e. reduced matrix elements of $\mathcal{M}_{\sigma\lambda}$.

From (6.4) to (6.7) we see that the units of the reduced transition probabilities are

$$[B(\mathrm{E}\lambda)] = e^2\mathrm{fm}^{2\lambda} , \quad [B(\mathrm{M}\lambda)] = (\mu_\mathrm{N}/c)^2\mathrm{fm}^{2\lambda-2} . \tag{6.12}$$

We note the handy relations for physical constants

$$\frac{1}{4\pi\epsilon_0}\frac{e^2}{\hbar c} \equiv \alpha = 1/137.04 , \quad \hbar c = 197.33 \text{ MeV fm} . \tag{6.13}$$

With use of these relations we can put the transition probabilities (6.3) into useful numerical forms:

$$\boxed{\begin{aligned} T_{fi}^{\mathrm{E}\lambda} &= 5.498 \times 10^{22} f(\lambda) \left(\frac{E_\gamma\,[\mathrm{MeV}]}{197.33}\right)^{2\lambda+1} B(\mathrm{E}\lambda)\,[e^2\mathrm{fm}^{2\lambda}]\,1/\mathrm{s} , \\ T_{fi}^{\mathrm{M}\lambda} &= 6.080 \times 10^{20} f(\lambda) \left(\frac{E_\gamma\,[\mathrm{MeV}]}{197.33}\right)^{2\lambda+1} B(\mathrm{M}\lambda)\,[(\mu_\mathrm{N}/c)^2\mathrm{fm}^{2\lambda-2}]\,1/\mathrm{s} , \\ f(\lambda) &\equiv \frac{\lambda+1}{\lambda[(2\lambda+1)!!]^2} . \end{aligned}}$$

$$\tag{6.14}$$

Reduced transition probabilities $B(\mathrm{E}2)$ are sometimes expressed in the literature in units of $e^2\mathrm{barn}^2$. The conversion, from $1\,\mathrm{barn} = 100\,\mathrm{fm}^2$, is

$$e^2\mathrm{barn}^2 = 10^4\,e^2\mathrm{fm}^4 . \tag{6.15}$$

This unit brings numerical values for collective nuclear excitations close to unity. In this sense it is a 'natural' unit for collective decays.

Equation (6.1) can be immediately generalized to the case where there are several final states available. The transition probabilities $T_{fi}$ are additive, so we sum them over the final states $f$, and (6.1) then gives for the (total) half-life of the initial state

$$\frac{1}{t_{1/2}} = \sum_f \frac{1}{t_{1/2}^{(f)}} . \tag{6.16}$$

Here $t_{1/2}^{(f)}$ is the *partial decay half-life* for decay to the final state $f$.

## 6.1.2 Selection Rules for Electromagnetic Transitions

The first selection rule is that *there are no E0 or M0 gamma transitions*. This is seen from the operators (6.6) and (6.7). With $\lambda = 0$ the first is a constant and the second vanishes. A constant cannot connect two different nuclear states. However, electromagnetic E0 transitions are possible via *internal conversion*, where the nucleus de-excites by ejecting an atomic electron; internal conversion is not discussed further in this book. The absence of all M0 transitions results fundamentally from the absence of magnetic monopoles in nature.

Electromagnetic transitions are classified according to their multipoles. To accomplish the classification let us look at the structure of the electric and magnetic $\lambda$-pole[7] operators in (6.6) and (6.7). We see that, for each $j$, the $\lambda$-pole operators have the structure

$$Q_{\lambda\mu} \propto r^\lambda Y_{\lambda\mu}\,, \quad M_{\lambda\mu} \propto \left.\begin{matrix} l \\ s \end{matrix}\right\} \cdot \boldsymbol{\nabla}(r^\lambda Y_{\lambda\mu})\,. \tag{6.17}$$

The parity of the spherical harmonic $Y_{\lambda\mu}$ is $(-1)^\lambda$ and that of the scalar $r^\lambda$ is $+1$. Hence the electric $\lambda$-pole operator has parity $\pi = (-1)^\lambda$. In the magnetic $\lambda$-pole operator the factors $l$ and $s$ are axial vectors, so their parity is $+1$. The parity of the vector operator $\boldsymbol{\nabla}$ is $-1$, and the parity of $r^\lambda Y_{\lambda\mu}$ is $(-1)^\lambda$. It follows that the parity of the magnetic $\lambda$-pole term can be written as $\pi = (-1)^{\lambda-1}$. Denoting the parity of the initial state by $\pi_i$ and that of the final state by $\pi_f$ we then have the *parity conservation* selection rule

$$\pi_i \pi_f = \begin{cases} (-1)^\lambda & \text{for E}\lambda\,, \\ (-1)^{\lambda-1} & \text{for M}\lambda\,. \end{cases} \tag{6.18}$$

The transition probability decreases drastically with increasing multipolarity. Therefore the likeliest transition is the one of the lowest multipolarity allowed by the angular momentum and parity selection rules. For a $\lambda$-pole transition between nuclear states of angular momenta $J_i$ and $J_f$ the angular momentum selection rule is the triangular condition $\Delta(J_f \lambda J_i)$ and the parity selection rule is (6.18). This leads to the hierarchical classification of electromagnetic decay transitions presented in Table 6.1.

The classification of Table 6.1 helps to track the leading electromagnetic multipole $\lambda$ responsible for the decay transition in question. Note that only the *lowest* multipoles are included. For example, for a $2^+ \rightarrow 2^+$ transition the table only gives M1. However, the selection rules allow all of M1, E2, M3 and E4. While M3 and E4 are without practical significance in this example, E2

---

[7] Traditionally the low multipoles are named according to Greek numbers for $2^\lambda$. Thus we have dipole, quadrupole, octupole, hexadecapole. This terminology was introduced in Subsect. 2.1.2.

**Table 6.1.** Lowest multipolarities for gamma transitions

| $\Delta J = \|J_f - J_i\|$ | $0^a$ | 1 | 2 | 3 | 4 | 5 |
|---|---|---|---|---|---|---|
| $\pi_i \pi_f = -1$ | E1 | E1 | M2 | E3 | M4 | E5 |
| $\pi_i \pi_f = +1$ | M1 | M1 | E2 | M3 | E4 | M5 |

$^a$Not $0 \to 0$.

often competes with M1. In certain collective transitions E2 in fact dominates over M1.

The ratio of the E2 and M1 reduced matrix elements is called the E2/M1 *mixing ratio*

$$\boxed{\Delta(\text{E2/M1}) = \frac{(\xi_f \, J_f \| \boldsymbol{Q}_2 \| \xi_i \, J_i)}{(\xi_f \, J_f \| \boldsymbol{M}_1 \| \xi_i \, J_i)}} \ . \tag{6.19}$$

Its units are given by (6.12) as

$$[\Delta(\text{E2/M1})] = \frac{e \, \text{fm}^2}{\mu_{\text{N}}/c} \ . \tag{6.20}$$

The mixing ratio, including its sign, is a measurable quantity. It serves as a sensitive test of nuclear models. A related, dimensionless quantity $\delta(\text{E2/M1})$ is commonly used in experimental reports. It is related to $\Delta(\text{E2/M1})$ according to

$$\delta(\text{E2/M1}) = 0.835 E_\gamma[\text{MeV}] \, \Delta(\text{E2/M1}) \left[ \frac{e \, \text{barn}}{\mu_{\text{N}}/c} \right] \ . \tag{6.21}$$

### 6.1.3 Single-Particle Matrix Elements of the Multipole Operators

As discussed in Chap. 4, the characteristics of operators are hidden inside their single-particle matrix elements. From (4.22) we can write the reduced matrix element of the electromagnetic operator as

$$(\xi_f \, J_f \| \boldsymbol{\mathcal{M}}_{\sigma\lambda} \| \xi_i \, J_i) = \widehat{\lambda}^{-1} \sum_{ab} (a \| \boldsymbol{\mathcal{M}}_{\sigma\lambda} \| b)(\xi_f \, J_f \| \left[ c_a^\dagger \tilde{c}_b \right]_\lambda \| \xi_i \, J_i) \ . \tag{6.22}$$

The reduced single-particle matrix element $(a \| \boldsymbol{\mathcal{M}}_{\sigma\lambda} \| b)$ can be easily written for electric transitions, $\sigma = \text{E}$, by using (6.6) and the reduced matrix element of the spherical harmonic $\boldsymbol{Y}_\lambda$ given in (2.57). The result is

$$
(a\|\boldsymbol{Q}_\lambda\|b) = \zeta_{ab}^{(\mathrm{E}\lambda)} \frac{e}{\sqrt{4\pi}}(-1)^{j_b+\lambda-\frac{1}{2}} \frac{1+(-1)^{l_a+l_b+\lambda}}{2}\, \widehat{\lambda}\widehat{j_a}\widehat{j_b} \begin{pmatrix} j_a & j_b & \lambda \\ \frac{1}{2} & -\frac{1}{2} & 0 \end{pmatrix} \mathcal{R}_{ab}^{(\lambda)} \,,
$$

$$
\mathcal{R}_{ab}^{(\lambda)} = \int_0^\infty g_{n_a l_a}(r) r^\lambda g_{n_b l_b}(r) r^2 \mathrm{d}r \,,
$$

$$
\zeta_{ab}^{(\mathrm{E}\lambda)} = \begin{cases} 1 & \text{Condon–Shortley phase convention}\,, \\ (-1)^{\frac{1}{2}(l_b-l_a+\lambda)} & \text{Biedenharn–Rose phase convention}\,. \end{cases}
$$

$$
(6.23)
$$

The fourth factor takes care of parity conservation according to (6.18): it is unity for allowed transitions and zero for forbidden ones. For the radial wave functions $g_{nl}(r)$ we choose to use the harmonic oscillator functions discussed in Subsect. 3.2.1.

The single-particle matrix element for magnetic transitions, $\sigma = \mathrm{M}$, is (see e.g. [16])[8]

$$
(a\|\boldsymbol{M}_\lambda\|b) = \zeta_{ab}^{(\mathrm{M}\lambda)} \frac{\mu_\mathrm{N}/c}{\sqrt{4\pi}}(-1)^{j_b+\lambda-\frac{1}{2}} \frac{1-(-1)^{l_a+l_b+\lambda}}{2}\, \widehat{\lambda}\widehat{j_a}\widehat{j_b} \begin{pmatrix} j_a & j_b & \lambda \\ \frac{1}{2} & -\frac{1}{2} & 0 \end{pmatrix}
$$

$$
\times (\lambda-\kappa)\left[ g_l\left(1+\frac{\kappa}{\lambda+1}\right) - \tfrac{1}{2}g_s\right]\mathcal{R}_{ab}^{(\lambda-1)}\,,
$$

$$
\kappa \equiv (-1)^{l_a+j_a+\frac{1}{2}}(j_a+\tfrac{1}{2}) + (-1)^{l_b+j_b+\frac{1}{2}}(j_b+\tfrac{1}{2})\,,
$$

$$
\zeta_{ab}^{(\mathrm{M}\lambda)} = \begin{cases} 1 & \text{Condon–Shortley phase convention}\,, \\ (-1)^{\frac{1}{2}(l_b-l_a+\lambda+1)} & \text{Biedenharn–Rose phase convention}\,. \end{cases}
$$

$$
(6.24)
$$

Again we note the fourth factor as the 'parity factor'.

The phase factors $\zeta$ in (6.23) and (6.24) are real, always $+1$ in the CS convention and $\pm 1$ in the BR convention. These phases arise from the definition of the single-particle wave functions in (3.63) and (3.67) and from the additional phase factors (6.10) and (6.11) defined in the operators (6.6) and (6.7).

The charges and $g$ factors in the single-particle matrix elements (6.23) and (6.24) are

$$
e(\mathrm{p}) = e \,, \quad e(\mathrm{n}) = 0 \,, \quad g_l(\mathrm{p}) = 1 \,, \quad g_l(\mathrm{n}) = 0 \,, \quad g_s(\mathrm{p}) = g_\mathrm{p} \,, \quad g_s(\mathrm{n}) = g_\mathrm{n}
$$

$$
(6.25)
$$

with $g_\mathrm{p}$ and $g_\mathrm{n}$ given in (6.8). These free-particle values are known as *bare* values. Instead of them, one can use *effective* values for one or more of them. Such effective values represent a summary way of taking into account effects not explicitly included in the description. Thus particle–hole excitations of the core cause effective charges. They can be represented as (see e.g. [12])

---

[8] One has to be careful when citing results from the literature: in [16] the coupling order for the single-particle states is $sl$, instead of our $ls$, which introduces a phase difference between our result and theirs.

$$e_{\text{eff}}^{\text{p}} = (1 + \chi)e , \quad e_{\text{eff}}^{\text{n}} = \chi e , \tag{6.26}$$

where $\chi$ is the electric *polarization constant*. The simplest example is that for electric dipole (E1) transitions it is customary to take $\chi = -Z/A$. This value relates to the spurious centre-of-mass motion discussed at the end of Sect. 6.4.1. The effective charges of (6.26) are adopted in this work. Also for the $g$ factors one can use effective values. They result from the inert core and meson exchange contributions. For a further discussion of the effects in the valence space induced by the chosen core, see [17].

We note the following symmetry properties of the reduced single-particle matrix elements, for both the Condon–Shortley (CS) and Biedenharn–Rose (BR) phases:

$$(b\|\boldsymbol{Q}_\lambda\|a)_{\text{CS}} = (-1)^{j_a+j_b+1}(a\|\boldsymbol{Q}_\lambda\|b)_{\text{CS}} , \tag{6.27}$$

$$(b\|\boldsymbol{M}_\lambda\|a)_{\text{CS}} = (-1)^{j_a+j_b+1}(a\|\boldsymbol{M}_\lambda\|b)_{\text{CS}} , \tag{6.28}$$

$$(b\|\boldsymbol{Q}_\lambda\|a)_{\text{BR}} = (-1)^{j_a+j_b+\lambda+1}(a\|\boldsymbol{Q}_\lambda\|b)_{\text{BR}} , \tag{6.29}$$

$$(b\|\boldsymbol{M}_\lambda\|a)_{\text{BR}} = (-1)^{j_a+j_b+\lambda}(a\|\boldsymbol{M}_\lambda\|b)_{\text{BR}} . \tag{6.30}$$

These symmetries reduce the effort when calculating single-particle matrix elements. A host of their numerical values are given in Sect. 6.1.5. We note that (6.27) and (6.28) are consistent with the general relation (2.32).

### 6.1.4 Properties of the Radial Integrals

To calculate numerical values of the reduced single-particle matrix elements (6.23) and (6.24) we need to evaluate the radial integrals $\mathcal{R}_{ab}^{(\lambda)}$ stated in (6.23). This can be done either by numerical integration or analytically by exploiting the properties of the radial harmonic oscillator eigenfunctions $g_{nl}(r)$.

The functions $g_{nl}(r)$ are scaled by the oscillator length $b$ given by (3.43). To obtain a table of universal values of the radial integrals we adopt a dimensionless variable $x \equiv r/b$ and include the parameter $b$ in the statement of the function, $g_{nl}(r, b)$. From (3.42) we see that

$$g_{nl}(r, b) = b^{-\frac{3}{2}} g_{nl}(x, b = 1) \equiv b^{-\frac{3}{2}} \tilde{g}_{nl}(x) , \quad x = \frac{r}{b} . \tag{6.31}$$

The radial integral can then be written as

$$\mathcal{R}_{ab}^{(\lambda)} = b^\lambda \int_0^\infty dx\, \tilde{g}_{n_a l_a}(x) x^{\lambda+2} \tilde{g}_{n_b l_b}(x) \equiv b^\lambda \tilde{\mathcal{R}}_{ab}^{(\lambda)} , \tag{6.32}$$

where $\tilde{\mathcal{R}}_{ab}^{(\lambda)}$ is independent of the oscillator length $b$. Values of the integrals $\tilde{\mathcal{R}}_{ab}^{(\lambda)}$ are given in Table 6.2. We note that $\mathcal{R}_{ab}^{(\lambda)} = \mathcal{R}_{ba}^{(\lambda)}$ and $\mathcal{R}_{ab}^{(0)} = \tilde{\mathcal{R}}_{ab}^{(0)}$.

For the analytical evaluation of radial integrals one can use the following mathematical expressions, which also provide for values beyond those in Table 6.2. The basic relations are

**Table 6.2.** Values of the radial integral $\tilde{\mathcal{R}}_{ab}^{(\lambda)} = \tilde{\mathcal{R}}_{ba}^{(\lambda)}$ given in (6.32) for $\lambda = 0, 1, 2, 3$ and for the 0s, 0p, 1s, 0d, 1p and 0f shells

| $a$ $b$ | $\lambda = 0$ | $\lambda = 1$ | $\lambda = 2$ | $\lambda = 3$ |
|---|---|---|---|---|
| 0s 0s | $1$ | $\frac{2}{\sqrt{\pi}} = 1.128$ | $\frac{3}{2} = 1.500$ | $\frac{4}{\sqrt{\pi}} = 2.257$ |
| 0s 0p | $2\sqrt{\frac{2}{3\pi}} = 0.921$ | $\sqrt{\frac{3}{2}} = 1.225$ | $4\sqrt{\frac{2}{3\pi}} = 1.843$ | $\frac{5}{2}\sqrt{\frac{3}{2}} = 3.062$ |
| 0s 1s | $0$ | $-\sqrt{\frac{2}{3\pi}} = -0.461$ | $-\sqrt{\frac{3}{2}} = -1.225$ | $-2\sqrt{\frac{6}{\pi}} = -2.764$ |
| 0s 0d | $\sqrt{\frac{3}{5}} = 0.775$ | $\frac{8}{\sqrt{15\pi}} = 1.165$ | $\frac{\sqrt{15}}{2} = 1.936$ | $8\sqrt{\frac{3}{5\pi}} = 3.496$ |
| 0s 0f | $8\sqrt{\frac{2}{105\pi}} = 0.623$ | $\sqrt{\frac{15}{14}} = 1.035$ | $8\sqrt{\frac{6}{35\pi}} = 1.869$ | $\frac{1}{2}\sqrt{\frac{105}{2}} = 3.623$ |
| 0s 1p | $\frac{2}{\sqrt{15\pi}} = 0.291$ | $0$ | $-\frac{4}{\sqrt{15\pi}} = -0.583$ | $-\frac{\sqrt{15}}{2} = -1.936$ |
| 0p 0p | $1$ | $\frac{8}{3\sqrt{\pi}} = 1.505$ | $\frac{5}{2} = 2.500$ | $\frac{8}{\sqrt{\pi}} = 4.514$ |
| 0p 1s | $-\frac{2}{3\sqrt{\pi}} = -0.376$ | $-1$ | $-\frac{4}{\sqrt{\pi}} = -2.257$ | $-5$ |
| 0p 0d | $\frac{8}{3}\sqrt{\frac{2}{5\pi}} = 0.951$ | $\sqrt{\frac{5}{2}} = 1.581$ | $8\sqrt{\frac{2}{5\pi}} = 2.855$ | $\frac{7}{2}\sqrt{\frac{5}{2}} = 5.534$ |
| 0p 0f | $\sqrt{\frac{5}{7}} = 0.845$ | $\frac{16}{\sqrt{35\pi}} = 1.526$ | $\frac{\sqrt{35}}{2} = 2.958$ | $\frac{64}{\sqrt{35\pi}} = 6.103$ |
| 0p 1p | $0$ | $-\frac{4}{3}\sqrt{\frac{2}{5\pi}} = -0.476$ | $-\sqrt{\frac{5}{2}} = -1.581$ | $-12\sqrt{\frac{2}{5\pi}} = -4.282$ |
| 1s 1s | $1$ | $\frac{3}{\sqrt{\pi}} = 1.693$ | $\frac{7}{2} = 3.500$ | $\frac{14}{\sqrt{\pi}} = 7.899$ |
| 1s 0d | $-\sqrt{\frac{2}{5}} = -0.632$ | $-4\sqrt{\frac{2}{5\pi}} = -1.427$ | $-\sqrt{10} = -3.162$ | $-4\sqrt{\frac{10}{\pi}} = -7.136$ |
| 1s 0f | $-\frac{8}{\sqrt{35\pi}} = -0.763$ | $-2\sqrt{\frac{5}{7}} = -1.690$ | $-8\sqrt{\frac{5}{7\pi}} = -3.815$ | $-\frac{3}{2}\sqrt{35} = -8.874$ |
| 1s 1p | $\frac{7}{3}\sqrt{\frac{2}{5\pi}} = 0.833$ | $\sqrt{\frac{5}{2}} = 1.581$ | $2\sqrt{\frac{10}{\pi}} = 3.568$ | $\frac{11}{2}\sqrt{\frac{5}{2}} = 8.696$ |
| 0d 0d | $1$ | $\frac{16}{5\sqrt{\pi}} = 1.805$ | $\frac{7}{2} = 3.500$ | $\frac{64}{5\sqrt{\pi}} = 7.222$ |
| 0d 0f | $\frac{16}{5}\sqrt{\frac{2}{7\pi}} = 0.965$ | $\sqrt{\frac{7}{2}} = 1.871$ | $\frac{64}{5}\sqrt{\frac{2}{7\pi}} = 3.860$ | $\frac{9}{2}\sqrt{\frac{7}{2}} = 8.419$ |
| 0d 1p | $-\frac{8}{15\sqrt{\pi}} = -0.301$ | $-1$ | $-\frac{24}{5\sqrt{\pi}} = -2.708$ | $-7$ |
| 0f 0f | $1$ | $\frac{128}{35\sqrt{\pi}} = 2.063$ | $\frac{9}{2} = 4.500$ | $\frac{128}{7\sqrt{\pi}} = 10.317$ |
| 0f 1p | $-\sqrt{\frac{2}{7}} = -0.534$ | $-\frac{24}{5}\sqrt{\frac{2}{7\pi}} = -1.448$ | $-\sqrt{14} = -3.742$ | $-32\sqrt{\frac{2}{7\pi}} = -9.650$ |
| 1p 1p | $1$ | $\frac{52}{15\sqrt{\pi}} = 1.956$ | $\frac{9}{2} = 4.500$ | $\frac{20}{\sqrt{\pi}} = 11.284$ |

$$\int_0^\infty r^m e^{-ar^2}\, \mathrm{d}r = \frac{\Gamma\left(\frac{m+1}{2}\right)}{2a^{\frac{m+1}{2}}} , \qquad a > 0 , \tag{6.33}$$

and

$$\Gamma\left(n + \tfrac{1}{2}\right) = \sqrt{\pi}\,\frac{(2n-1)!!}{2^n} , \qquad n = \text{integer} , \tag{6.34}$$

together with the expression (3.42) for the harmonic oscillator radial wave function. These lead to the useful formulas

$$\int_0^\infty g_{0l}(r) g_{0l'}(r) r^2\, \mathrm{d}r = \left(\frac{2}{\pi}\right)^{p/2} \frac{(l + l' + 1)!!}{\sqrt{(2l+1)!!(2l'+1)!!}} , \tag{6.35}$$

$$\int_0^\infty g_{1l}(r)g_{1l'}(r)r^2\mathrm{d}r = \left(\frac{2}{\pi}\right)^{p/2} \frac{(l+l'+1)!![3+l+l'-\frac{1}{2}(l-l')^2]}{\sqrt{(2l+3)!!(2l'+3)!!}} , \quad (6.36)$$

$$\int_0^\infty g_{0l}(r)g_{1l'}(r)r^2\mathrm{d}r = \left(\frac{2}{\pi}\right)^{p/2} \frac{(l'-l)(l+l'+1)!!}{\sqrt{2(2l+1)!!(2l'+3)!!}} , \quad (6.37)$$

where

$$p = \begin{cases} 0, & l+l' = \text{even}, \\ 1, & l+l' = \text{odd}. \end{cases} \quad (6.38)$$

Substituting the recursion relation (3.52) for associated Laguerre polynomials into the radial function (3.42) yields

$$rg_{nl}(r) = b\sqrt{n+l+\tfrac{3}{2}}\, g_{n,l+1}(r) - b\sqrt{n}\, g_{n-1,l+1}(r) . \quad (6.39)$$

Substituting this into (6.32) yields a relation for evaluating integrals with higher values of $\lambda$:

$$\boxed{\tilde{\mathcal{R}}^{(\lambda)}_{n_a l_a n_b l_b} = \sqrt{n_b + l_b + \tfrac{3}{2}}\, \tilde{\mathcal{R}}^{(\lambda-1)}_{n_a l_a n_b, l_b+1} - \sqrt{n_b}\, \tilde{\mathcal{R}}^{(\lambda-1)}_{n_a l_a, n_b-1, l_b+1} .} \quad (6.40)$$

The exact values in Table 6.2 were generated by using the relations (6.35)–(6.40).

As for numerical computing of the radial integrals, there is an elegant method [40] applicable to the case $l_a + l_b + \lambda =$ even, which is just the case for the electric and magnetic operators under discussion; see (6.23) and (6.24). The method is contained in the following general expression for the radial integral:

$$\tilde{\mathcal{R}}^{(\lambda)}_{ab} = (-1)^{n_a+n_b} \sqrt{\frac{n_a!n_b!}{\Gamma(n_a+l_a+\tfrac{3}{2})\Gamma(n_b+l_b+\tfrac{3}{2})}}\, \tau_a!\tau_b!$$

$$\times \sum_{\sigma=\sigma_{\min}}^{\sigma_{\max}} \frac{\Gamma[\tfrac{1}{2}(l_a+l_b+\lambda)+\sigma+\tfrac{3}{2}]}{\sigma!(n_a-\sigma)!(n_b-\sigma)!(\sigma+\tau_a-n_a)!(\sigma+\tau_b-n_b)!} , \quad (6.41)$$

where

$$\tau_a \equiv \tfrac{1}{2}(l_b-l_a+\lambda) \geq 0 , \quad \tau_b \equiv \tfrac{1}{2}(l_a-l_b+\lambda) \geq 0 , \quad (6.42)$$

$$\sigma_{\min} = \max\{0, n_a-\tau_a, n_b-\tau_b\} , \quad \sigma_{\max} = \min\{n_a, n_b\} . \quad (6.43)$$

The $\Gamma$ functions can be computed easily by starting from the initial value

$$\Gamma(\tfrac{3}{2}) = \tfrac{1}{2}\sqrt{\pi} \quad (6.44)$$

and using recursively the basic relation

$$\Gamma(m+1) = m\Gamma(m) . \quad (6.45)$$

### 6.1.5 Tables of Numerical Values of Single-Particle Matrix Elements

Tables 6.3–6.5 list numerical values of the electric single-particle reduced matrix elements in the CS phase convention. The tables include $\lambda = 1, 2, 3$ and the 0s, 0p, 0d-1s and 0f-1p major shells, which comprise the 10 lowest single-particle orbitals. The radial integral used in the tabulated matrix elements is the $b$-independent quantity $\tilde{\mathcal{R}}_{ab}^{(\lambda)}$ defined in (6.32). The charge $e$ is omitted from the tables. Thus, if we call the tabulated matrix elements $\overline{(a\|\boldsymbol{Q}_\lambda\|b)}_{\mathrm{CS}}$, we obtain the physical matrix elements (6.23) according to

**Table 6.3.** Electric dipole (E1) reduced matrix elements $\overline{(a\|\boldsymbol{Q}_1\|b)}_{\mathrm{CS}}$ for the 10 lowest single-particle orbitals

| $a\backslash b$ | $0s_{1/2}$ | $0p_{3/2}$ | $0p_{1/2}$ | $0d_{5/2}$ | $1s_{1/2}$ | $0d_{3/2}$ | $0f_{7/2}$ | $1p_{3/2}$ | $1p_{1/2}$ | $0f_{5/2}$ |
|---|---|---|---|---|---|---|---|---|---|---|
| $0s_{1/2}$ | 0 | −0.691 | −0.489 | 0 | 0 | 0 | 0 | 0 | 0 | 0 |
| $0p_{3/2}$ | 0.691 | 0 | 0 | −1.197 | −0.564 | −0.399 | 0 | 0 | 0 | 0 |
| $0p_{1/2}$ | −0.489 | 0 | 0 | 0 | 0.399 | −0.892 | 0 | 0 | 0 | 0 |
| $0d_{5/2}$ | 0 | 1.197 | 0 | 0 | 0 | 0 | −1.693 | −0.757 | 0 | −0.379 |
| $1s_{1/2}$ | 0 | 0.564 | 0.399 | 0 | 0 | 0 | 0 | −0.892 | −0.631 | 0 |
| $0d_{3/2}$ | 0 | −0.399 | 0.892 | 0 | 0 | 0 | 0 | 0.252 | −0.564 | −1.416 |
| $0f_{7/2}$ | 0 | 0 | 0 | 1.693 | 0 | 0 | 0 | 0 | 0 | 0 |
| $1p_{3/2}$ | 0 | 0 | 0 | 0.757 | 0.892 | 0.252 | 0 | 0 | 0 | 0 |
| $1p_{1/2}$ | 0 | 0 | 0 | 0 | −0.631 | 0.564 | 0 | 0 | 0 | 0 |
| $0f_{5/2}$ | 0 | 0 | 0 | −0.379 | 0 | 1.416 | 0 | 0 | 0 | 0 |

The physical matrix elements (6.23) are obtained by multiplying the tabulated numbers by $eb$

**Table 6.4.** Electric quadrupole (E2) reduced matrix elements $\overline{(a\|\boldsymbol{Q}_2\|b)}_{\mathrm{CS}}$ for the 10 lowest single-particle orbitals

| $a\backslash b$ | $0s_{1/2}$ | $0p_{3/2}$ | $0p_{1/2}$ | $0d_{5/2}$ | $1s_{1/2}$ | $0d_{3/2}$ | $0f_{7/2}$ | $1p_{3/2}$ | $1p_{1/2}$ | $0f_{5/2}$ |
|---|---|---|---|---|---|---|---|---|---|---|
| $0s_{1/2}$ | 0 | 0 | 0 | 1.338 | 0 | 1.092 | 0 | 0 | 0 | 0 |
| $0p_{3/2}$ | 0 | −1.410 | −1.410 | 0 | 0 | 0 | 2.676 | 0.892 | 0.892 | 1.093 |
| $0p_{1/2}$ | 0 | 1.410 | 0 | 0 | 0 | 0 | 0 | −0.892 | 0 | 2.044 |
| $0d_{5/2}$ | 1.338 | 0 | 0 | −2.585 | −2.185 | −1.293 | 0 | 0 | 0 | 0 |
| $1s_{1/2}$ | 0 | 0 | 0 | −2.185 | 0 | −1.784 | 0 | 0 | 0 | 0 |
| $0d_{3/2}$ | −1.092 | 0 | 0 | 1.293 | 1.784 | −1.975 | 0 | 0 | 0 | 0 |
| $0f_{7/2}$ | 0 | 2.676 | 0 | 0 | 0 | 0 | −3.918 | −3.385 | 0 | −1.357 |
| $1p_{3/2}$ | 0 | 0.892 | 0.892 | 0 | 0 | 0 | −3.385 | −2.539 | −2.539 | −1.382 |
| $1p_{1/2}$ | 0 | −0.892 | 0 | 0 | 0 | 0 | 0 | 2.539 | 0 | −2.586 |
| $0f_{5/2}$ | 0 | −1.093 | 2.044 | 0 | 0 | 0 | 1.357 | 1.382 | −2.586 | −3.324 |

The physical matrix elements (6.23) are obtained by multiplying the tabulated numbers by $eb^2$

**Table 6.5.** Electric octupole (E3) reduced matrix elements $\overline{(a\|Q_3\|b)}_{\mathrm{CS}}$ for the 10 lowest single-particle orbitals

| $a\backslash b$ | $0s_{1/2}$ | $0p_{3/2}$ | $0p_{1/2}$ | $0d_{5/2}$ | $1s_{1/2}$ | $0d_{3/2}$ | $0f_{7/2}$ | $1p_{3/2}$ | $1p_{1/2}$ | $0f_{5/2}$ |
|---|---|---|---|---|---|---|---|---|---|---|
| $0s_{1/2}$ | 0 | 0 | 0 | 0 | 0 | 0 | −2.891 | 0 | 0 | −2.503 |
| $0p_{3/2}$ | 0 | 0 | 0 | 3.420 | 0 | 4.189 | 0 | 0 | 0 | 0 |
| $0p_{1/2}$ | 0 | 0 | 0 | −3.824 | 0 | 0 | 0 | 0 | 0 | 0 |
| $0d_{5/2}$ | 0 | −3.420 | −3.824 | 0 | 0 | 0 | 6.717 | 4.326 | 4.837 | 4.248 |
| $1s_{1/2}$ | 0 | 0 | 0 | 0 | 0 | 0 | 7.080 | 0 | 0 | 6.132 |
| $0d_{3/2}$ | 0 | 4.189 | 0 | 0 | 0 | 0 | −3.878 | −5.299 | 0 | 5.203 |
| $0f_{7/2}$ | 2.891 | 0 | 0 | −6.717 | −7.080 | −3.878 | 0 | 0 | 0 | 0 |
| $1p_{3/2}$ | 0 | 0 | 0 | −4.326 | 0 | −5.299 | 0 | 0 | 0 | 0 |
| $1p_{1/2}$ | 0 | 0 | 0 | 4.837 | 0 | 0 | 0 | 0 | 0 | 0 |
| $0f_{5/2}$ | −2.503 | 0 | 0 | 4.248 | 6.132 | −5.203 | 0 | 0 | 0 | 0 |

The physical matrix elements (6.23) are obtained by multiplying the tabulated numbers by $eb^3$.

$$(a\|Q_\lambda\|b)_{\mathrm{CS}} = eb^\lambda \overline{(a\|Q_\lambda\|b)}_{\mathrm{CS}} . \tag{6.46}$$

The proper value of the oscillator length $b$ can be read from (3.43), combined with the Blomqvist–Molinari formula (3.45). The Biedenharn–Rose single-particle matrix elements can be obtained from our tabulated ones by inserting the phase factors given in (6.23).

Of the magnetic single-particle matrix elements (6.24) we only tabulate the most important multipolarity $\lambda = 1$. Because the orbital and spin parts are separate it is conveninent to tabulate the two terms separately,

$$(a\|M_1\|b)_{\mathrm{CS}} = g_l \mathcal{D}_{ab}^{(l)} + g_s \mathcal{D}_{ab}^{(s)} . \tag{6.47}$$

Tables 6.6 and 6.7 give the orbital-dipole term $\mathcal{D}_{ab}^{(l)}$ and the spin-dipole term $\mathcal{D}_{ab}^{(s)}$ in units of $\mu_{\mathrm{N}}/c$ for the 0s, 0p, 0d-1s and 0f-1p major shells. The radial integral $\mathcal{R}_{ab}^{(0)}$ entering the calculation is independent of the oscillator length $b$. The Biedenharn–Rose matrix elements can be obtained by attaching the phase factors given in (6.24) to the table entries.

### 6.1.6 Electromagnetic Multipole Moments

The static electric and magnetic multipole moments are important observables of nuclear structure. These moments are sensitive to details of the wave function used to compute them. Comparison of computed multipole moments with measured ones is one of the tests of a nuclear model. Apart from conventional constant factors to be introduced below, the $2^\lambda$-pole moment of a nucleus in a certain state is defined as

$$\boxed{\mathcal{M}(\sigma\lambda) \equiv \langle \xi\, J\, M = J | \mathcal{M}_{\sigma\lambda 0} | \xi\, J\, M = J \rangle ,} \tag{6.48}$$

**Table 6.6.** Magnetic dipole (M1) reduced matrix elements, orbital term $\mathcal{D}_{ab}^{(l)}$, for the 10 lowest single-particle orbitals in units of $\mu_N/c$

| $a\backslash b$ | $0s_{1/2}$ | $0p_{3/2}$ | $0p_{1/2}$ | $0d_{5/2}$ | $1s_{1/2}$ | $0d_{3/2}$ | $0f_{7/2}$ | $1p_{3/2}$ | $1p_{1/2}$ | $0f_{5/2}$ |
|---|---|---|---|---|---|---|---|---|---|---|
| $0s_{1/2}$ | 0 | 0 | 0 | 0 | 0 | 0 | 0 | 0 | 0 | 0 |
| $0p_{3/2}$ | 0 | 1.262 | 0.564 | 0 | 0 | 0 | 0 | 0 | 0 | 0 |
| $0p_{1/2}$ | 0 | −0.564 | 0.798 | 0 | 0 | 0 | 0 | 0 | 0 | 0 |
| $0d_{5/2}$ | 0 | 0 | 0 | 2.832 | 0 | 0.757 | 0 | 0 | 0 | 0 |
| $1s_{1/2}$ | 0 | 0 | 0 | 0 | 0 | 0 | 0 | 0 | 0 | 0 |
| $0d_{3/2}$ | 0 | 0 | 0 | −0.757 | 0 | 2.271 | 0 | 0 | 0 | 0 |
| $0f_{7/2}$ | 0 | 0 | 0 | 0 | 0 | 0 | 4.701 | 0 | 0 | 0.905 |
| $1p_{3/2}$ | 0 | 0 | 0 | 0 | 0 | 0 | 0 | 1.262 | 0.564 | 0 |
| $1p_{1/2}$ | 0 | 0 | 0 | 0 | 0 | 0 | 0 | −0.564 | 0.798 | 0 |
| $0f_{5/2}$ | 0 | 0 | 0 | 0 | 0 | 0 | −0.905 | 0 | 0 | 4.046 |

**Table 6.7.** Magnetic dipole (M1) reduced matrix elements, spin term $\mathcal{D}_{ab}^{(s)}$, for the 10 lowest single-particle orbitals in units of $\mu_N/c$

| $a\backslash b$ | $0s_{1/2}$ | $0p_{3/2}$ | $0p_{1/2}$ | $0d_{5/2}$ | $1s_{1/2}$ | $0d_{3/2}$ | $0f_{7/2}$ | $1p_{3/2}$ | $1p_{1/2}$ | $0f_{5/2}$ |
|---|---|---|---|---|---|---|---|---|---|---|
| $0s_{1/2}$ | 0.598 | 0 | 0 | 0 | 0 | 0 | 0 | 0 | 0 | 0 |
| $0p_{3/2}$ | 0 | 0.631 | −0.564 | 0 | 0 | 0 | 0 | 0 | 0 | 0 |
| $0p_{1/2}$ | 0 | 0.564 | −0.199 | 0 | 0 | 0 | 0. | 0 | 0 | 0 |
| $0d_{5/2}$ | 0 | 0 | 0 | 0.708 | 0 | −0.757 | 0 | 0 | 0 | 0 |
| $1s_{1/2}$ | 0 | 0 | 0 | 0 | 0.598 | 0 | 0 | 0 | 0 | 0 |
| $0d_{3/2}$ | 0 | 0 | 0 | 0.757 | 0 | −0.378 | 0 | 0 | 0 | 0 |
| $0f_{7/2}$ | 0 | 0 | 0 | 0 | 0 | 0 | 0.784 | 0 | 0 | −0.905 |
| $1p_{3/2}$ | 0 | 0 | 0 | 0 | 0 | 0 | 0 | 0.631 | −0.564 | 0 |
| $1p_{1/2}$ | 0 | 0 | 0 | 0 | 0 | 0 | 0 | 0.564 | −0.199 | 0 |
| $0f_{5/2}$ | 0 | 0 | 0 | 0 | 0 | 0 | 0.905 | 0 | 0 | −0.506 |

where $\xi$ carries all other quantum numbers than the angular momentum $J$ and its $z$ projection $M$ and the operator is the $\mu = 0$ component of the multipole tensor operator $\mathcal{M}_{\sigma\lambda}$ in (6.4). Applying the Wigner–Eckart theorem (2.27) yields

$$\mathcal{M}(\sigma\lambda) = \begin{pmatrix} J & \lambda & J \\ -J & 0 & J \end{pmatrix} (\xi J\|\mathcal{M}_{\sigma\lambda}\|\xi J) . \tag{6.49}$$

For the dipole ($\lambda = 1$) and quadrupole ($\lambda = 2$) cases the $3j$ symbol is given by (1.49) and (1.50) resulting in

$$\mathcal{M}(\sigma 1) = \sqrt{\frac{J}{(J+1)(2J+1)}} (\xi J\|\mathcal{M}_{\sigma 1}\|\xi J) , \tag{6.50}$$

$$\mathcal{M}(\sigma 2) = \sqrt{\frac{J(2J-1)}{(J+1)(2J+1)(2J+3)}} (\xi J\|\mathcal{M}_{\sigma 2}\|\xi J) . \tag{6.51}$$

The conventional *magnetic dipole moment* $\mu$ and *electric quadrupole moment* $Q$ are defined by the equations[9]

$$\frac{\mu}{c} \equiv \zeta\sqrt{\frac{4\pi}{3}}\mathcal{M}(\text{M1}) = \zeta\sqrt{\frac{4\pi}{3}}\sqrt{\frac{J}{(J+1)(2J+1)}}(\xi J\|\boldsymbol{M}_1\|\xi J)\,, \quad (6.52)$$

$$eQ \equiv \zeta\sqrt{\frac{16\pi}{5}}\mathcal{M}(\text{E2}) = \zeta\sqrt{\frac{16\pi}{5}}\sqrt{\frac{J(2J-1)}{(J+1)(2J+1)(2J+3)}}(\xi J\|\boldsymbol{Q}_2\|\xi J)\,,$$

$$(6.53)$$

where the operators are expressed in the conventional notation (6.5). These moments are defined so as to coincide with their classical analogues. The phase factors for both reduced matrix elements, given in (6.23) and (6.24), are $\zeta = +1$ for the Condon–Shortley convention and $\zeta = -1$ for the Biedenharn–Rose convention. These phase factors are included in (6.52) and (6.53) to cancel those in the matrix elements so that a given moment has the same sign in both conventions. The defining equations show explicitly the necessary conditions for a non-vanishing M1 and E2 moment, $J \geq \frac{1}{2}$ and $J \geq 1$, respectively. These conditions can already be read from the $3j$ symbol in (6.49).

The single-particle values of the magnetic dipole and electric quadrupole moments can be obtained from the single-particle matrix elements (6.24) and (6.23) respectively. The results are

$$\mu = \mu_N \frac{1-(-1)^{l+j+\frac{1}{2}}(2j+1)}{4(j+1)}\left\{g_s - g_l\left[2 + (-1)^{l+j+\frac{1}{2}}(2j+1)\right]\right\}\,, \quad (6.54)$$

$$Q = \frac{3-4j(j+1)}{2(j+1)(2j+3)}\mathcal{R}^{(2)}_{nlnl}\,. \quad (6.55)$$

Equation (6.54) shows directly that single-particle s states, $l = 0$, have no orbital magnetic moment, which is also clear from basic physics. Equation (6.55) shows that a non-vanishing single-particle quadrupole moment is always negative. This corresponds to the qualitative notion that in the defining state, with $M = J$, particles move around the nuclear equator and thus produce an oblate shape. The quadrupole moment has the unit of area, the oscillator length squared $b^2$, as is seen from the integral $\mathcal{R}^{(2)}_{nlnl}$.

### 6.1.7 Weisskopf Units and Transition Rates

There is a convenient simple estimate for the reduced transition probabilities $B(\sigma\lambda)$ defined in (6.4). It is derived by making some simplifying approximations in (6.23) and (6.24). To begin with, the radial wave function is assumed to be constant inside the nucleus and zero outside. Normalization then yields

---

[9] In Gaussian units there is no $c$ in (6.52). In both systems, SI and Gaussian, the unit of the magnetic dipole moment is the nuclear magneton; see (6.9).

$$g_{nl}(r) \approx \begin{cases} \sqrt{3/R^3}\,, & r \leq R\,, \\ 0\,, & r > R\,, \end{cases} \tag{6.56}$$

where $R$ is the nuclear radius. Then the radial integral becomes

$$\mathcal{R}_{ab}^{(\lambda)} \approx \frac{3}{R^3} \int_0^R r^{\lambda+2} \mathrm{d}r = \frac{3}{\lambda+3} R^\lambda\,. \tag{6.57}$$

For the angular momenta we take the 'stretched case' $j_a = \frac{1}{2}$ and $j_b = \lambda + \frac{1}{2}$, and assume such $l_a$ and $l_b$ that the parity factor in (6.23) and (6.24) is unity, i.e. that the transitions are parity allowed. Equation (1.45) gives the appropriate $3j$ symbol, leading to

$$\widehat{\lambda}\widehat{j_a}\widehat{j_b} \begin{pmatrix} j_a & j_b & \lambda \\ \frac{1}{2} & -\frac{1}{2} & 0 \end{pmatrix} = (-1)^\lambda \sqrt{2\lambda+2}\,. \tag{6.58}$$

Collecting the various factors in (6.23) and applying (6.4) we obtain

$$B(\mathrm{E}\lambda) \approx \frac{e^2}{4\pi} \left( \frac{3}{\lambda+3} \right)^2 R^{2\lambda}\,. \tag{6.59}$$

Finally, by using the relation $R = r_0 A^{1/3} = 1.2 A^{1/3}$ fm, we have the so-called *Weisskopf single-particle estimate* [41] or *Weisskopf unit* (W.u.)

$$\boxed{B_{\mathrm{W}}(\mathrm{E}\lambda) = \frac{1.2^{2\lambda}}{4\pi} \left( \frac{3}{\lambda+3} \right)^2 A^{2\lambda/3}\, e^2 \mathrm{fm}^{2\lambda}\,.} \tag{6.60}$$

For magnetic transitions one traditionally takes[10] [41]

$$\boxed{B_{\mathrm{W}}(\mathrm{M}\lambda) = \frac{10}{\pi} 1.2^{2\lambda-2} \left( \frac{3}{\lambda+3} \right)^2 A^{(2\lambda-2)/3} \left( \mu_{\mathrm{N}}/c \right)^2 \mathrm{fm}^{2\lambda-2}\,.} \tag{6.61}$$

By substituting the Weisskopf estimates for $B(\mathrm{E}\lambda)$ and $B(\mathrm{M}\lambda)$ in (6.14) we get Weisskopf estimates $T_{\mathrm{W}}(\mathrm{E}\lambda)$ and $T_{\mathrm{W}}(\mathrm{M}\lambda)$ for transition probabilities per unit time. Tables 6.8 and 6.9 list the relevant numerical expressions. The second columns of the tables are universally true, while the third columns contain the Weisskopf single-particle estimates for reduced transition probabilities. The fourth columns then restate the information in the third columns via the relations in the first coulumns. In all cases decay half-lives can be computed from the transition probabilities according to (6.1). The Weisskopf estimates $B_{\mathrm{W}}(\sigma\lambda)$ are very easy to compute since they only depend on the nuclear mass number $A$. However, one should be aware that they are essentially only order-of-magnitude approximations.

---

[10] There is no $c$ present in Gaussian units.

**Table 6.8.** Transition probabilities for $E\lambda$ transitions

| $E\lambda$ | $T(E\lambda)\,(\mathrm{s}^{-1})$ | $B_{\mathrm{W}}(E\lambda)\,(e^2\mathrm{fm}^{2\lambda})$ | $T_{\mathrm{W}}(E\lambda)\,(\mathrm{s}^{-1})$ |
|---|---|---|---|
| E1 | $1.587 \times 10^{15} E^3 B(\mathrm{E}1)$ | $6.446 \times 10^{-2} A^{2/3}$ | $1.023 \times 10^{14} E^3 A^{2/3}$ |
| E2 | $1.223 \times 10^{9} E^5 B(\mathrm{E}2)$ | $5.940 \times 10^{-2} A^{4/3}$ | $7.265 \times 10^{7} E^5 A^{4/3}$ |
| E3 | $5.698 \times 10^{2} E^7 B(\mathrm{E}3)$ | $5.940 \times 10^{-2} A^{2}$ | $3.385 \times 10^{1} E^7 A^{2}$ |
| E4 | $1.694 \times 10^{-4} E^9 B(\mathrm{E}4)$ | $6.285 \times 10^{-2} A^{8/3}$ | $1.065 \times 10^{-5} E^9 A^{8/3}$ |
| E5 | $3.451 \times 10^{-11} E^{11} B(\mathrm{E}5)$ | $6.928 \times 10^{-2} A^{10/3}$ | $2.391 \times 10^{-12} E^{11} A^{10/3}$ |

The transition energies $E$ are to be given in MeV and the reduced transition probabilities $B(E\lambda)$ in $e^2\mathrm{fm}^{2\lambda}$.

**Table 6.9.** Transition probabilities for $M\lambda$ transitions

| $M\lambda$ | $T(M\lambda)\,(\mathrm{s}^{-1})$ | $B_{\mathrm{W}}(M\lambda)\,((\mu_{\mathrm{N}}/c)^2\mathrm{fm}^{2\lambda-2})$ | $T_{\mathrm{W}}(M\lambda)\,(\mathrm{s}^{-1})$ |
|---|---|---|---|
| M1 | $1.779 \times 10^{13} E^3 B(\mathrm{M}1)$ | $1.790$ | $3.184 \times 10^{13} E^3$ |
| M2 | $1.371 \times 10^{7} E^5 B(\mathrm{M}2)$ | $1.650 A^{2/3}$ | $2.262 \times 10^{7} E^5 A^{2/3}$ |
| M3 | $6.387 \times 10^{0} E^7 B(\mathrm{M}3)$ | $1.650 A^{4/3}$ | $1.054 \times 10^{1} E^7 A^{4/3}$ |
| M4 | $1.899 \times 10^{-6} E^9 B(\mathrm{M}4)$ | $1.746 A^{2}$ | $3.316 \times 10^{-6} E^9 A^{2}$ |
| M5 | $3.868 \times 10^{-13} E^{11} B(\mathrm{M}5)$ | $1.924 A^{8/3}$ | $7.442 \times 10^{-13} E^{11} A^{8/3}$ |

The transition energies $E$ are to be given in MeV and the reduced transition probabilities $B(M\lambda)$ in $(\mu_{\mathrm{N}}/c)^2\mathrm{fm}^{2\lambda-2}$.

## 6.2 Electromagnetic Transitions in One-Particle and One-Hole Nuclei

The simplest possible nuclei are the one-particle and one-hole nuclei. Their states were discussed in Sect. 5.2. The simple wave functions of these states can be tested efficiently on electromagnetic observables.

### 6.2.1 Reduced Transition Probabilities

Wave functions of one-particle and one-hole nuclei were written down in Sect. 5.2. Let us now consider a *one-particle* nucleus and denote its initial and final states as

$$|\Psi_i\rangle = |n_i\, l_i\, j_i\, m_i\rangle = c_i^\dagger|\mathrm{CORE}\rangle\,, \tag{6.62}$$

$$|\Psi_f\rangle = |n_f\, l_f\, j_f\, m_f\rangle = c_f^\dagger|\mathrm{CORE}\rangle\,. \tag{6.63}$$

We calculate the reduced one-body transition density for insertion into (6.22). With use of (4.23) and (4.42) the one-body transition density becomes

$$\langle\Psi_f|\left[c_a^\dagger\tilde{c}_b\right]_{\lambda\mu}|\Psi_i\rangle = \sum_{m_\alpha m_\beta}(j_a\, m_\alpha\, j_b\, m_\beta|\lambda\,\mu)\langle\mathrm{CORE}|c_f c_\alpha^\dagger\tilde{c}_\beta c_i^\dagger|\mathrm{CORE}\rangle$$

$$= \delta_{af}\delta_{bi}(-1)^{j_i-m_i}(j_f\, m_f\, j_i\, -m_i|\lambda\,\mu)\,. \tag{6.64}$$

Application of (1.38) and the Wigner–Eckart theorem (2.27) gives the reduced one-body transition density

$$(\Psi_f \| [c_a^\dagger \tilde{c}_b]_\lambda \| \Psi_i) = \delta_{af} \delta_{bi} \hat{\lambda} \,. \tag{6.65}$$

From (6.22) we now have the reduced matrix element

$$(\Psi_f \| \mathcal{M}_{\sigma\lambda} \| \Psi_i) = (f \| \mathcal{M}_{\sigma\lambda} \| i) \,, \tag{6.66}$$

and (6.4) gives the reduced transition probability

$$\boxed{B(\sigma\lambda;\, \Psi_i \to \Psi_f) = \frac{1}{2j_i + 1} \left| (f \| \mathcal{M}_{\sigma\lambda} \| i) \right|^2 \,, \quad \sigma = \mathrm{E},\, \mathrm{M} \,.} \tag{6.67}$$

It should be noted that the reduced matrix element in (6.67) is between the *single-particle* states $|i\rangle$ and $|f\rangle$ whereas the physical states $|\Psi_i\rangle$ and $|\Psi_f\rangle$ include the core.

*One-hole* nuclei are treated similarly to one-particle nuclei. The initial and final states are

$$|\Phi_i\rangle = |(n_i\, l_i\, j_i\, m_i)^{-1}\rangle = h_i^\dagger |\mathrm{HF}\rangle \,, \tag{6.68}$$

$$|\Phi_f\rangle = |(n_f\, l_f\, j_f\, m_f)^{-1}\rangle = h_f^\dagger |\mathrm{HF}\rangle \,. \tag{6.69}$$

The one-body transition density is

$$\langle \Phi_f | [c_a^\dagger \tilde{c}_b]_{\lambda\mu} |\Phi_i\rangle = -\sum_{m_\alpha m_\beta} (j_a\, m_\alpha\, j_b\, m_\beta | \lambda\, \mu) \langle \mathrm{HF} | h_f \tilde{h}_\alpha h_\beta^\dagger h_i^\dagger |\mathrm{HF}\rangle$$

$$= -\delta_{fi} \sum_{m_\alpha} (-1)^{j_a + m_\alpha} (j_a\, m_\alpha\, j_a\, -m_\alpha | \lambda\, \mu)$$

$$+ \delta_{ai}\delta_{bf} (-1)^{j_i - m_i} (j_i\, -m_i\, j_f\, m_f | \lambda\, \mu) \,, \tag{6.70}$$

where we have used the conversion formulas (4.46) and (4.48) and the contractions (4.52). To perform the $m_\alpha$ summation in the first term, we note that

$$(-1)^{j_a + m_\alpha} = -\hat{j}_a (j_a\, m_\alpha\, j_a\, -m_\alpha | 0\, 0) \,. \tag{6.71}$$

Then the first term yields $\delta_{fi}\delta_{\lambda 0}\delta_{\mu 0}\hat{j}_a$, which can be omitted since for the electromagnetic operators always $\lambda \geq 1$. So we are left with

$$\langle \Phi_f | [c_a^\dagger \tilde{c}_b]_{\lambda\mu} |\Phi_i\rangle = \delta_{ai}\delta_{bf} (-1)^{j_i - m_i} (j_i\, -m_i\, j_f\, m_f | \lambda\, \mu) \,, \tag{6.72}$$

which via Wigner–Eckart leads to

$$(\Phi_f \| [c_a^\dagger \tilde{c}_b]_\lambda \| \Phi_i) = \delta_{ai}\delta_{bf} (-1)^{j_i + j_f + \lambda} \hat{\lambda} \,. \tag{6.73}$$

Substituting this into (6.22) yields

$$(\Phi_f\|\boldsymbol{\mathcal{M}}_{\sigma\lambda}\|\Phi_i) = (-1)^{j_i+j_f+\lambda}(f\|\boldsymbol{\mathcal{M}}_{\sigma\lambda}\|i) \ . \tag{6.74}$$

The reduced transition probability becomes

$$\boxed{B(\sigma\lambda\,;\,\Phi_i \to \Phi_f) = \frac{1}{2j_i+1}\left|(f\|\boldsymbol{\mathcal{M}}_{\sigma\lambda}\|i)\right|^2 \ , \quad \sigma = \mathrm{E, M} \ .} \tag{6.75}$$

The right-hand side is exactly the same as that of (6.67). The reduced matrix element is between the single-particle states $|i\rangle$ and $|f\rangle$ whereas the physical one-hole states $|\Phi_i\rangle$ and $|\Phi_f\rangle$ involve the Hartree–Fock vacuum $|\mathrm{HF}\rangle$. We have thus established that the reduced transition probability for a one-hole nucleus is the same as for the corresponding one-particle nucleus.

### 6.2.2 Example: Transitions in One-Hole Nuclei $^{15}\mathrm{N}$ and $^{15}\mathrm{O}$

Let us now discuss exhaustively an example of electromagnetic transitions in a pair of one-hole nuclei.

The lowest-lying one-hole states of the one-hole nuclei $^{15}_{7}\mathrm{N}_8$ and $^{15}_{8}\mathrm{O}_7$ are stated in (5.10)–(5.13) and depicted in Fig. 5.2. For both nuclei the states are $J_i^\pi = \frac{3}{2}^-$ and $J_f^\pi = \frac{1}{2}^-$, so $\Delta J = 1$ and $\pi_i\pi_f = +1$. Table 6.1 then gives M1 as the leading decay mode of the $3/2^-$ state. The triangular condition $\Delta(J_f\lambda J_i) = \Delta(\frac{1}{2}\lambda\frac{3}{2})$ also allows $\lambda = 2$, i.e. E2.

For the E2 transition equation (6.75) and Table 6.4 give

$$\begin{aligned} B\!\left(\mathrm{E2}\,;\,(0\mathrm{p}_{3/2})^{-1} \to (0\mathrm{p}_{1/2})^{-1}\right) &= \frac{1}{4}(0\mathrm{p}_{1/2}\|\boldsymbol{Q}_2\|0\mathrm{p}_{3/2})^2 \\ &= \tfrac{1}{4}(1.410eb^2)^2 = 4.270\,e^2\mathrm{fm}^4 \ . \end{aligned} \tag{6.76}$$

The last equality results from (3.43) and (3.45), which give $b = 1.712\,\mathrm{fm}$ for the oscillator length. It is of interest to express the result also in Weisskopf units. From Table 6.8 we calculate $B_\mathrm{W}(\mathrm{E2}) = 2.197\,e^2\mathrm{fm}^4$. So our result (6.76) is 1.943 W.u. Unless $e$ is replaced by a non-zero effective neutron charge, (6.76) applies only to the proton-hole nucleus $^{15}_{7}\mathrm{N}_8$.

For the M1 transition equations (6.47) and (6.75) and Tables 6.6 and 6.7 give

$$\begin{aligned} B\!\left(\mathrm{M1}\,;\,(0\mathrm{p}_{3/2})^{-1} \to (0\mathrm{p}_{1/2})^{-1}\right) &= \frac{1}{4}(0\mathrm{p}_{1/2}\|\boldsymbol{M}_1\|0\mathrm{p}_{3/2})^2 \\ &= \tfrac{1}{4}(-0.564g_l + 0.564g_s)^2\,(\mu_\mathrm{N}/c)^2 \ . \end{aligned} \tag{6.77}$$

Inserting the $g$ factors (6.8) we have for the proton-hole nucleus $^{15}_{7}\mathrm{N}_8$

$$\begin{aligned} B(\mathrm{M1}\,;\,{}^{15}\mathrm{N}) &= \tfrac{1}{4}(-0.564 \times 1 + 0.564 \times 5.586)^2\,(\mu_\mathrm{N}/c)^2 \\ &= 1.673\,(\mu_\mathrm{N}/c)^2 = 0.934\,\mathrm{W.u.} \end{aligned} \tag{6.78}$$

with the Weisskopf unit $1.790\,(\mu_\mathrm{N}/c)^2$ taken from Table 6.9. For the neutron-hole nucleus $^{15}_{8}\mathrm{O}_7$ we have similarly

$$B(M1\,;^{15}O) = \tfrac{1}{4}[-0.564 \times 0 + 0.564 \times (-3.826)]^2\,(\mu_N/c)^2$$
$$= 1.164\,(\mu_N/c)^2 = 0.650\,\text{W.u.} \qquad (6.79)$$

With the bare charges for the nucleons ($e_p = e$, $e_n = 0$) the E2 transition vanishes for $^{15}_{8}O_7$, so that M1 is the only possible decay mode. The formula in Table 6.9 with the experimental decay energy from Fig. 5.2 gives the transition probability

$$T(M1\,;^{15}O) = 1.779 \times 10^{13} \times 6.176^3 \times 1.164\,1/\text{s} = 4.878 \times 10^{15}\,1/\text{s}\,, \quad (6.80)$$

leading to the decay half-life (6.1)

$$t_{1/2}(M1\,;^{15}O) = \frac{\ln 2}{T(M1)} = 1.421 \times 10^{-16}\,\text{s} = 0.14\,\text{fs}\,. \qquad (6.81)$$

For the proton-hole nucleus $^{15}_{7}N_8$ the situation is more complicated because both E2 and M1 transitions are present. Using Tables 6.8 and 6.9 and taking the decay energy from Fig. 5.2, we obtain for the transition probabilities

$$T(E2\,;^{15}N) = 1.223 \times 10^9 \times 6.324^5 \times 4.270\,1/\text{s} = 5.282 \times 10^{13}\,1/\text{s}\,, \quad (6.82)$$
$$T(M1\,;^{15}N) = 1.779 \times 10^{13} \times 6.324^3 \times 1.673\,1/\text{s} = 7.527 \times 10^{15}\,1/\text{s}\,. \quad (6.83)$$

The decay is seen to be dominated by the M1 transition.

Transition probabilities are additive, so the total transition probability is

$$T(^{15}N) = T(E2 + M1) = T(E2) + T(M1) = 7.580 \times 10^{15}\,1/\text{s}\,. \qquad (6.84)$$

This gives the decay half-life

$$t_{1/2}(E2 + M1\,;^{15}N) = \frac{\ln 2}{T(E2 + M1)} = 9.145 \times 10^{-17}\,\text{s} = 0.09\,\text{fs}\,. \qquad (6.85)$$

This calculated value compares very well with the experimental result

$$t^{\text{exp}}_{1/2}(^{15}N\,;\,3/2^- \to 1/2^-) = 0.15\,\text{fs}\,. \qquad (6.86)$$

Since both M1 and E2 occur in the $3/2^- \to 1/2^-$ transition in $^{15}N$, one can calculate and measure their mixing ratio $\Delta(E2/M1)$, defined in (6.19). From (6.76) and (6.78) we have

$$\Delta(E2/M1) = \frac{+1.410\,eb^2}{+2.587\,\mu_N/c} = \frac{4.133\,e\,\text{fm}^2}{2.587\,\mu_N/c} = +0.01597\,\frac{e\,\text{barn}}{\mu_N/c}\,. \qquad (6.87)$$

In terms of the alternative quantity (6.21) we have

$$\delta(E2/M1) = 0.835 \times 6.324 \times 0.01597 = +0.084\,. \qquad (6.88)$$

This small value reflects the M1 dominance noted above. The sign of $\delta$ is important to note because it is measurable.

### 6.2.3 Magnetic Dipole Moments: Schmidt Lines

Equation (6.54) gives the magnetic moment of a single-particle state as

$$\mu_{sp} = \mu_N \frac{1-(-1)^{l+j+\frac{1}{2}}(2j+1)}{4(j+1)}\left\{g_s - g_l\left[2 + (-1)^{l+j+\frac{1}{2}}(2j+1)\right]\right\} . \quad (6.89)$$

We note that an equivalent expression can be derived by starting from the elementary definition of the single-particle magnetic moment operator, stated as (2.71) in Exercise 2.26,

$$\boldsymbol{\mu}_{sp} = \frac{\mu_N}{\hbar}(g_s \boldsymbol{s} + g_l \boldsymbol{l}) = \frac{\mu_N}{\hbar}\left[g_l \boldsymbol{j} + (g_s - g_l)\boldsymbol{s}\right] . \quad (6.90)$$

By means of (2.56) or the Landé formula (2.62) this yields

$$\mu_{sp} = \mu_N \left[g_l j + (g_s - g_l)\frac{j(j+1) - l(l+1) + \frac{3}{4}}{2j+2}\right] . \quad (6.91)$$

Written out explictly as a function of $j$ for $j = l \pm \frac{1}{2}$, either expression, (6.89) or (6.91), becomes a pair of simple equations,

$$\boxed{\begin{aligned} \mu_{sp} &= \mu_N\left[g_l j + \tfrac{1}{2}(g_s - g_l)\right] & \text{for } j = l + \tfrac{1}{2} , \\ \mu_{sp} &= \mu_N\left[g_l j - (g_s - g_l)\frac{j}{2j+2}\right] & \text{for } j = l - \tfrac{1}{2} . \end{aligned}} \quad (6.92)$$

The two equations (6.92) can be plotted as functions of $j$. The plots with the bare $g$ factors (6.25) are called the *Schmidt lines*.

**Table 6.10.** Theoretical and experimental proton and neutron single-particle magnetic dipole moments

| Nucleus | Active orbital | $\mu_{sp}$ ($\mu_N$) | $\mu_{exp}$ ($\mu_N$) |
|---|---|---|---|
| $^{15}$N | $(\pi 0p_{1/2})^{-1}$ | $-0.26$ | $-0.28$ |
| $^{15}$O | $(\nu 0p_{1/2})^{-1}$ | $0.64$ | $0.72$ |
| $^{17}$O | $\nu 0d_{5/2}$ | $-1.91$ | $-1.89$ |
| $^{17}$F | $\pi 0d_{5/2}$ | $4.79$ | $4.72$ |
| $^{39}$K | $(\pi 0d_{3/2})^{-1}$ | $0.12$ | $0.39$ |
| $^{41}$Ca | $\nu 0f_{7/2}$ | $-1.91$ | $-1.59$ |

The theoretical results are from the Schmidt formulas (6.92).

It follows from (6.66) and (6.74) that the theoretical magnetic moment of a one-particle or one-hole nucleus is precisely the single-particle quantity $\mu_{sp}$. The experimental dipole moments[11] of such nuclei are close to the Schmidt

---

[11] magnetic dipole and electric quadrupole moments of nuclei are tabulated in a compilation by Stone [42].

values. This is demonstrated in Table 6.10, which gives a sample of theoretical and experimental magnetic dipole moments of one-particle and one-hole nuclei discussed in Sect. 5.2.

When experimental dipole moments of *all* odd-*A* nuclei, not only those of the one-particle and one-hole type, are placed in a Schmidt diagram, most points are found to lie between the lines. This can be reproduced theoretically by using effective *g* factors, which reflect many-body effects. For compilations of data, see e.g. [9, 16].

## 6.3 Electromagnetic Transitions in Two-Particle and Two-Hole Nuclei

Two-particle and two-hole nuclei were discussed in Sect. 5.3. Consider a *two-proton* or *two-neutron* nucleus. Its initial and final states in an electromagnetic decay process can be written as

$$|\Psi_i\rangle = |a_i\, b_i\,;\, J_i\, M_i\rangle = \mathcal{N}_{a_i b_i}(J_i) \left[ c_{a_i}^\dagger c_{b_i}^\dagger \right]_{J_i M_i} |\text{CORE}\rangle\,, \tag{6.93}$$

$$|\Psi_f\rangle = |a_f\, b_f\,;\, J_f\, M_f\rangle = \mathcal{N}_{a_f b_f}(J_f) \left[ c_{a_f}^\dagger c_{b_f}^\dagger \right]_{J_f M_f} |\text{CORE}\rangle\,, \tag{6.94}$$

where $a_i, b_i$ as well as $a_f, b_f$ are all proton or neutron labels. The normalization factor is given in (5.21). We now derive in detail the amplitude for an electromagnetic transition from (6.93) to (6.94).

Using the Wigner–Eckart theorem (2.27), we can write the reduced one-body transition density as

$$
\begin{aligned}
(\Psi_f \| [c_a^\dagger \tilde{c}_b]_\lambda \| \Psi_i) &= (-1)^{J_f - M_f} \begin{pmatrix} J_f & \lambda & J_i \\ -M_f & \mu & M_i \end{pmatrix}^{-1} \mathcal{N}_{a_f b_f}(J_f) \mathcal{N}_{a_i b_i}(J_i) \\
&\quad \times \langle \text{CORE} | \left[ c_{a_f}^\dagger c_{b_f}^\dagger \right]_{J_f M_f}^\dagger \left[ c_a^\dagger \tilde{c}_b \right]_{\lambda\mu} \left[ c_{a_i}^\dagger c_{b_i}^\dagger \right]_{J_i M_i} | \text{CORE} \rangle \\
&= (-1)^{J_f - M_f} \begin{pmatrix} J_f & \lambda & J_i \\ -M_f & \mu & M_i \end{pmatrix}^{-1} \mathcal{N}_{a_f b_f}(J_f) \mathcal{N}_{a_i b_i}(J_i) \\
&\quad \times \sum_{\substack{m_{\alpha_f} m_{\beta_f} \\ m_\alpha m_\beta \\ m_{\alpha_i} m_{\beta_i}}} (j_{a_f}\, m_{\alpha_f}\, j_{b_f}\, m_{\beta_f} | J_f\, M_f)(j_a\, m_\alpha\, j_b\, m_\beta | \lambda\, \mu) \\
&\quad \times (j_{a_i}\, m_{\alpha_i}\, j_{b_i}\, m_{\beta_i} | J_i\, M_i)\langle \text{CORE} | c_{\beta_f} c_{\alpha_f} c_\alpha^\dagger \tilde{c}_\beta c_{\alpha_i}^\dagger c_{\beta_i}^\dagger | \text{CORE}\rangle\,.
\end{aligned} \tag{6.95}
$$

Performing the contractions in the core expectation value yields

$$
\begin{aligned}
\langle \text{CORE} | c_{\beta_f} c_{\alpha_f} c_\alpha^\dagger \tilde{c}_\beta c_{\alpha_i}^\dagger c_{\beta_i}^\dagger | \text{CORE}\rangle &= (-1)^{j_b + m_\beta}\big( \delta_{\beta_i \beta_f} \delta_{\alpha_f \alpha} \delta_{-\beta \alpha_i} \\
&\quad - \delta_{\beta_f \alpha_i} \delta_{\alpha_f \alpha} \delta_{-\beta \beta_i} + \delta_{\beta_f \alpha} \delta_{\alpha_i \alpha_f} \delta_{-\beta \beta_i} - \delta_{\beta_f \alpha} \delta_{\beta_i \alpha_f} \delta_{-\beta \alpha_i} \big)\,. \tag{6.96}
\end{aligned}
$$

This leads to

$$
(\Psi_f \| [c_a^\dagger \tilde{c}_b]_\lambda \| \Psi_i) = (-1)^{J_f - M_f} \begin{pmatrix} J_f & \lambda & J_i \\ -M_f & \mu & M_i \end{pmatrix}^{-1} \mathcal{N}_{a_f b_f}(J_f) \mathcal{N}_{a_i b_i}(J_i)
$$

$$
\times \Bigg[ \delta_{b_i b_f} \delta_{a a_f} \delta_{b a_i} \sum_{m_{\alpha_f} m_{\beta_f} m_{\alpha_i}} (-1)^{j_{a_i} - m_{\alpha_i}} (j_{a_f} \, m_{\alpha_f} \, j_{b_f} \, m_{\beta_f} | J_f \, M_f)
$$

$$
\times (j_{a_f} \, m_{\alpha_f} \, j_{a_i} \, -m_{\alpha_i} | \lambda \, \mu)(j_{a_i} \, m_{\alpha_i} \, j_{b_f} \, m_{\beta_f} | J_i \, M_i)
$$

$$
+ \text{ three similar terms} \Bigg] . \tag{6.97}
$$

The Clebsch–Gordan coefficients are converted into $3j$ symbols. The three $3j$ symbols can be summed into a $3j$ symbol times a $6j$ symbol. This sum can be obtained from (1.59) and is given in e.g. [2, 7, 22]. The resulting expression reads

$$
(\Psi_f \| [c_a^\dagger \tilde{c}_b]_\lambda \| \Psi_i) = (-1)^{J_f - M_f} \begin{pmatrix} J_f & \lambda & J_i \\ -M_f & \mu & M_i \end{pmatrix}^{-1} \mathcal{N}_{a_f b_f}(J_f) \mathcal{N}_{a_i b_i}(J_i)
$$

$$
\times \Bigg[ \delta_{b_i b_f} \delta_{a a_f} \delta_{b a_i} \widehat{\lambda}\widehat{J_i}\widehat{J_f} (-1)^{j_{a_f} + j_{b_f} - M_i - \mu} \begin{pmatrix} J_f & J_i & \lambda \\ -M_f & M_i & \mu \end{pmatrix} \begin{Bmatrix} J_f & J_i & \lambda \\ j_{a_i} & j_{a_f} & j_{b_f} \end{Bmatrix}
$$

$$
+ \text{ three similar terms} \Bigg] . \tag{6.98}
$$

When this reduced transition density is substituted into (6.22) we finally obtain the reduced matrix element for two-proton and two-neutron nuclei:

$$
\boxed{\begin{aligned}
(a_f b_f \, ; \, J_f \| \boldsymbol{\mathcal{M}}_{\sigma\lambda} \| a_i \, b_i \, ; \, J_i) &= \widehat{J_i}\widehat{J_f} \mathcal{N}_{a_i b_i}(J_i) \mathcal{N}_{a_f b_f}(J_f) \\
&\times \Bigg[ \delta_{b_i b_f} (-1)^{j_{a_f} + j_{b_f} + J_i + \lambda} \begin{Bmatrix} J_f & J_i & \lambda \\ j_{a_i} & j_{a_f} & j_{b_f} \end{Bmatrix} (a_f \| \boldsymbol{\mathcal{M}}_{\sigma\lambda} \| a_i) \\
&+ \delta_{a_i b_f} (-1)^{j_{a_f} + j_{b_i} + \lambda} \begin{Bmatrix} J_f & J_i & \lambda \\ j_{b_i} & j_{a_f} & j_{b_f} \end{Bmatrix} (a_f \| \boldsymbol{\mathcal{M}}_{\sigma\lambda} \| b_i) \\
&+ \delta_{a_i a_f} (-1)^{j_{a_i} + j_{b_i} + J_f + \lambda} \begin{Bmatrix} J_f & J_i & \lambda \\ j_{b_i} & j_{b_f} & j_{a_f} \end{Bmatrix} (b_f \| \boldsymbol{\mathcal{M}}_{\sigma\lambda} \| b_i) \\
&+ \delta_{b_i a_f} (-1)^{J_i + J_f + \lambda + 1} \begin{Bmatrix} J_f & J_i & \lambda \\ j_{a_i} & j_{b_f} & j_{a_f} \end{Bmatrix} (b_f \| \boldsymbol{\mathcal{M}}_{\sigma\lambda} \| a_i) \Bigg] .
\end{aligned}} \tag{6.99}
$$

The operators $\boldsymbol{\mathcal{M}}_{\sigma\lambda}$ are $\boldsymbol{\mathcal{M}}_{\mathrm{E}\lambda} = \boldsymbol{Q}_\lambda$ and $\boldsymbol{\mathcal{M}}_{\mathrm{M}\lambda} = \boldsymbol{M}_\lambda$, and their single-particle matrix elements are given by (6.23) and (6.24) respectively.

Consider next a *proton–neutron* nucleus. The initial and final states are

$$
|p_i \, n_i \, ; \, J_i \, M_i\rangle = [c_{p_i}^\dagger c_{n_i}^\dagger]_{J_i M_i} |\mathrm{CORE}\rangle , \tag{6.100}
$$

$$
|p_f \, n_f \, ; \, J_f \, M_f\rangle = [c_{p_f}^\dagger c_{n_f}^\dagger]_{J_f M_f} |\mathrm{CORE}\rangle . \tag{6.101}
$$

The decay amplitude for these states is obtained directly from (6.99) by setting $a_i = p_i$, $b_i = n_i$, $a_f = p_f$, $b_f = n_f$ and recognizing that $\delta_{pn} = 0$. Thus, with only the first and third terms contributing, the result is

$$
\begin{aligned}
(p_f\, n_f\, &;\, J_f \|\boldsymbol{\mathcal{M}}_{\sigma\lambda}\| p_i\, n_i\, ;\, J_i) \\
&= \widehat{J_i}\widehat{J_f}\left[ \delta_{n_i n_f}(-1)^{j_{p_f}+j_{n_f}+J_i+\lambda}
\begin{Bmatrix} J_f & J_i & \lambda \\ j_{p_i} & j_{p_f} & j_{n_f} \end{Bmatrix}
(p_f\|\boldsymbol{\mathcal{M}}_{\sigma\lambda}\|p_i) \right. \\
&\quad \left. + \delta_{p_i p_f}(-1)^{j_{p_i}+j_{n_i}+J_f+\lambda}
\begin{Bmatrix} J_f & J_i & \lambda \\ j_{n_i} & j_{n_f} & j_{p_f} \end{Bmatrix}
(n_f\|\boldsymbol{\mathcal{M}}_{\sigma\lambda}\|n_i) \right] .
\end{aligned}
\tag{6.102}
$$

A simple special case of the proton–proton or neutron–neutron formula (6.99) occurs when $J_f^\pi = 0^+$; this requires that $a_f = b_f$. The decay amplitude becomes

$$
\begin{aligned}
(a_f\, a_f\, ;\, 0^+ \|\boldsymbol{\mathcal{M}}_{\sigma\lambda}\| a_i\, b_i\, ;\, J_i) &= \sqrt{2}\, \delta_{\lambda J_i} \widehat{j}_{a_f}^{\,-1} \mathcal{N}_{a_i b_i}(J_i) \\
&\times \left[ \delta_{a_i a_f}(a_f\|\boldsymbol{\mathcal{M}}_{\sigma\lambda}\|b_i) - \delta_{b_i a_f}(-1)^{j_{a_i}+j_{a_f}+J_i}(a_f\|\boldsymbol{\mathcal{M}}_{\sigma\lambda}\|a_i) \right] .
\end{aligned}
\tag{6.103}
$$

A similar special case occurs for proton–neutron nuclei when $J_f = 0$ in (6.102):

$$
\begin{aligned}
(p_f\, n_f\, ;\, 0 \|\boldsymbol{\mathcal{M}}_{\sigma\lambda}\| p_i\, n_i\, ;\, J_i) &= \delta_{\lambda J_i} \delta_{j_{p_f} j_{n_f}}(-1)^{j_{p_i}+j_{p_f}+1} \widehat{j}_{p_f}^{\,-1} \\
&\times \left[ \delta_{n_i n_f}(-1)^{J_i}(p_f\|\boldsymbol{\mathcal{M}}_{\sigma\lambda}\|p_i) + \delta_{p_i p_f}(n_f\|\boldsymbol{\mathcal{M}}_{\sigma\lambda}\|n_i) \right] .
\end{aligned}
\tag{6.104}
$$

For *two-hole* nuclei one has to replace $c^\dagger$ with $h^\dagger$ in the wave functions (6.93), (6.94), (6.100) and (6.101). Calculations similar to those above give the results

$$
\begin{aligned}
(a_f^{-1}\, b_f^{-1}\, ;\, J_f \|\boldsymbol{\mathcal{M}}_{\sigma\lambda}\| a_i^{-1}\, b_i^{-1}\, ;\, J_i) &= (-1)^{\lambda+1}(a_f\, b_f\, ;\, J_f \|\boldsymbol{\mathcal{M}}_{\sigma\lambda}\| a_i\, b_i\, ;\, J_i) , \\
(p_f^{-1}\, n_f^{-1}\, ;\, J_f \|\boldsymbol{\mathcal{M}}_{\sigma\lambda}\| p_i^{-1}\, n_i^{-1}\, ;\, J_i) &= (-1)^{\lambda+1}(p_f\, n_f\, ;\, J_f \|\boldsymbol{\mathcal{M}}_{\sigma\lambda}\| p_i\, n_i\, ;\, J_i) .
\end{aligned}
\tag{6.105}
$$

It follows that a two-hole nucleus has the same $B(\sigma\lambda)$ value as the corresponding two-particle nucleus. This extends the principle established for one-hole and one-particle nuclei.

### 6.3.1 Example: Transitions in Two-Particle Nuclei $^{18}$O and $^{18}$Ne

Consider the electromagnetic decay of the first excited $2^+$ state in the two-particle nuclei $^{18}_{8}$O$_{10}$ and $^{18}_{10}$Ne$_8$. Their spectra are shown in Fig. 5.5. The $2^+_1 \to 0^+_{\text{gs}}$ decay is a pure E2 transition. The wave functions of the relevant states are, from (5.30) and (5.32),

$$
|^{18}\text{O}\, ;\, 0^+, 2^+\rangle = \frac{1}{\sqrt{2}} [c^\dagger_{\nu 0d_{5/2}} c^\dagger_{\nu 0d_{5/2}}]_{0^+,2^+} |\text{CORE}\rangle ,
\tag{6.106}
$$

$$
|^{18}\text{Ne}\, ;\, 0^+, 2^+\rangle = \frac{1}{\sqrt{2}} [c^\dagger_{\pi 0d_{5/2}} c^\dagger_{\pi 0d_{5/2}}]_{0^+,2^+} |\text{CORE}\rangle .
\tag{6.107}
$$

Because the final state is $J^\pi = 0^+$, we use (6.103) to compute the E2 decay amplitude. Equations (3.43) and (3.45) give an oscillator length of $b = 1.750\,\mathrm{fm}$. We adopt the effective charges (6.26). Substitution into (6.103) and use of Table 6.4 give

$$(0\mathrm{d}_{5/2}\,0\mathrm{d}_{5/2}\,;\,0^+ \| \boldsymbol{Q}_2 \| 0\mathrm{d}_{5/2}\,0\mathrm{d}_{5/2}\,;\,2^+)$$

$$= \sqrt{\frac{2}{6}}\,\frac{1}{\sqrt{2}}\left[(0\mathrm{d}_{5/2}\|\boldsymbol{Q}_2\|0\mathrm{d}_{5/2}) - (-1)^7(0\mathrm{d}_{5/2}\|\boldsymbol{Q}_2\|0\mathrm{d}_{5/2})\right]$$

$$= \sqrt{\tfrac{2}{3}}(0\mathrm{d}_{5/2}\|\boldsymbol{Q}_2\|0\mathrm{d}_{5/2}) = \sqrt{\tfrac{2}{3}}(-2.585)e_{\mathrm{eff}}b^2 = -6.464e_{\mathrm{eff}}\,\mathrm{fm}^2\ . \quad (6.108)$$

The reduced transition probabilities are

$$B(\mathrm{E2}\,;\,{}^{18}\mathrm{O}) = \frac{1}{5}(-6.464\chi e\,\mathrm{fm}^2)^2 = 8.357\chi^2 e^2 \mathrm{fm}^4 = 2.982\chi^2\ \mathrm{W.u.}\ , \quad (6.109)$$

$$B(\mathrm{E2}\,;\,{}^{18}\mathrm{Ne}) = \frac{1}{5}[-6.464(1+\chi)e\,\mathrm{fm}^2]^2 = 8.357(1+\chi)^2 e^2 \mathrm{fm}^4$$

$$= 2.982(1+\chi)^2\ \mathrm{W.u.}\ , \quad (6.110)$$

where we have used the Weisskopf unit from Table 6.8.

The experimental decay half-lives are $2.0\,\mathrm{ps}$ for ${}^{18}\mathrm{O}$, and $0.46\,\mathrm{ps}$ for ${}^{18}\mathrm{Ne}$. Using the formula in Table 6.8 with the energies from Fig. 5.5, we find

$$B(\mathrm{E2}\,;\,{}^{18}\mathrm{O})_{\mathrm{exp}} = 9.3\,e^2 \mathrm{fm}^4\ , \quad (6.111)$$

$$B(\mathrm{E2}\,;\,{}^{18}\mathrm{Ne})_{\mathrm{exp}} = 51.5\,e^2 \mathrm{fm}^4\ . \quad (6.112)$$

We deduce the following polarization constants $\chi$:

$$\chi = 1.1 \quad \text{for } {}^{18}\mathrm{O}\ , \quad (6.113)$$

$$\chi = 1.5 \quad \text{for } {}^{18}\mathrm{Ne}\ . \quad (6.114)$$

Two conclusions emerge. The first one is that the concept of effective charge is meaningful because the two nuclei are fitted with similar values of $\chi$; to a fair approximation we have for both nuclei

$$\chi \approx 1.3\ , \quad e_{\mathrm{eff}}^{\mathrm{p}} \approx 2.3e\ , \quad e_{\mathrm{eff}}^{\mathrm{n}} \approx 1.3e\ . \quad (6.115)$$

The second conclusion is that the present value of $\chi$ is very large, as one would expect it to be only a minor correction, $\chi \ll 1$. This means that configuration mixing in the 0d-1s shell is a significant effect. The present example is recalculated with configuration mixing in Subsect. 8.5.3.

## 6.4 Electromagnetic Transitions in Particle–Hole Nuclei

As discussed in Chap. 5, the particle–hole vacuum is the mean-field ground state of a doubly magic nucleus. On this ground state one can build excitations by lifting a nucleon across the Fermi surface. In a doubly magic nucleus excited states can be created through *charge-concerving* particle–hole excitations, which are proton-particle–proton-hole and neutron-particle–neutron-hole excitations. On the other hand, the ground and excited states of an odd–odd nucleus at a doubly magic shell closure are obtained through *charge-changing* particle–hole excitations of the particle–hole vacuum, namely proton-particle–neutron-hole and neutron-particle–proton-hole excitations. In the first kind a proton jumps from below the proton Fermi surface ending up a neutron above the neutron Fermi surface, and vice versa for the second kind. All these modes of particle–hole excitation are discussed below in the context of electromagnetic decay.

### 6.4.1 Transitions Involving Charge-Conserving Particle–Hole Excitations

Charge-conserving particle–hole excitations are excitations of even–even doubly magic nuclei. Electromagnetic transitions can operate either from a particle–hole excitation to the particle–hole vacuum or between two particle–hole excitations. In the first case the transition is to the ground state and in the second it is between two excited states.

### Decays to Particle–hole Vacuum

In the case of decays to the particle–hole vacuum the initial state is a charge-conserving particle–hole excitation, either of the proton or neutron type. Its wave function is

$$|a_i\, b_i^{-1}\,;\, J_i\, M_i\rangle = \left[c_{a_i}^\dagger h_{b_i}^\dagger\right]_{J_i M_i}|\mathrm{HF}\rangle\,, \qquad a_i, b_i \text{ are both } \pi \text{ or } \nu \text{ labels}\,. \quad (6.116)$$

The final state is the particle–hole vacuum $|\mathrm{HF}\rangle$. We use (4.46), (4.47), (4.53) and properties of the Clebsch–Gordan coefficients to calculate the transition density:

$$\langle \mathrm{HF}|\left[c_a^\dagger \tilde{c}_b\right]_{\lambda\mu}|a_i\, b_i^{-1}\,;\, J_i\, M_i\rangle$$

$$= \sum_{\substack{m_\alpha m_\beta \\ m_{\alpha_i} m_{\beta_i}}} (j_a\, m_\alpha\, j_b\, m_\beta|\lambda\,\mu)(j_{a_i}\, m_{\alpha_i}\, j_{b_i}\, m_{\beta_i}|J_i\, M_i)$$

$$\times (-1)^{j_b+m_\beta}\langle \mathrm{HF}|c_a^\dagger c_{-\beta} c_{\alpha_i}^\dagger h_{\beta_i}^\dagger|\mathrm{HF}\rangle$$

$$= \delta_{ab_i}\delta_{ba_i}\sum_{m_{\alpha_i} m_{\beta_i}} (-1)^{j_{a_i}-m_{\alpha_i}}(-1)^{j_{b_i}+m_{\beta_i}}(j_{b_i}\, -m_{\beta_i}\, j_{a_i}\, -m_{\alpha_i}|\lambda\,\mu)$$

$$\times (j_{a_i}\, m_{\alpha_i}\, j_{b_i}\, m_{\beta_i}|J_i\, M_i) = \delta_{ab_i}\delta_{ba_i}(-1)^{j_{a_i}-j_{b_i}+M_i}\delta_{\lambda J_i}\delta_{-\mu M_i}\,. \quad (6.117)$$

We put $\lambda = J_i$ and $\mu = -M_i$ in this equation and apply the Wigner–Eckart theorem (2.27) to the left-hand side. Inserting the simple $3j$ symbol (1.42) gives the reduced transition density immediately. However, we want to display the multipole degree $\lambda$. Its only value allowed by angular momentum conservation is $\lambda = J_i$, and we indicate this by including $\delta_{\lambda J_i}$ in the result:

$$(\text{HF}\| \left[c_a^\dagger \tilde{c}_b\right]_\lambda \| a_i \, b_i^{-1} \, ; \, J_i) = \delta_{\lambda J_i} \delta_{ab_i} \delta_{ba_i} (-1)^{j_{a_i} - j_{b_i} + J_i} \widehat{J}_i \, . \qquad (6.118)$$

Equation (6.22) gives finally for the decay amplitude

$$\boxed{(\text{HF}\| \mathcal{M}_{\sigma\lambda} \| a_i \, b_i^{-1} \, ; \, J_i) = \delta_{\lambda J_i} (-1)^{J_i} (a_i \| \mathcal{M}_{\sigma\lambda} \| b_i) \, ,} \qquad (6.119)$$

where we have used the symmetry relations (6.27) and (6.28) applicable to the CS phase convention. The single-particle matrix element needed on the right-hand side is given by (6.23) or (6.24).

### Transitions between Two Particle–hole States

Consider next electromagnetic transitions between two charge-conserving particle–hole states. These are always excited states of even–even doubly magic nuclei. The initial and final states are

$$|a_i \, b_i^{-1} \, ; \, J_i \, M_i\rangle = \left[c_{a_i}^\dagger h_{b_i}^\dagger\right]_{J_i M_i} |\text{HF}\rangle \, , \quad a_i, b_i \text{ are both } \pi \text{ or } \nu \text{ labels} \, , \tag{6.120}$$

$$|a_f \, b_f^{-1} \, ; \, J_f \, M_f\rangle = \left[c_{a_f}^\dagger h_{b_f}^\dagger\right]_{J_f M_f} |\text{HF}\rangle \, , \quad a_f, b_f \text{ are both } \pi \text{ or } \nu \text{ labels} \, . \tag{6.121}$$

Similarly to (6.95) in the treatment of two-particle nuclei we now have

$$(a_f \, b_f^{-1} \, ; \, J_f \| \left[c_a^\dagger \tilde{c}_b\right]_\lambda \| a_i \, b_i^{-1} \, ; \, J_i) = (-1)^{J_f - M_f} \begin{pmatrix} J_f & \lambda & J_i \\ -M_f & \mu & M_i \end{pmatrix}^{-1}$$

$$\times \sum_{\substack{m_{\alpha_f} m_{\beta_f} \\ m_\alpha m_\beta \\ m_{\alpha_i} m_{\beta_i}}} (j_{a_f} \, m_{\alpha_f} \, j_{b_f} \, m_{\beta_f} | J_f \, M_f)(j_a \, m_\alpha \, j_b \, m_\beta | \lambda \, \mu)(j_{a_i} \, m_{\alpha_i} \, j_{b_i} \, m_{\beta_i} | J_i \, M_i)$$

$$\times (-1)^{j_b + m_\beta} \langle \text{HF} | h_{\beta_f} c_{\alpha_f} c_\alpha^\dagger c_{-\beta} c_{\alpha_i}^\dagger h_{\beta_i}^\dagger | \text{HF} \rangle \, . \tag{6.122}$$

The contractions are done using (4.46), (4.47), (4.52) and (4.53), with the result

$$\langle \text{HF} | h_{\beta_f} c_{\alpha_f} c_\alpha^\dagger c_{-\beta} c_{\alpha_i}^\dagger h_{\beta_i}^\dagger | \text{HF} \rangle = \delta_{\beta_f \beta_i} \delta_{\alpha_f \alpha_i} \delta_{\alpha, -\beta} + \delta_{\beta_f \beta_i} \delta_{\alpha_f \alpha} \delta_{-\beta \alpha_i}$$

$$- (-1)^{j_{b_f} + m_{\beta_f} + j_{b_i} + m_{\beta_i}} \delta_{\beta_f \beta} \delta_{\alpha_f \alpha_i} \delta_{\alpha, -\beta_i} \, . \tag{6.123}$$

When we use (6.71), sum over $m_\alpha$ and $m_\beta$, and use (1.26), we obtain a factor $\delta_{\lambda 0} \delta_{\mu 0}$. This term is omitted because there is no $\lambda = 0$ multipole. Completing the derivation similarly to that of (6.99), we finally obtain

$$\begin{aligned}
(a_f\, b_f^{-1}\,;\, J_f \| \boldsymbol{\mathcal{M}}_{\sigma\lambda} \| a_i\, b_i^{-1}\,;\, J_i) &= (-1)^{j_{a_i}+j_{b_f}}\, \widehat{J_i}\widehat{J_f} \\
&\times \Bigg[ \delta_{b_i b_f}(-1)^{J_i+\lambda} \begin{Bmatrix} J_i & J_f & \lambda \\ j_{a_f} & j_{a_i} & j_{b_i} \end{Bmatrix} (a_i \| \boldsymbol{\mathcal{M}}_{\sigma\lambda} \| a_f) \\
&\quad + \delta_{a_i a_f}(-1)^{J_f+1} \begin{Bmatrix} J_i & J_f & \lambda \\ j_{b_f} & j_{b_i} & j_{a_i} \end{Bmatrix} (b_i \| \boldsymbol{\mathcal{M}}_{\sigma\lambda} \| b_f) \Bigg] ,
\end{aligned}$$

$$a_i, a_f, b_i, b_f \text{ are all either } \pi \text{ or } \nu \text{ labels.}$$

(6.124)

For the special case $J_f = 0$ we obtain the simplified expression

$$\begin{aligned}
(a_f\, b_f^{-1}\,;\, 0 \| \boldsymbol{\mathcal{M}}_{\sigma\lambda} \| a_i\, b_i^{-1}\,;\, J_i) &= \delta_{\lambda J_i}(-1)^{j_{b_i}+j_{b_f}+\lambda} \\
&\times \Big[ \delta_{b_i b_f}\delta_{j_{b_i} j_{a_f}}(-1)^{J_i+\lambda+1}\widehat{j_{a_f}}^{\,-1}(a_i \| \boldsymbol{\mathcal{M}}_{\sigma\lambda} \| a_f) \\
&\quad + \delta_{a_i a_f}\delta_{j_{a_i} j_{b_f}}\widehat{j_{b_f}}^{\,-1}(b_i \| \boldsymbol{\mathcal{M}}_{\sigma\lambda} \| b_f) \Big] .
\end{aligned}$$

(6.125)

## 6.4.2 Example: Doubly Magic Nucleus $^{16}$O

The doubly magic nucleus $^{16}_{8}\mathrm{O}_8$ provides an example of the application of the preceding formalism. The spectrum is shown in Fig. 5.10. Let us study the decays of the two lowest-lying $T = 0$ states, namely $3_1^-$ and $1_1^-$. The leading decay modes, from Table 6.1, are depicted in Fig. 6.1. The ground state is the particle–hole vacuum $|\mathrm{HF}\rangle$. The $3_1^-$ and $1_1^-$ states are given by (5.65) and (5.66), with the sign determined by (5.126), as

$$|^{16}\mathrm{O}\,;\, 3_1^-\rangle = \frac{1}{\sqrt{2}}\Big( \big[ c_{\pi 0d_{5/2}}^\dagger h_{\pi 0p_{1/2}}^\dagger \big]_{3^-} |\mathrm{HF}\rangle + \big[ c_{\nu 0d_{5/2}}^\dagger h_{\nu 0p_{1/2}}^\dagger \big]_{3^-} |\mathrm{HF}\rangle \Big) ,$$

(6.126)

$$|^{16}\mathrm{O}\,;\, 1_1^-\rangle = \frac{1}{\sqrt{2}}\Big( \big[ c_{\pi 1s_{1/2}}^\dagger h_{\pi 0p_{1/2}}^\dagger \big]_{1^-} |\mathrm{HF}\rangle + \big[ c_{\nu 1s_{1/2}}^\dagger h_{\nu 0p_{1/2}}^\dagger \big]_{1^-} |\mathrm{HF}\rangle \Big) .$$

(6.127)

For the decays to the ground state the single-particle matrix elements in (6.119) are

$$(0d_{5/2} \| \boldsymbol{Q}_3 \| 0p_{1/2}) = -3.824\,e_{\mathrm{eff}}b^3 , \quad (1s_{1/2} \| \boldsymbol{Q}_1 \| 0p_{1/2}) = 0.399\,e_{\mathrm{eff}}b \quad (6.128)$$

with the numerical values taken from Tables 6.3 and 6.5. This leads to

$$(\mathrm{HF} \| \boldsymbol{Q}_3 \| 3_1^-) = \frac{1}{\sqrt{2}}(-1)^3(-3.824)(e_{\mathrm{eff}}^{\mathrm{p}} + e_{\mathrm{eff}}^{\mathrm{n}})b^3 = 13.88(1 + 2\chi)e\,\mathrm{fm}^3 ,$$

(6.129)

$$(\mathrm{HF} \| \boldsymbol{Q}_1 \| 1_1^-) = \frac{1}{\sqrt{2}}(-1)^1(0.399)(e_{\mathrm{eff}}^{\mathrm{p}} + e_{\mathrm{eff}}^{\mathrm{n}})b = -0.487(1 + 2\chi)e\,\mathrm{fm} ,$$

(6.130)

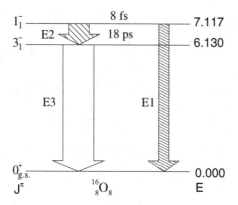

**Fig. 6.1.** The lowest $3^-$ and $1^-$ excited states and their decays in the nucleus $^{16}$O. The energies and decay half-lives are experimental data

where $b = 1.725\,\text{fm}$ from (3.43) and (3.45). The reduced transition probabilities become

$$B(\text{E3}) = 27.52(1 + 2\chi)^2\, e^2\text{fm}^6 , \tag{6.131}$$

$$B(\text{E1}) = 0.079(1 + 2\chi)^2\, e^2\text{fm}^2 . \tag{6.132}$$

For the E2 transition $1^-_1 \to 3^-_1$, which is between two excited states, (6.124) gives

$$\left(0\text{d}_{5/2}\,(0\text{p}_{1/2})^{-1} ; 3^- \|Q_2\| 1\text{s}_{1/2}\,(0\text{p}_{1/2})^{-1} ; 1^-\right)$$

$$= (-1)^{\frac{1}{2}+\frac{1}{2}}\sqrt{3}\sqrt{7}\left[(-1)^{1+2}\left\{\begin{matrix} 1 & 3 & 2 \\ \frac{5}{2} & \frac{1}{2} & \frac{1}{2} \end{matrix}\right\}(1\text{s}_{1/2}\|Q_2\|0\text{d}_{5/2}) + 0\right]$$

$$= \sqrt{21} \times \frac{1}{3\sqrt{2}}(-2.185 e_{\text{eff}}b^2) = -2.360 e_{\text{eff}}b^2 = -7.022 e_{\text{eff}}\,\text{fm}^2 , \tag{6.133}$$

where we have used Table 6.4 and (1.76). This leads to the reduced transition probability

$$B(\text{E2}; 1^-_1 \to 3^-_1) = \frac{1}{3}\left(\frac{1}{\sqrt{2}} \times \frac{1}{\sqrt{2}}\right)^2 (e^{\text{p}}_{\text{eff}} + e^{\text{n}}_{\text{eff}})^2 (-7.022)^2\,\text{fm}^4$$

$$= 4.109(1 + 2\chi)^2\, e^2\text{fm}^4 . \tag{6.134}$$

Figure 6.1 contains the experimental half-lives and energies of the states under study. Using (6.1) and Table 6.8 we find

$$B(\text{E3}; 3^-_1 \to 0^+_{\text{gs}})_{\text{exp}} = 208\, e^2\text{fm}^6 . \tag{6.135}$$

Comparing this with the theoretical result (6.131) gives

$$\chi = 0.87 \tag{6.136}$$

for the polarization constant. With this value of $\chi$ the other theoretical reduced transition probabilities become

$$B(E2) = 30.85\,e^2\text{fm}^4\,, \tag{6.137}$$

$$B(E1) = 0.593\,e^2\text{fm}^2\,, \tag{6.138}$$

which lead to the decay probabilities

$$T(E2) = 1.223 \times 10^9 \times 0.987^5 \times 30.85\,1/\text{s} = 3.53 \times 10^{10}\,1/\text{s}\,, \tag{6.139}$$

$$T(E1) = 1.587 \times 10^{15} \times 7.117^3 \times 0.593\,1/\text{s} = 3.39 \times 10^{17}\,1/\text{s}\,. \tag{6.140}$$

The E2 transition probability is negligible in comparison with the E1 transition probability, so the half-life of the $1_1^-$ state becomes

$$t_{1/2} = t_{1/2}(E1) = \frac{\ln 2}{T(E1)} \approx 2 \times 10^{-3}\,\text{fs}\,. \tag{6.141}$$

This is much too short compared with the experimental value of 8 fs. In this case our simplified approach is not sufficient to explain the very retarded experimental E1 transition.

The large $\chi$ value deduced from the E3 transition indicates that the transition is collective, i.e. many particle–hole pairs contribute to it. This can be described by means of the Tamm–Dancoff approximation (TDA) or random phase approximation (RPA), to be introduced later.

## E1 Transitions and Spurious Centre-of-mass Motion

The wide discrepancy between the calculated and experimental E1 decay rates in our $^{16}$O example arises from a kinematic source. The harmonic oscillator single-particle wave functions are not defined relative to the centre of mass of the nucleus, as they should be, but rather from the origin of a fixed external coordinate system. This gives rise to unphysical, or *spurious*, centre-of-mass contributions to computed nuclear observables. The consequences are particularly serious for the electric dipole operator. The simplest recipe to remove the spurious centre-of-mass contributions, on the average, is to adopt the effective charges

$$e_{\text{eff}}^{\text{p}} = \frac{N}{A}e\,, \quad e_{\text{eff}}^{\text{n}} = -\frac{Z}{A}e \tag{6.142}$$

for the E1 operator, as described in e.g. [16].

In our present example of a doubly magic nucleus the effective charges (6.142) would lead to a complete vanishing of the E1 decay probability of the lowest isoscalar $1^-$ excitation. This happens because the decay amplitude (6.130) is proportional to the sum of the proton and neutron effective charges and this sum is zero for $N = Z$.

One could think that the decay rate is determined by the E2 mode (6.134). However, (6.139) gives a half-life of 20 ps, which is far in excess of the experimental value of 8 fs. The E2 mode is thus no explanation, and the experimental decay rate must be understood in terms of E1. It has been concluded that incompleteness of isospin symmetry enables such E1 transitions to proceed [9].

This isospin breaking stems from the Coulomb energy between the protons. The Coulomb effect can be simulated by choosing slightly different relative single-particle energies for protons and neutrons.

This matter will be elaborated in Subsect. 9.4.6, where configuration mixing of particle–hole excitations is discussed in the Tamm–Dancoff approximation.

### 6.4.3 Transitions Between Charge-Changing Particle–Hole Excitations

Charge-changing particle–hole excitations of the particle–hole vacuum, either proton-to-neutron or neutron-to-proton, create states in odd–odd nuclei at doubly magic shells. The initial and final wave functions are

$$|p_i\,n_i^{-1}\,;\,J_i\,M_i\rangle = \left[c_{p_i}^\dagger h_{n_i}^\dagger\right]_{J_i M_i}|\mathrm{HF}\rangle\,, \tag{6.143}$$

$$|p_f\,n_f^{-1}\,;\,J_f\,M_f\rangle = \left[c_{p_f}^\dagger h_{n_f}^\dagger\right]_{J_f M_f}|\mathrm{HF}\rangle \tag{6.144}$$

for the proton-particle–neutron-hole states and

$$|n_i\,p_i^{-1}\,;\,J_i\,M_i\rangle = \left[c_{n_i}^\dagger h_{p_i}^\dagger\right]_{J_i M_i}|\mathrm{HF}\rangle\,, \tag{6.145}$$

$$|n_f\,p_f^{-1}\,;\,J_f\,M_f\rangle = \left[c_{n_f}^\dagger h_{p_f}^\dagger\right]_{J_f M_f}|\mathrm{HF}\rangle \tag{6.146}$$

for the neutron-particle–proton-hole states.

By methods similar to those used in the previous sections one can derive the following expressions for electromagnetic transition amplitudes involving charge-changing particle–hole excitations:

$$
\begin{aligned}
(p_f\,n_f^{-1}\,;\,J_f\|\boldsymbol{\mathcal{M}}_{\sigma\lambda}\|p_i\,n_i^{-1}\,;\,J_i) &= (-1)^{j_{p_i}+j_{n_f}}\,\widehat{J_i}\widehat{J_f} \\
&\times \Bigg[\delta_{n_i n_f}(-1)^{J_i+\lambda}\left\{\begin{matrix} J_i & J_f & \lambda \\ j_{p_f} & j_{p_i} & j_{n_i}\end{matrix}\right\}(p_i\|\boldsymbol{\mathcal{M}}_{\sigma\lambda}\|p_f) \\
&+ \delta_{p_i p_f}(-1)^{J_f+1}\left\{\begin{matrix} J_i & J_f & \lambda \\ j_{n_f} & j_{n_i} & j_{p_i}\end{matrix}\right\}(n_i\|\boldsymbol{\mathcal{M}}_{\sigma\lambda}\|n_f)\Bigg]
\end{aligned}
\tag{6.147}
$$

for proton-particle–neutron-hole transitions and

$$
\begin{aligned}
(n_f\,p_f^{-1}\,;\,J_f\|\boldsymbol{\mathcal{M}}_{\sigma\lambda}\|n_i\,p_i^{-1}\,;\,J_i) &= (-1)^{j_{n_i}+j_{p_f}}\,\widehat{J_i}\widehat{J_f} \\
&\times \Bigg[\delta_{p_i p_f}(-1)^{J_i+\lambda}\left\{\begin{matrix} J_i & J_f & \lambda \\ j_{n_f} & j_{n_i} & j_{p_i}\end{matrix}\right\}(n_i\|\boldsymbol{\mathcal{M}}_{\sigma\lambda}\|n_f) \\
&+ \delta_{n_i n_f}(-1)^{J_f+1}\left\{\begin{matrix} J_i & J_f & \lambda \\ j_{p_f} & j_{p_i} & j_{n_i}\end{matrix}\right\}(p_i\|\boldsymbol{\mathcal{M}}_{\sigma\lambda}\|p_f)\Bigg]
\end{aligned}
\tag{6.148}
$$

for neutron-particle–proton-hole transitions.

Equations (6.147) and (6.148) can be further simplified when the final state is $0^+$:

$$(p_f\, n_f^{-1}\,;\, 0\|\mathcal{M}_{\sigma\lambda}\|p_i\, n_i^{-1}\,;\, J_i) = \delta_{\lambda J_i}\delta_{j_{n_f}j_{p_f}}(-1)^{j_{n_i}+j_{n_f}+\lambda}\,\widehat{j_{p_f}}^{\,-1}$$
$$\times\,[\delta_{p_i p_f}(n_i\|\mathcal{M}_{\sigma\lambda}\|n_f) - \delta_{n_i n_f}(p_i\|\mathcal{M}_{\sigma\lambda}\|p_f)] \quad (6.149)$$

for proton-particle–neutron-hole transitions and

$$(n_f\, p_f^{-1}\,;\, 0\|\mathcal{M}_{\sigma\lambda}\|n_i\, p_i^{-1}\,;\, J_i) = \delta_{\lambda J_i}\delta_{j_{n_f}j_{p_f}}(-1)^{j_{p_i}+j_{p_f}+\lambda}\,\widehat{j_{p_f}}^{\,-1}$$
$$\times\,[\delta_{n_i n_f}(p_i\|\mathcal{M}_{\sigma\lambda}\|p_f) - \delta_{p_i p_f}(n_i\|\mathcal{M}_{\sigma\lambda}\|n_f)] \quad (6.150)$$

for neutron-particle–proton-hole transitions.

### 6.4.4 Example: Odd–Odd Nucleus $^{16}$N

The energy levels of the odd–odd nucleus $^{16}_{7}$N$_9$ are shown in Fig. 5.10. Consider the decay of the first excited state $0^-$ to the $2^-$ ground state. These states are, from (5.60) and (5.61),

$$|^{16}\mathrm{N}\,;\, 2^-_{\mathrm{gs}}\rangle = \left[c^\dagger_{\nu 0\mathrm{d}_{5/2}}h^\dagger_{\pi 0\mathrm{p}_{1/2}}\right]_{2^-}|\mathrm{HF}\rangle\,, \qquad (6.151)$$

$$|^{16}\mathrm{N}\,;\, 0^-_1\rangle = \left[c^\dagger_{\nu 1\mathrm{s}_{1/2}}h^\dagger_{\pi 0\mathrm{p}_{1/2}}\right]_{0^-}|\mathrm{HF}\rangle\,. \qquad (6.152)$$

We first evaluate the E2 transition amplitude $2^-_{\mathrm{gs}} \rightarrow 0^-_1$ using (6.150). The result is

$$\left(\nu 1\mathrm{s}_{1/2}\,(\pi 0\mathrm{p}_{1/2})^{-1}\,;\, 0^-\|Q_2\|\nu 0\mathrm{d}_{5/2}\,(\pi 0\mathrm{p}_{1/2})^{-1}\,;\, 2^-\right)$$
$$= (-1)^{\frac{1}{2}+\frac{1}{2}+2}\frac{1}{\sqrt{2}}[0 - (\nu 0\mathrm{d}_{5/2}\|Q_2\|\nu 1\mathrm{s}_{1/2})]$$
$$= \frac{1}{\sqrt{2}}(\nu 0\mathrm{d}_{5/2}\|Q_2\|\nu 1\mathrm{s}_{1/2}) = \frac{1}{\sqrt{2}}(-2.185 e^{\mathrm{n}}_{\mathrm{eff}}b^2) \qquad (6.153)$$

with the numerical value taken from Table 6.4 and the neutron effective charge inserted.

The reduced matrix element (6.153) is for the transition $2^- \rightarrow 0^-$, yet the transition concerned is $0^- \rightarrow 2^-$. By the symmetry relation (2.32) the matrix elements are the same, so (6.153) can be directly substituted into (6.4). We have $b = 1.725\,\mathrm{fm}$ for $A = 16$ from Subsect. 6.4.2, which gives the reduced transition probability

$$B(\mathrm{E2}\,;\, 0^-_1 \rightarrow 2^-_{\mathrm{gs}}) = \frac{1}{2} \times 2.185^2 b^4 \chi^2\, e^2 = 21.14\chi^2\, e^2\mathrm{fm}^4\,. \qquad (6.154)$$

The experimental half-life of the $0^-$ state is $5.3\,\mu$s, and Table 6.8 gives

$$B(E2; 0_1^- \to 2_{gs}^-)_{exp} = 4.3\,e^2\text{fm}^4 \,, \tag{6.155}$$

leading to

$$\chi = 0.45 \,. \tag{6.156}$$

This value of $\chi$ is not very large, so no major modifications are expected through configuration mixing as discussed in Sect. 10.2.

## 6.5 Isoscalar and Isovector Transitions

Sometimes it is convenient to decompose the electromagnetic operators into isoscalar and isovector parts. This decomposition leads to transparent selection rules for the electric and magnetic transitions in particle–hole and two–particle and two–hole nuclei.

### 6.5.1 Isospin Decomposition of the Electromagnetic Decay Operator

In this section we discuss the decomposition of the electric and magnetic operators and the associated transitions into isoscalar and isovector components. The key in the discussion is the isospin representation of the particle and hole operators introduced in Subsect. 5.5.3.

Consider the double-tensor operators $c_a^\dagger$ and $\hat{c}_a$. Equation (5.91) relates $\hat{c}_a$ to the spatial-only tensor operator $\tilde{c}_a$ as

$$\hat{c}_{am_{t\alpha}} = (-1)^{\frac{1}{2}+m_{t\alpha}}\tilde{c}_{a,-m_{t\alpha}} \,. \tag{6.157}$$

We thus identify

$$c_{a\frac{1}{2}}^\dagger = c_a^\dagger(\nu) \,, \qquad c_{a,-\frac{1}{2}}^\dagger = c_a^\dagger(\pi) \,, \tag{6.158}$$

$$\hat{c}_{a\frac{1}{2}} = -\tilde{c}_a(\pi) \,, \qquad \hat{c}_{a,-\frac{1}{2}} = \tilde{c}_a(\nu) \,. \tag{6.159}$$

Applying (1.29) in isospace gives

$$\left[c_{am_{t\alpha}}^\dagger \hat{c}_{bm_{t\beta}}\right]_{JM} = \sum_{TM_T} (\tfrac{1}{2}\,m_{t\alpha}\,\tfrac{1}{2}\,m_{t\beta}|T\,M_T)\left[c_a^\dagger \hat{c}_b\right]_{JM}^{TM_T} \,. \tag{6.160}$$

We can now write for neutrons

$$\left[c_a^\dagger(\nu)\tilde{c}_b(\nu)\right]_{JM} = \left[c_{a\frac{1}{2}}^\dagger \hat{c}_{a,-\frac{1}{2}}\right]_{JM} = \sum_T (\tfrac{1}{2}\,\tfrac{1}{2}\,\tfrac{1}{2}\,-\tfrac{1}{2}|T\,0)\left[c_a^\dagger \hat{c}_b\right]_{JM}^{T0}$$

$$= \frac{1}{\sqrt{2}}\left(\left[c_a^\dagger \hat{c}_b\right]_{JM}^{00} + \left[c_a^\dagger \hat{c}_b\right]_{JM}^{10}\right) \tag{6.161}$$

and for protons

$$\left[c_a^\dagger(\pi)\tilde{c}_b(\pi)\right]_{JM} = -\left[c_{a,-\frac{1}{2}}^\dagger \hat{c}_{a\frac{1}{2}}\right]_{JM} = -\sum_T (\tfrac{1}{2} - \tfrac{1}{2}\, \tfrac{1}{2}\, \tfrac{1}{2}|T\, 0)\left[c_a^\dagger \hat{c}_b\right]_{JM}^{T0}$$

$$= \frac{1}{\sqrt{2}}\left([c_a^\dagger \hat{c}_b]_{JM}^{00} - [c_a^\dagger \hat{c}_b]_{JM}^{10}\right) . \tag{6.162}$$

Separating (6.22) into sums over neutrons and protons we can write the electromagnetic operator as

$$\mathcal{M}_{\sigma\lambda\mu} = \hat{\lambda}^{-1}\sum_{ab}(a\|\mathcal{M}_{\sigma\lambda}\|b)\left[c_a^\dagger \tilde{c}_b\right]_{\lambda\mu}$$

$$= \hat{\lambda}^{-1}\Bigg\{ \sum_{\substack{ab \\ \text{neutrons}}} (a\|\mathcal{M}_{\sigma\lambda}\|b)_\nu \left[c_a^\dagger(\nu)\tilde{c}_b(\nu)\right]_{\lambda\mu}$$

$$+ \sum_{\substack{ab \\ \text{protons}}} (a\|\mathcal{M}_{\sigma\lambda}\|b)_\pi \left[c_a^\dagger(\pi)\tilde{c}_b(\pi)\right]_{\lambda\mu}\Bigg\} . \tag{6.163}$$

Substituting from (6.161) and (6.162) gives the final result

$$\begin{aligned}\mathcal{M}_{\sigma\lambda\mu} = \frac{1}{\sqrt{2}}\hat{\lambda}^{-1}\sum_{ab}\Big\{&[(a\|\mathcal{M}_{\sigma\lambda}\|b)_\nu + (a\|\mathcal{M}_{\sigma\lambda}\|b)_\pi]\left[c_a^\dagger \hat{c}_b\right]_{\lambda\mu}^{00}\\ &+ [(a\|\mathcal{M}_{\sigma\lambda}\|b)_\nu - (a\|\mathcal{M}_{\sigma\lambda}\|b)_\pi]\left[c_a^\dagger \hat{c}_b\right]_{\lambda\mu}^{10}\Big\} .\end{aligned} \tag{6.164}$$

This result means that the electromagnetic operator $\mathcal{M}_{\sigma\lambda\mu}$, and in fact any one-body tensor operator, can be decomposed into an isoscalar and an isovector part. Furthermore, the coupling to $M_T = 0$ shows that the operator cannot change the isospin projection, which in turn means that the electromagnetic operator cannot connect two different nuclei. As a special case of (6.164) the electric transition operator $Q_{\lambda\mu}$ can be expressed as

$$Q_{\lambda\mu} = \frac{1}{\sqrt{2}}\hat{\lambda}^{-1}\sum_{ab}\frac{(a\|Q_\lambda\|b)}{e}\Big\{(e_{\text{eff}}^{\text{n}} + e_{\text{eff}}^{\text{p}})\left[c_a^\dagger \hat{c}_b\right]_{\lambda\mu}^{00}$$

$$+ (e_{\text{eff}}^{\text{n}} - e_{\text{eff}}^{\text{p}})\left[c_a^\dagger \hat{c}_b\right]_{\lambda\mu}^{10}\Big\} . \tag{6.165}$$

### 6.5.2 Example: $3^-$ States in $^{16}$O

For charge-conserving particle–hole excitations in *doubly magic* nuclei the lowest states, including the ground state, are $T = 0$ states and the higher-lying states are mostly $T = 1$ states. An example is provided by the $3^-$ states in $^{16}$O as displayed in Fig. 5.10 and schematically represented in Fig. 6.2.

The transitions divide into isoscalar transitions ($\Delta T = 0$) and isovector transitions ($\Delta T = 1$). From (6.165) we see that

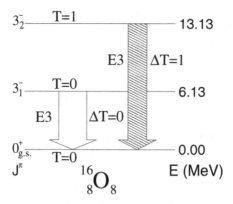

**Fig. 6.2.** The two lowest $3^-$ states and their electric decays via isoscalar and isovector E3 transitions in $^{16}$O. The energies are experimental data

$$(0^+_{\mathrm{gs}} \| \boldsymbol{Q}_\lambda \| \lambda^\pi ; T = 0) \propto e^{\mathrm{n}}_{\mathrm{eff}} + e^{\mathrm{p}}_{\mathrm{eff}} , \qquad (6.166)$$

$$(0^+_{\mathrm{gs}} \| \boldsymbol{Q}_\lambda \| \lambda^\pi ; T = 1) \propto e^{\mathrm{n}}_{\mathrm{eff}} - e^{\mathrm{p}}_{\mathrm{eff}} . \qquad (6.167)$$

This form of the transition amplitudes can also be directly read from the structure of a $T = 0$ and a $T = 1$ state given in (5.126) and (5.127). All electric transitions, of various multipolarities, between low-lying $T = 0$ states, like the three transitions in Fig. 6.1, are of isoscalar nature. From (6.165) we see that their amplitudes are proportional to the sum $e^{\mathrm{n}}_{\mathrm{eff}} + e^{\mathrm{p}}_{\mathrm{eff}}$. This is displayed explicitly in the results of Subsect. 6.4.2.

### 6.5.3 Isospin Selection Rules in Two-Particle and Two-Hole Nuclei

Let us discuss electromagnetic transitions in two-particle and two-hole nuclei from the isospin point of view. We start by examining transitions between two $T = 1$ states in two-particle nuclei. One can show that (Exercise 6.50)

$$\sum_{m_{ti} m_{tf} m_t} (-1)^{\frac{1}{2} - m_{ti}} (\tfrac{1}{2}\, m_{tf}\, \tfrac{1}{2}\, m_t | T_f\, m_{T_f})$$

$$\times (\tfrac{1}{2}\, m_{tf}\, \tfrac{1}{2} - m_{ti} | T\, m_T)(\tfrac{1}{2}\, m_{ti}\, \tfrac{1}{2}\, m_t | T_i\, m_{T_i})$$

$$= (-1)^{1+T+m_{T_i}} \widehat{T_i}\widehat{T_f}(T_f\, m_{T_f}\, T_i - m_{T_i} | T\, m_T) \begin{Bmatrix} T_f & T & T_i \\ \tfrac{1}{2} & \tfrac{1}{2} & \tfrac{1}{2} \end{Bmatrix} . \qquad (6.168)$$

By means of this result one can show that

$$\boxed{\begin{aligned} &\left(a_f\, b_f\, ; J_f\, ; T_f\, m_{T_f} \| \left[ c^\dagger_a \hat{c}_b \right]^{T M_T}_\lambda \| a_i\, b_i\, ; J_i\, ; T_i\, m_{T_i}\right) \\ &\quad = (-1)^{1+T+m_{T_i}} \widehat{T_i}\widehat{T_f}(T_f\, m_{T_f}\, T_i - m_{T_i} | T\, m_T) \begin{Bmatrix} T_f & T & T_i \\ \tfrac{1}{2} & \tfrac{1}{2} & \tfrac{1}{2} \end{Bmatrix} \\ &\quad \times (a_f\, b_f\, ; J_f \| \left[ c^\dagger_a \tilde{c}_b \right]_\lambda \| a_i\, b_i\, ; J_i) . \end{aligned}} \qquad (6.169)$$

For $T_i = T_f = 1$ this gives

$$\left(a_f\, b_f\, ;\, J_f\, ;\, 1\, m_{T_f}\, \middle\|\, \left[c_a^\dagger \hat{c}_b\right]_\lambda^{00}\, \middle\|\, a_i\, b_i\, ;\, J_i\, ;\, 1\, m_{T_i}\right)$$
$$= \frac{1}{\sqrt{2}}\, \delta_{m_{T_i} m_{T_f}}\, \left(a_f\, b_f\, ;\, J_f\, \middle\|\, \left[c_a^\dagger \tilde{c}_b\right]_\lambda\, \middle\|\, a_i\, b_i\, ;\, J_i\right)\, , \tag{6.170}$$

$$\left(a_f\, b_f\, ;\, J_f\, ;\, 1\, m_{T_f}\, \middle\|\, \left[c_a^\dagger \hat{c}_b\right]_\lambda^{10}\, \middle\|\, a_i\, b_i\, ;\, J_i\, ;\, 1\, m_{T_i}\right)$$
$$= \frac{1}{\sqrt{2}}(1 - \delta_{m_{T_i} 0})\delta_{m_{T_i} m_{T_f}}(-1)^{\frac{1}{2}(3+m_{T_i})}\left(a_f\, b_f\, ;\, J_f\, \middle\|\, \left[c_a^\dagger \tilde{c}_b\right]_\lambda\, \middle\|\, a_i\, b_i\, ;\, J_i\right)\, . \tag{6.171}$$

Consider the states with $m_{T_i} = 0 = m_{T_f}$, i.e. the states of proton–neutron two-particle nuclei. When we form the reduced matrix element of the operator (6.164) between these states, we see that only the isoscalar term (6.170) contributes, so that the result is

$$\boxed{\begin{aligned} &(p_f\, n_f\, ;\, J_f\, ;\, 1\, 0\|\boldsymbol{\mathcal{M}}_{\sigma\lambda}\|p_i\, n_i\, ;\, J_i\, ;\, 1\, 0) \\ &= \tfrac{1}{2}\widehat{\lambda}^{-1}\sum_{ab}[(a\|\boldsymbol{\mathcal{M}}_{\sigma\lambda}\|b)_\nu + (a\|\boldsymbol{\mathcal{M}}_{\sigma\lambda}\|b)_\pi](p_f\, n_f\, ;\, J_f\|\left[c_a^\dagger \tilde{c}_b\right]_\lambda\|p_i\, n_i\, ;\, J_i)\, . \end{aligned}}$$
$$\tag{6.172}$$

For electric transitions this becomes

$$(p_f\, n_f\, ;\, J_f\, ;\, 1\, 0\|\boldsymbol{Q}_\lambda\|p_i\, n_i\, ;\, J_i\, ;\, 1\, 0)$$
$$= \tfrac{1}{2}\widehat{\lambda}^{-1}(e_{\text{eff}}^{\text{p}} + e_{\text{eff}}^{\text{n}})\sum_{ab}\frac{(a\|\boldsymbol{Q}_\lambda\|b)}{e}(p_f\, n_f\, ;\, J_f\|\left[c_a^\dagger \tilde{c}_b\right]_\lambda\|p_i\, n_i\, ;\, J_i)\, , \tag{6.173}$$

as can be seen also directly from (6.165). Equation (6.172) contains the *selection rule* $\Delta T = 0$, valid for all electromagnetic transitions between $T = 1$ states in proton–neutron nuclei. Note that this is a special selection rule that transcends the general requirements of vector addition of isospin, which would also allow $\Delta T = 1$.

For two-neutron and two-proton nuclei, with $m_{T_i} = \pm 1$ respectively, we find similarly

$$\boxed{\begin{aligned} &(a_f\, b_f\, ;\, J_f\, ;\, 1\pm 1\|\boldsymbol{\mathcal{M}}_{\sigma\lambda}\|a_i\, b_i\, ;\, J_i\, ;\, 1\pm 1) \\ &= \tfrac{1}{2}\widehat{\lambda}^{-1}\sum_{ab}\big\{[(a\|\boldsymbol{\mathcal{M}}_{\sigma\lambda}\|b)_\nu + (a\|\boldsymbol{\mathcal{M}}_{\sigma\lambda}\|b)_\pi] \\ &\qquad\qquad \pm [(a\|\boldsymbol{\mathcal{M}}_{\sigma\lambda}\|b)_\nu - (a\|\boldsymbol{\mathcal{M}}_{\sigma\lambda}\|b)_\pi]\big\} \\ &\qquad\qquad \times (a_f\, b_f\, ;\, J_f\|\left[c_a^\dagger \tilde{c}_b\right]_\lambda\|a_i\, b_i\, ;\, J_i)\, , \end{aligned}}$$
$$\tag{6.174}$$

where the upper signs apply to neutron–neutron nuclei and the lower signs to proton–proton nuclei. Now the isoscalar term (6.170) and the isovector term (6.171) both contribute equal amounts to the transition amplitude. The corresponding selection rule is $\Delta T = 0, 1$. Again we state the special case of electric transitions:

$$(a_f b_f ; J_f ; 1 \pm 1 \| Q_\lambda \| a_i b_i ; J_i ; 1 \pm 1) = \tfrac{1}{2} \widehat{\lambda}^{-1} [(e_{\text{eff}}^{\text{n}} + e_{\text{eff}}^{\text{p}}) \pm (e_{\text{eff}}^{\text{n}} - e_{\text{eff}}^{\text{p}})]$$
$$\times \sum_{ab} \frac{(a \| Q_\lambda \| b)}{e} (a_f b_f ; J_f \| [c_a^\dagger \tilde{c}_b]_\lambda \| a_i b_i ; J_i) . \quad (6.175)$$

We thus have for a two-neutron (or two-neutron-hole) nucleus

$$(a_f b_f ; J_f ; 1 + 1 \| Q_\lambda \| a_i b_i ; J_i ; 1 + 1)$$
$$= \widehat{\lambda}^{-1} e_{\text{eff}}^{\text{n}} \sum_{ab} \frac{(a \| Q_\lambda \| b)}{e} (a_f b_f ; J_f \| [c_a^\dagger \tilde{c}_b]_\lambda \| a_i b_i ; J_i) \quad (6.176)$$

and for a two-proton (or two-proton-hole) nucleus

$$(a_f b_f ; J_f ; 1 - 1 \| Q_\lambda \| a_i b_i ; J_i ; 1 - 1)$$
$$= \widehat{\lambda}^{-1} e_{\text{eff}}^{\text{p}} \sum_{ab} \frac{(a \| Q_\lambda \| b)}{e} (a_f b_f ; J_f \| [c_a^\dagger \tilde{c}_b]_\lambda \| a_i b_i ; J_i) . \quad (6.177)$$

Figure 6.3 shows an example of the operation of the isospin selection rules.

In transitions between $T = 1$ states in particle–hole nuclei (with charge-conserving or charge-changing excitations) generally both the isoscalar ($\Delta T = 0$) and isovector ($\Delta T = 1$) parts of the electromagnetic operator contribute.

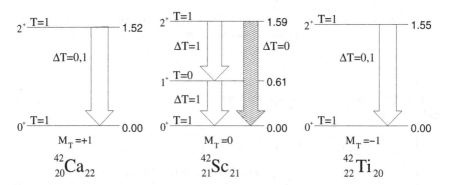

**Fig. 6.3.** Lowest-lying $2^+$ and $1^+$ excited states and their decays via isoscalar and isovector transitions in the $A = 42$ two-particle nuclei $^{42}$Ca, $^{42}$Sc and $^{42}$Ti. The experimental excitation energies (in MeV) and the isospin labels of the states are displayed. The transition governed by the special selection rule $\Delta T = 0$ is shown by a shaded arrow

# Epilogue

The most relevant features of the electromagnetic observables in nuclei were discussed in this chapter. The formalism developed concurrently was applied to describe the decay properties of nuclear systems with one or two active valence nucleons and simple mean-field wave functions. Having reached the end of this chapter, one should have become aware of the great sensitivity of electromagnetic processes to the details of the nuclear wave functions. The sensitivity was demonstrated by the inadequacy of effective charges as minor corrections to the bare charges. The necessity to go beyond the mean-field level became apparent. The quality of computed electromagnetic observables is decisively improved when allowing for the action of the residual nucleon–nucleon interaction, a subject to be discussed from Chap. 8 onwards.

# Exercises

**6.1.** Verify the values of the following useful quantities:

$$\hbar c = 197.33 \,\text{MeV fm} , \tag{6.178}$$

$$\frac{e^2}{4\pi\epsilon_0} = 1.440 \,\text{MeV fm} , \tag{6.179}$$

$$\frac{1}{4\pi\epsilon_0} \left( \frac{\mu_N}{c} \right)^2 = 0.0159 \,\text{MeV fm}^3 . \tag{6.180}$$

What are the corresponding relations in Gaussian units?

**6.2.** Derive (6.23).

**6.3.** Derive the symmetry properties of the reduced single-particle matrix elements of the electric and magnetic operators in (6.27)–(6.30).

**6.4.** Derive (6.35).

**6.5.** Derive (6.36).

**6.6.** Derive (6.37).

**6.7.** Derive (6.40) from the recursion relation (6.39).

**6.8.** By using the formulas (6.35)–(6.40) derive some of the values of the radial integral $\tilde{\mathcal{R}}_{ab}^{(\lambda)}$ given in Table 6.2.

**6.9.** By using (6.23) and Table 6.2 derive some of the values of the scaled single-particle matrix elements $\overline{(a\|Q_\lambda\|b)}_{\text{CS}}$, defined by (6.46), shown in Tables 6.3–6.5.

**6.10.** By using (6.24) and Table 6.2 derive some of the values of the orbital-dipole single-particle matrix elements $\mathcal{D}_{ab}^{(l)}$, defined by (6.47), shown in Table 6.6.

**6.11.** By using (6.24) and Table 6.2 derive some of the values of the spin-dipole single-particle matrix elements $\mathcal{D}_{ab}^{(s)}$, defined by (6.47), shown in Table 6.7.

**6.12.** Derive (6.50) and (6.51) from the definition (6.49) of a multipole moment.

**6.13.** Verify that

$$\sqrt{\frac{16\pi}{5}}\, r^2 Y_{20} = 3z^2 - r^2 . \tag{6.181}$$

This relation shows that the definition of the quantum-mechanical electric quadrupole moment (6.53) coincides with the classical one represented by the expression $3z^2 - r^2$.

**6.14.** Derive (6.54) and (6.55).

**6.15.** Derive the expression (6.56) for a step-function-like wave function, i.e. a wave function that is constant inside the nucleus and zero outside.

**6.16.** Give a detailed derivation of (6.67).

**6.17.** Give a detailed derivation of (6.75).

**6.18.** Show that the diagonal matrix element, with $m = j$, of the single-particle magnetic moment operator (6.90) coincides with the expression (6.89).

**6.19.** Show that (6.89) gives rise to the Schmidt lines defined by (6.92).

**6.20.** Derive the single-particle magnetic dipole moments of Table 6.10.

**6.21.** Evaluate the magnetic dipole and electric quadrupole moments of the ground states of $^{17}$F and $^{41}$Sc. Use the bare values (6.25) for the charges and gyromagnetic ratios. Compare the results with experimental data and comment.

**6.22.** Evaluate the effective charge $e_{\text{eff}}^{n}$ for $^{17}$O when experiment gives

$$B(\text{E2}; 1/2^+ \to 5/2^+)_{\text{exp}} = 6.3\, e^2 \text{fm}^4 \tag{6.182}$$

for this nucleus.

**6.23.** Evaluate the effective charge $e_{\text{eff}}^{p}$ for $^{17}$F when experiment gives

$$B(\text{E2}; 1/2^+ \to 5/2^+)_{\text{exp}} = 64\, e^2 \text{fm}^4 \tag{6.183}$$

for this nucleus.

**6.24.** Evaluate the electric quadrupole moments of the ground states of $^{17}$O and $^{17}$F by using the effective charges derived in Exercises 6.22 and 6.23. Compare the results with experimental data and comment.

**6.25.** Consider the decay of the $1/2^+$ first excited state to the $3/2^+$ ground state in $^{39}$K and $^{39}$Ca. Determine the values of the proton and neutron effective charges by comparing with experimental data.

**6.26.** Fill in the details for the derivation of (6.99).

**6.27.** Deduce (6.102) from (6.99).

**6.28.** Derive the special case (6.103) from (6.102).

**6.29.** Derive (6.104) starting from the very beginning, i.e. from (6.22).

**6.30.** Verify the symmetries (6.105).

**6.31.** Consider the decay of the first $2^+$ state in $^{38}$Ar and $^{38}$Ca. Compute the reduced transition probability $B(E2)$ for these decays. Determine the value of the polarization constant $\chi$ by comparing with experimental data.

**6.32.** By using the polarization constant found in Exercise 6.31 compute the value of $B(E2\,;\,2_1^+ \rightarrow 0_1^+)$ for $^{38}$K.

**6.33.** Compute the E2/M1 mixing ratios $\Delta$ and $\delta$ for the transition $2_1^+ \rightarrow 1_1^+$ in $^{38}$K. Use the bare values (6.25) for the charges and gyromagnetic ratios.

**6.34.** Compute the value of $B(E2\,;\,2_1^+ \rightarrow 0_{gs}^+)$ for $^{42}$Ca and $^{42}$Ti and determine the proton and neutron effective charges by comparison with experimental data. Compute $B(E2\,;\,2_1^+ \rightarrow 0_{gs}^+)$ for $^{42}$Sc by using the previously determined effective charges.

**6.35.** Consider the decay of the first $1^+$ state in $^{38}$K. Use the bare values (6.25) for the charges and gyromagnetic ratios.

(a) Compute the decay probability to the first $0^+$ state.
(b) Compute the decay probability to the $3^+$ ground state.
(c) Determine the total decay half-life of the $1^+$ state and compare it with experimental data.

**6.36.** Compute $B(E2\,;\,2_1^+ \rightarrow 0_{gs}^+)$ for $^{54}$Fe and determine the proton effective charge by comparison with experimental data.

**6.37.** Compute the electric quadrupole moment of the lowest $2^+$ state in $^{54}$Fe by using the proton effective charge extracted in Exercise 6.36. Compare with experimental data and comment.

**6.38.** Compute the electric quadrupole moment of the lowest $5^+$ state in $^{18}$F by using the electric polarization constant in (6.115). Compare with experimental data and comment.

**6.39.** The ground state of $^{18}$F is $1^+$.

(a) By using the electric polarization constant in (6.115) determine the half-lives of the $3_1^+$ and $5_1^+$ states.
(b) By adopting the bare values for the gyromagnetic ratios determine the half-life of the $0_1^+$ state.
(c) Compare the above results with experimental data and comment.

**6.40.** Give a detailed derivation of (6.124).

**6.41.** Derive the special case (6.125) from (6.124).

**6.42.** Give a detailed derivation of (6.148).

**6.43.** Derive the special case (6.150) from (6.148).

**6.44.** Consider the decays of the first $3^-$ and $5^-$ states in $^{40}$Ca. Determine the electric polarization constant $\chi$ from the available data. Compute the resulting decay half-lives of these states.

**6.45.** Consider the decays of the first $2^-$ and $5^-$ excited states in $^{40}$K. Use the polarization constant determined in Exercise 6.44 and the bare values of the gyromagnetic ratios to compute the following:

(a) the decay half-life of the $2^-$ state by considering its decay to the $3^-$ and $4^-$ states below it,
(b) the E2/M1 mixing ratio for the decay of the $5^-$ state to the $4^-$ state below it,
(c) the decay half-life of the $5^-$ state.

**6.46.** Compute $B(\text{E2}; 2_1^+ \rightarrow 0_{\text{gs}}^+)$ and $B(\text{E3}; 3_1^- \rightarrow 0_{\text{gs}}^+)$ for $^{48}$Ca. Compare the results with the experimental values

$$B(\text{E2}; 2^+ \rightarrow 0_{\text{gs}}^+)_{\text{exp}} = 1.58 \, \text{W.u.} \,, \tag{6.184}$$

$$B(\text{E3}; 3^- \rightarrow 0_{\text{gs}}^+)_{\text{exp}} = 6.8 \, \text{W.u.} \tag{6.185}$$

and determine the proton and neutron effective charges.

**6.47.** Compute $B(\text{E2}; 2_1^- \rightarrow 0_{\text{gs}}^-)$ for $^{16}$F by using the polarization constant (6.156).

**6.48.** Compute $B(\text{E2}; 2_1^+ \rightarrow 0_{\text{gs}}^+)$ for $^{56}$Ni by using the bare proton and neutron charges.

**6.49.** Derive the relation (6.168).

**6.50.** Derive (6.169) by using (6.168).

# 7

# Beta Decay

## Prologue

In the previous chapter a powerful method was introduced for probing the structure of nuclear states: computing electromagnetic decays and multipole moments, and comparing with experiment. In this chapter we introduce the various modes of nuclear beta decay and associated transitions. The basic theory of allowed beta decay is presented, but without a detailed derivation from hadronic and leptonic weak-interaction currents. In addition, the less frequently discussed forbidden unique beta-decay transitions are discussed in detail.

Comparing computed beta-decay rates with experiment probes the wave functions involved. Unlike electromagnetic decay, beta decay consists of charge-changing transitions leading from one nucleus to another. Studying simultaneously the electromagnetic transition probabilities between states in a nucleus and the beta-decay feeding of these states from the neighbouring nuclei offers a truly stringent test of nuclear models. In this chapter we test the simple wave functions of one- and two-particle and -hole nuclei of Chap. 5.

## 7.1 General Properties of Nuclear Beta Decay

In this section we discuss the general qualitative properties of beta decay without deriving them from the underlying formal theory framework of the *standard model of electroweak interactions* (Glashow, Weinberg and Salam [43–45]). In the following we call this model just the 'standard model' for brevity. In the standard model the possible beta-decay modes are determined by conservation of *electric charge, lepton number* and *baryon number*.

The charge can be zero, $+e$, $-e$ or some integral multiple of $\pm e$. The lepton number takes two values: $+1$ for leptons and $-1$ for antileptons. Each lepton *flavour*, electron, muon and tau, has its own lepton number that is conserved

**Table 7.1.** The electric charge $q$, baryon number $B$, lepton number $L$ and mass $m$ for the fermions involved in the beta-decay processes of this book

| Particle | $q$ | $B$ | $L$ | $m\,(\mathrm{MeV}/c^2)$ |
|---|---|---|---|---|
| electron (e$^-$) | $-e$ | 0 | $+1$ | 0.511 |
| positron (e$^+$) | $+e$ | 0 | $-1$ | 0.511 |
| electron neutrino ($\nu_e$) | 0 | 0 | $+1$ | 0 |
| electron antineutrino ($\overline{\nu}_e$) | 0 | 0 | $-1$ | 0 |
| proton (p) | $+e$ | $+1$ | 0 | 938.3 |
| neutron (n) | 0 | $+1$ | 0 | 939.6 |

In the standard model the lepton number is considered to be conserved separately for each lepton flavour: electron, muon and tau.

in the standard model.[1] Finally, the baryon number is $+1$ for baryons and $-1$ for antibaryons. Examples of baryons are the nucleons, protons and neutrons, that this book concentrates on.

The particles that take part in the beta-decay processes discussed in this book are the electron (e$^-$), the positron (e$^+$), the proton (p), the neutron (n), the electron neutrino ($\nu_e$) and the electron antineutrino ($\overline{\nu}_e$). These particles, along with their electric charge, baryon number, lepton number and rest-mass energy are listed in Table 7.1.

It is worth noting that in the standard model the neutrino and its antineutrino are considered to be different entities with zero mass; this is a property of the so-called *Dirac neutrino*. In some more elaborate theoretical particle-physics scenarios, such as grand-unified theories and supersymmetric extensions of the standard model, neutrinos can have a non-zero mass and the neutrino can be its own antiparticle ($\nu = \overline{\nu}$); such a neutrino is a so-called *Majorana particle*. In addition, lepton number conservation can be violated, which leads to lepton-flavour oscillations.[2]

The breaking of lepton number conservation also leads to exotic new decay modes, like neutrinoless double beta decay [51], which changes the electron lepton number by two units. Another example is muon-to-electron conversion

---

[1] Separate conservation of the electron, muon and tau lepton numbers guarantees also the conservation of lepton flavour. Lepton flavour conservation implies that one lepton flavour, say electron flavour, cannot convert into another one, say tau flavour. Hence lepton number conservation excludes the possibility of the recently discovered oscillations of neutrino flavour.

[2] At present we know from the large-scale neutrino experiments Super-Kamiokande [46], SNO [47], KamLAND [48] and CHOOZ [49] that lepton flavour conservation is indeed violated. This appears in lepton-flavour oscillations, first introduced by Pontecorvo [50]. In these oscillations the electron, muon and tau flavours convert into one another. This can happen only if at least one of the neutrino mass eigenstates describes a neutrino of non-zero mass. Thus we know now that the neutrino actually possesses a tiny mass.

[52], which violates lepton flavour conservation. Neither the lepton flavour violation nor the character of the neutrino (Dirac vs. Majorana) affects in a measurable way the results in this work.

The three processes of interest in this work, compatible with the conservation laws, are the following:

- $\beta^-$ *decay*

$$ n \xrightarrow{\beta^-} p + e^- + \bar{\nu}_e \,, \tag{7.1} $$

which describes the decay of a free neutron into a free proton, both being baryons. In addition, the final state contains a lepton and an antilepton, both of electron flavour. This decay is allowed by the mass difference between the neutron and the proton. The associated *decay energy*, i.e. the energy released as kinetic energy of the final-state particles, is

$$ Q_{\beta^-} = m_n c^2 - m_p c^2 - m_{e^-} c^2 > 0 \,. \tag{7.2} $$

The decay energy $Q$ is also called the $Q$ *value* of the decay.

- $\beta^+$ *decay*

$$ p \xrightarrow{\beta^+} n + e^+ + \nu_e \,, \tag{7.3} $$

which describes the decay of a proton into a neutron accompanied by an antilepton and a lepton, both of electron flavour. This decay mode is not allowed for a free proton. However, it is allowed in a nucleus, where the extra energy needed to create the neutron–proton mass difference and the positron mass $m_{e^+}$ can be available. The quantity $Q$ is in this case negative,

$$ Q_{\beta^+} = m_p c^2 - m_n c^2 - m_{e^+} c^2 < 0 \,. \tag{7.4} $$

- *Electron capture (EC)*

$$ p + e^- \xrightarrow{\text{EC}} n + \nu_e \,, \tag{7.5} $$

where a proton captures an electron and converts into a neutron and an electron neutrino. The $Q$ value of this process is

$$ Q_{\text{EC}} = m_p c^2 + m_{e^-} c^2 - m_n c^2 < 0 \,. \tag{7.6} $$

Hence electron capture can occur only if extra energy is supplied in a nuclear environment.

All these processes can occur in the many-body environment of a nucleus. In particular, the nuclear environment enables the $\beta^+$ and EC processes to proceed, which is not possible in free space because $Q < 0$.

The nuclear processes corresponding to the free-space processes (7.1), (7.3) and (7.5) are described below.

- *Nuclear $\beta^-$ decay*: A process involving two isobars where the nuclear charge number $Z$ increases by one unit,

$$(Z, N) \xrightarrow{\beta^-} (Z + 1, N - 1) + e^- + \overline{\nu}_e \,. \tag{7.7}$$

- *Nuclear $\beta^+$ decay*: A process involving two isobars where the nuclear charge number $Z$ decreases by one unit,

$$(Z, N) \xrightarrow{\beta^+} (Z - 1, N + 1) + e^+ + \nu_e \,. \tag{7.8}$$

- *Nuclear electron-capture (EC) decay*: A process involving two isobars where the nuclear charge number $Z$ decreases by one unit,

$$(Z, N) + e^- \xrightarrow{\text{EC}} (Z - 1, N + 1) + \nu_e \,. \tag{7.9}$$

The electron is captured from an atomic orbital, usually the s orbital whose wave function has its largest values in the region of the nucleus.

The $Q$ value of each of the processes (7.7)–(7.9) is defined as the total kinetic energy of the *final-state leptons*. The values depend on many-body aspects of the nuclei involved as reflected in their mass differences. This topic will be discussed in Subsect. 7.2.5.

The processes (7.7)–(7.9) are depicted by Feynman diagrams in Fig. 7.1. Their meaning is explained in the figure caption and in the text below.

In our treatment of nuclear beta decay, at the very moment of decay the decaying nucleon feels just the weak interaction and does not interact

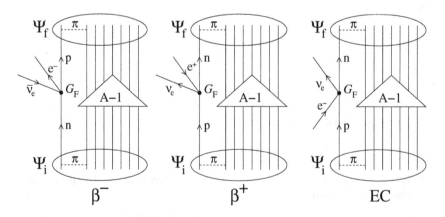

**Fig. 7.1.** Nuclear $\beta^-$, $\beta^+$ and EC decay in the impulse approximation, where only one nucleon takes part in the weak decay process and the remaining $A - 1$ nucleons are spectators. The initial and final states $\Psi_i$ and $\Psi_f$ are nuclear $A$-body states with strong two-nucleon interactions. At the weak-interaction vertices the antilepton lines are drawn as going backwards in time. The strength of the pointlike effective weak-interaction vertex is given by the Fermi constant $G_F$.

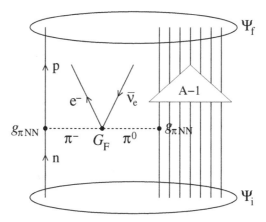

**Fig. 7.2.** A nuclear beta-decay process not included in the impulse approximation: two nucleons interact via pion exchange at the moment of the weak decay, $g_{\pi NN}$ being the coupling constant

via the strong force with the rest of the nucleons of the nucleus. Thus the $A - 1$ nucleons act as spectators with respect to the weak decay process. Only in the initial and final nuclear many-body states does the active nucleon interact strongly with the other $A - 1$ nucleons. This approximation is called the *impulse approximation*. A description of beta decay beyond the impulse approximation includes processes such as shown in Fig. 7.2.

In Fig. 7.1 the flow line of the nucleons is called the nucleon current, or more generally the *weak hadronic current*. Similarly, the flow line containing the leptons is called the *weak leptonic current*. The hadronic and leptonic currents interact at a *weak-interaction vertex*. The vertex can be described as pointlike in the energy range of nuclear beta decay. It incorporates the effect of the exchanged massive vector bosons $W^{\pm}$ into an effective decay strength constant $G_F$ named after Fermi.

A closer look at weak decay reveals, however, a more involved mechanism. This mechanism is shown in Fig. 7.3 for $\beta^-$ decay. The decay of a neutron to a proton proceeds via emission of a negatively charged W boson of mass $m_W \approx 80\,\text{GeV}/c^2$. The strength of interaction at this vertex is given by the weak-interaction coupling constant $g_W$. Due to its large mass the $W^-$ propagates a very short distance and then decays to an electron and its antineutrino with the coupling strength $g_W$.

This process is called the *current–current interaction*, where the interaction between the two currents is mediated by a massive charged vector boson. Since the bosons that mediate the interaction are charged, the associated weak currents are called *charged weak currents*. They are distinguished from *neutral weak currents*, where the mediator is the neutral massive $Z^0$ boson. Processes that proceed by charged weak currents involve exchange of charge, whereas

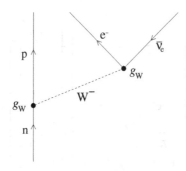

**Fig. 7.3.** Beta-minus decay of a neutron via $W^-$ boson coupling to the baryon and lepton vertices with weak-interaction coupling strength $g_W$

those proceeding by neutral weak currents do not. All modes of nuclear beta decay as depicted in Fig. 7.1 are charged-current processes.

It turns out that due to the large mass of the W boson and the small energy of nuclear beta decay the effective coupling constant $G_F$ can be written as (see e.g. [53])

$$\frac{G_F}{\sqrt{2}} = \frac{g_W^2}{8(m_W c^2)^2} .$$
(7.10)

To a good approximation one can replace the complicated decay pattern of Fig. 7.3 with the simple one occurring in Fig. 7.1. The two decay vertices of Fig. 7.3 are then replaced by one effective vertex with the effective coupling constant $G_F$. The effective vertex describes a pointlike current–current interaction.

## 7.2 Allowed Beta Decay

In this section we discuss the basic properties of so-called *allowed* beta decay. By definition, in allowed $\beta^\pm$ decay the final-state leptons are emitted in an s ($l = 0$) state relative to the nucleus. Similarly, in allowed electron capture, the initial electron is from an s shell and the final neutrino is in an s state relative to the nucleus. Thus the orbital angular momentum of the leptons cannot change the nuclear total angular momentum.

Other beta-decay processes involve higher values of lepton orbital angular momenta and are traditionally called *forbidden* beta transitions. This historical term is misleading since the transitions are not completely forbidden, only greatly hindered relative to allowed beta decays. Forbidden transitions will be discussed in Sect. 7.6.

In addition to any orbital angular momentum, each of the leptons involved has spin $s = \frac{1}{2}$. Thus in $\beta^\pm$ decay the final-state leptons can couple to total spin $S = 0$ or $S = 1$. In electron capture the initial proton and electron can couple to $j \pm \frac{1}{2}$ and the final neutron and neutrino can couple to $j \pm \frac{1}{2}$ or $j \mp \frac{1}{2}$,

**Table 7.2.** Selection rules for allowed beta-decay transitions

| Type of transition | $\Delta J = |J_f - J_i|$ | | $\pi_i \pi_f$ |
|---|---|---|---|
| Fermi | 0 | | +1 |
| Gamow–Teller | 1 | $(J_i = 0 \text{ or } J_f = 0)$ | +1 |
| Gamow–Teller | 0, 1 | $(J_i > 0, J_f > 0)$ | +1 |

Here $J_i$ ($J_f$) is the angular momentum of the initial (final) nuclear state and correspondingly for the parity $\pi$.

with coherent signs. Thus in all cases the lepton spins can change the nuclear total angular momentum $J$ by 0 or 1. In allowed beta decay, transitions with no angular momentum change are called *Fermi transitions* and those with an angular momentum change of one unit are called *Gamow–Teller transitions*. Note also that there is no source for a parity change in an allowed transition. The selection rules for allowed beta decay are collected in Table 7.2.

Derivation of beta-decay transition amplitudes from the interaction of leptonic and hadronic weak charged currents is far from trivial, as can be seen from e.g. [54, 55]. The relativistic quantum mechanics required is outside the scope of this book, and we cite various results without derivation.

### 7.2.1 Half-Lives, Reduced Transition Probabilities and $ft$ Values

As in the case of gamma decay in Subsect. 6.1.1, the transition probability $T_{fi}$ for beta decay is calculated by the 'golden rule' of time-dependent perturbation theory. It is related to the half-life as in (6.1),

$$t_{1/2} = \frac{\ln 2}{T_{fi}} \,. \tag{7.11}$$

The resulting expression is

$$t_{1/2} = \frac{\kappa}{f_0(B_F + B_{GT})} \,, \tag{7.12}$$

where the constant is [56]

$$\kappa \equiv \frac{2\pi^3 \hbar^7 \ln 2}{m_e^5 c^4 G_F^2} = 6147 \,\mathrm{s} \,, \tag{7.13}$$

$f_0$ is a phase-space integral that contains the lepton kinematics, and $B_F$ and $B_{GT}$ are the Fermi and Gamow–Teller *reduced transition probabilities*. They are conventionally broken up into factors as

$$B_F \equiv \frac{g_V^2}{2J_i + 1} |\mathcal{M}_F|^2 \,, \quad B_{GT} \equiv \frac{g_A^2}{2J_i + 1} |\mathcal{M}_{GT}|^2 \,, \tag{7.14}$$

where $J_i$ is the angular momentum of the initial nuclear state, the $g$ quantities are coupling constants to be discussed below, and the $\mathcal{M}$ quantities are matrix elements to be discussed in Subsect. 7.2.2.

The quantity $f_0 t_{1/2}$ is called the *ft value* of an allowed beta-decay transition.[3] It depends exclusively on nuclear structure, which is contained in the reduced matrix elements. In the literature it has also been called the comparative half-life [15] or the reduced half-life [54,55].

The factor $g_V = 1.0$ is the *vector coupling constant* of the weak interactions, and its value is determined by the CVC (conserved vector current) hypothesis of the standard model. The factor $g_A = 1.25$ is the *axial-vector coupling constant* of the weak interactions, and its value is determined by the PCAC (partially conserved axial-vector current) hypothesis of the standard model. In nuclei, the value of $g_A$ is affected by many-nucleon correlations; a value reduced by 20–30 % is sometimes used. For our purposes the free-nucleon value is accurate enough. Thus the values to be used throughout this book are

$$\boxed{g_V = 1.0 \, , \quad g_A = 1.25 \, .} \tag{7.15}$$

The presence of both the vector and axial-vector coupling constants in the half-life expression (7.12) reflects the *parity non-conserving* nature of the weak interactions. The vector and axial-vector parts have opposite space inversion symmetry, namely $\boldsymbol{V}(-\boldsymbol{r}) = -\boldsymbol{V}(\boldsymbol{r})$ for the vector part and $\boldsymbol{A}(-\boldsymbol{r}) = +\boldsymbol{A}(\boldsymbol{r})$ for the axial-vector part. For the lepton current the violation of parity conservation is maximal, and the weak-interaction amplitudes for the leptonic contribution contain the combination $\boldsymbol{V} - \boldsymbol{A}$, an equal division between vector and axial-vector contributions. The same happens at the quark level for hadrons. In the hadronic current the axial-vector contribution renormalizes due to the colour forces between the quarks, and a combination $\boldsymbol{V} - (g_A/g_V)\,\boldsymbol{A} = \boldsymbol{V} - 1.25\,\boldsymbol{A}$ is recovered. This 'vector-minus-axial-vector' structure of the weak charged currents is an indication of the 'left-handedness' of the weak interactions. For more discussion on this subject see e.g. [53].

Because the $ft$ values are usually large, it is normal to express them in terms of 'log $ft$ *values*'. The log $ft$ value is defined as

$$\log ft \equiv \log_{10}(f_0 t_{1/2}[\mathrm{s}]) \, . \tag{7.16}$$

For the logarithm it is essential that the half-life on the right-hand side is expressed as a dimensionless quantity because $f_0$ is dimensionless. Given the log $ft$ value, the half-life is

$$\boxed{t_{1/2} = 10^{\log ft - \log f_0} \, \mathrm{s} \, .} \tag{7.17}$$

---

[3] Modifications of the concept of $ft$ value are introduced for forbidden beta decay in Sect. 7.6.

### 7.2.2 Fermi and Gamow–Teller Matrix Elements

The reduced transition probabilities (7.14) contain the *Fermi matrix element* $\mathcal{M}_F$ [57] and the *Gamow–Teller matrix element* $\mathcal{M}_{GT}$ [58]. The initial and final nuclear wave functions in them carry the nuclear structure information. The Fermi operator is simply the unit operator 1 and the Gamow–Teller operator is the Pauli spin operator $\boldsymbol{\sigma}$. These operators are the simplest scalar and axial-vector operators that can be constructed, and they produce the selection rules of Table 7.2. Theoretically the operators can be derived as limiting expressions of a proper relativistic treatment.

In occupation number representation, following (4.25), the Fermi and Gamow–Teller nuclear matrix elements can be written as

$$\mathcal{M}_F \equiv (\xi_f \, J_f \|1\| \xi_i \, J_i) = \delta_{J_i J_f} \sum_{ab} \mathcal{M}_F(ab)(\xi_f \, J_f \| \left[ c_a^\dagger \tilde{c}_b \right]_0 \| \xi_i \, J_i) , \qquad (7.18)$$

$$\mathcal{M}_{GT} \equiv (\xi_f \, J_f \|\boldsymbol{\sigma}\| \xi_i \, J_i) = \sum_{ab} \mathcal{M}_{GT}(ab)(\xi_f \, J_f \| \left[ c_a^\dagger \tilde{c}_b \right]_1 \| \xi_i \, J_i) , \qquad (7.19)$$

where the reduced single-particle matrix elements are, from (2.33) and (2.56),[4]

$$\mathcal{M}_F(ab) = (a\|1\|b) = \delta_{ab} \widehat{j_a}$$
$$= (n_a \, l_a \, j_a \|1\| n_b \, l_b \, j_b) = \delta_{n_a n_b} \delta_{l_a l_b} \delta_{j_a j_b} \widehat{j_a} , \qquad (7.20)$$

$$\mathcal{M}_{GT}(ab) = \frac{1}{\sqrt{3}}(a\|\boldsymbol{\sigma}\|b) = \frac{1}{\sqrt{3}}(n_a \, l_a \, j_a \|\boldsymbol{\sigma}\| n_b \, l_b \, j_b)$$
$$= \sqrt{2}\, \delta_{n_a n_b} \delta_{l_a l_b} \widehat{j_a}\widehat{j_b}(-1)^{l_a + j_a + \frac{3}{2}} \left\{ \begin{array}{ccc} \frac{1}{2} & \frac{1}{2} & 1 \\ j_b & j_a & l_a \end{array} \right\} . \qquad (7.21)$$

Note that for $\beta^-$ decay $a$ is a proton index and $b$ is a neutron index, whereas for $\beta^+$ decay and electron capture $a$ is a neutron index and $b$ is a proton index.

The Fermi and Gamow–Teller single-particle matrix elements are the same in the CS and BR phase conventions, introduced in Sect. 3.3, since no orbital degrees of freedom are present in the transition operators. The symmetry properties of the single-particle matrix elements are

$$\mathcal{M}_F(ba) = \mathcal{M}_F(ab) , \qquad (7.22)$$

$$\mathcal{M}_{GT}(ba) = (-1)^{j_a + j_b + 1} \mathcal{M}_{GT}(ab) . \qquad (7.23)$$

The Gamow–Teller single-particle matrix elements for the lowest $lj$ combinations are tabulated in Table 7.3. As required by (7.21), they are independent of $n$ as long as $\Delta n = 0$ and they obey the selection rule $\Delta l = 0$.

---

[4] Following a convention in the literature, the right-hand side of (7.19) is defined differently from the general formula (4.25) in that the factor $1/\sqrt{3}$ is included in the single-particle matrix element (7.19). This convention is used also for the matrix elements of forbidden beta decay.

**Table 7.3.** Gamow–Teller single-particle matrix elements $\mathcal{M}_{GT}(ab)$

| $a\backslash b$ | $s_{1/2}$ | $p_{3/2}$ | $p_{1/2}$ | $d_{5/2}$ | $d_{3/2}$ | $f_{7/2}$ | $f_{5/2}$ | $g_{9/2}$ |
|---|---|---|---|---|---|---|---|---|
| $s_{1/2}$ | $\sqrt{2}$ | 0 | 0 | 0 | 0 | 0 | 0 | 0 |
| $p_{3/2}$ | 0 | $\frac{2\sqrt{5}}{3}$ | $-\frac{4}{3}$ | 0 | 0 | 0 | 0 | 0 |
| $p_{1/2}$ | 0 | $\frac{4}{3}$ | $-\frac{\sqrt{2}}{3}$ | 0 | 0 | 0 | 0 | 0 |
| $d_{5/2}$ | 0 | 0 | 0 | $\sqrt{\frac{14}{5}}$ | $-\frac{4}{\sqrt{5}}$ | 0 | 0 | 0 |
| $d_{3/2}$ | 0 | 0 | 0 | $\frac{4}{\sqrt{5}}$ | $-\frac{2}{\sqrt{5}}$ | 0 | 0 | 0 |
| $f_{7/2}$ | 0 | 0 | 0 | 0 | 0 | $2\sqrt{\frac{6}{7}}$ | $-4\sqrt{\frac{2}{7}}$ | 0 |
| $f_{5/2}$ | 0 | 0 | 0 | 0 | 0 | $4\sqrt{\frac{2}{7}}$ | $-\sqrt{\frac{10}{7}}$ | 0 |
| $g_{9/2}$ | 0 | 0 | 0 | 0 | 0 | 0 | 0 | $\frac{1}{3}\sqrt{\frac{110}{3}}$ |

### 7.2.3 Phase-Space Factors

The half-life (7.12) contains the integrated leptonic phase space in the form of a *phase-space factor*, sometimes called the Fermi integral. For $\beta^{\mp}$ decay the phase-space factor is

$$f_0^{(\mp)} = \int_1^{E_0} F_0(\pm Z_f, \varepsilon) p\varepsilon (E_0 - \varepsilon)^2 d\varepsilon , \qquad (7.24)$$

where $F_0$ is the so-called Fermi function to be discussed below and

$$\varepsilon \equiv \frac{E_e}{m_e c^2} , \quad E_0 \equiv \frac{E_i - E_f}{m_e c^2} , \quad p \equiv \sqrt{\varepsilon^2 - 1} , \qquad (7.25)$$

with $E_e$ the total energy of the emitted electron or positron, and $E_i$ and $E_f$ the energies of the initial and final *nuclear* states. For electron capture the phase-space factor is

$$f_0^{(EC)} = 2\pi (\alpha Z_i)^3 (\varepsilon_0 + E_0)^2 , \qquad (7.26)$$

where

$$\varepsilon_0 \equiv \frac{m_e c^2 - \mathcal{B}}{m_e c^2} \approx 1 - \tfrac{1}{2}(\alpha Z_i)^2 , \qquad (7.27)$$

where $\mathcal{B}$ is the binding energy of an electron in an atomic 1s orbital and $\alpha$ is the fine-structure constant, $\alpha \approx \frac{1}{137}$.

Note that (7.27) is generally not a good approximation because it assumes the simple non-relativistic s-electron wave function. The approximation is valid when $\alpha Z_i \ll 1$, which occurs for light nuclei; $Z_i < 40$ is a rule of thumb. For small decay energies additional corrections arise from the screening of the nuclear charge by atomic electrons and from the finite nuclear size. For the cases discussed in this work the decay energies are so large that the electron-capture branch is tiny relative to the $\beta^+$ branch, and the problems with the electron-capture phase-space factor do not affect our results. Accurate phase-space factors for electron capture are tabulated e.g. in [59].

The phase-space factors (7.24) and (7.26) are functions of the nuclear energy difference $E_0$. The final state of $\beta^{\mp}$ decay is a three-body state. Its complicated kinematics is reflected in the complicated $E_0$ dependence of $f^{\mp}$, explicitly displayed in (7.30). In electron capture the final state is a two-body state, and energy and momentum conservation result in a definite energy for the emitted neutrino. This is reflected in the simple phase-space factor $f^{(\mathrm{EC})}$ with parabolic dependence on $E_0$.

The *Fermi function* is a correction factor which approximately takes into account the Coulomb interaction between the emitted lepton and the final nucleus. It is the ratio of the absolute squares of the relativistic Coulomb wave function and the free lepton wave function at the nuclear radius $R$. In $\beta^-$ and $\beta^+$ decay the final state contains two leptons and the daughter nucleus. Because this is a three-body state, energy and momentum conservation do not uniquely determine the energy and momentum of the final-state leptons. The number $\mathrm{d}n_e$ of electrons in an energy interval $[\varepsilon, \varepsilon + \mathrm{d}\varepsilon]$ divided by $\mathrm{d}\varepsilon$ is plotted as a function of the electron energy $\varepsilon$. This function is given by

$$\frac{\mathrm{d}n_e}{\mathrm{d}\varepsilon} = F_0(\pm Z_f, \varepsilon) p\varepsilon (E_0 - \varepsilon)^2 , \tag{7.28}$$

and is called the *shape function* of allowed beta decay; it is the integrand of (7.24). The maximum energy $E_0$ of an electron in beta decay is called the *endpoint energy*.

Figure 7.4 shows the shape function for $E_0 = 6$ in three cases. In the case labelled as $Z = 0$ the Fermi function $F_0$ is omitted from (7.28). The other cases depict $\beta^-$ and $\beta^+$ decay for $Z_f = 20$. As can be seen, the charge of the final nucleus has an appreciable influence on the energy distribution of the emitted electrons and hence on the phase-space factor (7.24). In $\beta^-$ decay the positive nuclear charge decelerates the outgoing negative electrons thus shifting their energy distribution towards smaller energies, and oppositely in $\beta^+$ decay.

The Fermi function $F_0$ in (7.24) can be written analytically in a nonrelativistic approximation known as the *Primakoff–Rosen approximation* [60]:

$$\boxed{F_0(Z_f, \varepsilon) \approx \frac{\varepsilon}{p} F_0^{(\mathrm{PR})}(Z_f) , \quad F_0^{(\mathrm{PR})}(Z_f) = \frac{2\pi \alpha Z_f}{1 - \mathrm{e}^{-2\pi \alpha Z_f}} .} \tag{7.29}$$

The approximation, to be used throughout the book, is quite good unless the decay $Q$ value is very small. It leads to the phase-space factor

$$f_0^{(\mp)} \approx \frac{1}{30}(E_0^5 - 10E_0^2 + 15E_0 - 6)F_0^{(\mathrm{PR})}(\pm Z_f) . \tag{7.30}$$

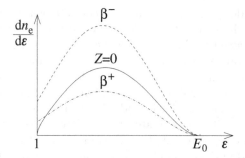

**Fig. 7.4.** Number of emitted electrons as a function of the electron energy $\varepsilon$ for $E_0 = 6$. For $\beta^\mp$ decay $Z = 20$; '$Z = 0$' marks the case with the Fermi function omitted

This expression is easy to use in pocket-calculator evaluations of beta-decay half-lives. More accurate phase-space factors are tabulated e.g. in [59].

### 7.2.4 Combined $\beta^+$ and Electron Capture Decays

Both $\beta^+$ decay and electron capture decrease the nuclear charge number by one. They can coexist and compete in the de-excitation of a nuclear state. The combined effect of these modes can be derived from the additivity of the decay rates $T_{fi}$ (transition probability per unit time),

$$T_{fi}^{(+)} = T_{fi}^{(\beta^+)} + T_{fi}^{(\mathrm{EC})} . \tag{7.31}$$

From (7.10) and (7.11) then follows that the total decay half-life of a combined $\beta^+$ and electron-capture transition, denoted by $\beta^+/\mathrm{EC}$, is given by

$$\boxed{f_0 t_{1/2} = \left[ f_0^{(+)} + f_0^{(\mathrm{EC})} \right] t_{1/2} = \frac{\kappa}{B_\mathrm{F} + B_\mathrm{GT}} .} \tag{7.32}$$

For energies $E_0 > 2$ the relation $f_0^{(+)} \gg f_0^{(\mathrm{EC})}$ is valid, and the half-life of a $\beta^+/\mathrm{EC}$ transition is determined by the $\beta^+$ decay. For small decay energies the electron-capture branch dominates, being even the only one possible for nuclear mass differences less than the positron mass (see Subsect. 7.2.5).

In summary, we have a full account of all allowed beta-decay transitions as follows:

$$\boxed{f_0 t_{1/2} = \frac{\kappa}{B_\mathrm{F} + B_\mathrm{GT}} , \quad f_0 = \begin{cases} f_0^{(-)} & \text{for } \beta^- \text{ decay} , \\ f_0^{(+)} + f_0^{(\mathrm{EC})} & \text{for } \beta^+/\mathrm{EC} \text{ decay} . \end{cases}} \tag{7.33}$$

## 7.2.5 Decay $Q$ Values

The $Q$ value of nuclear beta decay was defined in Sect. 1.1 as the total kinetic energy of the final-state leptons. The following useful relations connect the $Q$ value to the energy difference $\Delta E = E_i - E_f = E_0 m_e c^2$ of the initial and final nuclear states:

$$E_0 = \frac{Q_{\beta^-} + m_e c^2}{m_e c^2} \, , \tag{7.34}$$

$$E_0 = \frac{Q_{\beta^+} + m_e c^2}{m_e c^2} = \frac{Q_{EC} - m_e c^2}{m_e c^2} \, , \tag{7.35}$$

$$E_0 = \frac{Q_{EC} - m_e c^2}{m_e c^2} \, . \tag{7.36}$$

The $Q$ values $Q_{\beta^-}$ and $Q_{EC}$ are the ones tabulated in the *Table of Isotopes* [37] and elsewhere. Therefore the $\beta^+$ endpoint energy is here expressed also in terms of $Q_{EC} = Q_{\beta^+} + 2 m_e c^2$.

An endpoint energy $E_0$ extracted from (7.34)–(7.36) can be used in (7.24), (7.26) or (7.30) to compute the relevant phase-space factor. A beta-decay half-life can then be calculated in a straightforward manner once the one-body transition densities $(\xi_f J_f \| [c_a^\dagger \tilde{c}_b]_{0,1} \| \xi_i J_i)$ are known. Computation of these quantities is the subject of the following sections.

## 7.2.6 Partial and Total Decay Half-Lives; Decay Branchings

A nuclear state can generally beta decay to more than one final state. The transition probabilities are additive, as they are in electromagnetic decay. The beta-decay probability to a given final state $k$ corresponds to a *partial decay half-life* $t_{1/2}^{(k)}$. The total decay half-life $t_{1/2}$ is then given by

$$\boxed{\frac{1}{t_{1/2}} = \sum_k \frac{1}{t_{1/2}^{(k)}} \, .} \tag{7.37}$$

The partial half-life is obtained from the total half-life by using the so-called *branching probability*. This probability can be obtained from the measured decay branching by using the relation

$$B^{(k)} = (\text{experimental decay branching to final state } k \text{ in } \%)/100 \, . \tag{7.38}$$

The partial half-life is now obtained by dividing the total half-life by the branching probability, i.e.

$$\boxed{t_{1/2}^{(k)} = \frac{t_{1/2}}{B^{(k)}} \, .} \tag{7.39}$$

Examples of the use of (7.37)–(7.39) are given in the following sections.

## 7.2.7 Classification of Beta Decays

So far we have looked in any detail only into allowed beta decay. We now digress and consider the classification of all beta-decay transitions. The classification is done in terms of $\log ft$ values as shown in Table 7.4. For further detail see [54, 61].

**Table 7.4.** Classification of beta-decay transitions according to their $\log ft$ values

| Type of transition | $\log ft$ |
|---|---|
| superallowed | 2.9–3.7 |
| unfavoured allowed | 3.8–6.0 |
| $l$-forbidden allowed | $\geq 5.0$ |
| first-forbidden unique | 8–10 |
| first-forbidden non-unique | 6–9 |
| second-forbidden | 11–13 |
| third-forbidden | 17–19 |
| fourth-forbidden | $> 22$ |

In this table the allowed transitions are subdivided into three categories. We briefly discuss these categories in this subsection. The forbidden transitions shown in the table are the subject of Sect. 7.6. Note that the $\log ft$ boundaries are not sharp but rather enable a general grouping of transitions.

*Superallowed* transitions occur in light nuclei where the proton and neutron Fermi surfaces are roughly at the same position. This allows maximal overlap between the initial and final nuclear wave functions. The transitions are of the single-particle type and yield maximum values for the Fermi and Gamow–Teller matrix elements. The simplest transitions occur in light nuclei of one-particle or one-hole type. Examples of these are given in Sect. 7.3, with numerical values collected in Table 7.5.

Transitions of the *l-forbidden allowed* type occur in cases where the simple single-particle transition in the mean-field shell-model picture of Chap. 5 is forbidden by the $\Delta l = 0$ selection rule contained (7.20) and (7.21). The selection rules on nuclear angular momentum and parity, as stated in Table 7.2, are satisfied. Hence the forbiddenness is just a property of having a single configuration approximate each nuclear wave function. Introducing configuration mixing via the residual interaction lifts this forbiddenness and yields a finite magnitude for the computed $\log ft$ values. This mixing, however, is usually not strong enough to bring the $\log ft$ values below $\log ft \approx 5$.

*Unfavoured allowed* transitions are defined as those not belonging to either of the two types discussed above. They are allowed single-particle transitions in that there is no $l$ forbiddenness. However, the single-particle transitions are not pure but diluted in the initial and final many-nucleon wave functions. The contribution of the leading single-particle component is reduced by its redistribution among many nuclear states by the residual interaction. The ratio of the measured decay rate to the computed single-particle rate is called the *hindrance factor*. Its values range in the limits 0.004–0.01 in the 0d-1s and 0f-1p-0g$_{9/2}$ shells [9].

## 7.3 Beta-Decay Transitions in One-Particle and One-Hole Nuclei

In this section we discuss the beta decays of the simplest possible nuclei, namely the one-particle and one-hole nuclei. Examples of the application of the formalism are also given.

### 7.3.1 Matrix Elements and Reduced Transition Probabilities

The wave functions of one-particle and one-hole nuclei were discussed in Sects. 5.2 and 6.2. For the one-particle nuclei they were written as

$$|\Psi_i\rangle = |n_i\, l_i\, j_i\, m_i\rangle = c_i^\dagger|\text{CORE}\rangle \,, \tag{7.40}$$

$$|\Psi_f\rangle = |n_f\, l_f\, j_f\, m_f\rangle = c_f^\dagger|\text{CORE}\rangle \,, \tag{7.41}$$

and for the one-hole nuclei as

$$|\Phi_i\rangle = |(n_i\, l_i\, j_i\, m_i)^{-1}\rangle = h_i^\dagger|\text{HF}\rangle \,, \tag{7.42}$$

$$|\Phi_f\rangle = |(n_f\, l_f\, j_f\, m_f)^{-1}\rangle = h_f^\dagger|\text{HF}\rangle \,. \tag{7.43}$$

The following one-body transition densities were derived in Sect. 6.2:

$$(\Psi_f\|\left[c_a^\dagger\tilde{c}_b\right]_L\|\Psi_i) = \delta_{af}\delta_{bi}\widehat{L} \,, \tag{7.44}$$

$$(\Phi_f\|\left[c_a^\dagger\tilde{c}_b\right]_L\|\Phi_i) = \delta_{ai}\delta_{bf}(-1)^{j_i+j_f+L}\widehat{L} \,. \tag{7.45}$$

Subsituted in (7.18)–(7.20) these densities give the Fermi and Gamow–Teller matrix elements

$$\boxed{\mathcal{M}_\text{F}(\Psi_i \to \Psi_f) = -\mathcal{M}_\text{F}(\Phi_i \to \Phi_f) = \delta_{if}\widehat{j_i} \,,} \tag{7.46}$$

$$\boxed{\mathcal{M}_\text{GT}(\Psi_i \to \Psi_f) = \mathcal{M}_\text{GT}(\Phi_i \to \Phi_f) = \sqrt{3}\mathcal{M}_\text{GT}(fi) \,,} \tag{7.47}$$

where $\mathcal{M}_\text{GT}(fi)$ is the single-particle matrix element (7.21). Substituted into (7.14) these matrix elements lead to the reduced beta transition probabilities

$$\boxed{B_\text{F} = g_\text{V}^2\delta_{if} \,, \quad B_\text{GT} = g_\text{A}^2\frac{3}{2j_i + 1}|\mathcal{M}_\text{GT}(fi)|^2 \,.} \tag{7.48}$$

These reduced transition probabilities are valid for transitions between one-particle states and for transitions between one-hole states.

## 7.3.2 Application to Beta Decay of $^{15}$O; Other Examples

Consider the beta decay of $^{15}_8$O$_7$ to $^{15}_7$N$_8$ depicted in Fig. 7.5 with experimental data. In this case both $\beta^+$ decay and electron capture are active. With an experimental $\log ft$ value of 3.6 the transition is superallowed (see Table 7.4). It occurs between the two ground states with a branching of 100 %, i.e. all the decays go to the final ground state. The one-hole structure of the initial and final states, (5.12) and (5.10) respectively, is indicated in the figure. We proceed to calculate the reduced transition probabilities from these wave functions and compare subsequent results with the measured values.

Equation (7.48) and Table 7.3, with the coupling constants from Subsect. 7.2.1, give

$$B_F = g_V^2 = 1.0 \,, \tag{7.49}$$

$$B_{GT} = g_A^2 \frac{3}{2} |\mathcal{M}_{GT}(p_{1/2}\, p_{1/2})|^2 = 1.25^2 \times \frac{3}{2} \left( -\frac{\sqrt{2}}{3} \right)^2 = 0.521 \,. \tag{7.50}$$

From (7.32) we have

$$f_0 t_{1/2} = \frac{6147\,\text{s}}{1.0 + 0.521} = 4041\,\text{s} \,, \tag{7.51}$$

which gives $\log ft = 3.61$, in excellent agreement with experiment.

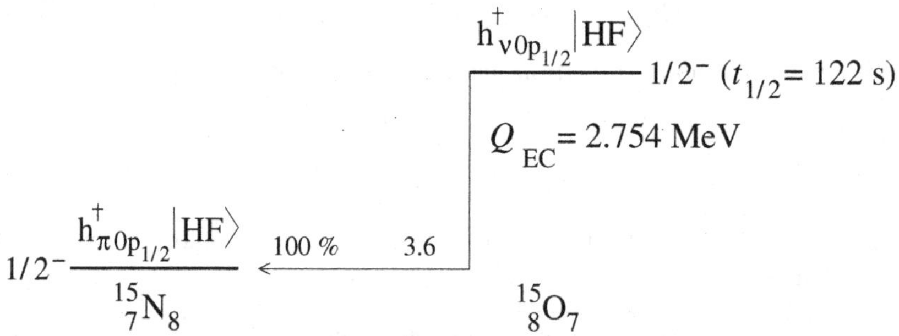

$$h^\dagger_{\nu 0 p_{1/2}} \big| \text{HF} \big\rangle$$

$$\underline{\hspace{3cm}} \ 1/2^-\ (t_{1/2} = 122\ \text{s})$$

$$Q_{EC} = 2.754\ \text{MeV}$$

$$h^\dagger_{\pi 0 p_{1/2}} \big| \text{HF} \big\rangle \qquad 100\ \% \qquad 3.6$$

$$1/2^- \frac{\phantom{xxxxx}}{^{15}_7\text{N}_8} \longleftarrow \qquad\qquad ^{15}_8\text{O}_7$$

**Fig. 7.5.** Superallowed beta decay of the $1/2^-$ ground state of $^{15}$O to the $1/2^-$ ground state of $^{15}$N. The decay proceeds via the $\beta^+$/EC decay mode. The experimental half-life, $Q$ value, branching and $\log ft$ value are given

We proceed to compute the half-life from (7.17). The phase-space factor needed is $f_0 = f_0^{(+)} + f_0^{(EC)}$, as is seen from (7.33). We calculate $f_0^{(+)}$ from (7.30) with input from (7.35),

$$E_0 = \frac{Q_{EC} - m_e c^2}{m_e c^2} = 4.389 \,, \tag{7.52}$$

and from (7.29) with $Z_f = 7$,

$$F_0^{(\mathrm{PR})}(-Z_f) = \frac{2\pi\alpha(-Z_f)}{1 - e^{2\pi\alpha Z_f}} = 0.848 . \tag{7.53}$$

Equation (7.30) now gives

$$f_0^{(+)} = 42.3 . \tag{7.54}$$

The phase-space factor $f_0^{(\mathrm{EC})}$ is calculated from (7.26) with input from (7.27),

$$\varepsilon_0 = 1 - \tfrac{1}{2}(\alpha \times 8)^2 = 0.998 . \tag{7.55}$$

Equation (7.36) gives the same $E_0$ as (7.52), so we have

$$f_0^{(\mathrm{EC})} = 2\pi(\alpha \times 8)^3 (0.998 + 4.389)^2 = 0.036 . \tag{7.56}$$

Since $f_0^{(\mathrm{EC})} \ll f_0^+$ the transition is dominated by $\beta^+$ decay, so that

$$f_0 \approx f_0^{(+)} = 42.3 . \tag{7.57}$$

This gives $\log f_0 = 1.63$, and substitution into (7.17) yields finally the theoretical half-life

$$t_{1/2} = 10^{3.61-1.63}\,\mathrm{s} = 95.5\,\mathrm{s} , \tag{7.58}$$

which is close to the experimental half-life of 122 s.

Other $\beta^+/\mathrm{EC}$ decays of one-particle and one-hole nuclei are calculated similarly. Table 7.5 summarizes the results of such calculations.

**Table 7.5.** Computed $\log f_0$ and $\log ft$ values, and resultant beta-decay half-lives, for $\beta^+/\mathrm{EC}$ transitions in one-particle and one-hole nuclei, together with experimental $Q_{\mathrm{EC}}$ values and half-lives

| Beta decay | $Q_{\mathrm{EC}}^{(\mathrm{exp})}$ (MeV) | $\log f_0$ | $\log ft$ | $t_{1/2}$ (s) | $t_{1/2}^{(\mathrm{exp})}$ (s) |
|---|---|---|---|---|---|
| $^{15}\mathrm{O}(1/2^-) \rightarrow {}^{15}\mathrm{N}(1/2^-)$ | 2.754 | 1.626 | 3.606 | 95.5 | 122 |
| $^{17}\mathrm{F}(5/2^+) \rightarrow {}^{17}\mathrm{O}(5/2^+)$ | 2.762 | 1.624 | 3.283 | 45.6 | 64.5 |
| $^{39}\mathrm{Ca}(3/2^+) \rightarrow {}^{39}\mathrm{K}(3/2^+)$ | 6.524 | 3.671 | 3.500 | 0.675 | 0.86 |
| $^{41}\mathrm{Sc}(7/2^-) \rightarrow {}^{41}\mathrm{Ca}(7/2^-)$ | 6.495 | 3.649 | 3.308 | 0.456 | 0.59 |

The computed half-lives agree well with the experimental ones. This means that the actual nuclear states involved are indeed rather pure one-particle or one-hole states. The transitions are thus single-particle transitions. According to Subsect. 7.2.7 such beta decays are superallowed. This is borne out by the $\log ft$ values which are in the range 2.9–3.7 stated in Table 7.4.

## 7.4 Beta-Decay Transitions in Particle–Hole Nuclei

In the following we discuss beta-decay transitions in particle–hole nuclei. As described in Sect. 5.4, there are two types of particle–hole excitation. The first, charge-conserving type consists of excited states in an even–even nucleus whose ground state is the vacuum of particle–hole excitations. Such an even–even nucleus can be called the *reference nucleus*. The second, charge-changing type consists of the ground and excited states in the odd–odd nuclei adjacent to the reference nucleus. Let us first discuss decays to the particle–hole vacuum of the reference nucleus. Depending on the energetics, the decay can also go in the opposite direction.

### 7.4.1 Beta Decay to and from the Even–Even Ground State

Charge-changing excitations of particle–hole nuclei can beta decay to the reference nucleus. The initial state is a state of an odd–odd nucleus, generated by making a charge-changing particle–hole excitation of the particle–hole vacuum. The final state is the particle–hole vacuum $|\text{HF}\rangle$, which is the ground state of the reference nucleus. The beta-decay matrix elements are constructed from the transition density (6.118),

$$(\text{HF}\| \left[ c_a^\dagger \tilde{c}_b \right]_L \| a_i \, b_i^{-1} \, ; \, J_i) = \delta_{L J_i} \delta_{ab_i} \delta_{ba_i} (-1)^{j_{a_i} - j_{b_i} + J_i} \widehat{J_i} \,. \tag{7.59}$$

Inserting this transition density into (7.18) and (7.19) yields

$$\boxed{\mathcal{M}_{\text{F}}(a_i b_i^{-1}) = \delta_{J_i 0} \delta_{a_i b_i} \widehat{j_{a_i}} \,,} \tag{7.60}$$

$$\boxed{\mathcal{M}_{\text{GT}}(a_i b_i^{-1}) = -\sqrt{3}\, \delta_{J_i 1} \mathcal{M}_{\text{GT}}(a_i b_i) \,,} \tag{7.61}$$

where the symmetry relation (7.23) has been used.

In the event the odd–odd nucleus has low-lying states below the particle–hole vacuum of the reference nucleus, beta decay can occur from the vacuum to the odd–odd nucleus. This is the situation in light nuclei. Equation (7.59) is now replaced by

$$(a_f \, b_f^{-1} \, ; \, J_f \| \left[ c_a^\dagger \tilde{c}_b \right]_L \| \text{HF}) = \delta_{L J_f} \delta_{aa_f} \delta_{bb_f} \widehat{J_f} \,. \tag{7.62}$$

Substituting this into (7.18) and (7.19) results in

$$\boxed{\mathcal{M}_{\text{F}}(a_f b_f^{-1}) = \delta_{J_f 0} \delta_{a_f b_f} \widehat{j_{a_f}} \,,} \tag{7.63}$$

$$\boxed{\mathcal{M}_{\text{GT}}(a_f b_f^{-1}) = \sqrt{3}\, \delta_{J_f 1} \mathcal{M}_{\text{GT}}(a_f b_f) \,.} \tag{7.64}$$

## 7.4.2 Application to Beta Decay of $^{56}$Ni

Consider the $\beta^+/EC$ decay depicted in Fig. 7.6, where $^{56}_{28}\mathrm{Ni}_{28}$ decays to $^{56}_{27}\mathrm{Co}_{29}$. This is a case described by (7.63) and (7.64). In our simple mean-field shell-model scheme the low-energy states of $^{56}\mathrm{Co}$ have the structure

$$|^{56}\mathrm{Co}\,;\,1^+,2^+,3^+,4^+,5^+,6^+\rangle$$
$$= \left[c^\dagger_{\nu 0f_{5/2}}\,h^\dagger_{\pi 0f_{7/2}}\right]_{1^+,2^+,3^+,4^+,5^+,6^+}|^{56}\mathrm{Ni}\,;\,0^+_{\mathrm{gs}}\rangle\,, \quad (7.65)$$

where $^{56}\mathrm{Ni}$ is the doubly magic particle–hole vacuum. Because of the angular momentum conditions in (7.63) and (7.64) the Fermi matrix element vanishes and the only possible final state for a Gamow–Teller matrix element is the $1^+$ state of $^{56}\mathrm{Co}$.

With the single-particle matrix element from Table 7.3, the Gamow–Teller matrix element (7.64) becomes

$$\mathcal{M}_{\mathrm{GT}} = \sqrt{3}\mathcal{M}_{\mathrm{GT}}(\mathrm{f}_{5/2}\,\mathrm{f}_{7/2}) = \sqrt{3}\times 4\sqrt{\frac{2}{7}} = 3.703\,, \quad (7.66)$$

leading to the reduced transition probability

$$B_{\mathrm{GT}} = g_A^2|\mathcal{M}_{\mathrm{GT}}|^2 = 1.25^2 \times 3.703^2 = 21.43\,. \quad (7.67)$$

Substituting the matrix element (7.67) into (7.32) gives the calculated $\log ft$ value

$$\log ft = 2.46\,. \quad (7.68)$$

This is far short of the experimental value $\log ft = 4.4$, which shows that the simple particle–hole description fails here. The transition is unfavoured

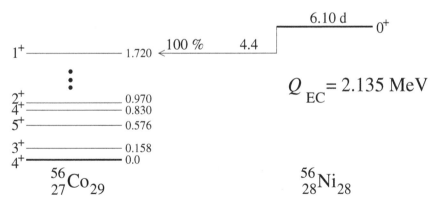

**Fig. 7.6.** Beta decay of the ground state of $^{56}$Ni to an excited state of $^{56}$Co. The decay proceeds via the $\beta^+/EC$ mode and is of the unfavoured allowed type, as seen from Table 7.4. The experimental half-life, $Q$ value, branching and $\log ft$ value are given

allowed, which means that the leading single-particle transition is diluted among several initial and final states by the residual two-body interaction. The hindrance factor is $10^{2.46-4.4} = 0.011$.

The coupling in (7.65) is the only way to produce a $1^+$ state by exciting a proton from the $0f_{7/2}$ shell to the rest of the fp shell for neutrons. Therefore the discrepancy suggests that more complicated configurations, two-particle–two-hole, etc., excitations, play an active part in the low-lying states of $^{56}$Co.

### 7.4.3 Beta-Decay Transitions Between Two Particle–Hole States

The tool to be used in this subsection is the general expression (6.124) for electromagnetic transitions between two arbitrary particle–hole states. The structure is just the same for any, not only electromagnetic, multipole operators. In particular we propose to apply (6.124) to the Fermi ($L = 0$) and Gamow–Teller ($L = 1$) operators. However, the results with a general value of $L$ turn out to be useful also for forbidden beta decay, where $L = 2, 3, \dots$.

Equation (6.124) carries the restriction that all single-particle orbitals involved are either proton or neutron orbitals. This restriction applies to electromagnetic decays, which are charge conserving. However, the same formula can be used for the charge-changing beta-decay transitions since no explicit use of charge conservation was made in deriving it.

We start from an even–even reference nucleus $(N, Z)$, which is the particle–hole vacuum. Its excited states are proton particle–hole $(pp^{-1})$ and neutron particle–hole $(nn^{-1})$ excitations. Consider $\beta^-$ decay of the neighbouring odd–odd nucleus $(N + 1, Z - 1)$. Its states are obtained from the $(N, Z)$ nucleus by neutron-particle–proton-hole $(np^{-1})$ excitations.

As a modified case of (4.22) the $\beta^-$ decay operator is

$$\beta^-_{LM} = \widehat{L}^{-1} \sum_{pn} (p\|\boldsymbol{\beta}_L\|n) \left[c_p^\dagger \tilde{c}_n\right]_{LM} , \qquad (7.69)$$

where the single-particle tensor operator is $\boldsymbol{\beta}_0 = \mathbf{1}$ or $\boldsymbol{\beta}_1 = \boldsymbol{\sigma}$, as introduced in (7.18) and (7.19). In (4.22) the summation indices $a$ and $b$ are equivalent, but here the proton index $p$ is distinct from the neutron index $n$. The transition amplitude for general states $|\Psi_i\rangle$ and $|\Psi_f\rangle$ is

$$(\Psi_f\|\beta^-_L\|\Psi_i) = \widehat{L}^{-1} \sum_{pn} (p\|\boldsymbol{\beta}_L\|n)(\Psi_f\| \left[c_p^\dagger \tilde{c}_n\right]_L \|\Psi_i) . \qquad (7.70)$$

The initial state is a neutron-particle–proton-hole state, and let us first take the final state to be a neutron particle–hole state:

$$|\Psi_i\rangle = \left[c_{n_i}^\dagger h_{p_i}^\dagger\right]_{J_i M_i} |\text{HF}\rangle , \quad |\Psi_f\rangle = \left[c_{n_f}^\dagger h_{n'_f}^\dagger\right]_{J_f M_f} |\text{HF}\rangle . \qquad (7.71)$$

Substituting these into (6.124) and taking account of the $p, n$ distinction $(\delta_{pn} = 0)$ we obtain

$$(n_f\, n_f'^{-1};\, J_f \|\beta_L^-\| n_i\, p_i^{-1};\, J_i) = \delta_{n_i n_f} (-1)^{j_{n_i}+j_{n_f'}+J_f+1}\, \widehat{J_i}\widehat{J_f}\widehat{L}$$
$$\times \left\{ \begin{matrix} J_i & J_f & L \\ j_{n_f'} & j_{p_i} & j_{n_i} \end{matrix} \right\} \mathcal{M}_L(p_i n_f')\,,$$
(7.72)

where $\mathcal{M}_L(p_i n_f')$ is either the Fermi or the Gamow–Teller single-particle matrix element, defined in (7.20) and (7.21),

$$\mathcal{M}_0(ab) \equiv \mathcal{M}_F(ab) = (a\|\beta_0\|b) = (a\|1\|b)\,,$$
(7.73)

$$\mathcal{M}_1(ab) \equiv \mathcal{M}_{GT}(ab) = \frac{1}{\sqrt{3}}(a\|\beta_1\|b) = \frac{1}{\sqrt{3}}(a\|\sigma\|b)\,.$$
(7.74)

In the single-particle matrix elements $\mathcal{M}_L$, proton and neutron labels are no longer considered as distinct.

Consider secondly the case where the final state is a proton particle–hole state,

$$|\Psi_f\rangle = \big[c_{p_f}^\dagger h_{p_f'}^\dagger\big]_{J_f M_f} |\mathrm{HF}\rangle\,,$$
(7.75)

Application of (6.124) now gives the transition amplitude as

$$(p_f\, p_f'^{-1};\, J_f \|\beta_L^-\| n_i\, p_i^{-1};\, J_i) = \delta_{p_i p_f'} (-1)^{j_{n_i}+j_{p_f'}+J_i+L}\, \widehat{J_i}\widehat{J_f}\widehat{L}$$
$$\times \left\{ \begin{matrix} J_i & J_f & L \\ j_{p_f} & j_{n_i} & j_{p_i} \end{matrix} \right\} \mathcal{M}_L(n_i p_f)\,.$$
(7.76)

Let us discuss next $\beta^+$ decay. The initial state in the neighbouring odd–odd $(N-1, Z+1)$ nucleus is generated by a proton-particle–neutron-hole $(pn^{-1})$ excitation of the $(N, Z)$ particle–hole vacuum. The final state is a charge-conserving particle–hole excitation of the even–even reference nucleus $(N, Z)$. Similarly to (7.70) the transition amplitude is

$$(\Psi_f\|\beta_L^+\|\Psi_i) = \widehat{L}^{-1} \sum_{np} (n\|\beta_L\|p)(\Psi_f\|\big[c_n^\dagger \tilde{c}_p\big]_L\|\Psi_i)\,.$$
(7.77)

The initial state is a proton-particle–neutron-hole state, and again we take first the case where the final state is a neutron particle–hole state:

$$|\Psi_i\rangle = \big[c_{p_i}^\dagger h_{n_i}^\dagger\big]_{J_i M_i} |\mathrm{HF}\rangle\,, \quad |\Psi_f\rangle = \big[c_{n_f}^\dagger h_{n_f'}^\dagger\big]_{J_f M_f} |\mathrm{HF}\rangle\,.$$
(7.78)

The transition amplitude now becomes

$$(n_f\, n_f'^{-1};\, J_f \|\beta_L^+\| p_i\, n_i^{-1};\, J_i) = \delta_{n_i n_f'} (-1)^{j_{p_i}+j_{n_f'}+J_i+L}\, \widehat{J_i}\widehat{J_f}\widehat{L}$$
$$\times \left\{ \begin{matrix} J_i & J_f & L \\ j_{n_f} & j_{p_i} & j_{n_i} \end{matrix} \right\} \mathcal{M}_L(p_i n_f)\,.$$
(7.79)

In the second case the final state is a proton particle–hole state, and we find

$$\boxed{\begin{aligned}
(p_f\,p_f'^{-1}\,;\,J_f\|\beta_L^+\|p_i\,n_i^{-1}\,;\,J_i) &= \delta_{p_i p_f}(-1)^{j_{p_i}+j_{p_f'}+J_f+1}\,\widehat{J_i}\widehat{J_f}\widehat{L} \\
&\times \left\{\begin{matrix} J_i & J_f & L \\ j_{p_f'} & j_{n_i} & j_{p_i} \end{matrix}\right\} \mathcal{M}_L(n_i p_f')\,.
\end{aligned}} \tag{7.80}$$

For clear emphasis on the Fermi and Gamow–Teller types of allowed transitions, we adopt the notation

$$\beta_0^{\mp} \equiv \beta_F^{\mp}\,, \quad \beta_1^{\mp} \equiv \beta_{GT}^{\mp}\,. \tag{7.81}$$

For Fermi transitions ($L = 0$) the simple $6j$ symbol is given by (1.65) and the transition amplitudes (7.72), (7.76), (7.79) and (7.80) reduce to

$$(n_f\,n_f'^{-1}\,;\,J_f\|\beta_F^-\|n_i\,p_i^{-1}\,;\,J_i) = -\delta_{J_iJ_f}\delta_{n_in_f}\delta_{p_in_f'}\Delta(j_{n_i}j_{n_f'}J_i)\widehat{J_i}\,, \tag{7.82}$$

$$(p_f\,p_f'^{-1}\,;\,J_f\|\beta_F^-\|n_i\,p_i^{-1}\,;\,J_i) = \delta_{J_iJ_f}\delta_{p_ip_f'}\delta_{n_ip_f}\Delta(j_{p_i}j_{p_f}J_i)\widehat{J_i}\,, \tag{7.83}$$

$$(n_f\,n_f'^{-1}\,;\,J_f\|\beta_F^+\|p_i\,n_i^{-1}\,;\,J_i) = \delta_{J_iJ_f}\delta_{n_in_f'}\delta_{p_in_f}\Delta(j_{n_i}j_{n_f}J_i)\widehat{J_i}\,, \tag{7.84}$$

$$(p_f\,p_f'^{-1}\,;\,J_f\|\beta_F^+\|p_i\,n_i^{-1}\,;\,J_i) = -\delta_{J_iJ_f}\delta_{p_ip_f}\delta_{n_ip_f'}\Delta(j_{p_i}j_{p_f'}J_i)\widehat{J_i}\,, \tag{7.85}$$

where the symbol $\delta_{pn}$ is understood so that the quantum numbers of the proton and neutron orbitals have to be the same.

### 7.4.4 Application to Beta Decay of $^{16}$N

Consider the Gamow–Teller $\beta^-$ decays of Fig. 7.7 from the $2^-$ ground state of $^{16}_7\mathrm{N}_9$ to the lowest negative-parity states in $^{16}_8\mathrm{O}_8$. The initial-state wave function, as given in (5.60), reads

$$|^{16}\mathrm{N}\,;\,2_{gs}^-\rangle = [c_{\nu 0d_{5/2}}^\dagger h_{\pi 0p_{1/2}}^\dagger]_{2^-}|\mathrm{HF}\rangle\,. \tag{7.86}$$

The simple mean-field wave functions for the $1_1^-$, $2_1^-$ and $3_1^-$ states of $^{16}$O, from (5.65) and (5.66), are

$$|^{16}\mathrm{O}\,;\,2_1^-,3_1^-\rangle = \frac{1}{\sqrt{2}}\Big([c_{\nu 0d_{5/2}}^\dagger h_{\nu 0p_{1/2}}^\dagger]_{2^-,3^-}|\mathrm{HF}\rangle$$
$$+ [c_{\pi 0d_{5/2}}^\dagger h_{\pi 0p_{1/2}}^\dagger]_{2^-,3^-}|\mathrm{HF}\rangle\Big)\,, \tag{7.87}$$

$$|^{16}\mathrm{O}\,;\,1_1^-\rangle = \frac{1}{\sqrt{2}}\Big([c_{\nu 1s_{1/2}}^\dagger h_{\nu 0p_{1/2}}^\dagger]_{1^-}|\mathrm{HF}\rangle + [c_{\pi 1s_{1/2}}^\dagger h_{\pi 0p_{1/2}}^\dagger]_{1^-}|\mathrm{HF}\rangle\Big)\,. \tag{7.88}$$

These are the $T = 0, M_T = 0$ states of the type (5.126).

These states are substituted into (7.72) and (7.76) to give the Gamow–Teller matrix elements, with the single-particle matrix elements taken from Table 7.3,

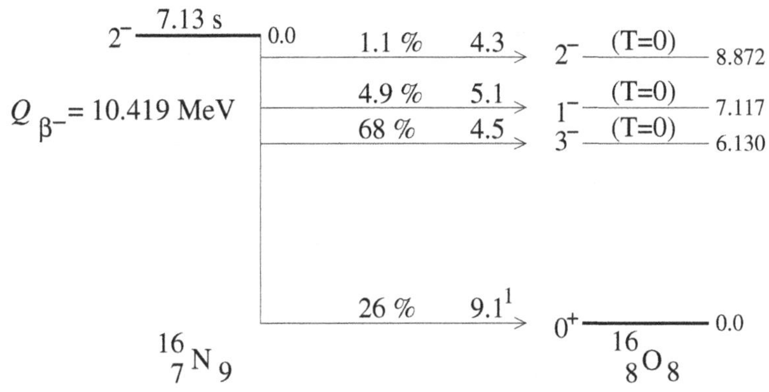

**Fig. 7.7.** Beta-minus decay of the $2^-$ ground state of $^{16}$N to the ground and excited states in $^{16}$O. The experimental half-life, $Q$ value, branchings and $\log ft$ values are shown. Gamow–Teller decay occurs to the $3^-$, $1^-$ and $2^-$ states. To the $2^-$ final state also Fermi decay is possible. The decay to the $0^+$ ground state is first-forbidden unique. The use of the phase-space factor $f_{1u}$ in the $\log ft$ value, as given in (7.165), is indicated by the superscript '1'

$$\left(\nu 0d_{5/2}\,(\nu 0p_{1/2})^{-1}\,;\,J_f\|\beta^-_{\mathrm{GT}}\|\nu 0d_{5/2}\,(\pi 0p_{1/2})^{-1}\,;\,2^-\right)$$

$$=\sqrt{\frac{10}{3}}(-1)^{J_f+1}\widehat{J_f}\left\{\begin{matrix}2 & J_f & 1\\ \frac{1}{2} & \frac{1}{2} & \frac{5}{2}\end{matrix}\right\}\equiv A(J_f)\,,\quad (7.89)$$

$$\left(\pi 0d_{5/2}\,(\pi 0p_{1/2})^{-1}\,;\,J_f\|\beta^-_{\mathrm{GT}}\|\nu 0d_{5/2}\,(\pi 0p_{1/2})^{-1}\,;\,2^-\right)$$

$$=\sqrt{42}\,\widehat{J_f}\left\{\begin{matrix}2 & J_f & 1\\ \frac{5}{2} & \frac{5}{2} & \frac{1}{2}\end{matrix}\right\}\equiv B(J_f)\,,\quad (7.90)$$

$$\left(\nu 1s_{1/2}\,(\nu 0p_{1/2})^{-1}\,;\,1^-\|\beta^-_{\mathrm{GT}}\|\nu 0d_{5/2}\,(\pi 0p_{1/2})^{-1}\,;\,2^-\right)=0\,,\quad (7.91)$$

$$\left(\pi 1s_{1/2}\,(\pi 0p_{1/2})^{-1}\,;\,1^-\|\beta^-_{\mathrm{GT}}\|\nu 0d_{5/2}\,(\pi 0p_{1/2})^{-1}\,;\,2^-\right)=0\,.\quad (7.92)$$

Equations (7.82) and (7.83) give the matrix elements of the Fermi operator. The only non-zero ones are those for $J_f = 2$, namely

$$\left(\nu 0d_{5/2}\,(\nu 0p_{1/2})^{-1}\,;\,2^-\|\beta^-_{\mathrm{F}}\|\nu 0d_{5/2}\,(\pi 0p_{1/2})^{-1}\,;\,2^-\right)=-\sqrt{5}\,,\quad (7.93)$$

$$\left(\pi 0d_{5/2}\,(\pi 0p_{1/2})^{-1}\,;\,2^-\|\beta^-_{\mathrm{F}}\|\nu 0d_{5/2}\,(\pi 0p_{1/2})^{-1}\,;\,2^-\right)=\sqrt{5}\,.\quad (7.94)$$

With the wave functions (7.86)–(7.88) our results (7.89)–(7.94) give the transition amplitudes

$$(1^-_1\|\beta^-_{\mathrm{GT}}\|2^-_{\mathrm{gs}})=0\,,\quad (7.95)$$

$$(2^-_1\|\beta^-_{\mathrm{GT}}\|2^-_{\mathrm{gs}})=\frac{1}{\sqrt{2}}[A(2)+B(2)]=\frac{1}{\sqrt{2}}\left(\frac{1}{3}\sqrt{\frac{10}{3}}+7\sqrt{\frac{2}{15}}\right)=2.238\,,$$

$$(7.96)$$

$$(3_1^- \|\beta_{\mathrm{GT}}^-\|2_{\mathrm{gs}}^-) = \frac{1}{\sqrt{2}}[A(3) + B(3)] = \frac{1}{\sqrt{2}}\left(\frac{1}{3}\sqrt{\frac{35}{3}} + \sqrt{\frac{7}{15}}\right) = 1.288 \,,$$

(7.97)

$$(2_1^- \|\beta_{\mathrm{F}}^-\|2_{\mathrm{gs}}^-) = \frac{1}{\sqrt{2}}\left(\sqrt{5} - \sqrt{5}\right) = 0 \,.$$

(7.98)

For our simple wave functions with no configuration mixing the decay rate to the $1_1^-$ state is exactly zero. This is an example of an $l$-forbidden allowed transition; see Subsect. 7.2.7. Also the $2^- \to 2^-$ Fermi transition vanishes for the simple wave functions.

Equation (7.14) gives the non-vanishing reduced transition probabilities as

$$B_{\mathrm{GT}}(2_{\mathrm{gs}}^- \to 2_1^-) = \frac{1.25^2}{5} \times 2.238^2 = 1.565 \,,$$

(7.99)

$$B_{\mathrm{GT}}(2_{\mathrm{gs}}^- \to 3_1^-) = \frac{1.25^2}{5} \times 1.288^2 = 0.518 \,,$$

(7.100)

whence (7.12) yields

$$\log ft(2_{\mathrm{gs}}^- \to 2_1^-) = 3.59 \,,$$

(7.101)

$$\log ft(2_{\mathrm{gs}}^- \to 3_1^-) = 4.07 \,.$$

(7.102)

These $\log ft$ values represent appreciably faster transitions than the corresponding experimental values 4.3 and 4.5. The partial half-lives, both theoretical and experimental, are stated in Table 10.1 relating to the example of Subsect. 10.3.4. That example is a continuation of the present one with simple configuration mixing included.

The preceding calculation can be repeated for the Gamow–Teller and Fermi $\beta^+$ decays of Fig. 7.12. In this case the $4^-$ ground state of $^{40}_{21}\mathrm{Sc}_{19}$ decays to negative-parity states in $^{40}_{20}\mathrm{Ca}_{20}$. Equations (7.79) and (7.80) serve to give the decay amplitudes.

## 7.5 Beta-Decay Transitions in Two-Particle and Two-Hole Nuclei

In the following we discuss beta-decay transitions between a pair of two-particle states and between a pair of two-hole states. The different combinations of initial and final configurations are summarized in the following subsection. Applications of the formalism are discussed in the subsequent subsections.

## 7.5.1 Transition Amplitudes

Equations (6.93), (6.94), (6.100) and (6.101) give the two-particle states needed for describing beta decay between two-particle states. They can be summarized as

$$|\Psi_i\rangle = |a_i\, b_i\,;\, J_i\, M_i\rangle = \mathcal{N}_{a_i b_i}(J_i)\big[c_{a_i}^\dagger c_{b_i}^\dagger\big]_{J_i M_i}|\text{CORE}\rangle\,, \tag{7.103}$$

$$|\Psi_f\rangle = |a_f\, b_f\,;\, J_f\, M_f\rangle = \mathcal{N}_{a_f b_f}(J_f)\big[c_{a_f}^\dagger c_{b_f}^\dagger\big]_{J_f M_f}|\text{CORE}\rangle\,, \tag{7.104}$$

where the labels $a_i$ and $b_i$ are either proton or neutron labels. The normalization factor $\mathcal{N}$ is given by (5.21) for the proton–proton and neutron–neutron states. For the proton–neutron states it is simply $\mathcal{N}_{pn} = 1$.

The method of Sect. 6.3 is used to derive the amplitudes for beta-decay transitions between proton–proton and proton–neutron excitations and between neutron–neutron and proton–neutron excitations. The resulting complete set is

$$\boxed{\begin{aligned}
\mathcal{M}_L^{(-)}(n_i\, n_i'\,;\, J_i \to p_f\, n_f\,;\, J_f) &= \widehat{L}\widehat{J_i}\widehat{J_f}\mathcal{N}_{n_i n_i'}(J_i) \\
&\times \Bigg[\delta_{n_i' n_f}(-1)^{j_{p_f}+j_{n_f}+J_i+L}\begin{Bmatrix} J_i & J_f & L \\ j_{p_f} & j_{n_i} & j_{n_f} \end{Bmatrix}\mathcal{M}_L(p_f n_i) \\
&+ \delta_{n_i n_f}(-1)^{j_{p_f}+j_{n_i'}+L}\begin{Bmatrix} J_i & J_f & L \\ j_{p_f} & j_{n_i'} & j_{n_f} \end{Bmatrix}\mathcal{M}_L(p_f n_i')\Bigg]\,,
\end{aligned}} \tag{7.105}$$

$$\boxed{\begin{aligned}
\mathcal{M}_L^{(-)}(p_i\, n_i\,;\, J_i \to p_f\, p_f'\,;\, J_f) &= \widehat{L}\widehat{J_i}\widehat{J_f}\mathcal{N}_{p_f p_f'}(J_f) \\
&\times \Bigg[\delta_{p_i p_f'}(-1)^{j_{p_f}+j_{n_i}+L}\begin{Bmatrix} J_i & J_f & L \\ j_{p_f} & j_{n_i} & j_{p_f'} \end{Bmatrix}\mathcal{M}_L(p_f n_i) \\
&+ \delta_{p_i p_f}(-1)^{j_{p_f}+j_{n_i}+J_f+L}\begin{Bmatrix} J_i & J_f & L \\ j_{p_f'} & j_{n_i} & j_{p_f} \end{Bmatrix}\mathcal{M}_L(p_f' n_i)\Bigg]\,,
\end{aligned}} \tag{7.106}$$

$$\boxed{\begin{aligned}
\mathcal{M}_L^{(+)}(p_i\, p_i'\,;\, J_i \to p_f\, n_f\,;\, J_f) &= \widehat{L}\widehat{J_i}\widehat{J_f}\mathcal{N}_{p_i p_i'}(J_i) \\
&\times \Bigg[\delta_{p_i p_f}(-1)^{j_{n_f}+j_{p_f}+J_f+L}\begin{Bmatrix} J_i & J_f & L \\ j_{n_f} & j_{p_i'} & j_{p_f} \end{Bmatrix}\mathcal{M}_L(p_i' n_f) \\
&+ \delta_{p_i' p_f}(-1)^{j_{p_i}+j_{n_f}+J_i+J_f+L}\begin{Bmatrix} J_i & J_f & L \\ j_{n_f} & j_{p_i} & j_{p_f} \end{Bmatrix}\mathcal{M}_L(p_i n_f)\Bigg] \\
&= (-1)^{J_i+J_f}\mathcal{M}_L^{(-)}(p_f\, n_f\,;\, J_f \to p_i\, p_i'\,;\, J_i)\,,
\end{aligned}} \tag{7.107}$$

$$
\begin{aligned}
\mathcal{M}_L^{(+)}(p_i\, n_i\,;\,J_i \to n_f\, n_f'\,;\,J_f) &= \widehat{L}\widehat{J_i}\widehat{J_f}\mathcal{N}_{n_f n_f'}(J_f) \\
&\times \Bigg[ \delta_{n_i n_f'}(-1)^{j_{p_i}+j_{n_i}+J_i+L} \begin{Bmatrix} J_i & J_f & L \\ j_{n_f} & j_{p_i} & j_{n_f'} \end{Bmatrix} \mathcal{M}_L(p_i n_f) \\
&\quad + \delta_{n_i n_f}(-1)^{j_{p_i}+j_{n_f'}+J_i+J_f+L} \begin{Bmatrix} J_i & J_f & L \\ j_{n_f'} & j_{p_i} & j_{n_f} \end{Bmatrix} \mathcal{M}_L(p_i n_f') \Bigg] \\
&= (-1)^{J_i+J_f}\mathcal{M}_L^{(-)}(n_i\, n_i'\,;\,J_i \to p_f\, n_f\,;\,J_f)\,.
\end{aligned}
\tag{7.108}
$$

Here the single-particle matrix elements are defined in (7.73) and (7.74).

In the important special cases $J_i = 0$ and $J_f = 0$, the expressions (7.105)–(7.108) reduce to

$$
\begin{aligned}
\mathcal{M}_L^{(-)}(n_i\, n_i'\,;\,J_i \to p_f\, n_f\,;\,J_f = 0) &= \delta_{J_i L}\delta_{j_{p_f} j_{n_f}}\widehat{J_i}\widehat{j_{p_f}}^{-1}\mathcal{N}_{n_i n_i'}(J_i) \\
&\times \left[ \delta_{n_i' n_f}(-1)^{j_{n_i}+j_{n_f}+J_i+1}\mathcal{M}_L(p_f n_i) + \delta_{n_i n_f}\mathcal{M}_L(p_f n_i') \right]\,,
\end{aligned}
\tag{7.109}
$$

$$
\begin{aligned}
\mathcal{M}_L^{(-)}(n_i\, n_i'\,;\,J_i = 0 \to p_f\, n_f\,;\,J_f) &= \delta_{J_f L}\delta_{j_{n_i} j_{n_i'}}\widehat{J_f}\widehat{j_{n_i}}^{-1}\mathcal{N}_{n_i n_i'}(0) \\
&\times \left[ \delta_{n_i' n_f}\mathcal{M}_L(p_f n_i) + \delta_{n_i n_f}\mathcal{M}_L(p_f n_i') \right]\,,
\end{aligned}
\tag{7.110}
$$

$$
\begin{aligned}
\mathcal{M}_L^{(-)}(p_i\, n_i\,;\,J_i \to p_f\, p_f'\,;\,J_f = 0) &= \delta_{J_i L}\delta_{j_{p_f} j_{p_f'}}\widehat{J_i}\widehat{j_{p_f}}^{-1}\mathcal{N}_{p_f p_f'}(0) \\
&\times \left[ \delta_{p_i p_f'}\mathcal{M}_L(p_f n_i) + \delta_{p_i p_f}\mathcal{M}_L(p_f' n_i) \right]\,,
\end{aligned}
\tag{7.111}
$$

$$
\begin{aligned}
\mathcal{M}_L^{(-)}(p_i\, n_i\,;\,J_i = 0 \to p_f\, p_f'\,;\,J_f) &= \delta_{J_f L}\delta_{j_{n_i} j_{p_i}}\widehat{J_f}\widehat{j_{n_i}}^{-1}\mathcal{N}_{p_f p_f'}(J_f) \\
&\times \left[ \delta_{p_i p_f'}\mathcal{M}_L(p_f n_i) + \delta_{p_i p_f}(-1)^{j_{p_f}+j_{p_f'}+J_f+1}\mathcal{M}_L(p_f' n_i) \right]\,,
\end{aligned}
\tag{7.112}
$$

$$
\begin{aligned}
\mathcal{M}_L^{(+)}(p_i\, p_i'\,;\,J_i \to p_f\, n_f\,;\,J_f = 0) &= \delta_{J_i L}\delta_{j_{n_f} j_{p_f}}\widehat{J_i}\widehat{j_{n_f}}^{-1}\mathcal{N}_{p_i p_i'}(J_i) \\
&\times \left[ \delta_{p_i p_f}(-1)^{j_{p_f}+j_{p_i'}+1}\mathcal{M}_L(p_i' n_f) + \delta_{p_i' p_f}(-1)^{J_i}\mathcal{M}_L(p_i n_f) \right]\,,
\end{aligned}
\tag{7.113}
$$

$$
\begin{aligned}
\mathcal{M}_L^{(+)}(p_i\, p_i'\,;\,J_i = 0 \to p_f\, n_f\,;\,J_f) &= \delta_{J_f L}\delta_{j_{p_i} j_{p_i'}}(-1)^{J_f}\widehat{J_f}\widehat{j_{p_i}}^{-1}\mathcal{N}_{p_i p_i'}(0) \\
&\times \left[ \delta_{p_i p_f}\mathcal{M}_L(p_i' n_f) + \delta_{p_i' p_f}\mathcal{M}_L(p_i n_f) \right]\,,
\end{aligned}
\tag{7.114}
$$

$$
\begin{aligned}
\mathcal{M}_L^{(+)}(p_i\, n_i\,;\,J_i \to n_f\, n_f'\,;\,J_f = 0) &= \delta_{J_i L}\delta_{j_{n_f} j_{n_f'}}(-1)^{J_i}\widehat{J_i}\widehat{j_{n_f}}^{-1}\mathcal{N}_{n_f n_f'}(0) \\
&\times \left[ \delta_{n_i n_f'}\mathcal{M}_L(p_i n_f) + \delta_{n_i n_f}\mathcal{M}_L(p_i n_f') \right]\,,
\end{aligned}
\tag{7.115}
$$

$$\mathcal{M}_L^{(+)}(p_i\,n_i\,;\,J_i=0\to n_f\,n'_f\,;\,J_f)=\delta_{J_f L}\delta_{j_{p_i}j_{n_i}}\widehat{J_f}\widehat{j_{p_i}}^{-1}\mathcal{N}_{n_f n'_f}(J_f)$$
$$\times\left[\delta_{n_i n'_f}(-1)^{j_{n_i}+j_{n_f}+1}\mathcal{M}_L(p_i n_f)+\delta_{n_i n_f}(-1)^{J_f}\mathcal{M}_L(p_i n'_f)\right].\quad(7.116)$$

One can show that for *two-hole nuclei* the formulas corresponding to (7.105)–(7.116) apply with the substitutions

$$\boxed{\mathcal{M}_L^{(\pm)}\xrightarrow{p\to h}\mathcal{M}_L^{(\mp)}\quad\text{and}\quad\mathcal{M}_L(pn)\xrightarrow{p\to h}(-1)^{L+1}\mathcal{M}_L(pn)\,.}\quad(7.117)$$

### 7.5.2 Application to Beta Decay of $^6$He

Consider the $\beta^-$ decay of $^6$He to $^6$Li, depicted in Fig. 7.8. With the experimental $\log ft$ value of 2.9 this is a superallowed transition according to Table 7.4. In our simple mean-field shell-model scheme the low-energy states of $^6$He and $^6$Li have the structure given by (5.26) and (5.27):

$$|^6\text{He}\,;\,0^+\rangle=\frac{1}{\sqrt{2}}\left[c^\dagger_{\nu 0p_{3/2}}c^\dagger_{\nu 0p_{3/2}}\right]_{0^+}|\text{CORE}\rangle\,,\quad(7.118)$$

$$|^6\text{Li}\,;\,1^+\rangle=\left[c^\dagger_{\pi 0p_{3/2}}c^\dagger_{\nu 0p_{3/2}}\right]_{1^+}|\text{CORE}\rangle\,.\quad(7.119)$$

The transition is from a two-neutron nucleus to a proton–neutron nucleus and the initial angular momentum is zero. Therefore a suitable formula for the calculation is (7.110), which gives

$$\mathcal{M}_{\text{GT}}^{(-)}(\nu 0p_{3/2}\,\nu 0p_{3/2}\,;\,J_i=0\to\pi 0p_{3/2}\,\nu 0p_{3/2}\,;\,J_f=1)$$
$$=\sqrt{3}\times\frac{1}{2}\times\frac{1}{\sqrt{2}}\left[\mathcal{M}_{\text{GT}}(p_{3/2}\,p_{3/2})+\mathcal{M}_{\text{GT}}(p_{3/2}\,p_{3/2})\right]=\sqrt{\frac{10}{3}}=1.826\,,$$
$$(7.120)$$

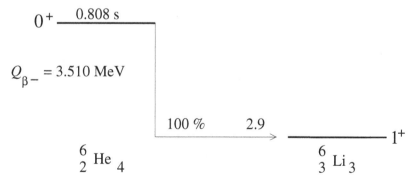

**Fig. 7.8.** Superallowed beta-minus decay of the $0^+$ ground state of $^6$He to the $1^+$ ground state of $^6$Li. The experimental half-life, $Q$ value, branching and $\log ft$ value are given

where $\mathcal{M}_{GT}(p_{3/2}\,p_{3/2}) = \frac{2\sqrt{5}}{3}$ was taken from Table 7.3. Via the equations of Subsect. 7.2.1 this leads to

$$B_{GT} = 5.208 , \quad \log ft = 3.07 . \tag{7.121}$$

This is in good agreement with the experimental value $\log ft = 2.9$. To calculate the half-life we first apply (7.25), (7.29), (7.30) and (7.34) to find the phase-space factor $f_0^{(-)}$. The intermediate and final results are

$$F_0^{(PR)}(Z_f = 3) = 1.070 , \quad E_0 = 7.869 , \quad f_0^{(-)} = 1058 . \tag{7.122}$$

Equation (7.17) now gives the half-life as

$$t_{1/2} = 1.11\,\text{s} , \tag{7.123}$$

which is close to the measured value of 0.808 s.

### 7.5.3 Application to the Beta-Decay Chain $^{18}$Ne $\rightarrow$ $^{18}$F $\rightarrow$ $^{18}$O

Our second example deals with the two-particle nuclei $^{18}$O, $^{18}$F and $^{18}$Ne. Their relevant energy levels and beta-decay transitions are presented in Fig. 7.9. Contrary to Fig. 5.5, here the Coulomb energy is retained. The beta decays contain both Fermi ($0^+ \rightarrow 0^+$) and Gamow–Teller ($0^+ \rightarrow 1^+$, $1^+ \rightarrow 0^+$) transitions.

The lowest $0^+$ states in $^{18}_{8}$O$_{10}$, $^{18}_{9}$F$_{9}$ and $^{18}_{10}$Ne$_{8}$ form an isospin triplet as shown in Fig. 5.5. Their structure is given by (5.30)–(5.32) as

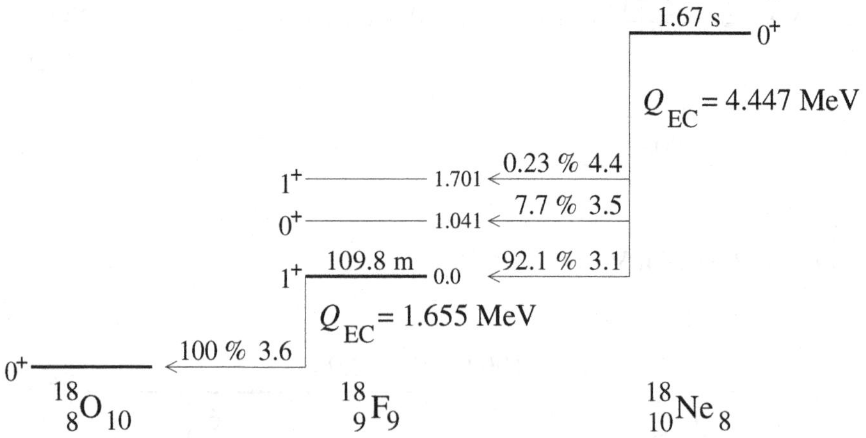

**Fig. 7.9.** Beta decay of the $0^+$ ground state of $^{18}$Ne to the $0^+$ and $1^+$ states of $^{18}$F, and the decay of the $1^+$ ground state of $^{18}$F to the $0^+$ ground state of $^{18}$O. The decays proceed via the $\beta^+$/EC mode. The experimental half-lives, $Q$ values, branchings and $\log ft$ values are given

$$|^{18}\mathrm{O}\,;\,0^+\rangle = \frac{1}{\sqrt{2}}\big[c^\dagger_{\nu 0d_{5/2}}c^\dagger_{\nu 0d_{5/2}}\big]_{0^+}|\mathrm{CORE}\rangle\,, \tag{7.124}$$

$$|^{18}\mathrm{F}\,;\,0^+\rangle = \big[c^\dagger_{\pi 0d_{5/2}}c^\dagger_{\nu 0d_{5/2}}\big]_{0^+}|\mathrm{CORE}\rangle\,, \tag{7.125}$$

$$|^{18}\mathrm{Ne}\,;\,0^+\rangle = \frac{1}{\sqrt{2}}\big[c^\dagger_{\pi 0d_{5/2}}c^\dagger_{\pi 0d_{5/2}}\big]_{0^+}|\mathrm{CORE}\rangle\,. \tag{7.126}$$

Consider the Fermi transition $^{18}\mathrm{Ne}(0^+) \to {}^{18}\mathrm{F}(0^+)$. Equation (7.113) with $J_i = 0$, or (7.114) with $J_f = 0$, yields

$$\mathcal{M}_{\mathrm{F}}^{(+)}(\pi 0d_{5/2}\,\pi 0d_{5/2}\,;\,J_i = 0 \to \pi 0d_{5/2}\,\nu 0d_{5/2}\,;\,J_f = 0)$$
$$= \frac{1}{\sqrt{6}} \times \frac{1}{\sqrt{2}} \times 2\mathcal{M}_{\mathrm{F}}(d_{5/2}\,d_{5/2}) = \sqrt{2}\,, \tag{7.127}$$

where $\mathcal{M}_{\mathrm{F}}(d_{5/2}\,d_{5/2}) = \sqrt{6}$ is from (7.20). It follows that

$$B_{\mathrm{F}} = 2\,,\quad \log ft = 3.49\,. \tag{7.128}$$

The perfect agreement with the experimental $\log ft$ value of 3.5 makes this a nice example of a superallowed transition.

Departing from the simple mean-field shell model, we now assume configuration mixing in the $1^+$ states of $^{18}\mathrm{F}$. Appreciable mixing can be expected between the states $[\pi 0d_{5/2}\nu 0d_{5/2}]_{1^+}|\mathrm{CORE}\rangle$ and $[\pi 1s_{1/2}\nu 1s_{1/2}]_{1^+}|\mathrm{CORE}\rangle$ because they are close in energy. We form two orthogonal combinations of these components,

$$|^{18}\mathrm{F}\,;\,1_1^+\rangle = \alpha\big[c^\dagger_{\pi 0d_{5/2}}c^\dagger_{\nu 0d_{5/2}}\big]_{1^+} + \beta\big[c^\dagger_{\pi 1s_{1/2}}c^\dagger_{\nu 1s_{1/2}}\big]_{1^+}|\mathrm{CORE}\rangle\,, \tag{7.129}$$

$$|^{18}\mathrm{F}\,;\,1_2^+\rangle = -\beta\big[c^\dagger_{\pi 0d_{5/2}}c^\dagger_{\nu 0d_{5/2}}\big]_{1^+} + \alpha\big[c^\dagger_{\pi 1s_{1/2}}c^\dagger_{\nu 1s_{1/2}}\big]_{1^+}|\mathrm{CORE}\rangle\,, \tag{7.130}$$

with the normalization condition

$$\alpha^2 + \beta^2 = 1\,. \tag{7.131}$$

From (5.31) we expect that the leading configuration of the $1_1^+$ state is $\pi 0d_{5/2}\,\nu 0d_{5/2}$, i.e. $\beta$ is appreciably smaller than $\alpha$.

Equations (7.114) and (7.115) and Table 7.3 give the relevant Gamow–Teller matrix elements

$$\mathcal{M}_{\mathrm{GT}}^{(+)}(\pi 0d_{5/2}\,\nu 0d_{5/2}\,;\,1 \to (\nu 0d_{5/2})^2\,;\,0) = -\sqrt{\tfrac{14}{5}}\,, \tag{7.132}$$

$$\mathcal{M}_{\mathrm{GT}}^{(+)}(\pi 1s_{1/2}\,\nu 1s_{1/2}\,;\,1 \to (\nu 0d_{5/2})^2\,;\,0) = 0\,, \tag{7.133}$$

$$\mathcal{M}_{\mathrm{GT}}^{(+)}((\pi 0d_{5/2})^2\,;\,0 \to \pi 0d_{5/2}\,\nu 0d_{5/2}\,;\,1) = -\sqrt{\tfrac{14}{5}}\,, \tag{7.134}$$

$$\mathcal{M}_{\mathrm{GT}}^{(+)}((\pi 0d_{5/2})^2\,;\,0 \to \pi 1s_{1/2}\,\nu 1s_{1/2}\,;\,1) = 0\,. \tag{7.135}$$

The final Gamow–Teller matrix elements for the wave functions (7.129) and (7.130) now become

$$\mathcal{M}_{GT}\left({}^{18}F(1_1^+) \rightarrow {}^{18}O(0_{gs}^+)\right) = -\sqrt{\tfrac{14}{5}}\alpha \,, \tag{7.136}$$

$$\mathcal{M}_{GT}\left({}^{18}Ne(0_{gs}^+) \rightarrow {}^{18}F(1_1^+)\right) = -\sqrt{\tfrac{14}{5}}\alpha \,, \tag{7.137}$$

$$\mathcal{M}_{GT}\left({}^{18}Ne(0_{gs}^+) \rightarrow {}^{18}F(1_2^+)\right) = \sqrt{\tfrac{14}{5}}\beta \,. \tag{7.138}$$

The reduced transition probabilities (7.14) are

$$B_{GT}(1_1^+ \rightarrow 0_{gs}^+) = \tfrac{14}{15}g_A^2\alpha^2 \,, \tag{7.139}$$

$$B_{GT}(0_{gs}^+ \rightarrow 1_1^+) = \tfrac{14}{5}g_A^2\alpha^2 \,, \tag{7.140}$$

$$B_{GT}(0_{gs}^+ \rightarrow 1_2^+) = \tfrac{14}{5}g_A^2\beta^2 \,. \tag{7.141}$$

We can use (7.140) and (7.141) to determine the ratio $\alpha^2/\beta^2$ from the experimental $\log ft$ values of Fig. 7.9. By means of (7.12) we find

$$B_{GT}(0_{gs}^+ \rightarrow 1_1^+)_{exp} = 4.88 \,, \tag{7.142}$$

$$B_{GT}(0_{gs}^+ \rightarrow 1_2^+)_{exp} = 0.245 \,. \tag{7.143}$$

Thus we have, with use of (7.131),

$$\frac{\alpha^2}{\beta^2} = \frac{4.88}{0.245} = 19.9 \,, \quad \beta^2 = \left(1 + \frac{\alpha^2}{\beta^2}\right)^{-1} = 0.048 \,, \quad \alpha^2 = 0.952 \,. \tag{7.144}$$

Substituting $\alpha^2$ and $\beta^2$ into (7.139)–(7.141) leads to the theoretical $\log ft$ values

$$\log ft\left({}^{18}F(1_1^+) \rightarrow {}^{18}O(0_{gs}^+)\right) = 3.65 \,, \tag{7.145}$$

$$\log ft\left({}^{18}Ne(0_{gs}^+) \rightarrow {}^{18}F(1_1^+)\right) = 3.17 \,, \tag{7.146}$$

$$\log ft\left({}^{18}Ne(0_{gs}^+) \rightarrow {}^{18}F(1_2^+)\right) = 4.47 \,. \tag{7.147}$$

The corresponding experimental values in Fig. 7.9 are 3.6, 3.1 and 4.4. The agreement between theory and experiment is excellent, but it must be remembered that the theoretical results were *fitted* to experiment to the extent that the ratio $\alpha^2/\beta^2$ was taken from the data.

An important qualitative result is that without the $\beta$ term in the $1^+$ wave functions there would be no transition ${}^{18}Ne(0_{gs}^+) \rightarrow {}^{18}F(1_2^+)$. Minor modifications to the present scheme can be expected from further configuration mixing, as will be discussed in later chapters. Our present conclusion is that the $1_1^+$ and $1_2^+$ states of ${}^{18}F$ are well described by the wave functions

$$|{}^{18}F;\, 1_1^+\rangle = 0.98\left[c_{\pi 0d_{5/2}}^\dagger c_{\nu 0d_{5/2}}^\dagger\right]_{1+}|CORE\rangle$$
$$+ 0.22\left[c_{\pi 1s_{1/2}}^\dagger c_{\nu 1s_{1/2}}^\dagger\right]_{1+}|CORE\rangle \,, \tag{7.148}$$

$$|{}^{18}F;\, 1_2^+\rangle = -0.22\left[c_{\pi 0d_{5/2}}^\dagger c_{\nu 0d_{5/2}}^\dagger\right]_{1+}|CORE\rangle$$
$$+ 0.98\left[c_{\pi 1s_{1/2}}^\dagger c_{\nu 1s_{1/2}}^\dagger\right]_{1+}|CORE\rangle \,. \tag{7.149}$$

### 7.5.4 Further Examples: Beta Decay in $A = 42$ and $A = 54$ Nuclei

In this subsection we proceed with further examples of beta decay in two-particle and two-hole nuclei. Figures 7.10 and 7.11 show beta-decay transitions in two-particle ($A = 42$) and two-hole ($A = 54$) nuclei in the $0f_{7/2}$ shell. All these decays are of the $\beta^+/EC$ type.

Every nucleus involved in these examples can be described as having either two particles in the $0f_{7/2}$ orbital ($^{42}$Ca, $^{42}$Sc and $^{42}$Ti) or two holes in the $0f_{7/2}$ orbital ($^{54}$Fe and $^{54}$Co). This simple description leads to good agreement between theoretical and experimental $\log ft$ values. In the figures the calculated $\log ft$ values are shown in parentheses next to the experimental ones. Only the $\log ft$ value for the $7^+ \rightarrow 6^+$ decay in the $A = 54$ nuclei is inadequately accounted for by the theory.

The computed $\log ft$ values can be verified, and half-lives can be computed, by calculations similar to those of the preceding examples. Comparison between theory and experiment can then be extended to half-lives.

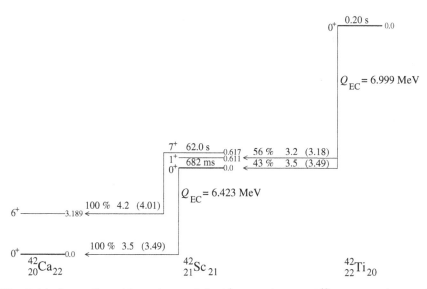

**Fig. 7.10.** Superallowed beta decay of the $0^+$ ground state of $^{42}$Ti to the $0^+$ and $1^+$ states of $^{42}$Sc. Also shown are decay of the $0^+$ ground state and the $7^+$ excited state of $^{42}$Sc to the $0^+$ ground state and $6^+$ excited state of $^{42}$Ca; the Fermi transition is superallowed. The decays proceed via the $\beta^+/EC$ mode. The experimental half-lives, $Q$ values, branchings and $\log ft$ values are also given. The numbers in parentheses are calculated $\log ft$ values

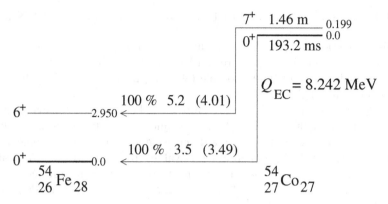

**Fig. 7.11.** Beta decay of the $0^+$ ground state and $7^+$ excited state of $^{54}$Co to the $0^+$ ground state and $6^+$ excited state of $^{54}$Fe. The Fermi transition is superallowed. The decays proceed via the $\beta^+$/EC mode. The experimental half-lives, $Q$ value, decay branchings and $\log ft$ values are given. The numbers in parentheses are calculated $\log ft$ values

## 7.6 Forbidden Unique Beta Decay

In this section forbidden unique beta-decay transitions are addressed in detail. These transitions are an important subgroup of forbidden decay transitions. For a given degree of forbiddenness they involve the maximum possible angular momentum difference between the initial and final states of decay. First we review some general aspects of first-forbidden beta decay and then concentrate on the exact formulation of first-forbidden and, subsequently, higher-forbidden unique beta decay.

### 7.6.1 General Aspects of First-Forbidden Beta Decay

So far in this chapter we have discussed allowed beta decay, with its operators and resultant matrix elements

$$\mathcal{O}_{\mathrm{F}} = 1 \quad \longrightarrow \quad \text{Fermi matrix element} , \tag{7.150}$$

$$\mathcal{O}_{\mathrm{GT}} = \sigma \quad \longrightarrow \quad \text{Gamow–Teller matrix element} . \tag{7.151}$$

The spectrum shape for allowed beta decay was given by (7.28) and depicted in Fig. 7.4. In the allowed decay modes leptons are emitted in an s state, as stated at the beginning of Sect. 7.2. The next step is to include the possibility of emission of p-wave leptons and the contributions coming from the small components of the relativistic Dirac wave functions. The p-wave contributions are suppressed relative to allowed decay by the factor

$$\left( \frac{Qb}{\hbar c} \right)^2 \lesssim \frac{1}{100} , \tag{7.152}$$

where $Q$ is the decay energy and $b$ is the oscillator length (3.43) calculated from (3.45). The relativistic effects from the small components of the Dirac spinors of the nucleons are suppressed relative to allowed decay by the factor

$$\left(\frac{\hbar c}{2m_N c^2 b}\right)^2 \lesssim \frac{1}{400} \, , \qquad (7.153)$$

where $m_N$ is the mass of the nucleon.

There are four types of nuclear matrix element that emerge from p-wave leptons, namely

$$\mathcal{O}_{SA} = \boldsymbol{\sigma} \cdot \boldsymbol{r} \quad \longrightarrow \quad \text{scalar–axial matrix element} , \qquad (7.154)$$

$$\mathcal{O}_{VA} = i\sqrt{\frac{3}{2}} \boldsymbol{\sigma} \times \boldsymbol{r} \quad \longrightarrow \quad \text{vector–axial matrix element} , \qquad (7.155)$$

$$\mathcal{O}_{TA} = \frac{\sqrt{3}}{2} [\boldsymbol{\sigma} \boldsymbol{r}]_2 \quad \longrightarrow \quad \text{tensor–axial matrix element} , \qquad (7.156)$$

$$\mathcal{O}_{VV} = -\sqrt{3} r \quad \longrightarrow \quad \text{vector–vector matrix element} . \qquad (7.157)$$

These are expressed in the CS phase convention introduced in (3.67) and in the so-called Cartesian notation [55]. The operator $\mathcal{O}_{SA}$ is a pseudoscalar, $\mathcal{O}_{VA}$ and $\mathcal{O}_{VV}$ are vectors, and $\mathcal{O}_{TA}$ is a rank-2 spherical pseudotensor.

There are two types of nuclear matrix element that emerge from the small components of the Dirac spinors, namely

$$\mathcal{O}_{RA} = -\gamma_5 \quad \longrightarrow \quad \text{recoil-axial matrix element} , \qquad (7.158)$$
$$\mathcal{O}_{RV} = \boldsymbol{\alpha} \quad \longrightarrow \quad \text{recoil-vector matrix element} . \qquad (7.159)$$

The Dirac matrices $\alpha_i$ and $\gamma_i$ are defined e.g. in [53]. The operator $\gamma_5$ is a pseudoscalar and the operator $\boldsymbol{\alpha}$ a vector.

The six matrix elements (7.154)–(7.159) contribute to *first-forbidden* nuclear beta decay. Each of the operators changes the parity (for more information see e.g. [54, 55, 62]). This means that the initial and final nuclear states must have opposite parities, i.e. $\pi_i \pi_f = -1$.

Similar to (7.24) and (7.26), the phase-space factors for first-forbidden beta decay are

$$f_1^{(\mp)} = \int_1^{E_0} S_1^{(\mp)}(Z_f, \varepsilon) p\varepsilon(E_0 - \varepsilon)^2 d\varepsilon \, , \qquad (7.160)$$

$$f_1^{(EC)} = 2\pi(\alpha Z_i)^3 S_1^{(EC)}(\varepsilon_0, E_0) \, , \qquad (7.161)$$

where the quantities $S_1^{(\mp)}$ and $S_1^{(EC)}$ are the shape functions of first-forbidden $\beta^{\mp}$ decay and electron capture. The shape functions contain all or part of the nuclear matrix elements (7.154)–(7.159) and the appropriate lepton kinematics.

The shape functions $S_1$ are much more complicated than their allowed-decay counterparts in (7.24) and (7.26). However, a considerable simplification is achieved when $\Delta J = 2$. This is the greatest value of $\Delta J$ achievable with the operators (7.154)–(7.159), and then only the tensor–axial matrix element (7.156) contributes. The associated lepton spectrum shape is simple. This decay mode is called *first-forbidden unique* decay. The transitions are analysed in detail in the following subsection.

### 7.6.2 First-Forbidden Unique Beta Decay

The shape function of a first-forbidden unique $\beta^{\mp}$ transition can be taken to be [59]

$$
\begin{aligned}
S_{1u}^{(\mp)}(Z_f, \varepsilon) &= F_{1u}(\pm Z_f, \varepsilon) p\varepsilon (E_0 - \varepsilon)^2 \\
&= \left[ F_0(\pm Z_f, \varepsilon)(E_0 - \varepsilon)^2 + F_1(\pm Z_f, \varepsilon)(\varepsilon^2 - 1) \right] p\varepsilon (E_0 - \varepsilon)^2 .
\end{aligned}
\tag{7.162}
$$

The functions $F_0(\pm Z_f, \varepsilon)$ and $F_1(\pm Z_f, \varepsilon)$ are Fermi functions to be discussed later.[5] The endpoint energy $E_0$ is given by (7.34) for a $\beta^-$ transition and by (7.35) for a $\beta^+$ transition. The phase-space factor is now

$$
f_{1u}^{(\mp)} = \int_1^{E_0} S_{1u}^{(\mp)}(Z_f, \varepsilon) d\varepsilon .
\tag{7.163}
$$

For first-forbidden unique electron capture the phase-space factor is

$$
f_{1u}^{(EC)} = \frac{2\pi}{9} (\alpha Z_i)^3 (\varepsilon_0 + E_0)^4 ,
\tag{7.164}
$$

where $\varepsilon_0$ is given by (7.25) and the endpoint energy $E_0$ by (7.36).

The $ft$ value for first-forbidden unique beta decay can be defined as [59]

$$
ft \equiv f_{1u} t_{1/2} = \frac{\kappa}{\frac{1}{12} B_{1u}} , \qquad B_{1u} = \frac{g_A^2}{2J_i + 1} |\mathcal{M}_{1u}|^2
\tag{7.165}
$$

with $\kappa$ given by (7.13) and

$$
f_{1u} = \begin{cases} f_{1u}^{(-)} & \text{for } \beta^- \text{ decay} , \\ f_{1u}^{(+)} + f_{1u}^{(EC)} & \text{for } \beta^+/EC \text{ decay} . \end{cases}
\tag{7.166}
$$

Like the $ft$ value (7.33) for allowed decay, the $ft$ value (7.165) is phase-space independent and depends only on nuclear structure. However, the difference between $f_0$ and $f_{1u}$ on the left-hand side means that allowed and

---

[5] The Fermi function $F_0$ appeared already in the phase-space factor (7.24), and an approximate expression for it was given in (7.29).

first-forbidden $ft$ values are not commensurate. This is recognized in the caption to Fig. 7.7.

The nuclear matrix element $\mathcal{M}_{1u}$ in (7.165) is

$$\mathcal{M}_{1u} = \frac{m_e c^2}{\sqrt{4\pi}} \zeta (\xi_f\, J_f \| [\boldsymbol{\sigma} r]_2 \| \xi_i\, J_i) = \sum_{ab} \mathcal{M}^{(1u)}(ab) (\xi_f\, J_f \| [c_a^\dagger \tilde{c}_b]_2 \| \xi_i\, J_i) ,$$

(7.167)

where $\zeta = 1$ for the CS and $\zeta = i$ for the BR phase convention. The single-particle matrix element is given by

$$\mathcal{M}^{(1u)}(ab) = 2.990 \times 10^{-3} \times b\,[\text{fm}] \times m^{(1u)}(ab) .$$

(7.168)

The number $2.990 \times 10^{-3}$ comes from various natural constants, and $m^{(1u)}(ab)$ is given by

$$
\begin{aligned}
m^{(1u)}(ab) &= \zeta_{ab}^{(1u)} \frac{1}{2\sqrt{10}} (-1)^{l_a + j_a + j_b} \frac{1 + (-1)^{l_a + l_b + 1}}{2} \hat{j}_a \hat{j}_b \begin{pmatrix} j_a & j_b & 2 \\ \tfrac{1}{2} & -\tfrac{1}{2} & 0 \end{pmatrix} \\
&\quad \times \left[ -\hat{j}_a^2 + (-1)^{j_a + j_b + 1} \hat{j}_b^2 + 4(-1)^{l_a + j_a + \frac{1}{2}} \right] \tilde{\mathcal{R}}_{ab}^{(1)} , \\
\zeta_{ab}^{(1u)} &= \begin{cases} 1 & \text{CS phase convention ,} \\ (-1)^{(l_b - l_a + 1)/2} & \text{BR phase convention ,} \end{cases}
\end{aligned}
$$

(7.169)

where the scaled radial integral is given in (6.32) and tabulated in Table 6.2. The scaled single-particle matrix elements $m^{(1u)}(ab)$ are given in Table 7.6 in the CS phase convention.

The symmetry relations for the basic single-particle matrix elements are

$$m_{\text{CS}}^{(1u)}(ba) = (-1)^{j_a + j_b + 1} m_{\text{CS}}^{(1u)}(ab) ,$$

(7.170)

$$m_{\text{BR}}^{(1u)}(ba) = (-1)^{j_a + j_b} m_{\text{BR}}^{(1u)}(ab)$$

(7.171)

for the CS and BR phase conventions.

The shape function $S_{1u}^{(\mp)}$ stated in (7.162) can be simplified by using the Primakoff–Rosen approximation (7.29) for the Fermi functions $F_0$ and $F_1$. In this approximation, valid for $\alpha Z \ll 1$,

$$F_1(Z, \varepsilon) \approx F_0(Z, \varepsilon) .$$

(7.172)

This leads to

$$
\begin{aligned}
S_{1u}^{(\mp)} &\approx F_0(\pm Z_f, \varepsilon)[(E_0 - \varepsilon)^2 + (\varepsilon^2 - 1)]p\varepsilon(E_0 - \varepsilon)^2 \\
&\approx F_0^{(\text{PR})}(\pm Z_f)[(E_0 - \varepsilon)^2 + (\varepsilon^2 - 1)]\varepsilon^2 (E_0 - \varepsilon)^2 .
\end{aligned}
$$

(7.173)

This, in turn, yields the phase-space factor (7.163) as

**Table 7.6.** Scaled nuclear single-particle matrix elements $m_{\text{CS}}^{(1u)}(ab)$ in the CS phase convention

| $a\backslash b$ | $0s_{1/2}$ | $0p_{3/2}$ | $0p_{1/2}$ | $0d_{5/2}$ | $1s_{1/2}$ | $0d_{3/2}$ | $0f_{7/2}$ | $1p_{3/2}$ | $0f_{5/2}$ | $1p_{1/2}$ | $0g_{9/2}$ |
|---|---|---|---|---|---|---|---|---|---|---|---|
| $0s_{1/2}$ | 0 | $\sqrt{3}$ | 0 | 0 | 0 | 0 | 0 | 0 | 0 | 0 | 0 |
| $0p_{3/2}$ | $-\sqrt{3}$ | 0 | 0 | $\sqrt{\frac{21}{5}}$ | $\sqrt{2}$ | $-\frac{2}{\sqrt{5}}$ | 0 | 0 | 0 | 0 | 0 |
| $0p_{1/2}$ | 0 | 0 | 0 | $2\sqrt{\frac{6}{5}}$ | 0 | $-\frac{1}{\sqrt{5}}$ | 0 | 0 | 0 | 0 | 0 |
| $0d_{5/2}$ | 0 | $-\sqrt{\frac{21}{5}}$ | $2\sqrt{\frac{6}{5}}$ | 0 | 0 | 0 | $\frac{9\sqrt{2}}{5}$ | $\frac{\sqrt{42}}{5}$ | $-\frac{4\sqrt{3}}{5}$ | $-\frac{4\sqrt{3}}{5}$ | 0 |
| $1s_{1/2}$ | 0 | $-\sqrt{2}$ | 0 | 0 | 0 | 0 | 0 | $\sqrt{5}$ | 0 | 0 | 0 |
| $0d_{3/2}$ | 0 | $-\frac{2}{\sqrt{5}}$ | $\frac{1}{\sqrt{5}}$ | 0 | 0 | 0 | $\frac{12\sqrt{2}}{5}$ | $\frac{2\sqrt{2}}{5}$ | $-\frac{3\sqrt{3}}{5}$ | $-\frac{\sqrt{2}}{5}$ | 0 |
| $0f_{7/2}$ | 0 | 0 | 0 | $-\frac{9\sqrt{2}}{5}$ | 0 | $\frac{12\sqrt{2}}{5}$ | 0 | 0 | 0 | 0 | $\sqrt{\frac{66}{7}}$ |
| $1p_{3/2}$ | 0 | 0 | 0 | $-\frac{\sqrt{42}}{5}$ | $-\sqrt{5}$ | $\frac{2\sqrt{2}}{5}$ | 0 | 0 | 0 | 0 | 0 |
| $0f_{5/2}$ | 0 | 0 | 0 | $-\frac{4\sqrt{3}}{5}$ | 0 | $\frac{3\sqrt{3}}{5}$ | 0 | 0 | 0 | 0 | $\frac{12}{\sqrt{7}}$ |
| $1p_{1/2}$ | 0 | 0 | 0 | $-\frac{4\sqrt{3}}{5}$ | 0 | $\frac{\sqrt{2}}{5}$ | 0 | 0 | 0 | 0 | 0 |
| $0g_{9/2}$ | 0 | 0 | 0 | 0 | 0 | 0 | $-\sqrt{\frac{66}{7}}$ | 0 | $\frac{12}{\sqrt{7}}$ | 0 | 0 |

$$f_{1u}^{(\mp)} \approx \frac{1}{30}\left(\frac{4}{7}E_0^7 - E_0^5 - 10E_0^4 + 30E_0^3 - 32E_0^2 + 15E_0 - \frac{18}{7}\right)F_0^{(\text{PR})}(\pm Z_f) \,. \tag{7.174}$$

### 7.6.3 Application to First-Forbidden Unique Beta Decay of $^{16}$N

Consider the $\beta^-$ decay of the $2^-$ ground state of $^{16}$N to the $0^+$ ground state of $^{16}$O, as depicted in Fig 7.7. This is a first-forbidden unique decay transition since $\pi_i\pi_f = -1$ and $\Delta J = 2$. The initial wave function, as given in (5.60), is

$$|^{16}\text{N}\,; 2^-\rangle = \left[c_{\nu 0d_{5/2}}^{\dagger}h_{\pi 0p_{1/2}}^{\dagger}\right]_{2^-}|\text{HF}\rangle \,. \tag{7.175}$$

The one-body transition density (7.59) gives for the nuclear matrix element (7.167) the expression

$$\mathcal{M}_{1u}(a_i b_i^{-1}\,; J_i \to 0_{\text{gs}}^+) = \sum_{ab}\mathcal{M}^{(1u)}(ab)\delta_{J_i 2}\delta_{ba_i}\delta_{ab_i}(-1)^{j_{a_i}+j_{b_i}+2+1}\sqrt{5}$$
$$= \delta_{J_i 2}(-1)^{j_{a_i}+j_{b_i}+1}\sqrt{5}\mathcal{M}^{(1u)}(b_i a_i) = \delta_{J_i 2}\sqrt{5}\mathcal{M}^{(1u)}(a_i b_i) \,, \tag{7.176}$$

where the symmetry relation (7.170) for the CS phase convention was used in the last step. This result,

$$\boxed{\mathcal{M}_{1u}(a_i b_i^{-1}\,; J_i \to 0_{\text{gs}}^+) = \delta_{J_i 2}\sqrt{5}\mathcal{M}^{(1u)}(a_i b_i) \,,} \tag{7.177}$$

applies to any decay transition from an odd–odd particle–hole nucleus to the particle–hole ground state of the adjacent even–even reference nucleus. The charge-changing particle–hole excitation can be either of the proton-particle–neutron-hole or neutron-particle–proton-hole type.

From (3.45) and (3.43) we find $b = 1.725\,\mathrm{fm}$. Equation (7.168) and Table 7.6 then yield for the nuclear matrix element (7.177) the value

$$\mathcal{M}_{1u} = \sqrt{5} \times 2.990 \times 10^{-3} \times 1.725 m^{(1u)}(0\mathrm{d}_{5/2}\,0\mathrm{p}_{1/2})$$

$$= 0.0115 \times 2\sqrt{\tfrac{6}{5}} = 0.0252 \ . \tag{7.178}$$

From (7.165) we find the reduced transition probability

$$B_{1u} = \frac{1.25^2}{5} \times 0.0252^2 = 1.98 \times 10^{-4} \tag{7.179}$$

and the log $ft$ value

$$\log f_{1u}t = 8.57 \ . \tag{7.180}$$

The experimental result from Fig. 7.7 is $\log f_{1u}t = 9.1$, so that the computed decay transition is slightly too fast.

With the $Q_{\beta-}$ value from Fig. 7.7, (7.34) gives $E_0 = 21.39$. Equation (7.174) then gives the phase-space factor

$$f_{1u}^{(-)} \approx 3.88 \times 10^7 \frac{2\pi\alpha \times 8}{1 - \exp(-2\pi\alpha \times 8)} = 3.88 \times 10^7 \times 1.19 = 4.62 \times 10^7 \ . \tag{7.181}$$

The resulting half-life from (7.17) is

$$t_{1/2} = 10^{\log f_{1u}t - \log f_{1u}}\,\mathrm{s} = 10^{8.57 - 7.66}\,\mathrm{s} = 8.13\,\mathrm{s} \ . \tag{7.182}$$

With the data from Fig. 7.7 the corresponding experimental half-life is given by (7.39) as

$$t_{1/2}^{(\mathrm{exp})}(2^- \to 0_{\mathrm{gs}}^+) = \frac{7.13\,\mathrm{s}}{0.26} = 27.4\,\mathrm{s} \ , \tag{7.183}$$

which is not very far from the theoretical value.

### 7.6.4 Higher-Forbidden Unique Beta Decay

In this subsection we discuss higher-forbidden, i.e. 2nd-, 3rd-, etc., forbidden, unique beta-decay transitions. Such a transition can be generally referred to as a $K$th-forbidden unique beta-decay transition.

As an extension of first-forbidden beta decay relating to p-wave leptons, $K$th-forbidden beta decay relates to lepton emission with higher angular momentum. Similarly to the first-forbidden case, $K$th-forbidden unique beta decay is characterized by a maximal change in nuclear angular momentum allowed by the appropriate four transition operators analogous to (7.155)–(7.157) and (7.159). That change, due to an operator analogous to (7.156), is $\Delta J = K + 1$.

Parity change alternates with successive levels of forbiddenness. Table 7.7 lists the angular-momentum and parity changes in $K$th-forbidden unique beta decay. It is to be noted that these are the *leading* decay modes; other theoretically possible modes are suppressed beyond observation.

**Table 7.7.** Identification of $K$th-forbidden unique beta-decay transitions

| $K$ | 1 | 2 | 3 | 4 | 5 | 6 |
|---|---|---|---|---|---|---|
| $\Delta J$ | 2 | 3 | 4 | 5 | 6 | 7 |
| $\pi_i \pi_f$ | $-1$ | $+1$ | $-1$ | $+1$ | $-1$ | $+1$ |

The shape function of $K$th-forbidden unique $\beta^{\mp}$ decay is [62] (see also the formulation of [55])

$$
\begin{aligned}
S_{Ku}^{(\mp)}(Z_f, \varepsilon) &\approx F_0(\pm Z_f, \varepsilon) p\varepsilon (E_0 - \varepsilon)^2 \\
&\times \sum_{k_e + k_\nu = K+2} \frac{(\varepsilon^2 - 1)^{k_e - 1}(E_0 - \varepsilon)^{2(k_\nu - 1)}}{(2k_e - 1)!(2k_\nu - 1)!},
\end{aligned}
\tag{7.184}
$$

where the approximation (7.172) has been used for all the Fermi functions $F_0, F_1, \ldots, F_K$ involved in the exact expression [55]. For light and medium-heavy nuclei this approximation is very good. The phase-space factor is

$$
f_{Ku}^{(\mp)} = \left(\frac{3}{4}\right)^K \frac{(2K)!!}{(2K+1)!!} \int_1^{E_0} S_{Ku}^{(\mp)}(Z_f, \varepsilon) d\varepsilon .
\tag{7.185}
$$

This phase-space factor is related to the one in (7.165) as

$$
f_{K=1,u}^{(\mp)} = \frac{1}{12} f_{1u}^{(\mp)} .
\tag{7.186}
$$

The nuclear matrix element for $K$th-forbidden unique beta decay is

$$
\mathcal{M}_{Ku} = \sum_{ab} \mathcal{M}^{(Ku)}(ab)(\xi_f J_f \| [c_a^\dagger \tilde{c}_b]_{K+1} \| \xi_i J_i) ,
\tag{7.187}
$$

The single-particle matrix element is given by

$$
\mathcal{M}^{(Ku)}(ab) = (2.990 \times 10^{-3} \times b \,[\text{fm}])^K m^{(Ku)}(ab)
\tag{7.188}
$$

with

$$m^{(K\mathrm{u})}(ab) = \zeta_{ab}^{(K\mathrm{u})} \frac{1}{2\sqrt{(K+1)(2K+3)}} (-1)^{l_a+j_a+j_b+K+1} \frac{1+(-1)^{l_a+l_b+K}}{2}$$

$$\times \hat{j}_a \hat{j}_b \begin{pmatrix} j_a & j_b & K+1 \\ \frac{1}{2} & -\frac{1}{2} & 0 \end{pmatrix} \left[ (-1)^K \hat{j}_a^{\,2} + (-1)^{j_a+j_b+1} \hat{j}_b^{\,2} \right.$$

$$\left. + 2(K+1)(-1)^{l_a+j_a+K-\frac{1}{2}} \right] \tilde{\mathcal{R}}_{ab}^{(K)} ,$$

$$\zeta_{ab}^{(K\mathrm{u})} = \begin{cases} 1 & \text{CS phase convention} , \\ (-1)^{(l_b-l_a+K)/2} & \text{BR phase convention} , \end{cases}$$

(7.189)

where $\tilde{\mathcal{R}}_{ab}^{(K)}$ is the scaled radial integral defined in (6.32). For the CS and BR phase conventions the symmetry properties are

$$m_{\mathrm{CS}}^{(K\mathrm{u})}(ba) = (-1)^{j_a+j_b+1} m_{\mathrm{CS}}^{(K\mathrm{u})}(ab) , \tag{7.190}$$

$$m_{\mathrm{BR}}^{(K\mathrm{u})}(ba) = (-1)^{j_a+j_b+K+1} m_{\mathrm{BR}}^{(K\mathrm{u})}(ab) . \tag{7.191}$$

The $K = 1$ equations (7.168) and (7.169) are special cases of (7.188) and (7.189). The matrix elements (7.189) with CS phases are given in Tables 7.8 and 7.9 for second- and third-forbidden beta transitions.

For $K$th-forbidden unique electron capture the phase-space factor is [55]

$$f_{K\mathrm{u}}^{(\mathrm{EC})} = \frac{2(2K)!!}{(2K+1)!!(2K+1)!} \pi (\alpha Z_i)^3 (\varepsilon_0 + E_0)^{2(K+1)} . \tag{7.192}$$

**Table 7.8.** Scaled nuclear single-particle matrix elements $m^{(2\mathrm{u})}(ab)$ in the CS phase convention

| $a\backslash b$ | $0\mathrm{s}_{1/2}$ | $0\mathrm{p}_{3/2}$ | $0\mathrm{p}_{1/2}$ | $0\mathrm{d}_{5/2}$ | $1\mathrm{s}_{1/2}$ | $0\mathrm{d}_{3/2}$ | $0\mathrm{f}_{7/2}$ | $1\mathrm{p}_{3/2}$ | $0\mathrm{f}_{5/2}$ | $1\mathrm{p}_{1/2}$ | $0\mathrm{g}_{9/2}$ |
|---|---|---|---|---|---|---|---|---|---|---|---|
| $0\mathrm{s}_{1/2}$ | 0 | 0 | 0 | $\sqrt{\frac{15}{2}}$ | 0 | 0 | 0 | 0 | 0 | 0 | 0 |
| $0\mathrm{p}_{3/2}$ | 0 | $-\sqrt{15}$ | 0 | 0 | 0 | 0 | $3\sqrt{\frac{10}{7}}$ | $\sqrt{6}$ | $-2\sqrt{\frac{2}{7}}$ | 0 | 0 |
| $0\mathrm{p}_{1/2}$ | 0 | 0 | 0 | 0 | 0 | 0 | $2\sqrt{\frac{30}{7}}$ | 0 | $-\sqrt{\frac{5}{14}}$ | 0 | 0 |
| $0\mathrm{d}_{5/2}$ | $\sqrt{\frac{15}{2}}$ | 0 | 0 | $-6\sqrt{\frac{3}{5}}$ | $-2\sqrt{5}$ | $4\sqrt{\frac{2}{5}}$ | 0 | 0 | 0 | 0 | $\sqrt{\frac{165}{7}}$ |
| $1\mathrm{s}_{1/2}$ | 0 | 0 | 0 | $-2\sqrt{5}$ | 0 | 0 | 0 | 0 | 0 | 0 | 0 |
| $0\mathrm{d}_{3/2}$ | 0 | 0 | 0 | $-4\sqrt{\frac{2}{5}}$ | 0 | $\sqrt{\frac{3}{5}}$ | 0 | 0 | 0 | 0 | $6\sqrt{\frac{10}{7}}$ |
| $0\mathrm{f}_{7/2}$ | 0 | $3\sqrt{\frac{10}{7}}$ | $-2\sqrt{\frac{30}{7}}$ | 0 | 0 | 0 | $-\frac{9}{7}\sqrt{22}$ | $-\frac{12}{\sqrt{7}}$ | $\frac{12}{7}\sqrt{6}$ | $8\sqrt{\frac{3}{7}}$ | 0 |
| $1\mathrm{p}_{3/2}$ | 0 | $\sqrt{6}$ | 0 | 0 | 0 | 0 | $-\frac{12}{\sqrt{7}}$ | $-9\sqrt{\frac{3}{5}}$ | $\frac{8}{\sqrt{35}}$ | 0 | 0 |
| $0\mathrm{f}_{5/2}$ | 0 | $2\sqrt{\frac{2}{7}}$ | $-\sqrt{\frac{5}{14}}$ | 0 | 0 | 0 | $-\frac{12}{7}\sqrt{6}$ | $-\frac{8}{\sqrt{35}}$ | $\frac{18}{7}\sqrt{\frac{3}{5}}$ | $\frac{2}{\sqrt{7}}$ | 0 |
| $1\mathrm{p}_{1/2}$ | 0 | 0 | 0 | 0 | 0 | 0 | $-8\sqrt{\frac{3}{7}}$ | 0 | $\frac{2}{\sqrt{7}}$ | 0 | 0 |
| $0\mathrm{g}_{9/2}$ | 0 | 0 | 0 | $\sqrt{\frac{165}{7}}$ | 0 | $-6\sqrt{\frac{10}{7}}$ | 0 | 0 | 0 | 0 | $-\frac{2}{7}\sqrt{715}$ |

**Table 7.9.** Scaled nuclear single-particle matrix elements $m^{(3u)}(ab)$ in the CS phase convention

| $a\backslash b$ | $0s_{1/2}$ | $0p_{3/2}$ | $0p_{1/2}$ | $0d_{5/2}$ | $1s_{1/2}$ | $0d_{3/2}$ | $0f_{7/2}$ | $1p_{3/2}$ | $0f_{5/2}$ | $1p_{1/2}$ | $0g_{9/2}$ |
|---|---|---|---|---|---|---|---|---|---|---|---|
| $0s_{1/2}$ | 0 | 0 | 0 | 0 | 0 | 0 | $\frac{\sqrt{105}}{2}$ | 0 | 0 | 0 | 0 |
| $0p_{3/2}$ | 0 | 0 | 0 | $-\frac{3\sqrt{35}}{2}$ | 0 | 0 | 0 | 0 | 0 | 0 | $\frac{5}{6}\sqrt{77}$ |
| $0p_{1/2}$ | 0 | 0 | 0 | 0 | 0 | 0 | 0 | 0 | 0 | 0 | $\frac{10}{3}\sqrt{7}$ |
| $0d_{5/2}$ | 0 | $\frac{3\sqrt{35}}{2}$ | 0 | 0 | 0 | 0 | $-\frac{3}{2}\sqrt{55}$ | $-3\sqrt{14}$ | $3\sqrt{2}$ | 0 | 0 |
| $1s_{1/2}$ | 0 | 0 | 0 | 0 | 0 | 0 | $-3\sqrt{\frac{35}{2}}$ | 0 | 0 | 0 | 0 |
| $0d_{3/2}$ | 0 | 0 | 0 | 0 | 0 | 0 | $-3\sqrt{5}$ | 0 | $\frac{3}{2}$ | 0 | 0 |
| $0f_{7/2}$ | $-\frac{\sqrt{105}}{2}$ | 0 | 0 | $\frac{3\sqrt{55}}{2}$ | $3\sqrt{\frac{35}{2}}$ | $-3\sqrt{5}$ | 0 | 0 | 0 | 0 | $-\frac{3}{2}\sqrt{\frac{715}{7}}$ |
| $1p_{3/2}$ | 0 | 0 | 0 | $3\sqrt{14}$ | 0 | 0 | 0 | 0 | 0 | 0 | $-\sqrt{\frac{385}{2}}$ |
| $0f_{5/2}$ | 0 | 0 | 0 | $3\sqrt{2}$ | 0 | $-\frac{3}{2}$ | 0 | 0 | 0 | 0 | $-3\sqrt{\frac{110}{7}}$ |
| $1p_{1/2}$ | 0 | 0 | 0 | 0 | 0 | 0 | 0 | 0 | 0 | 0 | $-2\sqrt{70}$ |
| $0g_{9/2}$ | 0 | $-\frac{5\sqrt{77}}{6}$ | $\frac{10\sqrt{7}}{3}$ | 0 | 0 | 0 | $\frac{3}{2}\sqrt{\frac{715}{7}}$ | $\sqrt{\frac{385}{2}}$ | $-3\sqrt{\frac{110}{7}}$ | $-2\sqrt{70}$ | 0 |

Note that the charge number is that of the initial nucleus. Here $\varepsilon_0$ is given by (7.27) and $E_0$ by (7.36).

Similarly to allowed beta decay, for $K$th-forbidden unique beta decay we have

$$f_{Ku}t_{1/2} = \frac{\kappa}{B_{Ku}}\,, \quad B_{Ku} = \frac{g_A^2}{2J_i+1}|\mathcal{M}_{Ku}|^2\,. \qquad (7.193)$$

The theoretical half-life can be computed from this. However, it is desirable to have $\log ft$ values that are directly comparable for different types of beta decay. In particular, all $\log ft$ values should be commensurate with those for allowed decay, where the phase-space factor is $f_0$. To that end the $ft$ value for $K$th-forbidden unique beta decay can be defined as [55]

$$\boxed{ft \equiv f_0^{(-)}t_{1/2} = \frac{f_0^{(-)}}{f_{Ku}}\frac{\kappa}{B_{Ku}}\,,} \qquad (7.194)$$

where

$$f_{Ku} = \begin{cases} f_{Ku}^{(-)} & \text{for } \beta^- \text{ decay}\,, \\ f_{Ku}^{(+)} + f_{Ku}^{(EC)} & \text{for } \beta^+/\text{EC decay}\,. \end{cases} \qquad (7.195)$$

Note that this $ft$ value differs from those for allowed and first-forbidden decay, (7.33) and (7.165), in that it depends not only on nuclear structure but also on the leptonic properties contained in the phase factors. Note also that the $ft$ value (7.194) is not completely commensurate with $ft$ values for allowed decay because it contains $f_0^{(-)}$ also for $\beta^+/\text{EC}$ decay.

The phase-space factors (7.185) can be integrated into explicit expressions when the Primakoff–Rosen approximation (7.29) is used. Up to $K=5$, the results are

$$
f_{2u}^{(\mp)} \approx \frac{3}{10} \left( \frac{E_0^9}{9072} - \frac{2E_0^7}{4725} - \frac{E_0^6}{360} + \frac{23E_0^5}{1800} - \frac{23E_0^4}{1080} + \frac{E_0^3}{54} - \frac{113E_0^2}{12\,600} + \frac{E_0}{432} \right.
$$
$$
\left. - \frac{29}{113\,400} \right) F_0^{(\mathrm{PR})}(\pm Z_f) \,, \tag{7.196}
$$

$$
f_{3u}^{(\mp)} \approx \frac{27}{140} \left( \frac{E_0^{11}}{498\,960} - \frac{13E_0^9}{1\,058\,400} - \frac{E_0^8}{15\,120} + \frac{11E_0^7}{26\,460} - \frac{E_0^6}{1080} + \frac{29E_0^5}{25\,200} \right.
$$
$$
\left. - \frac{17E_0^4}{18\,900} + \frac{E_0^3}{2160} - \frac{5E_0^2}{31\,752} + \frac{E_0}{30\,240} - \frac{19}{5\,821\,200} \right)
$$
$$
\times F_0^{(\mathrm{PR})}(\pm Z_f) \,, \tag{7.197}
$$

$$
f_{4u}^{(\mp)} \approx \frac{9}{700} \left( \frac{E_0^{13}}{4\,447\,872} - \frac{151E_0^{11}}{78\,586\,200} - \frac{E_0^{10}}{108\,864} + \frac{563E_0^9}{7\,620\,480} - \frac{37E_0^8}{181\,440} \right.
$$
$$
+ \frac{311E_0^7}{952\,560} - \frac{53E_0^6}{151\,200} + \frac{59E_0^5}{217\,728} - \frac{11E_0^4}{70\,560} + \frac{E_0^3}{15\,120}
$$
$$
\left. - \frac{2447E_0^2}{125\,737\,920} + \frac{19E_0}{5\,443\,200} - \frac{31}{108\,972\,864} \right) F_0^{(\mathrm{PR})}(\pm Z_f) \,, \tag{7.198}
$$

$$
f_{5u}^{(\mp)} \approx \frac{27}{30\,800} \left( \frac{E_0^{15}}{58\,378\,320} - \frac{20\,429E_0^{13}}{107\,883\,135\,360} - \frac{E_0^{12}}{1\,197\,504} + \frac{1627E_0^{11}}{197\,588\,160} \right.
$$
$$
- \frac{11E_0^{10}}{408\,240} + \frac{7927E_0^9}{150\,885\,504} - \frac{13E_0^8}{181\,440} + \frac{19E_0^7}{256\,608} - \frac{43E_0^6}{714\,420}
$$
$$
+ \frac{461E_0^5}{11\,975\,040} - \frac{29E_0^4}{1\,539\,648} + \frac{11E_0^3}{1\,632\,960} - \frac{1487E_0^2}{899\,026\,128}
$$
$$
\left. + \frac{E_0}{3\,991\,680} - \frac{137}{7\,705\,938\,240} \right) F_0^{(\mathrm{PR})}(\pm Z_f) \,. \tag{7.199}
$$

### 7.6.5 Application to Third-Forbidden Unique Beta Decay of $^{40}$K

Figure 7.12 shows the decays of the $4^-$ ground states of $^{40}_{19}$K$_{21}$ and $^{40}_{21}$Sc$_{19}$ to the ground and excited states of $^{40}_{20}$Ca$_{20}$. Let us calculate the $\beta^-$ decay $^{40}$K$(4^-_{\mathrm{gs}}) \to {}^{40}$Ca$(0^+_{\mathrm{gs}})$.
  The initial state of $^{40}$K is given in (5.67) as

$$
|^{40}\mathrm{K}\,;\, 4^-_{\mathrm{gs}} \rangle = \left[ c^\dagger_{\nu 0f_{7/2}} h^\dagger_{\pi 0d_{3/2}} \right]_{4-} |^{40}\mathrm{Ca}\rangle \,. \tag{7.200}
$$

The nuclear matrix element is obtained similarly to (7.176) leading to (7.177). In the CS phase convention it is

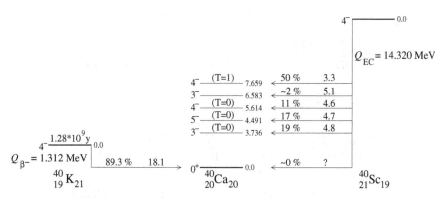

**Fig. 7.12.** Beta decays of the $4^-$ ground states of $^{40}$K and $^{40}$Sc to the $0^+$ ground state and five excited states of negative parity in $^{40}$Ca. The decays proceed via the $\beta^-$ and $\beta^+$/EC decay modes, respectively. The experimental half-lives, $Q$ values, branchings and log $ft$ values are given. Gamow–Teller decays go to the $3^-$, $4^-$ and $5^-$ states. Also Fermi decay goes to the $4^-$ final state. The decays to the final $0^+$ ground state are 3rd-forbidden unique

$$\boxed{\mathcal{M}_{Ku}(a_i\, b_i^{-1}\,;\, J_i \to 0_{gs}^+) = \delta_{J_i,K+1}(-1)^{K-1}\sqrt{2K+3}\mathcal{M}^{(Ku)}(a_ib_i)\,.}\quad (7.201)$$

With $K = 1$ this matrix element reduces to (7.177).

According to Table 7.9 the transition $^{40}$K$(4_{gs}^-) \to\ ^{40}$Ca$(0_{gs}^+)$ is a 3rd-forbidden unique transition, $K = 3$. Equation (7.201) gives the nuclear matrix element

$$\mathcal{M}_{3u}\left(\nu 0f_{7/2}\left(\pi 0d_{3/2}\right)^{-1}\,;\, 4^- \to 0_{gs}^+\right) = 3\mathcal{M}^{(3u)}(0f_{7/2}\,0d_{3/2})$$

$$= 3 \times (2.990 \times 10^{-3} \times 1.939)^3 m^{(3u)}(0f_{7/2}\,0d_{3/2})$$

$$= 5.846 \times 10^{-7} \times (-3\sqrt{5}) = -3.922 \times 10^{-6}\,,\qquad (7.202)$$

where we have used the oscillator length $b = 1.939\,\text{fm}$ from (3.45) and (3.43), and the value of $m^{(3u)}$ from Table 7.9. Equation (7.192) now gives the reduced transition probability

$$B_{3u} = \frac{1.25^2}{9}(-3.922 \times 10^{-6})^2 = 2.671 \times 10^{-12}\,.\qquad (7.203)$$

For input into the phase-space factors we have from (7.34) and (7.29)

$$E_0 = 3.568\,,\quad F_0^{(PR)}(20) = 1.528\,.\qquad (7.204)$$

Equations (7.30) and (7.197) then give

$$f_0^{(-)} = 25.39\,,\quad f_{3u}^{(-)} = 0.3512\,.\qquad (7.205)$$

The $ft$ value and its logarithm are now found from (7.194)

$$ft = \frac{25.39}{0.3512} \times \frac{6147}{2.671 \times 10^{-12}} = 1.664 \times 10^{17} , \quad \log ft = 17.22 . \quad (7.206)$$

The experimental $\log ft$ value in Fig. 7.12 is 18.1, so the agreement is not very good. This is not surprising in view of the simple wave functions used. Note, however, that our calculated value fits in the $\log ft$ range 17–19 given for third-forbidden transitions in Table 7.4.

The theoretical half-life for the $4^- \to 0^+_{gs}$ transition is, from (7.194),[6]

$$t_{1/2} = \frac{ft}{f_0^{(-)}} = \frac{1.664 \times 10^{17}}{25.39} = 6.554 \times 10^{15} \, \text{s} = 2.08 \times 10^8 \, \text{a} , \quad (7.207)$$

where we have used

$$1 \, \text{a} = 3.154 \times 10^7 \, \text{s} . \quad (7.208)$$

Figure 7.12 gives the experimental half-life as $1.28 \times 10^9$ a and the branching as 89.3 %. Thus the experimental partial half-life, from (7.39), is

$$t_{1/2}^{(exp)} = \frac{1.28 \times 10^9 \, \text{a}}{0.893} = 1.43 \times 10^9 \, \text{a} , \quad (7.209)$$

which is seven times longer than our calculated half-life (7.207).

### 7.6.6 Forbidden Unique Beta Decay in Few-Particle and Few-Hole Nuclei

In Sects. 7.3–7.5 we derived transition amplitudes for allowed beta decay in few-particle and few-hole nuclei. Many results were stated for a general $L$, although only $L = 0$ and $L = 1$ occur in allowed decay. The transition amplitude was defined in the same way for allowed, first-forbidden and $K$th-forbidden beta decay, in (7.18), (7.19), (7.167) and (7.187). Because the reduced transition density is the same for any one-body operator of rank $L$, the transition amplitudes for forbidden unique decay can be obtained from the results for allowed decay by an appropriate choice of $L$ and replacement of the single-particle matrix element.

**Beta-minus decay**

For $K$th-forbidden $(K \geq 1)$ unique $\beta^-$ decay the conversion procedure is

$$\boxed{\mathcal{M}_L(pn) \to \mathcal{M}^{(Ku)}(pn) , \quad L \to K + 1 .} \quad (7.210)$$

By this procedure all formulas for allowed $\beta^-$ decay, be they for particles or holes, can be converted into formulas for $K$th-forbidden $(K \geq 1)$ unique $\beta^-$ decay. Specifically the replacements work as follows.

---

[6] The SI symbol for year is 'a'. However, 'y', 'yr' and unabbreviated 'year' appear commonly in the literature.

- *One-particle and one-hole nuclei:* For transitions between one-particle nuclei and between one-hole nuclei, respectively, the transition density is given by (7.44) and (7.45). The resulting transition amplitudes for $K$th-forbidden unique transitions in such nuclei are

$$
\boxed{
\begin{aligned}
\mathcal{M}_{Ku}(i \to f) &= \sqrt{2K+3}\mathcal{M}^{(Ku)}(fi) \,, \\
\mathcal{M}_{Ku}(i^{-1} \to f^{-1}) &= \zeta^{(K)}(-1)^K \sqrt{2K+3}\mathcal{M}^{(Ku)}(fi) \,, \\
\zeta^{(K)} &= \begin{cases} 1 & \text{CS phase covention} \,, \\ (-1)^K & \text{BR phase covention} \,. \end{cases}
\end{aligned}
}
\tag{7.211}
$$

The symmetry relations (7.190) and (7.191) were used in the second relation.

- *Particle–hole nuclei:* For transitions from a particle–hole state to the even–even ground state of the reference nucleus, the result is given explicitly by (7.201). For transitions between two particle–hole states the substitutions (7.210) are made in (7.72) and (7.76).

- *Two-particle and two-hole nuclei:* For transitions between two two-particle or two-hole nuclei, the substitutions (7.210) are made in (7.105), (7.106) and (7.109)–(7.112).

## Beta-plus decay and electron capture

The $\beta^+$ operator can be obtained from the $\beta^-$ operator by Hermitian conjugation, i.e.

$$
\boxed{\beta^+_{K+1,M} = \zeta^{(K)}(-1)^M (\beta^-_{K+1,-M})^\dagger \,,}
\tag{7.212}
$$

where the spatial dependence of the $K$th-forbidden unique transition operator induces the phase factor defined in (7.211). It follows that a formula for allowed $\beta^+$/EC decay can be converted to one for $K$th-forbidden unique decay by the replacements

$$
\boxed{\mathcal{M}_L(pn) \to \zeta^{(K)}\mathcal{M}^{(Ku)}(pn) \,, \quad L \to K+1 \,.}
\tag{7.213}
$$

The difference between the CS and BR conventions is just the trivial overall phase factor $(-1)^K$.

The procedure of replacement is carried out the same way as for $\beta^-$ decay. For example, to find the $K$th-forbidden unique transition amplitude for $\beta^+$/EC decay between two-particle states, one has to make the substitution (7.213) in (7.107), (7.108) and (7.113)–(7.116).

### 7.6.7 Forbidden Non-Unique Beta Decays

As described in Subsect. 7.6.1, in first-forbidden beta decay there are six decay operators available to produce a given transition. For higher-forbidden transitions ($K \geq 2$) there are four decay operators available, as pointed out in Subsect. 7.6.4. The general case, with $K \geq 1$, of all these transitions can be called $K$th-forbidden *non-unique* beta decay.

The selection rules for forbidden non-unique beta decay are given in Table 7.10. Note that these selection rules do not include the forbidden unique ones in Table 7.7. Tables 7.7 and 7.10 together constitute the selection rules for forbidden beta decay.

**Table 7.10.** Identification of $K$th-forbidden non-unique beta-decay transitions

| $K$ | 1 | 2 | 3 | 4 | 5 | 6 |
|---|---|---|---|---|---|---|
| $\Delta J$ | 0,1 | 2 | 3 | 4 | 5 | 6 |
| $\pi_i \pi_f$ | $-1$ | $+1$ | $-1$ | $+1$ | $-1$ | $+1$ |

For forbidden non-unique beta decay the definition of the $ft$ value must be generalized beyond (7.194). The appropriate definition reads (see e.g. [55])

$$ft \equiv f_0^{(-)} t_{1/2} = f_0^{(-)} \frac{\kappa}{S_K} \,, \qquad K \geq 1 \,, \tag{7.214}$$

where

$$S_K = \begin{cases} S_K^{(-)} & \text{for } \beta^- \text{ decay} \,, \\ S_K^{(+)} + S_K^{(\text{EC})} & \text{for } \beta^+/\text{EC decay} \,. \end{cases} \tag{7.215}$$

The quantities $S_K^{(-)}$, $S_K^{(+)}$ and $S_K^{(\text{EC})}$ are the *shape functions* of $\beta^-$, $\beta^+$ and EC decay respectively. They contain up to six ($K = 1$, first forbidden) or four ($K \geq 2$) different nuclear matrix elements with the appropriate phase-space factors from the lepton kinematics.

As in the case of first-forbidden unique beta decay, the phase-space factor $f_0^{(-)}$ is used as a scaling quantity to normalize the integrated phase-space factors residing inside the shape functions of (7.215).

## Epilogue

The basic properties of beta-decay transitions were reviewed in this chapter. However, derivations from the interaction between hadronic and leptonic weak charged currents were omitted. The kinematics and nuclear matrix elements of allowed and unique forbidden beta decays were discussed in detail. The formalism was applied to the wave functions of few-particle and few-hole nuclei introduced in Chap. 5. In the following chapters the full power of beta-decay analysis is used to track down deficiencies in the wave functions of more sophisticated nuclear models that include configuration mixing.

## Exercises

**7.1.** Verify the symmetry relations (7.22) and (7.23) of the Fermi and Gamow–Teller single-particle matrix elements.

**7.2.** Compute a few non-zero entries of your choice in Table 7.3 by starting from the definition (7.21) of the Gamow–Teller single-particle matrix element.

**7.3.** Derive the analytical expression (7.30) for the phase-space factor of allowed beta decay by using the Primakoff–Rosen approximation (7.29) of (7.24).

**7.4.** Derive the one-body transition densities (7.44) and (7.45) and from them the matrix elements (7.46) and (7.47).

**7.5.** Derive the theoretical numbers listed in Table 7.5 for the decay of $^{17}$F. Start from the experimental electron-capture decay energy $Q$ listed in the second column of the table. Compute both the $\beta^+$ and EC phase-space factors and verify that the electron-capture contribution can be neglected.

**7.6.** Repeat Exercise 7.5 for the decay of $^{39}$Ca.

**7.7.** Repeat Exercise 7.5 for the decay of $^{41}$Sc.

**7.8.** Compute the log $ft$ value for the decay transition $^{11}$C$(3/2^-) \rightarrow {}^{11}$B$(3/2^-)$ and the corresponding half-life using the experimental $Q$ value. Take the filled orbitals of the particle–hole vacuum to be $0s_{1/2}$ and $0p_{3/2}$. Compare your results with experimental data and comment.

**7.9.** Repeat Exercise 7.8 for the decay transition $^{13}$N$(1/2^-) \rightarrow {}^{13}$C$(1/2^-)$.

**7.10.** Compute the log $ft$ value for the decay transition $^{33}$Cl$(3/2^+) \rightarrow {}^{33}$S$(3/2^+)$ and the corresponding half-life using the experimental $Q$ value. The Fermi levels of the particle–hole vacuum ($^{32}$S) can be taken to lie at $Z = 16$ and $N = 16$ in the 0d-1s shell. Compare your results with experimental data and comment.

**7.11.** Repeat Exercise 7.10 for the decay transition $^{31}$S$(1/2^+) \rightarrow {}^{31}$P$(1/2^+)$.

**7.12.** Derive Eqs. (7.62)–(7.64).

**7.13.** Consider the decay of the $1^+$ ground state of $^{12}$B to the $0^+$ ground state of $^{12}$C. The decay can be considered as a conversion of a neutron in the $0p_{1/2}$ orbital to a proton in the $0p_{3/2}$ orbital. Take the filled orbitals of the particle–hole vacuum to be of $0s_{1/2}$ and $0p_{3/2}$.

**7.14.** Repeat Exercise 7.13 for the decay of $^{12}$N to $^{12}$C.

**7.15.** The nucleus $^{32}$S can be considered as a doubly magic nucleus for which the Fermi levels of both protons and neutrons lie at the $1s_{1/2}$ orbital. The $1^+$ ground state of the adjacent nucleus $^{32}$P can be considered to be a linear combination of the particle–hole excitations $0d_{3/2}\text{-}(1s_{1/2})^{-1}$ and $0d_{3/2}\text{-}(0d_{5/2})^{-1}$. Determine the amplitudes of this linear combination by comparing the computed $\log ft$ value for the $^{32}$P($1^+$) $\rightarrow$ $^{32}$S($0^+$) transition with the experimental value 7.9. Evaluate the corresponding half-life by using the experimental $Q$ value.

**7.16.** Derive the formula (7.82) from (7.72)–(7.74).

**7.17.** Compute the $\log ft$ value for the decay transition $^{32}$Cl($1^+_{gs}$) $\rightarrow$ $^{32}$S($2^+_1$) by taking into account only the particle–hole excitation $0d_{3/2}\text{-}(1s_{1/2})^{-1}$. Compute the corresponding partial half-life of $^{32}$Cl. Compare the computed results with the experimental numbers and comment.

**7.18.** Compute the $\log ft$ values for the transitions $^{40}$Sc($4^-_{gs}$) $\rightarrow$ $^{40}$Ca($3^-_1$), $^{40}$Sc($4^-_{gs}$) $\rightarrow$ $^{40}$Ca($4^-_1$) and $^{40}$Sc($4^-_{gs}$) $\rightarrow$ $^{40}$Ca($5^-_1$) of Fig. 7.12. Compare with the data and comment.

**7.19.** Derive (7.109) and (7.110) from (7.105).

**7.20.** Verify by computation the $\log ft$ values inside parentheses in Fig. 7.10. The nuclei involved have two particles in the $0f_{7/2}$ orbital. Compute the half-lives of the ground state of $^{42}$Ti and the $7^+$ state of $^{42}$Sc. Compare the results with the data.

**7.21.** Verify by computation the $\log ft$ values in parentheses in Fig. 7.11. The nuclei involved have two holes in the $0f_{7/2}$ orbital.

**7.22.** Compute the $\log ft$ values for the decay of $^{38}$Ca to the first $0^+$ and $1^+$ states in $^{38}$K by assuming two holes in the $0d_{3/2}$ orbital. Compute the corresponding partial decay half-lives and compare with experiment. Comment on the result.

**7.23.** The $0^+$ ground state of $^{50}$Ca decays to the first $1^+$ state in $^{50}$Sc with experimental branching of 100 % and $\log ft = 4.1$. The $0^+$ ground state wave function can be guessed to consist of a leading component and some small components. The leading component is inactive in the decay transition. Compute an estimate for the amplitude of that component of the wave function which mediates the decay transition by reproducing the measured $\log ft$ value.

**7.24.** The $5^+$ ground state of $^{50}$Sc decays to the $4^+$ (branching 14 %, $\log ft = 6.4$) and $6^+$ (branching 86 %, $\log ft = 5.4$) states in $^{50}$Ti. The $5^+$ ground state can be guessed to consist of two components where the leading component is inactive in the transitions. Compute an estimate of the amplitudes of the two components by comparing the computed $\log ft$ values with the experimental ones.

**7.25.** Verify the symmetry relations (7.170) and (7.171) for the basic single-particle matrix elements of the first-forbidden unique transitions.

**7.26.** Derive the values of a few non-zero single-particle matrix elements of your choice in Table 7.6.

**7.27.** Derive the phase-space factor (7.174) by starting from (7.163) and using the result (7.173).

**7.28.** Verify the symmetry relations (7.190) and (7.191) for the basic single-particle matrix elements of $K$th-forbidden unique transitions.

**7.29.** Derive the values of a few non-zero single-particle matrix elements of your choice in Table 7.8.

**7.30.** Calculate the basic single-particle matrix element $m^{(3u)}(0f_{7/2}0d_{3/2})$.

**7.31.** Derive the values of a few single-particle matrix elements of your choice in Table 7.9.

**7.32.** Derive the expression (7.196) for the phase-space factor $f_{2u}^{(\pm)}$ by using the Primakoff–Rosen approximation in the shape function (7.184) in the integral of (7.185).

**7.33.** Evaluate the $\log ft$ value and partial half-life for the decay of the $3^+$ ground state of $^{38}$K to the $0^+$ ground state of $^{38}$Ar. Assume two holes in the $0d_{3/2}$ orbital. What are the chances of measuring this partial half-life?

**7.34.** Evaluate the $\log ft$ value and partial half-life for the decay of the $0^+$ ground state of $^{48}$Ca to the $5^+$ first excited state of $^{48}$Sc. For $^{48}$Sc assume a particle and a hole in the $0f_{7/2}$ orbital. The transition is 4th-forbidden unique and competes with the double beta decay of $^{48}$Ca to the $0^+$ ground state of $^{48}$Ti. Compare your result with the measured half-life [63, 64]

$$t_{1/2}(\beta\beta) = 4.2 \times 10^{19} \, \text{a} \tag{7.216}$$

for the double beta decay and comment. Hint: Compute the basic single-particle matrix element from (7.189) and evaluate the scaled radial integral by using the formalism of Subsect. 6.1.4.

# Nuclear Two-Body Interaction and Configuration Mixing

## Prologue

In previous chapters the nucleus was described as a collection of non-interacting nucleons in a mean-field potential. The wave function of a nuclear state was taken to be a Slater determinant corresponding to a definite way of placing the valence nucleons in the mean-field single-particle orbitals. In this way the energy of a nuclear state was fully determined by the energies of the occupied single-particle orbitals.

In this chapter we introduce the notion of configuration mixing, already alluded to several times previously. Configuration mixing leads to wave functions that consist of more than just one Slater determinant. Nucleon configurations are mixed by the residual interaction, the part of the nuclear Hamiltonian that was omitted in the mean-field description. Interactions between valence nucleons make them jump from one orbital to another, so that the wave function contains several configurations.

We start by studying configuration mixing in two-particle and two-hole nuclei, using a simple residual force, the so-called surface delta interaction. Despite its schematic nature it is realistic enough to yield results which can justifiably be compared with experiment.

## 8.1 General Properties of the Nuclear Two-Body Interaction

The basic properties of nuclear two-body interaction are reviewed in this section. Angular-momentum-coupled two-body interaction matrix elements are introduced and used to write down the angular-momentum-coupled form of the nuclear Hamiltonian. Important properties of the two-body matrix elements are discussed. The important special case of a separable interaction, the surface delta interaction (SDI) in particular, is taken up. The SDI will be used throughout this book to produce configuration mixing in various simple few-particle and, later, few-quasiparticle systems.

### 8.1.1 Coupled Two-Body Interaction Matrix Elements

Equation (4.28) gives the two-body part of the nuclear Hamiltonian as

$$V = \tfrac{1}{4} \sum_{\alpha\beta\gamma\delta} \bar{v}_{\alpha\beta\gamma\delta} c_\alpha^\dagger c_\beta^\dagger c_\delta c_\gamma \ , \tag{8.1}$$

where the

$$\bar{v}_{\alpha\beta\gamma\delta} = v_{\alpha\beta\gamma\delta} - v_{\alpha\beta\delta\gamma} \tag{8.2}$$

are the antisymmetrized two-nucleon interaction matrix elements (4.30). In (4.45) this matrix element was written as

$$\bar{v}_{\alpha\beta\gamma\delta} = \langle \alpha\beta|V|\gamma\delta \rangle \ , \tag{8.3}$$

where

$$|\alpha\beta\rangle = c_\alpha^\dagger c_\beta^\dagger|0\rangle \ , \quad |\gamma\delta\rangle = c_\gamma^\dagger c_\delta^\dagger|0\rangle \tag{8.4}$$

were the normalized and antisymmetrized two-nucleon states relative to the appropriate particle vacuum $|0\rangle$ (usually the state $|\text{CORE}\rangle$ introduced in Chap. 5).

Equations (5.21) and (5.23) give the normalized angular-momentum-coupled two-nucleon states

$$|a\,b\,;\,J\,M\rangle = \mathcal{N}_{ab}(J)\big[c_a^\dagger c_b^\dagger\big]_{JM}|0\rangle \ , \tag{8.5}$$

$$\mathcal{N}_{ab}(J) = \frac{\sqrt{1 + \delta_{ab}(-1)^J}}{1 + \delta_{ab}} \ . \tag{8.6}$$

Here the indices $a$ and $b$ may contain also the proton and neutron labels $\pi$ and $\nu$. Hence the wave function (8.5) can describe a proton–proton, neutron–neutron or proton–neutron two-particle system.

We now express the uncoupled wave function $|\alpha\beta\rangle$ in terms of the coupled wave functions (8.5). Equation (1.29) gives for the operators[1]

$$c_\alpha^\dagger c_\beta^\dagger = \sum_{JM}(j_a\, m_\alpha\, j_b\, m_\beta|J\,M)\big[c_a^\dagger c_b^\dagger\big]_{JM} \ . \tag{8.7}$$

Operating with this on the vacuum and taking into account the normalization contained in (8.5) we have

$$|\alpha\,\beta\rangle = \sum_{JM}(j_a\, m_\alpha\, j_b\, m_\beta|J\,M)[\mathcal{N}_{ab}(J)]^{-1}|a\,b\,;\,J\,M\rangle \ . \tag{8.8}$$

---

[1]  Because of the wave function normalization arising from fermion antisymmetry, the basically geometrical relation (1.29) cannot be applied directly.

Substituting (8.8) into the Hamiltonian (8.1) yields

$$V = \frac{1}{4} \sum_{\substack{\alpha\beta\gamma\delta \\ JMJ'M'}} [\mathcal{N}_{ab}(J)\mathcal{N}_{cd}(J')]^{-1}(j_a\, m_\alpha\, j_b\, m_\beta|J\,M)(j_c\, m_\gamma\, j_d\, m_\delta|J'\,M')$$
$$\times \langle a\,b;\, J\,M|V|c\,d;\, J'\,M'\rangle c_\alpha^\dagger c_\beta^\dagger c_\delta c_\gamma\,. \tag{8.9}$$

Converting the annihilation operators into spherical tensors (4.23) yields

$$c_\delta c_\gamma = (-1)^{-j_d+m_\delta-j_c+m_\gamma}\tilde{c}_{-\delta}\tilde{c}_{-\gamma} = (-1)^{j_c+j_d-M'+1}\tilde{c}_{-\gamma}\tilde{c}_{-\delta}\,, \tag{8.10}$$

where the property (1.22) of the second Clebsch–Gordan coefficient was used. By summing over the single-particle projection quantum numbers we obtain

$$V = \frac{1}{4} \sum_{\substack{abcd \\ JMJ'M'}} [\mathcal{N}_{ab}(J)\mathcal{N}_{cd}(J')]^{-1}\langle a\,b;\, J\,M|V|c\,d;\, J'\,M'\rangle$$
$$\times \left[c_a^\dagger c_b^\dagger\right]_{JM}(-1)^{J'+M'+1}\left[\tilde{c}_c\tilde{c}_d\right]_{J',-M'}\,. \tag{8.11}$$

The two-body interaction $V$ is a scalar, i.e. a spherical tensor of rank $\lambda = 0$. Applying the Wigner–Eckart theorem (2.27) to the two-body matrix element we obtain

$$\langle a\,b;\, J\,M|V|c\,d;\, J'\,M'\rangle = (-1)^{J-M}\begin{pmatrix} J & 0 & J' \\ -M & 0 & M' \end{pmatrix}(a\,b;\, J\|V\|c\,d;\, J')$$
$$= \delta_{JJ'}\delta_{MM'}\hat{J}^{-1}(a\,b;\, J\|V\|c\,d;\, J')\,, \tag{8.12}$$

where we used (1.39) and (1.42). Thus we see that the matrix element is diagonal in $J$ and $M$ and independent of the value of $M$, so that we can write

$$\langle a\,b;\, J\,M|V|c\,d;\, J'\,M'\rangle \equiv \delta_{JJ'}\delta_{MM'}\langle a\,b;\, J|V|c\,d;\, J\rangle\,. \tag{8.13}$$

The resulting expression for $V$ is

$$V = -\frac{1}{4}\sum_J \sum_{abcd}[\mathcal{N}_{ab}(J)\mathcal{N}_{cd}(J)]^{-1}\langle a\,b;\, J|V|c\,d;\, J\rangle$$
$$\times \sum_M (-1)^{J+M}\left[c_a^\dagger c_b^\dagger\right]_{JM}\left[\tilde{c}_c\tilde{c}_d\right]_{J,-M}\,. \tag{8.14}$$

With use of (2.51) this simplifies to

$$\boxed{V = -\frac{1}{4}\sum_J \sum_{abcd}[\mathcal{N}_{ab}(J)\mathcal{N}_{cd}(J)]^{-1}\hat{J}\langle a\,b;\, J|V|c\,d;\, J\rangle\left[\left[c_a^\dagger c_b^\dagger\right]_J\left[\tilde{c}_c\tilde{c}_d\right]_J\right]_{00}\,.}$$
$$\tag{8.15}$$

For completeness we note the relations, in both directions, between coupled and uncoupled two-nucleon interaction matrix elements:

$$\langle a\,b;\,J|V|c\,d;\,J\rangle$$

$$= \mathcal{N}_{ab}(J)\mathcal{N}_{cd}(J) \sum_{\substack{m_\alpha m_\beta \\ m_\gamma m_\delta}} (j_a\,m_\alpha\,j_b\,m_\beta|J\,M)(j_c\,m_\gamma\,j_d\,m_\delta|J\,M)\bar{v}_{\alpha\beta\gamma\delta}\,, \quad (8.16)$$

$$\bar{v}_{\alpha\beta\gamma\delta} = \sum_{JM} [\mathcal{N}_{ab}(J)\mathcal{N}_{cd}(J)]^{-1} (j_a\,m_\alpha\,j_b\,m_\beta|J\,M)(j_c\,m_\gamma\,j_d\,m_\delta|J\,M)$$

$$\times\,\langle a\,b;\,J|V|c\,d;\,J\rangle\,. \quad (8.17)$$

We now state the corresponding expressions in isospin representation. The normalized two-nucleon wave functions are given by (5.96) as

$$|\,a\,b;\,J\,M;\,T\,M_T\rangle = \mathcal{N}_{ab}(JT)\big[c_a^\dagger c_b^\dagger\big]_{JM}^{TM_T}|0\rangle\,, \quad (8.18)$$

$$\mathcal{N}_{ab}(JT) = \frac{\sqrt{1-\delta_{ab}(-1)^{J+T}}}{1+\delta_{ab}}\,, \quad (8.19)$$

where the labels $a$ and $b$ carry no proton or neutron labels.

The isospin counterpart of (8.8) is

$$|\alpha\,\beta\rangle = \sum_{\substack{JM \\ TM_T}} (j_a\,m_\alpha\,j_b\,m_\beta|J\,M)(\tfrac{1}{2}\,m_{t\alpha}\,\tfrac{1}{2}\,m_{t\beta}|T\,M_T)$$

$$\times\,[\mathcal{N}_{ab}(JT)]^{-1}|a\,b;\,J\,M;\,T\,M_T\rangle\,. \quad (8.20)$$

We assume that the two-nucleon interaction $V$ is not only a scalar but also an isoscalar. Applying the Wigner–Eckart theorem also in isospace allows us to write, as an extension of (8.13),

$$\langle a\,b;\,J\,M;\,T\,M_T|V|c\,d;\,J\,M;\,T'\,M_T'\rangle \equiv \delta_{TT'}\delta_{M_T M_T'}\langle a\,b;\,JT|V|c\,d;\,JT\rangle\,. \quad (8.21)$$

The expression corresponding to (8.15) becomes

$$\boxed{\begin{aligned} V = -\tfrac{1}{4} \sum_{\substack{J \\ T=0,1}} \sum_{abcd} &[\mathcal{N}_{ab}(JT)\mathcal{N}_{cd}(JT)]^{-1}\widehat{J}\widehat{T}\langle a\,b;\,JT|V|c\,d;\,JT\rangle \\ &\times \Big[\big[c_a^\dagger c_b^\dagger\big]_J^T\big[\hat{c}_c\hat{c}_d\big]_J^T\Big]_{00}^{00}\,. \end{aligned}} \quad (8.22)$$

The isospin extensions of (8.16) and (8.17) are

$$\langle a\,b;\,JT|V|c\,d;\,JT\rangle = \mathcal{N}_{ab}(JT)\mathcal{N}_{cd}(JT) \sum_{\substack{m_\alpha m_\beta m_{t\alpha} m_{t\beta} \\ m_\gamma m_\delta m_{t\gamma} m_{t\delta}}} (j_a\,m_\alpha\,j_b\,m_\beta|J\,M)$$

$$\times (\tfrac{1}{2}\,m_{t\alpha}\,\tfrac{1}{2}\,m_{t\beta}|T\,M_T)(j_c\,m_\gamma\,j_d\,m_\delta|J\,M)(\tfrac{1}{2}\,m_{t\gamma}\,\tfrac{1}{2}\,m_{t\delta}|T\,M_T)\bar{v}_{\alpha\beta\gamma\delta}\,, \quad (8.23)$$

$$\bar{v}_{\alpha\beta\gamma\delta} = \sum_{\substack{JM \\ TM_T}} [\mathcal{N}_{ab}(JT)\mathcal{N}_{cd}(JT)]^{-1}(j_a\,m_\alpha\,j_b\,m_\beta|J\,M)(\tfrac{1}{2}\,m_{t\alpha}\,\tfrac{1}{2}\,m_{t\beta}|T\,M_T)$$

$$\times\,(j_c\,m_\gamma\,j_d\,m_\delta|J\,M)(\tfrac{1}{2}\,m_{t\gamma}\,\tfrac{1}{2}\,m_{t\delta}|T\,M_T)\langle a\,b\,;\,JT|V|c\,d\,;\,JT\rangle\,. \tag{8.24}$$

Here the labels $\alpha$, $\beta$, $\gamma$ and $\delta$ contain also the isospin parts of (5.85).

## 8.1.2 Relations for Coupled Two-Body Matrix Elements

Matrix elements with angular momentum coupling and those with angular momentum and isospin coupling can be related to each other by applying the transformation formulas (5.107), (5.108), (5.114) and (5.115). The two-body interaction is assumed to be an isoscalar, i.e. to have the property (8.21).

For like-nucleon two-body matrix elements we have the simple relations

$$\boxed{\begin{aligned} \langle p_1\,p_2\,;\,J|V|p_3\,p_4\,;\,J\rangle &= \langle a_1\,a_2\,;\,J\,T=1|V|a_3\,a_4\,;\,J\,T=1\rangle\,, \\ \langle n_1\,n_2\,;\,J|V|n_3\,n_4\,;\,J\rangle &= \langle a_1\,a_2\,;\,J\,T=1|V|a_3\,a_4\,;\,J\,T=1\rangle\,. \end{aligned}} \tag{8.25}$$

The relations for proton–neutron and neutron–proton two-body matrix elements are

$$\boxed{\begin{aligned} \langle p_1\,n_2\,;\,J|V|p_3\,n_4\,;\,J\rangle &= \langle n_1\,p_2\,;\,J|V|n_3\,p_4\,;\,J\rangle \\ &= \frac{1}{2}\Big\{\sqrt{[1+(-1)^J\delta_{a_1a_2}][1+(-1)^J\delta_{a_3a_4}]} \\ &\qquad\qquad\times\,\langle a_1\,a_2\,;\,J\,T=1|V|a_3\,a_4\,;\,J\,T=1\rangle \\ &\qquad +\sqrt{[1-(-1)^J\delta_{a_1a_2}][1-(-1)^J\delta_{a_3a_4}]} \\ &\qquad\qquad\times\,\langle a_1\,a_2\,;\,J\,T=0|V|a_3\,a_4\,;\,J\,T=0\rangle\Big\}\,, \end{aligned}} \tag{8.26}$$

$$\boxed{\begin{aligned} \langle n_1\,p_2\,;\,J|V|p_3\,n_4\,;\,J\rangle &= \langle p_1\,n_2\,;\,J|V|n_3\,p_4\,;\,J\rangle \\ &= \frac{1}{2}\Big\{\sqrt{[1+(-1)^J\delta_{a_1a_2}][1+(-1)^J\delta_{a_3a_4}]} \\ &\qquad\qquad\times\,\langle a_1\,a_2\,;\,J\,T=1|V|a_3\,a_4\,;\,J\,T=1\rangle \\ &\qquad -\sqrt{[1-(-1)^J\delta_{a_1a_2}][1-(-1)^J\delta_{a_3a_4}]} \\ &\qquad\qquad\times\,\langle a_1\,a_2\,;\,J\,T=0|V|a_3\,a_4\,;\,J\,T=0\rangle\Big\}\,. \end{aligned}} \tag{8.27}$$

also note the simple symmetry

$$\boxed{\begin{aligned} \langle c\,d\,;\,J|V|a\,b\,;\,J\rangle &= \langle a\,b\,;\,J|V|c\,d\,;\,J\rangle\,, \\ \langle c\,d\,;\,JT|V|a\,b\,;\,JT\rangle &= \langle a\,b\,;\,JT|V|c\,d\,;\,JT\rangle\,. \end{aligned}} \tag{8.28}$$

It is necessary to use a consistent phase convention throughout a nuclear structure calculation. The single-particle wave functions (3.63) are defined according to (3.67) in either the CS or BR phase convention. The resulting two-body matrix elements are related as

$$
\begin{aligned}
\langle a\,b\,;\, J\,(T)|V|c\,d\,;\, J\,(T)\rangle_{\mathrm{BR}} \\
= (-1)^{\frac{1}{2}(l_c+l_d-l_a-l_b)}\langle a\,b\,;\, J\,(T)|V|c\,d\,;\, J\,(T)\rangle_{\mathrm{CS}} .
\end{aligned} \tag{8.29}
$$

This holds equally whether or not the isospin formalism is used. The quantity $\frac{1}{2}(l_c + l_d - l_a - l_b)$ is an integer since the initial and final two-nucleon states must have the same parity due to parity conservation of the nuclear force.

In both phase conventions the two-body matrix elements have the symmetry properties

$$
\begin{aligned}
\langle a\,b\,;\, J|V|c\,d\,;\, J\rangle &= (-1)^{j_a+j_b+J+1}\langle b\,a\,;\, J|V|c\,d\,;\, J\rangle \\
&= (-1)^{j_c+j_d+J+1}\langle a\,b\,;\, J|V|d\,c\,;\, J\rangle ,
\end{aligned} \tag{8.30}
$$

$$
\begin{aligned}
\langle a\,b\,;\, J\,T|V|c\,d\,;\, J\,T\rangle &= (-1)^{j_a+j_b+J+T}\langle b\,a\,;\, J\,T|V|c\,d\,;\, J\,T\rangle \\
&= (-1)^{j_c+j_d+J+T}\langle a\,b\,;\, J\,T|V|d\,c\,;\, J\,T\rangle .
\end{aligned} \tag{8.31}
$$

### 8.1.3 Different Types of Two-Body Interaction

In the following the nuclear two-body interactions are divided into four categories. These categories are only suggestive, other types being possible and existing in the literature. For a more exhaustive presentation see e.g. [16].

- *Realistic interactions: meson-exchange potentials:* Interactions based on meson-exchange potentials are considered as realistic interactions. They are developed starting from two interacting nucleons in free space. To make such a potential applicable in nuclear matter, namely to take into account the Pauli principle, one has to use the so-called $G$-matrix approach. The Pauli principle is very important since in nuclear matter the two interacting nucleons are surrounded by several other neighbouring nucleons.

- *Realistic interactions: fitted effective interactions:* In nuclear shell-model calculations the two-body matrix elements are often obtained by choosing them so that calculated energy levels and transition probabilities fit available data; further data can then be predicted. The collection of matrix elements can be considered as constituting a realistic interaction for the purpose at hand. Also other complicated interactions with a number of fitting parameters, like the different types of *Skyrme* and *Gogny* phenomenological forces, belong to the category of realistic interactions.

- *Schematic interactions:* Schematic interactions are designed to mimic salient features of realistic forces. One good example is the delta interaction of zero range.

- *Separable schematic interactions:* An interaction is called separable if it can be expressed as a sum of products where each factor contains labels of only one particle. Separable schematic interactions are discussed extensively in Sect. 8.2. Examples of this category are the SDI, to be discussed in Subsect. 8.2.2, all the simplified *multipole–multipole forces* like the 'pairing-plus-quadrupole interaction' and the simplified pure *pairing interaction* to be discussed in Chap. 12.

Nucleon–nucleon interactions can also be divided into two categories according to the direction of force. If the force is directed along the line connecting them, the force is a *central force.* Otherwise the force is non-central (tensor force). For a central force the mutual potential (energy) between the two interacting nucleons, located at $r_1$ and $r_2$, is a function of their relative distance only, so the potential simplifies to $v(r_1, r_2) = v(|r_1 - r_2|)$.

In the present work we discuss only central forces. The central two-body interaction can be conveniently handled by making a *multipole expansion* of it. In the multipole expansion the relative distance coordinate $|r_1 - r_2|$ is replaced by the coordinates of the individual nucleons in the following manner:

$$
\begin{aligned}
v(|r_1 - r_2|) &= \sum_\lambda v_\lambda(r_1, r_2) \sum_\mu Y^*_{\lambda\mu}(\Omega_1) Y_{\lambda\mu}(\Omega_2) \\
&= \sum_\lambda v_\lambda(r_1, r_2) Y_\lambda(\Omega_1) \cdot Y_\lambda(\Omega_2) , \\
v_\lambda(r_1, r_2) &= 2\pi \int_{-1}^{1} v(|r_1 - r_2|) P_\lambda(\cos\theta_{12}) \mathrm{d}(\cos\theta_{12}) ,
\end{aligned}
\tag{8.32}
$$

where $\theta_{12}$ is the angle between the coordinate vectors $r_1$ and $r_2$ and the scalar product is from (2.51). The result is derived by means of the addition theorem for Legendre polynomials. We see that separation always occurs in the angular coordinates, so separability depends on the radial part, i.e. on whether $v_\lambda(r_1, r_2)$ factorizes.

To clarify the use of the multipole-expansion technique we proceed to analyse two schematic zero-range forces. We start with the delta-function interaction

$$
v_\delta(r_1, r_2) = -V_\delta \delta(r_1 - r_2) .
\tag{8.33}
$$

The dimension of the strength constant $V_\delta$ is energy times volume. It follows from (8.32) that

$$
v_\lambda^{(\delta)}(r_1, r_2) = -V_\delta \frac{\delta(r_1 - r_2)}{r_1 r_2} .
\tag{8.34}
$$

This is not a product of a particle-1 term and a particle-2 term, so the delta interaction is not separable.

Consider next the *surface delta interaction* (SDI)

$$v_{\text{SDI}}(\boldsymbol{r}_1, \boldsymbol{r}_2) = -V_0 \delta(r_1 - R)\delta(r_2 - R)\delta(\Omega_1 - \Omega_2) \, . \qquad (8.35)$$

The dimension of $V_0$ is energy times area, and $R$ is the nuclear radius. The interaction thus occurs only when both nucleons are at the same point on the nuclear surface. Now (8.32) gives

$$v_\lambda^{(\text{SDI})}(r_1, r_2) = -V_0 \frac{\delta(r_1 - R)}{r_1} \frac{\delta(r_2 - R)}{r_2} \, . \qquad (8.36)$$

We see that the SDI is separable. Furthermore, it has the same radial term for all multipoles $\lambda$.

In fact, the SDI is surprisingly realistic since it reproduces the qualitative behaviour of nucleon–nucleon scattering data. The data show that the nucleon–nucleon interaction increases with decreasing relative kinetic energy of the two nucleons. In a nucleus the kinetic energies are the smallest in the surface region of the mean-field potential. This means that the nucleons interact most strongly in the surface region. An extreme limit of this would be the SDI with a sharp nuclear surface at $R$, where the relative kinetic energy of the interacting nucleons becomes zero, leading to an interaction of infinite strength. A more realistic interaction would contain a Gaussian peak at the nuclear surface, consistent with the diluted nuclear surface of thickness $a$ in the Woods–Saxon parametrization of the mean field in (3.20).

### 8.1.4 Central Forces with Spin and Isospin Dependendence

The central nucleon–nucleon potential $V(|\boldsymbol{r}_1 - \boldsymbol{r}_2|)$ was considered above. This coordinate-dependent potential can be complemented with spin and isospin dependence, so that it becomes

$$
\begin{aligned}
v^{(\text{spin–isospin})}(|\boldsymbol{r}_1 - \boldsymbol{r}_2|) &= f(|\boldsymbol{r}_1 - \boldsymbol{r}_2|) \\
&\times [A + B\boldsymbol{\sigma}_1 \cdot \boldsymbol{\sigma}_2 + C\boldsymbol{\tau}_1 \cdot \boldsymbol{\tau}_2 + D(\boldsymbol{\sigma}_1 \cdot \boldsymbol{\sigma}_2)(\boldsymbol{\tau}_1 \cdot \boldsymbol{\tau}_2)] \, , \quad (8.37)
\end{aligned}
$$

where $\boldsymbol{\sigma}_i$ and $\boldsymbol{\tau}_i$ are the spin and isospin operators respectively; see (2.37) and (5.76). The coefficients $A$, $B$, $C$ and $D$ are related to those of the historical *exchange forces*, defined in terms of the operators

$$P_\sigma = \tfrac{1}{2}(1 + \boldsymbol{\sigma}_1 \cdot \boldsymbol{\sigma}_2) \, , \quad P_\tau = \tfrac{1}{2}(1 + \boldsymbol{\tau}_1 \cdot \boldsymbol{\tau}_2) \, . \qquad (8.38)$$

These exchange forces are the *Bartlett* spin-exchange force $C_{\text{B}}P_\sigma$, the *Heisenberg* isospin-exchange force $C_{\text{H}}P_\tau$ and the *Majorana* spatial-exchange force $C_{\text{M}}P_\sigma P_\tau$.

Two frequently used special cases of (8.37) are the following.

- The *Rosenfeld* force, for which the radial dependence is given by

$$f(r) = V_0 e^{-r^2/\mu^2} , \quad r = |\boldsymbol{r}_1 - \boldsymbol{r}_2| \tag{8.39}$$

  with parameters $V_0 = -70.82\,\text{MeV}$ and $\mu = 1.48\,\text{fm}$. The spin–isospin parameters are $A = 0$, $B = 0$, $C = 0.1$ and $D = 0.233$.

- The *Serber* force, for which the radial dependence is given by (8.39), but with $V_0 = -35\,\text{MeV}$ and $\mu = 1.48\,\text{fm}$. The spin–isospin parameters are $A = \frac{3}{8}$, $B = -\frac{1}{8}$, $C = -\frac{1}{8}$ and $D = -\frac{1}{8}$.

After a lengthy but straightforward calculation the two-body interaction matrix elements of the general spin–isospin dependent residual interaction (8.37) can be written in the CS phase convention as

$$\langle a\,b\,;\,JT|V^{(\text{spin–isospin})}|c\,d\,;\,JT\rangle = \mathcal{N}_{ab}(JT)\mathcal{N}_{cd}(JT)$$
$$\times\,[V_{abcd}(JT) + (-1)^{j_c+j_d+J+T}V_{abdc}(JT)] \tag{8.40}$$

with the normalization constant $\mathcal{N}_{ab}(JT)$ stated in (8.19) and

$$V_{abcd}(JT) = \sum_\lambda \mathcal{R}^{(\lambda)}_{abcd} U^{JT\lambda}_{abcd} , \tag{8.41}$$

The quantities on the right-hand side are defined as

$$\mathcal{R}^{(\lambda)}_{abcd} = \int_0^\infty r_1^2 \mathrm{d}r_1 \int_0^\infty r_2^2 \mathrm{d}r_2\, g_{n_a l_a}(r_1) g_{n_b l_b}(r_2) v_\lambda(r_1, r_2) g_{n_c l_c}(r_1) g_{n_d l_d}(r_2) , \tag{8.42}$$

$$U^{JT\lambda}_{abcd} = [A + C(\delta_{T1} - 3\delta_{T0})]\Theta^{J\lambda}_{abcd} + [B + D(\delta_{T1} - 3\delta_{T0})]\Lambda^{J\lambda}_{abcd} , \tag{8.43}$$

where, in turn, we have the abbreviations

$$\Theta^{J\lambda}_{abcd} = (-1)^{j_b+j_c+J} \left\{ \begin{matrix} j_a & j_b & J \\ j_d & j_c & \lambda \end{matrix} \right\} \mathcal{Y}^{(\lambda)}_{ac} \mathcal{Y}^{(\lambda)}_{bd} , \tag{8.44}$$

$$\Lambda^{J\lambda}_{abcd} = (-1)^{j_b+j_c+J+\lambda+1} \sum_j (-1)^j \left\{ \begin{matrix} j_a & j_b & J \\ j_d & j_c & j \end{matrix} \right\} \mathcal{Z}^{(\lambda j)}_{ac} \mathcal{Z}^{(\lambda j)}_{bd} , \tag{8.45}$$

$$\mathcal{Y}^{(\lambda)}_{ac} = \frac{1}{4\sqrt{\pi}} (-1)^{j_c-\frac{1}{2}+\lambda} [1 + (-1)^{l_a+l_c+\lambda}] \hat{j}_a \hat{j}_c \begin{pmatrix} j_a & j_c & \lambda \\ \frac{1}{2} & -\frac{1}{2} & 0 \end{pmatrix} , \tag{8.46}$$

$$\mathcal{Z}^{(\lambda j)}_{ac} = \frac{1}{2\sqrt{\pi}} (-1)^{l_a} \hat{j}_a \hat{j}_c \hat{j} \, \hat{l}_a \hat{l}_c \hat{\lambda} \begin{pmatrix} l_a & \lambda & l_c \\ 0 & 0 & 0 \end{pmatrix} \left\{ \begin{matrix} l_a & \frac{1}{2} & j_a \\ l_c & \frac{1}{2} & j_c \\ \lambda & 1 & j \end{matrix} \right\} . \tag{8.47}$$

## 8.2 Separable Interactions; the Surface Delta Interaction

In the following we first address the properties of a general separable interaction and then specialize the results to our actual working mule, the SDI.

### 8.2.1 Multipole Decomposition of a General Separable Interaction

A separable interaction, by definition, possesses the convenient property that the radial part of each multipole term factorizes in (8.32). This property greatly simplifies the calculation of two-body matrix elements. The radial part of the $\lambda$ multipole term of a general separable interaction can be written in the form

$$v_\lambda(r_1, r_2) = f_\lambda(r_1) f_\lambda(r_2) \; . \tag{8.48}$$

In practice, the function $f$ is usually taken to be a power function,

$$f_\lambda(r) = C_\lambda r^\lambda \; . \tag{8.49}$$

This is the basic ingredient in a *multipole–multipole* force. For instance, the multipole–multipole interaction containing the monopole ($\lambda = 0$) and quadrupole ($\lambda = 2$) parts is called the pairing-plus-quadrupole interaction. For our separable interaction the expansion (8.32) is

$$\begin{aligned}
v(|\boldsymbol{r}_1 - \boldsymbol{r}_2|) &= \sum_\lambda f_\lambda(r_1) f_\lambda(r_2) \sum_\mu Y^*_{\lambda\mu}(\Omega_1) Y_{\lambda\mu}(\Omega_2) \\
&= \sum_{\lambda\mu} Q^*_{\lambda\mu}(\boldsymbol{r}_1) Q_{\lambda\mu}(\boldsymbol{r}_2) \; , \quad Q_{\lambda\mu}(\boldsymbol{r}) \equiv f_\lambda(r) Y_{\lambda\mu}(\Omega) \; . 
\end{aligned} \tag{8.50}$$

With the two-body operator expressed in the form (4.27), substitution of the separable interaction gives

$$V = \tfrac{1}{2} \sum_{\alpha\beta\gamma\delta} v_{\alpha\beta\gamma\delta} c^\dagger_\alpha c^\dagger_\beta c_\delta c_\gamma = \tfrac{1}{2} \sum_{\substack{\alpha\beta\gamma\delta \\ \lambda\mu}} \langle \alpha | Q^*_{\lambda\mu} | \gamma \rangle \langle \beta | Q_{\lambda\mu} | \delta \rangle c^\dagger_\alpha c^\dagger_\beta c_\delta c_\gamma \; . \tag{8.51}$$

Noting that $Q^*_{\lambda\mu} = (-1)^\mu Q_{\lambda,-\mu}$, we apply the Wigner–Eckart theorem (2.27) to the two matrix elements and after some algebra find

$$V = \tfrac{1}{2} \sum_{\substack{abcd \\ \lambda}} \hat{\lambda}^{-2} (a\|\boldsymbol{Q}_\lambda\|c)(b\|\boldsymbol{Q}_\lambda\|d) \left[ c^\dagger_a \tilde{c}_c \right]_\lambda \cdot \left[ c^\dagger_b \tilde{c}_d \right]_\lambda \; . \tag{8.52}$$

Except for the absence of the charge $e$ and the more general radial dependence, the operator $Q_{\lambda\mu}$ in (8.50) is the same as the electric operator (6.6). Therefore the reduced matrix elements appearing in (8.52) can be read off (6.23). In the CS phase convention the result is

$$(a\|\boldsymbol{Q}_\lambda\|c) = \frac{1}{\sqrt{4\pi}} (-1)^{j_c + \lambda - \frac{1}{2}} \frac{1 + (-1)^{l_a + l_c + \lambda}}{2} \hat{\lambda} \hat{j}_a \hat{j}_c \begin{pmatrix} j_a & j_c & \lambda \\ \frac{1}{2} & -\frac{1}{2} & 0 \end{pmatrix} \mathcal{R}^{(\lambda)}_{ac} \; , \tag{8.53}$$

where the radial integral, in general different from that in (6.23), is

$$\mathcal{R}_{ac}^{(\lambda)} = \int_0^\infty g_{n_a l_a}(r) f_\lambda(r) g_{n_c l_c}(r) r^2 dr \ . \tag{8.54}$$

We proceed to calculate the matrix element of the separable residual interaction (8.52) for angular-momentum-coupled two-nucleon states. The result for *like nucleons*, obtained after some contractions and angular momentum algebra, is

$$\langle a\,b\,;\,J|V|c\,d\,;\,J\rangle = \mathcal{N}_{ab}(J)\mathcal{N}_{cd}(J)[A_{abcd} - (-1)^{j_c+j_d+J} A_{abdc}] \ , \tag{8.55}$$

where

$$A_{abcd} \equiv \sum_\lambda (-1)^{j_a+j_b+J} \begin{Bmatrix} j_a & j_b & J \\ j_d & j_c & \lambda \end{Bmatrix} (c\|\boldsymbol{Q}_\lambda\|a)(b\|\boldsymbol{Q}_\lambda\|d) \ . \tag{8.56}$$

When $V$ is a separable *proton–neutron* interaction, (8.55) is replaced by

$$\langle p_1\,n_2\,;\,J|V|p_3\,n_4\,;\,J\rangle = A_{p_1 n_2 p_3 n_4} \ , \tag{8.57}$$

$$\langle p_1\,n_2\,;\,J|V|n_3\,p_4\,;\,J\rangle = (-1)^{j_{n_3}+j_{p_4}+J+1} A_{p_1 n_2 p_4 n_3} \ , \tag{8.58}$$

where we have used (1.30) and (4.15).

## 8.2.2 Two-Body Matrix Elements of the Surface Delta Interaction

Comparison of (8.36) and (8.48) shows that for the SDI

$$f_\lambda(r) = \sqrt{-V_0}\,\frac{\delta(r - R)}{r} \ . \tag{8.59}$$

Substituting this into (8.54) gives the simple radial integral

$$\begin{aligned}
\mathcal{R}_{ab}^{(\lambda)}(\text{SDI}) \Big/ \sqrt{-V_0} &= \int_0^\infty g_{n_a l_a}(r)\frac{\delta(r - R)}{r} g_{n_b l_b}(r) r^2 dr \\
&= g_{n_a l_a}(R) g_{n_b l_b}(R) R \equiv \kappa_{ab} = \kappa_{ba} \ .
\end{aligned} \tag{8.60}$$

We now calculate the auxiliary quantity $A_{abcd}$ defined in (8.56) by substituting from (8.53). After some simplification and rearranging we have

$$\begin{aligned}
A_{abcd} = K_{abcd}(-1)^{j_b+j_d+J}\widehat{j_a}\widehat{j_b}\widehat{j_c}\widehat{j_d}\big[1 + (-1)^{l_a+l_b+l_c+l_d}\big] \\
\times \sum_\lambda [1 + (-1)^{l_a+l_c+\lambda}]\widehat{\lambda}^2 \begin{pmatrix} j_c & j_a & \lambda \\ \tfrac{1}{2} & -\tfrac{1}{2} & 0 \end{pmatrix} \begin{pmatrix} j_b & j_d & \lambda \\ \tfrac{1}{2} & -\tfrac{1}{2} & 0 \end{pmatrix} \begin{Bmatrix} j_a & j_b & J \\ j_d & j_c & \lambda \end{Bmatrix} \ ,
\end{aligned} \tag{8.61}$$

where

$$K_{abcd} \equiv -\frac{V_0 \kappa_{ac}\kappa_{bd}}{16\pi} \ . \tag{8.62}$$

The sum over $\lambda$ is evaluated by means of the relations [7,22]

$$\sum_{\lambda}(-1)^{J+\lambda+1}\widehat{\lambda}^2 \begin{pmatrix} j_a & j_d & \lambda \\ \frac{1}{2} & -\frac{1}{2} & 0 \end{pmatrix} \begin{pmatrix} j_c & j_b & \lambda \\ \frac{1}{2} & -\frac{1}{2} & 0 \end{pmatrix} \begin{Bmatrix} j_a & j_d & \lambda \\ j_c & j_b & J \end{Bmatrix}$$

$$= \begin{pmatrix} j_a & j_b & J \\ \frac{1}{2} & -\frac{1}{2} & 0 \end{pmatrix} \begin{pmatrix} j_c & j_d & J \\ \frac{1}{2} & -\frac{1}{2} & 0 \end{pmatrix} \quad (8.63)$$

and

$$\sum_{\lambda}(-1)^{j_b+j_d+J}\widehat{\lambda}^2 \begin{pmatrix} j_c & j_a & \lambda \\ \frac{1}{2} & -\frac{1}{2} & 0 \end{pmatrix} \begin{pmatrix} j_b & j_d & \lambda \\ \frac{1}{2} & -\frac{1}{2} & 0 \end{pmatrix} \begin{Bmatrix} j_c & j_a & \lambda \\ j_b & j_d & J \end{Bmatrix}$$

$$= \begin{pmatrix} j_a & j_b & J \\ \frac{1}{2} & \frac{1}{2} & -1 \end{pmatrix} \begin{pmatrix} j_c & j_d & J \\ \frac{1}{2} & \frac{1}{2} & -1 \end{pmatrix} . \quad (8.64)$$

The result is

$$A_{abcd} = K_{abcd}\left[1 + (-1)^{l_a+l_b+l_c+l_d}\right]\widehat{j_a}\widehat{j_b}\widehat{j_c}\widehat{j_d}\left[\begin{pmatrix} j_a & j_b & J \\ \frac{1}{2} & \frac{1}{2} & -1 \end{pmatrix} \begin{pmatrix} j_c & j_d & J \\ \frac{1}{2} & \frac{1}{2} & -1 \end{pmatrix}\right.$$

$$\left. - (-1)^{l_a+l_c+j_b+j_d} \begin{pmatrix} j_a & j_b & J \\ \frac{1}{2} & -\frac{1}{2} & 0 \end{pmatrix} \begin{pmatrix} j_c & j_d & J \\ \frac{1}{2} & -\frac{1}{2} & 0 \end{pmatrix}\right]. \quad (8.65)$$

Substituted into (8.55) this gives the matrix element for like nucleons as

$$\boxed{\begin{aligned} \langle a\,b;\, J|V_{\mathrm{SDI}}|c\,d;\, J\rangle &= -K_{abcd}\mathcal{N}_{ab}(J)\mathcal{N}_{cd}(J)(-1)^{l_a+l_c+j_b+j_d} \\ &\times \left[1 + (-1)^{l_a+l_b+l_c+l_d}\right]\left[1 + (-1)^{l_c+l_d+J}\right] \\ &\times \widehat{j_a}\widehat{j_b}\widehat{j_c}\widehat{j_d} \begin{pmatrix} j_a & j_b & J \\ \frac{1}{2} & -\frac{1}{2} & 0 \end{pmatrix} \begin{pmatrix} j_c & j_d & J \\ \frac{1}{2} & -\frac{1}{2} & 0 \end{pmatrix} . \end{aligned}} \quad (8.66)$$

The proton–neutron matrix elements (8.57) and (8.58) are obtained directly from (8.65).

The two-body matrix element in isospin representation can be deduced immediately since there is no isospin-dependent term in the Hamiltonian.[2] Generalizing (8.55) we have

$$\langle a\,b;\, JT|V|c\,d;\, JT\rangle$$

$$= \mathcal{N}_{ab}(JT)\mathcal{N}_{cd}(JT)\left[A_{abcd} - (-1)^{j_c+j_d+J}(-1)^{\frac{1}{2}+\frac{1}{2}+T}A_{abdc}\right]$$

$$= \mathcal{N}_{ab}(JT)\mathcal{N}_{cd}(JT)\left[A_{abcd} + (-1)^{j_c+j_d+J+T}A_{abdc}\right], \quad (8.67)$$

which leads to

---

[2]  In general the nuclear two-body Hamiltonian need not be an isoscalar. The two creation operators, each an isospinor with $t = \frac{1}{2}$, can be coupled to $T_1 = 0$ or 1. Likewise the two annihilation operators can be coupled to $T_2 = 0$ or 1. Thus $T_1$ and $T_2$ can be coupled to $T = 0, 1, 2$. The Coulomb interaction contains such isoscalar, isovector and isotensor terms.

$$\langle a\,b;\, JT|V_{\mathrm{SDI}}|c\,d;\, JT\rangle$$

$$= K_{abcd}\mathcal{N}_{ab}(JT)\mathcal{N}_{cd}(JT)\left[1+(-1)^{l_a+l_b+l_c+l_d}\right]\hat{j}_a\hat{j}_b\hat{j}_c\hat{j}_d$$

$$\times\left\{\left[1+(-1)^T\right]\begin{pmatrix} j_a & j_b & J \\ \frac{1}{2} & \frac{1}{2} & -1 \end{pmatrix}\begin{pmatrix} j_c & j_d & J \\ \frac{1}{2} & \frac{1}{2} & -1 \end{pmatrix}\right.$$

$$\left.-(-1)^{l_a+l_c+j_b+j_d}\left[1-(-1)^{l_c+l_d+J+T}\right]\begin{pmatrix} j_a & j_b & J \\ \frac{1}{2} & -\frac{1}{2} & 0 \end{pmatrix}\begin{pmatrix} j_c & j_d & J \\ \frac{1}{2} & -\frac{1}{2} & 0 \end{pmatrix}\right\}.$$

$$(8.68)$$

Note that for $T=1$ this coincides with the like-nucleon result (8.66).

It turns out (see e.g. [28]) that the absolute value of the oscillator wave function $g_{nl}(r)$ at the nuclear surface, $r=R$, is nearly independent of $n$ and $l$. Accordingly the magnitude $|K_{abcd}|$ of the parameter (8.62) is practically independent of the indices $a$, $b$, $c$ and $d$. With the phasing of the $g_{nl}(r)$ given in (3.47), we can write

$$K_{abcd} = -|K_{abcd}|(-1)^{n_a+n_b+n_c+n_d}\ . \qquad (8.69)$$

Following [28] we now adopt a simplified interaction such that $K_{abcd}$ has a constant magnitude,

$$\boxed{K_{abcd}\to-\tfrac{1}{4}A_T(-1)^{n_a+n_b+n_c+n_d}\ ,\qquad T=0,1\ ,} \qquad (8.70)$$

where the strength constants $A_0$ and $A_1$ are treated as fitting parameters.

With explicit account of the normalizer (8.19) we find the notable special cases of (8.68)

$$\langle a^2;\, JT|V_{\mathrm{SDI}}|c^2;\, JT\rangle = -\tfrac{1}{4}A_T\left[1-(-1)^{J+T}\right]\hat{j}_a^{\,2}\hat{j}_c^{\,2}\left\{\tfrac{1}{2}\left[1+(-1)^T\right]\right.$$

$$\times\begin{pmatrix} j_a & j_a & J \\ \frac{1}{2} & \frac{1}{2} & -1 \end{pmatrix}\begin{pmatrix} j_c & j_c & J \\ \frac{1}{2} & \frac{1}{2} & -1 \end{pmatrix}-(-1)^{l_a+l_c+j_a+j_c}\begin{pmatrix} j_a & j_a & J \\ \frac{1}{2} & -\frac{1}{2} & 0 \end{pmatrix}\begin{pmatrix} j_c & j_c & J \\ \frac{1}{2} & -\frac{1}{2} & 0 \end{pmatrix}\right\},$$

$$(8.71)$$

$$\langle a^2;\, JT|V_{\mathrm{SDI}}|a^2;\, JT\rangle = -\tfrac{1}{4}A_T\left[1-(-1)^{J+T}\right]\hat{j}_a^{\,4}$$

$$\times\left\{\tfrac{1}{2}\left[1+(-1)^T\right]\begin{pmatrix} j_a & j_a & J \\ \frac{1}{2} & \frac{1}{2} & -1 \end{pmatrix}^2+\begin{pmatrix} j_a & j_a & J \\ \frac{1}{2} & -\frac{1}{2} & 0 \end{pmatrix}^2\right\}. \qquad (8.72)$$

In Tables 8.1 and 8.2 the (a priori) non-zero matrix elements of the SDI, given by (8.68) and (8.70) with $A_T=1$, are listed for the 0s-0p and 0d-1s shells . All the zeros appearing in these tables are non-trivial, i.e. not due to the angular momentum, parity or isospin selection rules. Tables 8.3 and 8.4 similarly list the matrix elements for interactions between the 0p and 0d-1s shells and between the 0d-1s and 0f$_{7/2}$ shells. Table 8.4 includes also the 0f$_{7/2}$ intra-shell matrix elements. The matrix elements in these tables

**Table 8.1.** Two-body matrix elements $\langle a\,b;\,JT|V_{\mathrm{SDI}}|c\,d;\,JT\rangle$ with $A_T = 1$ for the 0s-0p shells in the CS phase convention

| $abcd$ | $JT$ | $\langle V_{\mathrm{SDI}}\rangle$ | $JT$ | $\langle V_{\mathrm{SDI}}\rangle$ | $JT$ | $\langle V_{\mathrm{SDI}}\rangle$ | $JT$ | $\langle V_{\mathrm{SDI}}\rangle$ |
|---|---|---|---|---|---|---|---|---|
| 1111 | 01 | −1.0000 | 10 | −1.0000 | | | | |
| 1122 | 01 | 1.4142 | 10 | 1.0541 | | | | |
| 1123 | 10 | −1.3333 | | | | | | |
| 1133 | 01 | 1.0000 | 10 | −0.3333 | | | | |
| 1212 | 10 | −0.6667 | 11 | −1.3333 | 20 | −2.0000 | 21 | 0 |
| 1213 | 10 | 0.9428 | 11 | −0.9428 | | | | |
| 1313 | 00 | −2.0000 | 01 | 0 | 10 | −1.3333 | 11 | −0.6667 |
| 2222 | 01 | −2.0000 | 10 | −1.2000 | 21 | −0.4000 | 30 | −1.2000 |
| 2223 | 10 | 1.2649 | 21 | −0.5657 | | | | |
| 2233 | 01 | −1.4142 | 10 | 0.6325 | | | | |
| 2323 | 10 | −2.0000 | 11 | 0 | 20 | −1.2000 | 21 | −0.8000 |
| 2333 | 10 | 0 | | | | | | |
| 3333 | 01 | −1.0000 | 10 | −1.0000 | | | | |

The states are numbered $1 = 0\mathrm{s}_{1/2}$, $2 = 0\mathrm{p}_{3/2}$ and $3 = 0\mathrm{p}_{1/2}$. The first column gives the state labels, and the following columns give the $JT$ combinations and matrix elements.

serve as a reference and as a convenient source for small configuration mixing calculations.

It can be advantageous to add to the SDI the isospin-exchange contribution of the spin–isospin force (8.37). With $B = 0$ and $D = 0$ in (8.37), the resulting two-body interaction has the matrix elements

$$\langle a\,b;\,JT|V|c\,d;\,JT\rangle = \langle a\,b;\,JT|V_{\mathrm{SDI}}|c\,d;\,JT\rangle[A + C\,(\delta_{T1} - 3\delta_{T0})]\,, \quad (8.73)$$

as can be seen by an elementary calculation or from (8.43).

A modified version of the interaction in (8.73) is the so-called *modified surface delta interaction* (MSDI) [28], defined by its matrix elements

$$\langle a\,b;\,JT|V_{\mathrm{MSDI}}|c\,d;\,JT\rangle = \langle a\,b;\,JT\,|V_{\mathrm{SDI}}|\,c\,d;\,JT\rangle$$

$$+ (B_1\delta_{T1} + B_0\delta_{T0})\frac{\delta_{ac}\delta_{bd} + (-1)^{j_c+j_d+J+T}\delta_{ad}\delta_{bc}}{1 + \delta_{ab}}\,. \quad (8.74)$$

The added term, consistent with the form of (8.67), is diagonal in the nucleon labels. It thus effectively only changes the single-particle energies, differently for $T = 1$ and $T = 0$, and therefore does not affect the wave functions. The result is that the energies of all levels of a given $T$ shift by the constant amount $B_T$.

**Table 8.2.** Two-body matrix elements $\langle a\,b\,;\,J\,T|V_{\mathrm{SDI}}|c\,d\,;\,J\,T\rangle$ with $A_T = 1$ for the 0d-1s shells in the CS phase convention

| abcd | JT | $\langle V_{\mathrm{SDI}}\rangle$ | JT | $\langle V_{\mathrm{SDI}}\rangle$ | JT | $\langle V_{\mathrm{SDI}}\rangle$ | JT | $\langle V_{\mathrm{SDI}}\rangle$ | JT | $\langle V_{\mathrm{SDI}}\rangle$ |
|---|---|---|---|---|---|---|---|---|---|---|
| 1111 | 01 | −3.0000 | 10 | −1.6286 | 21 | −0.6857 | 30 | −0.9143 | 41 | −0.2857 |
| 1111 | 50 | −1.4286 | | | | | | | | |
| 1112 | 21 | −0.9071 | 30 | −1.3279 | | | | | | |
| 1113 | 10 | 1.8142 | 21 | −0.4849 | 30 | 0.5938 | 41 | −0.5714 | | |
| 1122 | 01 | −1.7320 | 10 | −1.1832 | | | | | | |
| 1123 | 10 | −0.6761 | 21 | −0.7407 | | | | | | |
| 1133 | 01 | −2.4495 | 10 | 1.1759 | 21 | −0.5237 | 30 | 0.3429 | | |
| 1212 | 20 | −0.8000 | 21 | −1.2000 | 30 | −2.0000 | 31 | 0 | | |
| 1213 | 20 | −1.0690 | 21 | −0.6414 | 30 | 1.0222 | 31 | 0 | | |
| 1223 | 20 | −0.9798 | 21 | −0.9798 | | | | | | |
| 1233 | 21 | −0.6928 | 30 | 0.2213 | | | | | | |
| 1313 | 10 | −3.6000 | 11 | 0 | 20 | −1.4286 | 21 | −0.3429 | 30 | −0.7429 |
| 1313 | 31 | 0 | 40 | −1.4286 | 41 | −1.1429 | | | | |
| 1322 | 10 | 1.7888 | | | | | | | | |
| 1323 | 10 | −0.8944 | 11 | 0 | 20 | −1.3093 | 21 | −0.5237 | | |
| 1333 | 10 | −0.5657 | 21 | −0.3703 | 30 | 0.3959 | | | | |
| 2222 | 01 | −1.0000 | 10 | −1.0000 | | | | | | |
| 2223 | 10 | 0 | | | | | | | | |
| 2233 | 01 | −1.4142 | 10 | 0.6325 | | | | | | |
| 2323 | 10 | −2.0000 | 11 | 0 | 20 | −1.2000 | 21 | −0.8000 | | |
| 2333 | 10 | 1.2649 | 21 | −0.5657 | | | | | | |
| 3333 | 01 | −2.0000 | 10 | −1.2000 | 21 | −0.4000 | 30 | −1.2000 | | |

The states are numbered $1 = 0\mathrm{d}_{5/2}$, $2 = 1\mathrm{s}_{1/2}$ and $3 = 0\mathrm{d}_{3/2}$. The first column gives the state labels, and the following columns give the $JT$ combinations and matrix elements.

# 8.3 Configuration Mixing in Two-Particle Nuclei

In the previous section we discussed the separable two-nucleon interaction, in particular the SDI. In this section we use the SDI to produce configuration mixing for states in two-particle nuclei.

### 8.3.1 Matrix Representation of an Eigenvalue Problem

The starting point in our discussion is the many-nucleon Schrödinger equation

$$H|\Psi_n\rangle = E_n|\Psi_n\rangle\,, \tag{8.75}$$

where $|\Psi_n\rangle$ is a many-nucleon state. Such a state can be expanded in any complete orthonormal basis $\{|\phi_k\rangle\}_{k=1}^{N}$ by writing

$$|\Psi_n\rangle = \sum_l a_l^{(n)}|\phi_l\rangle\,. \tag{8.76}$$

**Table 8.3.** Two-body matrix elements $\langle ab; JT|V_{\text{SDI}}|cd; JT\rangle$ with $A_T = 1$ between the 0p and 0d-1s shells in the CS phase convention

| abcd | JT | $\langle V_{\text{SDI}}\rangle$ | JT | $\langle V_{\text{SDI}}\rangle$ | JT | $\langle V_{\text{SDI}}\rangle$ | JT | $\langle V_{\text{SDI}}\rangle$ | JT | $\langle V_{\text{SDI}}\rangle$ |
|---|---|---|---|---|---|---|---|---|---|---|
| 1313 | 10 | −0.6667 | 11 | −1.3333 | 20 | −2.0000 | 21 | 0 | | |
| 1314 | 20 | 0 | 21 | 0 | | | | | | |
| 1315 | 10 | −0.9428 | 11 | 0.9428 | | | | | | |
| 1323 | 10 | −1.1926 | 11 | −0.5963 | 20 | 0.8000 | 21 | 0 | | |
| 1324 | 10 | 0.8944 | 11 | −1.7888 | 20 | 0.7856 | 21 | 0 | | |
| 1325 | 10 | −0.6667 | 11 | −1.3333 | 20 | 0.4000 | 21 | 0 | | |
| 1414 | 20 | −2.0000 | 21 | 0 | 30 | −1.1429 | 31 | −0.8571 | | |
| 1423 | 20 | 0.9798 | 21 | 0 | 30 | −1.2519 | 31 | 0.9389 | | |
| 1424 | 20 | −1.7105 | 21 | 0 | 30 | 0.2555 | 31 | 0.7666 | | |
| 1425 | 20 | −1.9596 | 21 | 0 | | | | | | |
| 1515 | 00 | −2.0000 | 01 | 0 | 10 | −1.3333 | 11 | −0.6667 | | |
| 1523 | 00 | 2.8284 | 01 | 0 | 10 | −1.6865 | 11 | 0.4216 | | |
| 1524 | 10 | 1.2649 | 11 | 1.2649 | | | | | | |
| 1525 | 10 | −0.9428 | 11 | 0.9428 | | | | | | |
| 2323 | 00 | −4.0000 | 01 | 0 | 10 | −2.1333 | 11 | −0.2667 | 20 | −0.8000 |
| 2323 | 21 | 0 | 30 | −1.3714 | 31 | −1.0286 | | | | |
| 2324 | 10 | 1.6000 | 11 | −0.8000 | 20 | 0.5237 | 21 | 0 | 30 | 0.2799 |
| 2324 | 31 | −0.8398 | | | | | | | | |
| 2325 | 10 | −1.1926 | 11 | −0.5963 | 20 | 0.8000 | 21 | 0 | | |
| 2424 | 10 | −1.2000 | 11 | −2.4000 | 20 | −1.7714 | 21 | 0 | 30 | −0.0571 |
| 2424 | 31 | −0.6857 | 40 | −2.5714 | 41 | 0 | | | | |
| 2425 | 10 | 0.8944 | 11 | −1.7888 | 20 | −1.8330 | 21 | 0 | | |
| 2525 | 10 | −0.6667 | 11 | −1.3333 | 20 | −2.0000 | 21 | 0 | | |

The first column gives the state labels, and the following columns give the $JT$ combinations and matrix elements. The states are numbered $1 = 0p_{1/2}$, $2 = 0p_{3/2}$, $3 = 0d_{3/2}$, $4 = 0d_{5/2}$ and $5 = 1s_{1/2}$.

Here we have assumed that the basis has $N < \infty$ basis states, an assumption valid for our purposes.

We first write the projection of the eigenvalue equation (8.75) onto the basis state $|\phi_k\rangle$, i.e.

$$\langle\phi_k|H|\Psi_n\rangle = E_n\langle\phi_k|\Psi_n\rangle \,, \tag{8.77}$$

and then expand it by using (8.76) to yield

$$\sum_l a_l^{(n)}\langle\phi_k|H|\phi_l\rangle = E_n a_k^{(n)} \tag{8.78}$$

Writing $\langle\phi_k|H|\phi_l\rangle \equiv H_{kl}$ this becomes

$$\sum_l H_{kl}a_l^{(n)} = E_n a_k^{(n)} \,, \tag{8.79}$$

which is the matrix equation

**Table 8.4.** Two-body matrix elements $\langle a\,b;\,J\,T|V_{\mathrm{SDI}}|c\,d;\,J\,T\rangle$ with $A_T = 1$ between the 0d-1s and $0f_{7/2}$ shells, and within the $0f_{7/2}$ shell, in the CS phase convention

| abcd | JT | $\langle V_{\mathrm{SDI}}\rangle$ | JT | $\langle V_{\mathrm{SDI}}\rangle$ | JT | $\langle V_{\mathrm{SDI}}\rangle$ | JT | $\langle V_{\mathrm{SDI}}\rangle$ | JT | $\langle V_{\mathrm{SDI}}\rangle$ |
|---|---|---|---|---|---|---|---|---|---|---|
| 1414 | 20 | −3.4286 | 21 | 0 | 30 | −1.1429 | 31 | −0.3809 | 40 | −0.7619 |
| 1414 | 41 | 0 | 50 | −1.4545 | 51 | −1.2121 | | | | |
| 1424 | 20 | −2.2857 | 21 | 0 | 30 | 0.3299 | 31 | 0.6598 | 40 | −0.9763 |
| 1424 | 41 | 0 | 50 | 0.1587 | 51 | 0.7935 | | | | |
| 1434 | 30 | 0.9897 | 31 | 0.6598 | 40 | −1.1269 | 41 | 0 | | |
| 2424 | 10 | −1.7143 | 11 | −3.4286 | 20 | −2.0952 | 21 | 0 | 30 | −0.0952 |
| 2424 | 31 | −1.1429 | 40 | −1.7922 | 41 | 0 | 50 | −0.0173 | 51 | −0.5195 |
| 2424 | 60 | −3.0303 | 61 | 0 | | | | | | |
| 2434 | 30 | −0.2857 | 31 | −1.1429 | 40 | −1.8687 | 41 | 0 | | |
| 3434 | 30 | −0.8571 | 31 | −1.1429 | 40 | −2.0000 | 41 | 0 | | |
| 4444 | 01 | −4.0000 | 10 | −2.0952 | 21 | −0.9524 | 30 | −0.9870 | 41 | −0.4675 |
| 4444 | 50 | −0.9391 | 61 | −0.2331 | 70 | −1.6317 | | | | |

The states are numbered $1 = 0d_{3/2}$, $2 = 0d_{5/2}$, $3 = 1s_{1/2}$ and $4 = 0f_{7/2}$. The first column gives the state labels, and the following columns give the $JT$ combinations and matrix elements.

$$\begin{pmatrix} H_{11} & H_{12} & \cdots & H_{1N} \\ H_{21} & H_{22} & \cdots & H_{2N} \\ \vdots & \vdots & \ddots & \vdots \\ H_{N1} & H_{N2} & \cdots & H_{NN} \end{pmatrix} \begin{pmatrix} a_1^{(n)} \\ a_2^{(n)} \\ \vdots \\ a_N^{(n)} \end{pmatrix} = E_n \begin{pmatrix} a_1^{(n)} \\ a_2^{(n)} \\ \vdots \\ a_N^{(n)} \end{pmatrix}. \tag{8.80}$$

So we have converted the Schrödinger equation (8.75) into an eigenvalue problem of the Hamiltonian matrix.

The eigenvalues and eigenstates of a general Hamiltonian are obtained by diagonalizing the Hamiltonian matrix. For two-particle nuclei the basis states are

$$|a\,b;\,J\,M;\,T\,M_T\rangle = \mathcal{N}_{ab}(JT)\left[c_a^\dagger c_b^\dagger\right]_{JM}^{TM_T}|\mathrm{CORE}\rangle \tag{8.81}$$

constructed within a chosen single-particle valence space. To solve for the eigenenergies and eigenstates we then need to diagonalize a Hamiltonian matrix with elements

$$\langle a\,b;\,J\,M;\,T\,M_T|H|c\,d;\,J\,M;\,T\,M_T\rangle. \tag{8.82}$$

### 8.3.2 Solving the Eigenenergies of a Two-by-Two Problem

The two-by-two eigenvalue problem is an analytically solvable special case of the general formalism introduced above. The matrix equation (8.80), to be satisfied by each $n = 1, 2$, becomes

$$\begin{pmatrix} H_{11} - E & H_{12} \\ H_{21} & H_{22} - E \end{pmatrix} \begin{pmatrix} a_1 \\ a_2 \end{pmatrix} = 0 . \tag{8.83}$$

A non-trivial solution of the two linear equations for $a_1$ and $a_2$ exists only if the determinant condition

$$0 = \begin{vmatrix} H_{11} - E & H_{12} \\ H_{21} & H_{22} - E \end{vmatrix} = (H_{11} - E)(H_{22} - E) - |H_{12}|^2 \tag{8.84}$$

is satisfied. Here we have used the fact that $H_{21} = H_{12}^*$ for a Hermitian matrix representing a Hermitian Hamiltonian. Solving for $E$ gives

$$\boxed{\begin{aligned} E_1 &= \frac{1}{2}\left(H_{11} + H_{22} - \sqrt{(H_{11} - H_{22})^2 + 4|H_{12}|^2}\right) , \\ E_2 &= \frac{1}{2}\left(H_{11} + H_{22} + \sqrt{(H_{11} - H_{22})^2 + 4|H_{12}|^2}\right) , \end{aligned}} \tag{8.85}$$

where we follow the usual labelling convention $E_1 < E_2$.

The eigenstates corresponding to $E_1$ and $E_2$ are, in the pattern of (8.76),

$$|\Psi_1\rangle = a_1^{(1)}|\phi_1\rangle + a_2^{(1)}|\phi_2\rangle , \tag{8.86}$$

$$|\Psi_2\rangle = a_1^{(2)}|\phi_1\rangle + a_2^{(2)}|\phi_2\rangle . \tag{8.87}$$

The ratio of the amplitudes $a_1^{(n)}$ and $a_2^{(n)}$ is found by substituting $E_n$ into (8.83). The result is

$$\frac{a_1^{(n)}}{a_2^{(n)}} = \frac{H_{12}}{E_n - H_{11}} = \frac{E_n - H_{22}}{H_{21}} , \tag{8.88}$$

which leads to

$$\frac{a_1^{(2)}}{a_2^{(2)}} = -\frac{a_2^{(1)*}}{a_1^{(1)*}} . \tag{8.89}$$

This allows us to write the eigenstates as

$$\begin{aligned} |\Psi_1\rangle &= a_1^{(1)}\left(|\phi_1\rangle + \frac{a_2^{(1)}}{a_1^{(1)}}|\phi_2\rangle\right) , \\ |\Psi_2\rangle &= a_2^{(2)}\left(\frac{a_1^{(2)}}{a_2^{(2)}}|\phi_1\rangle + |\phi_2\rangle\right) = a_2^{(2)}\left(-\frac{a_2^{(1)*}}{a_1^{(1)*}}|\phi_1\rangle + |\phi_2\rangle\right) . \end{aligned} \tag{8.90}$$

We note that the algebraic solution thus produces orthogonal eigenstates, $\langle\Psi_1|\Psi_2\rangle = 0$, as expected of the eigenstates of a Hermitian operator.

It remains to normalize the eigenstates $|\Psi_n\rangle$, so that $|a_1^{(n)}|^2 + |a_2^{(n)}|^2 = 1$. This is accomplished by inserting the amplitude ratio $a_2^{(1)}/a_1^{(1)}$ from (8.88) into (8.90) and determining $a_1^{(1)}$ and $a_2^{(2)}$ so that $\langle\Psi_1|\Psi_1\rangle = 1$ and $\langle\Psi_2|\Psi_2\rangle = 1$. This results in the magnitudes

$$|a_1^{(1)}| = \left(1 + |a_2^{(1)}/a_1^{(1)}|^2\right)^{-1/2} = |a_2^{(2)}| . \tag{8.91}$$

The phases of $a_1^{(1)}$ and $a_2^{(2)}$ remain undetermined. They are normally chosen to be real and such that the leading amplitude of each eigenstate is positive.

A simple analytic solution is generally not possible for three-by-three or larger problems. They are solved by readily available numerical methods and computer codes.

### 8.3.3 Matrix Elements of the Hamiltonian in the Two-Nucleon Basis

To carry out the evaluation of the matrix element (8.82) we need both the one-body and two-body parts of the nuclear Hamiltonian. The one-body part is given by (4.19), and we assume it has the diagonal form of (4.71),

$$T_\varepsilon = \sum_\alpha \varepsilon_\alpha c_\alpha^\dagger c_\alpha . \tag{8.92}$$

The matrix elements of this operator are evaluated in the basis (8.81). The operator $T_\varepsilon$ is a scalar and an isoscalar, so similarly to (8.21) we only need to consider matrix elements diagonal in angular momentum and isospin, and there is no dependence on the projection quantum numbers. Thus we have

$$\langle a\,b\,;\, J\,M\,;\, T\,M_T | T_\varepsilon | c\,d\,;\, J\,M\,;\, T\,M_T \rangle \equiv \langle a\,b\,;\, J\,T | T_\varepsilon | c\,d\,;\, J\,T \rangle$$

$$= \mathcal{N}_{ab}(JT)\mathcal{N}_{cd}(JT) \sum_{\alpha'} \varepsilon_{\alpha'} \langle \mathrm{CORE} | \left( \left[ c_a^\dagger c_b^\dagger \right]_{JM}^{TM_T} \right)^\dagger c_{\alpha'}^\dagger c_{\alpha'} \left[ c_c^\dagger c_d^\dagger \right]_{JM}^{TM_T} | \mathrm{CORE} \rangle$$

$$= \mathcal{N}_{ab}(JT)\mathcal{N}_{cd}(JT) \sum_{a'} \varepsilon_{a'} \sum_{m_{\alpha} m_{\beta}} \sum_{\substack{m_\alpha m_\beta \\ m_\gamma m_\delta}} \sum_{\substack{m_{t\alpha} m_{t\beta} \\ m_{t\gamma} m_{t\delta}}} (j_a\, m_\alpha\, j_b\, m_\beta | J\,M)$$

$$\times (\tfrac{1}{2}\, m_{t\alpha}\, \tfrac{1}{2}\, m_{t\beta} | T\,M_T)(j_c\, m_\gamma\, j_d\, m_\delta | J\,M)(\tfrac{1}{2}\, m_{t\gamma}\, \tfrac{1}{2}\, m_{t\delta} | T\,M_T)$$

$$\times \langle \mathrm{CORE} | c_\beta c_\alpha c_{\alpha'}^\dagger c_{\alpha'} c_\gamma^\dagger c_\delta^\dagger | \mathrm{CORE} \rangle . \tag{8.93}$$

Performing the contractions yields

$$\langle \mathrm{CORE} | c_\beta c_\alpha c_{\alpha'}^\dagger c_{\alpha'} c_\gamma^\dagger c_\delta^\dagger | \mathrm{CORE} \rangle$$

$$= \delta_{\beta\alpha'}\delta_{\alpha\gamma}\delta_{\alpha'\delta} - \delta_{\beta\alpha'}\delta_{\alpha\delta}\delta_{\alpha'\gamma} - \delta_{\beta\gamma}\delta_{\alpha\alpha'}\delta_{\alpha'\delta} + \delta_{\beta\delta}\delta_{\alpha\alpha'}\delta_{\alpha'\gamma} . \tag{8.94}$$

Properties of the Clebsch–Gordan coefficients lead to

$$\langle a\,b\,;\, J\,T | T_\varepsilon | c\,d\,;\, J\,T \rangle$$

$$= \frac{1 - \delta_{ab}(-1)^{J+T}}{(1 + \delta_{ab})^2} \left[ \delta_{ac}\delta_{bd} + (-1)^{j_a + j_b + J + T}\delta_{ad}\delta_{bc} \right](\varepsilon_a + \varepsilon_b) . \tag{8.95}$$

For our two-particle basis states we adopt the convention $a \leq b$ to avoid counting the same physical states twice. It follows that only the $\delta_{ac}\delta_{bd}$ term of (8.95) contributes when $a \neq b$, so that

$$\langle a\,b;\, JT|T_\varepsilon|c\,d;\, JT\rangle = \delta_{ac}\delta_{bd}(\varepsilon_a + \varepsilon_b) \quad \text{for } a < b. \qquad (8.96)$$

When $a = b$ both terms of (8.95) contribute, with the result

$$\langle a^2;\, JT|T_\varepsilon|c\,d;\, JT\rangle = \delta_{ac}\delta_{ad}\big[1 - (-1)^{J+T}\big]\varepsilon_a. \qquad (8.97)$$

Equations (8.96) and (8.97) can be combined into the single equation

$$\boxed{\langle a\,b;\, JT|T_\varepsilon|c\,d;\, JT\rangle = \delta_{ac}\delta_{bd}\frac{1 - \delta_{ab}(-1)^{J+T}}{1 + \delta_{ab}}(\varepsilon_a + \varepsilon_b), \quad a \leq b.} \qquad (8.98)$$

In the proton–neutron formalism the matrix element of $T_\varepsilon$ is derived similarly. The result is

$$\boxed{\langle a\,b;\, J|T_\varepsilon|c\,d;\, J\rangle = \delta_{ac}\delta_{bd}\frac{1 + \delta_{ab}(-1)^{J}}{1 + \delta_{ab}}(\varepsilon_a + \varepsilon_b), \quad a \leq b,} \qquad (8.99)$$

where the condition $a \leq b$ means the labelling order $p_1 \leq p_2$, $n_1 \leq n_2$, $p < n$. Note that (8.99) coincides with the $T = 1$ case of (8.98), as one would expect.

The two-body part $V$ of the nuclear Hamiltonian was expressed in terms of its two-body matrix elements in Subsect. 8.1.1. Thus the matrix elements of $V$ between wave functions of two nucleons are directly the two-body interaction matrix elements as expressed in (8.13) and (8.21).

### 8.3.4 Solving the Eigenvalue Problem for a Two-Particle Nucleus

The preceding discussion in this section can be condensed into the following three-part recipe to solve the eigenvalue problem for a two-particle nucleus.

#### Proton–neutron nucleus

Adopt a single-particle valence space according to the principles of Chap. 5. For a given angular momentum and parity $J^\pi$, form all possible proton–neutron states $|pn;\, J^\pi M\rangle$, with a common value of $M$. Because the matrix elements of the Hamiltonian are independent of $M$ (and $M_T$), we now omit projection quantum numbers in the notation for basic states: $|pn;\, J^\pi\rangle$.

Denoting the set of basis states as

$$\{|1\rangle,\, |2\rangle,\, \ldots,\, |N\rangle\} \qquad (8.100)$$

and the matrix elements as

$$T_{ij} \equiv \langle i|T_\varepsilon|j \rangle = \delta_{ij} T_{ii} , \quad V_{ij} \equiv \langle i|V|j \rangle , \tag{8.101}$$

we have the Hamiltonian matrix

$$\mathsf{H} = \begin{pmatrix} T_{11} + V_{11} & V_{12} & \cdots & V_{1N} \\ V_{21} & T_{22} + V_{22} & \cdots & V_{2N} \\ \vdots & \vdots & \ddots & \vdots \\ V_{N1} & V_{N2} & \cdots & T_{NN} + V_{NN} \end{pmatrix} . \tag{8.102}$$

Diagonalization of this matrix yields the eigenenergies $E_n$, $n = 1, 2, \ldots, N$, and the corresponding eigenstates

$$|\Psi_n \rangle = \sum_{i=1}^N a_i^{(n)} |i \rangle , \quad n = 1, 2, \ldots, N . \tag{8.103}$$

The Hamiltonian matrix is Hermitian, $\mathsf{H} = \mathsf{H}^\dagger$, i.e. its elements $H_{kl}$ have the property

$$H_{kl} = (\mathsf{H}^\dagger)_{kl} = H_{lk}^* . \tag{8.104}$$

It is therefore sufficient to compute just the diagonal and the upper or lower half of the Hamiltonian matrix. When the matrix elements are real, as they are in Tables 8.1–8.4 for the surface delta interaction, the relation is simply $H_{kl} = H_{lk}$.

**Proton–proton or neutron–neutron nucleus**

For a given $J^\pi$ form all possible proton–proton or neutron–neutron states $|a_1 a_2 \, (a_1 \le a_2) ; J^\pi \rangle$ in the adopted single-particle valence space and proceed as in the previous case.

**Isospin representation**

For a given $J^\pi T$ form all possible two-nucleon states $|a\,b\,(a \le b) ; J^\pi ; T \rangle$ in the adopted single-particle valence space and proceed as in the two previous cases. Note that the matrices to be diagonalized are usually smaller in isospin representation than they are in proton–neutron representation. Also note that isospin symmetry is realized only when the relative single-particle energies of protons and neutrons are the same.

### 8.3.5 Application to $A = 6$ Nuclei

The even–even nucleus $^6_2\text{He}_4$ offers an enlightening example of configuration mixing. The nucleus can be described as a two-particle nucleus with two neutrons in the $0p$ shell. All states of this model space have $T = 1$.

We take the energies of the orbitals to be $\varepsilon_{0p_{3/2}} = 0$ and $\varepsilon_{0p_{1/2}} \equiv \varepsilon = 6.0\,\mathrm{MeV}$ (see Fig. 3.3), for both protons and neutrons in order to have isospin symmetry. For the $0^+$ states the two-neutron basis states are given by

$$\{|1\rangle, |2\rangle\} = \{|(\nu 0p_{3/2})^2; 0^+\rangle, |(\nu 0p_{1/2})^2; 0^+\rangle\}. \tag{8.105}$$

We assume the SDI, $V = V_{\mathrm{SDI}}$. In the basis (8.105) the Hamiltonian matrix then becomes

$$\mathsf{H} = \begin{pmatrix} 2 \times 0 - 2.000A_1 & -1.414A_1 \\ -1.414A_1 & 2\varepsilon - 1.000A_1 \end{pmatrix}, \tag{8.106}$$

where the one-body part is from (8.99) and the two-body part from Table 8.1 with the strength constant $A_1$ included. Substituting the numerical value of $\varepsilon$ and taking[3] $A_1 = 1.0\,\mathrm{MeV}$ we have the matrix

$$\mathsf{H} = \begin{pmatrix} -2.000 & -1.414 \\ -1.414 & 11.000 \end{pmatrix} \tag{8.107}$$

with all elements in MeV. Equation (8.85) gives its eigenvalues as

$$E(0_1^+) = -2.152\,\mathrm{MeV}, \quad E(0_2^+) = 11.152\,\mathrm{MeV}. \tag{8.108}$$

The wave functions are found as in Subsection 8.3.2. The result is

$$|0_1^+\rangle \equiv |0_{\mathrm{gs}}^+\rangle = 0.994|1\rangle + 0.107|2\rangle$$
$$= 0.994|(\nu 0p_{3/2})^2; 0^+\rangle + 0.107|(\nu 0p_{1/2})^2; 0^+\rangle \tag{8.109}$$
$$|0_2^+\rangle = -0.107|1\rangle + 0.994|2\rangle$$
$$= -0.107|(\nu 0p_{3/2})^2; 0^+\rangle + 0.994|(\nu 0p_{1/2})^2; 0^+\rangle. \tag{8.110}$$

The result for the two-proton nucleus $^6_4\mathrm{Be}_2$ is the same except for the replacement $\nu \to \pi$. Also, because of isospin symmetry the same $0^+$ states, with appropriate $\pi\nu$ labels, occur in the proton–neutron nucleus $^6_3\mathrm{Li}_3$, e.g.

$$|^6\mathrm{Li}; 0_1^+\, T = 1\rangle = 0.994|\pi 0p_{3/2}\,\nu 0p_{3/2}; 0^+\rangle + 0.107|\pi 0p_{1/2}\,\nu 0p_{1/2}; 0^+\rangle. \tag{8.111}$$

Let us discuss next the $1^+$ states. For the two-proton and two-neutron nuclei, with $T = 1$, this problem is trivial since for odd angular momentum it is possible to construct only one basis state (remember the condition $a \leq b$), which is $|0p_{3/2}\,0p_{1/2}; 1^+\rangle$ with the energy

$$E(1^+\, T = 1) = T_{11} + V_{11} = \varepsilon_{0p_{3/2}} + \varepsilon_{0p_{1/2}} + 0 = 6.0\,\mathrm{MeV}. \tag{8.112}$$

Here the one-body part is from (8.99) and the two-body part from Table 8.1.

---

[3] It turns out that $A_0 = A_1 = 1\,\mathrm{MeV}$ is globally a reasonable first guess. In the text we present our calculations with that choice, but fitted values are used in figures containing comparison with experiment.

In the case of $^{6}_{3}\text{Li}_3$ the possible basis states for $1^+$ are

$$\{|1\rangle, |2\rangle, |3\rangle, |4\rangle\} = \{|\pi 0p_{3/2}\, \nu 0p_{3/2}\,;\, 1^+\rangle,\, |\pi 0p_{3/2}\, \nu 0p_{1/2}\,;\, 1^+\rangle,$$
$$|\pi 0p_{1/2}\, \nu 0p_{3/2}\,;\, 1^+\rangle,\, |\pi 0p_{1/2}\, \nu 0p_{1/2}\,;\, 1^+\rangle\}\,. \quad (8.113)$$

With (8.26), (8.99) and Table 8.1, the Hamiltonian matrix (8.102) becomes

$$\mathsf{H} = \begin{pmatrix} -1.200A_0 & 0.894A_0 & -0.894A_0 & 0.632A_0 \\ 0.894A_0 & -1.000A_0 + \varepsilon & 1.000A_0 & 0 \\ -0.894A_0 & 1.000A_0 & -1.000A_0 + \varepsilon & 0 \\ 0.632A_0 & 0 & 0 & -1.000A_0 + 2\varepsilon \end{pmatrix}\,. \quad (8.114)$$

We pick $A_0 = 1.0\,\text{MeV}$ and perform a numerical diagonalization. The eigenenergies come out as

$$E(1^+_1) = -1.522\,\text{MeV}\,, \quad E(1^+_2) = 4.288\,\text{MeV}\,,$$
$$E(1^+_3) = 6.000\,\text{MeV}\,, \quad E(1^+_4) = 11.033\,\text{MeV}\,. \quad (8.115)$$

The wave function of the lowest $1^+$ state comes out as

$$|1^+_1\rangle = 0.974|1\rangle - 0.158|2\rangle + 0.158|3\rangle - 0.049|4\rangle\,. \quad (8.116)$$

According to (8.112), we have only one $J^\pi = 1^+$ state with $T = 1$, namely the one at $6.0\,\text{MeV}$. This energy is identified with $E(1^+_3)$ in (8.115). We conclude that the other three $1^+$ states are $T = 0$ states. In this way it is possible to identify $T = 1$ and $T = 0$ states in two-nucleon nuclei without using the isospin formalism. Note also that if we did use the isospin formalism, the 4-by-4 matrix (8.114) would be replaced by a 1-by-1 matrix for $T = 1$ and a 3-by-3 matrix for $T = 0$.

For the $2^+$ states in $^{6}_{3}\text{Li}_3$ the basis is

$$\{|1\rangle, |2\rangle, |3\rangle\} = \{|\pi 0p_{3/2}\, \nu 0p_{3/2}\,;\, 2^+\rangle,\, |\pi 0p_{3/2}\, \nu 0p_{1/2}\,;\, 2^+\rangle,$$
$$|\pi 0p_{1/2}\, \nu 0p_{3/2}\,;\, 2^+\rangle\}\,. \quad (8.117)$$

The Hamiltonian matrix becomes

$$\mathsf{H} = \begin{pmatrix} -0.400A_1 & -0.400A_1 & 0.400A_1 \\ -0.400A_1 & \varepsilon - 0.600A_0 - 0.400A_1 & -0.600A_0 + 0.400A_1 \\ 0.400A_1 & -0.600A_0 + 0.400A_1 & \varepsilon - 0.600A_0 - 0.400A_1 \end{pmatrix}\,. \quad (8.118)$$

Here both the isoscalar and isovector interaction matrix elements are active, via the relation (8.26). Taking $A_0 = A_1 = 1.0\,\text{MeV}$ produces the eigenenergies

$$E(2^+_1) = -0.457\,\text{MeV}\,, \quad E(2^+_2) = 4.800\,\text{MeV}\,, \quad E(2^+_3) = 5.257\,\text{MeV}\,. \quad (8.119)$$

In $^{6}_{2}\text{He}_4$ the basis for $2^+$ states $(T = 1)$ is

$$\{|1\rangle, |2\rangle\} = \{|(\nu 0p_{3/2})^2\,;\, 2^+\rangle,\, |\nu 0p_{3/2}\, \nu 0p_{1/2}\,;\, 2^+\rangle\}\,. \quad (8.120)$$

The Hamiltonian matrix is

$$H = \begin{pmatrix} -0.400A_1 & -0.566A_1 \\ -0.566A_1 & \varepsilon - 0.800A_1 \end{pmatrix} . \tag{8.121}$$

With $A_1 = 1.0\,\text{MeV}$ the energy eigenvalues are

$$E(2_1^+\ T = 1) = -0.457\,\text{MeV} , \quad E(2_2^+\ T = 1) = 5.257\,\text{MeV} . \tag{8.122}$$

Comparison with (8.119) reveals that the $2^+$ state with the energy 4.800 MeV is a $T = 0$ state.

For $J^\pi = 3^+$ the only possible basis state is $|\pi 0p_{3/2}\, \nu 0p_{3/2}\, ; 3^+\rangle$. This state describes a $T = 0$ state in $^6_3\text{Li}_3$. With $A_0 = 1.0\,\text{MeV}$ its energy is

$$E(3^+\ T = 0) = -1.200\,\text{MeV} . \tag{8.123}$$

We have now constructed all the states of the $A = 6$ two-particle nuclei using the 0p shell as the valence space. The number of states in this space for the nucleus $^6_3\text{Li}_3$, classified by angular momentum and isospin, is given in Table 8.5. Of the ten states only the five $T = 1$ states are possible for the even–even $^6_2\text{He}_4$ and $^6_4\text{Be}_2$ nuclei.

**Table 8.5.** Numbers of 0p-shell states in $^6_3\text{Li}_3$ for different angular momenta and isospins

| Angular momentum | 0 | | 1 | | 2 | | 3 | |
|---|---|---|---|---|---|---|---|---|
| Isospin | 0 | 1 | 0 | 1 | 0 | 1 | 0 | 1 |
| Number of states | 0 | 2 | 3 | 1 | 1 | 2 | 1 | 0 |

The lowest experimental energy levels of the $A = 6$ nuclei are shown in Fig. 5.4. Our computed states with energies (8.108), (8.115), (8.119) and (8.123) can be identified with them, but the quantitative agreement is poor. In particular, the relative positions of the $T = 0$ and $T = 1$ states in $^6$Li are wrong: the ground state is $0^+, T = 1$ in theory while experimentally it is $1^+, T = 0$. However, our computations used the 'initial' parameter values $A_0 = A_1 = 1\,\text{MeV}$. With $A_0 = 2.4\,\text{MeV}$ and $A_1 = 0.5\,\text{MeV}$ a rather good fit is obtained, as shown in Fig. 8.1. Another way to improve the fit is to use the MSDI (8.74), which produces a shift between the $T = 0$ and $T = 1$ states.

## 8.3.6 Application to $A = 18$ Nuclei

In the 0d-1s shell the two-particle nuclei are $^{18}_{8}\text{O}_{10}$, $^{18}_{9}\text{F}_9$ and $^{18}_{10}\text{Ne}_8$. The relative single-particle energies

$$\varepsilon_{0d_{5/2}} = 0 , \quad \varepsilon_{1s_{1/2}} = 0.87\,\text{MeV} , \quad \varepsilon_{0d_{3/2}} = 5.08\,\text{MeV} \tag{8.124}$$

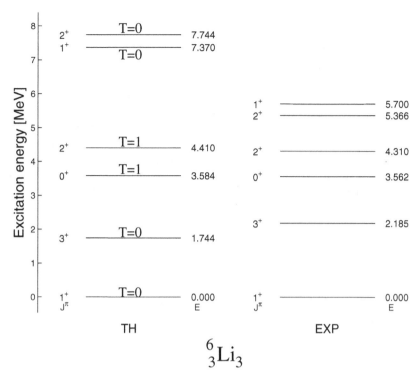

**Fig. 8.1.** Theoretical and experimental energy levels of $^6$Li. The calculation was done in the 0p valence space. The $0p_{1/2}$–$0p_{3/2}$ energy separation was taken to be 6.0 MeV and the SDI was used with the parameters $A_0 = 2.4\,$MeV and $A_1 = 0.5\,$MeV

are obtained from the experimental spectra of the one-particle nuclei $^{17}_{\ 8}O_9$ and $^{17}_{\ 9}F_8$ (see also Fig. 3.3). The simplest configuration mixing calculation is to restrict the valence space to the near-lying $0d_{5/2}$ and $1s_{1/2}$ orbitals. The matrices to be diagonalized are then no larger than 2-by-2, as seen from Table 8.6. In the valence space containing the complete 0d-1s shell the matrices are larger, as seen from the last row of the table.

**Table 8.6.** Numbers of states in $^{18}_{\ 9}F_9$ built from the $0d_{5/2}$-1s and 0d-1s shells for different angular momenta and isospins

| Angular momentum | 0 | | 1 | | 2 | | 3 | | 4 | | 5 | |
|---|---|---|---|---|---|---|---|---|---|---|---|---|
| Isospin | 0 | 1 | 0 | 1 | 0 | 1 | 0 | 1 | 0 | 1 | 0 | 1 |
| $0d_{5/2}$-1s | 0 | 2 | 2 | 0 | 1 | 2 | 2 | 1 | 0 | 1 | 1 | 0 |
| 0d-1s | 0 | 3 | 5 | 2 | 3 | 5 | 4 | 2 | 1 | 2 | 1 | 0 |

The spectra of the three two-particle nuclei, derived by using both the smaller and the larger valence space, are depicted in Figs. 8.2–8.4. The parameters $A_0 = 0.9$ MeV and $A_1 = 0.5$ MeV of the SDI were used throughout ($A_0$ affects only the proton–neutron nucleus $^{18}$F).

The calculated spectra for $^{18}$O and $^{18}$Ne are identical. Except for the unobserved $3^+$ state, Figs. 8.2 and 8.3 show that the theory predicts the correct low-lying states, but with poor energy agreement. We also see that the larger valence space brings no significant improvement. However, only the full 0d-1s space can produce the levels around 7 MeV. Contrary to the even–even nuclei, the description of the odd–odd nucleus $^{18}$F greatly improves when enlarging the valence space. In particular, the ground state is given correctly as $1^+$ only in the larger valence space. Furthermore, the observed $5^+$ state can be generated only in that space.

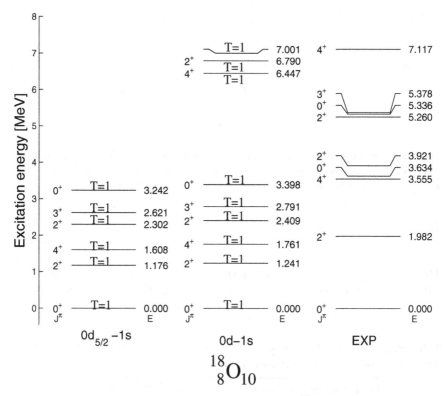

**Fig. 8.2.** Theoretical and experimental energy levels of $^{18}$O. The calculations were done in the $0d_{5/2}$-1s valence space and in the complete 0d-1s valence space. The single-particle energies (8.124) and the SDI, with $A_1 = 0.5$ MeV, were used

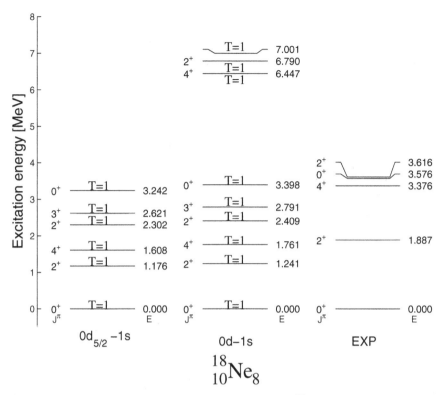

**Fig. 8.3.** Theoretical and experimental energy levels of $^{18}$Ne. The calculations were done in the $0d_{5/2}$-1s valence space and in the complete 0d-1s valence space. The single-particle energies (8.124) and the SDI, with $A_1 = 0.5$ MeV, were used

## 8.4 Configuration Mixing in Two-Hole Nuclei

The behaviour of two-hole nuclei is very much the same as that of two-particle nuclei. This results from the mirror kind of symmetry between particles and holes. First we survey the formalism of diagonalization of the residual interaction in the two-hole basis and then discuss some applications.

### 8.4.1 Diagonalization of the Residual Interaction in a Two-Hole Basis

The particle–hole symmetry shows up in the matrix elements of the nuclear Hamiltonian. When dealing with holes it is necessary to specify the one-body part as the Hartree–Fock Hamiltonian $H_{\mathrm{HF}}$ given in (4.72) and the two-body part as the residual interaction $V_{\mathrm{RES}}$ given in (4.73); the constant term of $V_{\mathrm{RES}}$ is here omitted because it has no effect on excitation energies.

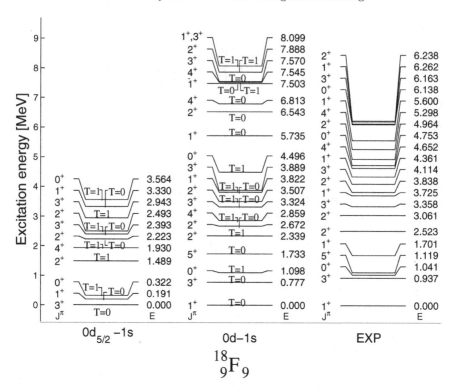

**Fig. 8.4.** Theoretical and experimental energy levels of $^{18}$F. The calculations were done in the $0d_{5/2}$-1s valence space and in the complete 0d-1s valence space. The single-particle energies (8.124) and the SDI, with $A_0 = 0.9$ MeV and $A_1 = 0.5$ MeV, were used

Proceeding as in Subsect. 8.3.3 one can calculate the matrix elements of $H_{\mathrm{HF}}$ and $V_{\mathrm{RES}}$ between two-hole states. The result for the two-body part is in proton–neutron representation

$$\langle a^{-1} b^{-1} ; J M | V_{\mathrm{RES}} | c^{-1} d^{-1} ; J M \rangle = \langle a\,b ; J | V | c\,d ; J \rangle \qquad (8.125)$$

and in isospin representation

$$\langle a^{-1} b^{-1} ; J M ; T M_T | V_{\mathrm{RES}} | c^{-1} d^{-1} ; J M ; T M_T \rangle = \langle a\,b ; J T | V | c\,d ; J T \rangle . \qquad (8.126)$$

So the two-body part of the Hamiltonian matrix in a two-hole problem is the same as in the corresponding two-particle problem.

The result for the one-body part is in proton–neutron representation

$$\langle a^{-1} b^{-1} ; J M | H_{\mathrm{HF}} | c^{-1} d^{-1} ; J M \rangle$$
$$= [\mathcal{N}_{ab}(J)]^2 \left[ \delta_{ac}\delta_{bd} + (-1)^{j_a+j_b+J+1}\delta_{ad}\delta_{bc} \right] (E - \varepsilon_a - \varepsilon_b) \qquad (8.127)$$

and in isospin representation

$$\langle a^{-1} b^{-1} ; JM ; TM_T | H_{HF} | c^{-1} d^{-1} ; JM ; TM_T \rangle$$
$$= [\mathcal{N}_{ab}(JT)]^2 [\delta_{ac}\delta_{bd} + (-1)^{j_a+j_b+J+T}\delta_{ad}\delta_{bc}] (E - \varepsilon_a - \varepsilon_b) , \quad (8.128)$$

where

$$E = \sum_\alpha \varepsilon_\alpha = \sum_a (2j_a + 1)\varepsilon_a . \quad (8.129)$$

Because the constant quantity $E$ appears on the diagonal of the Hamiltonian matrix and thus affects the energy eigenvalues by a common energy shift, it can be omitted when computing excitation energies. Thus the Hamiltonian matrix for a two-hole nucleus is the same as that for the corresponding two-particle nucleus within the same valence space except that the single-particle energies change sign,

$$\boxed{\varepsilon_a(\text{hole}) = -\varepsilon_a(\text{particle}) .} \quad (8.130)$$

To avoid confusion, we choose to list the non-negative numerical values of $\varepsilon_a$(particle) and insert the required minus signs in front of the $\varepsilon_a$ in the Hamiltonian matrix. This is illustrated in the examples below.

### 8.4.2 Application to $A = 14$ Nuclei

Consider the two-hole nuclei $^{14}_6C_8$, $^{14}_7N_7$ and $^{14}_8O_6$ in the valence space consisting of the 0p shell. We adopt the interaction parameters used for the two-particle nuclei in the same valence space, i.e. $A_0 = 2.4$ MeV and $A_1 = 0.5$ MeV (see Fig. 8.1). We also adopt the same single-particle energies, $\varepsilon_{0p_{3/2}} = 0$ and $\varepsilon_{0p_{1/2}} \equiv \varepsilon = 6.0$ MeV. This energy difference is supported by the spectra of the one-hole nuclei $^{15}_7N_8$ and $^{15}_8O_7$ in Fig. 5.2.

The Hamiltonian matrix for the $0^+$ states is the same as (8.106) except for $\varepsilon$ being replaced by $-\varepsilon$:

$$H = \begin{pmatrix} 2 \times 0 - 2.000A_1 & -1.414A_1 \\ -1.414A_1 & -2\varepsilon - 1.000A_1 \end{pmatrix} . \quad (8.131)$$

Likewise for all other angular momenta we have the same Hamiltonian matrices as in Subsect. 8.3.5 except for the replacements $\varepsilon_a \to -\varepsilon_a$. The isospin identification is the same as in the two-particle calculations.

Inserting the strength constants into (8.131) and diagonalizing the resulting matrix, we obtain the energy eigenvalues

$$E_1 = -12.543 \text{ MeV} , \quad E_2 = -0.957 \text{ MeV} . \quad (8.132)$$

The wave function of the first $0^+$ state becomes

$$|0_1^+\rangle = 0.998|(0p_{1/2})^2 ; 0^+\rangle + 0.061|(0p_{3/2})^2 ; 0^+\rangle , \quad (8.133)$$

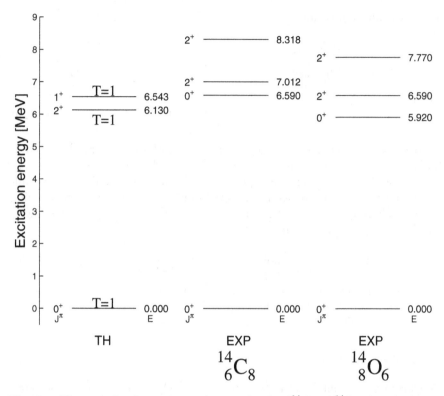

**Fig. 8.5.** Theoretical and experimental energy levels of $^{14}$C and $^{14}$O. The calculation was done in the 0p valence space. The difference between the $0p_{1/2}$ and $0p_{3/2}$ single-particle energies was taken to be 6.0 MeV and the SDI was used with $A_1 = 0.5$ MeV

where we have omitted the proton and neutron labels to cover the $0^+, T = 1$ states in all three nuclei at the same time. Comparison with (8.109) shows the expected reversal of the leading amplitudes.

Using the Hamiltonian matrices of Subsect. 8.3.5 with the change $\varepsilon \to -\varepsilon$ and the parameter values quoted above produces the energy spectra shown in Figs. 8.5 ($^{14}$C and $^{14}$O) and 8.6 ($^{14}$N). The agreement between the experimental and computed spectra is fair considering the simplicity of the theoretical approach.

### 8.4.3 Application to $A = 38$ Nuclei

The nuclei $^{38}_{18}\text{Ar}_{20}$, $^{38}_{19}\text{K}_{19}$ and $^{38}_{20}\text{Ca}_{18}$ have two holes in the 0d-1s shell. They can be treated in the same way as the two-hole $A = 14$ nuclei in the 0p shell.

The relative single-particle energies we adopt for these two-hole nuclei differ somewhat from those in (8.124) used for the two-particle nuclei. This difference can be understood when looking at the spectra of the one-particle

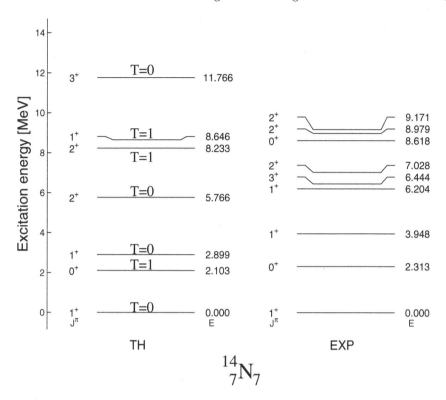

**Fig. 8.6.** Theoretical and experimental energy levels of $^{14}$N. The calculation was done in the 0p valence space. The difference between the $0p_{1/2}$ and $0p_{3/2}$ single-particle energies was taken to be 6.0 MeV and the SDI was used with $A_0 = 2.4$ MeV and $A_1 = 0.5$ MeV

and one-hole nuclei in the 0d-1s shell. In the beginning of the shell the single-particle energies can be deduced from the data on one-particle nuclei and at the end of the shell from the data on one-hole nuclei. From the one-hole nuclei $^{39}_{19}$K$_{20}$ and $^{39}_{20}$Ca$_{19}$ in Fig. 5.3 we deduce the relative single-particle energies

$$\varepsilon_{0d_{5/2}} = 0 \ , \quad \varepsilon_{1s_{1/2}} = 1.5 \,\text{MeV} \ , \quad \varepsilon_{0d_{3/2}} = 4.0 \,\text{MeV} \ . \tag{8.134}$$

In the two-hole calculations these are used with a minus sign according to (8.130).

The energy spectra of the two-hole nuclei are shown in Figs. 8.7 ($^{38}$Ar and $^{38}$Ca) and 8.8 ($^{38}$K). The SDI was used in the computation with $A_0 = A_1 = 0.9$ MeV. The experimental spectra are seen to contain some states not accounted for by the simple theory. They can be associated with configurations containing particle–hole excitations from the 0d-1s shell to the $0f_{7/2}$ shell. In addition, the relative positions of the $T = 0$ and $T = 1$ states in $^{38}$K are not correctly reproduced. Using the MSDI could improve this situation.

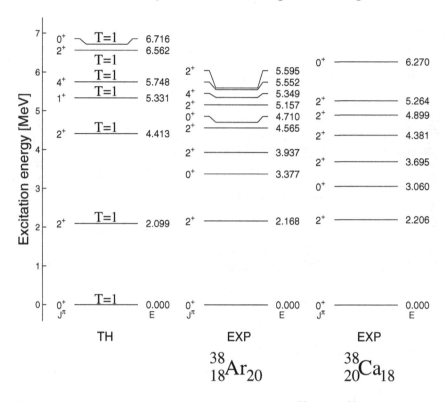

**Fig. 8.7.** Theoretical and experimental energy levels of $^{38}$Ar and $^{38}$Ca. The calculations were done in the 0d-1s valence space. The single-particle energies (8.134) and the SDI, with $A_0 = A_1 = 0.9$ MeV, were used

## 8.5 Electromagnetic and Beta-Decay Transitions in Two-Particle and Two-Hole Nuclei

In this section we compute electromagnetic and beta-decay transitions from and to nuclear states described by wave functions with configuration mixing. Configuration mixing is expected to provide a better description of these processes than does the simple single-configuration approach in Chaps. 6 and 7.

### 8.5.1 Transition Amplitudes With Configuration Mixing

Configuration mixing in two-particle and two-hole nuclei leads to a representation of the state vectors as linear combinations (8.103) of the basis states. For an electromagnetic or a beta-decay transition between an initial and a final state

$$|\Psi_i\rangle = \sum_k A_k |k\rangle \,, \quad |\Psi_f\rangle = \sum_l B_l |l\rangle \,, \tag{8.135}$$

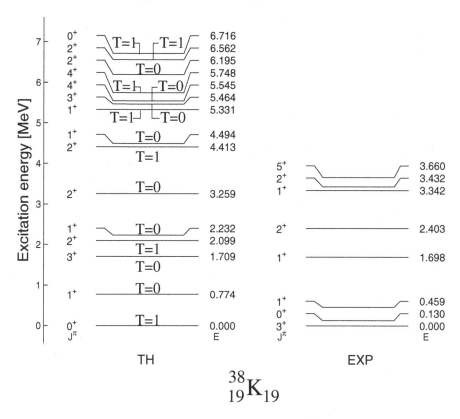

**Fig. 8.8.** Theoretical and experimental energy levels of $^{38}$K. The calculations were done in the 0d-1s valence space. The single-particle energies (8.134) and the SDI, with $A_0 = A_1 = 0.9\,\mathrm{MeV}$, were used

The transition amplitude (real $A_k, B_l$ assumed) becomes

$$(\Psi_f \| \boldsymbol{T}_\lambda \| \Psi_i) = \sum_{kl} A_k B_l (l \| \boldsymbol{T}_\lambda \| k) \; . \tag{8.136}$$

The tensor operator $\boldsymbol{T}_\lambda$ is either an electromagnetic operator (6.5) or a beta-decay operator appearing in (7.20), (7.21), (7.167) or (7.187). Consider $|k\rangle$ and $|l\rangle$ as two-particle or two-hole basis states. Below we discuss a few examples of the application of (8.136).

### 8.5.2 Application to Beta Decay of $^6$He

Consider the $\beta^-$ decay transition $^6\mathrm{He}(0^+_{\mathrm{gs}}) \to {}^6\mathrm{Li}(1^+_{\mathrm{gs}})$ shown in Fig. 7.8 and discussed in Subsect. 7.5.2. Our computation with the pure states (7.118) and (7.119) gave $\log ft = 3.07$ while the experimental value is $\log ft = 2.9$. We now propose to improve the agreement by including configuration mixing.

The configuration mixing calculation in Subsect. 8.3.5, with $A_0 = 2.4\,\mathrm{MeV}$ and $A_1 = 0.5\,\mathrm{MeV}$, gave $^6$Li the energy spectrum of Fig. 8.1, in rather nice agreement with experiment; note that with equal proton and neutron single-particle energies the unmixed states (7.118) and (7.119) are degenerate. The calculation gives the $^6$He ground-state wave function

$$|^6\mathrm{He}\,;\,0^+_{\mathrm{gs}}\,T = 1\rangle = 0.998|(\nu 0\mathrm{p}_{3/2})^2\,;\,0^+\rangle + 0.056|(\nu 0\mathrm{p}_{1/2})^2\,;\,0^+\rangle \quad (8.137)$$

and $^6$Li ground-state wave function[4]

$$|^6\mathrm{Li}\,;\,1^+_{\mathrm{gs}}\,T = 0\rangle = 0.883|\pi 0\mathrm{p}_{3/2}\,\nu 0\mathrm{p}_{3/2}\,;\,1^+\rangle - 0.325|\pi 0\mathrm{p}_{3/2}\,\nu 0\mathrm{p}_{1/2}\,;\,1^+\rangle$$
$$+ 0.325|\pi 0\mathrm{p}_{1/2}\,\nu 0\mathrm{p}_{3/2}\,;\,1^+\rangle - 0.094|\pi 0\mathrm{p}_{1/2}\,\nu 0\mathrm{p}_{1/2}\,;\,1^+\rangle\,. \quad (8.138)$$

Equations (8.137) and (8.138) give the amplitudes

$$A_1 = 0.998\,, \quad A_2 = 0.056\,, \tag{8.139}$$
$$B_1 = 0.883\,, \quad B_2 = -0.325\,, \quad B_3 = 0.325\,, \quad B_4 = -0.094\,. \tag{8.140}$$

The non-zero Gamow–Teller matrix elements are obtained from (7.110):

$$M_1 \equiv \mathcal{M}^{(-)}_{\mathrm{GT}}\big((\nu 0\mathrm{p}_{3/2})^2\,;\,0^+ \to \pi 0\mathrm{p}_{3/2}\,\nu 0\mathrm{p}_{3/2}\,;\,1^+\big) = \sqrt{\frac{10}{3}}\,, \tag{8.141}$$

$$M_2 \equiv \mathcal{M}^{(-)}_{\mathrm{GT}}\big((\nu 0\mathrm{p}_{3/2})^2\,;\,0^+ \to \pi 0\mathrm{p}_{1/2}\,\nu 0\mathrm{p}_{3/2}\,;\,1^+\big) = 2\sqrt{\frac{2}{3}}\,, \tag{8.142}$$

$$M_3 \equiv \mathcal{M}^{(-)}_{\mathrm{GT}}\big((\nu 0\mathrm{p}_{1/2})^2\,;\,0^+ \to \pi 0\mathrm{p}_{3/2}\,\nu 0\mathrm{p}_{1/2}\,;\,1^+\big) = -\frac{4}{\sqrt{3}}\,, \tag{8.143}$$

$$M_4 \equiv \mathcal{M}^{(-)}_{\mathrm{GT}}\big((\nu 0\mathrm{p}_{1/2})^2\,;\,0^+ \to \pi 0\mathrm{p}_{1/2}\,\nu 0\mathrm{p}_{1/2}\,;\,1^+\big) = -\sqrt{\frac{2}{3}}\,. \tag{8.144}$$

The transition amplitude is a linear combination of these matrix elements according to (8.136):

$$\mathcal{M}_{\mathrm{GT}}(0^+_{\mathrm{gs}} \to 1^+_{\mathrm{gs}}) = A_1 B_1 M_1 + A_1 B_3 M_2 + A_2 B_2 M_3 + A_2 B_4 M_4 = 2.185\,. \tag{8.145}$$

This should be compared with (7.120) or (8.141), which gives the result without configuration mixing as $\sqrt{10/3} = 1.826$. The value 2.185 leads to

$$\log ft = 2.91\,, \tag{8.146}$$

in perfect agreement with the experimental value. The immediate conclusion is that it is essential to consider configuration mixing when calculating nuclear decay transitions.

---

[4] These wave functions differ from (8.109) and (8.116) because those were calculated with $A_0 = A_1 = 1\,\mathrm{MeV}$.

### 8.5.3 Application to E2 Decays in $^{18}$O and $^{18}$Ne

Consider the nuclei $^{18}_{8}O_{10}$ and $^{18}_{10}Ne_8$ in the $0d_{5/2}$-$1s_{1/2}$ valence space. Their calculated energy levels are given on the left in Figs. 8.2 and 8.3. We set out to compute their reduced E2 transition probabilities $B(E2\,;\,2^+_1 \to 0^+_{gs})$. The original calculation yields the initial and final states

$$|2^+_1\rangle = 0.883|(0d_{5/2})^2\,;\,2^+\rangle + 0.470|0d_{5/2}\,1s_{1/2}\,;\,2^+\rangle\,, \qquad (8.147)$$

$$|0^+_{gs}\rangle = 0.961|(0d_{5/2})^2\,;\,0^+\rangle + 0.278|(1s_{1/2})^2\,;\,0^+\rangle\,. \qquad (8.148)$$

In $^{18}$O these states are two-neutron states, in $^{18}$Ne two-proton states, with their different effective charges.

In the notation of (8.136) the amplitudes in the wave functions (8.147) and (8.148) are

$$A_1 = 0.883\,, \quad A_2 = 0.470\,, \quad B_1 = 0.961\,, \quad B_2 = 0.278\,. \qquad (8.149)$$

The non-zero electric quadrupole transition matrix elements are, from (6.103) and Table 6.4,

$$\left((0d_{5/2})^2\,;\,0^+\|Q_2\|(0d_{5/2})^2\,;\,2^+\right) = -2.1106e_{\mathrm{eff}}b^2\,, \qquad (8.150)$$

$$\left((0d_{5/2})^2\,;\,0^+\|Q_2\|0d_{5/2}\,1s_{1/2}\,;\,2^+\right) = -1.2615e_{\mathrm{eff}}b^2\,, \qquad (8.151)$$

$$\left((1s_{1/2})^2\,;\,0^+\|Q_2\|0d_{5/2}\,1s_{1/2}\,;\,2^+\right) = -2.185e_{\mathrm{eff}}b^2\,. \qquad (8.152)$$

By using $b = 1.750\,\mathrm{fm}$ from Subsect. 6.3.1 we obtain the transition amplitude

$$(0^+_{gs}\|Q_2\|2^+_1) = [A_1B_1(-2.1106) + A_2B_1(-1.2615) + A_2B_2(-2.185)]e_{\mathrm{eff}}b^2$$
$$= -8.104e_{\mathrm{eff}}\,\mathrm{fm}^2\,, \qquad (8.153)$$

leading to the reduced transition probability

$$B(E2\,;\,2^+_1 \to 0^+_{gs}) = 13.13e^2_{\mathrm{eff}}\,\mathrm{fm}^4\,. \qquad (8.154)$$

The no-mixing calculation of Subsect. 6.3.1 gave $8.357e^2_{\mathrm{eff}}\,\mathrm{fm}^4$, so even our minimal configuration mixing calculation increased $B(E2)$ by nearly 60 %. The electric polarization constant now becomes

$$\chi \approx 0.9 \qquad (8.155)$$

instead of $\chi \approx 1.3$ in (6.115). Even the diminished value is so large that it evidently summarizes much configuration mixing not explicitly included.

To see whether further configuration mixing increases the $B(E2)$ value, we have repeated the present calculation in the complete 0d-1s valence space (middle spectra in Figs. 8.2 and 8.3). The result is

$$B(E2\,;\,2^+_1 \to 0^+_{gs})_{\mathrm{0d\text{-}1s}} = 14.32e^2_{\mathrm{eff}}\,\mathrm{fm}^4\,. \qquad (8.156)$$

This is indeed a further increase though not very large.

## Epilogue

In this chapter we have been able to extend the mean-field description of nuclear states. The handicap of single-configuration wave functions has been removed by the residual interaction. Within this scheme the states of two-particle and two-hole nuclei were allowed to include several configurations as demanded by the SDI. In the following three chapters that interaction is used to induce doubly magic nuclei to open their closed shells into their particle–hole excitation. In the latter half of this book the SDI serves to provide us with quasiparticles and their configuration mixing in open-shell nuclei.

## Exercises

**8.1.** Verify the relations (8.25).

**8.2.** Verify (8.26) and (8.27).

**8.3.** Derive (8.29).

**8.4.** Derive (8.32)

**8.5.** Derive the expression (8.34) for the radial part of the delta interaction by starting from (8.32).

**8.6.** Derive the expression (8.36) for the radial part of the SDI by starting from (8.32).

**8.7.** Derive the matrix element (8.55) by starting from (8.52).

**8.8.** Derive the expressions (8.57) and (8.58) by starting from (8.52).

**8.9.** Complete the details of the derivation of (8.66).

**8.10.** Complete the details of the derivation of (8.68).

**8.11.** Derive (8.71) and (8.72) by starting from (8.68).

**8.12.** Verify some numbers of your choice in Table 8.1.

**8.13.** Derive the relation (8.89).

**8.14.** Complete the details of the derivation of (8.98).

**8.15.** Complete the details of the derivation of (8.99).

**8.16.** Verify the matrix (8.114).

**8.17.** Write down all the wave functions corresponding to the energies of (8.115).

**8.18.** Verify the matrix of (8.118) and diagonalize it for $A_0 = 2.4\,\text{MeV}$ and $A_1 = 0.5\,\text{MeV}$. Compare with the theoretical energies of Fig. 8.1.

**8.19.** Verify the matrix (8.121). Compute the eigenenergies and eigenfunctions for $A_1 = 0.5\,\text{MeV}$.

**8.20.** Compute the eigenenergies and eigenvectors of the $1^+$, $2^+$, $3^+$ and $4^+$ states of $^{18}$O and $^{18}$Ne in the $0d_{5/2}$-$1s_{1/2}$ valence space. Use the energy difference $\varepsilon_{1s_{1/2}} - \varepsilon_{0d_{5/2}} = 0.87\,\text{MeV}$ and the parameter values $A_0 = 0.9\,\text{MeV}$ and $A_1 = 0.5\,\text{MeV}$ for the SDI. Draw the theoretical spectrum, compare it with experiment and comment.

**8.21.** The same as Exercise 8.20 but for $^{18}$F.

**8.22.** Compute the eigenenergies and eigenvectors of the $1^+$–$7^+$ states in $^{42}$Sc. Try to determine the parameters of the SDI in such a way that they best reproduce the experimental spectrum.

**8.23.** Compute the eigenenergies of the $0^+$, $1^+$, $2^+$ and $3^+$ states of $^6$Li in the $0p$ valence space by using $\varepsilon_{0p_{1/2}} - \varepsilon_{0p_{3/2}} = 6.0\,\text{MeV}$ and determining the parameters $A_0$, $A_1$, $B_0$ and $B_1$ of the MSDI (8.74) in such a way that they best reproduce the experimental spectrum. Draw the theoretical spectrum, compare it with the data and comment.

**8.24.** Derive (8.127).

**8.25.** Derive (8.128).

**8.26.** Compute the eigenenergies and eigenvectors of the $0^+$, $1^+$ and $2^+$ states of $^{14}$N in the $0p$ valence space. Use the energy difference $\varepsilon_{0p_{1/2}} - \varepsilon_{0p_{3/2}} = 6.0\,\text{MeV}$ and the parameter values $A_0 = 2.4\,\text{MeV}$ and $A_1 = 0.5\,\text{MeV}$ for the SDI. Draw the resulting theoretical spectrum, compare it with experimental data and comment.

**8.27.** Compute the eigenenergies and eigenvectors of the $0^+$, $1^+$, $2^+$ and $3^+$ states of $^{38}$Ar, $^{38}$K and $^{38}$Ca in the $1s_{1/2}$-$0d_{3/2}$ valence space. Take $A_0 = A_1 = 0.9\,\text{MeV}$ and use the energy difference $\varepsilon_{0d_{3/2}} - \varepsilon_{1s_{1/2}} = 2.5\,\text{MeV}$. Draw the resulting theoretical spectrum, compare it with experiment and comment.

**8.28.** For the nucleus $^{50}$Sc compute all the energies and wave functions that you can extract when using the $1p$-$0f_{5/2}$ valence space for neutrons. Take $A_0 = A_1 = 0.9\,\text{MeV}$ and use the single-particle energies $\varepsilon_{1p_{3/2}} = 0$, $\varepsilon_{1p_{1/2}} = 2.02\,\text{MeV}$ and $\varepsilon_{0f_{5/2}} = 3.9\,\text{MeV}$. Compare the structure of the ground-state wave function with that extracted in Exercise 7.24.

**8.29.** Compute the structure of the $0^+$ ground state of $^{50}$Ca by using the valence space and parameters of Exercise 8.28. Compare the result with that extracted in Exercise 7.23.

**8.30.** Compute the energies of the $2^+$ states of $^{50}$Ca by using the valence space and parameters of Exercise 8.28. Draw the spectrum, compare it with experiment and comment.

**8.31.** Compute the $\log ft$ values for the beta-decay transitions $^{18}$Ne$(0^+_{gs}) \rightarrow$ $^{18}$F$(1^+_{gs})$ and $^{18}$F$(1^+_{gs}) \rightarrow$ $^{18}$O$(0^+_{gs})$ with the wave functions computed in Exercises 8.20 and 8.21. Compare the wave functions and the $\log ft$ values with the results of the simplified calculation of Subsect. 7.5.3. Comment on the correspondence. Compare also with the experimental data and comment.

**8.32.** Calculate the $\log ft$ value of the transition $^6$He$(0^+_{gs}) \rightarrow$ $^6$Li$(1^+_{gs})$ by making a configuration mixing calculation in the 0p shell. Use $A_0 = A_1 = 1.0\,\text{MeV}$ and the energy difference $\varepsilon_{0p_{1/2}} - \varepsilon_{0p_{3/2}} = 6.0\,\text{MeV}$. Compare with the experimental data and the result of Subsect. 8.5.2.

**8.33.** Calculate the $\log ft$ values of the beta-decay transitions from the $0^+$ ground state of $^{14}$O to the $1^+_1$ and $1^+_2$ states in $^{14}$N by using the wave functions obtained in Exercise 8.26. Compare with the data and comment.

**8.34.** Using the wave functions of Exercise 8.27 calculate the $B(E2)$ values for the decay of the first $2^+$ state in $^{38}$Ar and $^{38}$Ca. Determine the electric polarization constant $\chi$, compare with the result of Exercise 6.31 and comment.

**8.35.** Using the wave functions of Exercise 8.27 calculate the $\log ft$ values for the decay of the $0^+$ ground state of $^{38}$Ca to the $0^+$ ground state and the first two excited $1^+$ states of $^{38}$K. Compare with the data and the results of Exercise 7.22 and comment.

# Particle–Hole Excitations and the Tamm–Dancoff Approximation

## Prologue

This chapter describes the configuration mixing of particle–hole excitations in doubly magic nuclei. The discussion is confined to one-particle–one-hole excitations within the simplest scheme of configuration mixing, namely the Tamm–Dancoff approximation (TDA). We show that the TDA arises from a variational principle and leads to diagonalization of the residual Hamiltonian in a basis of particle–hole excitations of the particle–hole vacuum.

We address the effects of particle–hole configuration mixing on electromagnetic observables. In particular, we introduce the notion of collectivity of a wave function and of an electromagnetic transition. We show that the general features of the TDA are conveniently represented by a schematic separable model with graphical solution.

## 9.1 The Tamm–Dancoff Approximation

In this section we derive the Tamm–Dancoff approximation (TDA) from a variational principle leading to a result known as Brillouin's theorem. This theorem justifies the diagonalization of the residual interaction within a one-particle–one-hole basis. We also write down the explicit form of the TDA matrix equations and tabulate the matrix elements for a large number of different particle–hole configurations.

### 9.1.1 Justification of the TDA: Brillouin's Theorem

Particle–hole excitations were introduced in Sect. 4.4 and extensively discussed in Sect. 5.4 and Subsect. 5.5.4 in the context of the mean-field shell model. In this subsection the residual interaction is allowed to mix particle–hole excitations to produce configuration-mixed nuclear states. The scheme is known as the TDA.

Consider the Hamiltonian $H$ of a nucleus and an arbitrary state vector $|\Psi\rangle$ in a *restricted part* of the nuclear Hilbert space. The expectation value of the energy in this state is

$$E[\Psi] \equiv \frac{\langle\Psi|H|\Psi\rangle}{\langle\Psi|\Psi\rangle} \, , \qquad (9.1)$$

where the denominator is included to guarantee normalization. The expectation value is written as a functional that appeared already in (3.14).

We wish to determine $|\Psi\rangle$ so that the energy (9.1) is minimized. That $|\Psi\rangle$ is then the best description of the state (usually the ground state) within the restricted space, i.e. the best approximation to the exact state realized in the full space. For the best $|\Psi\rangle$ the energy functional (9.1) must be stationary, $\delta E[\Psi] = 0$, under small variations of $|\Psi\rangle$.[1]

Let us vary $E[\Psi]$ so that $|\Psi\rangle \rightarrow |\Psi\rangle + |\delta\Psi\rangle$, where $|\delta\Psi\rangle$ is an arbitrary state of infinitesimal norm in the restricted space. By the elementary rules of differentiation and rearranging we obtain [4]

$$\langle\Psi|\Psi\rangle\delta E = \langle\delta\Psi|H - E|\Psi\rangle + \langle\Psi|H - E|\delta\Psi\rangle \, . \qquad (9.2)$$

The requirement $\delta E = 0$ now gives

$$\langle\delta\Psi|H - E|\Psi\rangle + \langle\Psi|H - E|\delta\Psi\rangle = 0 \, . \qquad (9.3)$$

Since $|\delta\Psi\rangle$ was chosen to be arbirary within the restricted space, we can do in that space a second variation where $|\delta\Psi\rangle$ is replaced by $\mathrm{i}|\delta\Psi\rangle$. Equation (9.3) is then replaced with

$$-\mathrm{i}\langle\delta\Psi|H - E|\Psi\rangle + \mathrm{i}\langle\Psi|H - E|\delta\Psi\rangle = 0 \, . \qquad (9.4)$$

If we now multiply (9.3) by $-\mathrm{i}$ and add it to (9.4), we find

$$\langle\delta\Psi|H - E|\Psi\rangle = 0 \, , \quad \langle\Psi|H - E|\delta\Psi\rangle = 0 \, . \qquad (9.5)$$

Let us now take the restricted space to be the space spanned by the $N$-particle Slater determinants (4.6) of single-particle states $\phi_\alpha$. The lowest-energy trial state in that space is the state $|\Phi_0\rangle$ which has the $N$ lowest levels occupied. A general variation of $|\Phi_0\rangle$ in the chosen space of Slater determinants is obtained by varying one of its single-particle states according to

$$\phi_\beta \rightarrow \phi_\beta + \eta\phi_\alpha \, , \qquad (9.6)$$

where $\eta$ is an arbitrary infinitesimal number. For this to be a true variation, i.e. an infinitesimal change of the state $|\Phi_0\rangle$, the single-particle state $\phi_\alpha$ has

---

[1] This is a necessary condition that does not guarantee that the extremum is a *minimum*. However, we may here safely assume that a minimum indeed results. For further discussion see [16].

to be one of the unoccupied states, so that $\varepsilon_\alpha > \varepsilon_F$. In occupation number representation we thus have

$$|\delta\Phi_0\rangle = \eta c_\alpha^\dagger c_\beta |\Phi_0\rangle \equiv \eta|\chi\rangle , \quad \varepsilon_\alpha > \varepsilon_F , \ \varepsilon_\beta \leq \varepsilon_F . \tag{9.7}$$

The state $|\chi\rangle$ is a particle–hole excitation of $|\Phi_0\rangle$ and clearly orthogonal to it, i.e. $\langle\chi|\Phi_0\rangle = 0$. The first equation (9.5) now gives

$$\boxed{\langle\chi|H|\Phi_0\rangle = 0 .} \tag{9.8}$$

This result, known as *Brillouin's theorem*, says that the nuclear Hamiltonian does not connect the particle–hole vacuum to its particle–hole excitations. Also, when the condition (9.8) is met, we have a formal proof that the state $|\Phi_0\rangle$ is indeed the ground state in the chosen space of Slater determinants.

Let us relate Brillouin's theorem to Hartree–Fock theory. From (4.62) and (4.63) we see that

$$\langle\chi|H|\Phi_0\rangle = t_{\alpha\beta} + \sum_{\substack{\gamma \\ \varepsilon_\gamma \leq \varepsilon_F}} \bar{v}_{\gamma\alpha\gamma\beta} = T_{\alpha\beta} . \tag{9.9}$$

According to Brillouin's theorem we have $T_{\alpha\beta} = 0$. This means that the effective one-body part of the nuclear Hamiltonian can be exactly diagonalized in a single-particle basis that only includes the orbitals below the Fermi level. After the diagonalization we have

$$T_{\beta\beta'} = \varepsilon_\beta \delta_{\beta\beta'} , \quad \varepsilon_\beta, \varepsilon_{\beta'} \leq \varepsilon_F . \tag{9.10}$$

This is the Hartree–Fock equation (4.68), with the additional information that only the states below the Fermi level contribute. Noting the unitary transformation (4.64) we identify the ground state $|\Phi_0\rangle$ as the Hartree–Fock vacuum $|HF\rangle$.

Since particle–hole excitations cannot mix into the Hartree–Fock vacuum, we conclude that the first possible perturbations of the vacuum come from two-particle–two-hole excitations. The resulting 'correlated ground state' appears in sophisticated particle–hole theories such as the RPA, to be discussed in Chap. 11.

When we include the residual interaction in the description of a doubly magic nucleus, in first approximation the ground state is the Hartree–Fock vacuum and the excited states are linear combinations of particle–hole excitations. The linear combinations are obtained by diagonalizing the nuclear Hamiltonian in the particle–hole basis; for a general description of the procedure see Subsect. 8.3.1. This first step in introducing configuration mixing of particle–hole excitations is called the TDA.

The TDA can only be used for doubly magic nuclei since they are the only ones that offer a converging hierarchy in $n$-particle–$n$-hole ($n = 1, 2, \ldots$)

excitations. This is seen from the fact that each new order $n$ of particle–hole excitation costs at least the energy needed to cross the magic gap. In open-shell nuclei particle–hole excitations do not cost much energy. In them the excitations for different $n$ coexist within the same energy range, so the hierarchical structure of perturbation theory is lost.

## 9.1.2 Derivation of Explicit Expressions for the TDA Matrix

In the proton–neutron formalism and angular-momentum-coupled particle–hole representation, the particle–hole basis states are $|a\,b^{-1};\,J\,M\rangle$. The Hamiltonian

$$H = H_{\mathrm{HF}} + V_{\mathrm{RES}} \tag{9.11}$$

is given in detail by (4.71) and (4.72). We now form the matrix elements of $H$ in the coupled basis.

The matrix element of $H_{\mathrm{HF}}$ is

$$\langle a\,b^{-1};\,J\,M|H_{\mathrm{HF}}|c\,d^{-1};\,J\,M\rangle = \sum_{\alpha'}\varepsilon_{\alpha'}\sum_{\substack{m_\alpha m_\beta \\ m_\gamma m_\delta}}(j_a\,m_\alpha\,j_b\,m_\beta|J\,M)$$

$$\times\,(j_c\,m_\gamma\,j_d\,m_\delta|J\,M)\langle\mathrm{HF}|h_\beta c_\alpha c_{\alpha'}^\dagger c_{\alpha'}c_\gamma^\dagger h_\delta^\dagger|\mathrm{HF}\rangle\,. \tag{9.12}$$

Using (4.46), (4.47), (4.52) and (4.53) we find

$$\langle\mathrm{HF}|h_\beta c_\alpha c_{\alpha'}^\dagger c_{\alpha'}c_\gamma^\dagger h_\delta^\dagger|\mathrm{HF}\rangle$$

$$= -\delta_{-\beta\alpha'}\delta_{\alpha\gamma}\delta_{\alpha',-\delta} + \delta_{\beta\delta}\delta_{\alpha\alpha'}\delta_{\alpha'\gamma} + \delta_{\beta\delta}\delta_{\alpha\gamma}\theta(\varepsilon_{\mathrm{F}} - \varepsilon_{\alpha'})\,, \tag{9.13}$$

where $\theta$ is the Heaviside step function defined as

$$\theta(x) = \begin{cases} 1\,, & x \geq 0\,, \\ 0\,, & x < 0\,. \end{cases} \tag{9.14}$$

Substitution into (9.12) results in

$$\langle a\,b^{-1};\,J\,M|H_{\mathrm{HF}}|c\,d^{-1};\,J\,M\rangle = \delta_{ac}\delta_{bd}\left(\varepsilon_a - \varepsilon_b + \sum_{\varepsilon_{\alpha'}\leq\varepsilon_{\mathrm{F}}}\varepsilon_{\alpha'}\right). \tag{9.15}$$

For *excitation energies* the last term in (9.15) can be omitted. This is because the ground-state energy (4.73) contains

$$\langle\mathrm{HF}|H_{\mathrm{HF}}|\mathrm{HF}\rangle = \sum_{\varepsilon_\alpha\leq\varepsilon_{\mathrm{F}}}\varepsilon_\alpha\,. \tag{9.16}$$

The contribution of the one-body part to the excitation energies is then

$$\boxed{\langle a\,b^{-1};\,J|H_{\mathrm{HF}}|c\,d^{-1};\,J\rangle = \delta_{ac}\delta_{bd}(\varepsilon_a - \varepsilon_b) \equiv \delta_{ac}\delta_{bd}\varepsilon_{ab}\,.} \tag{9.17}$$

The two-body part is

$$
\langle a\,b^{-1}\,;\,J\,M|V_{\mathrm{RES}}|c\,d^{-1}\,;\,J\,M\rangle
$$

$$
= \tfrac{1}{4} \sum_{\substack{m_\alpha m_\beta \\ m_\gamma m_\delta}} (j_a\,m_\alpha\,j_b\,m_\beta|J\,M)(j_c\,m_\gamma\,j_d\,m_\delta|J\,M)
$$

$$
\times \sum_{\alpha'\beta'\gamma'\delta'} \langle \alpha'\,\beta'|V|\gamma'\,\delta'\rangle \langle \mathrm{HF}|h_\beta c_\alpha \mathcal{N}\big[c_{\alpha'}^\dagger c_{\beta'}^\dagger c_{\delta'} c_{\gamma'}\big]c_\gamma^\dagger h_\delta^\dagger|\mathrm{HF}\rangle \quad (9.18)
$$

with the two-body matrix element (4.45). The normal order varies as the operators take on particle or hole character, and for a non-zero contribution there has to be four particle and four hole operators. With these observations the contractions yield

$$
\langle \mathrm{HF}|h_\beta c_\alpha \mathcal{N}\big[c_{\alpha'}^\dagger c_{\beta'}^\dagger c_{\delta'} c_{\gamma'}\big]c_\gamma^\dagger h_\delta^\dagger|\mathrm{HF}\rangle
$$

$$
= \delta_{\alpha\alpha'}\delta_{\beta,-\gamma'}\delta_{-\beta'\delta}\delta_{\delta'\gamma}(-1)^{j_{c'}+j_{b'}+m_{\gamma'}+m_{\beta'}}
$$

$$
- \delta_{\alpha\alpha'}\delta_{\beta,-\delta'}\delta_{\gamma\gamma'}\delta_{\delta,-\beta'}(-1)^{j_{d'}+j_{b'}+m_{\delta'}+m_{\beta'}}
$$

$$
- \delta_{\alpha\beta'}\delta_{\beta,-\gamma'}\delta_{\delta'\gamma}\delta_{-\alpha'\delta}(-1)^{j_{c'}+j_{a'}+m_{\alpha'}+m_{\gamma'}}
$$

$$
+ \delta_{\alpha\beta'}\delta_{\beta,-\delta'}\delta_{\gamma\gamma'}\delta_{-\alpha'\delta}(-1)^{j_{a'}+j_{d'}+m_{\alpha'}+m_{\delta'}} \,. \quad (9.19)
$$

After merging terms and simplifying this leads to

$$
\langle a\,b^{-1}\,;\,J\,M|V_{\mathrm{RES}}|c\,d^{-1}\,;\,J\,M\rangle = \sum_{\substack{m_\alpha m_\beta \\ m_\gamma m_\delta}} (-1)^{j_b+j_d+m_\beta+m_\delta}
$$

$$
\times (j_a\,m_\alpha\,j_b\,m_\beta|J\,M)(j_c\,m_\gamma\,j_d\,m_\delta|J\,M)\langle \alpha\,-\delta|V|-\beta\,\gamma\rangle \,. \quad (9.20)
$$

We now substitute (8.8) into (9.20), noticing that $a \neq b, c \neq d$. Since the particle–hole matrix element is independent of $M$, we can sum the equation over $M$ and divide the result by $2J+1$. Then, with formulas of Chap. 1, we convert the Clebsch–Gordan coefficients into $3j$ symbols and find

$$
\langle a\,b^{-1}\,;\,J|V_{\mathrm{RES}}|c\,d^{-1}\,;\,J\rangle
$$

$$
= \sum_{J'} \hat{J'}^2 \langle a\,d\,;\,J'|V|b\,c\,;\,J'\rangle \sum_{MM'} \sum_{\substack{m_\alpha m_\beta \\ m_\gamma m_\delta}} (-1)^{j_b+j_d+m_\beta+m_\delta}
$$

$$
\times \begin{pmatrix} j_a & j_b & J \\ m_\alpha & m_\beta & -M \end{pmatrix} \begin{pmatrix} j_a & j_d & J' \\ m_\alpha & -m_\delta & -M' \end{pmatrix} \begin{pmatrix} j_c & j_b & J' \\ -m_\gamma & m_\beta & M' \end{pmatrix} \begin{pmatrix} j_c & j_d & J \\ m_\gamma & m_\delta & -M \end{pmatrix}
$$

$$
= \sum_{J'} \hat{J'}^2 (-1)^{j_b+j_c+J'} \langle a\,d\,;\,J'|V|b\,c\,;\,J'\rangle \begin{Bmatrix} j_a & j_b & J \\ j_c & j_d & J' \end{Bmatrix} \,, \quad (9.21)
$$

where the explicit definition (1.59) of the $6j$ symbol was used in the last step. We use the symmetry property (8.30) and, to accommodate some of the literature and computer codes, express the result also in terms of the Racah symbol (1.63):

$$
\begin{aligned}
\langle a\,b^{-1}\,;\,J|V_{\text{RES}}|c\,d^{-1}\,;\,J\rangle &= -\sum_{J'} \widehat{J}'^{\,2} \begin{Bmatrix} j_a & j_b & J \\ j_c & j_d & J' \end{Bmatrix} \langle a\,d\,;\,J'|V|c\,b\,;\,J'\rangle \\
&= -\sum_{J'} \widehat{J}'^{\,2} W(j_a\,j_b\,j_d\,j_c;\,J\,J')\langle d\,a\,;\,J'|V|b\,c\,;\,J'\rangle\,.
\end{aligned}
$$

(9.22)

This result is known as the *Pandya transformation* of particle–hole matrix elements.

In isospin formalism an analogous derivation leads to the expression

$$
\begin{aligned}
\langle a\,b^{-1}&\,;\,J\,T|V_{\text{RES}}|c\,d^{-1}\,;\,J\,T\rangle \\
&= -\sum_{J'T'} \widehat{J}'^{\,2}\widehat{T}'^{\,2} \begin{Bmatrix} j_a & j_b & J \\ j_c & j_d & J' \end{Bmatrix} \begin{Bmatrix} \tfrac12 & \tfrac12 & T \\ \tfrac12 & \tfrac12 & T' \end{Bmatrix} \langle a\,d\,;\,J'\,T'|V|c\,b\,;\,J'\,T'\rangle \\
&= -\sum_{J'T'} \widehat{J}'^{\,2}\widehat{T}'^{\,2} W(j_a\,j_b\,j_d\,j_c;\,J\,J')W(\tfrac12\tfrac12\tfrac12\tfrac12;\,T\,T') \\
&\qquad\qquad \times \langle d\,a\,;\,J'\,T'|V|b\,c\,;\,J'\,T'\rangle\,.
\end{aligned}
$$

(9.23)

The particle–hole matrix elements (9.22) and (9.23) have the useful property

$$
\langle c\,d^{-1}\,;\,J\,(T)|V_{\text{RES}}|a\,b^{-1}\,;\,J\,(T)\rangle = \langle a\,b^{-1}\,;\,J\,(T)|V_{\text{RES}}|c\,d^{-1}\,;\,J\,(T)\rangle\,,
$$

(9.24)

which means that the TDA Hamiltonian matrix is *symmetric*. This was to be expected on general grounds: any Hamiltonian matrix is Hermitian and our particle–hole matrix elements are real.

By means of the unitarity relation (1.66) of $6j$ symbols we can invert (9.22) to yield

$$
\langle a\,d\,;\,J|V|c\,b\,;\,J\rangle = -\sum_{J'} \widehat{J}'^{\,2} \begin{Bmatrix} j_a & j_d & J \\ j_c & j_b & J' \end{Bmatrix} \langle a\,b^{-1}\,;\,J'|V_{\text{RES}}|c\,d^{-1}\,;\,J'\rangle\,. \quad (9.25)
$$

Substituting this into the right-hand side of (9.22) and applying (8.30) and (1.65) we can derive the relation

$$
\begin{aligned}
\langle p\,n^{-1}\,;\,J|V_{\text{RES}}|p'\,n'^{-1}\,;\,J\rangle &= (-1)^{j_n+j_{p'}+J+1} \sum_{J'} (-1)^{J'} \widehat{J}'^{\,2} \begin{Bmatrix} j_p & j_n & J \\ j_{n'} & j_{p'} & J' \end{Bmatrix} \\
&\qquad \times \langle p\,p'^{-1}\,;\,J'|V_{\text{RES}}|n\,n'^{-1}\,;\,J'\rangle\,.
\end{aligned}
$$

(9.26)

### 9.1.3 Tabulated Values of Particle–Hole Matrix Elements

Equation (9.22) or (9.23) gives the particle–hole matrix elements once the two-particle matrix elements are known. For the SDI the latter can be taken from Table 8.3 for excitations from the 0p shell to the 0d-1s shell, and from Table 8.4 for excitations from the 0d-1s shell to the $0f_{7/2}$ shell.

For more direct evaluation of particle–hole matrix elements Tables 9.1–9.4 list the quantities

$$\mathcal{M}_{abcd}(JT) \equiv -\sum_{J'} \hat{J'}^2 \begin{Bmatrix} j_a & j_b & J \\ j_c & j_d & J' \end{Bmatrix} \langle a\,d;\,J'\,T|V_{\text{SDI}}|c\,b;\,J'\,T\rangle_{A_T=1}\,. \quad (9.27)$$

This is the same as (9.22) for the SDI scaled to $A_T = 1$, except that here the two-particle matrix element is in isospin representation. Table 9.1 gives (9.27) for excitations from the 0s to the 0p shell, Table 9.2 from the 0p shell to the 0d-1s shells, Table 9.3 from the 0d-1s shells to the $0f_{7/2}$ shell and Table 9.4 from the $0f_{7/2}$ shell to the rest of the 0f-1p space. The CS phase convention is used throughout.

**Table 9.1.** Quantities $\mathcal{M}_{abcd}(JT)$ in the CS phase convention

| abcd | JT | $\mathcal{M}$ | JT | $\mathcal{M}$ | JT | $\mathcal{M}$ | JT | $\mathcal{M}$ |
|------|-----|---------|-----|---------|-----|---------|-----|---------|
| 1111 | 10 | 2.3333 | 11 | −0.3333 | 20 | 1.0000 | 21 | 1.0000 |
| 1121 | 10 | −0.9428 | 11 | 0.9428 | | | | |
| 2121 | 00 | 1.0000 | 01 | 1.0000 | 10 | 1.6667 | 11 | 0.3333 |

The particle states are numbered $1 = 0p_{3/2}$ and $2 = 0p_{1/2}$, and the hole state is numbered $1 = 0s_{1/2}$. The first column gives the particle-hole–particle-hole labels, and the following columns give the $JT$ combinations and values of $\mathcal{M}$.

By means of the relations (8.25) and (8.27) the particle–hole matrix elements (9.22) are obtained from the $\mathcal{M}_{abcd}(JT)$ according to

$$\langle p_1\,p_2^{-1};\,J|V_{\text{RES}}|p_3\,p_4^{-1};\,J\rangle = A_1\mathcal{M}_{a_1a_2a_3a_4}(J1)\,, \quad (9.28)$$

$$\langle n_1\,n_2^{-1};\,J|V_{\text{RES}}|n_3\,n_4^{-1};\,J\rangle = A_1\mathcal{M}_{a_1a_2a_3a_4}(J1)\,, \quad (9.29)$$

$$\langle p_1\,p_2^{-1};\,J|V_{\text{RES}}|n_3\,n_4^{-1};\,J\rangle$$
$$= \frac{1}{2}\Big\{ A_1\sqrt{[1+(-1)^J\delta_{a_1a_2}][1+(-1)^J\delta_{a_3a_4}]}\,\mathcal{M}_{a_1a_2a_3a_4}(J1)$$
$$- A_0\sqrt{[1-(-1)^J\delta_{a_1a_2}][1-(-1)^J\delta_{a_3a_4}]}\,\mathcal{M}_{a_1a_2a_3a_4}(J0) \Big\}\,. \quad (9.30)$$

The particle–hole matrix elements (9.23) are given directly by

$$\langle a\,b^{-1};\,J\,T = 0|V_{\text{RES}}|c\,d^{-1};\,J\,T = 0\rangle$$
$$= \tfrac{1}{2}[3A_1\mathcal{M}_{abcd}(J1) - A_0\mathcal{M}_{abcd}(J0)]\,, \quad (9.31)$$

**Table 9.2.** Quantities $\mathcal{M}_{abcd}(JT)$ in the CS phase convention

| abcd | JT | $\mathcal{M}$ | JT | $\mathcal{M}$ | JT | $\mathcal{M}$ | JT | $\mathcal{M}$ | JT | $\mathcal{M}$ |
|------|----|----|----|----|----|----|----|----|----|----|
| 1111 | 10 | 4.2000 | 11 | −0.6000 | 20 | 0.8857 | 21 | 0.8857 | 30 | 1.0571 |
| 1111 | 31 | −0.3143 | 40 | 1.2857 | 41 | 1.2857 | | | | |
| 1112 | 20 | −0.8552 | 21 | −0.8552 | 30 | 1.2777 | 31 | −0.2555 | | |
| 1121 | 10 | 2.2361 | 11 | −1.3416 | 20 | 0.9165 | 21 | 0.9165 | | |
| 1122 | 10 | 2.5298 | 11 | 0 | | | | | | |
| 1131 | 10 | −0.4000 | 11 | 1.2000 | 20 | 0.2619 | 21 | 0.2619 | 30 | −1.1198 |
| 1131 | 31 | 0.5599 | | | | | | | | |
| 1132 | 10 | 2.2361 | 11 | −1.3416 | 20 | −0.3928 | 21 | −0.3928 | | |
| 1212 | 20 | 1.0000 | 21 | 1.0000 | 30 | 1.8571 | 31 | 0.1429 | | |
| 1221 | 20 | −0.9798 | 21 | −0.9798 | | | | | | |
| 1231 | 20 | −0.4899 | 21 | −0.4899 | 30 | −0.7825 | 31 | 1.0954 | | |
| 1232 | 20 | 0 | 21 | 0 | | | | | | |
| 2121 | 10 | 2.3333 | 11 | −0.3333 | 20 | 1.0000 | 21 | 1.0000 | | |
| 2122 | 10 | 0.9428 | 11 | −0.9428 | | | | | | |
| 2131 | 10 | −1.4907 | 11 | −0.2981 | 20 | 0.4000 | 21 | 0.4000 | | |
| 2132 | 10 | 2.3333 | 11 | −0.3333 | 20 | −0.2000 | 21 | −0.2000 | | |
| 2222 | 00 | 1.0000 | 01 | 1.0000 | 10 | 1.6667 | 11 | 0.3333 | | |
| 2231 | 00 | −1.4142 | 01 | −1.4142 | 10 | 0.2108 | 11 | 1.0541 | | |
| 2232 | 10 | 0.9428 | 11 | −0.9428 | | | | | | |
| 3131 | 00 | 2.0000 | 01 | 2.0000 | 10 | 1.4667 | 11 | 0.9333 | 20 | 0.4000 |
| 3131 | 21 | 0.4000 | 30 | 2.2286 | 31 | 0.1714 | | | | |
| 3132 | 10 | −1.4907 | 11 | −0.2981 | 20 | 0.4000 | 21 | 0.4000 | | |
| 3232 | 10 | 2.3333 | 11 | −0.3333 | 20 | 1.0000 | 21 | 1.0000 | | |

The particle states are numbered $1 = 0d_{5/2}$, $2 = 1s_{1/2}$ and $3 = 0d_{3/2}$, and the hole states are numbered $1 = 0p_{3/2}$ and $2 = 0p_{1/2}$. The first column gives the particle–hole–particle-hole labels, and the following columns give the $JT$ combinations and values of $\mathcal{M}$. The zeros recorded do not result from conservation laws.

$$\langle a\,b^{-1}\,;\, J\,T = 1|V_{\mathrm{RES}}|c\,d^{-1}\,;\, J\,T = 1\rangle$$
$$= \tfrac{1}{2}[A_1\mathcal{M}_{abcd}(J1) + A_0\mathcal{M}_{abcd}(J0)]\,. \quad (9.32)$$

Particle–hole matrix elements in the BR phase convention are obtained from those in the CS convention according to

$$\langle a\,b^{-1}\,;\, J\,(T)|V_{\mathrm{RES}}|c\,d^{-1}\,;\, J\,(T)\rangle_{\mathrm{BR}}$$
$$= (-1)^{\frac{1}{2}(l_b+l_c-l_a-l_d)}\langle a\,b^{-1}\,;\, J\,(T)|V_{\mathrm{RES}}|c\,d^{-1}\,;\, J\,(T)\rangle_{\mathrm{CS}}\,. \quad (9.33)$$

The relations (9.24)–(9.26) are valid also in the BR convention.

**Table 9.3.** Quantities $\mathcal{M}_{abcd}(JT)$ in the CS phase convention

| abcd | JT | $\mathcal{M}$ | JT | $\mathcal{M}$ | JT | $\mathcal{M}$ | JT | $\mathcal{M}$ | JT | $\mathcal{M}$ |
|------|----|------|----|------|----|------|----|------|----|------|
| 1111 | 10 | 6.0000 | 11 | −0.8571 | 20 | 1.0476 | 21 | 1.0476 | 30 | 1.7619 |
| 1111 | 31 | −0.5238 | 40 | 0.8961 | 41 | 0.8961 | 50 | 0.7879 | 51 | −0.2511 |
| 1111 | 60 | 1.5151 | 61 | 1.5151 | | | | | | |
| 1112 | 30 | 1.8571 | 31 | −0.4286 | 40 | 0.9343 | 41 | 0.9343 | | |
| 1113 | 20 | −1.1429 | 21 | −1.1429 | 30 | 1.1547 | 31 | −0.1650 | 40 | −0.4882 |
| 1113 | 41 | −0.4882 | 50 | 1.2696 | 51 | −0.3174 | | | | |
| 1212 | 30 | 2.1429 | 31 | −0.1429 | 40 | 1.0000 | 41 | 1.0000 | | |
| 1213 | 30 | 1.4846 | 31 | 0.1650 | 40 | −0.5634 | 41 | −0.5634 | | |
| 1313 | 20 | 1.7143 | 21 | 1.7143 | 30 | 1.1429 | 31 | 0.3809 | 40 | 0.3809 |
| 1313 | 41 | 0.3809 | 50 | 2.5454 | 51 | 0.1212 | | | | |

The particle state is numbered $1 = 0\mathrm{f}_{7/2}$, and the hole states are numbered $1 = 0\mathrm{d}_{5/2}$, $2 = 1\mathrm{s}_{1/2}$ and $3 = 0\mathrm{d}_{3/2}$. The first column gives the particle-hole–particle-hole labels, and the following columns give the $JT$ combinations and values of $\mathcal{M}$.

**Table 9.4.** Quantities $\mathcal{M}_{abcd}(JT)$ in the CS phase convention

| abcd | JT | $\mathcal{M}$ | JT | $\mathcal{M}$ | JT | $\mathcal{M}$ | JT | $\mathcal{M}$ | JT | $\mathcal{M}$ |
|------|----|------|----|------|----|------|----|------|----|------|
| 1111 | 20 | 3.7714 | 21 | −0.3429 | 30 | 0.7619 | 31 | 0.7619 | 40 | 1.0159 |
| 1111 | 41 | −0.2540 | 50 | 1.3333 | 51 | 1.3333 | | | | |
| 1121 | 30 | 0.8248 | 31 | 0.8248 | 40 | −1.3147 | 41 | 0.1878 | | |
| 1131 | 20 | −1.8286 | 21 | −0.4571 | 30 | 0.4949 | 31 | 0.4949 | 40 | −1.0625 |
| 1131 | 41 | 0.0861 | 50 | 0.4761 | 51 | 0.4761 | | | | |
| 2121 | 30 | 1.0000 | 31 | 1.0000 | 40 | 1.8889 | 41 | 0.1111 | | |
| 2131 | 30 | 0.7143 | 31 | 0.7143 | 40 | 1.6139 | 41 | 0.2548 | | |
| 3131 | 10 | 2.5714 | 11 | 2.5714 | 20 | 1.2762 | 21 | 0.8190 | 30 | 0.6190 |
| 3131 | 31 | 0.6190 | 40 | 1.4156 | 41 | 0.3766 | 50 | 0.2684 | 51 | 0.2684 |
| 3131 | 60 | 2.9137 | 61 | 0.1165 | | | | | | |

The particle states are numbered $1 = 1\mathrm{p}_{3/2}$, $2 = 1\mathrm{p}_{1/2}$ and $3 = 0\mathrm{f}_{5/2}$, and the hole state is numbered $1 = 0\mathrm{f}_{7/2}$. The first column gives the particle-hole–particle-hole labels, and the following columns give the $JT$ combinations and values of $\mathcal{M}$.

### 9.1.4 TDA as an Eigenvalue Problem; Properties of the Solutions

The ground state of the TDA is the Hartree–Fock vacuum $|\mathrm{HF}\rangle$. We seek TDA excited states in the form

$$|\Psi_\omega^{\mathrm{TDA}}\rangle = \sum_{ab} X_{ab}^\omega |a\, b^{-1}\,;\, J^\pi\, M\rangle\,, \qquad (9.34)$$

where $\omega \equiv n J^\pi M$ and $n = 1, 2, 3, \ldots$ numbers the eigenvalues for a given multipolarity $J^\pi$. In the pattern of (8.78) we then have the matrix equation

$$\sum_{cd} \langle a\, b^{-1}\,;\, J^\pi |H| c\, d^{-1}\,;\, J^\pi \rangle X_{cd}^\omega = E_\omega X_{ab}^\omega\,. \qquad (9.35)$$

With (9.17) this becomes

$$\sum_{cd}(\delta_{ac}\delta_{bd}\varepsilon_{ab} + \langle a\,b^{-1}\,;\,J^\pi|V_{\mathrm{RES}}|c\,d^{-1}\,;\,J^\pi\rangle)X_{cd}^\omega = E_\omega X_{ab}^\omega\,. \qquad (9.36)$$

These, for various $\omega$, are the TDA equations to be solved.

We require orthonormality of the TDA eigenstates, i.e.

$$\langle\Psi_\omega^{\mathrm{TDA}}|\Psi_{\omega'}^{\mathrm{TDA}}\rangle = \delta_{\omega\omega'} = \delta_{nn'}\delta_{JJ'}\delta_{\pi\pi'}\delta_{MM'}\,. \qquad (9.37)$$

Substituting (9.34) in the left-hand side and using (5.45) we obtain

$$\sum_{abcd}X_{ab}^{\omega*}X_{cd}^{\omega'}\langle a\,b^{-1}\,;\,J^\pi\,M|c\,d^{-1}\,;\,J'^{\pi'}\,M'\rangle$$
$$= \sum_{abcd}X_{ab}^{\omega*}X_{cd}^{\omega'}\delta_{ac}\delta_{bd}\delta_{JJ'}\delta_{\pi\pi'}\delta_{MM'} = \delta_{JJ'}\delta_{\pi\pi'}\delta_{MM'}\sum_{ab}X_{ab}^{\omega*}X_{ab}^{\omega'}\,. \qquad (9.38)$$

Thus we see that orthogonality with respect to $J^\pi M$ is guaranteed by the basis. Since the amplitudes $X_{ab}^\omega$ are independent of $M$, we can write the remaining condition as

$$\sum_{ab}\left(X_{ab}^{nJ^\pi}\right)^*X_{ab}^{n'J^\pi} = \delta_{nn'} \quad \text{(TDA orthonormality)}\,. \qquad (9.39)$$

Normalization is carried by the special case

$$\sum_{ab}|X_{ab}^\omega|^2 = 1\,. \qquad (9.40)$$

The TDA states form a complete set in the particle–hole space, so that they satisfy

$$\sum_\omega|\Psi_\omega^{\mathrm{TDA}}\rangle\langle\Psi_\omega^{\mathrm{TDA}}| = 1\,. \qquad (9.41)$$

Inserting this relation into (5.45) gives

$$\delta_{ac}\delta_{bd} = \langle a\,b^{-1}\,;\,J^\pi\,M|\sum_\omega|\Psi_\omega^{\mathrm{TDA}}\rangle\langle\Psi_\omega^{\mathrm{TDA}}|c\,d^{-1}\,;\,J^\pi\,M\rangle = \sum_n X_{ab}^{nJ^\pi}\left(X_{cd}^{nJ^\pi}\right)^*\,, \qquad (9.42)$$

so we have

$$\sum_n X_{ab}^{nJ^\pi}\left(X_{cd}^{nJ^\pi}\right)^* = \delta_{ac}\delta_{bd} \quad \text{(TDA completeness)}\,. \qquad (9.43)$$

The calculation of the eigenenergies $E_\omega$ and the eigenfunctions $|\Psi_\omega^{\mathrm{TDA}}\rangle$ will be addressed in the following sections of this chapter.

## 9.2 TDA for General Separable Forces

In this section we present the particle–hole matrix element and simplify it by excluding part of it, which leads to a schematic model. Within that model, we derive a simple equation, a so-called dispersion relation, for approximate TDA energies. We solve the equation graphically for excitation energies in $^4$He.

### 9.2.1 Schematic Model; Dispersion Equation

In the following we develop a schematic model for solving the TDA equations (9.36) for separable forces. In spite of the radical assumptions required for its simplicity, the model gives qualitative and even quantitative insight into the basic principles and properties of the TDA and its solutions.

We write the Pandya transformation (9.22) and change the $cb$ coupling order in the two-particle matrix element by means of (8.30). We then insert the two-particle matrix element given by (8.55) and (8.56) for a general separable force. This gives

$$\langle a\,b^{-1}\,;\,J|V_{\mathrm{RES}}|c\,d^{-1}\,;\,J\rangle = -\sum_{J'}\widehat{J}'^{\,2}\left\{\begin{matrix} j_a & j_b & J \\ j_c & j_d & J' \end{matrix}\right\}\mathcal{N}_{ad}(J')\mathcal{N}_{bc}(J')$$

$$\times (-1)^{j_c+j_b+J'+1}\left[\sum_{\lambda}\chi_{\lambda}(-1)^{j_a+j_d+J'}\left\{\begin{matrix} j_a & j_d & J' \\ j_c & j_b & \lambda \end{matrix}\right\}(b\|\boldsymbol{Q}_{\lambda}\|a)(d\|\boldsymbol{Q}_{\lambda}\|c)\right.$$

$$\left. -(-1)^{j_b+j_c+J'}\sum_{\lambda}\chi_{\lambda}(-1)^{j_a+j_d+J'}\left\{\begin{matrix} j_a & j_d & J' \\ j_b & j_c & \lambda \end{matrix}\right\}(c\|\boldsymbol{Q}_{\lambda}\|a)(d\|\boldsymbol{Q}_{\lambda}\|b)\right]. \tag{9.44}$$

Here we have inserted interaction strengths $\chi_{\lambda}$, while in the original formulation of Subsect. 8.2.1 they were included in the reduced matrix elements of the $\boldsymbol{Q}_{\lambda}$. This changes the definition of the reduced matrix elements, as discussed in detail in Subsect. 9.2.3.

The sums over $J'$ can be carried out because $\mathcal{N}_{ad} = 1 = \mathcal{N}_{bc}$. The first term of (9.44) simplifies radically because of the $6j$ orthogonality relation (1.66). The second term also simplifies because of the relation (1.65), and (9.44) becomes

$$\langle a\,b^{-1}\,;\,J|V_{\mathrm{RES}}|c\,d^{-1}\,;\,J\rangle = (-1)^{j_a+j_b+j_c+j_d}\chi_J\widehat{J}^{-2}(b\|\boldsymbol{Q}_J\|a)(d\|\boldsymbol{Q}_J\|c)$$

$$-\sum_{\lambda}(-1)^{j_a+j_d+J+\lambda}\chi_{\lambda}\left\{\begin{matrix} j_a & j_b & J \\ j_d & j_c & \lambda \end{matrix}\right\}(c\|\boldsymbol{Q}_{\lambda}\|a)(d\|\boldsymbol{Q}_{\lambda}\|b). \tag{9.45}$$

The second, 'exchange' term is still complicated as it mixes different multipoles $\lambda$. We introduce now an approximate, schematic model that omits the explicit exchange term but takes it into account through the interaction parameters, as discussed in Subsect. 9.2.2.

254 9 Particle–Hole Excitations and the Tamm–Dancoff Approximation

From Subsect. 8.2.1 we know that the operator $\boldsymbol{Q}_\lambda$ in the separable force is essentially the same as the electric operator $\boldsymbol{Q}_\lambda$. We can therefore apply the symmetry property (6.27) to the reduced matrix elements in (9.45). The particle–hole matrix element in our *schematic model* then reads

$$\langle a\, b^{-1};\, J|V_{\mathrm{RES}}|c\, d^{-1};\, J\rangle = \chi_J \hat{J}^{-2}(a\|\boldsymbol{Q}_J\|b)(c\|\boldsymbol{Q}_J\|d) . \tag{9.46}$$

With the abbreviation

$$Q^J_{ab} \equiv \hat{J}^{-1}(a\|\boldsymbol{Q}_J\|b) \tag{9.47}$$

this becomes

$$\boxed{\langle a\, b^{-1};\, J|V_{\mathrm{RES}}|c\, d^{-1};\, J\rangle = \chi_J Q^J_{ab} Q^J_{cd} .} \tag{9.48}$$

Substituting (9.48) into the TDA eigenvalue equation (9.36) yields

$$\sum_{cd}(\delta_{ac}\delta_{bd}\varepsilon_{ab} + \chi_J Q^J_{ab}Q^J_{cd})X^\omega_{cd} = E_\omega X^\omega_{ab} , \tag{9.49}$$

which can be put in the form

$$(\varepsilon_{ab} - E_\omega)X^\omega_{ab} = Q^J_{ab}N_\omega , \quad N_\omega \equiv -\chi_J \sum_{cd} Q^J_{cd}X^\omega_{cd} . \tag{9.50}$$

Solving for the TDA amplitudes gives

$$X^\omega_{ab} = \frac{Q^J_{ab}}{\varepsilon_{ab} - E_\omega}N_\omega . \tag{9.51}$$

The magnitude of $N_\omega$ can be determined from the normalization condition (9.40):

$$1 = \sum_{ab} X^{\omega*}_{ab} X^\omega_{ab} = \sum_{ab} \frac{(Q^J_{ab})^2}{(\varepsilon_{ab} - E_\omega)^2}|N_\omega|^2 , \tag{9.52}$$

where we have noted that the $Q^J_{ab}$ are real but $N_\omega$ is complex if complex phases are used in the amplitudes $X^\omega_{ab}$. This leads to the result

$$|N_\omega|^{-2} = \sum_{ab} \frac{(Q^J_{ab})^2}{(\varepsilon_{ab} - E_\omega)^2} . \tag{9.53}$$

From the definition of $N_\omega$ in (9.50) and from the expression (9.51) for the TDA amplitude we obtain

$$N_\omega = -\chi_J \sum_{cd} Q^J_{cd}\frac{Q^J_{cd}}{\varepsilon_{cd} - E_\omega}N_\omega . \tag{9.54}$$

Dividing both sides by $N_\omega$, renaming the summation indices and rearranging, we have

$$\boxed{-\frac{1}{\chi_J} = \sum_{ab} \frac{(Q_{ab}^J)^2}{\varepsilon_{ab} - E_\omega}} \quad . \tag{9.55}$$

This equation gives the eigenenergies $E_\omega$, normally by numerical or graphical methods. Equations of this type in physics are known as *dispersion equations* or *dispersion relations*.

A graphical solution of a transcendental equation like (9.55) can give a panoramic view of the physics described by the equation. The basic method is to plot in the same coordinate system the left-hand and right-hand sides of the equation and then look for intersections of the curves. In the present case the curves are drawn as functions of the TDA eigenenergy $E_\omega$. This method is applied in Subsect. 9.2.4 to determine the TDA energies of $1^-$ states in $^4$He.

An interesting special case of the dispersion equation (9.55) occurs for degenerate particle–hole energies, i.e. $\varepsilon_{ab} = \varepsilon$ for all $ab$. Then all but one solution, for a given $\chi_J$, are trapped at the unperturbed energy $\varepsilon$. This can be seen from Fig. 9.1. For the untrapped solution equation (9.55) gives

$$-\frac{1}{\chi_J} = \frac{Q^2}{\varepsilon - E} , \quad Q^2 \equiv \sum_{ab}(Q_{ab}^J)^2 , \tag{9.56}$$

leading to

$$E = \varepsilon + \chi_J Q^2 \equiv E_{\text{coll}} . \tag{9.57}$$

The result (9.57) shows that the energy of the lowest solution decreases with increasing magnitude of $\chi_J$ for $\chi_J < 0$. The opposite happens to the

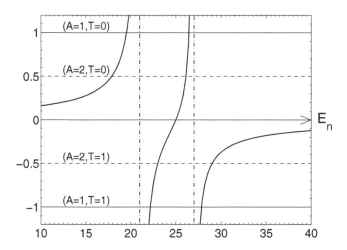

**Fig. 9.1.** Graphical solution of the transcendental equation (9.82) for the $1^-$ states in $^4$He. All energies are given in MeV. Solutions of the $T = 0$ and $T = 1$ eigenenergies are shown for two interaction strengths $A$. The unperturbed particle–hole energies are shown as vertical dash-dotted lines

highest solution if $\chi_J > 0$. This behaviour of the two untrapped TDA states displays their *collective* nature. In the following subsection we associate the attractive interaction with isoscalar states ($\chi_J < 0$, $T = 0$) and the repulsive interaction with isovector states ($\chi_J > 0$, $T = 1$).

### 9.2.2 The Schematic Model for $T = 0$ and $T = 1$

Let us now see how we can compensate for the omission of the exchange term of (9.45). To that end, we return to the complete equation (9.45) and designate its two terms as $D$ for 'direct' and $E$ for 'exchange':

$$\langle a\, b^{-1}\,;\, J | V_{\mathrm{RES}} | c\, d^{-1}\,;\, J \rangle = D - E \,. \tag{9.58}$$

This contains the proton and neutron terms

$$\langle p_1\, p_2^{-1}\,;\, J | V_{\mathrm{RES}} | p_3\, p_4^{-1}\,;\, J \rangle = D_{\pi\pi} - E_{\pi\pi} \,, \tag{9.59}$$

$$\langle n_1\, n_2^{-1}\,;\, J | V_{\mathrm{RES}} | n_3\, n_4^{-1}\,;\, J \rangle = D_{\nu\nu} - E_{\nu\nu} \,, \tag{9.60}$$

$$\langle p_1\, p_2^{-1}\,;\, J | V_{\mathrm{RES}} | n_3\, n_4^{-1}\,;\, J \rangle = D_{\pi\nu} \,, \tag{9.61}$$

$$\langle n_1\, n_2^{-1}\,;\, J | V_{\mathrm{RES}} | p_3\, p_4^{-1}\,;\, J \rangle = D_{\nu\pi} \,. \tag{9.62}$$

There cannot exist a term of the form $E_{\pi\nu}$ or $E_{\nu\pi}$ since it would contain factors like $(n_3 \| Q_\lambda \| p_1)$, in violation of charge conservation.

In the isospin formalism the particle–hole $T = 0$ and $T = 1$ states with $M_T = 0$ include both proton-particle–proton-hole and neutron-particle–neutron-hole excitations. These states, originally given by (5.126) and (5.127), are

$$|a_1\, a_2^{-1}\,;\, J M\,;\, 0\,0\rangle = \frac{1}{\sqrt{2}} \left( |n_1\, n_2^{-1}\,;\, J M\rangle + |p_1\, p_2^{-1}\,;\, J M\rangle \right) , \tag{9.63}$$

$$|a_1\, a_2^{-1}\,;\, J M\,;\, 1\,0\rangle = \frac{1}{\sqrt{2}} \left( |n_1\, n_2^{-1}\,;\, J M\rangle - |p_1\, p_2^{-1}\,;\, J M\rangle \right) . \tag{9.64}$$

Equations (9.59)–(9.62) now give

$$\langle a\, b^{-1}\,;\, J\,0 | V_{\mathrm{RES}} | c\, d^{-1}\,;\, J\,0 \rangle$$
$$= \tfrac{1}{2}[(D_{\nu\nu} - E_{\nu\nu}) + (D_{\pi\pi} - E_{\pi\pi}) + (D_{\pi\nu} + D_{\nu\pi})] \,, \tag{9.65}$$

$$\langle a\, b^{-1}\,;\, J\,1 | V_{\mathrm{RES}} | c\, d^{-1}\,;\, J\,1 \rangle$$
$$= \tfrac{1}{2}[(D_{\nu\nu} - E_{\nu\nu}) + (D_{\pi\pi} - E_{\pi\pi}) - (D_{\pi\nu} + D_{\nu\pi})] \,. \tag{9.66}$$

With isospin symmetry assumed, we have

$$D_{\pi\pi} = D_{\nu\nu} = D_{\pi\nu} = D_{\nu\pi} \equiv D \,, \quad E_{\pi\pi} = E_{\nu\nu} \equiv E \,, \tag{9.67}$$

so that

$$\langle a\,b^{-1}\,;\,J\,0|V_{\mathrm{RES}}|c\,d^{-1}\,;\,J\,0\rangle = 2D - E\,,\tag{9.68}$$

$$\langle a\,b^{-1}\,;\,J\,1|V_{\mathrm{RES}}|c\,d^{-1}\,;\,J\,1\rangle = -E\,.\tag{9.69}$$

For a typical set $abcd$ it turns out that $E \approx D$. This implies that the matrix elements (9.68) and (9.69) are roughly equal in magnitude and opposite in sign. On general experience from particle–hole spectra, as presented in Sect. 5.4, we expect the $T = 0$ matrix element to be smaller than the $T = 1$ matrix element, which implies

$$\langle a\,b^{-1}\,;\,J\,0|V_{\mathrm{RES}}|c\,d^{-1}\,;\,J\,0\rangle \approx D < 0\,,\tag{9.70}$$

$$\langle a\,b^{-1}\,;\,J\,1|V_{\mathrm{RES}}|c\,d^{-1}\,;\,J\,1\rangle \approx -D > 0\,.\tag{9.71}$$

For an attractive interaction in the direct term (9.48) we have $\chi_J < 0$. According to (9.70) and (9.71) we can then use the direct term with $\chi_J$ to calculate the $T = 0$ matrix element and with $-\chi_J$ to calculate the $T = 1$ matrix element. We adopt this approximation as an essential part of our schematic model.

The rather radical approximations introduced above find their practical justification in the application presented in Subsect. 9.2.4.

### 9.2.3 The Schematic Model with the Surface Delta Interaction

Equation (8.53) and the definition (9.47) provide an explicit expression for $Q_{ab}^{J}$. For the SDI, with details and notation from Subsect. 8.2.2, it becomes

$$Q_{ab}^{J} = \frac{|\kappa_{ab}|\sqrt{-V_0}}{4\sqrt{\pi}}(-1)^{n_a+n_b}(-1)^{j_b+J-\frac{1}{2}}\left[1 + (-1)^{l_a+l_b+J}\right]\hat{j}_a\hat{j}_b\begin{pmatrix} j_a & j_b & J \\ \frac{1}{2} & -\frac{1}{2} & 0 \end{pmatrix}\,.\tag{9.72}$$

If we formed $Q_{ab}^{J}Q_{cd}^{J}$ with (9.72), the constant in front would become the SDI constant $-\frac{1}{4}A_T$ according to the replacement (8.70). However, from Eq. (9.44) on we have removed the strength constant from the reduced matrix elements of $\boldsymbol{Q}_\lambda$ and included the external strength constant $\chi_\lambda$. Thus we now adopt

$$\boxed{Q_{ab}^{J}(\mathrm{SDI}) = (-1)^{n_a+n_b}(-1)^{j_b+J-\frac{1}{2}}\left[1 + (-1)^{l_a+l_b+J}\right]\hat{j}_a\hat{j}_b\begin{pmatrix} j_a & j_b & J \\ \frac{1}{2} & -\frac{1}{2} & 0 \end{pmatrix}\,.}\tag{9.73}$$

With $D$ given by (9.48), the particle–hole matrix elements (9.70) and (9.71) become

$$\langle a\,b^{-1}\,;\,J\,T|V_{\mathrm{RES}}|c\,d^{-1}\,;\,J\,T\rangle = -\tfrac{1}{4}A_T Q_{ab}^{J}(\mathrm{SDI})Q_{cd}^{J}(\mathrm{SDI})\,,\tag{9.74}$$

where

$$A_0 \equiv A > 0\,,\quad A_1 = -A\,.\tag{9.75}$$

Altogether the SDI matrix elements of our schematic model are thus

$$\boxed{\begin{aligned}\langle a\,b^{-1}\,;\,J\,0|V_{\text{RES}}|c\,d^{-1}\,;\,J\,0\rangle &= -\tfrac{1}{4}AQ_{ab}^{J}(\text{SDI})Q_{cd}^{J}(\text{SDI})\,,\\\langle a\,b^{-1}\,;\,J\,1|V_{\text{RES}}|c\,d^{-1}\,;\,J\,1\rangle &= +\tfrac{1}{4}AQ_{ab}^{J}(\text{SDI})Q_{cd}^{J}(\text{SDI})\,.\end{aligned}} \tag{9.76}$$

Comparison with (9.48) shows that we may write

$$\chi_J^{\text{SDI}}(T=0) = -\tfrac{1}{4}A\,, \qquad \chi_J^{\text{SDI}}(T=1) = +\tfrac{1}{4}A\,. \tag{9.77}$$

The formulas of this subsection are used in the following example to find the energies of $1^-$ states in $^4$He by means of the dispersion equation (9.55).

### 9.2.4 Application to $1^-$ Excitations in $^4$He

Consider the $1^-$ excitation spectrum of $^4_2\text{He}_2$ within the 0s-0p valence space. This nucleus is doubly magic, its ground state being the particle–hole vacuum $|\text{HF}\rangle$. Particle–hole nuclei were treated without configuration mixing in Sect. 5.4, with the experimental excitation spectrum of $^4$He shown in Fig. 5.9.

The energy gap between the two relevant major shells, 0s and 0p, can be obtained from the empirical formula

$$\Delta\varepsilon(\text{0p-0s}) \approx 33A^{-1/3}\,\text{MeV}\,. \tag{9.78}$$

From this we take $\Delta\varepsilon = 21.0\,\text{MeV}$. For the energy difference within the 0p shell we take $\varepsilon_{0p_{1/2}} - \varepsilon_{0p_{3/2}} = 6.0\,\text{MeV}$. The energies involved are also shown in the schematic drawing of Fig. 3.3.

The possible $1^-$ particle–hole excitations in our chosen valence space are

$$\{|1\rangle\,,\,|2\rangle\} = \{|0p_{3/2}\,(0s_{1/2})^{-1}\,;\,1^-\rangle\,,\,|0p_{1/2}\,(0s_{1/2})^{-1}\,;\,1^-\rangle\}\,, \tag{9.79}$$

for both protons and neutrons. The corresponding particle–hole energies $\varepsilon_{ab}$ are $\varepsilon_1 = 21.0\,\text{MeV}$ and $\varepsilon_2 = 27.0\,\text{MeV}$. Inserting the values of the $3j$ symbols into (9.73) gives

$$Q^1_{0p_{3/2}0s_{1/2}}(\text{SDI}) = \frac{4}{\sqrt{3}}\,, \qquad Q^1_{0p_{1/2}0s_{1/2}}(\text{SDI}) = -2\sqrt{\frac{2}{3}}\,. \tag{9.80}$$

Substituting into the dispersion equation (9.55) results in

$$-\frac{1}{\mp\tfrac{1}{4}A} = \frac{\tfrac{16}{3}}{21.0\,\text{MeV} - E} + \frac{\tfrac{8}{3}}{27.0\,\text{MeV} - E}\,, \tag{9.81}$$

where the upper sign is for $T = 0$ and the lower for $T = 1$. Simplifying, we obtain

$$\pm\frac{1}{A} = \frac{\tfrac{4}{3}}{21.0\,\text{MeV} - E} + \frac{\tfrac{2}{3}}{27.0\,\text{MeV} - E}\,. \tag{9.82}$$

**Table 9.5.** TDA eigenenergies $E_n$ of $J^\pi = 1^-$, $T = 0$ particle–hole states in $^4$He obtained by graphical solution of the schematic model and by exact diagonalization

| $A$ (MeV) | $E_1$ (MeV) | | $E_2$ (MeV) | |
|-----------|-------------|-------|-------------|-------|
| | Schematic | Exact | Schematic | Exact |
| 1.0 | 19.5 | 18.877 | 26.5 | 27.123 |
| 2.0 | 17.9 | 16.255 | 26.1 | 27.745 |

The SDI is used for both.

**Table 9.6.** TDA eigenenergies $E_n$ of $J^\pi = 1^-$, $T = 1$ particle–hole states in $^4$He obtained by graphical solution of the schematic model and by exact diagonalization

| $A$ (MeV) | $E_1$ (MeV) | | $E_2$ (MeV) | |
|-----------|-------------|-------|-------------|-------|
| | Schematic | Exact | Schematic | Exact |
| 1.0 | 22.2 | 22.000 | 27.8 | 28.000 |
| 2.0 | 23.0 | 23.000 | 29.0 | 29.000 |

The SDI is used for both.

The two equations (9.82) are solved graphically in Fig. 9.1. The left-hand sides (horizontal lines) and the common right-hand side are plotted as functions of $E$ for two values of $A$: $A = 1.0$ MeV and $A = 2.0$ MeV. The abscissas of the intersection points give the solutions $E_n$, separately for each $T$. The solutions read off the figure[2] are collected into Tables 9.5 and 9.6, and there compared with exact solutions obtained by numerical diagonalization in Subsect. 9.3.2.

Figure 9.1 confirms the statements about collectivity made in the last paragraph of Subsect. 9.2.1. Even in the realistic case where the single-particle energies are not degenerate, the solutions other than the lowest $T = 0$ and the highest $T = 1$ are confined between adjacent vertical lines and therefore do not change very much when $A$ is increased. In contrast, the lowest $T = 0$ solution is free to move to the left and the highest $T = 1$ solution to the right with increasing $A$. The collectivity appears not only in the energy spectrum but expressly in the electromagnetic properties, as discussed later in this chapter.

The collective $J^\pi = 1^-$, $T = 1$ state is known as the *giant dipole resonance*. It is well known experimentally, and realistic TDA calculations have been performed to reproduce its energy and dipole strength [14].

As can be seen from Tables 9.5 and 9.6, the solutions of the schematic model are surprisingly close to the exact solutions. However, varying the value of $A$ shows that the schematic description of the collective states deteriorates with increasing $A$.

---

[2] Because of the small valence space, equations (9.82) are of only second order in $E$ and can therefore be solved also analytically. The graphical solutions in Tables 9.5 and 9.6 have been thus checked.

## 9.3 Excitation Spectra of Doubly Magic Nuclei

In the following we apply the TDA method to calculate excitation energies of doubly magic nuclei by matrix diagonalization.

### 9.3.1 Block Decomposition of the TDA Matrix

As discussed in the previous section, particle–hole excitations have both proton and neutron contributions, i.e. we have both $pp^{-1}$ and $nn^{-1}$ excitations. Bearing this in mind, we can write the general form of a TDA matrix in a block form as

$$\mathsf{H}_{\mathrm{TDA}} = \begin{pmatrix} \mathsf{H}(pp^{-1} - pp^{-1}) & \mathsf{V}(pp^{-1} - nn^{-1}) \\ \mathsf{V}(nn^{-1} - pp^{-1}) & \mathsf{H}(nn^{-1} - nn^{-1}) \end{pmatrix} . \tag{9.83}$$

The proton–proton block $\mathsf{H}(pp^{-1} - pp^{-1})$ consists of the matrix elements

$$(\varepsilon_{p_1} - \varepsilon_{p_2})\delta_{p_1 p_3}\delta_{p_2 p_4} + \langle p_1\, p_2^{-1}\,;\, J | V_{\mathrm{RES}} | p_3\, p_4^{-1}\,;\, J \rangle . \tag{9.84}$$

The Pandya transformation (9.22) decomposes $\langle p_1\, p_2^{-1}\,;\, J | V_{\mathrm{RES}} | p_3\, p_4^{-1}\,;\, J \rangle$ into a sum of terms proportional to the two-particle matrix element

$$\langle p_1\, p_4\,;\, J' | V | p_3\, p_2\,;\, J' \rangle . \tag{9.85}$$

The neutron–neutron block is the same except that the proton labels are replaced with neutron labels. These two blocks have necessarily isospin $T = 1$.

The proton–neutron block $\mathsf{V}(pp^{-1} - nn^{-1})$ consists of the matrix elements

$$\langle p_1\, p_2^{-1}\,;\, J | V_{\mathrm{RES}} | n_3\, n_4^{-1}\,;\, J \rangle \tag{9.86}$$

and it decomposes into a sum of terms proportional to the two-particle matrix element

$$\langle p_1\, n_4\,;\, J' | V | n_3\, p_2\,;\, J' \rangle . \tag{9.87}$$

The neutron–proton block is the same as the proton–neutron block, only the labels $\pi$ and $\nu$ are interchanged. Equation (8.27) gives the matrix elements (9.87) in terms of the isospins $T = 1$ and $T = 0$.

Equations (9.28)–(9.30) give the particle–hole matrix elements in terms of the auxiliary quantity $\mathcal{M}_{abcd}(JT)$. This quantity is tabulated for various particle–hole valence spaces in Tables 9.1–9.4. Next we discuss some examples of the construction and diagonalization of TDA Hamiltonian matrices.

### 9.3.2 Application to $1^-$ States in $^4$He

Consider the $1^-$ excitations of the nucleus $^4_2\mathrm{He}_2$ in the particle–hole valence space $(0\mathrm{p}_{3/2}\text{-}0\mathrm{p}_{1/2})\text{-}(0\mathrm{s}_{1/2})^{-1}$ with the single-particle energies $\varepsilon_{0\mathrm{s}_{1/2}} = 0$,

$\varepsilon_{0p_{3/2}} = 21.0\,\mathrm{MeV}$ and $\varepsilon_{0p_{1/2}} = 27.0\,\mathrm{MeV}$. This is exactly the valence space of the example of Subsect. 9.2.4, where the same problem was discussed within the schematic model. The basis states are

$$\{|\pi_1\rangle, |\pi_2\rangle, |\nu_1\rangle, |\nu_2\rangle\}$$
$$= \{|\pi 0p_{3/2}\,(\pi 0s_{1/2})^{-1}\,;\,1^-\rangle,\ |\pi 0p_{1/2}\,(\pi 0s_{1/2})^{-1}\,;\,1^-\rangle,$$
$$|\nu 0p_{3/2}\,(\nu 0s_{1/2})^{-1}\,;\,1^-\rangle,\ |\nu 0p_{1/2}\,(\nu 0s_{1/2})^{-1}\,;\,1^-\rangle\}\,. \tag{9.88}$$

The TDA matrix (9.83) becomes

$$\mathsf{H}_{\mathrm{TDA}}(1^-) = \begin{pmatrix} \varepsilon_{\pi_1} + V_{\pi_1\pi_1} & V_{\pi_1\pi_2} & V_{\pi_1\nu_1} & V_{\pi_1\nu_2} \\ V_{\pi_2\pi_1} & \varepsilon_{\pi_2} + V_{\pi_2\pi_2} & V_{\pi_2\nu_1} & V_{\pi_2\nu_2} \\ V_{\nu_1\pi_1} & V_{\nu_1\pi_2} & \varepsilon_{\nu_1} + V_{\nu_1\nu_1} & V_{\nu_1\nu_2} \\ V_{\nu_2\pi_1} & V_{\nu_2\pi_2} & V_{\nu_2\nu_1} & \varepsilon_{\nu_2} + V_{\nu_2\nu_2} \end{pmatrix}, \tag{9.89}$$

where

$$\varepsilon_{\pi_1} = \varepsilon_{\pi 0p_{3/2}} - \varepsilon_{\pi 0s_{1/2}} = 21.0\,\mathrm{MeV}\,, \quad \varepsilon_{\pi_2} = \varepsilon_{\pi 0p_{1/2}} - \varepsilon_{\pi 0s_{1/2}} = 27.0\,\mathrm{MeV}\,,$$
$$\varepsilon_{\nu_1} = \varepsilon_{\nu 0p_{3/2}} - \varepsilon_{\nu 0s_{1/2}} = 21.0\,\mathrm{MeV}\,, \quad \varepsilon_{\nu_2} = \varepsilon_{\nu 0p_{1/2}} - \varepsilon_{\nu 0s_{1/2}} = 27.0\,\mathrm{MeV}\,.$$
$$\tag{9.90}$$

Using now Eqs. (9.28)–(9.30) and Table 9.1 we obtain, with single-particle energies in MeV and rounding off the two-body matrix elements to three decimals,

$$\mathsf{H}_{\mathrm{TDA}}(1^-) =$$
$$\begin{pmatrix} 21.0 - 0.333A_1 & 0.943A_1 & \frac{1}{2}(-0.333A_1 - 2.333A_0) & \frac{1}{2}(0.943A_1 + 0.943A_0) \\ 0.943A_1 & 27.0 + 0.333A_1 & \frac{1}{2}(0.943A_1 + 0.943A_0) & \frac{1}{2}(0.333A_1 - 1.667A_0) \\ \cdots & \cdots & 21.0 - 0.333A_1 & 0.943A_1 \\ \cdots & \cdots & 0.943A_1 & 27.0 + 0.333A_1 \end{pmatrix},$$
$$\tag{9.91}$$

where the dots represent the matrix elements symmetric with those of the proton–neutron block; the TDA matrix is symmetric, as stated in (9.24). By diagonalizing for $A_0 = A_1 = 1.0\,\mathrm{MeV}$ and $A_0 = A_1 = 2.0\,\mathrm{MeV}$ we obtain the eigenenergies quoted as 'exact' in Tables 9.5 and 9.6. The corresponding diagonal elements in the proton block and in the neutron block are the same because we have the same single-particle energies and the same interactions for both kinds of nucleon. The eigenstates have therefore good isospin, but the isospin assignment is not obvious.

The calculation of the $1^-$ states of $^4$He can be done also in the isospin formalism. This choice is preferable in general because it gives an immediate isospin identification and smaller matrices. For $^4$He we obtain two 2-by-2 matrices, one for $T = 0$ and the other for $T = 1$. The one-body part is given by (9.17) and the two-body part by the relations (9.31) and (9.32) and Table 9.1. The basis is

$$\{|1\rangle, |2\rangle\} = \{|0p_{3/2}(0s_{1/2})^{-1}; 1^- T\rangle, |0p_{1/2}(0s_{1/2})^{-1}; 1^- T\rangle\}. \quad (9.92)$$

The matrices become

$$H_{TDA}(1^-, T = 0) =$$
$$\begin{pmatrix} 21.0 + \frac{1}{2}[3(-0.3333A_1) - 2.3333A_0] & \frac{1}{2}[3(0.9428A_1) - (-0.9428A_0)] \\ \frac{1}{2}[3(0.9428A_1) - (-0.9428A_0)] & 27.0 + \frac{1}{2}[3(0.3333A_1) - 1.6667A_0] \end{pmatrix}, \quad (9.93)$$

$$H_{TDA}(1^-, T = 1) =$$
$$\begin{pmatrix} 21.0 + \frac{1}{2}(-0.3333A_1 + 2.3333A_0) & \frac{1}{2}(0.9428A_1 - 0.9428A_0) \\ \frac{1}{2}(0.9428A_1 - 0.9428A_0) & 27.0 + \frac{1}{2}(0.3333A_1 + 1.6667A_0) \end{pmatrix}. \quad (9.94)$$

The eigenvalues of the matrices (9.93) and (9.94) are quoted in Tables 9.5 and 9.6 respectively. We notice that for $A_0 = A_1 \equiv A$ the $T = 1$ matrix (9.94) happens to be diagonal, which results in the energies $E_1(T = 1) = 21.0\,\text{MeV} + A$ and $E_2(T = 1) = 27.0\,\text{MeV} + A$. This explains the 'round' figures in Table 9.6.

### 9.3.3 Application to Excited States in $^{16}$O

In our second example we consider particle–hole excitations across the $N = Z = 8$ magic shell gap in $^{16}_8$O$_8$. We take this gap from experiment as $11.6\,\text{MeV}$.[3] Our energies within the 0p and 0d-1s shells are those of Fig. 9.2 (a), for both protons and neutrons.

Let us compute the energies of the $2^-$ and $3^-$ states and compare them with experiment. The smallest possible particle–hole valence space for them is $0d_{5/2}\text{-}(0p_{1/2})^{-1}$. It contains the basis states

$$\{|\pi_1\rangle, |\nu_1\rangle\} = \{|\pi 0d_{5/2}(\pi 0p_{1/2})^{-1}; J^-\rangle, |\nu 0d_{5/2}(\nu 0p_{1/2})^{-1}; J^-\rangle\}. \quad (9.95)$$

Equations (9.28)–(9.30) and Table 9.2 give the following matrices:

$$H_{TDA}(2^-) = \begin{pmatrix} 11.6 + 1.0000A_1 & \frac{1}{2}(1.0000A_1 - 1.0000A_0) \\ \frac{1}{2}(1.0000A_1 - 1.0000A_0) & 11.6 + 1.0000A_1 \end{pmatrix}, \quad (9.96)$$

$$H_{TDA}(3^-) = \begin{pmatrix} 11.6 + 0.1429A_1 & \frac{1}{2}(0.1429A_1 - 1.8571A_0) \\ \frac{1}{2}(0.1429A_1 - 1.8571A_0) & 11.6 + 0.1429A_1 \end{pmatrix}. \quad (9.97)$$

Let us take $A_0 = A_1 = 1.0\,\text{MeV}$. The two $2^-$ states given by (9.96) are then degenerate at the energy $E(2^-) = 12.600\,\text{MeV}$. Diagonalizing the matrix (9.97) we obtain $E_1(3^-) = 10.886\,\text{MeV}$ and $E_2(3^-) = 12.600\,\text{MeV}$. The energies yielded by (9.96) and (9.97) are shown in the left-hand column of Fig. 9.3.

---

[3] The rule-of-thumb formula (9.78) gives $13.1\,\text{MeV}$, while Fig. 3.3 shows $\approx 12\,\text{MeV}$.

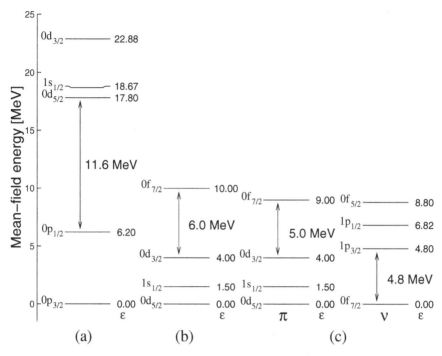

**Fig. 9.2.** Single-particle energies of the particle–hole valence spaces used to calculate spectra of the doubly magic nuclei $^{16}$O (a), $^{40}$Ca (b) and $^{48}$Ca (c). See also the rough scheme of Fig. 3.3

We have proceeded to calculate the energies of the 2⁻ and 3⁻ states in successively larger particle–hole valence spaces. The resulting spectra are shown in Fig. 9.3. In all these spectra every 2⁻ state is doubly degenerate.

Figure 9.3 shows that the TDA fails to bring the lowest 3⁻ state low enough for good agreement with experiment. This is understood to result from an insufficient collectivity of the lowest TDA 3⁻ state. The RPA introduced in Chap. 11 improves the description. However, the presence of collectivity in the lowest TDA states is borne out by their electromagnetic decay rates, to be discussed in the next section.

### 9.3.4 Further Examples: $^{40}$Ca and $^{48}$Ca

Similarly to the previous examples of $^{4}_{2}$He$_2$ and $^{16}_{8}$O$_8$, we have calculated TDA spectra for $^{40}_{20}$Ca$_{20}$ and $^{48}_{20}$Ca$_{28}$. Details of the calculations are omitted; only their results are displayed in Figs. 9.4 and 9.5. The single-particle energies used in these calculations are given in panels (b) and (c) of Fig. 9.2.

Figure 9.4 for $^{40}$Ca shows the TDA spectrum for the minimal hole space $(0d_{3/2})^{-1}$ and for the complete 0d-1s hole space. Use of the larger space greatly

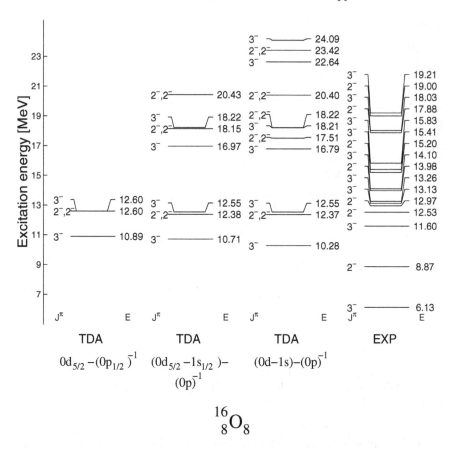

**Fig. 9.3.** TDA energies of the $2^-$ and $3^-$ states in $^{16}$O computed in successively larger valence spaces built from the single-particle levels of Fig. 9.2 (a), together with the experimental $2^-$ and $3^-$ levels. The SDI was used with parameters $A_0 = A_1 = 1.0$ MeV

improves the quality of the spectrum. The lowest $3^-$ and $5^-$ states drop considerably towards their experimental counterparts. However, only the RPA spectrum, shown for completeness, contains a good description of the $3^-$ and $5^-$ energies.

Figure 9.5 shows the calculated TDA and RPA spectra of $^{48}$Ca. The particle–hole valence space used in these calculations is shown in Fig. 9.2 (c). Here the protons have particle–hole excitations from the 0d-1s shell to the $0f_{7/2}$ orbital, across the $Z = 20$ magic gap, which produced the theoretical negative-parity states. The neutrons have particle–hole excitations from the $0f_{7/2}$ orbital to the 1p-$0f_{5/2}$ shell, across the $N = 28$ magic gap, which produced the theoretical positive-parity levels. There is no isospin symmetry

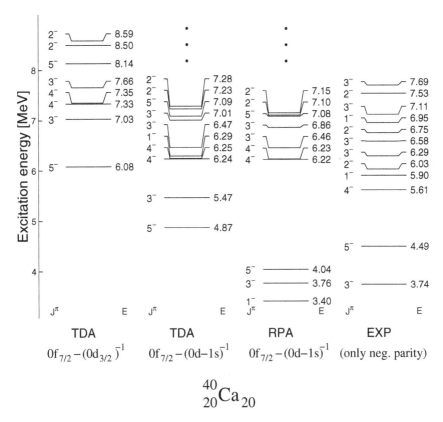

**Fig. 9.4.** TDA spectra of $^{40}$Ca in two different particle–hole valence spaces and the RPA spectrum in the larger valence space, together with the experimental spectrum for negative-parity states. The single-particle energies are those of Fig. 9.2 (b). The SDI was used with parameters $A_0 = 0.85\,\text{MeV}$ and $A_1 = 0.90\,\text{MeV}$

because of the completely different proton and neutron particle–hole spaces; this was noted at the end of Sect. 5.4.

The TDA and RPA spectra are not very different; both compare rather well with the experimental spectrum. No $0^+$ state can be produced in the chosen particle–hole valence space, so the $0^+$ states of the experimental spectrum must have another, more complicated structure.

## 9.4 Electromagnetic Transitions in Doubly Magic Nuclei

In this section we discuss the effects of configuration mixing on the electromagnetic decay properties of particle–hole states. These decays were discussed in Subsect. 6.4.1 with no configuration mixing. The transitions can be divided into two categories: ground-state transitions, involving only one TDA wave

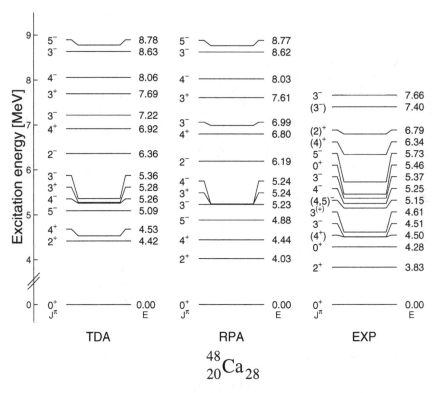

**Fig. 9.5.** TDA and RPA spectra of $^{48}$Ca in the particle–hole valence space of Fig. 9.2 (c), together with the experimental spectrum. The SDI was used with parameters $A_0 = A_1 = 1.0\,\mathrm{MeV}$

function, and transitions between two TDA states. We start by discussing transitions of the first category. The reduced transition probabilities and half-lives can be obtained with the formulas and tables of Sect. 6.1.

### 9.4.1 Transitions to the Particle–Hole Ground State

In Subsect. 6.4.1 we discussed the electromagnetic decay of a single particle–hole excitation to the particle–hole vacuum. In particular we established, as Eq. (6.119), that in the CS phase convention

$$(\mathrm{HF}\|\mathcal{M}_{\sigma\lambda}\|a_i\,b_i^{-1}\,;\,J_i) = \delta_{\lambda J_i}(-1)^{J_i}(a_i\|\mathcal{M}_{\sigma\lambda}\|b_i)\,,\qquad(9.98)$$

where $\mathcal{M}_{\sigma\lambda}$ is the electromagnetic multipole operator (6.5) of the electric, $\sigma = \mathrm{E}$, or magnetic, $\sigma = \mathrm{M}$, type. The particle–hole state in (9.98) is

$$|a_i\,b_i^{-1}\,;\,J_i\,M_i\rangle = \left[c_{a_i}^\dagger h_{b_i}^\dagger\right]_{J_i M_i}|\mathrm{HF}\rangle\,,\qquad(9.99)$$

where $a_i$ and $b_i$ are both either proton or neutron labels.

The TDA wave function is given by (9.34) as a linear combination of the particle–hole excitations (9.99). Thus we can immediately write the TDA decay amplitude as[4]

$$
(\mathrm{HF}\|\boldsymbol{\mathcal{M}}_{\sigma\lambda}\|\Psi_\omega^{(\mathrm{TDA})}) \equiv (\mathrm{HF}\|\boldsymbol{\mathcal{M}}_{\sigma\lambda}\|\omega) = \delta_{\lambda J}(-1)^J \sum_{ab} X_{ab}^\omega (a\|\boldsymbol{\mathcal{M}}_{\sigma\lambda}\|b) \,,
$$

$$(9.100)$$

where $\omega$ stands for $nJ^\pi$. The reduced transition probability is

$$
B(\sigma\lambda\,;\,\omega \to 0_{\mathrm{gs}}^+)_{\mathrm{TDA}} = \delta_{\lambda J}\widehat{J}^{-2}\big|(\mathrm{HF}\|\boldsymbol{\mathcal{M}}_{\sigma J}\|\omega)\big|^2 \,. \tag{9.101}
$$

For transitions in the opposite direction, the squared amplitude $\big|(\omega\|\boldsymbol{\mathcal{M}}_{\sigma\lambda}\|\mathrm{HF})\big|^2$ is called the *decay strength* to the final state $|\omega\rangle$.

### 9.4.2 Non-Energy-Weighted Sum Rule

The summed decay strength to all final states reached by the operator $\boldsymbol{\mathcal{M}}_{\sigma\lambda}$ from the particle–hole ground state is

$$
\sum_n \big|(\omega\|\boldsymbol{\mathcal{M}}_{\sigma\lambda}\|\mathrm{HF})\big|^2 = \sum_n \big|(\mathrm{HF}\|\boldsymbol{\mathcal{M}}_{\sigma\lambda}\|\omega)\big|^2 = \sum_n \Big|\sum_{ab} X_{ab}^\omega (a\|\boldsymbol{\mathcal{M}}_{\sigma\lambda}\|b)\Big|^2
$$

$$
= \sum_n \sum_{\substack{ab \\ a'b'}} X_{ab}^\omega (X_{a'b'}^\omega)^* (a\|\boldsymbol{\mathcal{M}}_{\sigma\lambda}\|b)(a'\|\boldsymbol{\mathcal{M}}_{\sigma\lambda}\|b')^*
$$

$$
= \sum_{ab} \big|(a\|\boldsymbol{\mathcal{M}}_{\sigma\lambda}\|b)\big|^2 \,, \tag{9.102}
$$

where the TDA completeness relation (9.43) was used in the last step. Spelling out $\omega$ and taking into account the Kronecker delta in (9.100) we then have the *TDA sum rule*

$$
\sum_n \big|(n\,\lambda^\pi\|\boldsymbol{\mathcal{M}}_{\sigma\lambda}\|\mathrm{HF})\big|^2 = \sum_{ab} \big|(a\|\boldsymbol{\mathcal{M}}_{\sigma\lambda}\|b)\big|^2 \,. \tag{9.103}
$$

The left-hand side of (9.103) depends on the TDA wave functions, whereas the right-hand side depends only on the chosen valence space and the characteristics of the transition operator. By computing each side of the equation we have a check on the correctness of our TDA wave functions. This sum rule is an example of the so-called *non-energy-weighted sum rule* (NEWSR).

For electric transitions with effective charges (6.26) we obtain from (9.103) the expression

---

[4] The phase factor $(-1)^J$ is replaced by $+1$ for $\sigma = \mathrm{E}$ and $-1$ for $\sigma = \mathrm{M}$ in the Biedenharn-Rose phase convention.

$$
\begin{aligned}
\sum_k \left| (k\,\lambda^\pi \| \boldsymbol{Q}_\lambda \| \mathrm{HF}) \right|^2 &= (e^{\mathrm{p}}_{\mathrm{eff}}/e)^2 \sum_{pp'} \left| (p\|\boldsymbol{Q}_\lambda\|p') \right|^2 \\
&\quad + (e^{\mathrm{n}}_{\mathrm{eff}}/e)^2 \sum_{nn'} \left| (n\|\boldsymbol{Q}_\lambda\|n') \right|^2 ,
\end{aligned}
\tag{9.104}
$$

where $p$ and $n$ are particle indices and $p'$ and $n'$ hole indices. The single-particle matrix elements are given by (6.23). With substitution from (9.100) the left-hand side of (9.104) becomes

$$
\begin{aligned}
\sum_k \left| (k\,\lambda^\pi \| \boldsymbol{Q}_\lambda \| \mathrm{HF}) \right|^2 &= \frac{1}{e^2} \sum_k \left| e^{\mathrm{p}}_{\mathrm{eff}} \sum_{pp'} X^{k\lambda^\pi}_{pp'} (p\|\boldsymbol{Q}_\lambda\|p') \right. \\
&\quad \left. + e^{\mathrm{n}}_{\mathrm{eff}} \sum_{nn'} X^{k\lambda^\pi}_{nn'} (n\|\boldsymbol{Q}_\lambda\|n') \right|^2 .
\end{aligned}
\tag{9.105}
$$

Next we discuss some examples to clarify the notion of a collective state and the related decay strength.

### 9.4.3 Application to Octupole Transitions in $^{16}$O

Consider the electric octupole (E3) transitions from the $3^-$ states of $^{16}_8$O$_8$ to the $0^+$ ground state. We calculate the transitions in the three successively larger particle–hole valence spaces of Fig. 9.3 with the single-particle energies of Fig. 9.2 (a). For the SDI we use $A_0 = A_1 = 1.0\,\mathrm{MeV}$ as in Fig. 9.3. The oscillator length for $A = 16$, $b = 1.725\,\mathrm{fm}$, we have from Subsect. 6.4.2.

### Valence space $0\mathrm{d}_{5/2}$-$(0\mathrm{p}_{1/2})^{-1}$

The smallest particle–hole valence space is $0\mathrm{d}_{5/2}$-$(0\mathrm{p}_{1/2})^{-1}$. Its basis states are given by (9.95) and the $3^-$ matrix by (9.97). The wave functions of the two $3^-$ states are[5]

$$
|^{16}\mathrm{O},\, 3^-_1\rangle = \frac{1}{\sqrt{2}}(|\pi_1\rangle + |\nu_1\rangle) ,
\tag{9.106}
$$

$$
|^{16}\mathrm{O},\, 3^-_2\rangle = \frac{1}{\sqrt{2}}(|\pi_1\rangle - |\nu_1\rangle) .
\tag{9.107}
$$

The state (9.106) can be identified as the $T = 0$ state (6.126), while (9.107) is its $T = 1$ companion.

We apply (9.100) with effective charges. The one single-particle matrix element needed, $(0\mathrm{d}_{5/2}\|\boldsymbol{Q}_3\|0\mathrm{p}_{1/2})$, is found from Table 6.5 and is directly quoted in (6.128). Thus we have

---

[5] It can be seen from the equations of Subsect. 8.3.2 that these wave functions are exact for the given space.

$$(\mathrm{HF}\|\boldsymbol{Q}_3\|3_1^-) = (-1)^3\frac{1}{\sqrt{2}}(-3.824)(e_{\mathrm{eff}}^{\mathrm{p}} + e_{\mathrm{eff}}^{\mathrm{n}})b^3 = 2.704e_+b^3 = 13.88e_+\ \mathrm{fm}^3\ ,$$

$$(9.108)$$

where we have abbreviated

$$e_\pm \equiv e_{\mathrm{eff}}^{\mathrm{p}} \pm e_{\mathrm{eff}}^{\mathrm{n}}\ .$$

$$(9.109)$$

Similarly we find

$$(\mathrm{HF}\|\boldsymbol{Q}_3\|3_2^-) = 13.88e_-\ \mathrm{fm}^3\ .$$

$$(9.110)$$

The left-hand side of the sum rule (9.104) is now

$$\sum_{k=1}^2 \left|(3_k^-\|\boldsymbol{Q}_3\|\mathrm{HF})\right|^2 = 13.88^2(e_+^2 + e_-^2)\ \mathrm{fm}^6 = 385.3[(e_{\mathrm{eff}}^{\mathrm{p}})^2 + (e_{\mathrm{eff}}^{\mathrm{n}})^2]\ \mathrm{fm}^6\ .$$

$$(9.111)$$

On the other hand, the right-hand side of (9.104) becomes

$$(e_{\mathrm{eff}}^{\mathrm{p}}/e)^2\left|(0\mathrm{d}_{5/2}\|\boldsymbol{Q}_3\|0\mathrm{p}_{1/2})\right|^2 + (e_{\mathrm{eff}}^{\mathrm{n}}/e)^2\left|(0\mathrm{d}_{5/2}\|\boldsymbol{Q}_3\|0\mathrm{p}_{1/2})\right|^2$$
$$= [(e_{\mathrm{eff}}^{\mathrm{p}})^2 + (e_{\mathrm{eff}}^{\mathrm{n}})^2](-3.824b^3)^2 = 385.3[(e_{\mathrm{eff}}^{\mathrm{p}})^2 + (e_{\mathrm{eff}}^{\mathrm{n}})^2]\ \mathrm{fm}^6\ . \quad (9.112)$$

Thus we see that the TDA sum rule is satisfied.

We finally note that this calculation substantially repeats the $3_1^-$ part of Subsect. 6.4.2 and the qualitative considerations of Subsect. 6.5.2, including Fig. 6.2.

## Valence space $(0\mathrm{d}_{5/2}\text{-}1\mathrm{s}_{1/2})\text{-}(0\mathrm{p})^{-1}$

We extend the previous example to the $(0\mathrm{d}_{5/2}\text{-}1\mathrm{s}_{1/2})\text{-}(0\mathrm{p})^{-1}$ particle–hole space. The $J^\pi = 3^-$ basis states are then[6]

$$\{|\pi_1\rangle,\ |\pi_2\rangle,\ |\nu_1\rangle,\ |\nu_2\rangle\}$$
$$= \{|\pi 0\mathrm{d}_{5/2}\,(\pi 0\mathrm{p}_{3/2})^{-1}\,;\,3^-\rangle,\ |\pi 0\mathrm{d}_{5/2}\,(\pi 0\mathrm{p}_{1/2})^{-1}\,;\,3^-\rangle,$$
$$|\nu 0\mathrm{d}_{5/2}\,(\nu 0\mathrm{p}_{3/2})^{-1}\,;\,3^-\rangle,\ |\nu 0\mathrm{d}_{5/2}\,(\nu 0\mathrm{p}_{1/2})^{-1}\,;\,3^-\rangle\}\ . \quad (9.113)$$

Proceeding as in Sect. 9.3 we form and diagonalize the Hamiltonian matrix in this basis. The lowest $3^-$ state becomes

$$|{}^{16}\mathrm{O}\,,\,3_1^-\rangle = 0.117|\pi_1\rangle + 0.697|\pi_2\rangle + 0.117|\nu_1\rangle + 0.697|\nu_2\rangle\ . \quad (9.114)$$

Note that the corresponding proton and neutron amplitudes are the same, as a result of the assumed proton–neutron symmetry and concomitant good isospin.

Using (9.100) and Table 6.5 we obtain

---

[6] Note that $1\mathrm{s}_{1/2}$ does not contribute.

$$(HF\|\mathbf{Q}_3\|3_1^-) = 3.065 e_+ b^3 = 15.73 e_+ \text{ fm}^3 . \qquad (9.115)$$

The decay strength to the $3_1^-$ state is thus

$$\left|(3_1^-\|\mathbf{Q}_3\|HF)\right|^2 = 247.4 e_+^2 \text{ fm}^6 . \qquad (9.116)$$

The four-dimensional basis (9.113) gives three more $3^-$ states. The results of the complete calculation, both the energies and the decay strengths, are given in Table 9.7.[7] The charge dependence of the transitions, as discussed in Subsect. 6.5.2, indicates that the $3_1^-$ and $3_3^-$ states have $T = 0$ while the $3_2^-$ and $3_4^-$ states have $T = 1$.

**Table 9.7.** TDA energies and octupole strengths of $3^-$ states in $^{16}$O for particle–hole valence space $(0d_{5/2}\text{-}1s_{1/2})\text{-}(0p)^{-1}$

| $k$ | 1 | 2 | 3 | 4 |
|---|---|---|---|---|
| $E_k$ (MeV) | 10.714 | 12.554 | 16.972 | 18.218 |
| $\left|(3_k^-\|\mathbf{Q}_3\|HF)\right|^2$ (fm$^6$) | $248.0 e_+^2$ | $161.3 e_-^2$ | $98.9 e_+^2$ | $185.6 e_-^2$ |

Note abbreviations $e_\pm \equiv e_{\text{eff}}^{\text{p}} \pm e_{\text{eff}}^{\text{n}}$.

Let us check whether the decay strengths in Table 9.7 satisfy the TDA sum relation (9.104). Its left-hand side gives

$$\sum_{k=1}^{4}\left|(3_k^-\|\mathbf{Q}_3\|HF)\right|^2 = (248.0 + 98.9) e_+^2 \text{ fm}^6 + (161.3 + 185.6) e_-^2 \text{ fm}^6$$

$$= (346.9 e_+^2 + 346.9 e_-^2) \text{ fm}^6 = 693.8[(e_{\text{eff}}^{\text{p}})^2 + (e_{\text{eff}}^{\text{n}})^2] \text{ fm}^6 . \qquad (9.117)$$

The right-hand side of (9.104) gives

$$(e_{\text{eff}}^{\text{p}})^2[(-3.420 b^3)^2 + (-3.824 b^3)^2] + (e_{\text{eff}}^{\text{n}})^2[(-3.420 b^3)^2 + (-3.824 b^3)^2]$$

$$= 693.4[(e_{\text{eff}}^{\text{p}})^2 + (e_{\text{eff}}^{\text{n}})^2] \text{ fm}^6 . \qquad (9.118)$$

So the sum rule is satisfied to within rounding-off errors resulting from the decimal display in Table 6.5.

**Valence space $(0d\text{-}1s)\text{-}(0p)^{-1}$**

Let us extend the previous analysis to the full $(0d\text{-}1s)\text{-}(0p)^{-1}$ particle–hole valence space. Then the basis becomes

---

[7] The table entries come from a complete computer calculation. They are therefore not subject to the three-decimal accuracy of Table 6.5 and of the wave functions like (9.114). This explains why the tabulated strength for $k = 1$ differs from our result (9.116).

$$\{|\pi_1\rangle,|\pi_2\rangle,|\pi_3\rangle,|\nu_1\rangle,|\nu_2\rangle,|\nu_3\rangle\}$$
$$= \{|\pi 0d_{5/2}(\pi 0p_{3/2})^{-1};3^-\rangle,|\pi 0d_{5/2}(\pi 0p_{1/2})^{-1};3^-\rangle,$$
$$|\pi 0d_{3/2}(\pi 0p_{3/2})^{-1};3^-\rangle,|\nu 0d_{5/2}(\nu 0p_{3/2})^{-1};3^-\rangle,$$
$$|\nu 0d_{5/2}(\nu 0p_{1/2})^{-1};3^-\rangle,|\nu 0d_{3/2}(\nu 0p_{3/2})^{-1};3^-\rangle\}. \quad (9.119)$$

The lowest $3^-$ state becomes

$$|^{16}O,3_1^-\rangle = 0.136|\pi_1\rangle + 0.681|\pi_2\rangle - 0.134|\pi_3\rangle$$
$$+ 0.136|\nu_1\rangle + 0.681|\nu_2\rangle - 0.134|\nu_3\rangle. \quad (9.120)$$

**Table 9.8.** TDA energies and octupole strengths of $3^-$ states in $^{16}O$ for particle–hole valence space $(0d\text{-}1s)\text{-}(0p)^{-1}$

| $k$ | 1 | 2 | 3 | 4 | 5 | 6 |
|---|---|---|---|---|---|---|
| $E_k$ (MeV) | 10.281 | 12.551 | 16.792 | 18.206 | 22.635 | 24.095 |
| $\left|(3_k^-\|Q_3\|HF)\right|^2$ (fm$^6$) | $347.2e_+^2$ | $167.1e_-^2$ | $129.4e_+^2$ | $167.5e_-^2$ | $101.6e_+^2$ | $243.4e_-^2$ |

Note abbreviations $e_\pm \equiv e_{\mathrm{eff}}^p \pm e_{\mathrm{eff}}^n$.

A hand calculation similar to the previous ones in this subsection gives

$$(HF\|Q_3\|3_1^-) = 3.631e_+ b^3 = 18.64e_+ \text{ fm}^3. \quad (9.121)$$

The results of a complete computer calculation are stated in Table 9.8. Again the transition strengths satisfy the sum rule (9.104): its both sides give $1156[(e_{\mathrm{eff}}^p)^2 + (e_{\mathrm{eff}}^n)^2]\,\text{fm}^6$.

### 9.4.4 Collective Transitions in the TDA

From the three examples of the previous subsection we can make the following observations.

- Enlarging the particle–hole valence space increases the transition strength from the first $3^-$ state. This is shown by the amplitudes (9.108), (9.115) and (9.121). Equation (9.101) gives the corresponding reduced transition probabilities as

$$B(E3;3_1^- \to 0_{gs}^+) = \begin{cases} 27.52e_+^2 \text{ fm}^6, & \text{space } 0d_{5/2}\text{-}(0p_{1/2})^{-1}, \\ 35.35e_+^2 \text{ fm}^6, & \text{space } (0d_{5/2}\text{-}1s_{1/2})\text{-}(0p)^{-1}, \\ 49.64e_+^2 \text{ fm}^6, & \text{space } (0d\text{-}1s)\text{-}(0p)^{-1}. \end{cases}$$
$$(9.122)$$

Inspection of the wave functions (9.106), (9.114) and (9.120) shows an increasing fragmentation into a larger number of more evenly sized amplitudes. This fragmentation is *coherent* and it amounts to *collectivity*: an increasing number of particle–hole components act in a coherent way to increase the decay probability.

- From Tables 9.7 and 9.8 we can access the distribution of the octupole strength among the $3^-$ final states. The trend is discernible that two of the states, namely the lowest and the highest, gather an increasing amount of the total strength as the space is enlarged. In Subsect. 9.2.4 these two states were characterized as collective states. The highest-lying collective state is in the E3 case known as the *giant octupole resonance*. The lowest and highest collective states are present even in the degenerate separable model as displayed in (9.57).

- When the proton and neutron single-particle excitation energies are the same, we have good isospin. As recorded in (6.166), a decay amplitude proportional to $e_+ = e^p_{eff} + e^n_{eff}$ connects a $T = 0$ state to the ground state. Likewise (6.167) tells us that a decay amplitude proportional to $e_- = e^p_{eff} - e^n_{eff}$ connects a $T = 1$ state to the ground state. Thus we see from Tables 9.7 and 9.8 that every other state $k$ has $T = 0$ and $T = 1$. The decays of such states are isoscalar and isovector decays respectively. In particular, the lowest collective particle–hole state decays by an isoscalar transition and the giant octupole state by an isovector transition.

The experimental reduced transition probability for the lowest $3^-$ state in $^{16}O$ is

$$B(E3\,;\, 3^-_1 \rightarrow 0^+_{gs})_{exp} = 205.3\, e^2 fm^6 = 13.5\, W.u. \tag{9.123}$$

This can be reproduced by fitting the effective charges. From (6.26) we have $e_+ = (1 + 2\chi)e$. The result $49.64 e^2_+ \, fm^6$ in (9.122) thus gives the electric polarization constant $\chi = 0.52$. The RPA yields more collectivity than does the TDA, so that the polarization constant is $\chi = 0.34$.

Based on the results of Subsect. 9.4.3, Fig. 9.6 shows the distribution of the $B(E3)$ values in $^{16}O$ for the three particle–hole valence spaces considered. The effective charges were chosen as $e^p_{eff} = 1.2e$ and $e^n_{eff} = 0.2e$, and the values are given in Weisskopf units. The successive panels display the evolution of the distribution of the octupole strength with the increasing size of the particle–hole valence space.

Successive enlargements of the valence space bring in new states at higher energies, while the old states remain roughly unmoved. Even the octupole strength of the low-lying states is not changed much except for the lowest state, whose strength increases appreciably when the valence space increases.

### 9.4.5 Application to Octupole Transitions in $^{40}Ca$

We calculate the reduced transition probability $B(E3\,;\, 3^-_1 \rightarrow 0^+_{gs})$ for $^{40}_{20}Ca_{20}$ in two particle–hole valence spaces. We start with the smallest possible space $0f_{7/2}$-$(0d_{3/2})^{-1}$. Proceeding as in the case of $^{16}O$, we find

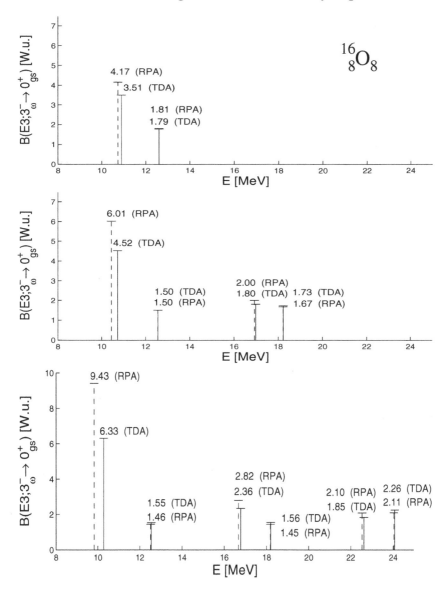

**Fig. 9.6.** Reduced E3 transition probabilities in Weisskopf units for transitions from the $3^-$ states to the ground state in $^{16}$O. The results were calculated with SDI parameters $A_0 = A_1 = 1.0$ MeV and effective charges $e_{\text{eff}}^{\text{p}} = 1.2e$ and $e_{\text{eff}}^{\text{n}} = 0.2e$. The top panel is for the particle–hole valence space $0d_{5/2}\text{-}(0p_{1/2})^{-1}$, the middle one for $(0d_{5/2}\text{-}1s_{1/2})\text{-}(0p)^{-1}$ and the bottom one for $(0d\text{-}1s)\text{-}(0p)^{-1}$. The horizontal axis gives the excitation energy

$$B(\text{E3}\,;\,3_1^- \rightarrow 0_{\text{gs}}^+) = 57.18(e_{\text{eff}}^{\text{p}} + e_{\text{eff}}^{\text{n}})^2\,\text{fm}^6 = 0.601(1 + 2\chi)^2\,\text{W.u.}, \quad (9.124)$$

where we used $b = 1.939\,\text{fm}$ as given by (3.45) and (3.43).

For the larger valence space $0f_{7/2}\text{-}(0d\text{-}1s)^{-1}$ the basis states are

$$\{|\pi_1\rangle, |\pi_2\rangle, |\pi_3\rangle, |\nu_1\rangle, |\nu_2\rangle, |\nu_3\rangle\}$$
$$= \{|\pi 0f_{7/2}\,(\pi 0d_{5/2})^{-1}\,;\,3^-\rangle,\ |\pi 0f_{7/2}\,(\pi 1s_{1/2})^{-1}\,;\,3^-\rangle,$$
$$|\pi 0f_{7/2}\,(\pi 0d_{3/2})^{-1}\,;\,3^-\rangle,\ |\nu 0f_{7/2}\,(\nu 0d_{5/2})^{-1}\,;\,3^-\rangle,$$
$$|\nu 0f_{7/2}\,(\nu 1s_{1/2})^{-1}\,;\,3^-\rangle,\ |\nu 0f_{7/2}\,(\nu 0d_{3/2})^{-1}\,;\,3^-\rangle\}\,. \tag{9.125}$$

With the the single-particle energies of Fig. 9.2 (b) and the SDI parameters $A_0 = 0.85\,\text{MeV}$ and $A_1 = 0.90\,\text{MeV}$, the lowest $3^-$ state becomes

$$|^{40}\text{Ca}, 3_1^-\rangle = 0.272|\pi_1\rangle + 0.314|\pi_2\rangle + 0.572|\pi_3\rangle$$
$$+ 0.272|\nu_1\rangle + 0.314|\nu_2\rangle + 0.572|\nu_3\rangle\,. \tag{9.126}$$

The wave function (9.126) leads to the theoretical value

$$B(\text{E3}\,;\,3_1^- \to 0_{\text{gs}}^+) = 298.3(e_{\text{eff}}^{\text{p}} + e_{\text{eff}}^{\text{n}})^2\,\text{fm}^6 = 3.138(1 + 2\chi)^2\,\text{W.u.} \tag{9.127}$$

This can be compared with the experimental value

$$B(\text{E3}\,;\,3_1^- \to 0_{\text{gs}}^+)_{\text{exp}} = 2504\,e^2\text{fm}^6 = 26.3\,\text{W.u.} \tag{9.128}$$

The theoretical octupole strength distribution is plotted in Fig. 9.7 for the $0f_{7/2}\text{-}(0d\text{-}1s)^{-1}$ particle–hole valence space and effective charges $e_{\text{eff}}^{\text{p}} = 1.3e$ and $e_{\text{eff}}^{\text{n}} = 0.3e$. The TDA value for the first $3^-$ state is far below the experimental one, whereas the RPA accurately reproduces the experimental value. The $3_1^-$ state is strongly collective in the RPA description.

### 9.4.6 E1 Transitions: Isospin Breaking in the Nuclear Mean Field

The E1 transition $1_1^- \to 0_{\text{gs}}^+$ in $^{16}\text{O}$ was studied in Subsect. 6.4.2. In that context, (6.142) gave a recipe for removing spurious centre-of-mass components from the dipole transition amplitude. Applied to a *self-conjugate* ($N = Z$) nucleus, the recipe gives

$$e_{\text{eff}}^{\text{p}} = \frac{1}{2}e\,,\quad e_{\text{eff}}^{\text{n}} = -\frac{1}{2}e\,. \tag{9.129}$$

According to (6.130), this leads to a vanishing E1 decay probability for $1^-$, $T = 0$ states. As long as the proton and neutron contributions are the same this behaviour persists even for larger spaces because the transition amplitude is proportional to $e_{\text{eff}}^{\text{p}} + e_{\text{eff}}^{\text{n}}$.

As was pointed out at the end of Subsect. 6.4.2, a measured finite E1 decay rate indicates a small isospin-breaking effect in the nuclear mean field. For $^{16}\text{O}$ this lack of complete isospin symmetry shows up as different single-particle energies of protons and neutrons in the 0d-1s shell. To make a quantitative

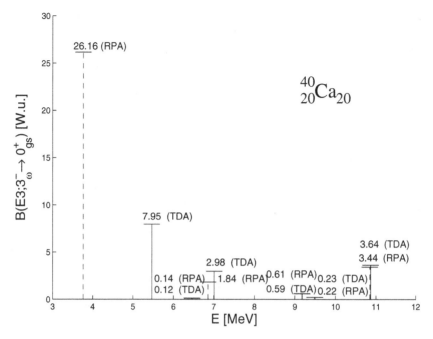

**Fig. 9.7.** Reduced E3 transition probabilities in Weisskopf units for transitions from the $3^-$ states to the ground state in $^{40}$Ca. The results were calculated with SDI parameters $A_0 = 0.85\,\mathrm{MeV}$ and $A_1 = 0.90\,\mathrm{MeV}$ and effective charges $e_{\mathrm{eff}}^{\mathrm{p}} = 1.3e$ and $e_{\mathrm{eff}}^{\mathrm{n}} = 0.3e$. The particle–hole valence space is $0\mathrm{f}_{7/2}$-$(0\mathrm{d}$-$1\mathrm{s})^{-1}$. The horizontal axis gives the excitation energy

estimate of this breaking one can examine the spectra of the one-neutron nucleus $^{17}_{8}\mathrm{O}_9$ and the one-proton nucleus $^{17}_{9}\mathrm{F}_8$. The extracted single-*proton* energies are

$$\varepsilon_{1\mathrm{s}_{1/2}} - \varepsilon_{0\mathrm{d}_{5/2}} = 0.49\,\mathrm{MeV}\,, \quad \varepsilon_{0\mathrm{d}_{3/2}} - \varepsilon_{0\mathrm{d}_{5/2}} = 5.10\,\mathrm{MeV}\,, \tag{9.130}$$

while the single-*neutron* energies are

$$\varepsilon_{1\mathrm{s}_{1/2}} - \varepsilon_{0\mathrm{d}_{5/2}} = 0.87\,\mathrm{MeV}\,, \quad \varepsilon_{0\mathrm{d}_{3/2}} - \varepsilon_{0\mathrm{d}_{5/2}} = 5.09\,\mathrm{MeV}\,. \tag{9.131}$$

We introduce isospin mixing within the TDA by using the single-particle energies (9.130) and (9.131) in the $(0\mathrm{d}$-$1\mathrm{s})$-$(0\mathrm{p}_{1/2})^{-1}$ particle–hole valence space. The energy gap between the $0\mathrm{p}$ shell and the $0\mathrm{d}$-$1\mathrm{s}$ shell is chosen as $11.6\,\mathrm{MeV}$ for both protons and neutrons. The basis states for $J^\pi = 1^-$ are

$$\{|\pi_1\rangle_1\,, |\pi_2\rangle_1\,, |\nu_1\rangle_1\,, |\nu_2\rangle_1\}$$
$$= \{|\pi 1\mathrm{s}_{1/2}\,(\pi 0\mathrm{p}_{1/2})^{-1}; 1^-\rangle\,, |\pi 0\mathrm{d}_{3/2}\,(\pi 0\mathrm{p}_{1/2})^{-1}; 1^-\rangle\,,$$
$$|\nu 1\mathrm{s}_{1/2}\,(\nu 0\mathrm{p}_{1/2})^{-1}; 1^-\rangle\,, |\nu 0\mathrm{d}_{3/2}\,(\nu 0\mathrm{p}_{1/2})^{-1}; 1^-\rangle\}\,, \tag{9.132}$$

where we have completed the notation on the left-hand side to distinguish this basis from the $J^\pi = 3^-$ basis (9.113) for $^{16}$O.

Equations (9.28)–(9.30) and Table 9.2 give the Hamiltonian matrix

$$H_{\mathrm{TDA}}(1^-) =$$

$$\begin{pmatrix} 12.09 + 0.333A_1 & -0.943A_1 & \frac{1}{2}(0.333A_1 - 1.667A_0) & \frac{1}{2}(-0.943A_1 - 0.943A_0) \\ -0.943A_1 & 16.68 - 0.333A_1 & \frac{1}{2}(-0.943A_1 - 0.943A_0) & \frac{1}{2}(-0.333A_1 - 2.333A_0) \\ \cdots & \cdots & 12.47 + 0.333A_1 & -0.943A_1 \\ \cdots & \cdots & -0.943A_1 & 16.67 - 0.333A_1 \end{pmatrix} .$$

$$(9.133)$$

Note that, in distinction to the matrix (9.91), the corresponding proton and neutron elements on the diagonal are only approximately equal. With $A_0 = A_1 = 1.0\,\mathrm{MeV}$ the lowest eigenstate becomes

$$|^{16}\mathrm{O}, 1_1^-\rangle = 0.691|\pi_1\rangle_1 + 0.302|\pi_2\rangle_1 + 0.583|\nu_1\rangle_1 + 0.303|\nu_2\rangle_1 . \qquad (9.134)$$

The symmetry break shows here in the inequality of the corresponding proton and neutron amplitudes.

With the wave function (9.134) the decay amplitude (9.100) becomes

$$(\mathrm{HF}\|Q_1\|1_1^-) = -[0.691(\pi 1s_{1/2}\|Q_1\|\pi 0p_{1/2}) + 0.302(\pi 0d_{3/2}\|Q_1\|\pi 0p_{1/2})$$
$$+ 0.583(\nu 1s_{1/2}\|Q_1\|\nu 0p_{1/2}) + 0.303(\nu 0d_{3/2}\|Q_1\|\nu 0p_{1/2})] . \qquad (9.135)$$

Taking the reduced matrix elements from Table 6.3 and using the effective charges (9.129) we obtain

$$(\mathrm{HF}\|Q_1\|1_1^-) = -0.0211b = -0.0364e\,\mathrm{fm} , \qquad (9.136)$$

where $b = 1.725\,\mathrm{fm}$.

The decay amplitude (9.136) gives the reduced transition probability

$$B(\mathrm{E1}; 1_1^- \to 0_{\mathrm{gs}}^+) = 4.42 \times 10^{-4}\,e^2\mathrm{fm}^2 . \qquad (9.137)$$

With the experimental transition energy $E = 7.117\,\mathrm{MeV}$ (see Fig. 6.1), Table 6.8 gives the decay probability

$$T(\mathrm{E1}) = 1.587 \times 10^{15} \times 7.117^3 \times 4.42 \times 10^{-4}\,1/\mathrm{s} = 2.53 \times 10^{14}\,1/\mathrm{s} , \qquad (9.138)$$

and the half-life becomes

$$t_{1/2} = \frac{\ln 2}{T(\mathrm{E1})} = 2.74\,\mathrm{fs} . \qquad (9.139)$$

This is of the same order as the experimental value of 8 fs.

This example shows that a small difference, in this case $\approx 0.4\,\mathrm{MeV}$, in the proton and neutron mean-field single-particle energies gives an isospin breaking sufficient to explain the observed E1 transition rate.

### 9.4.7 Transitions Between Two TDA Excitations

The matrix elements of the electromagnetic operator $\mathcal{M}_{\sigma\lambda}$ for transitions between particle–hole states were discussed in Subsect. 6.4.1. The initial and final TDA eigenfunctions are

$$|\Psi_{\omega_i}^{(\text{TDA})}\rangle = \sum_{a_i b_i} X_{a_i b_i}^{\omega_i} |a_i\, b_i^{-1}\,;\, J_i\, M_i\rangle\,, \tag{9.140}$$

$$|\Psi_{\omega_f}^{(\text{TDA})}\rangle = \sum_{a_f b_f} X_{a_f b_f}^{\omega_f} |a_f\, b_f^{-1}\,;\, J_f\, M_f\rangle\,. \tag{9.141}$$

The transition amplitude for these states is

$$\boxed{\left(\omega_f\|\mathcal{M}_{\sigma\lambda}\|\omega_i\right) = \sum_{\substack{a_i b_i \\ a_f b_f}} X_{a_f b_f}^{\omega_f *} X_{a_i b_i}^{\omega_i} \left(a_f\, b_f^{-1}\,;\, J_f\|\mathcal{M}_{\sigma\lambda}\|a_i\, b_i^{-1}\,;\, J_i\right).} \tag{9.142}$$

The particle–hole transition matrix element on the right-hand side is given by (6.124).

### 9.4.8 Application to the $5_1^- \to 3_1^-$ Transition in $^{40}$Ca

Consider the E2 transition between the first $3^-$ and $5^-$ states in $^{40}_{20}$Ca$_{20}$. These states are shown in the energy spectra, both theoretical and experimental, of Fig. 9.4. To calculate the transition rate, we use the particle–hole valence space $0f_{7/2}$-$(0d$-$1s)^{-1}$.

The $3^-$ basis states are given by (9.125), and the $5^-$ basis states are

$$\{|\pi_1\rangle_5\,,\, |\pi_2\rangle_5\,,\, |\nu_1\rangle_5\,,\, |\nu_2\rangle_5\}$$
$$= \{|\pi 0f_{7/2}\, (\pi 0d_{5/2})^{-1}\,;\, 5^-\rangle\,,\, |\pi 0f_{7/2}\, (\pi 0d_{3/2})^{-1}\,;\, 5^-\rangle\,,$$
$$\{|\nu 0f_{7/2}\, (\nu 0d_{5/2})^{-1}\,;\, 5^-\rangle\,,\, |\nu 0f_{7/2}\, (\nu 0d_{3/2})^{-1}\,;\, 5^-\rangle\}\,. \tag{9.143}$$

The wave function for the $3_1^-$ state is given by (9.126). For the $5_1^-$ state we obtain

$$|^{40}\text{Ca}\,,\, 5_1^-\rangle = 0.150|\pi_1\rangle_5 + 0.691|\pi_2\rangle_5 + 0.150|\nu_1\rangle_5 + 0.691|\nu_2\rangle_5\,. \tag{9.144}$$

We calculate the transition $5_1^- \to 3_1^-$, which is observed experimentally. The fact that the levels are in reverse order in the TDA does not matter. Substituting the wave-function amplitudes $X$ into (9.142) gives the E2 transition amplitude

$$(3_1^-\|\boldsymbol{Q}_2\|5_1^-) = [0.272 \times 0.150\,_3(1\|\boldsymbol{Q}_2\|1)_5 + 0.314 \times 0.150\,_3(2\|\boldsymbol{Q}_2\|1)_5$$
$$+ 0.572 \times 0.150\,_3(3\|\boldsymbol{Q}_2\|1)_5 + 0.272 \times 0.691\,_3(1\|\boldsymbol{Q}_2\|2)_5$$
$$+ 0.314 \times 0.691\,_3(2\|\boldsymbol{Q}_2\|2)_5 + 0.572 \times 0.691\,_3(3\|\boldsymbol{Q}_2\|2)_5]$$
$$\times (e_{\text{eff}}^{\text{p}} + e_{\text{eff}}^{\text{n}})/e\,. \tag{9.145}$$

It remains to calculate the reduced matrix elements (6.124) on the right-hand side. An example of such a calculation is provided by (6.133). The results are

$$_3(1\|\boldsymbol{Q}_2\|1)_5 = 0.4388eb^2 , \qquad _3(2\|\boldsymbol{Q}_2\|1)_5 = 1.6204eb^2 ,$$
$$_3(3\|\boldsymbol{Q}_2\|1)_5 = -1.1837eb^2 , \quad _3(1\|\boldsymbol{Q}_2\|2)_5 = 0.3915eb^2 ,$$
$$_3(2\|\boldsymbol{Q}_2\|2)_5 = 2.4752eb^2 , \qquad _3(3\|\boldsymbol{Q}_2\|2)_5 = 1.0978eb^2 . \tag{9.146}$$

Combining (9.145) and (9.146) we have

$$(3_1^-\|\boldsymbol{Q}_2\|5_1^-) = 1.037b^2(e_{\mathrm{eff}}^{\mathrm{p}} + e_{\mathrm{eff}}^{\mathrm{n}}) = 3.899(e_{\mathrm{eff}}^{\mathrm{p}} + e_{\mathrm{eff}}^{\mathrm{n}})\,\mathrm{fm}^2 , \tag{9.147}$$

where $b = 1.939\,\mathrm{fm}$. This gives the reduced transition probability

$$B(\mathrm{E2}\,;\, 5_1^- \to 3_1^-) = 1.382(e_{\mathrm{eff}}^{\mathrm{p}} + e_{\mathrm{eff}}^{\mathrm{n}})^2\,\mathrm{fm}^4 . \tag{9.148}$$

Based on the preceding examples we see that this is an isoscalar transition between $T = 0$ states.

When the same calculation is done in the minimal particle–hole valence space $0\mathrm{f}_{7/2}\text{-}(0\mathrm{d}_{3/2})^{-1}$, the result is

$$B(\mathrm{E2}\,;\, 5_1^- \to 3_1^-)_{\text{small space}} = 0.387(e_{\mathrm{eff}}^{\mathrm{p}} + e_{\mathrm{eff}}^{\mathrm{n}})^2\,\mathrm{fm}^4 . \tag{9.149}$$

This is much less than our 'large-space' result (9.148), and we may expect that the result would still increase appreciably with a further enlargement of the valence space.

From the experimental decay energy $0.75\,\mathrm{MeV}$ (see Fig. 9.4) and half-life $290\,\mathrm{ps}$, Table 6.8 gives the experimental value

$$B(\mathrm{E2}\,;\, 5_1^- \to 3_1^-)_{\mathrm{exp}} = 8.24\,e^2\mathrm{fm}^4 . \tag{9.150}$$

Comparing this with (9.148) gives the polarization constant

$$\chi = 0.72 . \tag{9.151}$$

It is of interest to see whether different transitions in the same nucleus lead to consistent polarization constant. Equations (9.127) and (9.128) give us the polarization constant for the $3_1^- \to 0_{\mathrm{gs}}^+$ transition as

$$\chi = 0.95 . \tag{9.152}$$

The values (9.151) and (9.152) are roughly compatible.

## 9.5 Electric Transitions on the Schematic Model

In this, the last section of the chapter we return to the schematic model introduced in Subsect. 9.2.1. The model was there considered only for TDA energies. We now extend it to transition probabilities.

### 9.5.1 Transition Amplitudes

The dispersion equation (9.55) of the schematic model of Subsect. 9.2.1 provides a rough estimate of TDA energies. These energies are easy to obtain by graphical means, as demonstrated in Subsect. 9.2.4, which makes the schematic model a valuable practical tool. The energies are calculated by taking the isoscalar and isovector strength constants of the SDI to be equal in magnitude and opposite in sign, as stated in (9.77).

For electric transitions, with the notation $\mathcal{M}_{\mathrm{E}\lambda} \equiv \boldsymbol{Q}_\lambda$, (9.100) becomes

$$(\mathrm{HF}\|\boldsymbol{Q}_\lambda\|\omega) = \delta_{\lambda J}(-1)^J \sum_{ab} X^\omega_{ab}(a\|\boldsymbol{Q}_\lambda\|b) , \qquad (9.153)$$

where $(a\|\boldsymbol{Q}_\lambda\|b)$ is the *electromagnetic* matrix element given by (6.23). It is to be distinguished from the identically denoted matrix element (8.53) used for a separable interaction. The relation between the electromagnetic matrix element and the quantity $Q^\lambda_{ab}(\mathrm{SDI})$ defined in (9.73) is

$$(a\|\boldsymbol{Q}_\lambda\|b) = (-1)^{n_a+n_b}\frac{e}{4\sqrt{\pi}}\widehat{\lambda}Q^\lambda_{ab}(\mathrm{SDI})\mathcal{R}^{(\lambda)}_{ab} . \qquad (9.154)$$

When we apply the schematic TDA with the SDI, as in Subsect. 9.2.4, the quantity $Q^\lambda_{ab}$ present in the equations of Subsect. 9.2.1 is precisely $Q^\lambda_{ab}(\mathrm{SDI})$.

Substituting (9.51) and (9.154) into (9.153) gives

$$(\mathrm{HF}\|\boldsymbol{Q}_\lambda\|\omega) = \delta_{\lambda J}(-1)^J \widehat{J}\frac{e_{\mathrm{eff}}}{4\sqrt{\pi}}N_\omega \sum_{ab}(-1)^{n_a+n_b}\frac{[Q^J_{ab}(\mathrm{SDI})]^2}{\varepsilon_{ab}-E_\omega}\mathcal{R}^{(J)}_{ab} ,$$

$$(9.155)$$

where $N_\omega$ is given by (9.53). For the overall effective charge we take

$$e_{\mathrm{eff}} = \begin{cases} \dfrac{e^{\mathrm{p}}_{\mathrm{eff}} + e^{\mathrm{n}}_{\mathrm{eff}}}{\sqrt{2}} = \dfrac{e_+}{\sqrt{2}} & \text{for } T = 0 , \\[3mm] \dfrac{e^{\mathrm{p}}_{\mathrm{eff}} - e^{\mathrm{n}}_{\mathrm{eff}}}{\sqrt{2}} = \dfrac{e_-}{\sqrt{2}} & \text{for } T = 1 . \end{cases} \qquad (9.156)$$

This choice is based on the anticipation that the amplitude will be squared and the protons and neutrons will contribute equally except for their different effective charges. The identification between charge dependence and isospin was established in Subsect. 6.5.2 and has been quoted throughout Sect. 9.4.

Before proceeding to an application of (9.155) we inspect electromagnetic transitions within the extremly simple degenerate model introduced at the end of Subsect. 9.2.1. In that model we obtain the collective TDA solution

$$E_{\mathrm{coll}} = \varepsilon + \chi_J Q^2 , \quad Q^2 = \sum_{ab}(Q^J_{ab})^2 . \qquad (9.157)$$

Equations (9.53) and (9.157) then yield

$$|N_{\text{coll}}|^{-2} = \frac{Q^2}{(\varepsilon - E_{\text{coll}})^2} = \frac{1}{\chi_J^2 Q^2} , \tag{9.158}$$

and we choose the phase so that

$$N_{\text{coll}} = \chi_J Q . \tag{9.159}$$

The wave-function amplitudes are given by (9.51) as

$$X_{ab}^{\text{coll}} = \frac{Q_{ab}^J}{\varepsilon - E_{\text{coll}}} N_{\text{coll}} = -\frac{Q_{ab}^J}{Q} . \tag{9.160}$$

Substituting these amplitudes into (9.153) we have

$$(\text{HF}\|\boldsymbol{Q}_J\|\text{coll}) = (-1)^{J+1} \sum_{ab} \frac{Q_{ab}^J}{Q} (a\|\boldsymbol{Q}_J\|b) . \tag{9.161}$$

Assuming that the radial integrals in (6.23) and (8.53) are the same, we may use (9.47) to write

$$(a\|\boldsymbol{Q}_J\|b) = e\widehat{J}Q_{ab}^J , \tag{9.162}$$

whereupon (9.161) becomes

$$(\text{HF}\|\boldsymbol{Q}_J\|\text{coll}) = (-1)^{J+1} \widehat{J}\frac{e}{Q} \sum_{ab} (Q_{ab}^J)^2 = (-1)^{J+1} \widehat{J}eQ . \tag{9.163}$$

Let us check whether the collective TDA state satisfies the sum rule (9.103). With (9.163) the left-hand side of (9.103) gives

$$\left|(\text{coll}\|\boldsymbol{Q}_J\|\text{HF})\right|^2 = \widehat{J}^2 e^2 Q^2 . \tag{9.164}$$

With (9.162) the right-hand side is

$$\sum_{ab} \left|(a\|\boldsymbol{Q}_J\|b)\right|^2 = e^2 \widehat{J}^2 \sum_{ab} (Q_{ab}^J)^2 = \widehat{J}^2 e^2 Q^2 , \tag{9.165}$$

so the sum rule is satisfied by the collective state alone; the trapped states make no contribution.

We have here considered the coupling constant of a given magnitude $|\chi_J|$ with a definite sign, say $\chi_J < 0$. In this case the collective state is the lowest state and has $T = 0$. In the case $\chi_J > 0$ the collective state is the highest state and has $T = 1$. In both cases all the multipole strength is carried by the collective state, as described by the extreme schematic model.

### 9.5.2 Application to Electric Dipole Transitions in $^4$He

In Subsect. 9.2.4 we found the energies of the $1^-$ states of $^4_2\text{He}_2$ by the schematic TDA and the SDI. Let us now continue the example and compute the reduced transition probabilities for the E1 transitions $1^- \rightarrow 0^+_{\text{gs}}$.

The squares of the $Q^1_{ab}(\text{SDI})$ are given in (9.81). Equation (9.53) then yields for the normalization

$$N_n^{-2} = \frac{\frac{16}{3}}{(21.0\,\text{MeV} - E_n)^2} + \frac{\frac{8}{3}}{(27.0\,\text{MeV} - E_n)^2} \,, \tag{9.166}$$

where we only quote the eigenvalue index $n$ since the multipolarity is fixed at $J^\pi = 1^-$. The energies $E_n$ of the schematic model are given in Tables 9.5 and 9.6 for two different interaction strengths, namely $A = 1.0\,\text{MeV}$ and $A = 2.0\,\text{MeV}$. Substituting into (9.155) we obtain

$$(\text{HF}\|Q_1\|1_n^-) = -\sqrt{3}\,\frac{e_{\text{eff}}}{4\sqrt{\pi}}\,N_n\left(\frac{\frac{16}{3}}{21.0\,\text{MeV} - E_n}\mathcal{R}^{(1)}_{0\text{p}0\text{s}}\right.$$
$$\left. + \frac{\frac{8}{3}}{27.0\,\text{MeV} - E_n}\mathcal{R}^{(1)}_{0\text{p}0\text{s}}\right). \tag{9.167}$$

Table 6.2 gives the radial integral as

$$\mathcal{R}^{(1)}_{0\text{p}0\text{s}} = b\tilde{\mathcal{R}}^{(1)}_{0\text{p}0\text{s}} = \sqrt{\tfrac{3}{2}}b\,. \tag{9.168}$$

We evaluate the decay amplitudes (9.155) with the $E_n$ for $A = 1.0\,\text{MeV}$. Equations (3.45) and (3.43) give $b = 1.499\,\text{fm}$. From the transition amplitudes we compute the reduced transition probabilities $B(\text{E1}\,;\,1_n^- \to 0^+_{\text{gs}})$. The results are stated in Table 9.9. For comparison, the table also gives the exact TDA values for $B(\text{E1})$, calculated as a follow-up to Subsect. 9.3.2.

**Table 9.9.** Reduced E1 matrix elements $(\text{HF}\|Q_1\|1_n^-)$ and transition probabilities $B(\text{E1}\,;\,1_n^- \to 0^+_{\text{gs}})$ for $^4_2\text{He}_2$ computed by the schematic and exact TDA

| | Schematic | | | Exact |
|---|---|---|---|---|
| $n$ | $E_n$ (MeV) | $(\|Q_1\|)$ (fm) | $B(\text{E1})$ (fm$^2$) | $B(\text{E1})$ (fm$^2$) |
| 1 | 19.5 | $-0.798e_+$ | $0.21e_+^2$ | $0.231e_+^2$ |
| 2 | 22.2 | $0.631e_-$ | $0.13e_-^2$ | $0.178e_-^2$ |
| 3 | 26.5 | $-0.420e_+$ | $0.06e_+^2$ | $0.036e_+^2$ |
| 4 | 27.8 | $0.631e_-$ | $0.13e_-^2$ | $0.089e_-^2$ |

The SDI parameters are $A = 1.0\,\text{MeV}$ (schematic) and $A_0 = A_1 = 1.0\,\text{MeV}$ (exact). Note abbreviations $e_\pm \equiv e^{\text{p}}_{\text{eff}} \pm e^{\text{n}}_{\text{eff}}$

We see that the schematic model produces results in surprisingly good agreement with the exact TDA, just as was the case for the energies in Tables 9.5 and 9.6. Based on this and other cases we can say that the schematic model reliably predicts the order of magnitude of both the excitation energies and the electric multipole transition rates in doubly magic $N = Z$ nuclei.

## Epilogue

In this chapter we have learned the first level of treating the mixing of one-particle–one-hole configurations. This level constitutes what is known as the TDA. In the TDA one diagonalizes the residual Hamiltonian in the combined proton-particle–proton-hole and neutron-particle–neutron-hole basis without touching the structure of the particle–hole ground state of the doubly magic nucleus under study.

In the following chapter the notion of configuration mixing of charge-conserving particle–hole excitations is extended to a description of charge-changing proton–neutron particle–hole excitations in the framework of the proton–neutron TDA. The TDA also paves the way to a more sophisticated particle–hole theory, the RPA. The RPA treats charge-conserving particle–hole excitations of a correlated ground state which itself contains complicated particle–hole configurations.

## Exercises

**9.1.** Derive the relation (9.9) by using the methods of Sect. 4.4.

**9.2.** Complete the details of the derivation of (9.17).

**9.3.** Verify the symmetry property (9.24).

**9.4.** Derive the relation (9.25).

**9.5.** Assume the result (9.25) known and derive the relation (9.26).

**9.6.** Consider the expression (9.45) for the two-body particle–hole matrix element in the form $D - E$ as stated in (9.58). Using the valence space (9.79) compute the direct terms $D$ and the exchange terms $E$. Compare the terms in light of the assertion in Subsect. 9.2.2 that $E \approx D$. Taking all multipole strengths $\chi_\lambda$ as equal to $-\frac{1}{4}A_1$, check your complete matrix elements $D - E$ against the entries in the matrix (9.91).

**9.7.** Solve analytically the second-order equations (9.82) and compare the results with the graphical solutions in Tables 9.5 and 9.6.

**9.8.** Verify the numbers in the matrix (9.91).

**9.9.** Verify the numbers in the isospin matrices (9.93) and (9.94).

**9.10.** Using an available computer routine, diagonalize the matrix (9.91). Check that you get the numbers quoted in Tables 9.5 and 9.6.

**9.11.** Verify the numbers in the matrix (9.97).

**9.12.** Calculate the approximate energies of the $3^-$ states in $^{16}$O by using the TDA dispersion equation (9.55) in the $(0d_{5/2}\text{-}1s_{1/2})\text{-}(0p)^{-1}$ particle–hole valence space. Take the single-particle energies from Fig. 9.2 (a) and choose the SDI parameter $A$ as
(a) $A = 1.0$ MeV,
(b) $A = 2.0$ MeV.

**9.13.** Calculate the energies of the $3^-$ states in $^{16}$O by exact diagonalization. Use the valence space of Exercise 9.12 and the SDI with parameters
(a) $A_0 = A_1 = 1.0$ MeV,
(b) $A_0 = A_1 = 2.0$ MeV.
Compare the results with those of Exercise 9.12 and comment.

**9.14.** Calculate the approximate energies of the $3^-$ states in $^{40}$Ca by using the TDA dispersion equation (9.55) in the $0f_{7/2}\text{-}(0d\text{-}1s)^{-1}$ particle–hole valence space. Take the single-particle energies from Fig. 9.2 (b) and choose the SDI parameter $A$ as
(a) $A = 1.0$ MeV,
(b) $A = 2.0$ MeV.

**9.15.** Calculate the energies of the $3^-$ states in $^{40}$Ca by exact diagonalization. Use the valence space of Exercise 9.14 and the SDI with parameters
(a) $A_0 = A_1 = 1.0$ MeV,
(b) $A_0 = A_1 = 2.0$ MeV.
Compare the results with those of Exercise 9.14 and comment.

**9.16.** Calculate the energies of the $2^+$ and $3^-$ states in $^{48}$Ca by exact diagonalization in the particle–hole valence space $0f_{7/2}\text{-}(0d_{3/2})^{-1}$ for the protons and $(1p\text{-}0f_{5/2})\text{-}(0f_{7/2})^{-1}$ for the neutrons. Take the single-particle energies from Fig. 9.2 (c) and use the SDI with parameters $A_0 = A_1 = 1.0$ MeV.

**9.17.** Verify the numbers in the matrix (9.133).

**9.18.** Verify the numbers in Table 9.7 by performing the diagonalization indicated and proceeding with the wave functions thus obtained.

**9.19.** Verify the numbers in Table 9.9.

**9.20.** Continue Exercise 9.12 and compute the reduced E3 transition probabilities for the ground-state decays of the $3^-$ states in $^{16}$O on the schematic model.

**9.21.** Continue Exercise 9.13 and compute the reduced E3 transition probabilities for the ground-state decays of the $3^-$ states in $^{16}$O by using the exact wave functions. Check that the TDA sum rule is satisfied. Compare the results with those of Exercise 9.20 and comment.

**9.22.** Continue Exercise 9.14 and compute the reduced E3 transition probabilities for the ground-state decays of the $3^-$ states in $^{40}$Ca on the schematic model.

**9.23.** Continue Exercise 9.15 and compute the reduced E3 transition probabilities for the ground-state decays of the $3^-$ states in $^{40}$Ca by using the exact wave functions. Check that the TDA sum rule is satisfied. Compare the results with those of Exercise 9.22 and comment.

**9.24.** Continue Exercise 9.16 and determine the electric polarization constant for neutrons, $\chi_{\rm n}$, by studying the E2 decay of the $2_1^+$ state of $^{48}$Ca.

**9.25.** By using the $\chi_{\rm n}$ of Exercise 9.24 and the experimental gamma energy calculate the decay half-life of the first $4^+$ state in $^{48}$Ca. Compare with experimental data and comment.

**9.26.** Calculate the energies of the $3^-$ and $5^-$ states of $^{40}$Ca in the $0f_{7/2}$-$(0d_{3/2})^{-1}$ particle–hole valence space by exact diagonalization. Take the single-particle energies from Fig. 9.2 (b) and use the SDI. Determine $A_0$ and $A_1$ such that they best reproduce the experimental energies.

**9.27.** The nucleus $^{12}_6$C$_6$ can be viewed as doubly magic in the $0p_{1/2}$-$(0p_{3/2})^{-1}$ particle–hole valence space. Set up a calculation of the eigenenergies by exact diagonalization. Take 6.0 MeV as the spin-orbit splitting and use the SDI. Determine the parameters $A_0$ and $A_1$ such that they best reproduce the experimental energies. How good is the approximation of double magicity?

**9.28.** The nucleus $^{32}_{16}$S$_{16}$ can be viewed as doubly magic in the $0d_{3/2}$-$(0d_{5/2}$-$1s_{1/2})^{-1}$ particle–hole valence space. Adopt the single-particle energies (8.124) and calculate the eigenenergies by exact diagonalization using the SDI with $A_0 = A_1 = 1.0$ MeV. Compare with the experimental spectrum. How good is the approximation of double magicity?

**9.29.** Continue Exercise 9.27 and compute the reduced E2 transition probabilities for the ground-state decays of the $2^+$ states in $^{12}$C. Check the TDA sum rule. Determine the electric polarization constant $\chi$ by using available experimental data.

**9.30.** Continue Exercise 9.28 and compute the reduced E2 transition probabilities for the ground-state decays of the $2^+$ states in $^{32}$S. Check the TDA sum rule. Determine the electric polarization constant $\chi$ by using available experimental data.

**9.31.** Continue Exercises 9.28 and 9.30. Compute the decay half-life of the $4_1^+$ state of $^{32}$S by using the experimental energies and taking the value of $\chi$ from Exercise 9.30. Compare your result with the experimental half-life.

**9.32.** Calculate the energies of the $4^-$ and $5^-$ states in $^{40}$Ca by exact diagonalization in the complete $0f_{7/2}$-$(0d$-$1s)^{-1}$ particle–hole valence space. Take the single-particle energies from Fig. 9.2 (b) and use the SDI with parameters $A_0 = A_1 = 1.0\,\text{MeV}$. Compute the decay half-life of the $4_1^-$ state by using the experimental energies and an electric polarization constant of the order of (9.151) and (9.152). Compare your results with experimental data.

**9.33.** Calculate the decay half-life of the $3_1^-$ state of $^{48}$Ca by assuming that it decays to the $2_1^+$ state and to the ground state (see the experimental level scheme in Fig. 9.5). Use the wave functions from Exercise 9.16, the electric polarization constant from Exercise 9.24 and the experimental energies. Compare with the experimental half-life.

# Charge-Changing Particle–Hole Excitations and the pnTDA

## Prologue

In this chapter we extend the Tamm–Dancoff approximation (TDA) to charge-changing particle–hole excitations. Such excitations consist of a proton particle and a neutron hole, or a neutron particle and a proton hole. These excitations of the doubly magic Hartree–Fock vacuum are nuclear states in the adjacent odd–odd nuclei. This formalism is well suited to describe beta-decay transitions from the states of one of the odd–odd nuclei to the ground and excited states of the even–even reference nucleus.

## 10.1 The Proton–Neutron Tamm–Dancoff Approximation

In this section we formulate the proton–neutron Tamm–Dancoff approximation (pnTDA) and cast it into an explicit matrix form. We then apply the pnTDA to a few selected examples.

### 10.1.1 Structure of the pnTDA Matrix

The pnTDA is based on neutron-to-proton or proton-to-neutron conversions. These are the proton-particle–neutron-hole and neutron-particle–proton-hole excitations discussed in Sect. 5.4. They are expressed as

$$|p\,n^{-1}; J\,M\rangle = \left[c_p^\dagger h_n^\dagger\right]_{JM}|\mathrm{HF}\rangle\,, \tag{10.1}$$

$$|n\,p^{-1}; J\,M\rangle = \left[c_n^\dagger h_p^\dagger\right]_{JM}|\mathrm{HF}\rangle\,, \tag{10.2}$$

where $|\mathrm{HF}\rangle$ is the ground state of the even–even reference nucleus $Z, N$. Equations (10.1) and (10.2), respectively, describe the generation of the states of the $Z + 1, N - 1$ nucleus and the $Z - 1, N + 1$ nucleus.

We can immediately extend (9.35) to the charge-changing case and write

$$\sum_{p'n'} \langle p\, n^{-1}\,;\, J^{\pi}|H|p'\, n'^{-1}\,;\, J^{\pi}\rangle X^{\omega}_{p'n'} = E_{\omega} X^{\omega}_{pn}\,, \qquad (10.3)$$

$$\sum_{n'p'} \langle n\, p^{-1}\,;\, J^{\pi}|H|n'\, p'^{-1}\,;\, J^{\pi}\rangle X^{\omega}_{n'p'} = E_{\omega} X^{\omega}_{np}\,, \qquad (10.4)$$

where $\omega = nJ^{\pi}M$. It remains to compute the matrix elements of the nuclear Hamiltonian $H = H_{\mathrm{HF}} + H_{\mathrm{RES}}$ in (10.3) and (10.4).

Proceeding as in Subsect. 9.1.2 we derive for the one-body part of the Hamiltonian the expressions

$$\boxed{\begin{aligned}
\langle p\, n^{-1}\,;\, J|H_{\mathrm{HF}}|p'\, n'^{-1}\,;\, J\rangle &= \delta_{pp'}\delta_{nn'}(\text{const.} + \varepsilon_p - \varepsilon_n)\,,\\
\langle n\, p^{-1}\,;\, J|H_{\mathrm{HF}}|n'\, p'^{-1}\,;\, J\rangle &= \delta_{nn'}\delta_{pp'}(\text{const.} + \varepsilon_n - \varepsilon_p)\,.
\end{aligned}} \qquad (10.5)$$

As in (9.15), the constants in these equations are irrelevant for energies relative to the particle–hole vacuum $|\mathrm{HF}\rangle$. They are therefore omitted in the subsequent discussion. Continuing as in Subsect. 9.1.2 we obtain for the two-body part the formula

$$\boxed{\begin{aligned}
&\langle p_1\, n_2^{-1}\,;\, J|V_{\mathrm{RES}}|p_3\, n_4^{-1}\,;\, J\rangle\\
&= -\tfrac{1}{2}\sum_{J'} \hat{J'}^2 \left\{\begin{matrix} j_{p_1} & j_{n_2} & J\\ j_{p_3} & j_{n_4} & J' \end{matrix}\right\} [\langle a_1\, a_4\,;\, J'\, T=1|V|a_3\, a_2\,;\, J'\, T=1\rangle\\
&\hspace{6cm} + \langle a_1\, a_4\,;\, J'\, T=0|V|a_3\, a_2\,;\, J'\, T=0\rangle]\\
&= \tfrac{1}{2}[A_1 \mathcal{M}_{a_1 a_2 a_3 a_4}(J1) + A_0 \mathcal{M}_{a_1 a_2 a_3 a_4}(J0)]\\
&= \langle n_1\, p_2^{-1}\,;\, J|V_{\mathrm{RES}}|n_3\, p_4^{-1}\,;\, J\rangle\,,
\end{aligned}} \qquad (10.6)$$

where the quantities $\mathcal{M}_{a_1 a_2 a_3 a_4}(JT)$ are defined in (9.27) and tabulated in Tables 9.1–9.4.

The pnTDA wave functions satisfy the orthonormality and completeness relations (9.39) and (9.43). Also the relation (9.33) between the CS and BR phases of particle–hole matrix elements is valid for the pnTDA.

Solving the proton-particle–neutron-hole eigenvalue problem (10.3) amounts to diagonalizing the matrix

$$\begin{pmatrix}
\varepsilon_1^{\mathrm{pn}} + V_{11}^{\mathrm{pn}} & V_{12}^{\mathrm{pn}} & \cdots & V_{1N}^{\mathrm{pn}}\\
V_{21}^{\mathrm{pn}} & \varepsilon_2^{\mathrm{pn}} + V_{22}^{\mathrm{pn}} & \cdots & V_{2N}^{\mathrm{pn}}\\
\vdots & \vdots & \ddots & \vdots\\
V_{N1}^{\mathrm{pn}} & V_{N2}^{\mathrm{pn}} & \cdots & \varepsilon_N^{\mathrm{pn}} + V_{NN}^{\mathrm{pn}}
\end{pmatrix}\,, \qquad (10.7)$$

where the particle–hole configurations are labelled with $i = 1, 2, \ldots, N$ and

$$\varepsilon_i^{\mathrm{pn}} \equiv \varepsilon_{p_i} - \varepsilon_{n_i}\,, \qquad V_{ij}^{\mathrm{pn}} \equiv \langle p_i\, n_i^{-1}\,;\, J|V_{\mathrm{RES}}|p_j\, n_j^{-1}\,;\, J\rangle\,. \qquad (10.8)$$

An analogous matrix can be built for the neutron-particle–proton-hole eigenvalue problem (10.4). The proton–neutron and neutron–proton modes of the pnTDA give the same eigenenergies and eigenfunctions of the same structure if the relative single-particle energies of protons and neutrons are equal, i.e. $\varepsilon_i^{pn} = \varepsilon_i^{np}$.

Examples of the application of the pnTDA are given in the following subsections.

## 10.1.2 Application to $^4_1\text{H}_3$ and $^4_3\text{Li}_1$

Consider the energy spectra of the nuclei $^4_1\text{H}_3$ and $^4_3\text{Li}_1$ in the $0p$-$(0s_{1/2})^{-1}$ particle–hole valence space. The basis states for $^4_1\text{H}_3$ are

$$|\nu_1\pi\rangle_{J=1,2} \equiv \left[c^\dagger_{\nu 0p_{3/2}} h^\dagger_{\pi 0s_{1/2}}\right]_{1-,2-} |\text{HF}\rangle , \qquad (10.9)$$

$$|\nu_2\pi\rangle_{J=0,1} \equiv \left[c^\dagger_{\nu 0p_{1/2}} h^\dagger_{\pi 0s_{1/2}}\right]_{0-,1-} |\text{HF}\rangle , \qquad (10.10)$$

and those for $^4_3\text{Li}_1$ are

$$|\pi_1\nu\rangle_{J=1,2} \equiv \left[c^\dagger_{\pi 0p_{3/2}} h^\dagger_{\nu 0s_{1/2}}\right]_{1-,2-} |\text{HF}\rangle , \qquad (10.11)$$

$$|\pi_2\nu\rangle_{J=0,1} \equiv \left[c^\dagger_{\pi 0p_{1/2}} h^\dagger_{\nu 0s_{1/2}}\right]_{0-,1-} |\text{HF}\rangle . \qquad (10.12)$$

The particle–hole vacuum $|\text{HF}\rangle$ is the ground state of $^4_2\text{He}_2$.

In the notation of (10.7) the neutron–proton single-particle energy differences are

$$\varepsilon_1^{np} \equiv \varepsilon_{0p_{3/2}} - \varepsilon_{0s_{1/2}} , \qquad \varepsilon_2^{np} \equiv \varepsilon_{0p_{1/2}} - \varepsilon_{0s_{1/2}} . \qquad (10.13)$$

We construct the Hamiltonian matrices of the type (10.7) for $^4_1\text{H}_3$. For the $J^\pi = 0^-$ and $2^-$ states Eqs. (10.5) and (10.6) and Table 9.1 give

$$\begin{aligned}
{}_0\langle\nu_2\pi|H|\nu_2\pi\rangle_0 \\
&= \varepsilon_2^{np} + \tfrac{1}{2}[A_1 \mathcal{M}_{p_{1/2}s_{1/2}p_{1/2}s_{1/2}}(01) + A_0\mathcal{M}_{p_{1/2}s_{1/2}p_{1/2}s_{1/2}}(00)] \\
&= \varepsilon_2^{np} + \tfrac{1}{2}(1.000A_1 + 1.000A_0) , \qquad (10.14)
\end{aligned}$$

$$\begin{aligned}
{}_2\langle\nu_1\pi|H|\nu_1\pi\rangle_2 \\
&= \varepsilon_1^{np} + \tfrac{1}{2}[A_1 \mathcal{M}_{p_{3/2}s_{1/2}p_{3/2}s_{1/2}}(21) + A_0\mathcal{M}_{p_{3/2}s_{1/2}p_{3/2}s_{1/2}}(20)] \\
&= \varepsilon_1^{np} + \tfrac{1}{2}(1.000A_1 + 1.000A_0) . \qquad (10.15)
\end{aligned}$$

These are the only elements of the $0^-$ and $2^-$ matrices, so the matrices are trivially diagonal. For the $1^-$ states we obtain

$$\begin{aligned}
\text{H}_{\text{pnTDA}}(1^-) &= \begin{pmatrix} {}_1\langle\nu_1\pi|H|\nu_1\pi\rangle_1 & {}_1\langle\nu_1\pi|H|\nu_2\pi\rangle_1 \\ {}_1\langle\nu_2\pi|H|\nu_1\pi\rangle_1 & {}_1\langle\nu_2\pi|H|\nu_2\pi\rangle_1 \end{pmatrix} \\
&= \begin{pmatrix} \varepsilon_2^{np} + \tfrac{1}{2}(-0.333A_1 + 2.333A_0) & \tfrac{1}{2}(0.943A_1 - 0.943A_0) \\ \tfrac{1}{2}(0.943A_1 - 0.943A_0) & \varepsilon_1^{np} + \tfrac{1}{2}(0.333A_1 + 1.667A_0) \end{pmatrix} . \qquad (10.16)
\end{aligned}$$

Generally the matrix (10.16) would have to be diagonalized, but with our standard choice of the SDI parameters, $A_0 = A_1$, the matrix is fortuitously diagonal. Then we can write all the eigenenergies immediately:

$$E(0^-) = \varepsilon_2^{np} + A_0 , \quad E(2^-) = \varepsilon_1^{np} + A_0 ,$$
$$E_1(1^-) = \varepsilon_1^{np} + A_0 , \quad E_2(1^-) = \varepsilon_2^{np} + A_0 , \tag{10.17}$$

where the decimal numbers have been replaced by the obvious exact values. As noted after (10.8), the eigenenergies are the same for $^4_3\text{Li}_1$ when we take $\varepsilon_i^{pn} = \varepsilon_i^{np}$. The degeneracies among the energies (10.17) are broken by taking $A_0 \neq A_1$.

Inspection of the matrix (9.94) shows that the $J^\pi = 1^-$, $T = 1$ states of $^4_2\text{He}_2$ given by the TDA are the same as the $1^-$ states obtained here for the odd–odd neighbours of $^4_2\text{He}_2$. This general feature of the TDA and pnTDA stems from isospin symmetry, which in turn results from the same proton and neutron single-particle energies and the same proton–proton, proton–neutron and neutron–neutron interactions. The nucleus $^4\text{He}$, with $M_T = 0$, has also $T = 0$ states, which were calculated in Subsect. 9.3.2. The neighbouring odd–odd nuclei have $M_T = \pm 1$ and thus cannot have states with $T = 0$.

To compute the energy spectra we need to choose the single-particle energy differences and the SDI strength parameters. On the basis of experimental data we take

$$\varepsilon_1^{np} = 24.0 \, \text{MeV} , \quad \varepsilon_2^{np} = 29.0 \, \text{MeV} . \tag{10.18}$$

These differ somewhat from the values used in Subsect. 9.2.4 and were chosen for an improved description of the experimental spectra. For the SDI parameters we use $A_0 = A_1 = 2.0 \, \text{MeV}$. The results of the calculations are shown in Fig. 10.1. Note that the same $1^-$ states are present here as in Tables 9.5 and 9.6, only with somewhat different energies.

In overall structure and energy scale, the computed spectra of Fig. 10.1 agree with the experimental ones in Fig. 5.9. The degeneracies due to our choice $A_0 = A_1$ are split by some 1–2 MeV in the experimental spectra. However, not all of our predicted states are seen experimentally, and experiment shows states not accounted for by our simple theory. In particular, the experimental spectrum of $^4\text{He}$ has two extra $0^-$ states around 20 MeV.

### 10.1.3 Further Examples: States of $^{16}_{7}\text{N}_9$ and $^{40}_{19}\text{K}_{21}$

Figure 10.2 shows the results of pnTDA calculations for $^{16}_{7}\text{N}_9$ in two particle–hole valence spaces, namely $(0d_{5/2}\text{-}1s_{1/2})\text{-}(0p_{1/2})^{-1}$ and $(0d\text{-}1s)\text{-}(0p)^{-1}$. The calculations used the single-particle energies of Fig. 9.2(a) and the SDI parameters $A_0 = A_1 = 1.0 \, \text{MeV}$.

On the whole, the computed states correspond very well with the experimental ones shown on the right in Fig. 10.2. In particular, as a group the first four levels are correctly described. The deviations in their detailed order

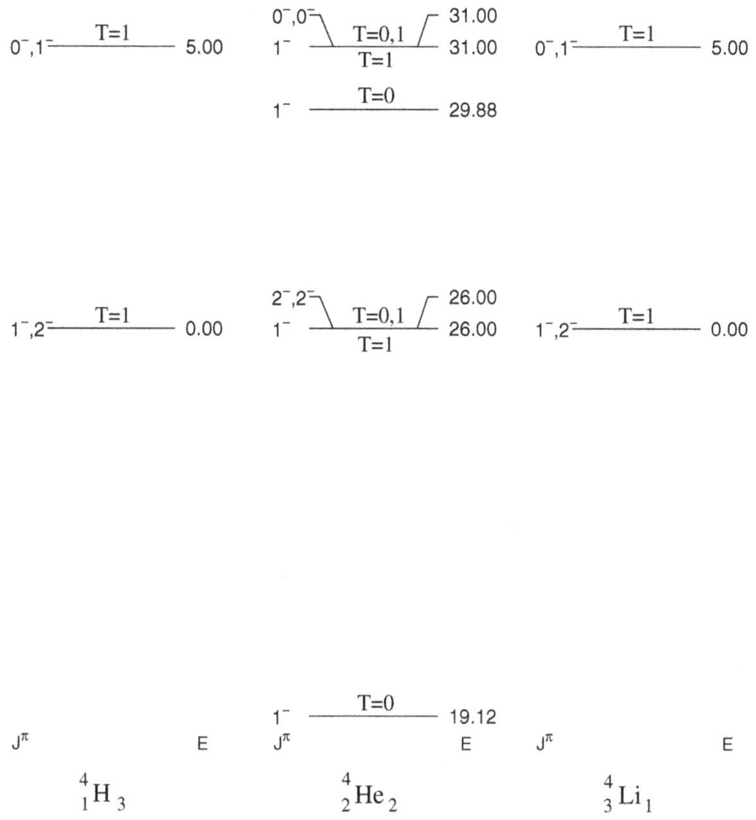

**Fig. 10.1.** Computed spectra of $^4$H, $^4$He and $^4$Li. The TDA and pnTDA were applied in the $0p$-$(0s_{1/2})^{-1}$ particle–hole valence space. The single-particle energies were $\varepsilon_{0s_{1/2}} = 0$, $\varepsilon_{0p_{3/2}} = 24.0\,\mathrm{MeV}$ and $\varepsilon_{0p_{1/2}} = 29.0\,\mathrm{MeV}$, and the SDI parameters $A_0 = A_1 = 2.0\,\mathrm{MeV}$

are a minor discrepancy because they lie so close to each other that small perturbations can change their relative positions. As expected from isospin symmetry, the same conclusions apply to $^{16}_9\mathrm{F}_7$, whose first four energies are $E(0^-) = 0$, $E(1^-) = 0.19\,\mathrm{MeV}$, $E(2^-) = 0.42\,\mathrm{MeV}$ and $E(3^-) = 0.72\,\mathrm{MeV}$.

Figure 10.3 shows the results of pnTDA calculations for $^{40}_{19}\mathrm{K}_{21}$ in the particle–hole valence spaces $0f_{7/2}$-$(0d_{3/2})^{-1}$ and $0f_{7/2}$-$(0d$-$1s)^{-1}$. The single-particle energies and SDI parameters were the same as for $^{40}_{20}\mathrm{Ca}_{20}$ in Fig. 9.4, namely from Fig. 9.2 (b), and $A_0 = 0.85\,\mathrm{MeV}$ and $A_1 = 0.90\,\mathrm{MeV}$.

The correspondence between the lowest four theoretical and experimental states is excellent. The higher-lying states, contained in the larger valence

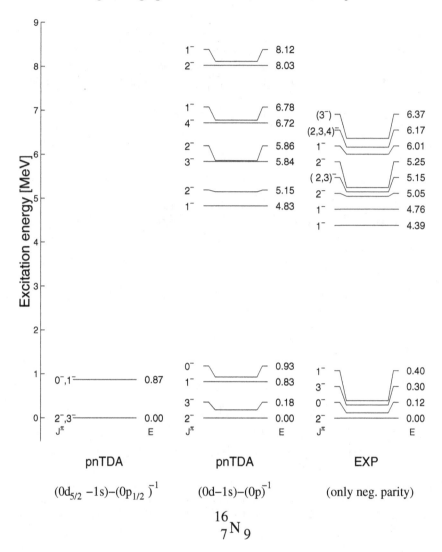

**Fig. 10.2.** Energy spectra of $^{16}$N computed in the pnTDA in the particle–hole valence spaces $(0d_{5/2}\text{-}1s_{1/2})\text{-}(0p_{1/2})^{-1}$ and $(0d\text{-}1s)\text{-}(0p)^{-1}$. The single-particle energies are from Fig. 9.2 (a) and the SDI parameters are $A_0 = A_1 = 1.0$ MeV. The experimental energy spectrum is shown on the right for comparison

space, are predicted somewhat too high. The computed spectra also apply to $^{40}_{21}\text{Sc}_{19}$, whose tentatively known first four experimental energies are virtually identical to those of $^{40}_{19}\text{K}_{21}$.

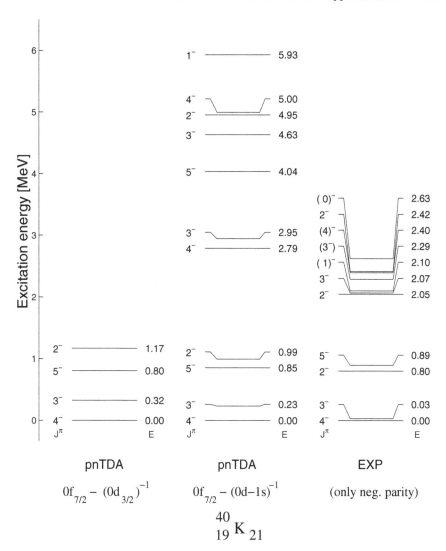

**Fig. 10.3.** Energy spectra of $^{40}$K computed in the pnTDA in the particle–hole valence spaces $0f_{7/2}$-$(0d_{3/2})^{-1}$ and $0f_{7/2}$-$(0d$-$1s)^{-1}$. The single-particle energies are from Fig. 9.2 (b) and the SDI parameters are $A_0 = 0.85\,\text{MeV}$ and $A_1 = 0.90\,\text{MeV}$. The experimental energy spectrum is shown on the right for comparison

## 10.2 Electromagnetic Transitions in the pnTDA

Electromagnetic decays within the pnTDA framework describe transitions between states of an odd–odd nucleus. Experimentally such transitions are not particularly well known. The reduced transition probabilities and half-lives can be obtained by using the formalism of Sect. 6.1.

### 10.2.1 Transition Amplitudes

We treat electromagnetic transitions within the pnTDA by extending the formalism of Subsect. 6.4.3 to include configuration mixing in the charge-changing particle–hole excitations. The configuration mixing in the TDA is described by the eigenvectors (9.34). In the pnTDA they take the form

$$
\begin{aligned}
|\omega\rangle &= \sum_{pn} X^{\omega}_{pn} |p\, n^{-1}\,;\, JM\rangle \quad \text{for } pn^{-1} \text{ nuclei}\,, \\
|\omega\rangle &= \sum_{np} X^{\omega}_{np} |n\, p^{-1}\,;\, JM\rangle \quad \text{for } np^{-1} \text{ nuclei}\,.
\end{aligned}
\tag{10.19}
$$

With the wave functions (10.19) the electromagnetic decay amplitudes can be expressed for the electric $(\mathcal{M}_{\sigma\lambda} = \boldsymbol{Q}_\lambda)$ and magnetic $(\mathcal{M}_{\sigma\lambda} = \boldsymbol{M}_\lambda)$ transition operators of Subsect. 6.1.3 as

$$
\begin{aligned}
(\omega_f\|\mathcal{M}_{\sigma\lambda}\|\omega_i) &= \sum_{\substack{p_i n_i \\ p_f n_f}} X^{\omega_f *}_{p_f n_f} X^{\omega_i}_{p_i n_i} (p_f\, n_f^{-1}\,;\, J_f\|\mathcal{M}_{\sigma\lambda}\|p_i\, n_i^{-1}\,;\, J_i)\,, \\
(\omega_f\|\mathcal{M}_{\sigma\lambda}\|\omega_i) &= \sum_{\substack{n_i p_i \\ n_f p_f}} X^{\omega_f *}_{n_f p_f} X^{\omega_i}_{n_i p_i} (n_f\, p_f^{-1}\,;\, J_f\|\mathcal{M}_{\sigma\lambda}\|n_i\, p_i^{-1}\,;\, J_i)\,,
\end{aligned}
\tag{10.20}
$$

where the first equation applies to a proton-particle–neutron-hole nucleus and the second equation to a neutron-particle–proton-hole nucleus. The particle–hole transition matrix elements are given in (6.147)–(6.150).

### 10.2.2 Application to the E2 Transition $0_1^- \to 2_{\mathrm{gs}}^-$ in $^{16}_{7}\mathrm{N}_9$

Consider the decay of the first $0^-$ state to the $2^-$ ground state in $^{16}_{7}\mathrm{N}_9$ by an E2 transition. This example was discussed in Subsect. 6.4.4 with the simple, unmixed wave functions (6.151) and (6.152). The relevant formula for the particle–hole transition matrix elements is (6.150).

We take the configuration-mixed $np^{-1}$ wave functions (10.19) from the large-space calculation of Fig. 10.2. With the $0^-$ particle–hole basis

$$
\{|1\rangle_0\,,\,|2\rangle_0\} = \{|\nu 1\mathrm{s}_{1/2}\,(\pi 0\mathrm{p}_{1/2})^{-1}\,;\, 0^-\rangle\,,\, |\nu 0\mathrm{d}_{3/2}\,(\pi 0\mathrm{p}_{3/2})^{-1}\,;\, 0^-\rangle\}\,,
\tag{10.21}
$$

the $0_1^-$ state is

$$|^{16}\text{N}\,;\,0_1^-\rangle = X_1^{0_1^-}|1\rangle_0 + X_2^{0_1^-}|2\rangle_0 = 0.993|1\rangle_0 + 0.121|2\rangle_0 \,. \tag{10.22}$$

The $2^-$ particle–hole basis states are

$$\{|1\rangle_2 , |2\rangle_2 , |3\rangle_2 , |4\rangle_2 , |5\rangle_2\}$$
$$= \{|\nu 0d_{5/2}\,(\pi 0p_{3/2})^{-1}\,;\,2^-\rangle, |\nu 0d_{5/2}\,(\pi 0p_{1/2})^{-1}\,;\,2^-\rangle,$$
$$|\nu 1s_{1/2}\,(\pi 0p_{3/2})^{-1}\,;\,2^-\rangle, |\nu 0d_{3/2}\,(\pi 0p_{3/2})^{-1}\,;\,2^-\rangle,$$
$$|\nu 0d_{3/2}\,(\pi 0p_{1/2})^{-1}\,;\,2^-\rangle\} \,. \tag{10.23}$$

The $2_1^-$ state is

$$|^{16}\text{N}\,;\,2_1^-\rangle = X_1^{2_1^-}|1\rangle_2 + X_2^{2_1^-}|2\rangle_2 + X_3^{2_1^-}|3\rangle_2 + X_4^{2_1^-}|4\rangle_2 + X_5^{2_1^-}|5\rangle_2$$
$$= 0.116|1\rangle_2 + 0.986|2\rangle_2 + 0.116|3\rangle_2 + 0.037|4\rangle_2 + 0.010|5\rangle_2 \,. \tag{10.24}$$

To be able to use (6.150) for the $np^{-1}$ matrix elements we first calculate the $2^- \to 0^-$ transition amplitude, as was done in Subsect. 6.4.4. Table 6.4 gives the single-particle matrix elements needed in (6.150); effective charges are included. The wave-function amplitudes (10.22) and (10.24) and the $np^{-1}$ matrix elements are then inserted into the second equation (10.20) to give the transition amplitude

$$(0_1^-\|\boldsymbol{Q}_2\|2_1^-) = \sum_{i=1}^{2}\sum_{j=1}^{5} X_i^{0_1^-} X_j^{2_1^-}\,_0(i\|\boldsymbol{Q}_2\|j)_2$$
$$= 0 \times X_1^{0_1^-} X_1^{2_1^-} - 1.545b^2 e_{\text{eff}}^{\text{n}} X_1^{0_1^-} X_2^{2_1^-} - 0.997b^2 e_{\text{eff}}^{\text{p}} X_1^{0_1^-} X_3^{2_1^-}$$
$$+ 0 \times X_1^{0_1^-} X_4^{2_1^-} + 1.261b^2 e_{\text{eff}}^{\text{n}} X_1^{0_1^-} X_5^{2_1^-} - 0.647b^2 e_{\text{eff}}^{\text{n}} X_2^{0_1^-} X_1^{2_1^-}$$
$$+ 0 \times X_2^{0_1^-} X_2^{2_1^-} - 0.892b^2 e_{\text{eff}}^{\text{n}} X_2^{0_1^-} X_3^{2_1^-}$$
$$+ (0.705e_{\text{eff}}^{\text{p}} - 0.988e_{\text{eff}}^{\text{n}})b^2 X_2^{0_1^-} X_4^{2_1^-} + 0.705b^2 e_{\text{eff}}^{\text{p}} X_2^{0_1^-} X_5^{2_1^-}$$
$$= -0.111b^2 e_{\text{eff}}^{\text{p}} - 1.526b^2 e_{\text{eff}}^{\text{n}} = -(0.330 + 4.871\chi)e\,\text{fm}^2 \,. \tag{10.25}$$

The oscillator length $b = 1.725\,\text{fm}$ from Subsect. 6.4.4 and the definition (6.26) of the electric polarization constant were used in the last step.

By the symmetry relation (2.32) the $2^- \to 0^-$ transition amplitude (10.25) is equal to the one for $0^- \to 2^-$. Equation (6.4) then gives the reduced transition probability

$$B(\text{E2}\,;\,0_1^- \to 2_{\text{gs}}^-) = (0.330 + 4.871\chi)^2\, e^2\text{fm}^4 \,. \tag{10.26}$$

Comparison with the experimental result $4.3\,e^2\text{fm}^4$ given in (6.155) leads to

$$\chi = 0.36 . \tag{10.27}$$

This is not far from the value $\chi = 0.45$ obtained in the unmixed case of Subsect. 6.4.4, which is not surprising because the leading amplitudes in the wave functions (10.22) and (10.24) are very near unity. Nevertheless, configuration mixing does reduce the polarization constant.

## 10.3 Beta-Decay Transitions in the pnTDA

The pnTDA states of an odd–odd nucleus allow for two kinds of beta-decay transition. The first kind consists of decays to the ground state of the reference nucleus. The second kind consists of transitions from a pnTDA state to a TDA state, i.e. from a state in an odd–odd nucleus to excited states of the reference nucleus.

### 10.3.1 Transitions to the Particle–Hole Vacuum

Consider beta decay to the particle–hole vacuum, i.e. the ground state of the even–even pnTDA reference nucleus. The initial state is in a neighbouring odd–odd nucleus and is described by one of the wave functions (10.19). The transition amplitudes are then

$$(\mathrm{HF}\|\beta_L^-\|\omega) = \sum_{np} X_{np}^\omega (\mathrm{HF}\|\beta_L^-\|n\,p^{-1}\,;\,J) , \tag{10.28}$$

$$(\mathrm{HF}\|\beta_L^+\|\omega) = \sum_{pn} X_{pn}^\omega (\mathrm{HF}\|\beta_L^+\|p\,n^{-1}\,;\,J) . \tag{10.29}$$

Equations (7.70) and (7.59) give the reduced matrix element on the right-hand side of (10.28) as

$$(\mathrm{HF}\|\beta_L^-\|n\,p^{-1}\,;\,J) = \widehat{L}^{-1} \sum_{p'n'} (p'\|\beta_L\|n')(\mathrm{HF}\| \left[ c_{p'}^\dagger \tilde{c}_{n'} \right]_L \|n\,p^{-1}\,;\,J)$$

$$= \delta_{LJ}(-1)^{j_n - j_p + J}(p\|\beta_L\|n) . \tag{10.30}$$

This relation is true for a generic tensor operator $\beta_L^-$ describing beta-minus decay.[1] For *allowed* decay we identify the single-particle matrix element as $(p\|\beta_L\|n) = \widehat{L}\mathcal{M}_L(pn)$ according to (7.73) and (7.74); the two values of $L$ are $L = 0$ for Fermi transitions and $L = 1$ for Gamow–Teller transitions. The single-particle symmetry relations (7.22) and (7.23) then simplify (10.30) to

$$(\mathrm{HF}\|\beta_L^-\|n\,p^{-1}\,;\,J) = \delta_{LJ}(-1)^J \widehat{J}\mathcal{M}_J(np) . \tag{10.31}$$

---

[1] The value of $L$ is not sufficient to identify the type of decay because of the possibility of forbidden non-unique decay; see Table 7.10. However, to keep the notation simple we ignore this possibility.

Similarly we obtain

$$(\text{HF}\|\beta_L^+\|p\,n^{-1};J) = \delta_{LJ}(-1)^J \hat{J} \mathcal{M}_J(pn) . \tag{10.32}$$

Substitution of (10.31) and (10.32) into (10.28) and (10.29) results in the formulas

$$\boxed{\begin{aligned}
(\text{HF}\|\beta_L^-\|\omega) &= \delta_{LJ}(-1)^J \hat{J} \sum_{np} X_{np}^{\omega} \mathcal{M}_J(np) , \\
(\text{HF}\|\beta_L^+\|\omega) &= \delta_{LJ}(-1)^J \hat{J} \sum_{pn} X_{pn}^{\omega} \mathcal{M}_J(pn) ,
\end{aligned}} \tag{10.33}$$

valid for allowed beta decay.

Let us next consider $K$*th-forbidden unique* beta decay ($K = 1, 2, 3, \ldots$). Equation (7.201), transcribed into the present notation, gives

$$(\text{HF}\|\beta_{K+1}^-\|n\,p^{-1};J) = \delta_{K+1,J}(-1)^J \hat{J} \mathcal{M}^{(Ku)}(np) . \tag{10.34}$$

The analogous relation for $\beta^+$ decay is

$$(\text{HF}\|\beta_{K+1}^+\|p\,n^{-1};J) = \delta_{K+1,J}(-1)^J \hat{J} \mathcal{M}^{(Ku)}(pn) . \tag{10.35}$$

For $K$th-forbidden unique beta decay the equations corresponding to (10.33) become

$$\boxed{\begin{aligned}
(\text{HF}\|\beta_{K+1}^-\|\omega) &= \delta_{K+1,J}(-1)^J \hat{J} \sum_{np} X_{np}^{\omega} \mathcal{M}^{(Ku)}(np) , \\
(\text{HF}\|\beta_{K+1}^+\|\omega) &= \delta_{K+1,J}(-1)^J \hat{J} \sum_{pn} X_{pn}^{\omega} \mathcal{M}^{(Ku)}(pn) .
\end{aligned}} \tag{10.36}$$

The following example illustrates application of the theory to first-forbidden unique beta decay.

## 10.3.2 First-Forbidden Unique Beta Decay of $^{16}_{7}\text{N}_9$

Subsection 7.6.3 treats the first-forbidden unique $\beta^-$ decay of the $2^-$ ground state of $^{16}_{7}\text{N}_9$ to the $0^+$ ground state of $^{16}_{8}\text{O}_8$ with unmixed states. We now extend the treatment to include configuration mixing in the initial $2^-$ state by using the pnTDA wave function given in (10.23) and (10.24).

We apply the first equation (10.36). The single-particle matrix elements $\mathcal{M}^{(1u)}(np)$ are given by (7.168) and (7.169). With our usual value $b = 1.725\,\text{fm}$ for $A = 16$, (7.168) gives

$$\mathcal{M}^{(1u)}(np) = 5.158 \times 10^{-3} m^{(1u)}(np) . \tag{10.37}$$

Using the labels $i = 1$–5 of (10.23) for the $np$ and taking the $m^{(1u)}(i)$ from Table 7.6, we have

$$(\mathrm{HF}\|\beta_2^-\|2_1^-) = \sqrt{5} \sum_i X_i^{2_1^-} \mathcal{M}^{(1\mathrm{u})}(i)$$

$$= \sqrt{5} \times 5.158 \times 10^{-3} \left( -\sqrt{\tfrac{21}{5}} X_1^{2_1^-} + 2\sqrt{\tfrac{6}{5}} X_2^{2_1^-} - \sqrt{2} X_3^{2_1^-} \right.$$

$$\left. - \tfrac{2}{\sqrt{5}} X_4^{2_1^-} + \tfrac{1}{\sqrt{5}} X_5^{2_1^-} \right) = 0.0200 . \tag{10.38}$$

With the relations (7.165) this gives

$$B_{1\mathrm{u}} = \frac{1.25^2}{5} \times 0.0200^2 = 1.25 \times 10^{-4} , \quad \log f_{1\mathrm{u}} t = 8.77 . \tag{10.39}$$

The new $\log ft$ value 8.8 is to be compared with the unmixed value 8.6 given in (7.180) and the experimental value 9.1. The improvement comes from the small negative contributions to the transition amplitude (10.38) dominated by the first positive term.

### 10.3.3 Transitions between Particle–Hole States

Beta-decay transitions between two particle–hole excitations were addressed in Subsect. 7.4.3 and all decay matrix elements were derived. When configuration mixing is included at the TDA level of approximation, the initial pnTDA state is given by (10.19) and the final TDA state by

$$|\omega_f\rangle = \sum_{p_f p_f'} X_{p_f p_f'}^{\omega_f} |p_f\, p_f'^{-1}; J_f\, M_f\rangle + \sum_{n_f n_f'} X_{n_f n_f'}^{\omega_f} |n_f\, n_f'^{-1}; J_f\, M_f\rangle , \tag{10.40}$$

where we have explicitly separated the proton and neutron particle–hole components.

For $\beta^-$ decay to the final state (10.40), the initial state is the neutron-particle–proton-hole state in (10.19). This gives the decay amplitude

$$\begin{aligned}
(\omega_f\|\beta_L^-\|\omega_i) &= \sum_{\substack{n_i p_i \\ p_f p_f'}} X_{p_f p_f'}^{\omega_f *} X_{n_i p_i}^{\omega_i} (p_f\, p_f'^{-1}; J_f\|\beta_L^-\|n_i\, p_i^{-1}; J_i) \\
&+ \sum_{\substack{n_i p_i \\ n_f n_f'}} X_{n_f n_f'}^{\omega_f *} X_{n_i p_i}^{\omega_i} (n_f\, n_f'^{-1}; J_f\|\beta_L^-\|n_i\, p_i^{-1}; J_i) ,
\end{aligned} \tag{10.41}$$

with the particle–hole transition matrix elements given by (7.76) and (7.72). In $\beta^+$ decay the initial state is the proton-particle–neutron-hole state in (10.19), and the decay amplitude becomes

$$\begin{aligned}
(\omega_f\|\beta_L^+\|\omega_i) &= \sum_{\substack{p_i n_i \\ p_f p_f'}} X_{p_f p_f'}^{\omega_f *} X_{p_i n_i}^{\omega_i} (p_f\, p_f'^{-1}; J_f\|\beta_L^+\|p_i\, n_i^{-1}; J_i) \\
&+ \sum_{\substack{p_i n_i \\ n_f n_f'}} X_{n_f n_f'}^{\omega_f *} X_{p_i n_i}^{\omega_i} (n_f\, n_f'^{-1}; J_f\|\beta_L^+\|p_i\, n_i^{-1}; J_i) ,
\end{aligned} \tag{10.42}$$

with the particle–hole transition matrix elements given by (7.80) and (7.79).

To illustrate the use of the formalism we revisit the example of Subsect. 7.4.4 below.

## 10.3.4 Allowed Beta Decay of $^{16}_{7}N_9$ to Excited States in $^{16}_{8}O_8$

Referring to Fig. 7.7, consider the Gamow–Teller and Fermi $\beta^-$ decay of the $2^-$ ground state of $^{16}_{7}N_9$ to the $1^-_1$, $2^-_1$ and $3^-_1$ states in $^{16}_{8}O_8$. The decay rates for these transitions were calculated in Subsect. 7.4.4 in the mean-field approximation, i.e. without configuration mixing. We now extend that treatment by assuming the particle–hole valence space $(0d-1s)-(0p_{1/2})^{-1}$.

The initial $2^-_{gs}$ state is to be constructed in the $2^-$ pnTDA basis

$$\{|\nu_1\pi\rangle_2\,,\,|\nu_2\pi\rangle_2\} = \{|\nu 0d_{5/2}\,(\pi 0p_{1/2})^{-1}\,;\,2^-\rangle\,,\,|\nu 0d_{3/2}\,(\pi 0p_{1/2})^{-1}\,;\,2^-\rangle\}\,. \tag{10.43}$$

Note that in the absence of the $0p_{3/2}$ proton-hole orbital this is a smaller basis than (10.23).

To find the wave function of the $2^-_{gs}$ state, we turn to Subsect. 9.1.3. From (5.128) we see that the state has $T = 1$, so the SDI matrix element is given by (9.32). The relevant entries in Table 9.2 are $\mathcal{M}_{1232}(21) = 0$ and $\mathcal{M}_{1232}(20) = 0$, whence

$$_2\langle\nu_1\pi|V_{\mathrm{RES}}|\nu_2\pi\rangle_2 = 0\,. \tag{10.44}$$

The initial state thus remains unmixed in this valence space, i.e. unchanged from (7.86).

The TDA basis states for $J = 3^-$ in $^{16}O$ are

$$\{|\pi\rangle_3\,,\,|\nu\rangle_3\} = \{|\pi 0d_{5/2}\,(\pi 0p_{1/2})^{-1}\,;\,3^-\rangle\,,\,|\nu 0d_{5/2}\,(\nu 0p_{1/2})^{-1}\,;\,3^-\rangle\}\,. \tag{10.45}$$

These are precisely the basis states appearing in the mean-field wave function (7.87), which then remains valid in the present basis. Since the $2^-_{gs}$ initial state also remained unchanged it follows that the decay amplitude for the $2^-_{gs} \to 3^-_1$ transition is given by (7.97).

For $2^-$ states the TDA basis becomes

$$\begin{aligned}
\{|\pi_1\rangle_2\,,\,|\pi_2\rangle_2\,,\,|\nu_1\rangle_2\,,\,|\nu_2\rangle_2\} \\
= \{|\pi 0d_{5/2}\,(\pi 0p_{1/2})^{-1}\,;\,2^-\rangle\,,\,|\pi 0d_{3/2}\,(\pi 0p_{1/2})^{-1}\,;\,2^-\rangle\,, \\
|\nu 0d_{5/2}\,(\nu 0p_{1/2})^{-1}\,;\,2^-\rangle\,,\,|\nu 0d_{3/2}\,(\nu 0p_{1/2})^{-1}\,;\,2^-\rangle\}\,.
\end{aligned} \tag{10.46}$$

However, for our standard choice $A_0 = A_1$ of the SDI parameters, the Hamiltonian matrix is diagonal because the off-diagonal matrix elements (9.28)–(9.30) with substitutions from Table 9.2 are zero. This means that the mean-field wave function (7.87) is recovered also for $2^-_1$. Consequently, the $2^-_{gs} \to 2^-_1$ Gamow–Teller transition amplitude (7.96) remains valid, and the Fermi amplitude (7.98) remains zero.

The mean-field $1_1^-$ state (7.88) gave a zero transition amplitude (7.95) for $2_{\mathrm{gs}}^- \to 1_1^-$. Configuration mixing is therefore needed for a non-vanishing decay rate. Our current particle–hole space provides the $1^-$ basis

$$\{|\pi_1\rangle_1 , |\pi_2\rangle_1 , |\nu_1\rangle_1 , |\nu_2\rangle_1\}$$
$$= \{|\pi 1s_{1/2} (\pi 0p_{1/2})^{-1} ; 1^-\rangle , |\pi 0d_{3/2} (\pi 0p_{1/2})^{-1} ; 1^-\rangle ,$$
$$|\nu 1s_{1/2} (\nu 0p_{1/2})^{-1} ; 1^-\rangle , |\nu 0d_{3/2} (\nu 0p_{1/2})^{-1} ; 1^-\rangle\} . \quad (10.47)$$

With the single-particle energies from Fig. 9.2 (a) the Hamiltonian matrix becomes

$$H_{\mathrm{TDA}}(1^-) =$$

$$\begin{pmatrix} 12.47 + 0.333A_1 & -0.943A_1 & \frac{1}{2}(0.333A_1 - 1.667A_0) & \frac{1}{2}(-0.943A_1 - 0.943A_0) \\ -0.943A_1 & 16.68 - 0.333A_1 & \frac{1}{2}(-0.943A_1 - 0.943A_0) & \frac{1}{2}(-0.333A_1 - 2.333A_0) \\ \cdots & \cdots & 12.47 + 0.333A_1 & -0.943A_1 \\ \cdots & \cdots & -0.943A_1 & 16.68 - 0.333A_1 \end{pmatrix} ,$$
$$(10.48)$$

Diagonalizing this matrix with $A_0 = A_1 = 1.0 \, \mathrm{MeV}$ yields

$$|^{16}\mathrm{O} ; 1_1^-\rangle = 0.634|\pi_1\rangle_1 + 0.314|\pi_2\rangle_1 + 0.634|\nu_1\rangle_1 + 0.314|\nu_2\rangle_1 . \quad (10.49)$$

As noted above, the leading components $|\pi_1\rangle_1$ and $|\nu_1\rangle_1$ do not connect to the initial $2^-$ ground state via Gamow–Teller single-particle matrix elements. Equations (7.76) and (7.72) with Table 7.3 give the matrix elements for the remaining components as

$$\left(\pi 0d_{3/2} (\pi 0p_{1/2})^{-1} ; 1^- \|\beta_{\mathrm{GT}}^-\| \nu 0d_{5/2} (\pi 0p_{1/2})^{-1} ; 2^-\right) = \frac{6}{\sqrt{5}} , \quad (10.50)$$

$$\left(\nu 0d_{3/2} (\nu 0p_{1/2})^{-1} ; 1^- \|\beta_{\mathrm{GT}}^-\| \nu 0d_{5/2} (\pi 0p_{1/2})^{-1} ; 2^-\right) = 0 . \quad (10.51)$$

Now from (10.41) we can obtain the decay amplitude

$$(1_1^- \|\beta_{\mathrm{GT}}^-\| 2_{\mathrm{gs}}^-) = 0.314 \times 1.0 \times \frac{6}{\sqrt{5}} = 0.843 . \quad (10.52)$$

By (7.14) and (7.33) the reduced transition probability and $\log ft$ value become

$$B_{\mathrm{GT}}(2_{\mathrm{gs}}^- \to 1_1^-) = \frac{1.25^2}{5} \times 0.843^2 = 0.222 , \quad \log ft = 4.44 . \quad (10.53)$$

Figure 7.7 gives the experimental $\log ft$ value as 5.1. We conclude that configuration mixing indeed produces a finite $2_{\mathrm{gs}}^- \to 1_1^-$ transition rate, which is a radical improvement over the mean-field description.

With the $\log ft$ values (7.101), (7.102) and (10.53) for the Gamow–Teller decays and (10.39) for the first-forbidden unique decay, and the vanishing

Fermi transition rate, we proceed to compute the partial and total decay half-lives. The phase-space factors for the Gamow–Teller decays are given by (7.30). The phase space factor (7.174) for the first-forbidden unique decay we already have from (7.181); its value $4.62 \times 10^7$ contains the Primakoff–Rosen factor 1.19 needed also for the Gamow–Teller decays.

Using (7.17) and extracting the experimental partial half-lives from the decay branchings by (7.39) we arrive at the numbers collected into Table 10.1. The computed partial half-lives are too short, especially for the transitions to the $1_1^-$ and $2_1^-$ states. The theoretical total half-life resulting from the partial half-lives of Table 10.1 by use of (7.37) is

$$t_{1/2}^{(\text{tot})}(\text{th}) = 2.75\,\text{s} . \tag{10.54}$$

In view of the wide range of beta-decay half-lives, this is not far from the experimental total half-life of 7.13 s shown in Fig. 7.7.

**Table 10.1.** Calculated and experimental data for the beta decay of the $2^-$ ground state of $^{16}_7\text{N}_9$ to the $^{16}_8\text{O}_8$ states $0_{\text{gs}}^+$, $3_1^-$, $1_1^-$ and $2_1^-$; see Fig. 7.7

| $J_f^\pi$ | $E_0$ | $\log f^{(-)}$ | $\log ft(\text{th})$ | $\log ft(\text{exp})$ | $t_{1/2}(\text{th})$ (s) | $t_{1/2}(\text{exp})$ (s) |
|---|---|---|---|---|---|---|
| $0_{\text{gs}}^+$ | 21.39 | 7.66 | 8.77 | 9.1 | 12.9 | 27.4 |
| $3_1^-$ | 9.39 | 3.46 | 4.07 | 4.5 | 4.1 | 10.5 |
| $1_1^-$ | 7.46 | 2.95 | 4.44 | 5.1 | 31 | 146 |
| $2_1^-$ | 4.03 | 1.58 | 3.59 | 4.3 | 102 | 648 |

The decay to $0_{\text{gs}}^+$ is first-forbidden unique; the other three are Gamow–Teller decays. The experimental quantity $E_0$ is the dimensionless endpoint energy used in calculating the phase-space factor $f^{(-)}$.

## Epilogue

In this chapter we have formulated the proton–neutron TDA. It serves to generate states in an odd–odd nucleus by starting from the adjacent doubly magic even–even reference nucleus. This formulation enabled us to calculate electromagnetic transitions in an odd–odd nucleus and beta-decay transitions from an odd–odd nucleus to its doubly even reference nucleus. In the following chapter we reach the climax of particle–hole theories when formulating the random phase approximation, RPA. The RPA produces a correlated ground state, which increases the collectivity of transitions beyond the TDA results.

## Exercises

**10.1.** Derive Eq. (10.5).

**10.2.** Verify the numbers in the matrix (10.16).

**10.3.** Verify the numbers in the matrix (10.48).

**10.4.** Verify the wave function (10.49).

**10.5.** Verify the value of the particle–hole transition matrix element (10.50).

**10.6.** Verify the numbers in Table 10.1.

**10.7.** Compute the energies of the $0^-$, $1^-$, $2^-$ and $3^-$ states of $^{16}_7N_9$ in the $(0d_{5/2}\text{-}1s_{1/2})\text{-}(0p)^{-1}$ particle–hole valence space. Use the single-particle energies of Fig. 9.2 (a) and the SDI. Try to fit the parameters $A_0$ and $A_1$ to reproduce the experimental energy levels.

**10.8.** Compute the energies of the $0^-$ and $3^-$ states of $^{16}_7N_9$ in the complete $(0d\text{-}1s)\text{-}(0p)^{-1}$ particle–hole valence space by using the single-particle energies of Fig. 9.2 (a) and the SDI with parameters $A_0 = A_1 = 1.0\,\text{MeV}$.

**10.9.** Compute the energies of the $2^-$ states of $^{16}_7N_9$ under the conditions of Exercise 10.8.

**10.10.** Compute the energies of the $2^-$, $3^-$, $4^-$ and $5^-$ states of $^{40}_{19}K_{21}$ in the $0f_{7/2}\text{-}(0d_{3/2})^{-1}$ particle–hole valence space. Assume an energy gap of $6.0\,\text{MeV}$ between the two single-particle levels and use the SDI. Try to fit the parameters $A_0$ and $A_1$ to reproduce the experimental energies.

**10.11.** Treat the nucleus $^{32}_{16}S_{16}$ as doubly magic in the $0d_{3/2}\text{-}(0d_{5/2}\text{-}1s_{1/2})^{-1}$ particle–hole valence space. Use it as the reference nucleus for $^{32}_{15}P_{17}$. Compute all states of $^{32}_{15}P_{17}$ in this valence space. Use the single-particle energies (8.124) and the SDI with parameters $A_0 = A_1 = 1.0\,\text{MeV}$. Compare with experimental data and comment.

**10.12.** Calculate the structure of the $4^-$ ground state of $^{40}_{21}Sc_{19}$ by diagonalizing the pnTDA matrix in the complete $0f_{7/2}\text{-}(0d\text{-}1s)^{-1}$ particle–hole valence space. Take the single-particle energies from Fig. 9.2 (b) and use the SDI with parameters $A_0 = A_1 = 1.0\,\text{MeV}$.

**10.13.** Calculate the $\log ft$ values of the $\beta^+$ transitions from the $4^-$ ground state of $^{40}Sc$ to the $5^-_1$, $4^-_1$ and $4^-_2$ states of $^{40}Ca$. Use the wave functions from Exercises 10.12 and 9.32. Compare with the experimental data shown in Fig. 7.12 and comment.

**10.14.** Using the wave function of Exercise 10.12 compute the $\log ft$ value of the third-forbidden unique beta-decay transition $4^-_{\text{gs}}(^{40}Sc) \rightarrow 0^+_{\text{gs}}(^{40}Ca)$.

**10.15.** Using the wave functions of Exercise 10.7 compute the reduced transition probability $B(E2\,;\,0^-_1 \rightarrow 2^-_1)$ in $^{16}N$. Deduce from the experimental half-life the value of the electric polarization constant $\chi$.

**10.16.** Using the wave functions of Exercise 10.11 compute the reduced transition probability $B(\text{M1}\,;2^+_1 \rightarrow 1^+_1)$ in $^{32}\text{P}$. Use the bare $g$ factors (6.25). Compute the half-life of the $2^+_1$ state by taking the gamma energy from experiment. Compare with the data and comment.

**10.17.** Using the wave functions of Exercise 10.10 compute the half-life of the $3^-_1$ state in $^{40}\text{K}$. Use the bare $g$ factors (6.25) and the experimental gamma energy. Compare with the data and comment.

**10.18.** Using the wave functions of Exercise 10.10 and the experimental gamma energies compute the half-lives of the $2^-_1$ and $5^-_1$ states in $^{40}\text{K}$. Assume the bare $g$ factors (6.25) and the value $\chi = 1$ for the electric polarization constant. Compare with the data and comment.

# 11

# The Random-Phase Approximation

## Prologue

In this chapter we extend the TDA particle–hole formalism of Chap. 9 to include correlations in the nuclear ground state. This sophisticated particle–hole formalism is called the random-phase approximation (RPA). In this description the simple Hartree–Fock particle–hole vacuum is replaced by a correlated ground state involving many-particle–many-hole excitations of the simple particle–hole vacuum. The resulting configuration mixing in excited states is more involved in the RPA than it is in the TDA. The ground-state correlations induce both particle–hole and hole–particle components in the RPA wave function.

The correlations of the RPA ground state can lead to strong collectivity of calculated electromagnetic decay rates. The traditional example is the electric octupole decay of the first $3^-$ state in $^{16}$O. As in the case of the TDA, a schematic separable model can be devised for the RPA. The schematic model is convenient for discussing general features of the RPA solutions of excitation energies and electromagnetic decay rates.

## 11.1 The Equations-of-Motion Method

An instructive way to arrive at a more sophisticated particle–hole theory than the TDA of Chap. 9 is to use the so-called *equations-of-motion* (EOM) method. This method can be used to derive the mean-field Hartree–Fock equation and the TDA, but also more sophisticated theories like the RPA, higher-RPA theories, etc. The basic idea behind the EOM is to avoid an explicit calculation of the ground state, which may be very complicated. Instead, the ground state is written implicitly and the excited states are obtained by performing simple operations on it. This enables a relatively easy calculation of the energy spectra and transition matrix elements between the ground and excited states.

In the previously discussed many-body approaches the ground state and the excited states were calculated separately, independently of each other. This is the thread of the shell-model philosophy. To simplify this scheme of calculation, the EOM lets the chosen form of excitations dictate the compatible form of the ground state. Hence the calculation of the ground state and that of the excited states are not independent. In fact, the relation between these two determines the excitation energies and the amplitudes for transitions between the ground and excited states. Operating with the relation between the ground state and the excited states enables one to avoid writing down the explicit form of the ground state.

The EOM method was introduced by Rowe [14]. Its derivation is analogous to the derivation of the method for solving the harmonic oscillator problem by using the so-called ladder operators (see, e.g. [3,6]). These operators create excited states from the oscillator vacuum, whose explicit form is not needed for obtaining the excitation spectrum or transition matrix elements.

### 11.1.1 Derivation of the Equations of Motion

By analogy to the harmonic oscillator problem, we can start from a postulated basic relation between the ground state and an excited state, written as

$$|\omega\rangle = Q_\omega^\dagger |0\rangle \ . \tag{11.1}$$

It is required that the annihilation operator $Q_\omega$, which corresponds to the creation operator $Q_\omega^\dagger$, deletes the vacuum $|0\rangle$, i.e.

$$Q_\omega |0\rangle = 0 \quad \text{for all } \omega \ . \tag{11.2}$$

Furthermore, the excitation is assumed to satisfy the Schrödinger equation

$$H|\omega\rangle = E_\omega |\omega\rangle \ . \tag{11.3}$$

The above relations allow us to write

$$[H, Q_\omega^\dagger]|0\rangle = (E_\omega - E_0)Q_\omega^\dagger|0\rangle \ , \tag{11.4}$$

where $E_0$ is the energy of the ground state, i.e.

$$H|0\rangle = E_0|0\rangle \ . \tag{11.5}$$

Equation (11.4) constitutes the equation of motion of the excitation creation operator $Q_\omega^\dagger$.

Consider the operator $Q_\omega^\dagger$ as composed of a set of blocks $\delta Q^\dagger$, such that all $\delta Q|0\rangle = 0$, with certain amplitudes. In what follows, we refer to $\delta Q^\dagger$ as the variation of the basic excitation $Q_\omega^\dagger$. To compute the amplitudes we form the overlap between $\delta Q^\dagger|0\rangle$ and the relation (11.4):

$$\langle 0|\delta Q[H, Q_\omega^\dagger]|0\rangle = (E_\omega - E_0)\langle 0|\delta Q\, Q_\omega^\dagger|0\rangle \ . \tag{11.6}$$

The evaluation of (11.6) can be simplified by writing it in the commutator form

$$\langle 0| [\delta Q, [H, Q_\omega^\dagger]] |0\rangle = (E_\omega - E_0) \langle 0|[\delta Q, Q_\omega^\dagger]|0\rangle \ . \tag{11.7}$$

The simplification results from the fact that the commutator of any two operators is of a lower particle rank than is their product. What this means is, for example, that a commutator of the type $[cc, c^\dagger c^\dagger]$ consists of terms of the type $cc^\dagger$. We can simplify (11.7) further by letting $E_\omega$ denote the excitation energy.

Equation (11.7) is derived from a *variational principle*, namely $\delta Q|0\rangle = 0$, and is *exact*. However, the vacuum state $|0\rangle$ is usually not known exactly, so we are led to replace it with some approximate vacuum state $|\Psi_0\rangle$. Even with an approximate vacuum state it is necessary to guarantee the orthogonality of the basic excitations, i.e.

$$\langle \Psi_0|Q_\omega Q_{\omega'}^\dagger|\Psi_0\rangle = \delta_{\omega\omega'} \ . \tag{11.8}$$

It turns out that to satisfy the condition (11.8) we have to symmetrize the double commutator in (11.7), as discussed in [14]. Furthermore, we must then consider two different situations, namely where the basic excitations $Q_\omega^\dagger$ are either Fermi-like or Bose-like. In the Fermi case, $Q_\omega^\dagger$ contains an odd number of the particle operators $c^\dagger$ and $c$, and in the Bose case an even number of them. This is schematically represented as

$$\mathcal{O}_F^\dagger = c^\dagger + c^\dagger c^\dagger c^\dagger + c^\dagger c^\dagger c + \dots \ , \tag{11.9}$$

$$\mathcal{O}_B^\dagger = c^\dagger c + c^\dagger c^\dagger + c^\dagger c^\dagger cc + \dots \ . \tag{11.10}$$

The Fermi-like excitations can be imagined to behave roughly as fermions and the Bose-like roughly as bosons.

The resultant final form of the equations of motion can be written as

$$\boxed{\langle \Psi_0|[\delta Q, H, Q_\omega^\dagger]_\pm|\Psi_0\rangle = E_\omega \langle \Psi_0|[\delta Q, Q_\omega^\dagger]_\pm|\Psi_0\rangle \ ,} \tag{11.11}$$

where the symmetrized *double commutator* is defined as

$$[A, B, C]_\pm \equiv \frac{1}{2}\Big( [A, [B, C]]_\pm + [[A, B], C]_\pm \Big) \ . \tag{11.12}$$

Here the notation is

$$[A, B]_\pm \equiv AB \pm BA \ , \tag{11.13}$$

which relates to the standard notation for commutators and anticommutators as

$$[A, B]_- = [A, B] \ , \quad [A, B]_+ = \{A, B\} \ . \tag{11.14}$$

If the basic excitations are Fermi-like (Bose-like), anticommutators (commutators) are used in (11.11). Unless $|\Psi_0\rangle = |0\rangle$, the equation of motion (11.11) is *not* derived from a variational principle and is *not* exact.

The motivation for the commutator formalism is to reduce the dependence of (11.11) on the structure of the ground state as much as possible. This is achieved by reducing the particle rank of the operators through commutation, because the smaller the particle rank the more accurate is the replacement $|0\rangle \rightarrow |\Psi_0\rangle$. In particular, if the double commutator yields a c-number, the EOM is *exact* because normalization is assumed preserved in the replacement. If the double commutator yields a one-body operator then the accuracy of the EOM depends on the goodness of the approximation

$$\langle \Psi_0 | cc^\dagger | \Psi_0 \rangle \approx \langle 0 | cc^\dagger | 0 \rangle . \qquad (11.15)$$

The EOM is a very general and useful method. It can be used to derive all nuclear many-body theories in common use. This situation is schematically depicted in Fig. 11.1. Many of these theories are listed also in Table 11.1, where the structure of the basic excitation and its variation(s) is given. Also the approximate ground state used in deriving the equations of motion is stated.

Table 11.1 contains the particle–hole and pair creation operators

$$\mathcal{A}^\dagger_{ab}(JM) \equiv \left[ c^\dagger_a h^\dagger_b \right]_{JM} , \qquad (11.16)$$

$$A^\dagger_{ab}(JM) \equiv \mathcal{N}_{ab}(J) \left[ a^\dagger_a a^\dagger_b \right]_{JM} \qquad (11.17)$$

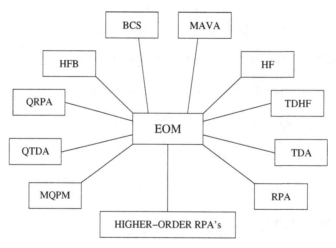

**Fig. 11.1.** The genealogy of many-body theories derivable by the equations-of-motion (EOM) method. The abbreviations are: HF = Hartree–Fock, TDHF = time-dependent Hartree–Fock, TDA = Tamm–Dancoff approximation, RPA = random-phase approximation, BCS = Bardeen–Cooper–Schrieffer theory of nucleon pairing, HFB = Hartree–Fock–Bogoliubov self-consistent pairing theory, QTDA = quasiparticle TDA, QRPA = quasiparticle RPA, MQPM = microscopic quasiparticle–phonon model [65], MAVA = microscopic anharmonic vibrator approach [66]

**Table 11.1.** The basic excitation and its variation(s) for a number of widely used nuclear many-body theories listed in Fig. 11.1. (Also the approximate ground state $|\Psi_0\rangle$ is given)

| Theory | $Q_\omega^\dagger$ | $\delta Q$ | $|\Psi_0\rangle$ |
|---|---|---|---|
| HF | $c_\omega^\dagger = \sum_l U_{\omega l} b_l^\dagger$ | $b_l$ | $|\text{HF}\rangle$ |
| TDA | $Q_\omega^\dagger = \sum_{ab} X_{ab}^\omega A_{ab}^\dagger(JM)$ | $A_{ab}(JM)$ | $|\text{HF}\rangle$ |
| RPA | $Q_\omega^\dagger = \sum_{ab}[X_{ab}^\omega A_{ab}^\dagger(JM)$ | $A_{ab}(JM)$ | |
| | $\qquad - Y_{ab}^\omega \tilde{A}_{ab}(JM)]$ | $\tilde{A}_{ab}^\dagger(JM)$ | $|\text{HF}\rangle$ |
| BCS | $a_\omega^\dagger = u_\omega c_\omega^\dagger + v_\omega \tilde{c}_\omega$ | $c_\omega, \tilde{c}_\omega^\dagger$ | $|\text{BCS}\rangle$ |
| HFB | $a_\omega^\dagger = \sum_l (U_{\omega l} c_l^\dagger + V_{\omega l} \tilde{c}_l)$ | $c_l, \tilde{c}_l^\dagger$ | $|\text{HFB}\rangle$ |
| QTDA | $Q_\omega^\dagger = \sum_{ab} X_{ab}^\omega A_{ab}^\dagger(JM)$ | $A_{ab}(JM)$ | $|\text{BCS}\rangle$ |
| QRPA | $Q_\omega^\dagger = \sum_{ab}[X_{ab}^\omega A_{ab}^\dagger(JM)$ | $A_{ab}(JM)$ | |
| | $\qquad - Y_{ab}^\omega \tilde{A}_{ab}(JM)]$ | $\tilde{A}_{ab}^\dagger(JM)$ | $|\text{BCS}\rangle$ |

with the normalization constant $\mathcal{N}_{ab}(J)$ given by (5.21). The operators $a^\dagger$ and $a$ are *quasiparticle* creation and annihilation operators defined either through the Bogoliubov–Valatin or the Hartree–Fock–Bogoliubov transformation (see, e.g. [16]). The former transformation is taken up when dealing with the BCS theory in Part II of this book. The adjoint tensor operators are defined as

$$\tilde{\mathcal{A}}_{ab}(JM) \equiv (-1)^{J+M}\big(\mathcal{A}_{ab}^\dagger(J-M)\big)^\dagger$$
$$= (-1)^{J+M}\mathcal{A}_{ab}(J-M) = -\big[\tilde{c}_a \tilde{h}_b\big]_{JM}, \qquad (11.18)$$

$$\tilde{A}_{ab}(JM) \equiv (-1)^{J+M}\big(A_{ab}^\dagger(J-M)\big)^\dagger$$
$$= (-1)^{J+M}A_{ab}(J-M) = -\mathcal{N}_{ab}(J)\big[\tilde{a}_a \tilde{a}_b\big]_{JM}. \qquad (11.19)$$

Let us review the general principle of approximation in the EOM. The state $|\Psi_0\rangle$ in (11.11) is intended to be an approximation of the exact ground state $|0\rangle$. In comparison with (11.2), $Q_\omega|0\rangle = 0$, this means that $Q_\omega|\Psi_0\rangle \neq 0$. Therefore (11.11) does *not* satisfy a variational principle, and the calculated ground state can lie below the exact one. If, however, a theory is formulated from (11.11) with $|0\rangle$ in place of $|\Psi_0\rangle$, the theory does satisfy a variational principle and the calculated ground state is exact.

Of the theories in Fig. 11.1 and Table 11.1 that are discussed in this book, the HF, TDA, BCS and QTDA approaches emerge from a variational principle; this is implied by the last column of the table. The other theories discussed in the book, namely the RPA and QRPA, do not satisfy variational principles. However, this flaw does not significantly detract from the usefulness of these theories.

To end this subsection, we list some commutator and anticommutator relations that are useful when evaluating (11.11). They read

$$[AB, C] = A[B, C] + [A, C]B = A\{B, C\} - \{A, C\}B \ , \tag{11.20}$$

$$[A, BC] = [A, B]C + B[A, C] = \{A, B\}C - B\{A, C\} \ , \tag{11.21}$$

$$[AB, CD] = A[B, C]D + AC[B, D] + [A, C]DB + C[A, D]B$$
$$= A\{B, C\}D - AC\{B, D\} + \{A, C\}DB - C\{A, D\}B \ , \tag{11.22}$$

$$\{A, BC\} = \{A, B\}C - B[A, C] = [A, B]C + B\{A, C\} \ . \tag{11.23}$$

Note that the compound commutators are stated in terms either of the basic commutators or the basic anticommutators. If the basic operators $A$, $B$, $C$ and $D$ obey commutation (anticommutation) rules then one should use the expansion given in terms of commutators (anticommutators).

We discuss next a non-trivial example illustrating how to use the EOM when starting from the postulated form of a basic excitation.

### 11.1.2 Derivation of the Hartree–Fock Equations by the EOM

To illustrate the EOM, we derive the Hartree–Fock equations by it. We start from the basic excitation and its variation

$$c_\omega^\dagger = \sum_l U_{\omega l} b_l^\dagger \ , \qquad \delta c = b_l \tag{11.24}$$

given in Table 11.1. In this case the label $\omega$ carries the quantum numbers of a single-particle orbital. The basic excitation postulated in (11.24) is a linear combination of operators each of which creates a particle into a single-particle state $\phi_l(\boldsymbol{x})$. The amplitudes $U$ of the basic excitation can be viewed as elements of a unitary transformation between the optimal particle creation operators $c_\omega^\dagger$ of the Hartree–Fock single-particle basis and the initial, arbitrary particle creation operators $b_l^\dagger$.

The basic excitations are now Fermi-like, so that anticommutators are used in (11.11). Expressed in the initial basis, the nuclear Hamiltonian is

$$H = T + V = \sum_{\alpha\beta} t_{\alpha\beta} b_\alpha^\dagger b_\beta + \tfrac{1}{4} \sum_{\alpha\beta\gamma\delta} \bar{v}_{\alpha\beta\gamma\delta} b_\alpha^\dagger b_\beta^\dagger b_\delta b_\gamma \ . \tag{11.25}$$

This gives for the commutators

$$[T, c_\omega^\dagger] = \sum_{\alpha\beta} t_{\alpha\beta} U_{\omega\beta} b_\alpha^\dagger \ , \tag{11.26}$$

$$[V, c_\omega^\dagger] = \tfrac{1}{4} \sum_{\substack{\alpha\beta\gamma\delta \\ l}} \bar{v}_{\alpha\beta\gamma\delta} b_\alpha^\dagger b_\beta^\dagger (U_{\omega\gamma} b_\delta - U_{\omega\delta} b_\gamma)$$

$$= -\tfrac{1}{2} \sum_{\alpha\beta\gamma\delta} \bar{v}_{\alpha\beta\gamma\delta} U_{\omega\delta} b_\alpha^\dagger b_\beta^\dagger b_\gamma \ , \tag{11.27}$$

$$[\delta c, T] = \sum_\beta t_{l\beta} b_\beta \ , \tag{11.28}$$

$$[\delta c, V] = \frac{1}{2} \sum_{\beta\gamma\delta} \bar{v}_{l\beta\gamma\delta} b_\beta^\dagger b_\delta b_\gamma \ , \tag{11.29}$$

where we have used the commutator formulas (11.20) and (11.21) and the symmetry relations (4.29).

We proceed to compute the anticommutators and write down the HF matrix elements:

$$\langle HF|\{\delta c, [T, c_\omega^\dagger]\}|HF\rangle = \sum_\beta t_{l\beta} U_{\omega\beta} \ , \tag{11.30}$$

$$\langle HF|\{\delta c, [V, c_\omega^\dagger]\}|HF\rangle = \sum_{\alpha\gamma\delta} \bar{v}_{\alpha l\gamma\delta} U_{\omega\delta} \langle HF|b_\alpha^\dagger b_\gamma|HF\rangle \ , \tag{11.31}$$

$$\langle HF|\{[\delta c, T], c_\omega^\dagger\}|HF\rangle = \sum_\beta t_{l\beta} U_{\omega\beta} \ , \tag{11.32}$$

$$\langle HF|\{[\delta c, V], c_\omega^\dagger\}|HF\rangle = \sum_{\alpha\gamma\delta} \bar{v}_{\alpha l\gamma\delta} U_{\omega\delta} \langle HF|b_\alpha^\dagger b_\gamma|HF\rangle \ , \tag{11.33}$$

where we have used (11.21), (11.23) and (4.29). Collecting the terms (11.30)–(11.33) into the left-hand side of (11.11) and inserting the matrix element

$$\langle HF|\{\delta c, c_\omega^\dagger\}|HF\rangle = U_{\omega l} \tag{11.34}$$

into the right-hand side, we finally obtain the equations of motion

$$\sum_\beta U_{\omega\beta}\left(t_{l\beta} + \sum_{\alpha\gamma} \bar{v}_{\alpha l\gamma\beta}\langle HF|b_\alpha^\dagger b_\gamma|HF\rangle\right) = E_\omega U_{\omega l} \ . \tag{11.35}$$

Since $|HF\rangle$ is the particle–hole vacuum of the $b$ operators (11.35) becomes

$$\sum_\beta U_{\omega\beta}\left(t_{l\beta} + \sum_{\substack{\alpha \\ \varepsilon_\alpha \leq \varepsilon_F}} \bar{v}_{\alpha l\alpha\beta}\right) = \sum_\beta U_{\omega\beta} T_{l\beta} = E_\omega U_{\omega l} \ , \tag{11.36}$$

where the two-body operator $T_{l\beta}$ is the same as that defined in (4.63). The one-body and two-body parts of the Hamiltonian are separately Hermitian, so we have $T_{l\beta} = (T^\dagger)_{l\beta} = T_{\beta l}^*$. In the normal case that the $T_{l\beta}$ are real we then have $T_{l\beta} = T_{\beta l}$. Substituting this into (11.36), multiplying its both sides from the right by $U_{\omega' l}^*$ and summing over $l$, we obtain

$$\sum_{\beta l} U_{\omega\beta} T_{\beta l} U_{\omega' l}^* = E_\omega \sum_l U_{\omega l} U_{\omega' l}^* \ . \tag{11.37}$$

In matrix notation this reads

$$(\mathsf{U}\mathsf{T}\mathsf{U}^\dagger)_{\omega\omega'} = E_\omega (\mathsf{U}\mathsf{U}^\dagger)_{\omega\omega'} = E_\omega \delta_{\omega\omega'} \ , \tag{11.38}$$

where we have used the unitarity of U. The left-hand side represents the diagonalized matrix T′ of (4.67). Equation (11.38) gives directly the Hartree–Fock equation (4.69), in the current notation

$$t_{\omega\omega'} + \sum_{\substack{\alpha \\ \varepsilon_\alpha \leq \varepsilon_F}} \bar{v}_{\alpha\omega\alpha\omega'} = E_\omega \delta_{\omega\omega'} . \tag{11.39}$$

## 11.2 Sophisticated Particle–Hole Theories: The RPA

The basic difference between the relatively simple TDA and the more sophisticated RPA is the replacement of the simple particle–hole vacuum $|HF\rangle$ of the TDA by the *correlated ground state* in the RPA. This correlated ground state consists of the particle–hole vacuum and components with two-particle–two-hole, four-particle–four-hole, etc., correlations. Simple particle–hole correlations cannot occur in the RPA ground state, as is stated by Brillouin's theorem (9.8).

The RPA equations can be derived in numerous ways:

- by the so-called quasiboson approximation, where the commutators of particle–hole operators are replaced by their vacuum expectation values in the uncorrelated particle–hole state;

- by linearization of two-particle–two-hole excitations, with substitutions like $c^\dagger c^\dagger cc \rightarrow c^\dagger c \langle HF|c^\dagger c|HF\rangle$;

- by time-dependent Hartree–Fock theory;

- by the Green function formalism;

- by the EOM.

In the following we use the EOM to derive the RPA equations.

### 11.2.1 Derivation of the RPA Equations by the EOM

We choose to use the EOM to derive the RPA equations because it can also be readily applied to derive higher-RPA theories. In this case Table 11.1 gives the basic excitation as

$$Q_\omega^\dagger = \sum_{ab} \left[ X_{ab}^\omega \mathcal{A}_{ab}^\dagger(JM) - Y_{ab}^\omega \tilde{\mathcal{A}}_{ab}(JM) \right] , \tag{11.40}$$

where $\omega = nJ^\pi M$. From (11.40) we obtain by Hermitian conjugation

$$Q_\omega = \sum_{ab} \left[ X_{ab}^{\omega*} \mathcal{A}_{ab}(JM) - Y_{ab}^{\omega*} \tilde{\mathcal{A}}_{ab}^\dagger(JM) \right] \tag{11.41}$$

with the particle–hole operator and its adjoint defined in (11.16) and (11.18).

The variations of the first and the second term, respectively, of the basic excitation are given by

$$\delta Q = \mathcal{A}_{ab}(JM) , \quad \delta Q = \tilde{\mathcal{A}}^{\dagger}_{ab}(JM) . \tag{11.42}$$

They are Bose-like, so (11.11) gives the equation of motion[1]

$$\langle \mathrm{HF} | [\delta Q, H, Q^{\dagger}_{\omega}] | \mathrm{HF} \rangle = E_{\omega} \langle \mathrm{HF} | [\delta Q, Q^{\dagger}_{\omega}] | \mathrm{HF} \rangle . \tag{11.43}$$

Note that we have here $|\Psi_0\rangle = |\mathrm{HF}\rangle$, which is not the true ground state. As discussed in Subsect. 11.1.1, this means that the RPA does not satisfy a variational principle.

To evaluate the right-hand side of (11.43) we need commutator relations for the particle–hole operators (11.16) and (11.18). The identity (11.22) and various relations of Chap. 4 give

$$\langle \mathrm{HF} | [\mathcal{A}_{ab}(JM), \mathcal{A}^{\dagger}_{cd}(J'M')] | \mathrm{HF} \rangle = \delta_{ac}\delta_{bd}\delta_{JJ'}\delta_{MM'} , \tag{11.44}$$

$$\langle \mathrm{HF} | [\tilde{\mathcal{A}}_{ab}(JM), \tilde{\mathcal{A}}^{\dagger}_{cd}(J'M')] | \mathrm{HF} \rangle = \delta_{ac}\delta_{bd}\delta_{JJ'}\delta_{MM'} , \tag{11.45}$$

while the expectation values of all other commutators vanish. Note that in these expectation values the Bose-like particle–hole operators behave as exact boson operators.

We substitute (11.40) and the variation $\delta Q = \mathcal{A}_{ab}(JM)$ into (11.43). Use of (11.44) then gives

$$\sum_{cd} \langle \mathrm{HF} | [\mathcal{A}_{ab}, H, \mathcal{A}^{\dagger}_{cd}] | \mathrm{HF} \rangle X^{\omega}_{cd} - \sum_{cd} \langle \mathrm{HF} | [\mathcal{A}_{ab}, H, \tilde{\mathcal{A}}_{cd}] | \mathrm{HF} \rangle Y^{\omega}_{cd} = E_{\omega} X^{\omega}_{ab} . \tag{11.46}$$

The variation $\delta Q = \tilde{\mathcal{A}}^{\dagger}_{ab}(JM)$ yields similarly

$$\sum_{cd} \langle \mathrm{HF} | [\tilde{\mathcal{A}}^{\dagger}_{ab}, H, \mathcal{A}^{\dagger}_{cd}] | \mathrm{HF} \rangle X^{\omega}_{cd} - \sum_{cd} \langle \mathrm{HF} | [\tilde{\mathcal{A}}^{\dagger}_{ab}, H, \tilde{\mathcal{A}}_{cd}] | \mathrm{HF} \rangle Y^{\omega}_{cd} = E_{\omega} Y^{\omega}_{ab} . \tag{11.47}$$

To deal with the left-hand side of (11.46) in short-hand notation we define matrices $\mathsf{A}(J)$ and $\mathsf{B}(J)$ with elements

$$\boxed{\begin{aligned} A_{ab,cd}(J) &\equiv \langle \mathrm{HF} | [\mathcal{A}_{ab}(JM), H, \mathcal{A}^{\dagger}_{cd}(JM)] | \mathrm{HF} \rangle , \\ B_{ab,cd}(J) &\equiv -\langle \mathrm{HF} | [\mathcal{A}_{ab}(JM), H, \tilde{\mathcal{A}}_{cd}(JM)] | \mathrm{HF} \rangle . \end{aligned}} \tag{11.48}$$

An examination of the tensor rank resulting from the relations (11.16) and (11.18) shows that the matrix elements (11.48) do not depend on $M$. The defining equation (11.12) of the double commutator gives the identities

$$[A, B, C]^{\dagger} = [C^{\dagger}, B^{\dagger}, A^{\dagger}] , \quad [A, B, C] = [C, B, A] . \tag{11.49}$$

---

[1] All the following double commutators involve Bose-like excitations, so the subscript '−' in the defining equation (11.12) is omitted.

With (11.18) and (11.49), and noting that $H^\dagger = H$, the double commutators in (11.47) can be related to those in (11.46) according to

$$\left[\tilde{\mathcal{A}}_{ab}^\dagger(JM), H, \mathcal{A}_{cd}^\dagger(JM)\right] = \left[\mathcal{A}_{cd}(JM), H, \tilde{\mathcal{A}}_{ab}(JM)\right]^\dagger , \qquad (11.50)$$

$$\left[\tilde{\mathcal{A}}_{ab}^\dagger(JM), H, \tilde{\mathcal{A}}_{cd}(JM)\right] = \left[\mathcal{A}_{cd}(J-M), H, \mathcal{A}_{ab}^\dagger(J-M)\right] . \qquad (11.51)$$

To simplify the notation we omit $JM$ for the time being and identify

$$\langle \mathrm{HF}| \left[\tilde{\mathcal{A}}_{ab}^\dagger, H, \mathcal{A}_{cd}^\dagger\right] |\mathrm{HF}\rangle = -B_{cd,ab}^* = -\left(\mathsf{B}^\dagger\right)_{ab,cd} , \qquad (11.52)$$

$$\langle \mathrm{HF}| \left[\tilde{\mathcal{A}}_{ab}^\dagger, H, \tilde{\mathcal{A}}_{cd}\right] |\mathrm{HF}\rangle = A_{cd,ab} = \left(\mathsf{A}^\mathrm{T}\right)_{ab,cd} . \qquad (11.53)$$

By means of (11.48), (11.52) and (11.53), Eqs. (11.46) and (11.47) become

$$\sum_{cd} A_{ab,cd} X_{cd}^\omega + \sum_{cd} B_{ab,cd} Y_{cd}^\omega = E_\omega X_{ab}^\omega , \qquad (11.54)$$

$$-\sum_{cd} \left(\mathsf{B}^\dagger\right)_{ab,cd} X_{cd}^\omega - \sum_{cd} \left(\mathsf{A}^\mathrm{T}\right)_{ab,cd} Y_{cd}^\omega = E_\omega Y_{ab}^\omega , \qquad (11.55)$$

or in matrix form

$$\mathsf{A}\mathsf{X}^\omega + \mathsf{B}\mathsf{Y}^\omega = E_\omega \mathsf{X}^\omega , \qquad (11.56)$$

$$-\mathsf{B}^\dagger \mathsf{X}^\omega - \mathsf{A}^\mathrm{T}\mathsf{Y}^\omega = E_\omega \mathsf{Y}^\omega . \qquad (11.57)$$

From the defining equations (11.48) and the identities (11.49) one can show that the matrices $\mathsf{A}$ and $\mathsf{B}$ have the properties

$$\mathsf{A}^\dagger = \mathsf{A} , \quad \mathsf{B}^\mathrm{T} = \mathsf{B} , \qquad (11.58)$$

i.e. $\mathsf{A}$ is Hermitian and $\mathsf{B}$ is symmetric. The square matrices in (11.57) are thus complex conjugate matrices of $\mathsf{A}$ and $\mathsf{B}$: $\mathsf{A}^\mathrm{T} = \mathsf{A}^*$, $\mathsf{B}^\dagger = \mathsf{B}^*$. Equation (11.57) now becomes

$$-\mathsf{B}^* \mathsf{X}^\omega - \mathsf{A}^* \mathsf{Y}^\omega = E_\omega \mathsf{Y}^\omega . \qquad (11.59)$$

The two matrix equations (11.56) and (11.59) can be combined into a single matrix equation whose elements themselves are matrices:

$$\boxed{\begin{pmatrix} \mathsf{A} & \mathsf{B} \\ -\mathsf{B}^* & -\mathsf{A}^* \end{pmatrix} \begin{pmatrix} \mathsf{X}^\omega \\ \mathsf{Y}^\omega \end{pmatrix} = E_\omega \begin{pmatrix} \mathsf{X}^\omega \\ \mathsf{Y}^\omega \end{pmatrix} .} \qquad (11.60)$$

This is the *RPA matrix equation*. The matrix containing $\mathsf{A}$ and $\mathsf{B}$ is not Hermitian, and the RPA eigenvalue problem is indeed *non-Hermitian*.

To study the contents of the $\mathsf{A}$ matrix we expand the commutators in the defining expression in (11.48) and obtain

$$A_{ab,cd} = \tfrac{1}{2}\big(2\langle\mathrm{HF}|\mathcal{A}_{ab}H\mathcal{A}^{\dagger}_{cd}|\mathrm{HF}\rangle - \langle\mathrm{HF}|\mathcal{A}_{ab}\mathcal{A}^{\dagger}_{cd}H|\mathrm{HF}\rangle - \langle\mathrm{HF}|H\mathcal{A}^{\dagger}_{cd}\mathcal{A}_{ab}|\mathrm{HF}\rangle$$
$$+ 2\langle\mathrm{HF}|\mathcal{A}^{\dagger}_{cd}H\mathcal{A}_{ad}|\mathrm{HF}\rangle - \langle\mathrm{HF}|H\mathcal{A}_{ab}\mathcal{A}^{\dagger}_{cd}|\mathrm{HF}\rangle - \langle\mathrm{HF}|\mathcal{A}^{\dagger}_{cd}\mathcal{A}_{ab}H|\mathrm{HF}\rangle\big)\,.$$
$$(11.61)$$

In the third, fourth and sixth terms an annihilation operator operates on the particle–hole vacuum, so the terms vanish. The second and fifth terms are diagonal and give a constant contribution of $-E_{\mathrm{HF}}$; see (4.73). Without changing the notation, we drop these terms since we are interested in the excitation energies. This leaves us with the first term, which gives

$$A_{ab,cd}(J) = \langle\mathrm{HF}|\mathcal{A}_{ab}(JM)H\mathcal{A}^{\dagger}_{cd}(JM)|\mathrm{HF}\rangle = \langle a\,b^{-1}\,;\,J\,M|H|c\,d^{-1}\,;\,J\,M\rangle\,.$$
$$(11.62)$$

Comparison with (9.35) shows that the matrix $\mathsf{A}$ is nothing but the TDA matrix given by (9.17) and (9.22). In our current matrix notation, (9.35) reads $\mathsf{A}\mathsf{X}^{\omega} = E_{\omega}\mathsf{X}^{\omega}$. From Chap. 9 we know how to construct matrix $\mathsf{A}$ in detail.

The new matrix $\mathsf{B}$ is called the *correlation matrix*, and it is discussed in detail in the following subsection.

### 11.2.2 Explicit Form of the Correlation Matrix

Here we derive the correlation matrix $\mathsf{B}$ explicitly. We expand the double commutator in the defining equation in (11.48). The result is similar to (11.61) except that now there are no diagonal terms. In four terms an annihilation operator operates on the particle–hole vacuum, so they are zero. The remaining two terms combine into

$$B_{ab,cd}(J) = \langle\mathrm{HF}|\mathcal{A}_{ab}(JM)\tilde{\mathcal{A}}_{cd}(JM)H|\mathrm{HF}\rangle\,.$$
$$(11.63)$$

Expanding the angular momentum couplings in (11.63) results in

$$B_{ab,cd}(J) = (-1)^{J+M}\sum_{\substack{m_\alpha m_\beta \\ m_\gamma m_\delta}}(j_a\,m_\alpha\,j_b\,m_\beta|J\,M)(j_c\,m_\gamma\,j_d\,m_\delta|J\,-M)$$
$$\times\,\langle\mathrm{HF}|h_\beta c_\alpha h_\delta c_\gamma H|\mathrm{HF}\rangle\,.$$
$$(11.64)$$

The Hamiltonian is given by (4.70)–(4.72), and with the constant term omitted the matrix element in (11.64) becomes

$$\langle\mathrm{HF}|h_\beta c_\alpha h_\delta c_\gamma H|\mathrm{HF}\rangle = \sum_{\alpha'}\varepsilon_{\alpha'}\langle\mathrm{HF}|h_\beta c_\alpha h_\delta c_\gamma c^{\dagger}_{\alpha'}c_{\alpha'}|\mathrm{HF}\rangle$$
$$+\tfrac{1}{4}\sum_{\alpha'\beta'\gamma'\delta'}\bar{v}_{\alpha'\beta'\gamma'\delta'}\langle\mathrm{HF}|h_\beta c_\alpha h_\delta c_\gamma\mathcal{N}\big[c^{\dagger}_{\alpha'}c^{\dagger}_{\beta'}c_{\delta'}c_{\gamma'}\big]|\mathrm{HF}\rangle\,.\quad(11.65)$$

From (4.72) we see that $c^{\dagger}_{\alpha'}c_{\alpha'}$ in the one-body term is diagonal. The remaining string of operators $h_\beta c_\alpha h_\delta c_\gamma$ makes the term vanish. The normal-ordered product in the two-body term has to contain two particle-creation and two hole-creation operators for the matrix element not to vanish. From (4.46) and (4.47) we have $c_\beta = (-1)^{j_b - m_\beta} h^\dagger_{-\beta}$. Making this substitution and performing the contractions, we obtain

$$(-1)^{j_{d'} - m_{\delta'} + j_{c'} - m_{\gamma'}} \langle \mathrm{HF} | h_\beta c_\alpha h_\delta c_\gamma c^\dagger_{\alpha'} c^\dagger_{\beta'} h^\dagger_{-\delta'} h^\dagger_{-\gamma'} | \mathrm{HF} \rangle$$

$$= (-1)^{j_{d'} - m_{\delta'} + j_{c'} - m_{\gamma'}} (\delta_{\alpha\alpha'} \delta_{\gamma\beta'} \delta_{\beta, -\gamma'} \delta_{\delta, -\delta'} - \delta_{\alpha\alpha'} \delta_{\gamma\beta'} \delta_{\beta, -\delta'} \delta_{\delta, -\gamma'}$$

$$+ \delta_{\gamma\alpha'} \delta_{\alpha\beta'} \delta_{\beta, -\delta'} \delta_{\delta, -\gamma'} - \delta_{\gamma\alpha'} \delta_{\alpha\beta'} \delta_{\beta, -\gamma'} \delta_{\delta, -\delta'}) . \quad (11.66)$$

Substituting this into (11.65) and using the symmetry properties (4.29) of $\bar{v}$, we arrive at the simple expression

$$\langle \mathrm{HF} | h_\beta c_\alpha h_\delta c_\gamma H | \mathrm{HF} \rangle = (-1)^{j_b + m_\beta + j_d + m_\delta} \bar{v}_{\alpha\gamma, -\beta, -\delta} . \quad (11.67)$$

To express (11.67) in terms of coupled two-body matrix elements we use (8.17) and the Clebsch–Gordan symmetry (1.31). The result is

$$\langle \mathrm{HF} | h_\beta c_\alpha h_\delta c_\gamma H | \mathrm{HF} \rangle = \sum_{J'M'} (-1)^{J'+M'} [\mathcal{N}_{ac}(J') \mathcal{N}_{bd}(J')]^{-1}$$

$$\times (j_a \, m_\alpha \, j_c \, m_\gamma | J' \, M')(j_b \, m_\beta \, j_d \, m_\delta | J' \, -M') \langle a\, c; \, J' | V | b\, d; \, J' \rangle . \quad (11.68)$$

We substitute this into (11.64), convert all the Clebsch–Gordan coefficients to $3j$ symbols, and sum over $M$ and divide by $2J + 1$. This yields

$$B_{ab,cd}(J) = \sum_{J'} \hat{J'}^2 [\mathcal{N}_{ac}(J') \mathcal{N}_{bd}(J')]^{-1} \langle a\, c; \, J' | V | b\, d; \, J' \rangle$$

$$\times \sum_{\substack{m_\alpha m_\beta M \\ m_\gamma m_\delta M'}} (-1)^{J+M+J'+M'} \begin{pmatrix} j_a & j_b & J \\ m_\alpha & m_\beta & -M \end{pmatrix} \begin{pmatrix} j_c & j_d & J \\ m_\gamma & m_\delta & M \end{pmatrix}$$

$$\times \begin{pmatrix} j_a & j_c & J' \\ m_\alpha & m_\gamma & -M' \end{pmatrix} \begin{pmatrix} j_b & j_d & J' \\ m_\beta & m_\delta & M' \end{pmatrix} . \quad (11.69)$$

By means of (1.39) and (1.40) the $3j$ symbols can be rearranged so that they sum into a $6j$ symbol according to (1.59). In addition, the two-body matrix elements take care of the selection of $J'$ values, so that the normalization factors, in explicit form, can be brought outside the sum over $J'$. The final result is

$$\boxed{\begin{aligned} B_{ab,cd}(J) &= (-1)^{j_b + j_c + J} \sqrt{(1 + \delta_{ac})(1 + \delta_{bd})} \\ &\times \sum_{J'} (-1)^{J'} \hat{J'}^2 \begin{Bmatrix} j_a & j_b & J \\ j_d & j_c & J' \end{Bmatrix} \langle a\, c; \, J' | V | b\, d; \, J' \rangle . \end{aligned}} \quad (11.70)$$

An analogous expression applies in the isospin formalism:

$$B_{ab,cd}(JT) = (-1)^{j_b+j_c+J+1+T}\sqrt{(1+\delta_{ac})(1+\delta_{bd})}$$
$$\times \sum_{J'T'}(-1)^{J'+T'}\widehat{J'}^2\widehat{T'}^2 \begin{Bmatrix} j_a & j_b & J \\ j_d & j_c & J' \end{Bmatrix}\begin{Bmatrix} \frac{1}{2} & \frac{1}{2} & T \\ \frac{1}{2} & \frac{1}{2} & T' \end{Bmatrix} \qquad (11.71)$$
$$\times \langle a\,c;\,J'\,T'|V|b\,d;\,J'\,T'\rangle\,.$$

### 11.2.3 Numerical Tables of Correlation Matrix Elements

For SDI, the elements of the A matrix are given in a convenient form by Eqs. (9.28)–(9.32) and Tables 9.1–9.4. We now state the elements of the B matrix similarly through Eqs. (11.74)–(11.78) and Tables 11.2–11.5. The auxiliary matrix elements analogous to those defined by (9.27) are

$$\tilde{\mathcal{M}}_{abcd}(JT) \equiv (-1)^{j_b+j_c+J}\sqrt{(1+\delta_{ac})(1+\delta_{bd})}$$
$$\times \sum_{J'}(-1)^{J'}\widehat{J'}^2 \begin{Bmatrix} j_a & j_b & J \\ j_d & j_c & J' \end{Bmatrix}\langle a\,c;\,J'\,T|V_{\text{SDI}}|b\,d;\,J'\,T\rangle_{A_T=1}\,. \qquad (11.72)$$

The chosen SDI strength parameters $A_T$ are to be inserted as shown in Eqs. (11.74)–(11.78). The CS phase convention is used in the tables. The BR matrix elements can be obtained from the CS ones by the relation

$$B_{ab,cd}^{(\text{BR})}(J(T)) = (-1)^{\frac{1}{2}(l_b+l_d-l_a-l_c)}B_{ab,cd}^{(\text{CS})}(J(T))\,. \qquad (11.73)$$

This is identical in form to (9.33) applicable to the A matrix elements.

In proton–neutron representation the correlation matrix elements are given by (11.70). Via (8.25) and (8.26) they are expressed in terms of the isospin-dependent auxiliary matrix elements (11.72). The result is

$$B_{p_1p_2,p_3p_4}(J) = A_1\tilde{\mathcal{M}}_{a_1a_2a_3a_4}(J1)\,, \qquad (11.74)$$
$$B_{n_1n_2,n_3n_4}(J) = A_1\tilde{\mathcal{M}}_{a_1a_2a_3a_4}(J1)\,, \qquad (11.75)$$

$$B_{p_1p_2,n_3n_4}(J) = \frac{1}{2}\Big\{A_1\sqrt{[1+(-1)^J\delta_{a_1a_2}][1+(-1)^J\delta_{a_3a_4}]}\,\tilde{\mathcal{M}}_{a_1a_2a_3a_4}(J1)$$
$$+ A_0\sqrt{[1-(-1)^J\delta_{a_1a_2}][1-(-1)^J\delta_{a_3a_4}]}\,\tilde{\mathcal{M}}_{a_1a_2a_3a_4}(J0)\Big\}\,. \qquad (11.76)$$

These matrix elements are analogous to (9.28)–(9.30).

The matrix elements $B_{ab,cd}(JT)$, analogous to (9.31) and (9.32), are obtained directly from (11.71) and (11.72) as

$$B_{ab,cd}(J\,T=0) = \tfrac{1}{2}[3A_1\tilde{\mathcal{M}}_{abcd}(J1) + A_0\tilde{\mathcal{M}}_{abcd}(J0)]\,, \qquad (11.77)$$
$$B_{ab,cd}(J\,T=1) = \tfrac{1}{2}[-A_1\tilde{\mathcal{M}}_{abcd}(J1) + A_0\tilde{\mathcal{M}}_{abcd}(J0)]\,. \qquad (11.78)$$

Table 11.2 lists the auxiliary particle–hole matrix elements (11.72) for excitations from the 0s to the 0p shell; the notation is essentially the same as in Table 9.1. Table 11.3 gives them for excitations from the 0p shell to the 0d-1s shell, Table 11.4 from the 0d-1s shells to the $0f_{7/2}$ shell and Table 11.5 from the $0f_{7/2}$ shell to the $0p-0f_{5/2}$ shells.

**Table 11.2.** Quantities $\tilde{\mathcal{M}}_{abcd}(JT)$ in the CS phase convention

| abcd | JT | $\tilde{\mathcal{M}}$ | JT | $\tilde{\mathcal{M}}$ | JT | $\tilde{\mathcal{M}}$ | JT | $\tilde{\mathcal{M}}$ |
|------|----|----|----|----|----|----|----|----|
| 1111 | 10 | 1 .6667 | 11 | 1 .0000 | 20 | −1 .0000 | 21 | 1 .0000 |
| 1121 | 10 | −1 .8856 | 11 | 0 | | | | |
| 2121 | 00 | −1 .0000 | 01 | 1 .0000 | 10 | 0 .3333 | 11 | 1 .0000 |

The particle states are numbered $1 = 0p_{3/2}$ and $2 = 0p_{1/2}$, and the hole state is numbered $1 = 0s_{1/2}$. The first column gives the particle-hole–particle-hole labels, and the following columns give the $JT$ combinations and values of $\tilde{\mathcal{M}}$. The zero recorded does not result from conservation laws.

## 11.3 Properties of the RPA Solutions

The solutions of the RPA equations are more intricate than those of the TDA equations. This is due to the additional complication arising from the replacement of the simple particle–hole vacuum of the TDA by a correlated ground state in the RPA. In addition, the non-Hermiticity of the RPA eigenvalue problem produces an overcomplete set of solutions. The overcompleteness leads to positive- and negative-energy solutions as discussed below.

### 11.3.1 RPA Energies and Amplitudes

RPA excitations are created by the operator (11.40) and annihilated by its Hermitian conjugate (11.41). Thus the ground state satisfies, for all $\omega$,

$$Q_\omega|\text{RPA}\rangle = 0 . \tag{11.79}$$

Note that we started the RPA derivation from (11.11) with $|\Psi_0\rangle = |\text{HF}\rangle$ but $Q_\omega|\text{HF}\rangle \neq 0$. This signifies the non-variational nature of the RPA, as discussed earlier in this chapter. The wave function of a single excitation with its set of quantum numbers labelled by $\omega$ is

$$|\Psi_\omega^{(\text{RPA})}\rangle \equiv |\omega\rangle = Q_\omega^\dagger|\text{RPA}\rangle , \tag{11.80}$$

Such an excitation is called a *phonon* of type $\omega$.

**Table 11.3.** Quantities $\tilde{\mathcal{M}}_{abcd}(JT)$ in the CS phase convention

| abcd | JT | $\tilde{\mathcal{M}}$ | JT | $\tilde{\mathcal{M}}$ | JT | $\tilde{\mathcal{M}}$ | JT | $\tilde{\mathcal{M}}$ | JT | $\tilde{\mathcal{M}}$ |
|------|----|------|----|------|----|------|----|------|----|------|
| 1111 | 10 | 3.0000 | 11 | 1.8000 | 20 | −0.8857 | 21 | 0.8857 | 30 | 1.0000 |
| 1111 | 31 | 0.3714 | 40 | −1.2857 | 41 | 1.2857 | | | | |
| 1112 | 20 | 0.8552 | 21 | −0.8552 | 30 | 1.0222 | 31 | 0.5111 | | |
| 1121 | 10 | 3.1305 | 11 | 0.4472 | 20 | −0.9165 | 21 | 0.9165 | | |
| 1122 | 10 | 1.2649 | 11 | 1.2649 | | | | | | |
| 1131 | 10 | −2.0000 | 11 | 0.4000 | 20 | −0.2619 | 21 | 0.2619 | 30 | −1.3997 |
| 1131 | 31 | −0.2799 | | | | | | | | |
| 1132 | 10 | 3.1305 | 11 | 0.4472 | 20 | 0.3928 | 21 | −0.3928 | | |
| 1212 | 20 | −1.0000 | 21 | 1.0000 | 30 | 0.7143 | 31 | 1.0000 | | |
| 1221 | 20 | 0.9798 | 21 | −0.9798 | | | | | | |
| 1231 | 20 | 0.4899 | 21 | −0.4899 | 30 | −2.0344 | 31 | 0.1565 | | |
| 1232 | 20 | 0 | 21 | 0 | | | | | | |
| 2121 | 10 | 1.6667 | 11 | 1.0000 | 20 | −1.0000 | 21 | 1.0000 | | |
| 2122 | 10 | 1.8856 | 11 | 0 | | | | | | |
| 2131 | 10 | −0.2981 | 11 | −0.8944 | 20 | −0.4000 | 21 | 0.4000 | | |
| 2132 | 10 | 1.6667 | 11 | 1.0000 | 20 | 0.2000 | 21 | −0.2000 | | |
| 2222 | 00 | −1.0000 | 01 | 1.0000 | 10 | 0.3333 | 11 | 1.0000 | | |
| 2231 | 00 | 1.4142 | 01 | −1.4142 | 10 | −1.4757 | 11 | 0.6325 | | |
| 2232 | 10 | 1.8856 | 11 | 0 | | | | | | |
| 3131 | 00 | −2.0000 | 01 | 2.0000 | 10 | −0.6667 | 11 | 1.2000 | 20 | −0.4000 |
| 3131 | 21 | 0.4000 | 30 | 0.8571 | 31 | 1.2000 | | | | |
| 3132 | 10 | −0.2981 | 11 | −0.8944 | 20 | −0.4000 | 21 | 0.4000 | | |
| 3232 | 10 | 1.6667 | 11 | 1.0000 | 20 | −1.0000 | 21 | 1.0000 | | |

The particle states are numbered $1 = 0d_{5/2}$, $2 = 1s_{1/2}$ and $3 = 0d_{3/2}$ and the hole states are numbered $1 = 0p_{3/2}$ and $2 = 0p_{1/2}$. The first column gives the particle–hole–particle-hole labels, and the following columns give the $JT$ combinations and values of $\tilde{\mathcal{M}}$. The zeros recorded do not result from conservation laws.

The phonon creation operator $Q_\omega^\dagger$ contains the TDA type of components with amplitudes $X_{ab}^\omega$, generated by the TDA part A of the RPA equation (11.60). It also contains the terms with amplitudes $Y_{ab}^\omega$, describing the ground-state correlations generated by the B matrix in (11.60).

The correlation part of (11.40) yields zero when it acts on the particle–hole vacuum $|\mathrm{HF}\rangle$. However, as shown below, the correlated ground state $|\mathrm{RPA}\rangle$ has $2n$-particle–$2n$-hole components with $n = 1, 2, 3, \ldots$. It follows that action of the correlation part on $|\mathrm{RPA}\rangle$ produces non-zero contributions. For example, action by $\tilde{\mathcal{A}}$ on the two-particle–two-hole components produces one-particle–one-hole terms. They are of the same particle–hole order as the terms produced by $\mathcal{A}^\dagger$ acting on the particle–hole vacuum. In this way the excited RPA states have one-particle–one-hole, three-particle–three-hole, etc., components. The magnitudes of the $Y^\omega$ amplitudes are a measure of the amount of correlation in $|\mathrm{RPA}\rangle$.

**Table 11.4.** Quantities $\tilde{\mathcal{M}}_{abcd}(JT)$ in the CS phase convention

| abcd | JT | $\tilde{\mathcal{M}}$ | JT | $\tilde{\mathcal{M}}$ | JT | $\tilde{\mathcal{M}}$ | JT | $\tilde{\mathcal{M}}$ | JT | $\tilde{\mathcal{M}}$ |
|------|----|------|----|------|----|------|----|------|----|------|
| 1111 | 10 | 4.2857 | 11 | 2.5714 | 20 | −1.0476 | 21 | 1.0476 | 30 | 1.6667 |
| 1111 | 31 | 0.6190 | 40 | −0.8961 | 41 | 0.8961 | 50 | 0.7706 | 51 | 0.2684 |
| 1111 | 60 | −1.5151 | 61 | 1.5151 | | | | | | |
| 1112 | 30 | 1.5714 | 31 | 0.7143 | 40 | −0.9343 | 41 | 0.9343 | | |
| 1113 | 20 | 1.1429 | 21 | −1.1429 | 30 | 0.8248 | 31 | 0.4949 | 40 | 0.4882 |
| 1113 | 41 | −0.4882 | 50 | 1.1109 | 51 | 0.4761 | | | | |
| 1212 | 30 | 1.2857 | 31 | 1.0000 | 40 | −1.0000 | 41 | 1.0000 | | |
| 1213 | 30 | 0.4949 | 31 | 0.8248 | 40 | 0.5634 | 41 | −0.5634 | | |
| 1313 | 20 | −1.7143 | 21 | 1.7143 | 30 | 0 | 31 | 0.7619 | 40 | −0.3809 |
| 1313 | 41 | 0.3809 | 50 | 1.0909 | 51 | 1.3333 | | | | |

The particle state is numbered $1 = 0f_{7/2}$, and the hole states are numbered $1 = 0d_{5/2}$, $2 = 1s_{1/2}$ and $3 = 0d_{3/2}$. The first column gives the particle-hole–particle-hole labels, and the following columns give the $JT$ combinations and values of $\tilde{\mathcal{M}}$. The zero recorded does not result from conservation laws

**Table 11.5.** Quantities $\tilde{\mathcal{M}}_{abcd}(JT)$ in the CS phase convention

| abcd | JT | $\tilde{\mathcal{M}}$ | JT | $\tilde{\mathcal{M}}$ | JT | $\tilde{\mathcal{M}}$ | JT | $\tilde{\mathcal{M}}$ | JT | $\tilde{\mathcal{M}}$ |
|------|----|------|----|------|----|------|----|------|----|------|
| 1111 | 20 | −2.4000 | 21 | −1.7143 | 30 | 0.7619 | 31 | −0.7619 | 40 | −0.8889 |
| 1111 | 41 | −0.3809 | 50 | 1.3333 | 51 | −1.3333 | | | | |
| 1121 | 30 | 0.8248 | 31 | −0.8248 | 40 | 0.9391 | 41 | 0.5634 | | |
| 1131 | 20 | 0.2286 | 21 | 1.1429 | 30 | 0.4949 | 31 | −0.4949 | 40 | 0.6604 |
| 1131 | 41 | 0.4882 | 50 | 0.4761 | 51 | −0.4761 | | | | |
| 2121 | 30 | 1.0000 | 31 | −1.0000 | 40 | −0.7778 | 41 | −1.0000 | | |
| 2131 | 30 | 0.7143 | 31 | −0.7143 | 40 | −0.4247 | 41 | −0.9343 | | |
| 3131 | 10 | 2.5714 | 11 | −2.5714 | 20 | 0.5905 | 21 | −1.0476 | 30 | 0.6190 |
| 3131 | 31 | −0.6190 | 40 | −0.1429 | 41 | −0.8961 | 50 | 0.2684 | 51 | −0.2684 |
| 3131 | 60 | −1.2820 | 61 | −1.5151 | | | | | | |

The particle states are numbered $1 = 1p_{3/2}$, $2 = 1p_{1/2}$ and $3 = 0f_{5/2}$, and the hole state is numbered $1 = 0f_{7/2}$. The first column gives the particle-hole–particle-hole labels, and the following columns give the $JT$ combinations and values of $\tilde{\mathcal{M}}$

## Normalization and orthogonality

We require orthonormality of the RPA states $|\omega\rangle$. With the phonon operator (11.40) we then have

$$\delta_{\omega\omega'} = \langle\omega|\omega'\rangle = \langle\text{RPA}|Q_\omega Q_{\omega'}^\dagger|\text{RPA}\rangle = \langle\text{RPA}|[Q_\omega, Q_{\omega'}^\dagger]|\text{RPA}\rangle$$

$$\overset{\text{QBA}}{\approx} \langle\text{HF}|[Q_\omega, Q_{\omega'}^\dagger]|\text{HF}\rangle = \sum_{\substack{ab\\cd}} \left(X_{ab}^{\omega*} X_{cd}^{\omega'}\langle\text{HF}|[\mathcal{A}_{ab}, \mathcal{A}_{cd}^\dagger]|\text{HF}\rangle\right.$$

$$\left. + Y_{ab}^{\omega*} Y_{cd}^{\omega'}\langle\text{HF}|[\tilde{\mathcal{A}}_{ab}^\dagger, \tilde{\mathcal{A}}_{cd}]|\text{HF}\rangle\right).$$
(11.81)

The approximation designated as QBA is the *quasiboson approximation*. It involves the replacement of the expectation value of a commutator in the true ground state $|\text{RPA}\rangle$ by its expectation value in the uncorrelated ground state $|\text{HF}\rangle$. Our application of the EOM in fact relies on the QBA through the presence of $|\text{HF}\rangle$ in the equation of motion (11.43).

Equation (11.81) also demonstrates the endeavour in the RPA (and its extensions) to replace products of operators with their commutators wherever possible. This technique guarantees that the $Y$ amplitudes do not disappear from the expectation values in the particle–hole vacuum $|\text{HF}\rangle$.

Substituting (11.44) and (11.45) into (11.81) we find

$$\boxed{\sum_{ab} \left(X_{ab}^{nJ^\pi*} X_{ab}^{n'J^\pi} - Y_{ab}^{nJ^\pi*} Y_{ab}^{n'J^\pi}\right) = \delta_{nn'} \quad \text{(RPA orthonormality)}.}$$ (11.82)

This contains the normalization condition

$$\langle\omega|\omega\rangle = \sum_{ab} \left(|X_{ab}^\omega|^2 - |Y_{ab}^\omega|^2\right) = 1.$$ (11.83)

This normalization applies only to the physical, positive-energy solutions of the RPA equations (11.60). The meaning of the positive- and negative-energy solutions is discussed below.

## Completeness

Within given values of the quantum numbers $J^\pi M$, the RPA wave functions satisfy the completeness, or closure, relations

$$\boxed{\begin{aligned}\sum_{\substack{n\\E_n>0}} \left(X_{ab}^{nJ^\pi} X_{cd}^{nJ^\pi*} - Y_{ab}^{nJ^\pi*} Y_{cd}^{nJ^\pi}\right) &= \delta_{ac}\delta_{bd}, \\ \sum_{\substack{n\\E_n>0}} \left(X_{ab}^{nJ^\pi} Y_{cd}^{nJ^\pi*} - Y_{ab}^{nJ^\pi*} X_{cd}^{nJ^\pi}\right) &= 0.\end{aligned}}$$ (11.84)

The derivation of these relations is similar to that of the corresponding quasiparticle-RPA (QRPA) relations in Subsect. 18.2.1.

The relations (11.84) can be written in matrix form as

$$\sum_{\substack{n \\ E_n > 0}} \left[ \begin{pmatrix} X^\omega \\ Y^\omega \end{pmatrix} (X^{\omega\dagger}, -Y^{\omega\dagger}) - \begin{pmatrix} Y^{\omega*} \\ X^{\omega*} \end{pmatrix} (Y^{\omega T}, -X^{\omega T}) \right] = \begin{pmatrix} 1 & 0 \\ 0 & 1 \end{pmatrix} . \qquad (11.85)$$

This form turns out to be handy when deriving the energy-weighted sum rule (EWSR) of the RPA in Subsect. 11.6.3.

**Positive- and negative-energy states**

Inspection of (11.60) shows that if the triplet $E_\omega$, $X^\omega$, $Y^\omega$ of energy and amplitudes constitutes a solution of the RPA equations, then also

$$E_{\omega'} = -E_\omega , \quad X^{\omega'} = Y^{\omega*} , \quad Y^{\omega'} = X^{\omega*} \qquad (11.86)$$

is a solution. This means that every solution with energy $E_\omega$ has a partner with energy $-E_\omega$.

Let us compute the squared norm of the negative-energy solution $|\omega_-\rangle$ corresponding to the positive-energy solution $|\omega_+\rangle$. Equation (11.83) gives

$$\left\| |\omega_-\rangle \right\|^2 = \langle \omega_- | \omega_- \rangle = \sum_{ab} \left( |X_{ab}^{\omega-}|^2 - |Y_{ab}^{\omega-}|^2 \right)$$

$$= \sum_{ab} \left( |Y_{ab}^{\omega+}|^2 - |X_{ab}^{\omega+}|^2 \right) = -\langle \omega_+ | \omega_+ \rangle = -\left\| |\omega_+\rangle \right\|^2 = -1 . \qquad (11.87)$$

An absolute value cannot be negative, so this is an obvious contradiction. Furthermore, according to (11.84) the positive-energy solutions constitute a complete set of eigenstates. We conclude that only the positive-energy solutions have physical meaning and the negative-energy solutions are to be discarded as unphysical. The sign of the squared norm serves to distinguish between physical and unphysical solutions in practical applications. The doubling of solutions is a special property of the non-Hermiticity of the RPA 'supermatrix' (11.60).

The RPA represents a refinement of the TDA. Thus it is to be expected that

$$|Y_{ab}^\omega| \ll 1 \quad \text{for all } \omega, ab . \qquad (11.88)$$

This expectation is borne out by the $Y_{ab}^{\omega+}$ but not by the $Y_{ab}^{\omega-}$, which is another consequence of the unphysical nature of the negative-energy states.

**11.3.2 The RPA Ground State**

The RPA ground state $|\text{RPA}\rangle$ is derived from its defining condition (11.79),

$$Q_\omega |\text{RPA}\rangle = 0 \qquad (11.89)$$

with $Q_\omega$ given by (11.41) as

$$Q_\omega = \sum_{ab} \left[ X_{ab}^{\omega*} \mathcal{A}_{ab}(JM) - Y_{ab}^{\omega*} \tilde{\mathcal{A}}_{ab}^\dagger(JM) \right] . \qquad (11.90)$$

The derivation is carried out by means of the *Thouless theorem*.[2] The theorem states that the RPA ground state is given by

$$|\text{RPA}\rangle = \mathcal{N}_0 e^S |\text{HF}\rangle , \qquad (11.91)$$

where $\mathcal{N}_0$ is a normalization factor and

$$S = \tfrac{1}{2} \sum_{\substack{abcd \\ JM}} C_{abcd}(J) \mathcal{A}_{ab}^\dagger(JM) \tilde{\mathcal{A}}_{cd}^\dagger(JM) \qquad (11.92)$$

with certain amplitudes $C_{abcd}(J)$ having the symmetry property

$$C_{cdab}(J) = C_{abcd}(J) . \qquad (11.93)$$

We proceed to determine the $C_{abcd}(J)$. For any operators $O$ and $S$ one can show that

$$\mathcal{O}e^S = e^S \left( \mathcal{O} + [\mathcal{O}, S] + \frac{1}{2!}[[\mathcal{O}, S], S] + \dots \right) . \qquad (11.94)$$

With $O = Q_\omega$ and $S$ given by (11.92), Eqs. (11.89) and (11.91) imply

$$\left( Q_\omega + [Q_\omega, S] + \frac{1}{2!}[[Q_\omega, S], S] + \dots \right)|\text{HF}\rangle = 0 . \qquad (11.95)$$

The second commutator gives

$$[Q_\omega, S] = \tfrac{1}{2} \sum_{\substack{ab \\ cdc'd' \\ J'M'}} C_{cdc'd'}(J')$$

$$\times \left[ X_{ab}^{\omega*} \mathcal{A}_{ab}(JM) - Y_{ab}^{\omega*} \tilde{\mathcal{A}}_{ab}^\dagger(JM), \mathcal{A}_{cd}^\dagger(J'M') \tilde{\mathcal{A}}_{c'd'}^\dagger(J'M') \right]$$

$$= \tfrac{1}{2} \sum_{\substack{ab \\ cdc'd' \\ J'M'}} C_{cdc'd'}(J') X_{ab}^{\omega*} \left( \left[ \mathcal{A}_{ab}(JM), \mathcal{A}_{cd}^\dagger(J'M') \right] \tilde{\mathcal{A}}_{c'd'}^\dagger(J'M') \right.$$

$$\left. + \mathcal{A}_{cd}^\dagger(J'M') \left[ \mathcal{A}_{ab}(JM), \tilde{\mathcal{A}}_{c'd'}^\dagger(J'M') \right] \right) . \qquad (11.96)$$

To simplify (11.96) we turn to the QBA, which we already used in (11.81). We now apply another aspect of the QBA, namely replacing the basic $[\mathcal{A}, \mathcal{A}^\dagger]$ commutator by its HF expectation value stated in (11.44):

$$[\mathcal{A}_{ab}(JM), \mathcal{A}_{cd}^\dagger(J'M')] \xrightarrow{\text{QBA}} \langle\text{HF}|[\mathcal{A}_{ab}(JM), \mathcal{A}_{cd}^\dagger(J'M')]|\text{HF}\rangle$$

$$= \delta_{ac}\delta_{bd}\delta_{JJ'}\delta_{MM'} . \qquad (11.97)$$

---

[2] For further details see [16]

An analogous relation applies to the commutator in (11.45).

By using (11.18), renaming summation indices and applying (11.93), we can combine the two terms of (11.96) to read

$$[Q_\omega, S] \overset{\text{QBA}}{\approx} \sum_{abcd} X^{\omega*}_{ab} C_{abcd}(J) \tilde{A}^\dagger_{cd}(JM) . \tag{11.98}$$

Substitution of the QBA result (11.98) into the double commutator in (11.95) results in an outer commutator involving creation operators only. The double commutator thus vanishes, and consequently all later terms in the expansion (11.95) also vanish. Equation (11.95) then reduces to

$$\left(Q_\omega + [Q_\omega, S]\right)|\text{HF}\rangle = 0 , \tag{11.99}$$

and substitution from (11.90) and (11.98) yields

$$\sum_{cd} \left[ -Y^{\omega*}_{cd} + \sum_{ab} X^{\omega*}_{ab} C_{abcd}(J) \right] \tilde{A}^\dagger_{cd}(JM)|\text{HF}\rangle = 0 . \tag{11.100}$$

This can be satisfied only if

$$\sum_{ab} X^{\omega*}_{ab} C_{abcd}(J) = Y^{\omega*}_{cd} \quad \text{for all } \omega, cd . \tag{11.101}$$

The amplitudes $X^\omega_{ab}$ and $Y^\omega_{ab}$ are solved from (11.60); the ground state is not needed at that stage. Equations (11.101) are thus a set of linear equations with known coefficients from which the unknown quantities $C_{abcd}(J)$ can be solved to constitute the RPA ground state according to (11.91).

The QBA is an essential part of the RPA. Therefore approximate equalities resulting from the QBA are denoted as equalities after the introductory relation such as (11.98).

As was noted already in connection with (11.44) and (11.45), the QBA makes the particle–hole operators $\mathcal{A}^\dagger, \mathcal{A}$ behave as genuine boson operators. The resultant commutation relations for the RPA phonon operators $Q^\dagger, Q$ similarly obey boson commutation relations within the QBA. Calling them 'phonon' operators comes from the macroscopic interpretation of the excitation $Q^\dagger|\text{RPA}\rangle$ as a quantum of vibration of the nuclear surface.

### 11.3.3 RPA One-Particle Densities

We next derive expressions, in terms of the RPA amplitudes $X$ and $Y$, for the *one-particle densities*

$$\langle \text{RPA}|c^\dagger_\alpha c_\alpha|\text{RPA}\rangle , \quad \langle \text{RPA}|h^\dagger_\beta h_\beta|\text{RPA}\rangle , \tag{11.102}$$

where the $c$'s operate on particle states and the $h$'s on hole states. These densities are useful because they form the basis for extensions of the RPA.

Using (11.22) we find

$$\sum_{m_{\alpha'}} \left[ c_{\alpha'}^\dagger c_{\alpha'}, \mathcal{A}_{ab}^\dagger(JM) \right] = \delta_{a'a} \mathcal{A}_{ab}^\dagger(JM) \, . \tag{11.103}$$

This relation, together with (11.18) and (11.93), leads to

$$\sum_{m_{\alpha'}} \left[ c_{\alpha'}^\dagger c_{\alpha'}, S \right] = \tfrac{1}{2} \sum_{\substack{m_{\alpha'} \\ abcd \\ JM}} C_{abcd}(J) \left[ c_{\alpha'}^\dagger c_{\alpha'}, \mathcal{A}_{ab}^\dagger(JM) \tilde{\mathcal{A}}_{cd}^\dagger(JM) \right]$$

$$= \tfrac{1}{2} \sum_{\substack{abcd \\ JM}} C_{abcd}(J) \left[ \delta_{a'a} \mathcal{A}_{ab}^\dagger(JM) \tilde{\mathcal{A}}_{cd}^\dagger(JM) \right.$$

$$\left. + \, \delta_{a'c} \tilde{\mathcal{A}}_{ab}^\dagger(J-M) \mathcal{A}_{cd}^\dagger(J-M) \right]$$

$$= \sum_{\substack{bcd \\ JM}} C_{a'bcd}(J) \mathcal{A}_{a'b}^\dagger(JM) \tilde{\mathcal{A}}_{cd}^\dagger(JM) \, . \tag{11.104}$$

Because all $m$ substates are occupied with equal weight, we can write

$$(2j_{a'} + 1)\langle \mathrm{RPA}|c_{\alpha'}^\dagger c_{\alpha'}|\mathrm{RPA}\rangle = \sum_{m_{\alpha'}} \langle \mathrm{RPA}|c_{\alpha'}^\dagger c_{\alpha'}|\mathrm{RPA}\rangle$$

$$= \mathcal{N}_0 \sum_{m_{\alpha'}} \langle \mathrm{RPA}|c_{\alpha'}^\dagger c_{\alpha'} e^S|\mathrm{HF}\rangle$$

$$= \mathcal{N}_0 \sum_{m_{\alpha'}} \langle \mathrm{RPA}|e^S \left( c_{\alpha'}^\dagger c_{\alpha'} + [c_{\alpha'}^\dagger c_{\alpha'}, S] + \frac{1}{2!} \left[ [c_{\alpha'}^\dagger c_{\alpha'}, S], S \right] + \dots \right)|\mathrm{HF}\rangle$$

$$= \sum_{\substack{bcd \\ JM}} C_{a'bcd}(J)\langle \mathrm{RPA}|\mathcal{A}_{a'b}^\dagger(JM) \tilde{\mathcal{A}}_{cd}^\dagger(JM)|\mathrm{RPA}\rangle \, , \tag{11.105}$$

where we have used (11.92) and (11.94). Only the second term in the expansion contributes: the first term gives zero when acting on $|\mathrm{HF}\rangle$, and the outer commutator in the third term is seen to vanish when (11.104) is inserted into the inner commutator.

In basic form, the RPA completeness relations (11.84), for fixed $J^\pi M$, are

$$\sum_{\substack{n \\ E_n>0}} |nJ^\pi M\rangle\langle nJ^\pi M| = \sum_{\substack{n \\ E_n>0}} Q_{nJ^\pi M}^\dagger|\mathrm{RPA}\rangle\langle \mathrm{RPA}|Q_{nJ^\pi M} = 1 \, . \tag{11.106}$$

We insert this into the matrix element in (11.105) and use (11.40), (11.41), (11.44), (11.45) and (11.84):

$$\langle \text{RPA}|\mathcal{A}^\dagger_{a'b}(JM)\tilde{\mathcal{A}}^\dagger_{cd}(JM)|\text{RPA}\rangle = \langle \text{RPA}|\tilde{\mathcal{A}}^\dagger_{cd}(JM)\mathcal{A}^\dagger_{a'b}(JM)|\text{RPA}\rangle$$

$$= \sum_{\substack{n \\ E_n > 0}} \langle \text{RPA}|\tilde{\mathcal{A}}^\dagger_{cd}(JM)Q^\dagger_{nJ^\pi M}|\text{RPA}\rangle \langle \text{RPA}|Q_{nJ^\pi M}\mathcal{A}^\dagger_{a'b}(JM)|\text{RPA}\rangle$$

$$= \sum_{\substack{n \\ E_n > 0}} \langle \text{RPA}|\left[\tilde{\mathcal{A}}^\dagger_{cd}(JM), Q^\dagger_{nJ^\pi M}\right]|\text{RPA}\rangle \langle \text{RPA}|\left[Q_{nJ^\pi M}, \mathcal{A}^\dagger_{a'b}(JM)\right]|\text{RPA}\rangle$$

$$\overset{\text{QBA}}{\approx} \sum_{\substack{n \\ E_n > 0}} \langle \text{HF}|\left[\tilde{\mathcal{A}}^\dagger_{cd}(JM), Q^\dagger_{nJ^\pi M}\right]|\text{HF}\rangle \langle \text{HF}|\left[Q_{nJ^\pi M}, \mathcal{A}^\dagger_{a'b}(JM)\right]|\text{HF}\rangle$$

$$= \sum_{\substack{n \\ E_n > 0}} Y^{nJ^\pi}_{cd} X^{nJ^\pi *}_{a'b} = \sum_{\substack{n \\ E_n > 0}} X^{nJ^\pi *}_{cd} Y^{nJ^\pi}_{a'b} . \tag{11.107}$$

Equation (11.105) now becomes

$$\widehat{j_{a'}}^2 \langle \text{RPA}|c^\dagger_{\alpha'} c_{\alpha'}|\text{RPA}\rangle = \sum_{\substack{bcd \\ JM}} C_{a'bcd}(J) \sum_{\substack{n \\ E_n > 0}} X^{nJ^\pi *}_{cd} Y^{nJ^\pi}_{a'b}$$

$$= \sum_{\substack{bJn \\ E_n > 0}} \hat{J}^2 Y^{nJ^\pi}_{a'b} \sum_{cd} X^{nJ^\pi *}_{cd} C_{cda'b}(J) = \sum_{\substack{bJn \\ E_n > 0}} \hat{J}^2 \left|Y^{nJ^\pi}_{a'b}\right|^2 , \tag{11.108}$$

where (11.101) was used in the final step.

A derivation analogous to that of (11.108) gives for hole operators the result

$$\widehat{j_{a'}}^2 \langle \text{RPA}|h^\dagger_{\alpha'} h_{\alpha'}|\text{RPA}\rangle = \sum_{\substack{aJn \\ E_n > 0}} \hat{J}^2 \left|Y^{nJ^\pi}_{aa'}\right|^2 . \tag{11.109}$$

In summary, we rewrite the one-particle densities in (11.108) and (11.109) as

$$\boxed{\begin{aligned} \langle \text{RPA}|c^\dagger_\alpha c_\alpha|\text{RPA}\rangle &= \widehat{j_a}^{-2} \sum_{\substack{bnJ \\ E_n > 0}} \hat{J}^2 \left|Y^{nJ^\pi}_{ab}\right|^2 , \\ \langle \text{RPA}|h^\dagger_\beta h_\beta|\text{RPA}\rangle &= \widehat{j_b}^{-2} \sum_{\substack{anJ \\ E_n > 0}} \hat{J}^2 \left|Y^{nJ^\pi}_{ab}\right|^2 . \end{aligned}} \tag{11.110}$$

If the structure of the RPA ground state $|\text{RPA}\rangle$ is to be dominated by the structure of the particle–hole ground state $|\text{HF}\rangle$, the sums in (11.110) have to be small. In that case the implication is that

$$\left|Y^{nJ^\pi}_{ab}\right| \ll 1 \quad \text{for all } nJ, ab . \tag{11.111}$$

This repeats the statement in (11.88). If the condition (11.111) applies, we can expect the quasiboson approximation to be good.

The expressions (11.110) for the one-particle densities open up a way to improve the RPA. In particular they help to better take into account the

Pauli principle, which is partly lost in the quasiboson approximation. The deterioration of the Pauli principle can in fact cause excessive, unphysical correlations in the RPA ground state (see Subsect. 11.4.2). To go beyond the RPA level of approximation means that we must upgrade the definitions (11.48) of the A and B matrices to

$$A_{ab,cd} \equiv \langle \text{RPA} | \left[ \mathcal{A}_{ab}, H, \mathcal{A}^\dagger_{cd} \right] | \text{RPA} \rangle , \tag{11.112}$$

$$B_{ab,cd} \equiv -\langle \text{RPA} | \left[ \mathcal{A}_{ab}, H, \tilde{\mathcal{A}}_{cd} \right] | \text{RPA} \rangle . \tag{11.113}$$

When evaluating these expressions we encounter the one-particle densities (11.110) in addition to the standard terms in (11.60). The matrix equation replacing (11.60) has the structure

$$\begin{pmatrix} \mathsf{A}(\mathsf{X}^\omega, \mathsf{Y}^\omega) & \mathsf{B}(\mathsf{X}^\omega, \mathsf{Y}^\omega) \\ -\mathsf{B}^*(\mathsf{X}^\omega, \mathsf{Y}^\omega) & -\mathsf{A}^*(\mathsf{X}^\omega, \mathsf{Y}^\omega) \end{pmatrix} \begin{pmatrix} \mathsf{X}^\omega \\ \mathsf{Y}^\omega \end{pmatrix} = E_\omega \begin{pmatrix} \mathsf{X}^\omega \\ \mathsf{Y}^\omega \end{pmatrix} . \tag{11.114}$$

Unlike the linear matrix equation (11.60), this equation is nonlinear, which means that it has to be solved iteratively.

The equations contained in (11.114) can be solved to various degrees of accuracy. The solution methods constitute variations of what is generally known as *higher RPA*. Examples are the *self-consistent RPA* (SCRPA) and the *renormalized RPA* (RRPA). For more information on this subject see, e.g. [14, 16].

## 11.4 RPA Solutions of the Schematic Separable Model

We can solve the RPA equations with a general separable force similar to the treatment of the TDA equations in Sect. 9.2. The schematic solutions allow an assessment of the similarities and differences between the TDA and RPA, both for the energy spectrum and for the electric decay rates. In particular, the schematic model provides a straightforward comparison between the collective features of the two models.

### 11.4.1 The RPA Dispersion Equation

The general separable force is treated within the schematic model similar to the TDA study in Sect. 9.2. The schematic A matrix of the RPA can be read off the TDA equation (9.49) as

$$A_{ab,cd} = \delta_{ac}\delta_{bd}\varepsilon_{ab} + \chi_J Q^J_{ab} Q^J_{cd} . \tag{11.115}$$

To derive the schematic B matrix we start from the expression (11.70) and substitute into it the matrix element of the separable force as given by (8.55) and (8.56). This leads to

$$B_{ab,cd}(J) = (-1)^{j_b+j_c+J} \sqrt{(1+\delta_{ac})(1+\delta_{bd})} \sum_{J'}(-1)^{J'}\widehat{J'}^2 \begin{Bmatrix} j_a & j_b & J \\ j_d & j_c & J' \end{Bmatrix}$$

$$\times \mathcal{N}_{ac}(J')\mathcal{N}_{bd}(J')\Bigg[\sum_\lambda \chi_\lambda(-1)^{j_a+j_c+J'}\begin{Bmatrix} j_a & j_c & J' \\ j_d & j_b & \lambda \end{Bmatrix}(b\|\boldsymbol{Q}_\lambda\|a)(c\|\boldsymbol{Q}_\lambda\|d)$$

$$- (-1)^{j_b+j_d+J'}\sum_\lambda \chi_\lambda(-1)^{j_a+j_c+J'}\begin{Bmatrix} j_a & j_c & J' \\ j_b & j_d & \lambda \end{Bmatrix}(d\|\boldsymbol{Q}_\lambda\|a)(c\|\boldsymbol{Q}_\lambda\|b)\Bigg].$$

$$(11.116)$$

The interaction strengths $\chi_\lambda$ are included here similarly to (9.44). As in the case of (11.69), the normalization factors can be taken outside the $J'$ sum. They then cancel against the square-root factor in the first line of (11.116). Using the properties (1.66) and (1.67) of the $6j$ symbols we obtain

$$B_{ab,cd}(J) = (-1)^{j_a+j_b+J+1}\chi_J \widehat{J}^{-2}(b\|\boldsymbol{Q}_J\|a)(c\|\boldsymbol{Q}_J\|d)$$

$$+ (-1)^{j_a+j_d+1}\sum_\lambda(-1)^\lambda \chi_\lambda \begin{Bmatrix} j_a & j_b & J \\ j_c & j_d & \lambda \end{Bmatrix}(d\|\boldsymbol{Q}_\lambda\|a)(c\|\boldsymbol{Q}_\lambda\|b)$$

$$\approx (-1)^J \chi_J Q^J_{ab}Q^J_{cd}. \tag{11.117}$$

In the final step the second, exchange term was neglected as in the schematic TDA, and the symmetry property (6.27) and the definition (9.47) were used to simplify the expression.

Written out, (11.60) gives the RPA matrix equations

$$\mathsf{A}\mathsf{X}^\omega + \mathsf{B}\mathsf{Y}^\omega = E_\omega \mathsf{X}^\omega, \tag{11.118}$$

$$\mathsf{B}^*\mathsf{X}^\omega + \mathsf{A}^*\mathsf{Y}^\omega = -E_\omega \mathsf{Y}^\omega. \tag{11.119}$$

Substituting the $\mathsf{A}$ matrix from (11.115) and the $\mathsf{B}$ matrix from (11.117) we have, with the $Q^J_{ab}$ being real,

$$\sum_{cd}(\delta_{ac}\delta_{bd}\varepsilon_{ab} + \chi_J Q^J_{ab}Q^J_{cd})X^\omega_{cd} + \sum_{cd}(-1)^J \chi_J Q^J_{ab}Q^J_{cd}Y^\omega_{cd} = E_\omega X^\omega_{ab},$$

$$(11.120)$$

$$\sum_{cd}(-1)^J \chi_J Q^J_{ab}Q^J_{cd}X^\omega_{cd} + \sum_{cd}(\delta_{ac}\delta_{bd}\varepsilon_{ab} + \chi_J Q^J_{ab}Q^J_{cd})Y^\omega_{cd} = -E_\omega Y^\omega_{ab}.$$

$$(11.121)$$

We define, similarly to the TDA definition in (9.50),

$$N_\omega \equiv -\chi_J \sum_{cd}\left[Q^J_{cd}X^\omega_{cd} + (-1)^J Q^J_{cd}Y^\omega_{cd}\right]. \tag{11.122}$$

Equations (11.120) and (11.121) then give the RPA amplitudes as

$$\boxed{X^\omega_{ab} = \frac{Q^J_{ab}}{\varepsilon_{ab} - E_\omega}N_\omega, \quad Y^\omega_{ab} = (-1)^J\frac{Q^J_{ab}}{\varepsilon_{ab} + E_\omega}N_\omega.} \tag{11.123}$$

Like the TDA case, Eq. (9.52), $N_\omega$ is evaluated from the normalization condition (11.83) of the RPA solutions.

Substituting from (11.123) into (11.122) results in

$$N_\omega = -\chi_J \sum_{cd} \left[ Q_{cd}^J \frac{Q_{cd}^J}{\varepsilon_{cd} - E_\omega} N_\omega + (-1)^J Q_{cd}^J (-1)^J \frac{Q_{cd}^J}{\varepsilon_{cd} + E_\omega} N_\omega \right]. \quad (11.124)$$

The quantity $N_\omega$ cancels out, and the final result is the *RPA dispersion equation*

$$\boxed{-\frac{1}{\chi_J} = 2 \sum_{ab} \frac{\left(Q_{ab}^J\right)^2 \varepsilon_{ab}}{\varepsilon_{ab}^2 - E_\omega^2}.} \quad (11.125)$$

This is a transcendental equation, similar to the TDA dispersion equation (9.55), to be solved numerically or graphically. The graphical method was presented in Subsect. 9.2.4. For the SDI the interaction strength $\chi_J$ is given by (9.77). To illustrate the RPA and its difference from the TDA, we repeat the TDA example of Subsect. 9.2.4 now in the RPA.

### 11.4.2 Application to $1^-$ Excitations in $^4$He

Consider the $1^-$ excitation spectrum of $^4_2\text{He}_2$ within the 0s-0p valence space. These excitations were treated in Subsect. 9.2.4 within the schematic TDA using the SDI. Here we take the coupling constant as in Subsect. 9.2.3 and use it in the RPA dispersion equation (11.125).

As in the TDA case, we take the energy difference between the 0s and 0p shells as $\Delta E(\text{0p-0s}) = 21.0\,\text{MeV}$ and within the 0p shell as $\varepsilon_{0p_{1/2}} - \varepsilon_{0p_{3/2}} = 6.0\,\text{MeV}$. The matrix elements $Q_{ab}^J$ are given by (9.80), and substituting them and the energies into (11.125) gives

$$-\frac{1}{\mp\frac{1}{4}A} = 2 \left[ \frac{\frac{16}{3} \times 21.0\,\text{MeV}}{(21.0\,\text{MeV})^2 - E^2} + \frac{\frac{8}{3} \times 27.0\,\text{MeV}}{(27.0\,\text{MeV})^2 - E^2} \right], \quad (11.126)$$

where the upper sign is for $T = 0$ and the lower for $T = 1$. This simplifies to

$$\pm\frac{1}{A} = \frac{4}{3} \left[ \frac{42.0\,\text{MeV}}{(21.0\,\text{MeV})^2 - E^2} + \frac{27.0\,\text{MeV}}{(27.0\,\text{MeV})^2 - E^2} \right]. \quad (11.127)$$

Graphical solutions of the two equations (11.127) are shown in Fig. 11.2. There the common right-hand side and the left-hand sides (horizontal lines) are plotted as functions of $E$ for two values of $A$, namely $A = 1.0\,\text{MeV}$ and $A = 2.0\,\text{MeV}$. The abscissas of the intersection points give the solutions $E_n$, separately for each $T$. The solutions read off the figure are collected into Tables 11.6 and 11.7, and there compared with non-schematic, numerical solutions from Subsect. 11.5.4. The solutions of the schematic model are seen to be rather close to the exact solutions, but the difference increases with increasing interaction strength.

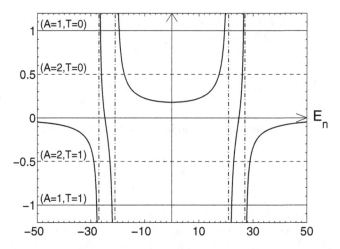

**Fig. 11.2.** Graphical solution of the transcendental equation (11.127) for the $1^-$ states in $^4$He. All energies are given in Mega-electron volts. Solutions of the $T = 0$ and $T = 1$ eigenenergies are shown for two interaction strengths $A$. The unperturbed particle–hole energies are shown as vertical dash-dotted lines

**Table 11.6.** RPA eigenenergies $E_n$ of $J^\pi = 1^-$, $T = 0$ particle–hole states in $^4$He obtained by graphical solution of the schematic model and by exact diagonalization

| $A$ (MeV) | $E_1$ (MeV) | | $E_2$ (MeV) | |
|---|---|---|---|---|
| | Schematic | Exact | Schematic | Exact |
| 1.0 | 19.4 | 18.668 | 26.4 | 27.083 |
| 2.0 | 17.5 | 15.180 | 26.0 | 27.633 |

The SDI is used for both

**Table 11.7.** RPA eigenenergies $E_n$ of $J^\pi = 1^-$, $T = 1$ particle–hole states in $^4$He obtained by graphical solution of the schematic model and by exact diagonalization

| $A$ (MeV) | $E_1$ (MeV) | | $E_2$ (MeV) | |
|---|---|---|---|---|
| | Schematic | Exact | Schematic | Exact |
| 1.0 | 22.1 | 21.980 | 27.8 | 27.980 |
| 2.0 | 22.9 | 22.922 | 28.9 | 28.924 |

The SDI is used for both

The lowest $T = 0$ solution, interpreted as collective, is typical of TDA and RPA theory. The collectivity is seen as a lowering of the energy $E_1$ from the non-interacting particle–hole energy with increasing strength of the two-body interaction. Figure 11.2 shows that the lowering occurs without bound to zero value. The lowering is related to the attractive nature of the $T = 0$ particle–

hole interaction. The collectivity of the first $T = 0$ state is displayed also by its electromagnetic decay properties, to be discussed in Sect. 11.6.

The highest $T = 1$ solution also behaves collectively. Its energy increases without bound with increasing interaction strength. The energy $E_2$ in Table 11.7 does not show a conspicuous increase but Fig. 11.2 does. This behaviour is due to the repulsive nature of the $T = 1$ particle–hole interaction. As noted in Subsect. 9.2.4, the collective $T = 1$ state is known as the giant dipole resonance. The rest of the RPA states, other than the two collective ones, are bound in energy by the unperturbed particle–hole energies $\varepsilon_{ab}$.

Comparison of Tables 11.6 and 11.7 with the corresponding TDA Tables 9.5 and 9.6 reveals that the RPA and TDA results are quite close to each other, both for the schematic model and for exact diagonalization. Only the lowest TDA and RPA solutions differ notably with increasing interaction strength. Although the RPA and TDA solutions are rather similar, they have the following conspicuous differences:

- The solutions of the TDA are doubled by the RPA. The negative-energy solutions of the RPA are, however, unphysical. As seen in Subsect. 11.3.1, the positive-energy RPA solutions form a complete set of states.

- Figure 11.2 shows that the lowest RPA energy approaches zero with increasing interaction strength. Beyond a certain *critical* value the horizontal line no longer intersects the curve; formally the energy becomes imaginary. This never happens in the TDA and is related to the non-Hermitian nature of the RPA supermatrix. Since the RPA is not derived from a variational principle, the ground-state correlations increasing with the interaction strength can push the first excited state below the ground state. This means that the excessive ground-state correlations make the particle–hole ground state $|\mathrm{HF}\rangle$ unstable. The critical value of the interaction strength corresponds to the *breaking point* of the RPA. As it turns out, this instability can only be avoided by letting the ground state build a static deformation.

### 11.4.3 The Degenerate Model

Same as the discussion of the TDA at the end of Subsect. 9.2.1, let us consider the RPA in the degenerate limit $\varepsilon_{ab} \to \varepsilon$ for all $ab$. For the TDA and RPA alike, there are two untrapped collective solutions. In the RPA their energy is obtained from (11.125), which now becomes

$$-\frac{1}{\chi_J} = \frac{2\varepsilon Q^2}{\varepsilon^2 - E_{\mathrm{coll}}^2}, \quad Q^2 \equiv \sum_{ab} \left(Q_{ab}^J\right)^2 . \tag{11.128}$$

Solving for $E_{\mathrm{coll}}$ gives

$$E_{\mathrm{coll}} = \pm\sqrt{\varepsilon(\varepsilon + 2\chi_J Q^2)} . \tag{11.129}$$

Taking the physical positive-energy solution and using (9.77) for the SDI gives finally

$$E_{\text{coll}}(T = 0) = \sqrt{\varepsilon\left(\varepsilon - \tfrac{1}{2}AQ^2\right)} \,, \quad E_{\text{coll}}(T = 1) = \sqrt{\varepsilon\left(\varepsilon + \tfrac{1}{2}AQ^2\right)} \,. \quad (11.130)$$

Let us examine limiting values of the interaction strength. For small values of $|\chi_J|$ the positive solution (11.129) becomes $E_{\text{coll}} \approx \varepsilon + \chi_J Q^2$, which is the TDA result (9.57). This indicates that for $|\chi_J| \ll \varepsilon/Q^2$ the RPA does not differ significantly from the TDA. In addition we observe that

$$\left(E_{\text{coll}}^{\text{RPA}}\right)^2 = \varepsilon^2 + 2\varepsilon\chi_J Q^2 = \left(E_{\text{coll}}^{\text{TDA}}\right)^2 - \chi_J^2 Q^4 < \left(E_{\text{coll}}^{\text{TDA}}\right)^2 \,, \quad (11.131)$$

so that for the two untrapped solutions the RPA energy is always lower than the corresponding TDA energy.

The result (11.130) implies that the lowest $T = 0$ solution breaks down at the point where the interaction strength $A$ exceeds $2\varepsilon/Q^2$. For the energy of the giant resonance this critical value yields an upper limit of $\sqrt{2}\,\varepsilon$.

## 11.5 RPA Description of Doubly Magic Nuclei

In this section we solve the RPA matrix equations starting from the two-body interaction matrix elements of the SDI. First we have to construct the RPA matrices A and B in a way similar to that used for the A matrix of the TDA. What is more elaborate in the RPA than in the TDA is the diagonalization of the RPA supermatrix. This is due to the fact that the associated eigenvalue problem is non-Hermitian and half of the solutions are unphysical. One special method to deal with this problem is exhaustively discussed in Subsect. 11.5.2.

### 11.5.1 Examples of the RPA Matrices

Let us first discuss two simple examples of applying the RPA formalism to the excitation spectra of doubly magic nuclei.

#### Lowest $3^-$ state in $^{48}$Ca solved

Consider the lowest $3^-$ state in $^{48}_{20}$Ca$_{28}$ in the proton $0f_{7/2}$-$(0d_{3/2})^{-1}$ particle–hole valence space. This is the simplest particle–hole valence space where a $3^-$ state can be constructed; the neutron particle–hole space $(1p-0f_{5/2})$-$(0f_{7/2})^{-1}$ produces no negative-parity states. In our valence space the matrix A in the RPA matrix equation (11.60) only consists of one element,

$$A_{0f_{7/2}0d_{3/2},0f_{7/2}0d_{3/2}}(3) \equiv a \,. \quad (11.132)$$

We take the single-particle energies from Fig. 9.2 (c) and the SDI matrix element from (9.28) and Table 9.3, with the result

$$a = \varepsilon_{0f_{7/2}} - \varepsilon_{0d_{3/2}} + 0.3809A_1 = 5.0\,\text{MeV} + 0.3809A_1 \,. \tag{11.133}$$

Similarly, the only element of the B matrix in the RPA matrix equation (11.60) is

$$B_{0f_{7/2}0d_{3/2},0f_{7/2}0d_{3/2}}(3) \equiv b \,. \tag{11.134}$$

Equation (11.74) and Table 11.4 give

$$b = 0.7619A_1 \,. \tag{11.135}$$

Equation (11.60) now becomes

$$\begin{pmatrix} a & b \\ -b & -a \end{pmatrix} \begin{pmatrix} X^\omega \\ Y^\omega \end{pmatrix} = E_\omega \begin{pmatrix} X^\omega \\ Y^\omega \end{pmatrix} \,. \tag{11.136}$$

We solve this equation for the eigenenergies and the corresponding wave functions. The energy eigenvalues are given by the secular equation

$$0 = \begin{vmatrix} a - E_\omega & b \\ -b & -a - E_\omega \end{vmatrix} = -(a^2 - E_\omega^2) + b^2 \,. \tag{11.137}$$

Solving for $E_\omega$ we obtain

$$E_\omega = \pm\sqrt{a^2 - b^2} = \pm\sqrt{(5.0\,\text{MeV} + 0.3809A_1)^2 - (0.7619A_1)^2} \,. \tag{11.138}$$

Keeping only the physical positive-energy solution and inserting the default SDI strength $A_1 = 1.0\,\text{MeV}$, we have for the energy of the lowest $3^-$ state

$$E(3_1^-) = 5.33\,\text{MeV} \,. \tag{11.139}$$

For the eigenvector of the $3_1^-$ state we obtain from (11.136)

$$Y = \frac{E - a}{b}X = \frac{\sqrt{a^2 - b^2} - a}{b}X \,. \tag{11.140}$$

We assume real amplitudes $X$ and $Y$. The normalization condition (11.83) then reduces to $X^2 - Y^2 = 1$, and we find

$$X = \frac{b}{\sqrt{2}}\left(b^2 - a^2 + a\sqrt{a^2 - b^2}\right)^{-1/2} \,, \tag{11.141}$$

$$Y = \frac{\sqrt{a^2 - b^2} - a}{\sqrt{2}}\left(b^2 - a^2 + a\sqrt{a^2 - b^2}\right)^{-1/2} \,. \tag{11.142}$$

Equations (11.133) and (11.135) show that $b \ll a$. Expanding the square roots we find to order $b^2/a^2$

$$X \approx 1 + \frac{b^2}{8a^2} \,, \quad Y \approx -\frac{b}{2a} \,. \tag{11.143}$$

This shows that values of $X$ can exceed unity and that the ground-state correlations, related to the $Y$ amplitudes, arise from the elements of the B matrix. We also see that as $b \to 0$ we recover the TDA: $X = 1$, $Y = 0$.

In the present case we have $b/a = 0.142$, which gives the amplitudes (11.143) as

$$X \approx 1.0025 , \quad Y \approx -0.0710 . \tag{11.144}$$

These are excellent approximations; the exact equations (11.141) and (11.142) give $X = 1.0025$ and $Y = -0.0713$.

### Setting up the problem for $1^-$ excitations in $^4$He

Consider the $1^-$ excitations of $^4_2$He$_2$ in the particle–hole valence space $0\mathrm{p}_{3/2}$-$(0\mathrm{s}_{1/2})^{-1}$, with the single-particle energy difference $\varepsilon_{0\mathrm{p}_{3/2}} - \varepsilon_{0\mathrm{s}_{1/2}} = 21.0\,\mathrm{MeV}$ used in Chap. 9. The basis states are

$$\{|\pi_1\rangle, |\nu_1\rangle\} = \{|\pi 0\mathrm{p}_{3/2}\,(\pi 0\mathrm{s}_{1/2})^{-1}; 1^-\rangle, |\nu 0\mathrm{p}_{3/2}\,(\nu 0\mathrm{s}_{1/2})^{-1}; 1^-\rangle\} . \tag{11.145}$$

The TDA matrix A can be extracted from the matrix (9.91). The rows and columns in the matrix (9.89) containing the proton or neutron label 2 are discarded. From the first and third rows and columns of (9.91) we then have

$$\mathsf{A}(1^-) = \begin{pmatrix} 21.0\,\mathrm{MeV} - 0.333A_1 & \frac{1}{2}(-0.333A_1 - 2.333A_0) \\ \frac{1}{2}(-0.333A_1 - 2.333A_0) & 21.0\,\mathrm{MeV} - 0.333A_1 \end{pmatrix} . \tag{11.146}$$

The correlation matrix B of the RPA is constructed by using the relations (11.74)–(11.76) and Table 11.2. The result, with three decimal places, is

$$\mathsf{B}(1^-) = \begin{pmatrix} 1.000A_1 & \frac{1}{2}(A_1 + 1.667A_0) \\ \frac{1}{2}(A_1 + 1.667A_0) & 1.000A_1 \end{pmatrix} . \tag{11.147}$$

In the previous example the matrices A and B were trivial in that their dimension was 1-by-1. In the present case we have the non-trivial 2-by-2 matrices A and B stated in (11.146) and (11.147). The task is to diagonalize the supermatrix (11.60). This diagonalization has to be carried out in a special way because the matrix is non-Hermitian, or non-symmetric in the case of a real matrix.

In addition, we have to take care of the normalization condition (11.83), which is not the usual Euclidean dot product of two vectors. There are several ways to perform the diagonalization. For example, we can use a suitable commercial code designed to diagonalize non-Hermitian matrices with general orthogonality criteria. We can also use methods by which a non-Hermitian eigenvalue problem is converted to a Hermitian one.

In the following we discuss a diagonalization method due to Ullah and Rowe [67]. The method is suitable for real RPA matrices with real eigenvalues, or with at most one pair of imaginary eigenvalues. It is based on similarity transformations which turn the original problem into a symmetric one. The resulting matrices are of the TDA type and can be diagonalized by the usual methods.

## 11.5.2 Diagonalization of the RPA Supermatrix by Similarity Transformations

To be able to proceed with the $^4$He example mentioned above, we digress to develop a general matrix method of solving the RPA equations. We assume that the A and B matrices are *real*. According to the relations (11.58) both A and B are then *symmetric*: $A^T = A$, $B^T = B$. We can now write the RPA supermatrix equation (11.60) as the pair of matrix equations

$$AX^\omega + BY^\omega = E_\omega X^\omega , \qquad (11.148)$$

$$-BX^\omega - AY^\omega = E_\omega Y^\omega . \qquad (11.149)$$

We form the difference and sum of these equations:

$$M_+ U_+^\omega = E_\omega U_-^\omega , \qquad (11.150)$$

$$M_- U_-^\omega = E_\omega U_+^\omega , \qquad (11.151)$$

where we have defined the new matrices and amplitudes as

$$M_\pm \equiv A \pm B , \quad U_\pm^\omega \equiv X^\omega \pm Y^\omega . \qquad (11.152)$$

To solve Eqs. (11.150) and (11.151) we can first diagonalize the matrix $M_-$ by an orthogonal similarity transformation T, i.e. $T^{-1} = T^T$. The result of the diagonalization is

$$D_- = T^T M_- T , \qquad (11.153)$$

where $D_-$ is a diagonal matrix. In this transformation, the symmetric matrix $M_+$ is carried over to the symmetric matrix

$$M'_+ = T^T M_+ T \qquad (11.154)$$

and Eqs. (11.150) and (11.151) become

$$M'_+ V_+^\omega = E_\omega V_-^\omega , \qquad (11.155)$$

$$D_- V_-^\omega = E_\omega V_+^\omega , \qquad (11.156)$$

where

$$V_\pm^\omega \equiv T^T U_\pm^\omega . \qquad (11.157)$$

By substituting from (11.156) into (11.155) we can eliminate $V_+^\omega$ and find for $V_-^\omega$ the equation

$$M'_+ D_- V_-^\omega = E_\omega^2 V_-^\omega . \qquad (11.158)$$

The matrix $M'_+ D_-$ is not symmetric. Equation (11.158) is therefore not an eigenvalue problem for a symmetric matrix. However, it can be turned into one by defining a square-root matrix $D_-^{1/2}$ through

$$D_-^{1/2} D_-^{1/2} = D_- . \qquad (11.159)$$

The matrix $D_-^{1/2}$ is diagonal and its elements are the square roots of the corresponding elements of $D_-$. Inserting (11.159) into (11.158) and multiplying by $D_-^{1/2}$ from the left we create a symmetric matrix $D_-^{1/2}M'_+D_-^{1/2}$. With the notation

$$M''_+ \equiv D_-^{1/2}M'_+D_-^{1/2} , \quad R_-^\omega \equiv D_-^{1/2}V_-^\omega \tag{11.160}$$

we now have the real symmetric eigenvalue problem

$$\boxed{M''_+R_-^\omega = E_\omega^2 R_-^\omega .} \tag{11.161}$$

Instead of proceeding from (11.153) we could have started by diagonalizing $M_+$ by an orthogonal transformation $T$ (not the same as for $M_-$). Then all the $+$ and $-$ subscripts are reversed. The result analogous to (11.161) is

$$\boxed{M''_-R_+^\omega = E_\omega^2 R_+^\omega .} \tag{11.162}$$

Suppose we have solved (11.161) and found the eigenenergies $E_\omega$ and eigenvectors $R_-^\omega$. We proceed to find the RPA amplitudes $X^\omega$ and $Y^\omega$. Equations (11.152), (11.156) and (11.157) give

$$X^\omega = \tfrac{1}{2}T \left( \frac{D_-}{E_\omega} + 1 \right) V_-^\omega , \quad Y^\omega = \tfrac{1}{2}T \left( \frac{D_-}{E_\omega} - 1 \right) V_-^\omega . \tag{11.163}$$

From (11.160) we can write

$$V_-^\omega = D_-^{-1/2}R_-^\omega , \tag{11.164}$$

where the diagonal matrix $D_-^{-1/2}$ is defined through $D_-^{1/2}D_-^{-1/2} = 1$, i.e. the elements of $D_-^{-1/2}$ are the inverses of the corresponding elements of $D_-^{1/2}$. The solutions for $X^\omega$ and $Y^\omega$ can now be written as

$$\boxed{X^\omega = G_+^\omega R_-^\omega , \quad Y^\omega = G_-^\omega R_-^\omega ,} \tag{11.165}$$

where

$$G_\pm^\omega \equiv \tfrac{1}{2}T \left( \frac{1}{E_\omega}D_-^{1/2} \pm D_-^{-1/2} \right) . \tag{11.166}$$

Similarly, when we have the eigenvalues $E_\omega$ and eigenvectors $R_+^\omega$ from (11.162) we obtain

$$\boxed{X^\omega = F_+^\omega R_+^\omega , \quad Y^\omega = F_-^\omega R_+^\omega ,} \tag{11.167}$$

where

$$F_\pm^\omega \equiv \tfrac{1}{2}T \left( D_+^{-1/2} \pm \frac{1}{E_\omega}D_+^{1/2} \right) . \tag{11.168}$$

Let us discuss next the implications of the normalization condition (11.83) for the transformed equations (11.161) and (11.162). In matrix form the condition is

$$\mathsf{X}^{\omega\mathrm{T}}\mathsf{X}^{\omega} - \mathsf{Y}^{\omega\mathrm{T}}\mathsf{Y}^{\omega} = 1 \ . \tag{11.169}$$

Inverting Eq. (11.152) for $\mathsf{X}^{\omega}$ and $\mathsf{Y}^{\omega}$ transforms the RPA normalization condition into

$$\begin{aligned}
1 &= \mathsf{X}^{\omega\mathrm{T}}\mathsf{X}^{\omega} - \mathsf{Y}^{\omega\mathrm{T}}\mathsf{Y}^{\omega} \\
&= \tfrac{1}{4}\left[\left(\mathsf{U}_{+}^{\omega\,\mathrm{T}} + \mathsf{U}_{-}^{\omega\,\mathrm{T}}\right)\left(\mathsf{U}_{+}^{\omega} + \mathsf{U}_{-}^{\omega}\right) - \left(\mathsf{U}_{+}^{\omega\,\mathrm{T}} - \mathsf{U}_{-}^{\omega\,\mathrm{T}}\right)\left(\mathsf{U}_{+}^{\omega} - \mathsf{U}_{-}^{\omega}\right)\right] \\
&= \mathsf{U}_{+}^{\omega\,\mathrm{T}}\mathsf{U}_{-}^{\omega} \ . \tag{11.170}
\end{aligned}$$

We solve $\mathsf{U}_{+}^{\omega}$ from (11.151) and substitute the result into (11.170), which leads to the normalization condition

$$1 = \frac{1}{E_{\omega}}\mathsf{U}_{-}^{\omega\,\mathrm{T}}\mathsf{M}_{-}^{\mathrm{T}}\mathsf{U}_{-}^{\omega} \ . \tag{11.171}$$

Equation (11.153) yields

$$\mathsf{M}_{-} = \mathsf{T}\mathsf{D}_{-}\mathsf{T}^{\mathrm{T}} \ , \tag{11.172}$$

and $\mathsf{M}_{-} = \mathsf{M}_{-}^{\mathrm{T}}$. Substituting for $\mathsf{M}_{-}^{\mathrm{T}}$ in (11.171) and using (11.157), (11.159) and (11.160), we have

$$1 = \frac{1}{E_{\omega}}\mathsf{U}_{-}^{\omega\,\mathrm{T}}\mathsf{T}\mathsf{D}_{-}\mathsf{T}^{\mathrm{T}}\mathsf{U}_{-}^{\omega} = \frac{1}{E_{\omega}}\mathsf{V}_{-}^{\omega\,\mathrm{T}}\mathsf{D}_{-}\mathsf{V}_{-}^{\omega} = \frac{1}{E_{\omega}}\mathsf{R}_{-}^{\omega\,\mathrm{T}}\mathsf{R}_{-}^{\omega} \ . \tag{11.173}$$

The normalization condition for the eigenvectors $\mathsf{R}_{-}^{\omega}$ is thus

$$\mathsf{R}_{-}^{\omega\,\mathrm{T}}\mathsf{R}_{-}^{\omega} = E_{\omega} \ . \tag{11.174}$$

The normalization condition for the eigenvectors $\mathsf{R}_{+}^{\omega}$ of (11.162) is derived in the same way as (11.174). The analogous result is

$$\mathsf{R}_{+}^{\omega\,\mathrm{T}}\mathsf{R}_{+}^{\omega} = E_{\omega} \ . \tag{11.175}$$

### 11.5.3 Application to 1⁻ Excitations in ⁴He Carried Through

Let us apply the method derived in the preceding subsection to solve the RPA equations for the $1^{-}$ excitations in $^{4}_{2}\mathrm{He}_{2}$. This is continuation of the example set up at the end of Subsect. 11.5.1. The $\mathsf{A}$ and $\mathsf{B}$ matrices were constructed in the proton–neutron formalism. They can also be constructed in the isospin formalism, whereupon the subsequent diagonalization becomes straightforward, as shown at the end of this subsection.

## Diagonalization in the proton–neutron formalism

We start from the A matrix (11.146) and the B matrix (11.147), constructed in the proton–proton and neutron–neutron basis (11.145). With $A_0 = A_1 = 1.0\,\text{MeV}$ these matrices become[3]

$$A(1^-) = \begin{pmatrix} 20.667 & -1.333 \\ -1.333 & 20.667 \end{pmatrix} \text{MeV}\ , \quad B(1^-) = \begin{pmatrix} 1.000 & 1.333 \\ 1.333 & 1.000 \end{pmatrix} \text{MeV}\ .$$

(11.176)

The sum and difference matrices (11.152) become

$$M_+ = \begin{pmatrix} 21.667 & 0 \\ 0 & 21.667 \end{pmatrix} \text{MeV}\ , \quad M_- = \begin{pmatrix} 19.667 & -2.667 \\ -2.667 & 19.667 \end{pmatrix} \text{MeV}\ . \quad (11.177)$$

We observe that $M_+$ is already in diagonal form (and in this special case even proportional to the unit matrix $1_2$). This means that the relevant equations are (11.153)–(11.160) with reversed $\pm$ subscripts. It follows that $D_+ = M_+$ and $T = 1_2$, i.e. the 2-by-2 unit matrix. The square-root matrix becomes

$$D_+^{1/2} = \begin{pmatrix} 4.655 & 0 \\ 0 & 4.655 \end{pmatrix} \sqrt{\text{MeV}}\ . \tag{11.178}$$

The matrix to be diagonalized according to (11.162) is

$$M_-'' = D_+^{1/2} M_- D_+^{1/2} = \begin{pmatrix} 426.11 & -57.778 \\ -57.778 & 426.11 \end{pmatrix} \text{MeV}^2\ . \tag{11.179}$$

The eigenvalues $E_\omega^2$ of $M_-''$ are

$$E_1^2 = 368.33\,\text{MeV}^2\ , \quad E_2^2 = 483.89\,\text{MeV}^2\ , \tag{11.180}$$

and their positive square roots are

$$E_1 = 19.192\,\text{MeV}\ , \quad E_2 = 21.997\,\text{MeV}\ . \tag{11.181}$$

Here the label $\omega = n$ only enumerates the solutions, while it is understood that the states in question are $1^-$.

We proceed to find the amplitudes $X^1$, $Y^1$ by means of Eqs. (11.167) and (11.168). To do so, we need first the vector $R_+^1$. To within normalization it is obtained from (11.162) as

$$(M_-'' - E_1^2 1)R_+^1 = 0\ . \tag{11.182}$$

Substituting the value of $E_1^2$ from (11.180) into (11.182) with the matrix $M_-''$ given by (11.179), we find[4]

---

[3] The calculation is carried out by greater accuracy than is displayed.

[4] This result is exact because the diagonal elements of the matrix (11.179) are equal.

$$R^1_+ = \mathcal{N}\begin{pmatrix} 1 \\ 1 \end{pmatrix}. \tag{11.183}$$

The normalization constant $\mathcal{N}$ is found from (11.175) as $\mathcal{N} = \sqrt{E_1/2}$, so that

$$R^1_+ = \sqrt{\frac{E_1}{2}}\begin{pmatrix} 1 \\ 1 \end{pmatrix} = \begin{pmatrix} 3.098 \\ 3.098 \end{pmatrix}\sqrt{\mathrm{MeV}}. \tag{11.184}$$

Next we evaluate the matrices (11.168). The matrix $\mathsf{T}$ is simply the unit matrix and $\mathsf{D}^{1/2}_+$ is given by (11.178). The inverse matrix $\mathsf{D}^{-1/2}_+$ is

$$\mathsf{D}^{-1/2}_+ = \begin{pmatrix} 0.215 & 0 \\ 0 & 0.215 \end{pmatrix}\frac{1}{\sqrt{\mathrm{MeV}}}. \tag{11.185}$$

With this input the matrices (11.168) become

$$\mathsf{F}^1_+ = \begin{pmatrix} 0.229 & 0 \\ 0 & 0.229 \end{pmatrix}\frac{1}{\sqrt{\mathrm{MeV}}}, \quad \mathsf{F}^1_- = \begin{pmatrix} -0.014 & 0 \\ 0 & -0.014 \end{pmatrix}\frac{1}{\sqrt{\mathrm{MeV}}}. \tag{11.186}$$

Equations (11.167), with $\mathsf{R}^1_+$ given by (11.184), now give the amplitudes

$$\mathsf{X}^1 = \begin{pmatrix} 0.708 \\ 0.708 \end{pmatrix}, \quad \mathsf{Y}^1 = \begin{pmatrix} -0.043 \\ -0.043 \end{pmatrix}. \tag{11.187}$$

As a check we note that these amplitudes satisfy the normalization condition (11.169).

Finding the amplitudes $\mathsf{X}^2$ and $\mathsf{Y}^2$ is relegated to Exercise 11.33.

### Diagonalization in the isospin formalism

The above mentioned problem can also be solved in the isospin formalism. The $\mathsf{A}$ matrix is given by the single-particle energy 21.0 MeV, Eqs. (9.31) and (9.32) and Table 9.1. The $\mathsf{B}$ matrix is given by Eqs. (11.77) and (11.78) and Table 11.2. Now the matrices are 1-by-1 in dimension, and we denote

$$A_{0\mathrm{p}_{3/2}0\mathrm{s}_{1/2},0\mathrm{p}_{3/2}0\mathrm{s}_{1/2}}(1T) \equiv a_T, \quad B_{0\mathrm{p}_{3/2}0\mathrm{s}_{1/2},0\mathrm{p}_{3/2}0\mathrm{s}_{1/2}}(1T) \equiv b_T. \tag{11.188}$$

The values of these are

$$a_0 = 21.0\,\mathrm{MeV} + \tfrac{1}{2}[3(-0.333A_1) - 2.333A_0] = 19.334\,\mathrm{MeV}, \tag{11.189}$$
$$b_0 = \tfrac{1}{2}[3(1.000A_1) + 1.667A_0] = 2.334\,\mathrm{MeV}, \tag{11.190}$$
$$a_1 = 21.0\,\mathrm{MeV} + \tfrac{1}{2}[-0.333A_1 + 2.333A_0] = 22.000\,\mathrm{MeV}, \tag{11.191}$$
$$b_1 = \tfrac{1}{2}[-(1.000A_1) + 1.667A_0] = 0.334\,\mathrm{MeV}, \tag{11.192}$$

where we have inserted $A_0 = A_1 = 1.0\,\mathrm{MeV}$.

Because the matrices A and B are now single numbers, the RPA super-matrix equation (11.60) reduces to a pair of normal matrix equations, one for $T = 0$ and one for $T = 1$. For distinction from the proton–neutron case, we now call the amplitudes $X_T, Y_T$ and the energies $E^T$. Substituting from (11.189)–(11.192) we then have

$$\begin{pmatrix} 19.334 & 2.334 \\ -2.334 & -19.334 \end{pmatrix} \begin{pmatrix} X_0 \\ Y_0 \end{pmatrix} \text{MeV} = E^0 \begin{pmatrix} X_0 \\ Y_0 \end{pmatrix} , \tag{11.193}$$

$$\begin{pmatrix} 22.000 & 0.334 \\ -0.334 & -22.000 \end{pmatrix} \begin{pmatrix} X_1 \\ Y_1 \end{pmatrix} \text{MeV} = E^1 \begin{pmatrix} X_1 \\ Y_1 \end{pmatrix} . \tag{11.194}$$

These matrix equations have the form of (11.136), and the resulting secular equations are like (11.137) with solutions (11.138). The positive, physical solutions are

$$E^0 = \sqrt{19.334^2 - 2.334^2}\,\text{MeV} = 19.193\,\text{MeV} , \tag{11.195}$$

$$E^1 = \sqrt{22.000^2 - 0.334^2}\,\text{MeV} = 21.997\,\text{MeV} . \tag{11.196}$$

These eigenenergies are in agreement with the energies (11.181), and we can identify the proton–neutron states $n = 1$ and $n = 2$ with the isospin states $T = 0$ and $T = 1$, respectively.

Equations (11.141) and (11.142) serve to give the amplitudes $X_T$ and $Y_T$. The results are

$$X_0 = 1.002 , \quad Y_0 = -0.061 , \tag{11.197}$$

$$X_1 = 1.000 , \quad Y_1 = -0.008 . \tag{11.198}$$

In conclusion we note that the solution in the isospin formalism is much easier than in the proton–neutron formalism. This kind of simplification results whenever a symmetry reduces matrix sizes.

### 11.5.4 The $1^-$ Excitations of $^4$He Revisited

Let us extend the example of the $1^-$ states in $^4$He to the $0p$-$(0s_{1/2})^{-1}$ valence space. This particular example was discussed at the TDA level of approximation in Subsect. 9.3.2. We use the single-particle energies of that example and the particle–hole basis (9.88). Equations (11.74)–(11.76) and Table 11.2 give the symmetric B matrix as

$$\mathsf{B}(1^-) = \begin{pmatrix} 1.000A_1 & 0A_1 & \frac{1}{2}(1.000A_1 + 1.667A_0) & \frac{1}{2}(0A_1 - 1.886A_0) \\ 0A_1 & 1.000A_1 & \frac{1}{2}(0A_1 - 1.886A_0) & \frac{1}{2}(1.000A_1 + 0.333A_0) \\ \cdots & \cdots & 1.000A_1 & 0A_1 \\ \cdots & \cdots & 0A_1 & 1.000A_1 \end{pmatrix} . \tag{11.199}$$

The A matrix is given by (9.91).

Inserting the interaction strengths $A_0 = A_1 = 1.0\,\text{MeV}$ into the A and B matrices, we obtain the difference and sum matrices $\mathsf{M}_{\mp} \equiv \mathsf{A} \mp \mathsf{B}$ as

$$\mathsf{M}_- = \begin{pmatrix} 19.667 & 0.943 & -2.667 & 1.886 \\ 0.943 & 26.333 & 1.886 & -1.333 \\ -2.667 & 1.886 & 19.667 & 0.943 \\ 1.886 & -1.333 & 0.943 & 26.333 \end{pmatrix} \text{MeV} \,, \tag{11.200}$$

$$\mathsf{M}_+ = \begin{pmatrix} 21.667 & 0.943 & 0 & 0 \\ 0.943 & 28.333 & 0 & 0 \\ 0 & 0 & 21.667 & 0.943 \\ 0 & 0 & 0.943 & 28.333 \end{pmatrix} \text{MeV} \,. \tag{11.201}$$

We diagonalize $\mathsf{M}_-$, with the resulting diagonal matrix

$$\mathsf{D}_- = \text{diag}(16.101, 22.172, 25.899, 27.828)\,\text{MeV} \,. \tag{11.202}$$

The matrix T that affects the orthogonal similarity transformation according to (11.153) is composed of the normalized column eigenvectors of $\mathsf{M}_-$ as

$$\mathsf{T} = \begin{pmatrix} 0.674 & 0.697 & -0.214 & 0.120 \\ -0.214 & 0.120 & -0.674 & -0.697 \\ 0.674 & -0.697 & -0.214 & -0.120 \\ -0.214 & -0.120 & -0.674 & 0.697 \end{pmatrix} \,. \tag{11.203}$$

The transformed sum matrix (11.154) then becomes

$$\mathsf{M}'_+ = \mathsf{T}^T \mathsf{M}_+ \mathsf{T} = \begin{pmatrix} 21.734 & 0 & 1.155 & 0 \\ 0 & 22.172 & 0 & -2.000 \\ 1.155 & 0 & 28.266 & 0 \\ 0 & -2.000 & 0 & 27.828 \end{pmatrix} \text{MeV} \,. \tag{11.204}$$

The square roots of the elements in (11.202) constitute the matrix $\mathsf{D}_-^{1/2}$ defined by (11.159). With $\mathsf{D}_-^{1/2}$ we build the matrix in (11.160):

$$\mathsf{M}''_+ = \mathsf{D}_-^{1/2} \mathsf{M}'_+ \mathsf{D}_-^{1/2} = \begin{pmatrix} 349.940 & 0 & 23.580 & 0 \\ 0 & 491.579 & 0 & -49.679 \\ 23.580 & 0 & 732.060 & 0 \\ 0 & -49.679 & 0 & 774.422 \end{pmatrix} \text{MeV}^2 \,. \tag{11.205}$$

The eigenvalues of this matrix are

$$E_n^2 = (348.490, 483.107, 733.510, 782.893)\,\text{MeV}^2 \,, \tag{11.206}$$

and their square roots are

$$E_n = (18.668, 21.980, 27.083, 27.980)\,\text{MeV} \,. \tag{11.207}$$

These are the exact energy eigenvalues quoted in Tables 11.6 and 11.7.

To obtain the eigenvector belonging to the energy $E_1 = 18.668\,\text{MeV}$ we construct the matrices $\mathsf{G}^1_\pm$ defined in (11.166). The result is

$$
\mathsf{G}^1_+ = \begin{pmatrix}
0.156 & 0.162 & -0.050 & 0.028 \\
-0.050 & 0.028 & -0.158 & -0.165 \\
0.156 & -0.162 & -0.050 & -0.028 \\
-0.050 & -0.028 & -0.158 & 0.165
\end{pmatrix} \frac{1}{\sqrt{\text{MeV}}} ,
\tag{11.208}
$$

$$
\mathsf{G}^1_- = \begin{pmatrix}
-0.012 & 0.014 & -0.008 & 0.006 \\
0.004 & 0.002 & -0.026 & -0.032 \\
-0.012 & -0.014 & -0.008 & -0.006 \\
0.004 & -0.002 & -0.026 & 0.032
\end{pmatrix} \frac{1}{\sqrt{\text{MeV}}} .
\tag{11.209}
$$

The diagonalization of (11.205) gives the eigenvectors $\mathsf{R}^n_-$ according to (11.162). With the normalization (11.174), the $\mathsf{R}^1_-$ vector becomes

$$
\mathsf{R}^1_- = \sqrt{18.668} \begin{pmatrix}
0.998 \\
0.000 \\
-0.061 \\
0.000
\end{pmatrix} \sqrt{\text{MeV}} .
\tag{11.210}
$$

Equations (11.165) give the RPA amplitude vectors as

$$
\mathsf{X}^1 = \mathsf{G}^1_+ \mathsf{R}^1_- = \begin{pmatrix}
-0.688 \\
0.172 \\
-0.688 \\
0.172
\end{pmatrix} , \quad
\mathsf{Y}^1 = \mathsf{G}^1_- \mathsf{R}^1_- = \begin{pmatrix}
0.048 \\
-0.023 \\
0.048 \\
-0.023
\end{pmatrix} .
\tag{11.211}
$$

These are seen to satisfy the normalization condition (11.169).

### 11.5.5 Further Examples

Let us compute the RPA energies of the $3^-$ states of $^{16}\text{O}$ in the $0\text{d}_{5/2}\text{-}(0\text{p}_{1/2})^{-1}$ particle–hole valence space (9.95) and with the single-particle energies of Fig. 9.2(a). The TDA matrix $\mathsf{A}$ is stated in (9.97). The RPA correlation matrix $\mathsf{B}$ is given by the relations (11.74)–(11.76) and Table 11.3 as

$$
\mathsf{B}(3^-) = \begin{pmatrix}
1.000 A_1 & \frac{1}{2}(1.000 A_1 + 0.714 A_0) \\
\frac{1}{2}(1.000 A_1 + 0.714 A_0) & 1.000 A_1
\end{pmatrix} .
\tag{11.212}
$$

The calculation proceeds as in the previous examples. With $A_0 = A_1 = 1.0\,\text{MeV}$ we obtain the eigenenergies

$$
E_1(3^-) = 10.726\,\text{MeV} , \quad E_2(3^-) = 12.599\,\text{MeV} .
\tag{11.213}
$$

We can continue the RPA calculations in successively larger particle–hole valence spaces, and also include the $2^-$ states, as was done in the TDA and shown in Fig. 9.3. Figure 11.3 shows the corresponding RPA results.

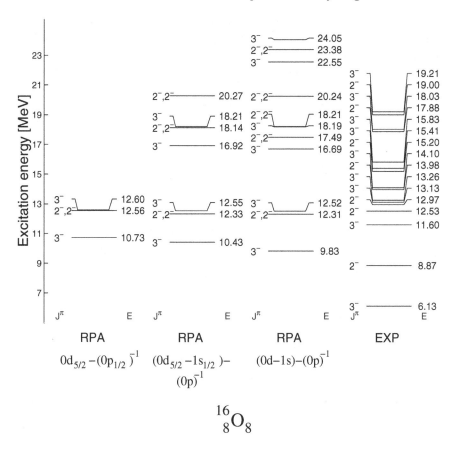

**Fig. 11.3.** RPA energies of the $2^-$ and $3^-$ states in $^{16}$O, computed in the particle–hole valence spaces indicated in Fig. 9.2 and compared with experiment. The SDI parameters are $A_0 = A_1 = 1.0\,\mathrm{MeV}$

Comparing Figs. 9.3 and 11.3 we observe that the RPA energies are only slightly lower than their TDA counterparts. In fact, the only appreciable differences occur for the lowest $3^-$ states. The differences increase with the increasing valence space, yet the largest of them is only 0.45 MeV. The best $3_1^-$ prediction is still 3.7 MeV above the experimental energy, but the SDI parameters have not been optimized. In general, an increase in the interaction strength increases the differences between the TDA and RPA energies.

Further comparison between the TDA and RPA energies can be seen in Figs. 9.4 and 9.5 for $^{40}$Ca and $^{48}$Ca, respectively. The differences in $^{40}$Ca are large for the lowest $3^-$ and $5^-$ states, which become quite collective in the RPA description. In addition, there is a dramatic decrease in the energy of the first $1^-$ state. A small increase in SDI interaction strength would lead to an imaginary $1^-$ RPA energy. The steep descent stems from the spurious centre-of-mass contributions to the wave function of the $1_1^-$ state and is therefore completely unphysical.

In $^{48}$Ca the differences between the TDA and RPA energies are moderate. However, the collectivity of the lowest states can be quite different depending on the description. Striking differences appear in the electromagnetic decay rates, as discussed in the following section.

## 11.6 Electromagnetic Transitions in the RPA Framework

Following the pattern of Sect. 9.4, we divide the electromagnetic transitions within the RPA framework into transitions to the ground state and transitions between two excited states. Transitions to the ground state are discussed first. As in the previous chapters, we take the oscillator parameter $b$ from the relations (3.43) and (3.45) and use the formalism of Sect. 6.1 to deduce reduced transition probabilities and half-lives.

### 11.6.1 Transitions to the RPA Ground State

Our aim is to evaluate the electromagnetic transition amplitude

$$\langle \text{RPA} | \mathcal{M}_{\sigma\lambda\mu} | \omega \rangle \,, \tag{11.214}$$

where $|\text{RPA}\rangle$ is the correlated RPA ground state, $\mathcal{M}_{\sigma\lambda}$ is the electromagnetic operator (6.5) and $\omega$ represents the quantum numbers $nJ^\pi(M)$. The excited initial state is

$$|\omega\rangle = Q_\omega^\dagger |\text{RPA}\rangle = \sum_{ab} \left[ X_{ab}^\omega \mathcal{A}_{ab}^\dagger(JM) - Y_{ab}^\omega \tilde{\mathcal{A}}_{ab}(JM) \right] |\text{RPA}\rangle \,, \tag{11.215}$$

as given by (11.40).

To evaluate the amplitude (11.214) we start by writing (11.214) as

$$\langle \text{RPA} | \mathcal{M}_{\sigma\lambda\mu} | \omega \rangle = \langle \text{RPA} | \mathcal{M}_{\sigma\lambda\mu} Q_\omega^\dagger | \text{RPA} \rangle = \langle \text{RPA} | [\mathcal{M}_{\sigma\lambda\mu}, Q_\omega^\dagger] | \text{RPA} \rangle \,, \tag{11.216}$$

where the commutator could be introduced since $Q_\omega | \text{RPA} \rangle = 0$, as stated in (11.89). The method of introducing commutators on every possible occasion in the RPA is general and comes from the EOM background of the RPA. This matter was already discussed when introducing the quasiboson approximation in the derivation of the RPA orthonormality condition (11.82).

With the commutator present in (11.216), we can immediately make the QBA by replacing the RPA ground state $|\text{RPA}\rangle$ by the particle–hole ground state $|\text{HF}\rangle$ as we did in (11.81):

$$\langle \text{RPA} | [\mathcal{M}_{\sigma\lambda\mu}, Q_\omega^\dagger] | \text{RPA} \rangle \overset{\text{QBA}}{\approx} \langle \text{HF} | [\mathcal{M}_{\sigma\lambda\mu}, Q_\omega^\dagger] | \text{HF} \rangle \,. \tag{11.217}$$

We calculate the contributions of the $\mathcal{A}^\dagger$ and $\tilde{\mathcal{A}}$ terms of $Q_\omega^\dagger$ separately. With the electromagnetic operator given by (4.22) as

$$\mathcal{M}_{\sigma\lambda\mu} = \hat{\lambda}^{-1} \sum_{ab} (a\|\mathcal{M}_{\sigma\lambda}\|b) \left[c_a^\dagger \tilde{c}_b\right]_{\lambda\mu} , \tag{11.218}$$

we first calculate the commutator $\left[\mathcal{M}_{\sigma\lambda\mu}, \mathcal{A}_{cd}^\dagger(JM)\right]$ and then take its expectation value in the state $|\text{HF}\rangle$. The result, obtained with operator relations from Chap. 4 and angular momentum relations from Chap. 1, is

$$\langle \text{HF}|\left[\mathcal{M}_{\sigma\lambda\mu}, \mathcal{A}_{cd}^\dagger(JM)\right]|\text{HF}\rangle$$
$$= \hat{\lambda}^{-1} \sum_{ab} (a\|\mathcal{M}_{\sigma\lambda}\|b)\delta_{\lambda J}\delta_{\mu,-M}(-1)^{j_a+j_b+M+1}\delta_{bc}\delta_{ad} . \tag{11.219}$$

The $\tilde{\mathcal{A}}$ contribution is calculated similarly, with the result

$$\langle \text{HF}|\left[\mathcal{M}_{\sigma\lambda\mu}, \tilde{\mathcal{A}}_{cd}(JM)\right]|\text{HF}\rangle$$
$$= \hat{\lambda}^{-1} \sum_{ab} (a\|\mathcal{M}_{\sigma\lambda}\|b)\delta_{\lambda J}\delta_{\mu,-M}(-1)^{J+M+1}\delta_{ac}\delta_{db} . \tag{11.220}$$

Substituting (11.219) and (11.220) into (11.217) yields for (11.216) in the quasiboson approximation

$$\langle \text{RPA}|\mathcal{M}_{\sigma\lambda\mu}|\omega\rangle = \hat{\lambda}^{-1} \sum_{ab} (a\|\mathcal{M}_{\sigma\lambda}\|b)\delta_{\lambda J}\delta_{\mu,-M}$$
$$\times \left[X_{ba}^\omega(-1)^{j_a+j_b+M+1} - Y_{ab}^\omega(-1)^{J+M+1}\right] . \tag{11.221}$$

We now apply the Wigner–Eckart theorem (2.27) to the left-hand side of (11.216) and find

$$(\text{RPA}\|\mathcal{M}_{\sigma\lambda}\|\omega) = \delta_{\lambda J} \sum_{ab} (a\|\mathcal{M}_{\sigma\lambda}\|b)\left[(-1)^{j_a+j_b+J+1}X_{ba}^\omega + Y_{ab}^\omega\right] . \tag{11.222}$$

With use of the symmetry properties (6.27)–(6.30) we can write this as

$$\boxed{\begin{aligned} (\text{RPA}\|\mathcal{M}_{\sigma\lambda}\|\omega) &= \delta_{\lambda J} \sum_{ab} (a\|\mathcal{M}_{\sigma\lambda}\|b)\left[\zeta^{(\lambda)} X_{ab}^\omega + Y_{ab}^\omega\right] , \\ \zeta^{(\lambda)} &= \begin{cases} (-1)^\lambda & \text{CS phase convention ;} \\ \pm 1 & \text{BR phase convention,} \\ & + \text{ for } \sigma = \text{E} , \ - \text{ for } \sigma = \text{M} . \end{cases} \end{aligned}} \tag{11.223}$$

The reduced transition probability (6.4) becomes

$$B(\sigma\lambda ; \omega \to 0_{\text{gs}}^+)_{\text{RPA}} = \hat{J}^{-2}\left|(\text{RPA}\|\mathcal{M}_{\sigma\lambda}\|\omega)\right|^2 . \tag{11.224}$$

For $Y_{ab}^{\omega} = 0$ the RPA transition amplitude (11.223) reduces to the TDA result (9.100). In the RPA also the $Y$ amplitudes contribute to the decay strength. They are small relative to the $X$ amplitudes; see (11.88) and (11.111). Nevertheless, the $Y$ amplitudes can enhance collective transitions beyond the enhancement created by the TDA. This is demonstrated by the 'extreme collective model' of the following subsection.

## 11.6.2 Extreme Collective Model

The RPA amplitude (11.223) can produce considerable enhancement of the decay strength of a collective state $|\omega\rangle$. This happens when the products of wave-function amplitudes and single-particle matrix elements sum coherently. To demonstrate this we assume that

$$\left|X_{ab}^{\omega}\right| = X , \quad \left|Y_{ab}^{\omega}\right| = Y \quad \text{for all } ab . \tag{11.225}$$

If the number of contributing particle–hole configurations $ab$ is $N$, the normalization condition (11.83) yields

$$X^2 - Y^2 = \frac{1}{N} , \quad \text{whence} \quad X^2 = \frac{1}{N} + Y^2 . \tag{11.226}$$

Equation (11.226) implies

$$X \gtrsim \frac{1}{\sqrt{N}} . \tag{11.227}$$

We assume that for the collective state $|\omega\rangle \equiv |\text{coll}\rangle$ the transition amplitude (11.223) sums coherently over the particle–hole contributions. It follows that

$$\left|(\text{RPA}\|\boldsymbol{\mathcal{M}}_{\sigma\lambda}\|\text{coll})\right| \approx (X + Y) \sum_{ab} \left|(a\|\boldsymbol{\mathcal{M}}_{\sigma\lambda}\|b)\right|$$

$$\gtrsim \left(\frac{1}{\sqrt{N}} + Y\right) \sum_{ab} \left|(a\|\boldsymbol{\mathcal{M}}_{\sigma\lambda}\|b)\right| . \tag{11.228}$$

The corresponding expression for the TDA reads

$$\left|(\text{HF}\|\boldsymbol{\mathcal{M}}_{\sigma\lambda}\|\text{coll})\right| \approx \frac{1}{\sqrt{N}} \sum_{ab} \left|(a\|\boldsymbol{\mathcal{M}}_{\sigma\lambda}\|b)\right| . \tag{11.229}$$

The schematic model giving rise to the estimates (11.228) and (11.229) can be called the *extreme collective model*. The ratio of the reduced transition probabilities is

$$\frac{B(\sigma\lambda\,;\,\text{coll} \to \text{gs})_{\text{RPA}}}{B(\sigma\lambda\,;\,\text{coll} \to \text{gs})_{\text{TDA}}} \gtrsim \left(\frac{\frac{1}{\sqrt{N}} + Y}{\frac{1}{\sqrt{N}}}\right)^2 \underset{N\to\infty}{\sim} NY^2 . \tag{11.230}$$

This shows that very large RPA transition amplitudes can be obtained when the number $N$ of contributing particle–hole excitations increases. A concrete example of this enhancement is given in the following subsection.

### 11.6.3 Octupole Decay in $^{16}$O

Let us consider the electric octupole decay of the lowest $3^-$ state in $^{16}$O in two different particle–hole valence spaces. This decay was discussed within the TDA in Subsect. 9.4.3. We now quote the earlier results and compare them with their RPA counterparts.

The two particle–hole valence spaces are $(0d_{5/2}\text{-}1s_{1/2})\text{-}(0p)^{-1}$ and $(0d\text{-}1s)\text{-}(0p)^{-1}$. In each space the proton and neutron contributions are proportional to the respective effective charges $e^p_{\text{eff}}$ and $e^n_{\text{eff}}$, and it is convenient to use the abbreviations (9.109)

$$e_\pm = e^p_{\text{eff}} \pm e^n_{\text{eff}} . \tag{11.231}$$

The TDA wave functions, i.e. the amplitudes $X^{(\text{TDA})}$, are given by (9.114) and (9.120) for the two spaces. The RPA wave functions, with amplitudes $X^{(\text{RPA})}$ and $Y^{(\text{RPA})}$, come from the calculations whose energy spectra are shown in Fig. 11.3. The single-particle energies of Fig. 9.2 (a) and the SDI parameters $A_0 = A_1 = 1.0\,\text{MeV}$ were used in all the calculations.

The numbers substituted into the right-hand sides of (9.100) and (11.223) for each kind of nucleon and the resulting total decay amplitudes are listed in Tables 11.8 and 11.9, respectively, for the smaller and the larger valence space.[5] The single-particle matrix elements are from Table 6.5. From the table entries we obtain, according to (6.46),

$$(a\|\mathbf{Q}_3\|b) = eb^3\overline{(a\|\mathbf{Q}_3\|b)} \tag{11.232}$$

with our usual $A = 16$ oscillator length $b = 1.725\,\text{fm}$.

We note that the TDA parts of Tables 11.8 and 11.9 merely repeat the work in Subsect. 9.4.3 leading to the results (9.115) and (9.121).

**Table 11.8.** Electric octupole decay amplitude of the $3^-_1$ state in $^{16}$O calculated in the particle–hole valence space $(0d_{5/2}\text{-}1s_{1/2})\text{-}(0p)^{-1}$

| $a$ | $b$ | $\overline{(a\|\mathbf{Q}_3\|b)}$ | $X^{(\text{TDA})}_{ab}$ | $(\mathcal{M}_{\text{E3}})_{\text{TDA}}$ | $X^{(\text{RPA})}_{ab}$ | $Y^{(\text{RPA})}_{ab}$ | $(\mathcal{M}_{\text{E3}})_{\text{RPA}}$ |
|---|---|---|---|---|---|---|---|
| $0d_{5/2}$ | $0p_{3/2}$ | $-3.420$ | $0.117$ | $0.400e_+b^3$ | $0.133$ | $-0.041$ | $0.594e_+b^3$ |
| $0d_{5/2}$ | $0p_{1/2}$ | $-3.824$ | $0.697$ | $2.665e_+b^3$ | $0.699$ | $-0.071$ | $2.944e_+b^3$ |
| Total | | | | $3.065e_+b^3$ | | | $3.538e_+b^3$ |

The contributions of particle–hole configurations $ab$ and the total are listed for TDA and RPA.

Equations (9.101) and (11.224) give the TDA and RPA reduced transition probabilities $B(\text{E3})$ for the decay of the $3^-_1$ state to the $0^+$ ground state. We have already calculated the transitions amplitudes, and they are stated as the totals in Tables 11.8 and 11.9. Thus we find immediately

---

[5] The numerical accuracy exceeded that recorded.

**Table 11.9.** The same as Table 11.8 except that the particle–hole valence space is $(0\mathrm{d}\text{-}1\mathrm{s})\text{-}(0\mathrm{p})^{-1}$

| $a$ | $b$ | $\overline{(a\|Q_3\|b)}$ | $X_{ab}^{\mathrm{(TDA)}}$ | $(\mathcal{M}_{\mathrm{E3}})_{\mathrm{TDA}}$ | $X_{ab}^{\mathrm{(RPA)}}$ | $Y_{ab}^{\mathrm{(RPA)}}$ | $(\mathcal{M}_{\mathrm{E3}})_{\mathrm{RPA}}$ |
|---|---|---|---|---|---|---|---|
| $0\mathrm{d}_{5/2}$ | $0\mathrm{p}_{3/2}$ | $-3.420$ | $0.136$ | $0.465e_+b^3$ | $0.159$ | $-0.051$ | $0.718e_+b^3$ |
| $0\mathrm{d}_{5/2}$ | $0\mathrm{p}_{1/2}$ | $-3.824$ | $0.681$ | $2.604e_+b^3$ | $0.681$ | $-0.083$ | $2.922e_+b^3$ |
| $0\mathrm{d}_{3/2}$ | $0\mathrm{p}_{3/2}$ | $4.189$ | $-0.134$ | $0.561e_+b^3$ | $-0.149$ | $0.040$ | $0.792e_+b^3$ |
| Total | | | | $3.631e_+b^3$ | | | $4.431e_+b^3$ |

$$B(\mathrm{E3}; 3^-_1 \to 0^+_{\mathrm{gs}})_{\mathrm{TDA}} = 35.4e^2_+\,\mathrm{fm}^6\,, \quad \text{space } (0\mathrm{d}_{5/2}\text{-}1\mathrm{s}_{1/2})\text{-}(0\mathrm{p})^{-1}\,, \tag{11.233}$$

$$B(\mathrm{E3}; 3^-_1 \to 0^+_{\mathrm{gs}})_{\mathrm{RPA}} = 47.2e^2_+\,\mathrm{fm}^6\,, \quad \text{space } (0\mathrm{d}_{5/2}\text{-}1\mathrm{s}_{1/2})\text{-}(0\mathrm{p})^{-1}\,, \tag{11.234}$$

$$B(\mathrm{E3}; 3^-_1 \to 0^+_{\mathrm{gs}})_{\mathrm{TDA}} = 49.6e^2_+\,\mathrm{fm}^6\,, \quad \text{space } (0\mathrm{d}\text{-}1\mathrm{s})\text{-}(0\mathrm{p})^{-1}\,, \tag{11.235}$$

$$B(\mathrm{E3}; 3^-_1 \to 0^+_{\mathrm{gs}})_{\mathrm{RPA}} = 73.9e^2_+\,\mathrm{fm}^6\,, \quad \text{space } (0\mathrm{d}\text{-}1\mathrm{s})\text{-}(0\mathrm{p})^{-1}\,. \tag{11.236}$$

From these $B(\mathrm{E3})$ values and from Tables 11.8 and 11.9 we can draw the following conclusions about electromagnetic transitions in the RPA. It turns out that these conclusions have general validity beyond the example studied.

- In a given particle–hole valence space the RPA develops more collectivity than does the TDA for the lowest-lying collective excitation, as was also shown by the result (11.230) of the extreme collective model. Furthermore, the collectivity increases with an increasing number of contributing particle–hole configurations, i.e. with the size of the particle–hole valence space, again in agreement with (11.230).

- When the protons and neutrons have the same relative single-particle energies, electromagnetic transitions can be classified according to isospin. As in the TDA, transitions are then of isoscalar or isovector character, signified by the charge $e_+$, or $e_-$, respectively. The lowest, collective state decays in isoscalar mode. Extending our RPA example to include all $3^-$ states, we obtain the results shown in Tables 11.10 and 11.11. From them we see that isoscalar and isovector decays alternate.

**Table 11.10.** RPA energies and octupole strengths of $3^-$ states in $^{16}\mathrm{O}$ for particle–hole valence space $(0\mathrm{d}_{5/2}\text{-}1\mathrm{s}_{1/2})\text{-}(0\mathrm{p})^{-1}$

| $n$ | 1 | 2 | 3 | 4 |
|---|---|---|---|---|
| $E_n$ (MeV) | 10.429 | 12.551 | 16.919 | 18.213 |
| $\left\|(3^-_n\|Q_3\|\mathrm{RPA})\right\|^2$ (fm$^6$) | $330.2e^2_+$ | $161.4e^2_-$ | $109.6e^2_+$ | $179.6e^2_-$ |

**Table 11.11.** RPA energies and octupole strengths of $3^-$ states in $^{16}$O for particle–hole valence space $(0d\text{-}1s)\text{-}(0p)^{-1}$

| $n$ | 1 | 2 | 3 | 4 | 5 | 6 |
|---|---|---|---|---|---|---|
| $E_n$ (MeV) | 9.833 | 12.519 | 16.686 | 18.192 | 22.547 | 24.053 |
| $\left\|(3_n^-\|\boldsymbol{Q}_3\|\text{RPA})\right\|^2$ (fm$^6$) | $517.7e_+^2$ | $156.9e_-^2$ | $154.6e_+^2$ | $156.3e_-^2$ | $115.3e_+^2$ | $227.1e_-^2$ |

The energies $E_n$ in Tables 11.10 and 11.11 can be recognized in Fig. 11.3. For the decay strengths, we compare these tables with their TDA counterparts, Tables 9.7 and 9.8. For the $3_1^-$ state the RPA provides 30–50 % more enhancement than does the TDA; this repeats our previous conclusion from the $B$(E3) values in (11.233)–(11.236). However, for the other E3 decays there is little difference between the TDA and the RPA. The TDA–RPA difference is further demonstrated in Figs. 9.6 and 9.7. Figure 9.6 shows that the TDA–RPA difference in $B$(E3) values in $^{16}$O clearly increases with increasing size of the particle–hole valence space, most notably for the lowest $3^-$ excitation.

Figure 9.7 shows the $B$(E3) values in $^{40}$Ca for the $0f_{7/2}\text{-}(0d\text{-}1s)^{-1}$ particle–hole valence space. Here the TDA–RPA difference is truly striking for the first $3^-$ excitation, both in energy and $B$(E3) value. The differences are minor for the other states.

### 11.6.4 The Energy-Weighted Sum Rule

In Subsect. 9.4.2 it was shown that the TDA satisfies the NEWSR. For the RPA there exists no NEWSR, but rather an *energy-weighted sum rule* (EWSR), which can be stated as

$$\sum_n E_\omega \left|(n\,\lambda^\pi\|\boldsymbol{\mathcal{M}}_{\sigma\lambda}\|\text{RPA})\right|^2 = \tfrac{1}{2}\widehat{\lambda}^2\langle\text{HF}|[\mathcal{M}_{\sigma\lambda\mu}^\dagger, H, \mathcal{M}_{\sigma\lambda\mu}]|\text{HF}\rangle\,, \quad (11.237)$$

where $\omega = n\lambda^\pi$ (since $\lambda = J$) and $\mathcal{M}_{\sigma\lambda\mu}^\dagger$ is the Hermitian conjugate of the electromagnetic multipole operator $\mathcal{M}_{\sigma\lambda\mu}$. The double commutator is defined in (11.12), and we omit the subscript '$-$' from the notation.

We proceed to prove (11.237). In the CS phase convention $\boldsymbol{\mathcal{M}}_{\sigma\lambda}$ is a Hermitian tensor operator as defined and discussed in Subsect. 2.2.1. In the BR convention it is not, and an extra phase factor enters. Extended to both conventions (see Exercise 11.50), the symmetry relation (2.32) gives

$$(n\,\lambda^\pi\|\boldsymbol{\mathcal{M}}_{\sigma\lambda}\|\text{RPA}) = \zeta^{(\lambda)}(\text{RPA}\|\boldsymbol{\mathcal{M}}_{\sigma\lambda}\|n\,\lambda^\pi)^* \quad (11.238)$$

with the phase factor $\zeta^{(\lambda)}$ stated in (11.223). For the single-particle matrix elements (6.23) and (6.24) we adopt the abbreviations

$$p_{ab}^\lambda \equiv \zeta^{(\lambda)}(a\|\boldsymbol{\mathcal{M}}_{\sigma\lambda}\|b)\,, \quad q_{ab}^\lambda \equiv (a\|\boldsymbol{\mathcal{M}}_{\sigma\lambda}\|b) = \zeta^{(\lambda)}p_{ab}^\lambda\,, \quad (11.239)$$

where effective charges are included. The quantities $p_{ab}^\lambda$ and $q_{ab}^\lambda$ are real (see Subsect. 6.1.3). Substituting them into (11.223) and using (11.238) we can write

$$(n\,\lambda^\pi\|\mathcal{M}_{\sigma\lambda}\|\mathrm{RPA}) = \zeta^{(\lambda)} \sum_{ab}(X_{ab}^{\omega*}p_{ab}^\lambda + Y_{ab}^{\omega*}q_{ab}^\lambda)$$

$$= \zeta^{(\lambda)}\left(X^{\omega\dagger}, -Y^{\omega\dagger}\right)\begin{pmatrix} p^\lambda \\ -q^\lambda \end{pmatrix}. \qquad (11.240)$$

The complex conjugate of this is

$$(n\,\lambda^\pi\|\mathcal{M}_{\sigma\lambda}\|\mathrm{RPA})^* = (n\,\lambda^\pi\|\mathcal{M}_{\sigma\lambda}\|\mathrm{RPA})^\dagger = \zeta^{(\lambda)}\left(p^{\lambda\,\mathrm{T}}, -q^{\lambda\,\mathrm{T}}\right)\begin{pmatrix} X^\omega \\ -Y^\omega \end{pmatrix}, \qquad (11.241)$$

and we form the left-hand side of (11.237):

$$\sum_n E_\omega\left|(n\,\lambda^\pi\|\mathcal{M}_{\sigma\lambda}\|\mathrm{RPA})\right|^2$$

$$= \sum_n E_\omega\left(p^{\lambda\,\mathrm{T}}, -q^{\lambda\,\mathrm{T}}\right)\begin{pmatrix} X^\omega \\ -Y^\omega \end{pmatrix}\left(X^{\omega\dagger}, -Y^{\omega\dagger}\right)\begin{pmatrix} p^\lambda \\ -q^\lambda \end{pmatrix}. \qquad (11.242)$$

Applying the RPA equation (11.60) to this we obtain

$$\sum_n E_\omega\left|(n\,\lambda^\pi\|\mathcal{M}_{\sigma\lambda}\|\mathrm{RPA})\right|^2$$

$$= \left(p^{\lambda\,\mathrm{T}}, -q^{\lambda\,\mathrm{T}}\right)\begin{pmatrix} A & B \\ B^* & A^* \end{pmatrix}\sum_n\begin{pmatrix} X^\omega \\ Y^\omega \end{pmatrix}\left(X^{\omega\dagger}, -Y^{\omega\dagger}\right)\begin{pmatrix} p^\lambda \\ -q^\lambda \end{pmatrix}. \qquad (11.243)$$

The complex conjugate of (11.242) is

$$\sum_n E_\omega\left|(n\,\lambda^\pi\|\mathcal{M}_{\sigma\lambda}\|\mathrm{RPA})\right|^2$$

$$= \sum_n E_\omega\left(p^{\lambda\,\mathrm{T}}, -q^{\lambda\,\mathrm{T}}\right)\begin{pmatrix} X^{\omega*} \\ -Y^{\omega*} \end{pmatrix}\left(X^{\omega\mathrm{T}}, -Y^{\omega\mathrm{T}}\right)\begin{pmatrix} p^\lambda \\ -q^\lambda \end{pmatrix}. \qquad (11.244)$$

Exchanging the $p$ and $q$ quantities according to (11.239) gives

$$\sum_n E_\omega\left|(n\,\lambda^\pi\|\mathcal{M}_{\sigma\lambda}\|\mathrm{RPA})\right|^2$$

$$= \sum_n E_\omega\left(q^{\lambda\,\mathrm{T}}, -p^{\lambda\,\mathrm{T}}\right)\begin{pmatrix} X^{\omega*} \\ -Y^{\omega*} \end{pmatrix}\left(X^{\omega\mathrm{T}}, -Y^{\omega\mathrm{T}}\right)\begin{pmatrix} q^\lambda \\ -p^\lambda \end{pmatrix}$$

$$= \sum_n E_\omega\left(p^{\lambda\,\mathrm{T}}, -q^{\lambda\,\mathrm{T}}\right)\begin{pmatrix} Y^{\omega*} \\ -X^{\omega*} \end{pmatrix}\left(Y^{\omega\mathrm{T}}, -X^{\omega\mathrm{T}}\right)\begin{pmatrix} p^\lambda \\ -q^\lambda \end{pmatrix}. \qquad (11.245)$$

Similar to the step from (11.242) to (11.243), we again apply the RPA equation and find

$$\sum_n E_\omega \left| (n\,\lambda^\pi \| \mathcal{M}_{\sigma\lambda} \| \mathrm{RPA}) \right|^2$$

$$= - \left( p^{\lambda\,\mathrm{T}}, -q^{\lambda\,\mathrm{T}} \right) \begin{pmatrix} A & B \\ B^* & A^* \end{pmatrix} \sum_n \begin{pmatrix} Y^{\omega*} \\ X^{\omega*} \end{pmatrix} \left( Y^{\omega\,\mathrm{T}}, -X^{\omega\,\mathrm{T}} \right) \begin{pmatrix} p^\lambda \\ -q^\lambda \end{pmatrix} . \quad (11.246)$$

We can combine (11.243) and (11.246) to read

$$\sum_n E_\omega \left| (n\,\lambda^\pi \| \mathcal{M}_{\sigma\lambda} \| \mathrm{RPA}) \right|^2 = \frac{1}{2} \left( p^{\lambda\,\mathrm{T}}, -q^{\lambda\,\mathrm{T}} \right) \begin{pmatrix} A & B \\ B^* & A^* \end{pmatrix}$$

$$\times \sum_n \left[ \begin{pmatrix} X^\omega \\ Y^\omega \end{pmatrix} \left( X^{\omega\dagger}, -Y^{\omega\dagger} \right) - \begin{pmatrix} Y^{\omega*} \\ X^{\omega*} \end{pmatrix} \left( Y^{\omega\,\mathrm{T}}, -X^{\omega\,\mathrm{T}} \right) \right] \begin{pmatrix} p^\lambda \\ -q^\lambda \end{pmatrix} . \quad (11.247)$$

The quantity in the square brackets is precisely the left-hand side of the RPA completeness relation (11.85), which gives

$$\sum_n E_\omega \left| (n\,\lambda^\pi \| \mathcal{M}_{\sigma\lambda} \| \mathrm{RPA}) \right|^2 = \frac{1}{2} \left( p^{\lambda\,\mathrm{T}}, -q^{\lambda\,\mathrm{T}} \right) \begin{pmatrix} A & B \\ B^* & A^* \end{pmatrix} \begin{pmatrix} p^\lambda \\ -q^\lambda \end{pmatrix}$$

$$= \tfrac{1}{2} \left( p^{\lambda\,\mathrm{T}} A p^\lambda - p^{\lambda\,\mathrm{T}} B q^\lambda - q^{\lambda\,\mathrm{T}} B^* p^\lambda + q^{\lambda\,\mathrm{T}} A^* q^\lambda \right) . \quad (11.248)$$

Let us now evaluate the right-hand side of (11.237). The particle–hole part of the operator (11.218) can be expressed in terms of the RPA operators $\mathcal{A}^\dagger_{ab}(\lambda\mu)$ and $\tilde{\mathcal{A}}_{ab}(\lambda\mu)$ defined in (11.16) and (11.18), with $a$ the particle index and $b$ the hole index. To proceed, we need the relations (4.46)–(4.48) between particle and hole creation and annihilation operators. Expressing everything in terms of the $c^\dagger, c$ operators we then obtain

$$\mathcal{A}^\dagger_{ab}(\lambda\mu) = \left[ c^\dagger_a \tilde{c}_b \right]_{\lambda\mu} , \quad \tilde{\mathcal{A}}_{ab}(\lambda\mu) = (-1)^{j_a + j_b - \lambda + 1} \left[ c^\dagger_b \tilde{c}_a \right]_{\lambda\mu} . \quad (11.249)$$

Multiplied by the appropriate single-particle matrix element to enter into (11.218), the coupled operators become

$$\zeta^{(\lambda)} p^\lambda_{ab} \left[ c^\dagger_a \tilde{c}_b \right]_{\lambda\mu} = \zeta^{(\lambda)} p^\lambda_{ab} \mathcal{A}^\dagger_{ab}(\lambda\mu) , \quad (11.250)$$

$$q^\lambda_{ba} \left[ c^\dagger_b \tilde{c}_a \right]_{\lambda\mu} = \zeta^{(\lambda)} q^\lambda_{ab} \tilde{\mathcal{A}}_{ab}(\lambda\mu) \quad (11.251)$$

with use of the abbreviations (11.239) and the symmetry relations (6.27)–(6.30). The multipole operator now becomes

$$\mathcal{M}_{\sigma\lambda\mu} = \zeta^{(\lambda)} \hat{\lambda}^{-1} \sum_{ab} \left[ p^\lambda_{ab} \mathcal{A}^\dagger_{ab}(\lambda\mu) + q^\lambda_{ab} \tilde{\mathcal{A}}_{ab}(\lambda\mu) \right]$$

$$+ \text{particle–particle and hole–hole terms} . \quad (11.252)$$

Only the particle–hole terms explicitly stated in (11.252) contribute to the right-hand side of (11.237). Substitution then yields

$$\frac{1}{2}\widehat{\lambda}^2\langle\mathrm{HF}|[\mathcal{M}^\dagger_{\sigma\lambda\mu}, H, \mathcal{M}_{\sigma\lambda\mu}]|\mathrm{HF}\rangle$$

$$= \frac{1}{2}\sum_{\substack{ab\\cd}}\langle\mathrm{HF}|\,[p^\lambda_{ab}A_{ab} + q^\lambda_{ab}\tilde{A}^\dagger_{ab}, H, p^\lambda_{cd}A^\dagger_{cd} + q^\lambda_{cd}\tilde{A}_{cd}]\,|\mathrm{HF}\rangle$$

$$= \frac{1}{2}\sum_{\substack{ab\\cd}}(p^\lambda_{ab}A_{ab,cd}p^\lambda_{cd} - p^\lambda_{ab}B_{ab,cd}q^\lambda_{cd} - q^\lambda_{ab}B^*_{cd,ab}p^\lambda_{cd} + q^\lambda_{ab}A_{cd,ab}q^\lambda_{cd})$$

$$= \frac{1}{2}(\mathsf{p}^{\lambda\mathrm{T}}\mathsf{A}\mathsf{p}^\lambda - \mathsf{p}^{\lambda\mathrm{T}}\mathsf{B}\mathsf{q}^\lambda - \mathsf{q}^{\lambda\mathrm{T}}\mathsf{B}^*\mathsf{p}^\lambda + \mathsf{q}^{\lambda\mathrm{T}}\mathsf{A}^*\mathsf{q}^\lambda)\,, \tag{11.253}$$

where we have used (11.48), (11.52), (11.53) and (11.58). This is equal to the expression (11.248) for the left-hand side of (11.237), which concludes the proof of the EWSR (11.237).

As a by-product of the proof, (11.248) provides a convenient expression for the EWSR. Assuming that the A and B matrices are real and using (11.239), we have

$$\boxed{\sum_n E_\omega\left|(n\,\lambda^\pi\|\mathcal{M}_{\sigma\lambda}\|\mathrm{RPA})\right|^2 = \mathsf{p}^{\lambda\mathrm{T}}\big[\mathsf{A} - \zeta^{(\lambda)}\mathsf{B}\big]\mathsf{p}^\lambda\,.} \tag{11.254}$$

As in the case of the NEWSR of the TDA, the left-hand side of the sum rule involves the details of the energies and wave functions. The right-hand side involves just the single-particle matrix elements of the electromagnetic operator and the effect of the nuclear Hamiltonian through the A and B matrices.

### 11.6.5 Sum Rule for the Octupole Transitions in $^{16}$O

We continue the example of Subsect. 11.6.3 in the smaller basis (9.113) to check whether the values in Table 11.10 satisfy the EWSR (11.254). In this case the vector $\mathsf{p}^\lambda = \mathsf{p}^3$ reads (see Table 11.8)

$$\mathsf{p}^3 = \begin{pmatrix} 3.420e^{\mathrm{p}}_{\mathrm{eff}} \\ 3.824e^{\mathrm{p}}_{\mathrm{eff}} \\ 3.420e^{\mathrm{n}}_{\mathrm{eff}} \\ 3.824e^{\mathrm{n}}_{\mathrm{eff}} \end{pmatrix} b^3\,. \tag{11.255}$$

The A and B matrices, from the calculation quoted in Subsect. 11.6.3, are

$$\mathsf{A}(3) = \begin{pmatrix} 17.486 & -0.256 & -0.686 & -0.767 \\ -0.256 & 11.743 & -0.767 & -0.857 \\ -0.686 & -0.767 & 17.486 & -0.256 \\ -0.767 & -0.857 & -0.256 & 11.743 \end{pmatrix} \mathrm{MeV}\,, \tag{11.256}$$

$$\mathsf{B}(3) = \begin{pmatrix} 0.371 & 0.511 & 0.686 & 0.767 \\ 0.511 & 1.000 & 0.767 & 0.857 \\ 0.686 & 0.767 & 0.371 & 0.511 \\ 0.767 & 0.857 & 0.511 & 1.000 \end{pmatrix} \mathrm{MeV}\,. \tag{11.257}$$

Evaluating the right-hand side of (11.254) gives

$$p^{3\dagger}(A+B)p^3 = 402\left[(e_{\text{eff}}^{\text{p}})^2 + (e_{\text{eff}}^{\text{n}})^2\right]b^6 \text{ MeV}$$
$$= 1.06 \times 10^4 \left[(e_{\text{eff}}^{\text{p}})^2 + (e_{\text{eff}}^{\text{n}})^2\right] \text{ MeV fm}^6 . \qquad (11.258)$$

For the left-hand side of (11.254) Table 11.10 gives

$$\sum_n E_n \left|(3_n^-\|Q_3\|\text{RPA})\right|^2 = (3443e_+^2 + 2026e_-^2 + 1854e_+^2 + 3271e_-^2) \text{ MeV fm}^6$$
$$= 5297(e_+^2 + e_-^2) \text{ MeV fm}^6$$
$$= 1.06 \times 10^4 \left[(e_{\text{eff}}^{\text{p}})^2 + (e_{\text{eff}}^{\text{n}})^2\right] \text{ MeV fm}^6 . \qquad (11.259)$$

This agrees with (11.258), so the EWSR is indeed obeyed by our calculated octupole transitions in $^{16}$O.

### 11.6.6 Electric Transitions to the RPA Ground State on the Schematic Model

The schematic separable model was applied to the RPA in Sect. 11.4. We now extend the description to electromagnetic transitions in a way parallel to the TDA treatment in Subsect. 9.5.1. The scheme is to use (11.223) for an E$\lambda$ transition with the single-particle matrix elements given by (9.154) as

$$(a\|Q_\lambda\|b) = (-1)^{n_a+n_b}\,\hat{\lambda}\frac{e_{\text{eff}}}{4\sqrt{\pi}}Q_{ab}^\lambda(\text{SDI})\mathcal{R}_{ab}^{(\lambda)} \qquad (11.260)$$

and the $X$ and $Y$ amplitudes given by (11.123). The effective charge to be used here is defined by (9.156) for each isospin mode.

Substituting (11.260) and (11.123) into (11.223), with CS phases, gives

$$(\text{RPA}\|Q_\lambda\|\omega) = \delta_{\lambda J}\hat{\lambda}\frac{e_{\text{eff}}}{4\sqrt{\pi}}\sum_{ab}(-1)^{n_a+n_b}Q_{ab}^\lambda(\text{SDI})\mathcal{R}_{ab}^{(\lambda)}$$
$$\times \left[(-1)^\lambda\frac{Q_{ab}^J(\text{SDI})}{\varepsilon_{ab}-E_\omega}N_\omega + (-1)^J\frac{Q_{ab}^J(\text{SDI})}{\varepsilon_{ab}+E_\omega}N_\omega\right] \qquad (11.261)$$

with $\omega = nJ^\pi$. This simplifies to

$$\boxed{(\text{RPA}\|Q_\lambda\|\omega) = \delta_{\lambda J}(-1)^J\hat{J}\frac{e_{\text{eff}}}{2\sqrt{\pi}}N_\omega\sum_{ab}(-1)^{n_a+n_b}\frac{[Q_{ab}^J(\text{SDI})]^2\varepsilon_{ab}}{\varepsilon_{ab}^2-E_\omega^2}\mathcal{R}_{ab}^{(J)} .}$$
$$(11.262)$$

In the absence of the second term in (11.261) we have the TDA result (9.155).

To find the normalization constant $N_\omega$ we have from (11.83) and (11.123)

$$1 = \sum_{ab}\left(|X_{ab}^\omega|^2 - |Y_{ab}^\omega|^2\right) = \sum_{ab}\left\{\frac{[Q_{ab}^J(\text{SDI})]^2}{(\varepsilon_{ab}-E_\omega)^2} - \frac{[Q_{ab}^J(\text{SDI})]^2}{(\varepsilon_{ab}+E_\omega)^2}\right\}|N_\omega|^2 ,$$
$$(11.263)$$

leading to

$$|N_\omega|^{-2} = 4E_\omega \sum_{ab} \frac{[Q_{ab}^J(\text{SDI})]^2 \varepsilon_{ab}}{(\varepsilon_{ab}^2 - E_\omega^2)^2} . \qquad (11.264)$$

The relations (11.262) and (11.264) are simple to use to obtain first approximations for electric decays to the ground state.

## 11.6.7 Electric Dipole Transitions in ⁴He on the Schematic Model

To demonstrate the above formalism we revisit the TDA example of Subsect. 9.5.2 and compute the reduced E1 transition probabilities from the $1^-$ states to the ground state in $^4_2\text{He}_2$.
    Equation (11.264) gives for the normalization constant

$$N_n^{-2} = 4E_n \left\{ \frac{\frac{16}{3} \times 21.0\,\text{MeV}}{[(21.0\,\text{MeV})^2 - E_n^2]^2} + \frac{\frac{8}{3} \times 27.0\,\text{MeV}}{[(27.0\,\text{MeV})^2 - E_n^2]^2} \right\} . \qquad (11.265)$$

We apply the schematic model consistently and use the 'schematic' energies $E_n$ for $A_0 = A_1 = 1.0\,\text{MeV}$ from Tables 11.6 and 11.7. The energies and the normalization constants computed from (11.265) are recorded in Table 11.12.
    Equation (11.262) gives the transition amplitude

$$(\text{RPA}\|Q_1\|1_n^-) = -\sqrt{3}\frac{e_\text{eff}}{2\sqrt{\pi}}N_\omega \left[ \frac{\frac{16}{3} \times 21.0\,\text{MeV}}{(21.0\,\text{MeV})^2 - E_n^2}\mathcal{R}_{0p0s}^{(1)} \right.$$
$$\left. + \frac{\frac{8}{3} \times 27.0\,\text{MeV}}{(27.0\,\text{MeV})^2 - E_n^2}\mathcal{R}_{0p0s}^{(1)} \right] . \qquad (11.266)$$

Inserting $\mathcal{R}_{0p0s}^{(1)} = \sqrt{\frac{3}{2}}b$ with $b = 1.499\,\text{fm}$, as in the TDA example, yields the reduced transition probabilities in Table 11.12.
    Table 11.12 shows good agreement between the $B(\text{E1})$ values from the schematic model and those from the exact diagonalization. This agreement

**Table 11.12.** Reduced transition probabilities $B(\text{E1}\,;\,1_n^- \to 0_\text{gs}^+)$ for $^4_2\text{He}_2$ computed by the schematic and exact RPA

| | | Schematic | | Exact |
|---|---|---|---|---|
| $n$ | $E_n$ (MeV) | $N_n$ (MeV) | $B(\text{E1})$ (fm$^2$) | $B(\text{E1})$ (fm$^2$) |
| 1 | 19.4 | 0.686 | $0.24e_+^2$ | $0.272e_+^2$ |
| 2 | 22.1 | 0.471 | $0.13e_-^2$ | $0.171e_-^2$ |
| 3 | 26.4 | 0.363 | $0.06e_+^2$ | $0.040e_+^2$ |
| 4 | 27.8 | 0.483 | $0.12e_-^2$ | $0.086e_-^2$ |

The SDI parameters are $A = 1.0\,\text{MeV}$ (schematic) and $A_0 = A_1 = 1.0\,\text{MeV}$ (exact).

supports the fact that the schematic model is rather reliable both for excitation energies and for E$\lambda$ transitions in doubly magic nuclei with good isospin.

Comparing the energies and $B(E1)$ values in Tables 11.12 and 9.9 shows hardly any difference between the RPA and TDA, schematic or exact. As regards the $B(E1)$ values, this is at odds with the general qualitative conclusion drawn from the extreme collective model in Subsect. 11.6.2 and the degenerate model in Subsect. 11.6.8, as well as with the detailed E3 example in Subsect. 11.6.3. The reason is that in the present case the particle–hole space is small and the interaction is not very strong.

### 11.6.8 The Degenerate Model

In the limit of degenerate particle–hole energies, the lowest and highest, untrapped solutions of the RPA equations are given by (11.129) as

$$E_{\text{coll}} = \sqrt{\varepsilon(\varepsilon + 2\chi_J Q^2)}, \quad Q^2 = \sum_{ab} \left(Q_{ab}^J\right)^2. \tag{11.267}$$

Equation (11.264) in this case becomes

$$|N_{\text{coll}}|^{-2} = \frac{\sqrt{\varepsilon(\varepsilon + 2\chi_J Q^2)}}{\varepsilon \chi_J^2 Q^2} = \frac{E_{\text{coll}}}{\varepsilon \chi_J^2 Q^2}. \tag{11.268}$$

With the $X$ and $Y$ amplitudes from (11.123) and with the abbreviation (9.162), the transition matrix element (11.223) in the CS phase convention becomes

$$(\text{RPA}\|Q_J\|\text{coll}) = \sum_{ab} e\hat{J}Q_{ab}^J \left[(-1)^J \frac{Q_{ab}^J}{\varepsilon - E_{\text{coll}}} + (-1)^J \frac{Q_{ab}^J}{\varepsilon + E_{\text{coll}}}\right] N_{\text{coll}}$$

$$= (-1)^J \hat{J}e \frac{2\varepsilon Q^2}{\varepsilon^2 - E_{\text{coll}}^2} N_{\text{coll}}. \tag{11.269}$$

In the degenerate case the RPA dispersion equation (11.125) is reduced to

$$-\frac{1}{\chi_J} = 2\frac{Q^2 \varepsilon}{\varepsilon^2 - E_{\text{coll}}^2}. \tag{11.270}$$

Substituting this and the normalization constant from (11.268) into (11.269) we find

$$(\text{RPA}\|Q_J\|\text{coll}) = (-1)^{J+1} \hat{J}e\sqrt{\frac{\varepsilon}{E_{\text{coll}}}} Q. \tag{11.271}$$

Comparison with the corresponding TDA result (9.163) shows that

$$\boxed{(\text{RPA}\|Q_J\|\text{coll}) = \sqrt{\frac{\varepsilon}{E_{\text{coll}}}}(\text{HF}\|Q_J\|\text{coll}).} \tag{11.272}$$

The result (11.272) indicates that the RPA predicts a notable enhancement of electric transitions over the predictions of the TDA. The enhancement factor $\sqrt{\varepsilon/E_{\text{coll}}}$ can be large for the lowest collective RPA solution since $E_{\text{coll}} \to 0$ for large interaction strengths. The same conclusion was reached in (11.230) by using the extreme collective model.

### 11.6.9 Electromagnetic Transitions Between Two RPA Excitations

Electromagnetic transitions between two TDA excitations were discussed in Subsect. 9.4.7, with the general result stated in (9.142). We now extend that formalism to the RPA, where commutator techniques are important for including the ground-state correlations via the $Y$ amplitudes.

In the present case both the initial and final states are of the form $Q_\omega^\dagger |\text{RPA}\rangle$, with $Q_\omega^\dagger$ given by (11.40). The electromagnetic decay amplitude is then

$$\langle n_f\, J_f\, M_f | \mathcal{M}_{\sigma\lambda\mu} | n_i\, J_i\, M_i \rangle = \langle \text{RPA} | Q_{\omega_f} \mathcal{M}_{\sigma\lambda\mu} Q_{\omega_i}^\dagger | \text{RPA} \rangle \,. \tag{11.273}$$

To evaluate the right-hand side we form the commutator

$$\begin{aligned}
[Q_\omega, Q_{\omega'}^\dagger] = \sum_{\substack{ab\\cd}} &\left\{ X_{ab}^{\omega*} X_{cd}^{\omega'} \left[ A_{ab}(JM), A_{cd}^\dagger(J'M') \right] \right.\\
&\left. + Y_{ab}^{\omega*} Y_{cd}^{\omega'} \left[ \tilde{A}_{ab}^\dagger(JM), \tilde{A}_{cd}(J'M') \right] \right\}\\
\approx \delta_{JJ'}\delta_{MM'} &\sum_{ab} \left( X_{ab}^{\omega*} X_{ab}^{\omega'} - Y_{ab}^{\omega*} Y_{ab}^{\omega'} \right) = \delta_{JJ'}\delta_{MM'}\delta_{nn'}\delta_{\pi\pi'} \,,
\end{aligned} \tag{11.274}$$

where we have made the quasiboson approximation (11.97) and used the RPA orthogonality relation (11.82). We have thus established the approximate commutation relation

$$\boxed{[Q_\omega, Q_{\omega'}^\dagger] \overset{\text{QBA}}{\approx} \delta_{\omega\omega'} \,.} \tag{11.275}$$

As our next step we calculate

$$\begin{aligned}
&\langle \text{RPA} | \left[ Q_{\omega_f}, \mathcal{M}_{\sigma\lambda\mu}, Q_{\omega_i}^\dagger \right] | \text{RPA} \rangle\\
&= \tfrac{1}{2} \langle \text{RPA} | Q_{\omega_f} \left[ \mathcal{M}_{\sigma\lambda\mu}, Q_{\omega_i}^\dagger \right] - \left[ \mathcal{M}_{\sigma\lambda\mu}, Q_{\omega_i}^\dagger \right] Q_{\omega_f}\\
&\qquad + \left[ Q_{\omega_f}, \mathcal{M}_{\sigma\lambda\mu} \right] Q_{\omega_i}^\dagger - Q_{\omega_i}^\dagger \left[ Q_{\omega_f}, \mathcal{M}_{\sigma\lambda\mu} \right] | \text{RPA} \rangle\\
&= \tfrac{1}{2} \langle \text{RPA} | Q_{\omega_f} \left[ \mathcal{M}_{\sigma\lambda\mu}, Q_{\omega_i}^\dagger \right] + \left[ Q_{\omega_f}, \mathcal{M}_{\sigma\lambda\mu} \right] Q_{\omega_i}^\dagger | \text{RPA} \rangle\\
&= \langle \text{RPA} | Q_{\omega_f} \mathcal{M}_{\sigma\lambda\mu} Q_{\omega_i}^\dagger - \tfrac{1}{2} Q_{\omega_f} Q_{\omega_i}^\dagger \mathcal{M}_{\sigma\lambda\mu} - \tfrac{1}{2} \mathcal{M}_{\sigma\lambda\mu} Q_{\omega_f} Q_{\omega_i}^\dagger | \text{RPA} \rangle \,.
\end{aligned} \tag{11.276}$$

From (11.275) we have

$$Q_{\omega_f} Q_{\omega_i}^{\dagger} \approx Q_{\omega_i}^{\dagger} Q_{\omega_f} + \delta_{\omega_i \omega_f} , \qquad (11.277)$$

so the second and third terms vanish in the QBA. Consequently the matrix element (11.273) becomes

$$\langle n_f J_f M_f | \mathcal{M}_{\sigma\lambda\mu} | n_i J_i M_i \rangle \approx \langle \mathrm{RPA} | \big[ Q_{\omega_f}, \mathcal{M}_{\sigma\lambda\mu}, Q_{\omega_i}^{\dagger} \big] | \mathrm{RPA} \rangle$$
$$\approx \langle \mathrm{HF} | \big[ Q_{\omega_f}, \mathcal{M}_{\sigma\lambda\mu}, Q_{\omega_i}^{\dagger} \big] | \mathrm{HF} \rangle , \qquad (11.278)$$

where the second approximate equality results from the QBA in the form it appears in (11.81). The continuation is understood to occur within the QBA, so approximate equality will be denoted as equality.

With the electromagnetic operator (11.218) the matrix element (11.278) becomes

$$\langle n_f J_f M_f | \mathcal{M}_{\sigma\lambda\mu} | n_i J_i M_i \rangle$$
$$= \widehat{\lambda}^{-1} \sum_{ab} (a\|\boldsymbol{\mathcal{M}}_{\sigma\lambda}\|b) \langle \mathrm{HF} | \big[ Q_{\omega_f}, [c_a^{\dagger} \tilde{c}_b]_{\lambda\mu}, Q_{\omega_i}^{\dagger} \big] | \mathrm{HF} \rangle . \qquad (11.279)$$

In the standard RPA way, the calculation would continue with a straightforward evaluation of the double commutator. However, that would in effect include a rederivation of the particle–hole matrix element (6.124). It is therefore more economical to keep the particle–hole creation and annihilation operators $\mathcal{A}^{\dagger}, \mathcal{A}$ intact and exploit (6.124). We proceed as follows.

To evaluate the expectation value of the double commutator (11.279) we first expand the commutator:

$$\big[ Q_{\omega_f}, [c_a^{\dagger} \tilde{c}_b]_{\lambda\mu}, Q_{\omega_i}^{\dagger} \big] = Q_{\omega_f} [c_a^{\dagger} \tilde{c}_b]_{\lambda\mu} Q_{\omega_i}^{\dagger} + Q_{\omega_i}^{\dagger} [c_a^{\dagger} \tilde{c}_b]_{\lambda\mu} Q_{\omega_f}$$
$$- \tfrac{1}{2} \big( Q_{\omega_f} Q_{\omega_i}^{\dagger} [c_a^{\dagger} \tilde{c}_b]_{\lambda\mu} + [c_a^{\dagger} \tilde{c}_b]_{\lambda\mu} Q_{\omega_i}^{\dagger} Q_{\omega_f}$$
$$+ [c_a^{\dagger} \tilde{c}_b]_{\lambda\mu} Q_{\omega_f} Q_{\omega_i}^{\dagger} + Q_{\omega_i}^{\dagger} Q_{\omega_f} [c_a^{\dagger} \tilde{c}_b]_{\lambda\mu} \big) . \qquad (11.280)$$

When $Q_{\omega_i}^{\dagger}$ and $Q_{\omega_f}$ are substituted here from (11.40) and (11.41), most terms disappear because $\mathcal{A}|\mathrm{HF}\rangle = 0$ and $\langle \mathrm{HF}|\mathcal{A}^{\dagger} = 0$ and because two-particle–two-hole states cannot connect to the particle–hole vacuum. We then have

$$\langle \mathrm{HF} | \big[ Q_{\omega_f}, [c_a^{\dagger} \tilde{c}_b]_{\lambda\mu}, Q_{\omega_i}^{\dagger} \big] | \mathrm{HF} \rangle$$
$$= \sum_{\substack{a_i b_i \\ a_f b_f}} \Big\{ X_{a_f b_f}^{\omega_f *} X_{a_i b_i}^{\omega_i} \langle \mathrm{HF} | \mathcal{A}_{a_f b_f} [c_a^{\dagger} \tilde{c}_b]_{\lambda\mu} \mathcal{A}_{a_i b_i}^{\dagger} | \mathrm{HF} \rangle$$
$$+ Y_{a_i b_i}^{\omega_i} Y_{a_f b_f}^{\omega_f *} \langle \mathrm{HF} | \tilde{\mathcal{A}}_{a_i b_i} [c_a^{\dagger} \tilde{c}_b]_{\lambda\mu} \tilde{\mathcal{A}}_{a_f b_f}^{\dagger} | \mathrm{HF} \rangle$$
$$- \tfrac{1}{2} \big( X_{a_f b_f}^{\omega_f *} X_{a_i b_i}^{\omega_i} \langle \mathrm{HF} | \mathcal{A}_{a_f b_f} \mathcal{A}_{a_i b_i}^{\dagger} [c_a^{\dagger} \tilde{c}_b]_{\lambda\mu} | \mathrm{HF} \rangle$$
$$+ Y_{a_i b_i}^{\omega_i} Y_{a_f b_f}^{\omega_f *} \langle \mathrm{HF} | [c_a^{\dagger} \tilde{c}_b]_{\lambda\mu} \tilde{\mathcal{A}}_{a_i b_i} \tilde{\mathcal{A}}_{a_f b_f}^{\dagger} | \mathrm{HF} \rangle$$
$$+ X_{a_f b_f}^{\omega_f *} X_{a_i b_i}^{\omega_i} \langle \mathrm{HF} | [c_a^{\dagger} \tilde{c}_b]_{\lambda\mu} \mathcal{A}_{a_f b_f} \mathcal{A}_{a_i b_i}^{\dagger} | \mathrm{HF} \rangle$$
$$+ Y_{a_i b_i}^{\omega_i} Y_{a_f b_f}^{\omega_f *} \langle \mathrm{HF} | \tilde{\mathcal{A}}_{a_i b_i} \tilde{\mathcal{A}}_{a_f b_f}^{\dagger} [c_a^{\dagger} \tilde{c}_b]_{\lambda\mu} | \mathrm{HF} \rangle \big) \Big\} . \qquad (11.281)$$

The last four terms of (11.281) turn out to be zero. To see this, we calculate the commutator $[\mathcal{A}, \mathcal{A}^\dagger]$. The result is

$$
\begin{aligned}
[\mathcal{A}_{ab}(JM), \mathcal{A}_{cd}^\dagger(J'M')] = &\, \delta_{ac}\delta_{bd}\delta_{JJ'}\delta_{MM'} \\
&- \delta_{ac} \sum_{m_\alpha m_\beta m_\delta} (j_a\, m_\alpha\, j_b\, m_\beta | J\, M)(j_a\, m_\alpha\, j_d\, m_\delta | J'M')h_\delta^\dagger h_\beta \\
&- \delta_{bd} \sum_{m_\alpha m_\beta m_\gamma} (j_a\, m_\alpha\, j_b\, m_\beta | J\, M)(j_c\, m_\gamma\, j_b\, m_\beta | J'M')c_\gamma^\dagger c_\alpha \, . \quad (11.282)
\end{aligned}
$$

From (11.18) we also have

$$
[\tilde{\mathcal{A}}_{ab}(JM), \tilde{\mathcal{A}}_{cd}^\dagger(J'M')] = (-1)^{J+M+J'+M'} [\mathcal{A}_{ab}(J\,-M), \mathcal{A}_{cd}^\dagger(J'\,-M')] \, . \quad (11.283)
$$

Substituting $\mathcal{A}_{a_f b_f}\mathcal{A}_{a_i b_i}^\dagger$ and $\tilde{\mathcal{A}}_{a_i b_i}\tilde{\mathcal{A}}_{a_f b_f}^\dagger$ from here into (11.281) shows that the last four terms of (11.281) indeed vanish. Note that the first term in (11.282) also gives a vanishing contribution because $\lambda \geq 1$ for a gamma transition.

The first matrix element on the right-hand side of (11.281) is recognized as the one-body transition density for particle–hole states. With use of (11.218), it leads to the electromagnetic matrix element

$$
\begin{aligned}
\hat{\lambda}^{-1} \sum_{ab} &(a\|\mathcal{M}_{\sigma\lambda}\|b)\langle \mathrm{HF}|\mathcal{A}_{a_f b_f}(J_f M_f)\left[c_a^\dagger \tilde{c}_b\right]_{\lambda\mu} \mathcal{A}_{a_i b_i}^\dagger(J_i M_i)|\mathrm{HF}\rangle \\
&= \langle a_f\, b_f^{-1}\,;\, J_f\, M_f | \mathcal{M}_{\sigma\lambda\mu} | a_i\, b_i^{-1}\,;\, J_i\, M_i\rangle \\
&= (-1)^{J_f - M_f} \begin{pmatrix} J_f & \lambda & J_i \\ -M_f & \mu & M_i \end{pmatrix} (a_f\, b_f^{-1}\,;\, J_f \|\mathcal{M}_{\sigma\lambda}\| a_i\, b_i^{-1}\,;\, J_i) \, , \quad (11.284)
\end{aligned}
$$

where the Wigner–Eckart theorem (2.27) was used in the second step. The second term leads similarly to

$$
\begin{aligned}
\hat{\lambda}^{-1} \sum_{ab} &(a\|\mathcal{M}_{\sigma\lambda}\|b)\langle \mathrm{HF}|\tilde{\mathcal{A}}_{a_i b_i}(J_i M_i)\left[c_a^\dagger \tilde{c}_b\right]_{\lambda\mu} \tilde{\mathcal{A}}_{a_f b_f}^\dagger(J_f M_f)|\mathrm{HF}\rangle \\
&= (-1)^{J_i + M_i + J_f + M_f} \langle a_i\, b_i^{-1}\,;\, J_i\, -M_i | \mathcal{M}_{\sigma\lambda\mu} | a_f\, b_f^{-1}\,;\, J_f\, -M_f\rangle \\
&= (-1)^{J_f + M_f} \begin{pmatrix} J_i & \lambda & J_f \\ M_i & \mu & -M_f \end{pmatrix} (a_i\, b_i^{-1}\,;\, J_i \|\mathcal{M}_{\sigma\lambda}\| a_f\, b_f^{-1}\,;\, J_f) \\
&= (-1)^{J_f - M_f} \begin{pmatrix} J_f & \lambda & J_i \\ -M_f & \mu & M_i \end{pmatrix} \\
&\quad \times (-1)^{J_i - J_f}\left(\zeta^{(\sigma\lambda)}\right)^2 (a_f\, b_f^{-1}\,;\, J_f \|\mathcal{M}_{\sigma\lambda}\| a_i\, b_i^{-1}\,;\, J_i)^* \, . \quad (11.285)
\end{aligned}
$$

The last step is due to the general symmetry relation (11.290) with the phase factors $\zeta^{(\sigma\lambda)}$ defined in equations (6.10) and (6.11). The squared phase factor $(\zeta^{(\sigma\lambda)})^2$ is unity in the CS phase convention; in the BR convention it is $(-1)^\lambda$ for $\sigma = \mathrm{E}$ and $(-1)^{\lambda+1}$ for $\sigma = \mathrm{M}$. The complete phase factor in front of the reduced matrix element is thus equal to $(-1)^{\Delta J+\lambda}\zeta^{(\lambda)}$, where $\Delta J = |J_f - J_i|$ and $\zeta^{(\lambda)}$ is given by (11.223).

The electromagnetic single-particle matrix elements are real in both phase conventions (see Subsect. 6.1.3). It follows that the particle–hole matrix element (6.124) that appears in (11.284) and (11.285) is also real. The asterisk for complex conjugation in (11.285) can therefore be removed.

When the Wigner–Eckart theorem is applied to the left-hand side of (11.279), the same phase factor and $3j$ symbol come in front of the reduced matrix element as in (11.284) and (11.285). The reduced matrix element for an electromagnetic transition between two excited RPA states is thus

$$
(\omega_f \| \mathcal{M}_{\sigma\lambda} \| \omega_i)_{\text{RPA}} = \sum_{\substack{a_i b_i \\ a_f b_f}} \left[ X_{a_f b_f}^{\omega_f *} X_{a_i b_i}^{\omega_i} + (-1)^{\Delta J+\lambda} \zeta^{(\lambda)} Y_{a_f b_f}^{\omega_f *} Y_{a_i b_i}^{\omega_i} \right]
$$
$$
\times (a_f\, b_f^{-1}\,;\, J_f \| \mathcal{M}_{\sigma\lambda} \| a_i\, b_i^{-1}\,;\, J_i)\,.
$$

(11.286)

The convention-dependent phase factor is given through (11.223).

We can compare the transition amplitude (11.286) with the TDA result (9.142). For vanishing $Y$ amplitudes the RPA result is reduced to the TDA expression. Since the $Y$ amplitudes are generally small, we expect that the RPA strength of an electromagnetic transition between two excited states only moderately exceeds the TDA strength; there is no such relation as (11.272). Application of the result (11.286) is illustrated in the following subsection.

### 11.6.10 The E2 Transition $5_1^- \rightarrow 3_1^-$ in $^{40}$Ca

The example of Subsect. 9.4.8 presents a TDA calculation of the E2 transition $5_1^- \rightarrow 3_1^-$ in $^{40}_{20}\text{Ca}_{20}$. The calculation is now repeated in the RPA, using the same particle–hole valence space $0f_{7/2}\text{-}(0d\text{-}1s)^{-1}$ and the same single-particle energies from Fig. 9.2 (b). We use the CS phase convention.

We compare the TDA and RPA results by listing the TDA $(X)$ and RPA $(X$ and $Y)$ amplitudes of the initial and final states in Table 11.13; the $X$ values of the TDA are copied from (9.145). The $X$ and $Y$ amplitude products appearing in the E2 decay amplitudes (9.142) and (11.286) are displayed in Table 11.14. Note that the amplitudes in the tables are for one kind of nucleon, so they are normalized to 0.5 rather than 1.

Comparison between the first and last lines of Table 11.14 shows that an RPA contribution to the decay amplitude can be roughly twice the TDA value. The enhancement comes from the increased values of the $X$ amplitudes and the finite $Y$ amplitudes. The increased $X$ amplitudes are allowed by the RPA normalization condition (11.83), where they are compensated by the non-zero $Y$ values.

Equation (9.146) gives the E2 particle–hole matrix elements needed in (11.280), and we insert the effective charges. Combining them with the numbers in the last line of Table 11.14 leads to the decay amplitude

**Table 11.13.** TDA and RPA amplitudes for the $3_1^-$ and $5_1^-$ states in $^{40}$Ca in the particle–hole valence space $0f_{7/2}\text{-}(0d\text{-}1s)^{-1}$

| Model | $J_1^-$ | $X_1$ | $Y_1$ | $X_2$ | $Y_2$ | $X_3$ | $Y_3$ |
|-------|---------|-------|-------|-------|-------|-------|-------|
| TDA | $3_1^-$ | 0.272 | | 0.314 | | 0.572 | |
| RPA | $3_1^-$ | 0.384 | −0.187 | 0.467 | −0.224 | 0.502 | −0.179 |
| TDA | $5_1^-$ | 0.150 | | 0.691 | | | |
| RPA | $5_1^-$ | 0.187 | −0.085 | 0.719 | −0.210 | | |

The SDI parameters are $A_0 = 0.85\,\text{MeV}$, $A_1 = 0.90\,\text{MeV}$.

**Table 11.14.** TDA and RPA amplitude products in (9.142) and (11.280), with amplitudes from Table 11.13

| Model | Product | 11 | 21 | 31 | 12 | 22 | 32 |
|-------|---------|----|----|----|----|----|-----|
| | | | | | $ij$ | | |
| TDA | $X_i^{3_1^-} X_j^{5_1^-}$ | 0.041 | 0.047 | 0.086 | 0.188 | 0.217 | 0.395 |
| RPA | $X_i^{3_1^-} X_j^{5_1^-}$ | 0.072 | 0.087 | 0.094 | 0.276 | 0.336 | 0.361 |
| RPA | $Y_i^{3_1^-} Y_j^{5_1^-}$ | 0.016 | 0.019 | 0.015 | 0.039 | 0.047 | 0.038 |
| RPA | Total | 0.088 | 0.106 | 0.109 | 0.315 | 0.383 | 0.399 |

The last line gives the sum of the two previous lines, i.e. the term inside the square brackets in (11.280).

$$
\begin{aligned}
(3_1^- \| Q_2 \| 5_1^-) &= (0.4388 \times 0.088 + 1.6204 \times 0.106 - 1.1837 \times 0.109 \\
&\quad + 0.3915 \times 0.315 + 2.4752 \times 0.383 + 1.0978 \times 0.399) \\
&\quad \times (e_{\text{eff}}^{\text{p}} + e_{\text{eff}}^{\text{n}}) b^2 = 1.59(e_{\text{eff}}^{\text{p}} + e_{\text{eff}}^{\text{n}}) b^2 \\
&= 5.98(e_{\text{eff}}^{\text{p}} + e_{\text{eff}}^{\text{n}})\,\text{fm}^2
\end{aligned}
\tag{11.287}
$$

with the oscillator length $b = 1.939\,\text{fm}$ from Subsect. 9.4.8.

For the reduced transition probability, we then obtain

$$
B(\text{E2};\, 5_1^- \to 3_1^-)_{\text{RPA}} = 3.25(e_{\text{eff}}^{\text{p}} + e_{\text{eff}}^{\text{n}})^2\,\text{fm}^4 .
\tag{11.288}
$$

This is 2.35 times the TDA value (9.148). From (6.26) and the experimental $B(\text{E2})$ value (9.150) we can determine the electric polarization constant $\chi$. The result is

$$
\chi = 0.30 ,
\tag{11.289}
$$

to be compared with the TDA value $\chi = 0.72$ given in (9.151). Our RPA result (11.289) agrees with the value $\chi = 0.3$ deduced by comparing experimental and RPA values of $B(\text{E3};\, 3_1^- \to 0_{\text{gs}}^+)$. The E3 result is indicated in the caption to Fig. 9.7.

# Epilogue

In this chapter we have developed a sophisticated theoretical framework to describe a correlated ground state and excited states with configuration mixing consisting of particle–hole excitations. This RPA theory was found capable of producing strong collective effects, in particular for the energy and decay rate of the first RPA excitation. Having introduced the RPA, our sometimes hard and weary but in the end rewarding path in the particle–hole territory has ended. In Part II of this book we are challenged by the BCS-based quasiparticles to explore the wealth of new possibilities offered by them. The neighbourhoods of closed shells, discussed thus far, will be extended further towards the open-shell region to describe open-shell vibrational nuclei.

# Exercises

**11.1.** Derive the relations (11.18) and (11.19).

**11.2.** Verify the commutator relations (11.20)–(11.23).

**11.3.** Verify the commutators (11.26)–(11.29).

**11.4.** Verify the relations (11.30)–(11.33).

**11.5.** Verify the ground-state expectation values (11.44) and (11.45).

**11.6.** Show that the quantities (11.48) do not depend on the projection quantum number $M$.

**11.7.** Verify the double commutator identities (11.49).

**11.8.** Verify the relations (11.50) and (11.51).

**11.9.** Derive the matrix relations (11.58).

**11.10.** Derive the expression (11.63) for the matrix B.

**11.11.** Give a detailed derivation of (11.66) starting from (11.65).

**11.12.** Give a detailed derivation of (11.67) starting from (11.66).

**11.13.** Give a detailed derivation of (11.70) starting from (11.64) and (11.67).

**11.14.** Derive the result (11.71).

**11.15.** Verify the numbers in Table 11.2.

**11.16.** Derive the completeness relations (11.84). Hint: See Sect. 18.2.

**11.17.** Show that if $E_\omega$, $X^\omega$ and $Y^\omega$ form a solution of the RPA equations, then also the quantities (11.86) constitute a solution.

**11.18.** Consider the operator $e^{-S}Oe^S$, where $S$ and $O$ are arbitrary operators, and derive (11.94).

**11.19.** Derive the relation (11.109).

**11.20.** Calculate the approximate energies of the $3^-$ states in $^{16}$O by using the RPA dispersion equation in the $(0d_{5/2}\text{-}1s_{1/2})\text{-}(0p)^{-1}$ particle–hole valence space. Take the single-particle energies from Fig. 9.2 (a) and use the SDI with parameter values
(a) $A = 1.0\,\text{MeV}$,
(b) $A = 2.0\,\text{MeV}$.
Compare with the TDA results of Exercise 9.12 and comment.

**11.21.** Calculate the approximate energies of the $3^-$ states in $^{40}$Ca by using the RPA dispersion equation in the $0f_{7/2}\text{-}(0d\text{-}1s)^{-1}$ particle–hole valence space. Take the single-particle energies from Fig. 9.2 (b) and use the SDI with parameter values
(a) $A = 1.0\,\text{MeV}$,
(b) $A = 2.0\,\text{MeV}$.
Compare with the TDA results of Exercise 9.14 and comment.

**11.22.** Solve the RPA equations for the energy of the $3^-$ state in $^{48}$Ca. Use the $0f_{7/2}\text{-}(0d_{3/2})^{-1}$ particle–hole valence space, the single-particle energy difference $\varepsilon_{0f_{7/2}} - \varepsilon_{0d_{3/2}} = 5.0\,\text{MeV}$ and the SDI with $A_0 = A_1 = 1.0\,\text{MeV}$.

**11.23.** Verify the numbers in the matrix (11.147).

**11.24.** Form the RPA matrices for
(a) the $2^+$ states,
(b) the $3^-$ states
in $^{48}$Ca. Use the particle–hole valence space $0f_{7/2}\text{-}(0d_{3/2})^{-1}$ for protons and $(1p\text{-}0f_{5/2})\text{-}(0f_{7/2})^{-1}$ for neutrons, the single-particle energies of Fig. 9.2 (c) and the SDI parameters $A_0 = A_1 = 1.0\,\text{MeV}$.

**11.25.** Show that the following statements are true.

(a) The matrix $\mathsf{M}''_+$ defined in (11.160) is symmetric, i.e. $\mathsf{M}''^{\mathrm{T}}_+ = \mathsf{M}''_+$.
(b) The column matrix $\mathsf{X}^\omega$ is obtained as $\mathsf{X}^\omega = \mathsf{G}^\omega_+ \mathsf{R}^\omega_-$ with $\mathsf{G}^\omega_+$ defined in (11.166).

**11.26.** Compute the energies of the $2^+$ states in $^{48}$Ca by diagonalizing the RPA matrix constructed in Exercise 11.24 (a). Compare with the TDA results of Exercise 9.16 and comment.

**11.27.** Compute the energies of the $3^-$ states in $^{48}$Ca by diagonalizing the RPA matrix constructed in Exercise 11.24 (b). Compare with the TDA results of Exercise 9.16 and comment.

**11.28.** Compute the wave functions of the $2^+$ states in $^{48}$Ca by using the results of Exercise 11.26.

**11.29.** Compute the wave functions of the $3^-$ states in $^{48}$Ca by using the results of Exercise 11.27.

**11.30.** Consider $4^+$ excitations in $^{48}$Ca.

(a) Form the RPA matrix in the neutron particle–hole valence space (1p-$0f_{5/2}$)-$(0f_{7/2})^{-1}$. Use the single-particle energies of Fig. 9.2 (c) and the SDI with parameters $A_0 = A_1 = 1.0\,$MeV.
(b) Diagonalize the RPA matrix to find the wave function of the lowest state. Compare the eigenenergies with experiment and comment.

**11.31.** Derive the matrix equation (11.162).

**11.32.** Derive the normalization condition (11.175).

**11.33.** Derive the X and Y vectors corresponding to the energy $E_2$ in (11.181).

**11.34.** Verify the numbers in the matrix (11.199).

**11.35.** Verify the numbers in the matrix (11.203).

**11.36.** Verify the numbers in the matrices (11.204) and (11.205).

**11.37.** Verify Eqs. (11.208)–(11.211).

**11.38.** Derive the X and Y vectors corresponding to the energy $E_2 = 21.980\,$MeV in (11.207).

**11.39.** Verify the numbers in the matrix (11.212).

**11.40.** Form the RPA supermatrix in (11.60) for the $3^-$ states in $^{40}$Ca for the two cases in Exercise 11.21.

**11.41.** Diagonalize the RPA matrices of Exercise 11.40 to find the eigenenergies and eigenfunctions of the $3^-$ states in $^{40}$Ca. Compare with the results of the schematic model of Exercise 11.21 and with the TDA results of Exercise 9.15, and comment.

**11.42.** The same task as in Exercise 9.27 but by using the RPA.

**11.43.** The same task as in Exercise 9.28 but by using the RPA.

**11.44.** Verify the numbers in Table 11.8.

**11.45.** Verify the numbers in the results (11.233) and (11.234).

**11.46.** Verify the numbers in Table 11.9.

**11.47.** Verify the numbers in the results (11.235) and (11.236).

**11.48.** Verify the numbers of Table 11.10.

**11.49.** Verify the numbers of Table 11.11.

**11.50.** Consider a tensor operator $O_L = \zeta_L T_L$, where $T_L$ is a Hermitian tensor operator, as defined in (2.31), and $\zeta_L$ is a phase factor. Show that

$$(\xi\, j\|O_L\|\xi'\, j') = (-1)^{j-j'}(\zeta_L)^2(\xi'\, j'\|O_L\|\xi\, j)^* \,. \qquad (11.290)$$

Apply this relation to the electromagnetic multipole operator $\mathcal{M}_{\sigma\lambda}$ as defined in the CS and BR phase conventions through (6.10) and (6.11), and derive (11.238).

**11.51.** Carry out the application of the RPA supermatrix equation (11.60) to go from (11.242) to (11.243), and from (11.245) to (11.246).

**11.52.** Verify the numbers in the matrices (11.256) and (11.257).

**11.53.** Check that the EWSR (11.254) is satisfied for the $2^+$ and $3^-$ states computed in Exercises 11.28 and 11.29.

**11.54.** Verify the numbers in Table 11.12.

**11.55.** Compute the reduced E3 transition probabilities for the ground-state decays of the $3^-$ states in $^{16}$O within the schematic model by using the approximate energies from Exercise 11.20(b). Compare with the results of the TDA calculation in Exercise 9.20 and comment.

**11.56.** Compute the reduced E3 transition probabilities for the ground-state decays of the $3^-$ states in $^{40}$Ca within the schematic model by using the approximate energies from Exercise 11.21. Compare with the results of the TDA calculation in Exercise 9.22 and comment.

**11.57.** Compute the reduced E3 transition probabilities for the ground-state decays of the $3^-$ states in $^{40}$Ca by using the exact wave functions from Exercise 11.41. Compare with the results of Exercise 11.56 and with the TDA results of Exercise 9.23. Comment on the similarities and differences. Check that the RPA sum rule is satisfied.

**11.58.** Derive the result (11.281).

**11.59.** Derive the commutation relation (11.282).

**11.60.** Fill in the details in the development of (11.285).

**11.61.** Verify the numbers in Tables 11.13 and 11.14.

**11.62.** Continue Exercise 11.42 and compute the reduced E2 transition probabilities for the ground-state decays of the $2^+$ states in $^{12}$C. Check the RPA sum rule. Determine the electric polarization constant $\chi$ by using available experimental data. Compare with the TDA result of Exercise 9.29 and comment.

**11.63.** Continue Exercise 11.43 and compute the reduced E2 transition probabilities for the ground-state decays of the $2^+$ states in $^{32}$S. Check the RPA sum rule. Determine the electric polarization constant $\chi$ by using available experimental data. Compare with the TDA result of Exercise 9.30 and comment.

**11.64.** Take the wave function of the $2^+_1$ state in $^{48}$Ca from Exercise 11.28. Compute the reduced E2 transition probability for the decay of this state to the ground state. Determine the electric polarization constant $\chi$ from the experimental half-life of the $2^+_1$ state. Compare with the TDA result of Exercise 9.24 and comment.

**11.65.** Take the wave functions of the $2^+_1$ and $3^-_1$ states in $^{48}$Ca from Exercises 11.28 and 11.29. Compute the decay half-life of the $3^-_1$ state by assuming that it decays to the $2^+_1$ state and to the ground state (see the experimental level scheme in Fig. 9.5). Take the value of $\chi$ from Exercise 11.64 and use the experimental energies. Compare with the TDA result of Exercise 9.33 and with experiment, and comment.

**11.66.** Compute the decay half-life of the $4^+_1$ state in $^{48}$Ca by using the results of Exercises 11.28 and 11.30. Take the value of $\chi$ from Exercise 11.64 and use the experimental energies. Compare with the TDA result of Exercise 9.25 and with experiment, and comment.

**11.67.** The same task as in Exercise 9.31 but by using the RPA. Take the value of $\chi$ from Exercise 11.63. Compare with experiment and with the TDA result of Exercise 9.31, and comment.

**11.68.** The same task as in Exercise 9.32 but by using the RPA. Take the value of $\chi$ from (11.289). Compare with experiment and with the TDA result of Exercise 9.32, and comment.

# Part II

Quasiparticles

# Nucleon Pairing and Seniority

## Prologue

Until now we have been dealing with particle and hole aspects of nuclear structure. In this second part of the book we go farther away from a closed major shell. Still near the beginning or end of a major shell we encounter vibrational, spherical open-shell nuclei that cannot be described in terms of a few particles or holes. Farther towards the middle of the shell the spherical shape will give way to permanent deformation signalled by rotational bands analogous to those of diatomic molecules. Microscopic description of such nuclei requires a deformed mean field as the starting point. Deformed nuclei will not be considered in any detail in this book.

Even the description of vibrational nuclei requires an extension of the concepts of particle and hole. It is found expedient to introduce a special type of quasiparticle which mixes the particle and hole degrees of freedom. The essense of the quasiparticle method is contained in the simple quasiparticle shell model which consists of a quasiparticle vacuum and non-interacting quasiparticles. These quasiparticles absorb a good part of the short-range residual interaction.

The approximation of non-interacting quasiparticles constitutes the quasiparticle mean-field approach. We use this level of approximation to access electromagnetic and beta-decay properties of vibrational nuclei. The long-range part of the residual interaction can be implemented as a residual force between the quasiparticles. This leads to nuclear states which, at their simplest, are built from one or two quasiparticles. Quasiparticles interact via the residual Hamiltonian, which leads to quasiparticle configuration mixing.

In this chapter we discuss experimental observations that provide justification for the quasiparticle approach. We discuss short-range nuclear correlations and how they can be mimicked by an extremely simple nucleon–nucleon force, the pure pairing force of zero range. This interaction is the foundation of the seniority model and other simplified models dealing with correlated nucleon pairs. In these models a simple Hamiltonian is diagonalized exactly.

The models can be used as test benches for more sophisticated approaches such as the BCS or the Lipkin–Nogami BCS.

## 12.1 Evidence of Nucleon Pairing

Nuclei with several nucleons outside closed major shells are called *open-shell nuclei*. For them the particle–hole methods, like the TDA and RPA, become inapplicable. This follows from the fact that inside a major shell the single-particle energy differences and two-body matrix elements are of the same order of magnitude. The residual interaction is then able to scatter nucleons in such a way that the particle–hole hierarchy is lost. In other words, the particle–hole energies inside a major shell are so small that one-particle–one-hole excitations have roughly the same energies as two-particle–two-hole, etc., excitations.

At the same time, the residual interaction becomes more important than in the pure particle–hole picture. This is especially true of the short-range interaction, which is not taken into account in the Hartree–Fock mean field. The short-range part of the residual interaction shows as *pairing correlations* between the nucleons, also known as nucleon *pairing*. The *pairing interaction* is a short-range attraction between two nucleons. It lowers their total energy by an amount $2\Delta$, where $\Delta$ is called the *pairing gap*.

The effects of nucleon pairing can be seen experimentally in the following ways.

- The pairing interaction reduces the energy of nucleon pairs, most efficiently in a state of angular momentum zero, as shown in Sect. 12.3. Hence the nucleons in a major shell tend to form $J = 0$ pairs, which leads to a condensate of paired nucleons in a collective state with total angular momentum zero. This condensate is seen experimentally in all even–even nuclei, whose ground states are without exception $0^+$.

  The pair condensate is a kind of *superfluid*, analogous to that formed by correlated electron pairs in a metallic superconductor. The fundamental difference between the two systems, however, is that there is a direct attraction between nucleons while there is a (screened) Coulomb repulsion between electrons and the pairing is mediated by lattice phonons.

- Excitations of a nucleus can be produced by breaking one or more of the pairs forming the condensate. The lowest excitations have one broken pair; their energy is the *pairing energy* $2\Delta$ above the ground state. As in the case of the superconductor, the pair condensate in the nuclear ground state can be described by a BCS many-body wave function. A broken pair forms a pair of BCS *quasiparticles*. The lowest excitations in an even–even nucleus are *two-quasiparticle excitations*, a gap $2\Delta$ above the condensate which is the quasiparticle vacuum. This pattern is seen experimentally in even–even nuclei.

Experimental data show *collective* excitations, of vibrational or rotational nature, below the pairing gap. *Vibrational* states can be viewed as waves on the nuclear surface, microscopically described as coherent combinations of two-quasiparticle excitations. This is analogous to the vibrations as coherent combinations of particle–hole excitations in closed-shell nuclei, discussed in Chaps. 9 and 11 for the TDA and RPA.

The energy gap is a feature of even–even nuclei that cannot be observed in odd-$A$ or odd–odd nuclei. In the latter kinds of nuclei the low-energy spectra are complicated and have a much larger level density than adjacent even-even nuclei. In odd-$A$ nuclei the level density is increased by one- and three-quasiparticle excitations. In odd–odd nuclei two-quasiparticle excitations form the ground and low-lying excited states.

- Pairing can also be seen as an *odd–even effect*, relating to the masses of even–even and odd-$A$ nuclei. Experiment has established that the total binding energy of an odd-$A$ nucleus is less than the average of the total binding energies of the two neighbouring even–even nuclei. We assign the difference to the pairing gap $\Delta$ and write the mass of the odd-$A$ nucleus as

$$\boxed{M_A = \tfrac{1}{2}(M_{A-1} + M_{A+1}) + \Delta/c^2 \,.}\tag{12.1}$$

The odd–even effect is demonstrated in Fig. 12.1 for the odd-$A$ nucleus $^{105}_{46}\mathrm{Pd}_{59}$. In this figure the masses $m$ are the *atomic masses*, related to the *nuclear masses* $M$ by

$$M = m - Zm_\mathrm{e} + B_\mathrm{e}(Z)/c^2 \,,\tag{12.2}$$

where $B_\mathrm{e}(Z)$ is the total binding energy of the atomic electrons. For the palladium isotopes the latter two terms in (12.2) are common, so the differences in the ground-state energies come out the same whether one uses atomic or nuclear masses. Figure 12.1 shows that the deviation from the average is possible to measure and that it is very small compared to the nuclear mass differences. The experimental pairing gaps are reproduced by the formula

$$\boxed{\Delta \approx 12A^{-1/2}\,\mathrm{MeV} \,.}\tag{12.3}$$

Apart from being a fit to experimental data, this formula is supported by the liquid-drop model of the nucleus.

- Pairing effects show up in the moments of inertia of deformed nuclei. Moments of inertia calculated without pairing are two to three times larger than the measured ones. This is interpreted so that only part of the nucleus participates in the rotation, namely the paired superfluid of the valence space. The macroscopic motion is constituted by a tidal wave rotating around the inert core.

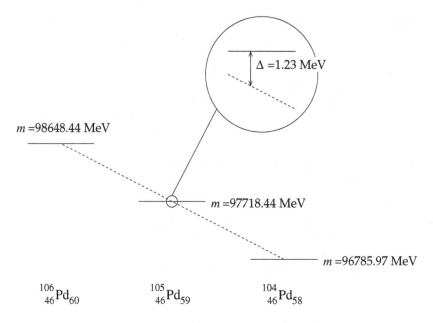

**Fig. 12.1.** Atomic masses of $^{106}_{46}\text{Pd}_{60}$, $^{105}_{46}\text{Pd}_{59}$ and $^{104}_{46}\text{Pd}_{58}$. The magnified insert shows the pairing gap $\Delta$ extracted from the masses. The semi-empirical formula (12.3) gives $\Delta \approx 1.2\,\text{MeV}$

- A Hartree–Fock calculation always predicts a deformed ground state for a nucleus with an open major shell. However, experimentally most open-shell nuclei are spherical in their low-lying states. The explanation is that the short-range pairing interaction favours spherical symmetry and maintains the spherical shape rather far into the open shell. Adding more nucleons will eventually cause a rapid transition from spherical to deformed shape. In microscopic theory this can be described by replacing the Hartree–Fock calculation by a Hartree–Fock–Bogoliubov calculation that includes the pairing interaction in a self-consistent variational scheme.

## 12.2 The Pure Pairing Force

As an example, consider the two-particle interaction of like nucleons ($T = 1$) in the $0f_{7/2}$ shell. The two-nucleon interaction matrix element for the SDI is given by (8.72) as

$$\langle (0f_{7/2})^2\,;\, J\,T = 1|V_{\text{SDI}}|(0f_{7/2})^2\,;\, J\,T = 1\rangle = -32A_1 \begin{pmatrix} \frac{7}{2} & \frac{7}{2} & J \\ \frac{1}{2} & -\frac{1}{2} & 0 \end{pmatrix}^2 . \quad (12.4)$$

Looking up the $3j$ coefficients or taking the values directly from Table 8.4 produces the numbers in Table 12.1.

**Table 12.1.** The two-body interaction matrix elements (12.4) for different values of the total angular momentum $J$

| $J$ | 0 | 2 | 4 | 6 |
|---|---|---|---|---|
| $\langle V_{\text{SDI}} \rangle (A_1)$ | $-4.000$ | $-0.952$ | $-0.468$ | $-0.233$ |

The matrix elements are given in units of the isovector strength parameter $A_1$ of the SDI.

As can be seen from Table 12.1, the interaction for the $J = 0$ pair is much stronger than for the $J \neq 0$ pairs. This suggests that we may simplify the interaction in a single $j$ shell by making the approximation that only the $J = 0$ channel contributes. Starting from the general form (8.15) of the two-body interaction Hamiltonian, we have

$$V_{\text{RES}} = -\frac{1}{2} \sum_J \widehat{J} \langle j\,j;\, J\,|V|\,j\,j;\, J\rangle \left[ \left[ c_j^\dagger c_j^\dagger \right]_J \left[ \tilde{c}_j \tilde{c}_j \right]_J \right]_{00}$$

$$\overset{\text{pairing}}{\approx} -\frac{1}{2} \langle j\,j;\, 0\,|V|\,j\,j;\, 0\rangle \left[ c_j^\dagger c_j^\dagger \right]_0 \left[ \tilde{c}_j \tilde{c}_j \right]_0$$

$$= -\frac{1}{2} \langle j\,j;\, 0\,|V|\,j\,j;\, 0\rangle \widehat{j}^{-2} \sum_{mm'} (-1)^{j-m+j-m'} c_{jm}^\dagger c_{j,-m}^\dagger \tilde{c}_{jm'} \tilde{c}_{j,-m'}$$

$$= \frac{1}{2} \langle j\,j;\, 0\,|V|\,j\,j;\, 0\rangle \widehat{j}^{-2} \sum_{mm'} c_{jm}^\dagger \tilde{c}_{jm}^\dagger \tilde{c}_{jm'} c_{jm'}$$

$$= 2 \langle j\,j;\, 0\,|V|\,j\,j;\, 0\rangle \widehat{j}^{-2} \sum_{\substack{m>0 \\ m'>0}} c_{jm}^\dagger \tilde{c}_{jm}^\dagger \tilde{c}_{jm'} c_{jm'} \;, \tag{12.5}$$

where we have used (1.34), (4.23) and (4.9). With the abbreviation

$$2 \langle j\,j;\, 0\,|V|\,j\,j;\, 0\rangle \widehat{j}^{-2} \equiv -G \tag{12.6}$$

we have the *pure pairing interaction*, or just the pairing interaction, $V_{\text{PAIR}}$ for a *single $j$ shell* as

$$\boxed{V_{\text{PAIR}} = -G \sum_{mm'>0} c_{jm}^\dagger \tilde{c}_{jm}^\dagger \tilde{c}_{jm'} c_{jm'}\;.} \tag{12.7}$$

The pairing interaction (12.7) can be immediately generalized to *several $j$ shells*, to read

$$\boxed{V_{\text{PAIR}} = -G \sum_{jj'} \sum_{mm'>0} c_{jm}^\dagger \tilde{c}_{jm}^\dagger \tilde{c}_{j'm'} c_{j'm'}\;.} \tag{12.8}$$

The idea of the pairing interaction is that it is attractive and of short range. Therefore $G$ has to be a positive constant. In the single-shell case the two-body matrix element in (12.6) is obviously negative for an attractive force $V$, which guarantees $G > 0$. In the multishell case we must specifically require that all the two-body matrix elements of the valence space satisfy

$$\langle j\,j\,;\,0\,|V|\,j'\,j'\,;\,0\rangle < 0 \tag{12.9}$$

so that it is possible to make the replacement

$$\langle j\,j\,;\,0\,|V|\,j'\,j'\,;\,0\rangle \rightarrow -\tfrac{1}{2}\widehat{j}\widehat{j'}G < 0 \,. \tag{12.10}$$

It turns out that this condition is satisfied in the BR phase convention but not in all cases in the CS convention (see Exercise 12.2).

The relation between the strength $G$ of the pairing interaction and the general two-body $J = 0$, or 'monopole', matrix element in the CS phase convention, is obtained directly from (8.29). Our result for both phase conventions is then

$$\boxed{\begin{aligned}\langle j\,j\,;\,0\,|V|\,j'\,j'\,;\,0\rangle &= -\tfrac{1}{2}\zeta^{(ll')}\widehat{j}\widehat{j'}G \,, \quad G > 0 \,,\\[4pt]\zeta^{(ll')} &= \begin{cases}(-1)^{l+l'} & \text{CS phase convention}\,,\\ 1 & \text{BR phase convention}\,.\end{cases}\end{aligned}} \tag{12.11}$$

## 12.3 Two-Particle Spectrum of the Pure Pairing Force

For a single $j$ shell the pairing Hamiltonian is given by (12.7). We proceed to diagonalize it. Let us choose the single-particle energy as $\varepsilon_j = 0$ and the two-particle basis states as

$$A_{jm}^\dagger|0\rangle \equiv c_{jm}^\dagger \tilde{c}_{jm}^\dagger|0\rangle \,, \quad m > 0 \,. \tag{12.12}$$

The elements of the Hamiltonian matrix are

$$\begin{aligned}\langle 0|A_{jm}V_{\text{PAIR}}A_{jm'}^\dagger|0\rangle &= -G \sum_{m_1 m_2 > 0} \langle 0|\tilde{c}_{jm}c_{jm}c_{jm_1}^\dagger \tilde{c}_{jm_1}^\dagger \tilde{c}_{jm_2} c_{jm_2} c_{jm'}^\dagger \tilde{c}_{jm'}^\dagger|0\rangle \\[4pt]&= -G \sum_{m_1 m_2 > 0} \tilde{c}_{jm}c_{jm}c_{jm_1}^\dagger \tilde{c}_{jm_1}^\dagger \tilde{c}_{jm_2} c_{jm_2} c_{jm'}^\dagger \tilde{c}_{jm'}^\dagger \\[4pt]&= -G \,.\end{aligned} \tag{12.13}$$

The Hamiltonian matrix is then

$$H_{\text{PAIR}} = -G \begin{pmatrix} 1 & 1 & \cdots & 1 \\ 1 & 1 & \cdots & 1 \\ \vdots & \vdots & \ddots & \vdots \\ 1 & 1 & \cdots & 1 \end{pmatrix} \,. \tag{12.14}$$

It is an $\Omega \times \Omega$ matrix, where

$$\Omega \equiv \frac{1}{2}(2j+1) \tag{12.15}$$

is the pair degeneracy.

We diagonalize the matrix (12.14). Calculating the characteristic determinant and setting it equal to zero we have

$$(-G\Omega - E)(-E)^{\Omega-1} = 0 . \tag{12.16}$$

This gives the eigenvalues

$$E_1 = -\Omega G , \quad E_i = 0 \text{ for } i = 2, 3, \dots, \Omega . \tag{12.17}$$

The eigenstate corresponding to the lowest eigenvalue can be written as

$$\Psi_1 = \frac{1}{\sqrt{\Omega}} \begin{pmatrix} 1 \\ 1 \\ \vdots \\ 1 \end{pmatrix} . \tag{12.18}$$

In occupation number representation the state (12.18) is

$$|\Psi_1\rangle = \frac{1}{\sqrt{\Omega}} \sum_{m>0} A_{jm}^\dagger |0\rangle = \frac{1}{\sqrt{2}} \sum_m \frac{(-1)^{j+m}}{\hat{j}} c_{jm}^\dagger c_{j,-m}^\dagger |0\rangle$$

$$= -\frac{1}{\sqrt{2}} [c_j^\dagger c_j^\dagger]_{00} |0\rangle = -|j^2 ; J = 0 , M = 0\rangle , \tag{12.19}$$

where the Clebsch–Gordan coefficient (1.34) was recognized. The result (12.19) shows that diagonalization of the pairing Hamiltonian in the uncoupled basis indeed gives a zero-coupled pair as the lowest eigenstate.

The simple two-nucleon spectrum (12.17) for the $0f_{7/2}$ shell is shown on the right in Fig. 12.2. For comparison, the experimental spectrum of $^{42}$Ca is shown on the left and a theoretical SDI spectrum in the middle. The pairing strength parameter $G = 0.750$ MeV has been fitted to a summary $J \neq 0$ excitation energy of 3.00 MeV and the SDI strength parameter $A_1 = 0.847$ MeV to $E(6^+) = 3.19$ MeV.

Figure 12.2 shows that the zero-range surface delta force is not able to account sufficiently for the spread of the $2^+$, $4^+$ and $6^+$ levels. A more realistic finite-range force would do better. For the pairing force there is no spread between the $J = 2, 4, 6$ states; they are degenerate. The situation is sometimes characterized by saying that 'the pairing force is shorter-ranged than the delta force'.

The behaviour of the pairing force is a direct consequence of its definition in Sect. 12.2, where the two-body matrix elements with $J \geq 2$ were omitted. Then the $J \neq 0$ pairs do not interact at all and remain at their unperturbed energy $2\varepsilon_j$, whereas the $J = 0$ pair interacts and is lowered in energy by the attractive force.

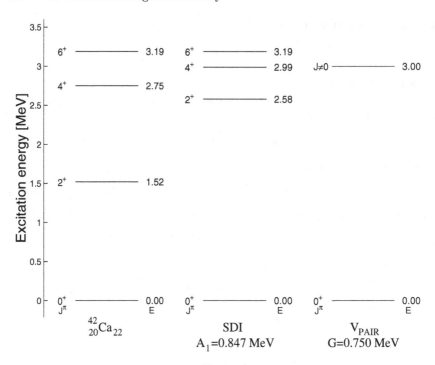

**Fig. 12.2.** Experimental spectrum of $^{42}$Ca compared with calculated SDI and pairing spectra

## 12.4 Seniority Model of the Pure Pairing Force

In this section we assume that there are $N$ nucleons that occupy a single $j$ shell, with a single-particle energy $\varepsilon_j = 0$. First we derive analytical expressions for the energies and degeneracies of nuclear states with good seniority. We then apply the formalism to nuclei whose valence nucleons occupy the $0f_{7/2}$ shell.

### 12.4.1 Derivation of the Seniority-Zero Spectrum

Here we derive the excitation spectrum for states with seniority zero. The derivation is based on the commutation relations between the pair operators and the pure pairing Hamiltonian (12.7). With reference to (12.12), we define a *pair creation operator* as

$$A^\dagger \equiv \frac{1}{\sqrt{\Omega}} \sum_{m>0} A^\dagger_{jm} = \frac{1}{\sqrt{\Omega}} \sum_{m>0} c^\dagger_{jm} \tilde{c}^\dagger_{jm} \ . \tag{12.20}$$

As seen from (12.19), this operator creates a zero-coupled pair. It can be used to write the pairing interaction (12.7) in the form

$$V_{\text{PAIR}} = -G\Omega A^\dagger A . \tag{12.21}$$

The *particle number operator* $\hat{n}$ is defined as

$$\hat{n} \equiv \sum_m c^\dagger_{jm} c_{jm} = \sum_{m>0} \left( c^\dagger_{jm} c_{jm} + \tilde{c}^\dagger_{jm} \tilde{c}_{jm} \right) . \tag{12.22}$$

Straightforward derivations give the commutation relations

$$
\begin{array}{|l|}
\hline
[A, A^\dagger] = 1 - \hat{n}/\Omega , \\
[A^\dagger, \hat{n}] = -2A^\dagger , \\
[V_{\text{PAIR}}, A^\dagger] = -GA^\dagger(\Omega - \hat{n}) = -G(\Omega - \hat{n} + 2)A^\dagger . \\
\hline
\end{array}
\tag{12.23}
$$

The commutator of the pair operator $A^\dagger$ with the pairing interaction $V_{\text{PAIR}}$ gives

$$
\begin{aligned}
V_{\text{PAIR}} A^\dagger |0\rangle &= [V_{\text{PAIR}}, A^\dagger]|0\rangle + A^\dagger V_{\text{PAIR}}|0\rangle \\
&= -G(\Omega - \hat{n} + 2)A^\dagger|0\rangle = -G\Omega A^\dagger|0\rangle .
\end{aligned}
\tag{12.24}
$$

Action of the pairing potential on a state of two pairs yields in the same way

$$
\begin{aligned}
V_{\text{PAIR}} (A^\dagger)^2|0\rangle &= [V_{\text{PAIR}}, A^\dagger]A^\dagger|0\rangle + A^\dagger V_{\text{PAIR}} A^\dagger|0\rangle \\
&= -G(\Omega - \hat{n} + 2 + \Omega)(A^\dagger)^2|0\rangle = -2G(\Omega - 1)(A^\dagger)^2|0\rangle .
\end{aligned}
\tag{12.25}
$$

From (12.24) and (12.25) we see how the procedure continues for any number of zero-coupled pairs. By induction one can prove the general result

$$V_{\text{PAIR}} (A^\dagger)^{N/2}|0\rangle = -\tfrac{1}{4}GN(2\Omega - N + 2)(A^\dagger)^{N/2}|0\rangle . \tag{12.26}$$

We now define a new quantum number, the *seniority* $v$, which is the number of nucleons not pairwise coupled to angular momentum zero, in short the number of unpaired nucleons. The fully paired state of $N$ nucleons is then

$$(A^\dagger)^{N/2}|0\rangle = |N, v = 0\rangle , \tag{12.27}$$

and (12.26) gives its energy

$$E_{v=0}(N) = -\tfrac{1}{4}GN(2\Omega - N + 2) . \tag{12.28}$$

### 12.4.2 Spectra of Seniority-One and Seniority-Two States

To extend the previous discussion of seniority-zero states to non-zero seniorities we define the $\Omega - 1$ operators $B^\dagger_J$, $J \neq 0$

$$B^\dagger_J \equiv \sqrt{2} \sum_{m>0} (-1)^{j+m}(j\,m\,j\,-m|J\,0)c^\dagger_{jm}\tilde{c}^\dagger_{jm} = \frac{1}{\sqrt{2}}[c^\dagger_j c^\dagger_j]_{J0} . \tag{12.29}$$

The operators $A^\dagger$ and $B_J^\dagger$ commute trivially, i.e.

$$[A^\dagger, B_J^\dagger] = 0 . \tag{12.30}$$

A further trivial relation is

$$V_{\text{PAIR}} B_J^\dagger |0\rangle = V_{\text{PAIR}} |N = 2, \, v = 2; \, J \, 0\rangle = 0 , \tag{12.31}$$

the zero coming from (12.17). Adding one paired couple yields, with use of (12.23),

$$V_{\text{PAIR}} A^\dagger B_J^\dagger |0\rangle = V_{\text{PAIR}} |N = 4, \, v = 2; \, J \, 0\rangle = -G A^\dagger (\Omega - \hat{n}) B_J^\dagger |0\rangle$$
$$= -G(\Omega - 2) A^\dagger B_J^\dagger |0\rangle = -G(\Omega - 2)|N = 4, \, v = 2; \, J \, 0\rangle . \tag{12.32}$$

By induction we obtain the general result

$$V_{\text{PAIR}} \left(A^\dagger\right)^{(N-2)/2} B_J^\dagger |0\rangle = V_{\text{PAIR}} |N, \, v = 2; \, J\rangle$$
$$= -\tfrac{1}{4} G(N-2)(2\Omega - N)|N, \, v = 2; \, J \, 0\rangle . \tag{12.33}$$

From this we read the energy of the seniority-two state as

$$E_{v=2}(N) = -\tfrac{1}{4} G(N-2)(2\Omega - N) . \tag{12.34}$$

Note that this energy is degenerate with respect to $J$.

For odd-mass nuclei we obtain in an analogous way

$$V_{\text{PAIR}} c_{jm}^\dagger |0\rangle = V_{\text{PAIR}} |N = 1, \, v = 1; \, j\,m\rangle = 0 , \tag{12.35}$$

$$V_{\text{PAIR}} A^\dagger c_{jm}^\dagger |0\rangle = V_{\text{PAIR}} |N = 3, \, v = 1; \, j\,m\rangle$$
$$= -G(\Omega - 1)|N = 3, \, v = 1; \, j\,m\rangle , \tag{12.36}$$

$$V_{\text{PAIR}} \left(A^\dagger\right)^{(N-1)/2} c_{jm}^\dagger |0\rangle = V_{\text{PAIR}} |N, \, v = 1; \, j\,m\rangle$$
$$= -\tfrac{1}{4} G(N-1)(2\Omega - N + 1)|N, \, v = 1; \, j\,m\rangle . \tag{12.37}$$

The $j$-degenerate energy of the seniority-one state is thus

$$E_{v=1}(N) = -\tfrac{1}{4} G(N-1)(2\Omega - N + 1) . \tag{12.38}$$

We have now constructed a full set of $v = 1$ and $v = 2$ states. States with $v > 2$ cannot be directly constructed by repeatedly operating with $B_J^\dagger$. The states so obtained would form an overcomplete and non-orthogonal set that would have to be separately orthogonalized. A more straightforward method is introduced in the following subsection.

## 12.4.3 States of Higher Seniority

We can circumvent the difficulty of overcompleteness and non-orthogonality by starting from the state of maximal seniority and proceeding towards lower seniority. Because all the unpaired couples are in the zero-energy state, as indicated by (12.17), we can write

$$V_{\text{PAIR}}|N, v = N\rangle = 0 . \tag{12.39}$$

With use of (12.23) this gives

$$
\begin{aligned}
V_{\text{PAIR}}|N, v = N - 2\rangle &= V_{\text{PAIR}} A^\dagger |N - 2, v = N - 2\rangle \\
&= A^\dagger V_{\text{PAIR}}|N - 2, v = N - 2\rangle - G A^\dagger (\Omega - \hat{n})|N - 2, v = N - 2\rangle \\
&= -G(\Omega - N + 2)|N, v = N - 2\rangle = -G(\Omega - v)|N, v = N - 2\rangle .
\end{aligned}
\tag{12.40}
$$

By induction we find the general expression

$$V_{\text{PAIR}}|N\,v\rangle = V_{\text{PAIR}}\left(A^\dagger\right)^{(N-v)/2}|v\,v\rangle = -\tfrac{1}{4}G(N - v)(2\Omega - N - v + 2)|N\,v\rangle . \tag{12.41}$$

The energy of a seniority-$v$ state is thus

$$\boxed{E_v(N) = -\tfrac{1}{4}G(N - v)(2\Omega - N - v + 2) .} \tag{12.42}$$

Our previous results for $v = 0, 1, 2$, equations (12.28), (12.38) and (12.34) respectively, are reproduced by this general formula.

Equation (12.42) provides us with the excitation energies, for even–even and odd-mass nuclei separately. For $N$ even, the excitation energy is

$$\boxed{E_v(N) - E_{v=0}(N) = \tfrac{1}{4}Gv(2\Omega - v + 2) ,} \tag{12.43}$$

while for $N$ odd it is

$$\boxed{E_v(N) - E_{v=1}(N) = \tfrac{1}{4}G\left[v(2\Omega - v + 2) - 2\Omega - 1\right] .} \tag{12.44}$$

It is interesting that these excitation energies are independent of the number $N$ of valence particles. This property of the seniority scheme suggests that theories that mix particle numbers, like the BCS theory of the following chapter, do not lose much in accuracy because of the mixing.

The states of the seniority model are highly degenerate. The degeneracy $D(v)$, stated here without proof, is

$$D(v) = \begin{pmatrix} 2\Omega \\ v \end{pmatrix} - \begin{pmatrix} 2\Omega \\ v - 2 \end{pmatrix} \theta(v - 2) , \quad N \leq \Omega , \tag{12.45}$$

where the column symbols are the usual binomial coefficients,

$$\binom{n}{m} = \frac{n!}{m!(n-m)!} \, , \tag{12.46}$$

and $\theta(x)$ is the Heaviside step function defined in (9.14). Note that the degeneracy is independent of the particle number $N$. For states with $N > \Omega$ the degeneracy is obtained by counting holes instead of particles, according to $N_{\text{hole}} = 2\Omega - N$. The same applies to the determination of the possible seniorities and angular momenta.

### 12.4.4 Application of the Seniority Model to $0f_{7/2}$-Shell Nuclei

Figure 12.3 shows the results for different particle numbers in the $0f_{7/2}$ shell within the seniority scheme. Up to midshell, $\Omega = 4$, we count particles and thereafter holes. The energies of the different seniority states are given by (12.43) and (12.44).

In the simple cases $N = 1, 2$ the connection between angular momentum and seniority is seen immediately. For $N = 1$ the only choice is trivially $v = 1$. For $N = 2$ the definition of seniority in Subsect. 12.4.1 means that the $J = 0$ state has $v = 0$ and the $J \neq 0$ states have $v = 2$. For a particle number $N > 2$ we must first find the possible angular momenta $J$. This can be done by the $m$-table technique mentioned in Sect. 1.3. Seniorities can then be assigned to angular momentum states by comparing the seniority degeneracies $D(v)$ and the angular momentum degeneracies $2J + 1$.

In the example of Fig. 12.3 the configuration $(\frac{7}{2})^3$ contains the angular momenta $J = \frac{3}{2}, \frac{5}{2}, \frac{7}{2}, \frac{9}{2}, \frac{11}{2}, \frac{15}{2}$. The degeneracy of the $J = \frac{7}{2}$ state is 8; on the other hand (12.45) gives $D(1) = 8$. The summed degeneracy of the remaining $J$ states is 48; on the other hand (12.45) gives $D(3) = 48$. We conclude that

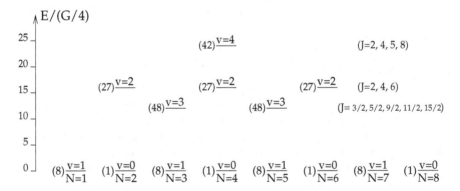

**Fig. 12.3.** Excitation spectra in the seniority scheme for different numbers $N$ of particles occupying the $0f_{7/2}$ shell. The seniority $v$ is indicated for each level. The numbers in parentheses to the left of the levels give the degeneracies. The angular momentum content of the levels is given on the far right

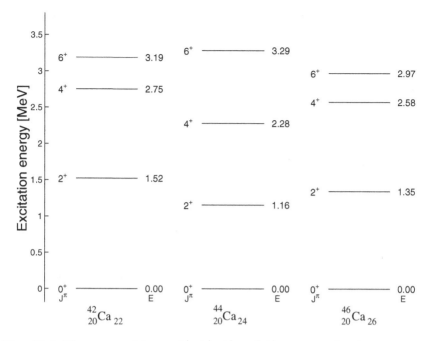

**Fig. 12.4.** The measured lowest $0^+$, $2^+$, $4^+$ and $6^+$ states in the three even–even calcium nuclei of the $0f_{7/2}$ shell

the $J = \frac{7}{2}$ state has seniority $v = 1$ while the other states have $v = 3$. The configuration $(\frac{7}{2})^4$ contains the angular momentum states $J = 0, 2^2, 4^2, 5, 6, 8$. Again by comparing degeneracies we find that the $J = 0$ state has $v = 0$, the $J = 2_1, 4_1, 6$ states have $v = 2$, and the $J = 2_2, 4_2, 5, 8$ states have $v = 4$.

The seniority scheme in the $j = 0f_{7/2}$ shell is partly realized in nature. This is seen from Fig. 12.4, where the ground state can be considered as a $v = 0$ state and the $J = 2, 4, 6$ states as $v = 2$ states. The excitation energies are roughly independent of the neutron number along the $0f_{7/2}$ shell, as predicted by the seniority model (see Fig. 12.3). On the other hand, the predicted $J = 2, 4, 6$ degeneracy is far from being realized. The degeneracy is broken by the true residual interaction which contains all multipole components, not only the monopole component of the pairing force. There are more convincing examples of the validity of the (generalized) seniority scheme [17]. A parade example are the even–even tin isotopes in the mass range $102 \le A \le 130$ [18].

## 12.5 The Two-Level Model

In this section we diagonalize the pure pairing Hamiltonian in a valence space of two $j$ shells, $j_1$ and $j_2$, for the seniority $v = 0$ states. To this end, we define for each of the two shells its own pair creation operator of the form (12.20). The two-level model is a nice test bench for approximate many-body theories. It will be used for this purpose in the chapters to come.

### 12.5.1 The Pair Basis

In the two-level model the starting point are the pair-creation operators for the two $j$ shells, $j_1$ and $j_2$. They are defined, in the pattern of (12.20), as

$$A_j^\dagger \equiv \frac{1}{\sqrt{2}} \left[ c_j^\dagger c_j^\dagger \right]_{00} = \frac{1}{\sqrt{\Omega_j}} \sum_{m>0} c_{jm}^\dagger \tilde{c}_{jm}^\dagger , \qquad \Omega_j \equiv \frac{1}{2}(2j+1) . \qquad (12.47)$$

The number operators for the two levels are, in the pattern of (12.22),

$$\hat{n}_j \equiv \sum_m c_{jm}^\dagger c_{jm} = \sum_{m>0} \left( c_{jm}^\dagger c_{jm} + \tilde{c}_{jm}^\dagger \tilde{c}_{jm} \right) . \qquad (12.48)$$

The first two commutation relations (12.23) become now generalized to

$$\boxed{\begin{aligned} \left[ A_{j_1}, A_{j_2}^\dagger \right] &= \delta_{j_1 j_2}(1 - \hat{n}_{j_1}/\Omega_{j_1}) , \\ \left[ A_{j_1}^\dagger, \hat{n}_{j_2} \right] &= -2\delta_{j_1 j_2} A_{j_1}^\dagger . \end{aligned}} \qquad (12.49)$$

For subsequent use we derive a number of further relations for the pair operators. By induction the commutation relations (12.49) lead to

$$\left[ A_j, \left( A_j^\dagger \right)^k \right] = k \left( A_j^\dagger \right)^{k-1} \left( 1 - \frac{k-1+\hat{n}_j}{\Omega_j} \right) , \qquad (12.50)$$

$$\left[ \left( A_j^\dagger \right)^k, \hat{n}_j \right] = -2k \left( A_j^\dagger \right)^k , \qquad (12.51)$$

Hermitian conjugation of these yields

$$\left[ \left( A_j \right)^k, A_j^\dagger \right] = k \left( 1 - \frac{k-1+\hat{n}_j}{\Omega_j} \right) \left( A_j \right)^{k-1} , \qquad (12.52)$$

$$\left[ \left( A_j \right)^k, \hat{n}_j \right] = 2k \left( A_j \right)^k . \qquad (12.53)$$

By means of (12.50) we have

$$A_j^\dagger A_j \left( A_j^\dagger \right)^k |0\rangle = A_j^\dagger \left[ A_j, \left( A_j^\dagger \right)^k \right] |0\rangle = k \left( 1 - \frac{k-1}{\Omega_j} \right) \left( A_j^\dagger \right)^k |0\rangle . \qquad (12.54)$$

We will use this result in deriving the matrix element of $V_{\text{PAIR}}$ in the next subsection. Apart from that, we note that the operator $A_j^\dagger A_j$ is an approximate number operator for zero-coupled $j$-type pairs. For $k = 0$ and $k = 1$ it is exact. For large $\Omega_j$ and small $k$ it is a good approximation; then the Pauli principle is not very effective and the pairs behave approximately as bosons.

Consider the $n$-pair state $\left(A_j^\dagger\right)^n|0\rangle$. It is obviously orthogonal to $\left(A_{j'}^\dagger\right)^k|0\rangle$ with $j' \neq j$ or $k \neq n$,

$$\langle 0|\left(A_{j'}\right)^k\left(A_j^\dagger\right)^n|0\rangle \propto \delta_{jj'}\delta_{kn} . \tag{12.55}$$

To calculate the norm of $\left(A_j^\dagger\right)^n|0\rangle$, we apply (12.50) repeatedly and find

$$\langle 0|\left(A_j\right)^n\left(A_j^\dagger\right)^n|0\rangle = \langle 0|\left[\left(A_j\right)^n, \left(A_j^\dagger\right)^n\right]|0\rangle$$
$$= [n]_j[n-1]_j \ldots [2]_j[1]_j = [n]_j! , \tag{12.56}$$

where

$$[n]_j \equiv n\left(1 - \frac{n-1}{\Omega_j}\right) . \tag{12.57}$$

The normalized $n$-pair state of $j$-shell nucleons is thus

$$|n\rangle = \frac{1}{\sqrt{[n]_j!}}\left(A_j^\dagger\right)^n|0\rangle . \tag{12.58}$$

Since all $j_1$ pair operators commute with all $j_2$ pair operators, we can write the normalized two-level basis states as

$$\boxed{|n\,m\rangle = \frac{1}{\sqrt{[n]_{j_1}![m]_{j_2}!}}\left(A_{j_1}^\dagger\right)^n\left(A_{j_2}^\dagger\right)^m|0\rangle .} \tag{12.59}$$

### 12.5.2 Matrix Elements of the Pairing Hamiltonian

The Hamiltonian of our two-level model consists of a one-body mean-field term and the pure pairing potential (12.8),

$$H_{\mathrm{PAIR}} = \varepsilon_{j_1}\hat{n}_{j_1} + \varepsilon_{j_2}\hat{n}_{j_2} + V_{\mathrm{PAIR}} , \quad \varepsilon_{j_1} < \varepsilon_{j_2} . \tag{12.60}$$

With (12.47) it becomes

$$H_{\mathrm{PAIR}} = \varepsilon_{j_1}\hat{n}_{j_1} + \varepsilon_{j_2}\hat{n}_{j_2} - G\sum_{jj'}\sqrt{\Omega_j\Omega_{j'}}\,A_j^\dagger A_{j'} , \tag{12.61}$$

where the summation indices take on the values $j_1, j_2$.

We are now in a position to derive the matrix element of $V_{\mathrm{PAIR}}$ in the basis (12.59):

$$\langle n\,m|V_{\mathrm{PAIR}}|n'\,m'\rangle = \frac{-G}{\sqrt{[n]_{j_1}![m]_{j_2}![n']_{j_1}![m']_{j_2}!}}\sum_{jj'}\sqrt{\Omega_j\Omega_{j'}}$$
$$\times \langle 0|\left(A_{j_2}\right)^m\left(A_{j_1}\right)^n A_j^\dagger A_{j'}\left(A_{j_1}^\dagger\right)^{n'}\left(A_{j_2}^\dagger\right)^{m'}|0\rangle \equiv V_{j_1j_1} + V_{j_1j_2} + V_{j_2j_1} + V_{j_2j_2} , \tag{12.62}$$

where the four terms label the contributions of the $jj'$ sum. The first term is

$$V_{j_1 j_1} = \frac{-G\Omega_{j_1}}{\sqrt{[n]_{j_1}! [m]_{j_2}! [n']_{j_1}! [m']_{j_2}!}} \langle 0| \left(A_{j_2}\right)^m \left(A_{j_1}\right)^n A_{j_1}^\dagger A_{j_1} \left(A_{j_1}^\dagger\right)^{n'} \left(A_{j_2}^\dagger\right)^{m'} |0\rangle .$$
(12.63)

Because the $j_1$ and $j_2$ operators commute, we can think of the vacuum as $|0\rangle = |0_{j_1}\rangle |0_{j_2}\rangle$. This justifies factorizing the matrix element as

$$\langle 0| \left(A_{j_2}\right)^m \left(A_{j_1}\right)^n A_{j_1}^\dagger A_{j_1} \left(A_{j_1}^\dagger\right)^{n'} \left(A_{j_2}^\dagger\right)^{m'} |0\rangle$$
$$= \langle 0| \left(A_{j_1}\right)^n A_{j_1}^\dagger A_{j_1} \left(A_{j_1}^\dagger\right)^{n'} |0\rangle \langle 0| \left(A_{j_2}\right)^m \left(A_{j_2}^\dagger\right)^{m'} |0\rangle .$$
(12.64)

With use of (12.54)–(12.57) the right-hand side becomes

$$[n']_{j_1} \langle 0| \left(A_{j_1}\right)^n \left(A_{j_1}^\dagger\right)^{n'} |0\rangle \langle 0| \left(A_{j_2}\right)^m \left(A_{j_2}^\dagger\right)^{m'} |0\rangle = [n']_{j_1} \delta_{nn'} [n]! \delta_{mm'} [m]! ,$$
(12.65)

whence (12.63) becomes

$$V_{j_1 j_1} = -G\Omega_{j_1} \delta_{nn'} \delta_{mm'} [n]_{j_1} .$$
(12.66)

In complete analogy we find

$$V_{j_2 j_2} = -G\Omega_{j_2} \delta_{nn'} \delta_{mm'} [m]_{j_2} .$$
(12.67)

Next we calculate the term

$$V_{j_1 j_2} = \frac{-G\sqrt{\Omega_{j_1} \Omega_{j_2}}}{\sqrt{[n]_{j_1}! [m]_{j_2}! [n']_{j_1}! [m']_{j_2}!}} \langle 0| \left(A_{j_2}\right)^m \left(A_{j_1}\right)^n A_{j_1}^\dagger A_{j_2} \left(A_{j_1}^\dagger\right)^{n'} \left(A_{j_2}^\dagger\right)^{m'} |0\rangle .$$
(12.68)

Factorizing the matrix element as in (12.64) and using (12.50), (12.52) and (12.55)–(12.57) we obtain

$$\langle 0| \left(A_{j_2}\right)^m \left(A_{j_1}\right)^n A_{j_1}^\dagger A_{j_2} \left(A_{j_1}^\dagger\right)^{n'} \left(A_{j_2}^\dagger\right)^{m'} |0\rangle$$
$$= \langle 0| \left(A_{j_1}\right)^n A_{j_1}^\dagger \left(A_{j_1}^\dagger\right)^{n'} |0\rangle \langle 0| \left(A_{j_2}\right)^m A_{j_2} \left(A_{j_2}^\dagger\right)^{m'} |0\rangle$$
$$= [n]_{j_1} \delta_{n-1,n'} [n-1]_{j_1}! [m']_{j_2} \delta_{m,m'-1} [m]_{j_2}! .$$
(12.69)

Substituting this into (12.68) yields

$$V_{j_1 j_2} = -G\sqrt{\Omega_{j_1} \Omega_{j_2}} \delta_{n',n-1} \delta_{m',m+1} \sqrt{[n]_{j_1} [m+1]_{j_2}} .$$
(12.70)

Similarly we find

$$V_{j_2 j_1} = -G\sqrt{\Omega_{j_1} \Omega_{j_2}} \delta_{n',n+1} \delta_{m',m-1} \sqrt{[n+1]_{j_1} [m]_{j_2}} .$$
(12.71)

The interaction $V_{\text{PAIR}}$ cannot change the total number of pairs, so within a calculation that number is a constant

$$\mathcal{N} = n + m \; . \tag{12.72}$$

Substituting $m = \mathcal{N} - n$ and collecting the terms (12.66), (12.67), (12.70) and (12.71), we have

$$
\begin{aligned}
\langle n\, m | V_{\text{PAIR}} | n'\, m' \rangle &= \langle n\; \mathcal{N} - n | V_{\text{PAIR}} | n'\; \mathcal{N} - n' \rangle \\
&= -G \delta_{nn'} \left( \Omega_{j_1}[n]_{j_1} + \Omega_{j_2}[\mathcal{N} - n]_{j_2} \right) \\
&\quad - G \sqrt{\Omega_{j_1} \Omega_{j_2}} \Big( \delta_{n',n-1} \sqrt{[n]_{j_1}[\mathcal{N} - n + 1]_{j_2}} \\
&\qquad\qquad + \delta_{n',n+1} \sqrt{[n+1]_{j_1}[\mathcal{N} - n]_{j_2}} \; \Big) \; .
\end{aligned}
\tag{12.73}
$$

The matrix element of the mean-field part of $H_{\text{PAIR}}$ in (12.60) is diagonal, and we can write directly

$$\langle n\; \mathcal{N} - n | \varepsilon_{j_1} \hat{n}_{j_1} + \varepsilon_{j_2} \hat{n}_{j_2} | n'\; \mathcal{N} - n' \rangle = \delta_{nn'} \left[ 2n\varepsilon_{j_1} + 2(\mathcal{N} - n)\varepsilon_{j_2} \right] \; . \tag{12.74}$$

The matrix element of $H_{\text{PAIR}}$ now becomes, with the quantities (12.57) expanded,

$$
\boxed{
\begin{aligned}
\langle n\; \mathcal{N} - n | H_{\text{PAIR}} | n'\; \mathcal{N} - n' \rangle &= \delta_{nn'} \Big\{ 2n\varepsilon_{j_1} + 2(\mathcal{N} - n)\varepsilon_{j_2} \\
&\quad - G\big[ n(\Omega_{j_1} - n + 1) + (\mathcal{N} - n)(\Omega_{j_2} - \mathcal{N} + n + 1) \big] \Big\} \\
&\quad - G\delta_{n',n-1} \sqrt{n(\Omega_{j_1} - n + 1)(\mathcal{N} - n + 1)(\Omega_{j_2} - \mathcal{N} + n)} \\
&\quad - G\delta_{n',n+1} \sqrt{(n+1)(\Omega_{j_1} - n)(\mathcal{N} - n)(\Omega_{j_2} - \mathcal{N} + n + 1)} \; .
\end{aligned}
}
\tag{12.75}
$$

When using this formula one must remember that $\mathcal{N}$ is the number of nucleon pairs, i.e. $\mathcal{N} = N/2$, where $N$ is the number of nucleons occupying the two-level system. The quantity $n$ stands for the number of pairs occupying the lower level of degeneracy $2\Omega_{j_1}$.

Equation (12.75) gives the matrix elements of the pairing Hamiltonian in the seniority $v = 0$ basis

$$\{ |\mathcal{N}\; 0\rangle, \; |\mathcal{N} - 1\; 1\rangle, \; |\mathcal{N} - 2\; 2\rangle, \; \ldots, \; |0\; \mathcal{N}\rangle \} \; , \tag{12.76}$$

listed from lowest to highest energy ($\varepsilon_{j_1} < \varepsilon_{j_2}$). Diagonalization in this basis results in a total of $\mathcal{N}$ seniority $v = 0$ states. The one with the lowest energy is the ground state

$$|0^+_{\text{gs}}\rangle = \sum_n X_n^{(\text{gs})} |n\; \mathcal{N} - n\rangle \; , \tag{12.77}$$

which can be compared e.g. with the BCS ground state, to be discussed in the next chapter.

A two-level model where

$$\Omega_{j_1} = \Omega_{j_2} \equiv \Omega , \quad \varepsilon_{j_1} = -\tfrac{1}{2}\varepsilon , \quad \varepsilon_{j_2} = +\tfrac{1}{2}\varepsilon , \quad N = 2\Omega \qquad (12.78)$$

is called the Lipkin–Meshkov–Glick model, or simply the *Lipkin model* [68]. The Lipkin model is exactly solvable for many kinds of Hamiltonian, which makes it a popular test bench for various nuclear many-body theories.

A broken-pair operator of the type (12.29) can be added to the two-level model in a straightforward way. The broken pair can occupy either the lower or the upper level. We can form the Hamiltonian matrix containing one broken pair and any number of zero-coupled pairs by using the methods of this and the previous section. Diagonalization of the matrix will then yield the spectrum of the seniority $v = 2$ states, analogously to the seniority model for a single $j$ orbital.

### 12.5.3 Application to a Two-Particle System

Consider a very simple application of the two-level formalism derived in the previous subsection. We take just two particles that occupy the two-level system. Then $\mathcal{N} = 1$ and the basis states (12.76) are

$$\{|1\,0\rangle , |0\,1\rangle\} . \qquad (12.79)$$

Equation (12.75) gives the Hamiltonian matrix elements

$$\langle 1\,0|H_{\mathrm{PAIR}}|1\,0\rangle = 2\varepsilon_{j_1} - G\Omega_{j_1} , \qquad (12.80)$$

$$\langle 1\,0|H_{\mathrm{PAIR}}|0\,1\rangle = -G\sqrt{\Omega_{j_1}\Omega_{j_2}} , \qquad (12.81)$$

$$\langle 0\,1|H_{\mathrm{PAIR}}|1\,0\rangle = -G\sqrt{\Omega_{j_1}\Omega_{j_2}} , \qquad (12.82)$$

$$\langle 0\,1|H_{\mathrm{PAIR}}|0\,1\rangle = 2\varepsilon_{j_2} - G\Omega_{j_2} . \qquad (12.83)$$

Taking for simplicity $j_1 = j_2$ (but $\varepsilon_{j_1} < \varepsilon_{j_2}$) and defining $\Omega_{j_1} = \Omega_{j_2} \equiv \Omega$, we have the Hamiltonian matrix

$$H_{\mathrm{PAIR}} = \begin{pmatrix} 2\varepsilon_{j_1} - g & -g \\ -g & 2\varepsilon_{j_2} - g \end{pmatrix} , \qquad (12.84)$$

where $g \equiv G\Omega$. The eigenvalues are, from (8.85),

$$E_{\mp} = \varepsilon_{j_1} + \varepsilon_{j_2} - g \mp \sqrt{(\varepsilon_{j_2} - \varepsilon_{j_1})^2 + g^2} . \qquad (12.85)$$

We lose no generality by adopting the values[1] $\varepsilon_{j_1} = -\tfrac{1}{2}\varepsilon$ and $\varepsilon_{j_2} = +\tfrac{1}{2}\varepsilon$. This leads to an extremely simple expression for the energies of the $v = 0$ states in the two-particle system, namely

$$E_{\mp} = -g \mp \sqrt{\varepsilon^2 + g^2} . \qquad (12.86)$$

---

[1] The use of symmetric shell energies is the traditional way to discuss two-level systems with the upper and lower levels having the same degeneracy.

## 12.6 Two Particles in a Valence Space of Many $j$ Shells

In this section we consider a pair of particles occupying several $j$ shells with single-particle energies $\varepsilon_j$ and interacting through the pairing force (12.8).

### 12.6.1 Dispersion Equation

The complete Hamiltonian (12.61) is now generalized to

$$H_{\mathrm{PAIR}} = \sum_j \varepsilon_j \hat{n}_j - G \sum_{jj'} \sqrt{\Omega_j \Omega_{j'}} A_j^\dagger A_{j'} . \tag{12.87}$$

With the pair structure given by (12.12), the matrix element of the pairing potential (12.8) becomes

$$\langle 0|A_{jm} V_{\mathrm{PAIR}} A_{j'm'}^\dagger|0\rangle$$
$$= -G \sum_{\substack{j_1 j_2 \\ m_1 m_2 > 0}} \langle 0|\tilde{c}_{jm} c_{jm} c_{j_1 m_1}^\dagger \tilde{c}_{j_1 m_1}^\dagger \tilde{c}_{j_2 m_2} c_{j_2 m_2} c_{j'm'}^\dagger \tilde{c}_{j'm'}^\dagger|0\rangle = -G . \tag{12.88}$$

This is an immediate extension of the single-$j$ shell result (12.13). The matrix element of the complete Hamiltonian (12.87) is

$$\langle 0|A_{jm} H_{\mathrm{PAIR}} A_{j'm'}^\dagger|0\rangle = 2\varepsilon_j \delta_{jj'} \delta_{mm'} - G \equiv H_{jm,j'm'} , \quad m, m' > 0 . \tag{12.89}$$

For the Hamiltonian matrix elements (12.89) the eigenvalue equation

$$\sum_{j',m'>0} H_{jm,j'm'} X_{j'm'}^\omega = E_\omega X_{jm}^\omega \tag{12.90}$$

becomes

$$2\varepsilon_j X_{jm}^\omega - G \sum_{j',m'>0} X_{j'm'}^\omega = E_\omega X_{jm}^\omega . \tag{12.91}$$

This can be rewritten as

$$X_{jm}^\omega = \frac{G}{2\varepsilon_j - E_\omega} N_\omega , \tag{12.92}$$

where

$$\sum_{j',m'>0} X_{j'm'}^\omega \equiv N_\omega . \tag{12.93}$$

When we substitute $X_{jm}^\omega$ from (12.92) into (12.93), the constant $N_\omega$ appears on both sides and can be cancelled, and we are left with

$$1 = \sum_{j',m'>0} \frac{G}{2\varepsilon_{j'} - E_\omega} = \sum_{j'} \frac{G\Omega_{j'}}{2\varepsilon_{j'} - E_\omega} . \tag{12.94}$$

The final result is thus the *dispersion equation*

$$\frac{1}{G} = \sum_j \frac{\Omega_j}{2\varepsilon_j - E_\omega} \cdot$$

(12.95)

The solutions of (12.95) are the energies of the seniority $v = 0$ states for two nucleons in a multi-$j$ valence space. The seniority $v = 2$ states are left at their unperturbed energies $2\varepsilon_j$. Equation (12.95) is similar to the TDA dispersion equation (9.55) and the RPA dispersion equation (11.125). Its solutions can be obtained by the same graphical method as was used in the previous cases.

The simplest example of the use of the dispersion equation (12.95) is the case of two levels with equal degeneracy. Denoting $\Omega_{j_1} = \Omega_{j_2} \equiv \Omega$ we have

$$\frac{1}{G} = \frac{\Omega}{2\varepsilon_{j_1} - E_\omega} + \frac{\Omega}{2\varepsilon_{j_2} - E_\omega} \cdot$$

(12.96)

This is a second-degree equation, so we have an exact solution. As one would expect, the resulting energies $E_\omega$ are the same as those given by (12.85).

### 12.6.2 The Three-Level Case

Let us apply the dispersion equation (12.95) to a three-level case. We can take this case to be d-s shell with single-particle energies

$$\varepsilon_1 = \varepsilon_{0d_{5/2}} = 0 , \quad \varepsilon_2 = \varepsilon_{1s_{1/2}} = 1.0 \,\text{MeV} , \quad \varepsilon_3 = \varepsilon_{0d_{3/2}} = 5.0 \,\text{MeV} .$$

(12.97)

We then obtain from (12.95) the dispersion equation

$$\frac{1}{G} = \frac{3}{0 - E_\omega} + \frac{1}{2.0 \,\text{MeV} - E_\omega} + \frac{2}{10.0 \,\text{MeV} - E_\omega} \cdot$$

(12.98)

The graphical solution of (12.98) is shown in Fig. 12.5 for $G = 1.0 \,\text{MeV}$ and $G = 2.0 \,\text{MeV}$. The energies $E_\omega$ are also listed in Table 12.2, with an accuracy obtained from a numerical solution. Note that the broken-pair states stay at their unperturbed energies; they are the $J = 2, 4$ states at zero excitation energy and the $J = 2$ state at 10.0 MeV.

**Table 12.2.** Numerical solutions $E_\omega$ of the dispersion equation (12.98) for two different values of the pairing strength $G$

| $G$ (MeV) | $E_1$ (MeV) | $E_2$ (MeV) | $E_3$ (MeV) |
|---|---|---|---|
| 1.0 | −4.281 | 1.618 | 8.663 |
| 2.0 | −9.617 | 1.546 | 8.070 |

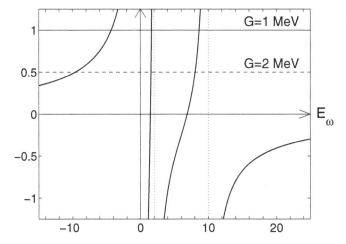

**Fig. 12.5.** Graphical solution of the dispersion equation (12.98) for pairing strengths $G = 1.0\,\mathrm{MeV}$ and $G = 2.0\,\mathrm{MeV}$. The energy $E_\omega$ is in MeV

## Epilogue

In this chapter we have tried to justify the use of a simple pure pairing force in the description of open-shell vibrational nuclei. The justification came in two ways: first by observing experimental evidence of nucleon pairing, secondly by being able to develop exactly solvable models for testing nuclear many-body approaches. In the following chapters these simple models are used to compare their exact solutions with the approximate solutions provided by nuclear models of various degrees of complexity.

## Exercises

**12.1.** Work out the details of the derivation (12.5).

**12.2.** Using the SDI expression (8.71), which is in the CS phase convention, compute the two-body matrix element in (12.9) and note that its sign depends on the $l$ values involved.

**12.3.** Derive the characteristic equation (12.16).

**12.4.** Verify the form of the eigenstate in (12.18) and (12.19).

**12.5.** Derive the commutation relations (12.23) by starting from (12.20)–(12.22).

**12.6.** Prove by induction the general expression (12.26) for the action of the pair potential.

**12.7.** Prove by induction the general expression (12.33) for the action of the pair potential.

**12.8.** Prove by induction the general expression (12.37) for the action of the pair potential.

**12.9.** Prove by induction the general expression (12.41) for the action of the pair potential.

**12.10.** Draw the excitation spectra for the $0g_{9/2}$ shell in the seniority scheme on the pattern of Fig. 12.3. You need not indicate the angular momentum content of the states.

**12.11.** Determine the angular momentum content of the seniority states of Exercise 12.10 by using the $m$-table technique and reasoning as in Subsect. 12.4.4.

**12.12.** Derive the commutation relation (12.50).

**12.13.** Derive the commutation relation (12.51).

**12.14.** Derive the commutation relation (12.52).

**12.15.** Derive the relation (12.54).

**12.16.** Derive the ground-state expectation value (12.56).

**12.17.** Work out the details of the derivation of (12.67).

**12.18.** Work out the details of the derivation of (12.71).

**12.19.** Compute the energy spectrum of the Lipkin model for two $j = \frac{3}{2}$ shells. Use the values
(a) $\varepsilon = G$,
(b) $\varepsilon = 2G$.

**12.20.** Compute the energies of the seniority-zero states for different numbers of particle pairs in a valence space consisting of the $0p_{3/2}$ and $0p_{1/2}$ orbitals. Take $\varepsilon_{0p_{3/2}} = 0$ and $\varepsilon_{0p_{1/2}} = 6.0\,\mathrm{MeV}$ and
(a) $G = 1.0\,\mathrm{MeV}$,
(b) $G = 2.0\,\mathrm{MeV}$.

**12.21.** Compute the energies of the seniority-zero states for different numbers of particle pairs in a valence space consisting of the $0d_{3/2}$ and $0f_{7/2}$ orbitals. Take $\varepsilon_{0d_{3/2}} = 0$ and $\varepsilon_{0f_{7/2}} = 4.0\,\mathrm{MeV}$ and
(a) $G = 1.0\,\mathrm{MeV}$,
(b) $G = 2.0\,\mathrm{MeV}$.

**12.22.** Consider the $1p$-$0f_{5/2}$-$0g_{9/2}$ shells with the energies $\varepsilon_{1p_{3/2}} = 0$, $\varepsilon_{1p_{1/2}} = 2.02\,\mathrm{MeV}$, $\varepsilon_{0f_{5/2}} = 3.60\,\mathrm{MeV}$ and $\varepsilon_{0g_{9/2}} = 4.00\,\mathrm{MeV}$. Find the energies of the $0^+$ states of two particles distributed on these orbitals when the two-body force is the pure pairing force with strength
(a) $G = 1.0\,\mathrm{MeV}$,
(b) $G = 2.0\,\mathrm{MeV}$.

# 13

# BCS Theory

## Prologue

Up to now we have treated vibrational open-shell nuclei within simplified schemes using the pure pairing Hamiltonian. This Hamiltonian is far too simple for a realistic description of excitation spectra, but it has the advantage of providing exact solutions within simplified approaches such as the seniority scheme and the two-level model discussed in the preceding chapter.

In this section the concept of quasiparticle becomes an active working tool. We introduce BCS quasiparticles through the transformation leading to them. These quasiparticles are made out of particle and hole components with certain occupation amplitudes. BCS theory is an advanced many-body theory, but it can be used with the simple pairing force. However, the framework provides for a generalization to any two-body interaction. The meaning of the quasiparticle mean field and single-quasiparticle energies will be addressed. BCS solutions will be discussed both generally and within simple, exactly solvable models.

## 13.1 BCS Quasiparticles and Their Vacuum

The seniority model of Sect. 12.4 can be generalized to a model where $N$ particles can occupy several $j$ shells. This model is called the *generalized seniority* model [18]. Here we do not want to discuss that model, but instead introduce a realistic microscopic model known as the BCS model or BCS theory.

The BCS theory was first introduced by Bardeen, Cooper and Schrieffer in 1957 [69] for microscopic description of the superconductivity of metals. In superconducting metals the long-range effective attraction between two electrons is mediated by quantized lattice vibrations, lattice phonons. The correlated electrons form pairs with total spin zero. These effective bosonic pairs condense to the ground state, with a resultant energy gap to the excited states. At low temperatures, electrons scattering from the lattice cannot overcome this energy gap. They stay in the ground state and traverse the bulk metal without losing energy in collisions with the lattice. Therefore an electric current can flow with no resistance.

As discussed in Sect. 12.1, there is experimental evidence for the presence of a similar collective condensate in atomic nuclei. The valence nucleons of a nucleus feel a strong attractive force which stems from the short-range component of the nucleon–nucleon interaction. This short-range attraction was mimicked by the pure pairing force in Chap. 12. Having recognized the pairing phenomenon in nuclei, Bohr, Mottelson and Pines [70] and Belyaev [71] in 1958–59 proposed to apply BCS theory to nuclei. The theory has become a standard part of the description of nuclear structure.

### 13.1.1 The BCS Ground State

The BCS theory can be viewed as a Rayleigh–Ritz variational problem with suitably parametrized ansatz wave functions. Then the best wave function is found by varying the parameters to achieve minimum energy. We thus write an ansatz for the *BCS ground state* as

$$|\text{BCS}\rangle = \prod_{\alpha>0} \left( u_a - v_a A_\alpha^\dagger \right) |\text{CORE}\rangle \,, \tag{13.1}$$

where $u_a$ and $v_a$ are variational parameters and the operator (12.12),

$$A_\alpha^\dagger = c_\alpha^\dagger \tilde{c}_\alpha^\dagger \,, \tag{13.2}$$

creates a pair of like nucleons. Here we use the Baranger notation, defined in (3.62),

$$\alpha = (a, m_\alpha) \,, \quad a = (n_a, l_a, j_a) \,. \tag{13.3}$$

As denoted, we have made the natural choice for spherical nuclei that the $u$ and $v$ parameters are independent of the projection quantum number $m_\alpha$.

The BCS state (13.1) does not have a good particle number. However, we can decompose it into a sum of eigenstates $|N\rangle$. To that end, consider an exponential operator function and expand it in a Taylor series:

$$\exp\left( -\frac{v_a}{u_a} A_\alpha^\dagger \right) = \exp\left( -\frac{v_a}{u_a} c_\alpha^\dagger \tilde{c}_\alpha^\dagger \right) = 1 - \frac{v_a}{u_a} c_\alpha^\dagger \tilde{c}_\alpha^\dagger + 0 = 1 - \frac{v_a}{u_a} A_\alpha^\dagger \,, \tag{13.4}$$

where the zero results from the anticommutation relations (4.9). By using (13.4) we can rewrite the BCS ground state (13.1) as

$$|\text{BCS}\rangle = \prod_{\alpha>0} u_a \left(1 - \frac{v_a}{u_a}A_\alpha^\dagger\right)|\text{CORE}\rangle = \prod_{\alpha>0} u_a \exp\left(-\frac{v_a}{u_a}A_\alpha^\dagger\right)|\text{CORE}\rangle$$

$$= \prod_{\beta>0} u_b \prod_{\alpha>0} \exp\left(-\frac{v_a}{u_a}A_\alpha^\dagger\right)|\text{CORE}\rangle = \prod_{\beta>0} u_b \exp\left(-\sum_{\alpha>0}\frac{v_a}{u_a}A_\alpha^\dagger\right)|\text{CORE}\rangle$$

$$= \prod_{\beta>0} u_b \sum_n \frac{1}{n!}\left(-\sum_{\alpha>0}\frac{v_a}{u_a}A_\alpha^\dagger\right)^n|\text{CORE}\rangle . \tag{13.5}$$

To highlight the dependence on nucleon number $N$ we write the final result in the form

$$|\text{BCS}\rangle = \prod_{\beta>0} u_b \sum_{N=\text{even}} \frac{1}{(N/2)!}|N\rangle , \tag{13.6}$$

where

$$|N\rangle \equiv \left(-\sum_{\alpha>0}\frac{v_a}{u_a}A_\alpha^\dagger\right)^{N/2}|\text{CORE}\rangle \tag{13.7}$$

is an unnormalized eigenstate of the nucleon number. The seniority model of Sect. 12.4 showed only weak dependence on particle number. This suggests that the lack of good nucleon number in the state (13.6) may not be a severe shortcoming.

We require the state (13.1) to be normalized. It follows that for all $a$

$$|u_a|^2 + |v_a|^2 = 1 . \tag{13.8}$$

The amplitudes $u_a$ and $v_a$ are chosen to be real, so the normalization condition is

$$\boxed{u_a^2 + v_a^2 = 1 \quad \text{for all } a .} \tag{13.9}$$

### 13.1.2 BCS Quasiparticles

The BCS state (13.1) is much easier to work with than the number-conserving state $|N\rangle$. This is because $|\text{BCS}\rangle$ is the vacuum for *BCS quasiparticles*. We define $a_\alpha^\dagger$ as the operator that creates a quasiparticle in orbital $\alpha$. The corresponding annihilation operator is $a_\alpha$, and $\tilde{a}_\alpha = (-1)^{j_a+m_\alpha}a_{-\alpha}$ is its companion with good tensorial properties according to (4.23). The quasiparticle operators are linear combinations of particle operators via the *Bogoliubov–Valatin transformation*

$$\boxed{\begin{aligned} a_\alpha^\dagger &= u_a c_\alpha^\dagger + v_a \tilde{c}_\alpha , \\ \tilde{a}_\alpha &= u_a \tilde{c}_\alpha - v_a c_\alpha^\dagger , \end{aligned}} \tag{13.10}$$

introduced separately c. 1960 by Bogoliubov [72] and Valatin [73].

The Hermitian conjugates of the relations (13.10) are

$$a_\alpha = u_a c_\alpha + v_a \tilde{c}_\alpha^\dagger , \quad \tilde{a}_\alpha^\dagger = u_a \tilde{c}_\alpha^\dagger - v_a c_\alpha . \tag{13.11}$$

We invert the relations (13.10) and use the normalization condition (13.9). The result is

$$c_\alpha^\dagger = u_a a_\alpha^\dagger - v_a \tilde{a}_\alpha \ , \quad \tilde{c}_\alpha = u_a \tilde{a}_\alpha + v_a a_\alpha^\dagger \ . \tag{13.12}$$

The Hermitian conjugates of these relations are

$$c_\alpha = u_a a_\alpha - v_a \tilde{a}_\alpha^\dagger \ , \quad \tilde{c}_\alpha^\dagger = u_a \tilde{a}_\alpha^\dagger + v_a a_\alpha \ . \tag{13.13}$$

From the anticommutation relations (4.9), the transformation equations (13.10) and (13.11) and the normalization condition (13.9) we obtain for BCS quasiparticles the anticommutation relations

$$\boxed{\left\{a_\alpha^\dagger, a_\beta^\dagger\right\} = 0 \ , \quad \left\{a_\alpha, a_\beta\right\} = 0 \ , \quad \left\{a_\alpha, a_\beta^\dagger\right\} = \delta_{\alpha\beta} \ .} \tag{13.14}$$

Thus the quasiparticles are fermions just like the particles they are built from. A transformation that preserves the form of the basic commutation relations is known as a *quantum-mechanical canonical transformation*.

Operating with $a_\alpha$ or $\tilde{a}_\alpha$ on the BCS ground state (13.1) gives

$$\boxed{a_\alpha|\text{BCS}\rangle = 0 \ , \quad \tilde{a}_\alpha|\text{BCS}\rangle = 0 \ ,} \tag{13.15}$$

so $|\text{BCS}\rangle$ can be appropriately called the *BCS vacuum*.

The case of particles and holes, as introduced in Sect. 4.4, is recovered when certain coefficients are $u_a, v_a = 1, 0$ and others are $u_b, v_b = 0, 1$. In this case we have

$$a_\alpha^\dagger = c_\alpha^\dagger \ , \ a_\alpha = c_\alpha \ ; \quad a_\beta^\dagger = \tilde{c}_\beta \ , \ a_\beta = \tilde{c}_\beta^\dagger \ . \tag{13.16}$$

The vacuum $|\text{BCS}\rangle$ is now the particle–hole vacuum $|\text{HF}\rangle$, and the quasiparticle creation operator $a_\alpha^\dagger$ creates a particle above the Fermi surface and the quasiparticle creation operator $a_\beta^\dagger$ creates a hole below the Fermi surface.

In the normal BCS case each operator $a_\alpha^\dagger$ creates a quasiparticle that is a particle with probability amplitude $u_a$ and a hole with probability amplitude $v_a$. This is understood so that the single-particle orbital $\alpha$ is empty with a probability $u_a^2$ and occupied with a probability $v_a^2$, with the proper probability normalization (13.9). Therefore $v_a$ is called the *occupation amplitude* and $u_a$ the *unoccupation amplitude* of the orbital $\alpha$; generically both are called occupation amplitudes. In all, a $j$ orbital is thus occupied by $(2j+1)v_j^2$ particles and $(2j+1)u_j^2$ holes.

## 13.2 Occupation Number Representation for BCS Quasiparticles

In Sect. 4.1 we discussed occupation number representation in particle space, with the vacuum $|0\rangle$, and in Sect. 4.4 we discussed the particle–hole representation, with the vacuum $|\text{HF}\rangle$. We now proceed in a similar way with quasiparticle operators and define their normal ordering and contractions relative to the BCS vacuum $|\text{BCS}\rangle$.

## 13.2.1 Contraction Properties

For BCS quasiparticles, normal ordering and contractions of operators are done, by definition, relative to the BCS ground state $|\mathrm{BCS}\rangle$. The contractions of quasiparticle creation and annihilation operators can be written immediately as

$$\boxed{\;\contraction{}{a}{_\alpha}{a} a_\alpha a_\beta^\dagger = \langle \mathrm{BCS}|a_\alpha a_\beta^\dagger|\mathrm{BCS}\rangle = \delta_{\alpha\beta}\;, \quad \text{other contractions} = 0\;.\;}$$
(13.17)

With these quasiparticle contractions and the transformation equations (13.12) and (13.13) we can find the contractions of the particle operators with respect to the BCS vacuum, with the result

$$\contraction{}{c}{_\alpha}{c} c_\alpha c_\beta^\dagger = u_a^2 \delta_{\alpha\beta}\;, \quad \contraction{}{c}{_\alpha^\dagger}{c} c_\alpha^\dagger c_\beta = v_a^2 \delta_{\alpha\beta}\;,$$
(13.18)

$$\contraction{}{c}{_\alpha^\dagger}{c} c_\alpha^\dagger c_\beta^\dagger = u_a v_a (-1)^{j_a - m_\alpha} \delta_{\alpha,-\beta}\;,$$
(13.19)

$$\contraction{}{c}{_\alpha}{c} c_\alpha c_\beta = u_a v_a (-1)^{j_a + m_\alpha} \delta_{\alpha,-\beta}\;.$$
(13.20)

These contractions are sometimes useful, as in the derivation of the quasiparticle representation of the nuclear Hamiltonian in the next subsection.

## 13.2.2 Quasiparticle Representation of the Nuclear Hamiltonian

Let us now apply Wick's theorem (4.41) to the nuclear Hamiltonian $H = T + V$ as was done in Sect. 4.5. There we managed to decompose the Hamiltonian into a one-body part representing the Hartree–Fock mean field and a residual interaction, as shown in (4.70)–(4.72). Here we set out to do likewise, namely to write the Hamiltonian as a sum of a one-quasiparticle term, describing the quasiparticle mean field, and a residual interaction.

For the two-body part, Wick's theorem gives

$$4V = \sum_{\alpha\beta\gamma\delta} \bar{v}_{\alpha\beta\gamma\delta} c_\alpha^\dagger c_\beta^\dagger c_\delta c_\gamma = \sum_{\alpha\beta\gamma\delta} \bar{v}_{\alpha\beta\gamma\delta} \mathcal{N}\big[c_\alpha^\dagger c_\beta^\dagger c_\delta c_\gamma\big]$$

$$+ \sum_{\alpha\beta\gamma\delta} \bar{v}_{\alpha\beta\gamma\delta} \langle \mathrm{BCS}|c_\alpha^\dagger c_\beta^\dagger|\mathrm{BCS}\rangle \mathcal{N}\big[c_\delta c_\gamma\big] - \sum_{\alpha\beta\gamma\delta} \bar{v}_{\alpha\beta\gamma\delta} \langle \mathrm{BCS}|c_\alpha^\dagger c_\delta|\mathrm{BCS}\rangle \mathcal{N}\big[c_\beta^\dagger c_\gamma\big]$$

$$+ \sum_{\alpha\beta\gamma\delta} \bar{v}_{\alpha\beta\gamma\delta} \langle \mathrm{BCS}|c_\alpha^\dagger c_\gamma|\mathrm{BCS}\rangle \mathcal{N}\big[c_\beta^\dagger c_\delta\big] + \sum_{\alpha\beta\gamma\delta} \bar{v}_{\alpha\beta\gamma\delta} \langle \mathrm{BCS}|c_\beta^\dagger c_\delta|\mathrm{BCS}\rangle \mathcal{N}\big[c_\alpha^\dagger c_\gamma\big]$$

$$- \sum_{\alpha\beta\gamma\delta} \bar{v}_{\alpha\beta\gamma\delta} \langle \mathrm{BCS}|c_\beta^\dagger c_\gamma|\mathrm{BCS}\rangle \mathcal{N}\big[c_\alpha^\dagger c_\delta\big] + \sum_{\alpha\beta\gamma\delta} \bar{v}_{\alpha\beta\gamma\delta} \langle \mathrm{BCS}|c_\delta c_\gamma|\mathrm{BCS}\rangle \mathcal{N}\big[c_\alpha^\dagger c_\beta^\dagger\big]$$

$$+ \sum_{\alpha\beta\gamma\delta} \bar{v}_{\alpha\beta\gamma\delta} \langle \mathrm{BCS}|c_\alpha^\dagger c_\beta^\dagger|\mathrm{BCS}\rangle \langle \mathrm{BCS}|c_\delta c_\gamma|\mathrm{BCS}\rangle$$

$$- \sum_{\alpha\beta\gamma\delta} \bar{v}_{\alpha\beta\gamma\delta} \langle \mathrm{BCS}|c_\alpha^\dagger c_\delta|\mathrm{BCS}\rangle \langle \mathrm{BCS}|c_\beta^\dagger c_\gamma|\mathrm{BCS}\rangle$$

$$+ \sum_{\alpha\beta\gamma\delta} \bar{v}_{\alpha\beta\gamma\delta} \langle \mathrm{BCS}|c_\alpha^\dagger c_\gamma|\mathrm{BCS}\rangle \langle \mathrm{BCS}|c_\beta^\dagger c_\delta|\mathrm{BCS}\rangle \ . \tag{13.21}$$

Next we use (13.12) and (13.13), (13.18)–(13.20) and the definition of normal-ordered product (4.31). Terms are combined by means of the symmetry properties (4.29) and the relation (see Exercise 13.6)

$$\bar{v}_{-\alpha,-\beta,-\gamma,-\delta} = (-1)^{j_a-m_\alpha+j_b-m_\beta+j_c-m_\gamma+j_d-m_\delta} \bar{v}_{\alpha\beta\gamma\delta} \ . \tag{13.22}$$

This relation enables us to express all operators without tildes. After a long but straightforward calculation we obtain

$$
\begin{aligned}
4V = & \sum_{\alpha\beta\gamma\delta} \bar{v}_{\alpha\beta\gamma\delta} \mathcal{N}\left[c_\alpha^\dagger c_\beta^\dagger c_\delta c_\gamma\right] + 2\sum_{\alpha\beta} \bar{v}_{\alpha\beta\alpha\beta} v_a^2 v_b^2 \\
& + \sum_{\alpha\beta} \bar{v}_{\alpha,-\alpha\beta,-\beta} (-1)^{j_a-m_\alpha+j_b-m_\beta} u_a v_a u_b v_b \\
& + 2\sum_{\alpha\gamma\delta} \bar{v}_{\alpha,-\alpha,-\gamma\delta} (-1)^{j_a-m_\alpha+j_c-m_\gamma} u_a v_a (u_d v_c + v_d u_c) a_\gamma^\dagger a_\delta \\
& + 4\sum_{\alpha\gamma\delta} \bar{v}_{\alpha\gamma\alpha\delta} v_a^2 (u_c u_d - v_d v_c) a_\gamma^\dagger a_\delta \\
& + 4\sum_{\alpha\gamma\delta} \bar{v}_{\alpha\delta\alpha,-\gamma} (-1)^{j_c-m_\gamma} v_a^2 u_d v_c (a_\gamma^\dagger a_\delta^\dagger + a_\delta a_\gamma) \\
& + \sum_{\alpha\gamma\delta} \bar{v}_{\alpha,-\alpha\gamma\delta} (-1)^{j_a+m_\alpha} u_a v_a (v_c v_d - u_c u_d)(a_\gamma^\dagger a_\delta^\dagger + a_\delta a_\gamma) \ . \tag{13.23}
\end{aligned}
$$

By means of (8.17) and various Clebsch–Gordan relations from Chap. 1 we re-express the contracted terms of (13.23) in terms of coupled two-body matrix elements and one-body operators. The results are

$$2\sum_{\alpha\beta} \bar{v}_{\alpha\beta\alpha\beta} v_a^2 v_b^2 = 2\sum_{ab} v_a^2 v_b^2 \sum_J \hat{J}^2 [\mathcal{N}_{ab}(J)]^{-2} \langle a\,b\,;\,J|V|a\,b\,;\,J\rangle \ , \tag{13.24}$$

$$
\sum_{\alpha\beta} \bar{v}_{\alpha,-\alpha\beta,-\beta} (-1)^{j_a-m_\alpha+j_b-m_\beta} u_a v_a u_b v_b
$$
$$
= 2\sum_{ab} \hat{j}_a \hat{j}_b u_a v_a u_b v_b \langle a\,a\,;\,0|V|b\,b\,;\,0\rangle \ , \tag{13.25}
$$

$$
2\sum_{\alpha\gamma\delta} \bar{v}_{\alpha,-\alpha,-\gamma\delta} (-1)^{j_a-m_\alpha+j_c-m_\gamma} u_a v_a (u_d v_c + v_d u_c) a_\gamma^\dagger a_\delta
$$
$$
= -2\sqrt{2}\sum_{acd} \hat{j}_a u_a v_a (u_d v_c + v_d u_c)[\mathcal{N}_{cd}(0)]^{-1}\delta_{j_c j_d}\langle a\,a\,;0\,|V|\,c\,d\,;0\rangle \left[a_c^\dagger \tilde{a}_d\right]_{00} \ ,
$$
$$
\tag{13.26}
$$

$$4\sum_{\alpha\gamma\delta}\bar{v}_{\alpha\gamma\alpha\delta}v_a^2(u_cu_d-v_dv_c)a_\gamma^\dagger a_\delta = 4\sum_{acd}\hat{j}_c^{-1}v_a^2(u_cu_d-v_dv_c)$$

$$\times\sum_J\hat{J}^2[\mathcal{N}_{ac}(J)\mathcal{N}_{ad}(J)]^{-1}\delta_{j_cj_d}\langle a\,c\,;\,J|V|a\,d\,;\,J\rangle\big[a_c^\dagger\tilde{a}_d\big]_{00}\,,\quad (13.27)$$

$$4\sum_{\alpha\gamma\delta}\bar{v}_{\alpha\delta\alpha,-\gamma}(-1)^{j_c-m_\gamma}v_a^2u_du_c(a_\gamma^\dagger a_\delta^\dagger + a_\delta a_\gamma) = 4\sum_{acd}\hat{j}_c^{-1}v_a^2u_du_c$$

$$\times\sum_J\hat{J}^2[\mathcal{N}_{ad}(J)\mathcal{N}_{ac}(J)]^{-1}\delta_{j_dj_c}\langle a\,d\,;\,J|V|a\,c\,;\,J\rangle\Big(\big[a_c^\dagger a_d^\dagger\big]_{00}-\big[\tilde{a}_d\tilde{a}_c\big]_{00}\Big)\,,$$

$$(13.28)$$

$$\sum_{\alpha\gamma\delta}\bar{v}_{\alpha,-\alpha\gamma\delta}(-1)^{j_a+m_\alpha}u_av_a(v_cv_d-u_cu_d)(a_\gamma^\dagger a_\delta^\dagger + a_\delta a_\gamma) = \sqrt{2}\sum_{acd}\hat{j}_au_av_a$$

$$\times (u_cu_d-v_cv_d)[\mathcal{N}_{cd}(0)]^{-1}\delta_{j_cj_d}\langle a\,a\,;\,0|V|c\,d\,;\,0\rangle\Big(\big[a_c^\dagger a_d^\dagger\big]_{00}-\big[\tilde{a}_d\tilde{a}_c\big]_{00}\Big)\,.$$

$$(13.29)$$

The one-body part $T$ of the Hamiltonian has the simple form (4.71). Substitution from (13.12) and (13.13) gives after a short calculation

$$T = \sum_\alpha \varepsilon_\alpha c_\alpha^\dagger c_\alpha = \sum_a \varepsilon_a\hat{j}_a^2 v_a^2 + \sum_a \varepsilon_a\hat{j}_a(u_a^2-v_a^2)\big[a_a^\dagger\tilde{a}_a\big]_{00}$$

$$+ \sum_a \varepsilon_a\hat{j}_a u_av_a\Big(\big[a_a^\dagger a_a^\dagger\big]_{00}-\big[\tilde{a}_a\tilde{a}_a\big]_{00}\Big)\,.\qquad (13.30)$$

We are now ready to write the full Hamiltonian explicitly in quasiparticle representation. The non-constant two-body terms (13.26)–(13.29) contain the Kronecker delta $\delta_{j_cj_d}$. While it excludes coupling between different $j$ values, it allows coupling between different orbital quantum numbers $n$; the parity $\pi = (-1)^l$ cannot change because the Hamiltonian is a scalar. However, pairs of such single-particle states occur in harmonic oscillator shells $2\hbar\omega$ apart, and their coupling can be neglected to a good approximation. We thus make the replacement

$$\delta_{j_cj_d} \to \delta_{cd}\,,\qquad (13.31)$$

which brings about an appreciable simplification.

Collecting the terms (13.24)–(13.30) in the approximation (13.31) results in

$$\boxed{H = H_0 + \sum_b H_{11}(b)\big[a_b^\dagger\tilde{a}_b\big]_{00} + \sum_b H_{20}(b)\Big(\big[a_b^\dagger a_b^\dagger\big]_{00}-\big[\tilde{a}_b\tilde{a}_b\big]_{00}\Big) + V_{\text{RES}}\,.}$$

$$(13.32)$$

The last term is the residual interaction

$$V_{\text{RES}} = \tfrac{1}{4}\sum_{\alpha\beta\gamma\delta}\bar{v}_{\alpha\beta\gamma\delta}\mathcal{N}\big[c_\alpha^\dagger c_\beta^\dagger c_\delta c_\gamma\big]_{\text{BCS}}\,,\qquad (13.33)$$

where $\mathcal{N}[\ldots]_{\mathrm{BCS}}$ means normal ordering with respect to the BCS vacuum $|\mathrm{BCS}\rangle$. The term $H_0$ is a c-number given by

$$H_0 = \sum_a \varepsilon_a \widehat{j_a}^2 v_a^2 + \tfrac{1}{2} \sum_{abJ} v_a^2 v_b^2 \widehat{J}^2 [\mathcal{N}_{ab}(J)]^{-2} \langle a\,b\,;\,J|V|a\,b\,;\,J\rangle$$

$$+ \tfrac{1}{2} \sum_{ab} \widehat{j_a}\widehat{j_b} u_a v_a u_b v_b \langle a\,a\,;\,0|V|b\,b\,;\,0\rangle \;. \tag{13.34}$$

The coefficients of the $H_{11}$ and $H_{20}$ parts are

$$H_{11}(b) = \varepsilon_b \widehat{j_b}(u_b^2 - v_b^2) - 2 u_b v_b \sum_a \widehat{j_a} u_a v_a \langle a\,a\,;\,0|V|b\,b\,;\,0\rangle$$

$$+ \widehat{j_b}^{-1}(u_b^2 - v_b^2) \sum_{aJ} v_a^2 \widehat{J}^2 [\mathcal{N}_{ab}(J)]^{-2} \langle a\,b\,;\,J|V|a\,b\,;\,J\rangle \;, \tag{13.35}$$

$$H_{20}(b) = \varepsilon_b \widehat{j_b} u_b v_b + \widehat{j_b}^{-1} u_b v_b \sum_{aJ} v_a^2 \widehat{J}^2 [\mathcal{N}_{ab}(J)]^{-2} \langle a\,b\,;\,J|V|a\,b\,;\,J\rangle$$

$$+ \tfrac{1}{2}(u_b^2 - v_b^2) \sum_a \widehat{j_a} u_a v_a \langle a\,a\,;\,0|V|b\,b\,;\,0\rangle \;. \tag{13.36}$$

The expressions (13.32)–(13.36) constitute the *quasiparticle representation of the nuclear Hamiltonian*.

## 13.3 Derivation of the BCS Equations

The BCS equations can be derived in at least two very different ways. Both methods amount to minimizing the BCS ground-state expectation value $\langle \mathrm{BCS}|H|\mathrm{BCS}\rangle$. Our choice is a Rayleigh–Ritz variational treatment with respect to the occupation amplitudes $u_a$ and $v_a$.

The BCS vacuum does not possess good particle number, as is expressly displayed in (13.6). Various procedures have been devised to make up for this deficiency. The most accurate way is to project good particle number, but the procedure is tedious and will not be considered here. In standard BCS theory the ground-state expectation value $\bar{n}$ of the particle number operator $\hat{n}$ is constrained to be the desired number of particles. The Lipkin–Nogami extension of BCS theory, discussed in Chap. 14, takes into account not only $\hat{n}$ but also $\hat{n}^2$. We derive the standard BCS equations in the next two subsections.

### 13.3.1 BCS as a Constrained Variational Problem

The simplest way to reduce uncertainties arising from the non-conservation of particle number in BCS theory is to constrain the variational problem to yield a good *average* particle number $\bar{n}$. With $\hat{n}$, the particle number operator, the constraint reads

$$\langle \mathrm{BCS}|\hat{n}|\mathrm{BCS}\rangle = \bar{n} \ . \tag{13.37}$$

The operator $\hat{n}$ can be written down immediately in the pattern of (13.30):

$$\hat{n} = \sum_\alpha c_\alpha^\dagger c_\alpha = \sum_a \hat{j_a}^2 v_a^2 + \sum_a \hat{j_a}(u_a^2 - v_a^2)\left[a_a^\dagger \tilde{a}_a\right]_{00}$$

$$+ \sum_a \hat{j_a} u_a v_a \left( \left[a_a^\dagger a_a^\dagger\right]_{00} - \left[\tilde{a}_a \tilde{a}_a\right]_{00} \right) \ . \tag{13.38}$$

Only the constant term of $\hat{n}$ contributes to $\bar{n}$, so we have

$$\bar{n} = \sum_a \hat{j_a}^2 v_a^2 \ . \tag{13.39}$$

In actual BCS calculations, protons and neutrons are treated separately. We require the average numbers $\bar{n}_\mathrm{p}$ and $\bar{n}_\mathrm{n}$ of active protons and neutrons, respectively, to be the numbers $Z_\mathrm{act}$ and $N_\mathrm{act}$ of protons and neutrons in the valence space.

Imposing the constraint on the average particle number leads to a *constrained variational problem*. This problem can be solved by the method of *Lagrange undetermined multipliers*, where a parameter $\lambda$ is introduced to yield an *unconstrained variational problem*. In the present case we define an auxiliary Hamiltonian

$$\mathcal{H} \equiv H - \lambda \hat{n} \tag{13.40}$$

and pose the variational problem as

$$\delta\langle \mathrm{BCS}|\mathcal{H}|\mathrm{BCS}\rangle = 0 \ . \tag{13.41}$$

From (13.32) and (13.38) we see that

$$\langle \mathrm{BCS}|\mathcal{H}|\mathrm{BCS}\rangle = H_0 - \lambda \sum_a \hat{j_a}^2 v_a^2 \equiv \mathcal{H}_0 \ . \tag{13.42}$$

With the expression (13.34) for $H_0$ this becomes

$$\mathcal{H}_0 = \sum_a (\varepsilon_a - \lambda)\hat{j_a}^2 v_a^2 + \tfrac{1}{2}\sum_{abJ} v_a^2 v_b^2 \hat{J}^2 [\mathcal{N}_{ab}(J)]^{-2}\langle a\,b\,;\,J|V|a\,b\,;\,J\rangle$$

$$+ \tfrac{1}{2}\sum_{ab} \hat{j_a}\hat{j_b} u_a v_a u_b v_b \langle a\,a\,;\,0|V|b\,b\,;\,0\rangle \ . \tag{13.43}$$

The change from $H_0$ to $\mathcal{H}_0$ amounts to replacing the single-particle energies $\varepsilon_a$ in $H_0$ with $\varepsilon_a - \lambda$. The same change occurs in (13.35) and (13.36):

$$\mathcal{H}_{11}(b)|_\varepsilon = H_{11}(b)|_{\varepsilon-\lambda} \ , \quad \mathcal{H}_{20}(b)|_\varepsilon = H_{20}(b)|_{\varepsilon-\lambda} \ . \tag{13.44}$$

To prepare the Hamiltonian for the variational treatment, and subsequent physical interpretation, we adopt the notation

$$\Delta_b \equiv -\widehat{j_b}^{-1} \sum_a \widehat{j_a} u_a v_a \langle a\,a\,;\,0|V|b\,b\,;\,0\rangle \,,$$

$$\mu_b \equiv -\widehat{j_b}^{-2} \sum_{aJ} v_a^2 \widehat{J}^2 [\mathcal{N}_{ab}(J)]^{-2} \langle a\,b\,;\,J|V|a\,b\,;\,J\rangle \,, \qquad (13.45)$$

$$\eta_b \equiv \varepsilon_b - \lambda - \mu_b \,.$$

These abbreviations result in the concise expressions

$$\mathcal{H}_0 = \sum_b \widehat{j_b}^2 \left[ v_b^2 (\eta_b + \tfrac{1}{2}\mu_b) - \tfrac{1}{2} u_b v_b \Delta_b \right] \,, \qquad (13.46)$$

$$\mathcal{H}_{11}(b) = \widehat{j_b} \left[ (u_b^2 - v_b^2)\eta_b + 2 u_b v_b \Delta_b \right] \,, \qquad (13.47)$$

$$\mathcal{H}_{20}(b) = \widehat{j_b} \left[ u_b v_b \eta_b - \tfrac{1}{2}(u_b^2 - v_b^2)\Delta_b \right] \,. \qquad (13.48)$$

### 13.3.2 The Gap Equation and the Quasiparticle Mean Field

We are now in a position to perform the variational calculation that leads to the BCS equations. It is sufficient to require that

$$\frac{\partial}{\partial v_c} \mathcal{H}_0 = 0 \quad \text{for all } c \qquad (13.49)$$

because the normalization condition (13.9) makes each $u_c$ dependent on $v_c$. That condition gives

$$\frac{\partial u_b}{\partial v_c} = -\delta_{bc} \frac{v_c}{u_c} \,. \qquad (13.50)$$

In preparation for the differentiation we note from the last relation in (13.45) that

$$\frac{\partial}{\partial v_c}(\eta_b + \tfrac{1}{2}\mu_b) = -\frac{1}{2}\frac{\partial \mu_b}{\partial v_c} \,. \qquad (13.51)$$

We also form the derivatives

$$\frac{\partial \mu_b}{\partial v_c} = -2\widehat{j_b}^{-2} v_c \sum_J \widehat{J}^2 [\mathcal{N}_{bc}(J)]^{-2} \langle b\,c\,;\,J|V|b\,c\,;\,J\rangle \,,$$

$$\frac{\partial \Delta_b}{\partial v_c} = -\widehat{j_b}^{-1} \widehat{j_c} \frac{u_c^2 - v_c^2}{u_c} \langle b\,b\,;\,0|V|c\,c\,;\,0\rangle \,, \qquad (13.52)$$

where we have used the symmetry properties (4.29) so that $\mu_c$ and $\Delta_c$ can be recognized in the sums over $b$ below. The requirement (13.49) now gives

$$0 = \frac{\partial}{\partial v_c} \mathcal{H}_0 = \sum_b \widehat{j_b}^2 \left[ -\tfrac{1}{2} v_b^2 \frac{\partial \mu_b}{\partial v_c} + 2 v_b \delta_{bc}(\eta_b + \tfrac{1}{2}\mu_b) \right.$$

$$\left. -\tfrac{1}{2} u_b v_b \frac{\partial \Delta_b}{\partial v_c} - \tfrac{1}{2} u_b \delta_{bc} \Delta_b + \tfrac{1}{2} \delta_{bc} \frac{v_c}{u_c} v_b \Delta_b \right]$$

$$= v_c \sum_{bJ} v_b^2 \hat{J}^2 [\mathcal{N}_{bc}(J)]^{-2} \langle bc; J|V|bc; J\rangle + 2\hat{j}_c^2 v_c(\eta_c + \tfrac{1}{2}\mu_c)$$

$$+ \hat{j}_c \frac{u_c^2 - v_c^2}{2u_c} \sum_b \hat{j}_b u_b v_b \langle bb; 0|V|cc; 0\rangle - \tfrac{1}{2}\hat{j}_c^2 u_c \Delta_c + \tfrac{1}{2}\hat{j}_c^2 \frac{v_c^2}{u_c}\Delta_c$$

$$= 2\hat{j}_c^2 v_c \eta_c + \hat{j}_c^2 \frac{v_c^2 - u_c^2}{u_c}\Delta_c \ . \tag{13.53}$$

This gives us the equation

$$(u_c^2 - v_c^2)\Delta_c = 2u_c v_c \eta_c \ . \tag{13.54}$$

Squaring both sides of (13.54) and using (13.9) we find

$$u_c^2 v_c^2 = \frac{\Delta_c^2}{4(\eta_c^2 + \Delta_c^2)} \ . \tag{13.55}$$

From this we solve the occupation amplitudes

$$\boxed{u_c = \theta^{(l_c)} \frac{1}{\sqrt{2}}\sqrt{1 + \frac{\eta_c}{E_c}} \ , \quad v_c = \frac{1}{\sqrt{2}}\sqrt{1 - \frac{\eta_c}{E_c}} \ ,} \tag{13.56}$$

where $\theta^{(l_c)}$ is a phase factor and

$$\boxed{E_c \equiv \sqrt{\eta_c^2 + \Delta_c^2}} \tag{13.57}$$

is the *quasiparticle energy* as becomes clear below.

The phases are chosen according to

$$\theta^{(l_c)} = \begin{cases} (-1)^{l_c} & \text{CS phase convention} , \\ 1 & \text{BR phase convention} . \end{cases} \tag{13.58}$$

The phases have to do with the time-reversal properties of the single-particle orbitals. The time-reversal operator $\mathcal{T}$ relates the orbitals $\alpha$ and $-\alpha$ according to

$$\mathcal{T}c_\alpha^\dagger \mathcal{T}^{-1} = \zeta^{(l_a)}\tilde{c}_\alpha^\dagger \ . \tag{13.59}$$

In the BCS ground state (13.1) the paired nucleons occupy time-reversed orbitals as seen from the structure of the pair-creation operator (13.2). A simple interpretation is that a nucleon in orbital $\alpha$ and another in $-\alpha$ revolve along the same path in opposite directions. For more discussion about time-reversal and other symmetries see, e.g. [12, 17].

The quantity $\Delta_b$, introduced in (13.45) as an abbreviation for a certain block of terms, is called the *pairing gap*, for reasons that become evident in

Subsect. 13.4.1. We can derive an equation for determining the pairing gaps $\Delta_b$. The positive square root of (13.55) is

$$u_c v_c = \frac{\Delta_c}{2E_c} \; . \tag{13.60}$$

Substituting this into the expression for $\Delta_b$ in (13.45) gives the so-called *gap equation*

$$2\widehat{j_b}\Delta_b = -\sum_a \frac{\widehat{j_a}\Delta_a}{\sqrt{\eta_a^2 + \Delta_a^2}} \langle a\,a\,;\,0|V|b\,b\,;\,0\rangle \; . \tag{13.61}$$

The conditions for a minimal value of $\mathcal{H}_0$ appear in the equations above. Applying those equations, in particular (13.54), (13.56), (13.57) and (13.60), to (13.46)–(13.48), we find

$$\mathcal{H}_0 = \tfrac{1}{2}\sum_b \frac{\widehat{j_b}^2}{E_b}\left[(E_b - \eta_b)(\eta_b + \tfrac{1}{2}\mu_b) - \tfrac{1}{2}\Delta_b^2\right] , \tag{13.62}$$

$$\mathcal{H}_{11}(b) = \widehat{j_b}E_b \; , \tag{13.63}$$

$$\mathcal{H}_{20}(b) = 0 \; . \tag{13.64}$$

The auxiliary Hamiltonian thus becomes

$$\mathcal{H} = \mathcal{H}_0 + \sum_b \widehat{j_b}E_b\left[a_b^\dagger \tilde{a}_b\right]_{00} + V_{\mathrm{RES}} \; . \tag{13.65}$$

Writing the middle term in uncoupled form as $\sum_\beta E_b a_\beta^\dagger a_\beta$ makes clear the interpretation of $E_b$ as the quasiparticle energy.

According to the defining relation (13.42) the constant term $H_0$ in the original Hamiltonian (13.32) is

$$H_0 = \mathcal{H}_0 + \lambda \sum_a \widehat{j_b}^2 v_b^2 = \mathcal{H}_0 + \lambda\bar{n} \; . \tag{13.66}$$

Substituting for $v_b^2$ from (13.56) and then using the last definition in (13.45) we find

$$H_0 = \tfrac{1}{2}\sum_b \frac{\widehat{j_b}^2}{E_b}\left[(E_b - \eta_b)(\varepsilon_b - \tfrac{1}{2}\mu_b) - \tfrac{1}{2}\Delta_b^2\right] . \tag{13.67}$$

Equations (13.56), (13.57) and (13.61) constitute what are known as the *BCS equations*. Another popular way to derive these equations is to require the 'bad term' $\mathcal{H}_{20}(b)$ to vanish, without resorting to a Rayleigh–Ritz variation of the occupation amplitudes. Equation (13.64) shows that these two BCS derivations are equivalent.

The first two terms of the Hamiltonian (13.65) carry a large part of the original residual interaction, i.e. the interaction remaining after the subtraction of the nuclear mean field. They describe *non-interacting quasiparticles* with energies $E_b$. This approximation can be called the *quasiparticle mean field*, in analogy to the Hartree–Fock mean field of non-interacting particles. The residual interaction $V_{\mathrm{RES}}$ in (13.65) is one between the quasiparticles. It produces configuration mixing between many-quasiparticle states. Methods to handle this mixing, such as the QTDA and the QRPA, will be discussed later in this book.

## 13.4 Properties of the BCS Solutions

### 13.4.1 Physical Meaning of the Basic Parameters

In this subsection we address the physical interpretation of three key parameters of the BCS theory, namely the pairing gap, self-energy and chemical potential.

### Pairing Gap and Pairing Energy

The excited states of an even–even nucleus are created by breaking one or more pairs in the superfluid ground state |BCS⟩. The least energy is needed to break just one pair. Then an extra energy equal to the binding energy of the pair has to be supplied from the outside. The broken pair is interpreted as two quasiparticles, referred to as a two-quasiparticle configuration or a *two-quasiparticle excitation*. This excitation corresponds to a seniority $v = 2$ state of the seniority model, discussed in Subsect. 12.4.2. The energy needed to break a pair is the excitation energy of the two-quasiparticle configuration.

The fact that the quasiparticle energies (13.57) satisfy

$$E_a = \sqrt{\eta_a^2 + \Delta_a^2} \geq \Delta_a \tag{13.68}$$

implies for the energies of two-quasiparticle excitations

$$E_{2\mathrm{qp}} \geq 2\Delta_{\mathrm{smallest}} \ . \tag{13.69}$$

In the terminology introduced in Sect. 12.1 the quantities $\Delta_a$ are pairing gaps while the quantities $2\Delta_a$ are pairing energies. In Sect. 12.3 we treated two particles in a single $j$ shell and interacting through the pure pairing force. According to (12.17) the pairs not coupled to angular momentum zero lie an energy $\Omega G$ above the zero-coupled pair, where $G$ is the pairing strength and $\Omega = \frac{1}{2}(2j+1)$ is the degeneracy. We can thus identify the quasiparticle energy with half the pairing energy, $E_a = \frac{1}{2}\Omega_a G$.

**Self-energy**

The quantity $\mu_a$, defined in (13.45), is called the *self-energy*. It describes a renormalization of the single-particle energy $\varepsilon_a$ due to the fact that the energy of a nucleon in orbital $a$ gets additional contributions from its interactions with the other nucleons.

**Chemical Potential**

The quantity $\lambda$ is called the *chemical potential*. To study its meaning we write, from (13.40) and (13.42),

$$\mathcal{H}_0 = \langle \text{BCS}|H|\text{BCS} \rangle - \lambda n \ . \tag{13.70}$$

The number constraint in the variational problem means that the derivative of $\mathcal{H}_0$ with respect to the particle number $n$ has to vanish at the correct required particle number $n = \bar{n}$. Taking the derivative of (13.70) with respect to $n$ and setting it to zero yields

$$\boxed{\lambda = \frac{\partial}{\partial n} \langle \text{BCS}|H|\text{BCS} \rangle \Big|_{n=\bar{n}} \ .} \tag{13.71}$$

According to this equation $\lambda$ tells us how much the energy of the BCS ground state grows when one particle is added to it. This is nothing but the standard definition of chemical potential in statistical mechanics.

## 13.4.2 Particle Number and Its Fluctuations

The particle number and its fluctuations are important considerations in BCS theory. This is due to the fact that good particle number was lost in the Bogoliubov–Valatin transformation (13.10) replacing particles with quasiparticles. Particle number issues are addressed in this subsection.

**Number Constraint on the BCS Vacuum**

The particle number constraint (13.37) led to the condition (13.39) on the occupation probabilities $v_a^2$. Expressing the $v_a^2$ through (13.56) gives

$$\boxed{\bar{n} = \sum_a \hat{j}_a^{\,2} v_a^2 = \tfrac{1}{2} \sum_a \hat{j}_a^{\,2} \left(1 - \frac{\eta_a}{E_a}\right) ,} \tag{13.72}$$

where $\bar{n}$ is the number of protons or neutrons in the valence space adopted. When solving the BCS problem, this equation and the BCS equations (13.56), (13.57) and (13.61) have to be solved simultaneously. The solving tactics of these equations are demonstrated in the following chapter.

**Particle Number Fluctuations**

Since the BCS ground state has only a correct average number of nucleons, it contains a spread in particle number, for both protons and neutrons. This uncertainty in the particle number shows up as *particle number fluctuations*.

A measure of the fluctuations is provided by the mean-square deviation from the mean value $\bar{n}$,

$$(\Delta n)^2 \equiv \langle \text{BCS}|(\hat{n} - \bar{n})^2|\text{BCS}\rangle = \langle \text{BCS}|\hat{n}^2|\text{BCS}\rangle - \bar{n}^2 . \tag{13.73}$$

Using (13.38) we can write

$$\hat{n}|\text{BCS}\rangle = \left[\sum_a \hat{j_a}^2 v_a^2 + \sum_\alpha u_a v_a (-1)^{j_a - m_\alpha} a_\alpha^\dagger a_{-\alpha}^\dagger\right]|\text{BCS}\rangle . \tag{13.74}$$

Multiplying this by the corresponding ket vector $\langle \text{BCS}|\hat{n}$ gives

$$\langle \text{BCS}|\hat{n}^2|\text{BCS}\rangle = \sum_{ab} \hat{j_a}^2 \hat{j_b}^2 v_a^2 v_b^2$$

$$+ \sum_{\alpha\beta} u_a v_a u_b v_b (-1)^{j_a - m_\alpha + j_b - m_\beta} \langle \text{BCS}|a_{-\alpha} a_\alpha a_\beta^\dagger a_{-\beta}^\dagger|\text{BCS}\rangle$$

$$= \sum_{ab} \hat{j_a}^2 \hat{j_b}^2 v_a^2 v_b^2 + 2 \sum_a \hat{j_a}^2 u_a^2 v_a^2 . \tag{13.75}$$

The first term is recognized from (13.72) as $\bar{n}^2$, so the mean-square deviation (13.73) becomes

$$(\Delta n)^2 = 2 \sum_a \hat{j_a}^2 u_a^2 v_a^2 = \tfrac{1}{2} \sum_a \hat{j_a}^2 \frac{\Delta_a^2}{E_a^2} \tag{13.76}$$

with (13.60) used in the last step. This deviation will be discussed in a transparent way in the context of exactly solvable models in the next section.

**Number Parity**

The BCS states are not eigenstates of the number operator. However, the ground state and the two-quasiparticle states contain even particle numbers while the one-quasiparticle states contain odd particle numbers. According to this evenness or oddness these states are said to have even or odd *number parity*.

In general, excitations containing an even number of like quasiparticles also contain an even number particles of that type. Hence, an excitation containing even numbers of proton and neutron quasiparticles belongs to an even–even nucleus. Excitations that contain an odd number of either proton or neutron quasiparticles (not both) also contain an odd number of protons or neutrons and thus belong to odd–even or even–odd nuclei. Finally, if an excitation contains odd numbers of proton and neutron quasiparticles, it belongs to an odd–odd nucleus.

### 13.4.3 Odd–Even Effect

Consider an even–even nucleus with mass number $A - 1$ in its ground state $|BCS\rangle$. According to (13.42) the energy of the state is $\langle BCS|H|BCS\rangle = H_0$. Operating on $|BCS\rangle$ with the quasiparticle creation operator $a_\alpha^\dagger$ results in state $|\alpha = j_a n_a l_a m_a\rangle$ of an odd-$A$ nucleus with mass number $A$.[1] What is the energy of the state $|\alpha\rangle$?

From (13.71) we can write

$$d\langle BCS|H|BCS\rangle = \lambda dn \ . \tag{13.77}$$

The change from $A - 1$ to $A$ means $dn = 1$, and (13.77) gives the increment to $H_0$ as

$$d\langle BCS|H|BCS\rangle = \lambda \ . \tag{13.78}$$

The energy of the mass-$A$ nucleus in the state $|\alpha\rangle$ contains additionally the quasiparticle energy $E_a$, so it is

$$\langle BCS|a_\alpha H a_\alpha^\dagger|BCS\rangle = H_0 + \lambda + E_a \ . \tag{13.79}$$

Equations (13.78) and (13.79) lead to relations for nuclear masses. When $A$ is odd we have

$$M_{A-1}c^2 = H_0 \ , \tag{13.80}$$

$$M_A c^2 = H_0 + \lambda + \Delta \ , \tag{13.81}$$

$$M_{A+1}c^2 = H_0 + 2\lambda \ . \tag{13.82}$$

Here $\Delta$ is the lowest quasiparticle energy, $\Delta \equiv \min\{E_a\}$. Equations (13.80) and (13.82) give

$$\tfrac{1}{2}(M_{A-1} + M_{A+1})c^2 = H_0 + \lambda \ . \tag{13.83}$$

From (13.81) we then notice that

$$M_A = \tfrac{1}{2}(M_{A-1} + M_{A+1}) + \Delta/c^2 \ . \tag{13.84}$$

This is exactly the same as the empirical relation (12.1). Hence the BCS theory successfully reproduces the odd–even effect.

## 13.5 Solution of the BCS Equations for Simple Models

In this section we solve the BCS equations for a single $j$ shell and for the two-level Lipkin model. These simple examples reveal a great deal about the properties of BCS solutions since exact solutions are available for comparison.

---

[1] Because of the particle–hole symmetry of BCS quasiparticles the state $|\alpha\rangle$ exists equally in the adjacent nucleus with mass number $A - 2$.

### 13.5.1 Single $j$ Shell

Consider the case of just one $j$ shell and take its energy as $\varepsilon_j = 0$. This simplifies the BCS approach considerably. The results can be compared with those of the seniority model if we use the pure pairing force as the source of the pairing matrix elements. Equations (13.45) and (13.57) give now

$$\Delta = -u_j v_j \langle j^2 ; 0 | V | j^2 ; 0 \rangle = -u_j v_j G_0 \,, \tag{13.85}$$

$$\mu = -\hat{j}^{-2} \sum_J \hat{J}^2 v_j^2 [N_{jj}(J)]^{-2} \langle j^2 ; J | V | j^2 ; J \rangle$$

$$= -2\hat{j}^{-2} v_j^2 \sum_{J=\text{even}} \hat{J}^2 G_J \,, \tag{13.86}$$

$$\eta = -\lambda - \mu \,, \quad E = \sqrt{\eta^2 + \Delta^2} \,, \tag{13.87}$$

where we have defined

$$G_J \equiv \langle j^2 ; J | V | j^2 ; J \rangle \,. \tag{13.88}$$

The gap equation (13.61) then yields, together with (13.87), for the quasiparticle energy

$$E = -\tfrac{1}{2} G_0 \,. \tag{13.89}$$

The particle number constraint (13.72) gives

$$\bar{n} = \tfrac{1}{2} \hat{j}^2 \left( 1 - \frac{\eta}{E} \right) \,, \tag{13.90}$$

whence

$$\eta = \left( 1 - \frac{2\bar{n}}{2j+1} \right) E \,. \tag{13.91}$$

With this $\eta$ the occupation amplitudes (13.56) become

$$u_j = \theta^{(l)} \sqrt{1 - \frac{\bar{n}}{2j+1}} \,, \quad v_j = \sqrt{\frac{\bar{n}}{2j+1}} \,. \tag{13.92}$$

Let us now specify the matrix elements (13.88) to be of the pure pairing type as given by (12.11). Since the pure pairing force only affects $J = 0$ pairs, we can write

$$G_J = -\tfrac{1}{2} \delta_{J0} \hat{j}^2 G \,, \quad G > 0 \,. \tag{13.93}$$

The quasiparticle energy (13.89) now becomes, in the notation of Chap. 12,

$$E = \tfrac{1}{2} \Omega G \,, \quad \Omega = \tfrac{1}{2} (2j+1) \,. \tag{13.94}$$

Denoting the number of quasiparticles by $n_{\text{qp}}$ we have the quasiparticle spectrum

$$E(n_{\text{qp}}) = n_{\text{qp}} \times \tfrac{1}{2} \Omega G \,. \tag{13.95}$$

Let us compare this quasiparticle spectrum for an even–even nucleus with that of the seniority model given by (12.43). Since the number of quasiparticles is the number of valence particles not in pairs of zero angular momentum, we identify $n_{\mathrm{qp}}$ with the seniority $v$. Equation (12.43) then shows that the BCS solution is exact for $n_{\mathrm{qp}} = v = 2$. For seniorities $v \geq 4$ the relative deviation from the exact solution is of the order of $(v - 2)/2\Omega$. The deviation is the smaller the larger is the $j$ shell.

We proceed to calculate the ground-state energy $H_0$ from (13.67) in the pure pairing case. For that we need the quantities (13.85)–(13.87). Equations (13.91)–(13.94) yield for them

$$\Delta = \theta^{(l)} \sqrt{\frac{\bar{n}}{2}\left(\Omega - \frac{\bar{n}}{2}\right)} \, G \,, \tag{13.96}$$

$$\mu = \frac{\bar{n}}{2\Omega} G \,, \tag{13.97}$$

$$\eta = \tfrac{1}{2}(\Omega - \bar{n}) G \,, \tag{13.98}$$

$$\lambda = -\eta - \mu = \tfrac{1}{2}\left(\bar{n} - \Omega - \frac{\bar{n}}{\Omega}\right) G \,. \tag{13.99}$$

Substituting these into (13.67) gives

$$H_0 = -\tfrac{1}{4} G\bar{n}\left(2\Omega - \bar{n} + \frac{\bar{n}}{\Omega}\right). \tag{13.100}$$

This BCS result is close to the exact solution (12.28) of the seniority model when we identify the average particle number $\bar{n}$ with the precise number $N$. The expressions are identical except for the last term.

We finally calculate the mean-square deviation (13.76). With (13.94) and (13.96) we obtain

$$(\Delta n)^2 = 2\bar{n}\left(1 - \frac{\bar{n}}{2\Omega}\right), \tag{13.101}$$

whence

$$\left|\frac{\Delta n}{\bar{n}}\right| \lesssim \sqrt{\frac{2}{\bar{n}}}. \tag{13.102}$$

This indicates that the spread or fluctuation in particle number decreases with increasing number of particles in the shell.

### 13.5.2 The Lipkin Model

In this subsection we discuss the BCS within the Lipkin model, which is a special case of the general two-level model of Sect. 12.5. As in the previous subsection, we use the pure pairing interaction. The Lipkin model has two $j$ shells with $j_1 l_1 = j_2 l_2 = jl$ and energies $\varepsilon_1 = -\tfrac{1}{2}\varepsilon$ and $\varepsilon_2 = \tfrac{1}{2}\varepsilon$. By definition of the model, the average particle number is $\bar{n} = 2\Omega = 2j+1$. Thus the ground state of non-interacting particles has the lower $j$ shell completely filled and the upper shell completely empty. Turning on the interaction will scatter particles from the lower shell to the upper one. Hence the occupation numbers of the shells depend on the strength of the interaction.

In the present case the relation (12.11) between the pairing matrix elements and the interaction strength of the pure pairing force becomes

$$\langle a\,b\,;\,0|V|c\,d\,;\,0\rangle = -\Omega G \,, \tag{13.103}$$

where $a, b, c, d$ are any of the two orbitals $1, 2$. From (13.45) we now obtain

$$\Delta_1 = \Delta_2 \equiv \Delta = (u_1 v_1 + u_2 v_2)\Omega G \,, \tag{13.104}$$

$$\mu_1 = (v_1^2 + \tfrac{1}{2}v_2^2)G \,, \quad \mu_2 = (\tfrac{1}{2}v_1^2 + v_2^2)G \,, \tag{13.105}$$

The gap equation (13.61) gives

$$\frac{2}{\Omega G} = \frac{1}{E_1} + \frac{1}{E_2} \,. \tag{13.106}$$

The particle number constraint (13.72) yields for $\bar{n} = 2\Omega$

$$\bar{n} = 2\Omega(v_1^2 + v_2^2) = 2\Omega \,, \tag{13.107}$$

whence $v_1^2 + v_2^2 = 1$. The normalization condition (13.9) then implies that $v_2^2 = u_1^2$ and $v_1^2 = u_2^2$. Assuming that $l$ is even, so that according to (13.56) the $u$ coefficient is positive also in the CS phase convention, we have in either phase convention

$$v_2 = u_1 \,, \quad v_1 = u_2 \,. \tag{13.108}$$

The result describes a mirror symmetry between the occupation and unoccupation amplitudes of the upper and lower shells.

Next we set out to find the occupation amplitudes $v_1$ and $v_2$. Equation (13.60) gives

$$u_1 v_1 = \frac{\Delta}{2E_1} \,, \quad u_2 v_2 = \frac{\Delta}{2E_2} \,. \tag{13.109}$$

Equation (13.108) implies that $u_1 v_1 = u_2 v_2$, so that $E_1 = E_2 \equiv E$. It follows from (13.57) that $\eta_1 = \pm\eta_2$. From (13.106) we now obtain

$$E = \Omega G \,. \tag{13.110}$$

The equations (13.56) together with (13.108) give

$$v_1 = \frac{1}{\sqrt{2}}\sqrt{1 - \frac{\eta_1}{E}} = u_2 = \frac{1}{\sqrt{2}}\sqrt{1 + \frac{\eta_2}{E}} \,, \tag{13.111}$$

which implies $\eta_1 = -\eta_2$. Using the definition of $\eta_b$ in (13.45) and the relations (13.105) and (13.107) we obtain for the chemical potential the expression

$$\lambda = -\tfrac{1}{2}(\mu_1 + \mu_2) = -\tfrac{3}{4}G \,. \tag{13.112}$$

Substituting into the definition of $\eta_1$ from (13.112) and from (13.105) and (13.107) gives the expression

$$\eta_1 = -\tfrac{1}{2}\varepsilon + \tfrac{3}{4}G - \tfrac{1}{2}(1 + v_1^2)G . \tag{13.113}$$

The last term contains the unknown $\eta_1$ through (13.111). Making the substitution, using (13.110) and solving for $\eta_1$ we find

$$\eta_1 = -\frac{2\Omega\varepsilon}{4\Omega - 1} = -\eta_2 \equiv -\eta . \tag{13.114}$$

From (13.111) we finally obtain the desired occupation amplitudes

$$v_1 = \frac{1}{\sqrt{2}}\sqrt{1 + \frac{2\varepsilon}{(4\Omega - 1)G}} = u_2 , \tag{13.115}$$

$$u_1 = \frac{1}{\sqrt{2}}\sqrt{1 - \frac{2\varepsilon}{(4\Omega - 1)G}} = v_2 . \tag{13.116}$$

Having found the occupation amplitudes we can readily write down final forms for the pairing gap (13.104) and the self-energies (13.105):

$$\Delta = \sqrt{\Omega^2 G^2 - \left(\frac{2\Omega\varepsilon}{4\Omega - 1}\right)^2} , \tag{13.117}$$

$$\mu_1 = \tfrac{1}{2}(1 + v_1^2)G , \quad \mu_2 = \tfrac{1}{2}(1 + v_2^2)G . \tag{13.118}$$

For the ground-state energy (13.67), we obtain

$$\begin{aligned}
H_0 &= \frac{\Omega}{E}\left[(E + \eta)(-\tfrac{1}{2}\varepsilon - \tfrac{1}{2}\mu_1) + (E - \eta)(\tfrac{1}{2}\varepsilon - \tfrac{1}{2}\mu_2) - \Delta^2\right] \\
&= \frac{1}{G}\left[-\tfrac{3}{4}\Omega G^2 - \tfrac{1}{2}\varepsilon\eta\frac{8\Omega - 1}{4\Omega - 1} - \Omega^2 G^2 + \left(\frac{2\Omega\varepsilon}{4\Omega - 1}\right)^2\right] . 
\end{aligned} \tag{13.119}$$

From (13.117) we notice that there exists a pairing gap only if the expression under the square root is positive, which occurs for

$$G > \frac{2\varepsilon}{4\Omega - 1} \equiv G_{\text{crit}} , \tag{13.120}$$

the *critical strength*. This is interpreted to mean that there is no BCS solution when the condition is not satisfied. For $G \leq G_{\text{crit}}$ the pairing gap vanishes and the occupation amplitudes attain the trivial values 1 or 0. In this case the BCS ground state $|\text{BCS}\rangle$ reduces to the particle–hole vacuum $|\text{HF}\rangle$.

The present results for the pairing gap, occupation amplitudes and ground-state energy assume remarkably simple forms when we define

$$x \equiv \frac{4\Omega - 1}{2}\frac{G}{\varepsilon} . \tag{13.121}$$

In terms of $x$ we have

$$\Delta = \Omega G \sqrt{1 - \frac{1}{x^2}} \, , \tag{13.122}$$

$$v_1 = \frac{1}{\sqrt{2}} \sqrt{1 + \frac{1}{x}} \, , \quad v_2 = \frac{1}{\sqrt{2}} \sqrt{1 - \frac{1}{x}} \, , \tag{13.123}$$

$$H_0 = -\frac{1}{2} \Omega \varepsilon \left( \frac{4\Omega + 3}{4\Omega - 1} x + \frac{1}{x} \right) \, . \tag{13.124}$$

In the following example we discuss the physical contents of these expressions through comparison with exact results of the Lipkin model of Sect. 12.5.

### 13.5.3 Example: The Lipkin Model for Two $j = \frac{7}{2}$ Shells

As an illustrative application of the BCS formalism we discuss here the Lipkin model with two $j = \frac{7}{2}$ shells. In this case $\Omega = 4$ and $\bar{n} = 8$. To begin with, we plot in Fig. 13.1 the quantity $\Delta/G$ as given by (13.122), with the parameter $x$ defined in (13.121). The figure shows that there exists a pairing gap only for $x > 1$, which corresponds to the criterion (13.120). At the critical strength $G_{\text{crit}}$ the pairing gap vanishes and the BCS theory breaks down, producing the trivial values 0 or 1 for the occupation amplitudes. The critical strength is proportional to the width $\varepsilon$ of the gap between the two $j$ shells: the wider the gap, the larger the value of $G_{\text{crit}}$ required for a BCS solution. At large values of $x \propto G/\varepsilon$ the gap is seen to saturate to the value $\Omega G = 4G$.

The occupation amplitudes $v_1$ and $v_2$ given in (13.123) are shown in Fig. 13.2, together with the results of an exact solution of the Lipkin model.

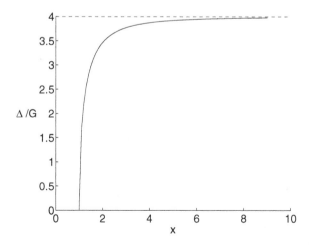

**Fig. 13.1.** The quantity $\Delta/G$ plotted as a function of the parameter $x$, defined in (13.121), for the Lipkin model with two $j = \frac{7}{2}$ shells

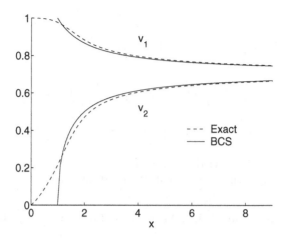

**Fig. 13.2.** The occupation amplitudes $v_1$ and $v_2$ of the lower and upper $j$ shells of the Lipkin model for the $j = \frac{7}{2}$ case as functions of the parameter $x$ defined in (13.121). The BCS results (13.123) are compared with the exact solutions (13.125) and (13.126)

These exact amplitudes can be obtained from the definition of the occupation amplitudes for a *general two-level model*:

$$v_1^2 = \frac{1}{2\Omega_{j_1}} \langle 0_{\mathrm{gs}}^+ | \hat{n}_{j_1} | 0_{\mathrm{gs}}^+ \rangle = \frac{1}{\Omega_{j_1}} \sum_m (\mathcal{N} - m) \left( X_m^{(\mathrm{gs})} \right)^2 , \qquad (13.125)$$

$$v_2^2 = \frac{1}{2\Omega_{j_2}} \langle 0_{\mathrm{gs}}^+ | \hat{n}_{j_2} | 0_{\mathrm{gs}}^+ \rangle = \frac{1}{\Omega_{j_2}} \sum_m m \left( X_m^{(\mathrm{gs})} \right)^2 . \qquad (13.126)$$

These expressions are obtained by using the ground-state wave function (12.77) of the two-level model. The label $m$ denotes the number of nucleon pairs occupying the upper shell and $\mathcal{N} - m$ denotes those in the lower shell. The amplitudes $X_m^{(\mathrm{gs})}$ for the wave function $|0_{\mathrm{gs}}^+\rangle$ are obtained by diagonalizing the pairing Hamiltonian in the basis (12.76). Figure 13.2 shows that where the BCS solutions exist, i.e. for $x > 1$, they are close to the exact ones.

Figure 13.3 shows the BCS ground-state energy (13.124) as a function of $x$. Comparison with the exact result indicates that the BCS is able to reproduce the ground-state energy very well. The BCS energy lies somewhat above the exact result because the ansatz for the BCS ground state does not contain enough degrees of freedom to allow the variation to reach the exact solution.

### 13.5.4 The Two-Level Model for Two $j = \frac{7}{2}$ Shells

The comparison between the BCS approach and the exact solution can be extended to the general two-level model. This is done in Chap. 14, which

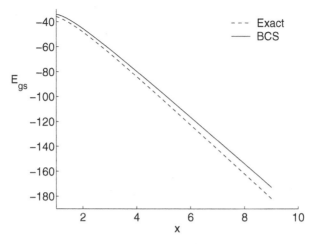

**Fig. 13.3.** The BCS ground-state energy (13.124) and the exact result for the $j = \frac{7}{2}$ Lipkin model plotted as functions of the parameter $x$ defined in (13.121). In this plot we have taken $\varepsilon = 7.5\,\mathrm{MeV}$, which leads to the simple relation $x = G\,[\mathrm{MeV}]$

presents the BCS solution for the general case of many $j$ shells. In anticipation, we discuss here the case of two $j = \frac{7}{2}$ shells with the number of particles $\bar{n} = 2$–14. The occupation amplitudes are quoted in Table 13.1 for two different ratios $\varepsilon/G$. The ratio represents the ability of the interaction to scatter particles from the lower level to the higher one: the smaller the ratio the greater the scattering. The table shows excellent agreement between the BCS solutions and the exact ones.

**Table 13.1.** Exact and BCS occupation amplitudes $v_1$ and $v_2$ for $\bar{n}$ particles occupying two $j = \frac{7}{2}$ shells an energy $\varepsilon$ apart

| | $\dfrac{\varepsilon}{G} = 1.0$ | | | | | $\dfrac{\varepsilon}{G} = 6.0$ | | | | |
|---|---|---|---|---|---|---|---|---|---|---|
| | Exact | | BCS | | | Exact | | BCS | | |
| $\bar{n}$ | $v_1$ | $v_2$ | $v_1$ | $v_2$ | $\lambda\,(\mathrm{MeV})$ | $v_1$ | $v_2$ | $v_1$ | $v_2$ | $\lambda\,(\mathrm{MeV})$ |
| 2 | 0.394 | 0.308 | 0.391 | 0.312 | $-3.177$ | 0.479 | 0.145 | 0.479 | 0.144 | $-4.672$ |
| 4 | 0.550 | 0.444 | 0.547 | 0.449 | $-2.288$ | 0.676 | 0.206 | 0.677 | 0.203 | $-3.674$ |
| 6 | 0.665 | 0.555 | 0.661 | 0.559 | $-1.395$ | 0.828 | 0.254 | 0.830 | 0.246 | $-2.524$ |
| 8 | 0.756 | 0.655 | 0.753 | 0.658 | $-0.500$ | 0.957 | 0.291 | 0.949 | 0.316 | $-0.500$ |
| 10 | 0.832 | 0.747 | 0.829 | 0.750 | 0.395 | 0.967 | 0.561 | 0.969 | 0.557 | 1.524 |
| 12 | 0.896 | 0.835 | 0.894 | 0.837 | 1.288 | 0.979 | 0.737 | 0.979 | 0.736 | 2.674 |
| 14 | 0.952 | 0.919 | 0.950 | 0.921 | 2.177 | 0.989 | 0.878 | 0.990 | 0.878 | 3.672 |

The results are given for two interaction strengths, stated as the ratio of $\varepsilon$ to pairing strength $G$. Also the chemical potential $\lambda$ is given.

Table 13.1 also gives the chemical potential $\lambda$ as a function of the number of particles in the valence space. The table demonstrates that $\lambda$ grows from negative to zero to positive as the two-level space is being filled. This means that in the beginning the ground-state energy becomes more negative when a pair of particles is added to the system, i.e. the system becomes more bound when adding particle pairs. Once the space is more than half full, $\lambda$ is positive, so adding more particle pairs reduces the total binding energy although the interactions are attractive.

The behaviour of the chemical potential can be understood in terms of the Pauli principle. As a result of their interaction, particles will scatter into available states. The more particles in the valence space, the fewer available states. Hence the normal attractive interactions are increasingly frustrated as the space fills.

The evolution of the chemical potential can also be seen in a single $j$ shell. Equation (13.99) gives values of $\lambda$ from $-\frac{1}{2}\Omega G$ to $\frac{1}{2}(\Omega - 2)G$, with zero at $\bar{n} = \Omega^2/(\Omega - 1)$. The ground-state energy (13.100) has its minimum at this point. The same feature shows up in the exact solution of the single-$j$ case, namely in the seniority model of Sect. 12.4. Equation (12.28) for the ground-state energy has its minimum at $N = \Omega + 1$.

We have seen that our simple solvable models exhibit striking features of many-fermion systems. In particular, these models demonstrate the competition between the attractive two-nucleon interaction and the effective repulsion caused by the Pauli principle. The relation of the chemical potential to the total binding energy of the system is further discussed in Sect. 14.2, where the Lipkin–Nogami pairing theory is introduced.

## Epilogue

In this chapter we have encountered a quasiparticle in the true meaning of the word. The BCS quasiparticles emerge from the Bogoliubov–Valatin transformation and they can be viewed as being partly particles, partly holes. This interesting property of theirs makes the notion of a precise particle number disappear for a quasiparticle theory of the nucleus. Consequences of this were seen and will be seen in the computed wave functions of the ground and excited states of nuclei. Theories based on quasiparticles and their configuration mixing will be developed further in the following chapters.

## Exercises

**13.1.** Derive the normalization condition (13.8).

**13.2.** Derive the transformation equations (13.12).

**13.3.** Derive the anticommutation relations (13.14).

**13.4.** Verify the relations (13.15).

**13.5.** Derive the contractions (13.18)–(13.20).

**13.6.** Starting from (8.17) show that

$$\bar{v}_{-\alpha,-\beta,-\gamma,-\delta} = (-1)^{j_a+j_b+j_c+j_d}\bar{v}_{\alpha\beta\gamma\delta}$$
$$= (-1)^{j_a-m_a+j_b-m_\beta+j_c-m_\gamma+j_d-m_\delta}\bar{v}_{\alpha\beta\gamma\delta} . \qquad (13.127)$$

**13.7.** Derive the relations (13.24) and (13.29).

**13.8.** Derive the quasiparticle representation (13.30) of the one-body part of the nuclear Hamiltonian.

**13.9.** Complete the details leading to the representation (13.32) of the nuclear Hamiltonian.

**13.10.** Derive the expressions (13.56) of the BCS occupation amplitudes by starting from (13.9) and (13.55).

**13.11.** Complete the details of the derivation of the ground-state expectation value (13.75).

**13.12.** Apply the calculations of Subsect. 13.5.1 to the case $j = \frac{7}{2}$ and $\varepsilon_{7/2} = 0$. Use the pure pairing force with $G = 1.0\,\text{MeV}$.

**13.13.** Plot the BCS ground-state energy of Exercise 13.12 as a function of the mean particle number $\bar{n}$ and compare with the corresponding plot of the seniority model.

**13.14.** Consider the $j = \frac{9}{2}$ case of the single $j$-shell model of Subsect. 13.5.1 for $\varepsilon_{9/2} = 0$. Use the pure pairing force with $G = 1.0\,\text{MeV}$. Plot the BCS ground-state energy as a function of the mean particle number $\bar{n}$ and compare with the corresponding plot of the seniority model.

**13.15.** Derive the expression (13.114) for $\eta_1$.

**13.16.** Plot the pairing gap $\Delta$ as a function of the interaction strength $G$ of the pure pairing force for the $j = \frac{13}{2}$ Lipkin model with a single-particle energy gap of $\varepsilon = 5.0\,\text{MeV}$.

**13.17.** Consider the $j = \frac{1}{2}$ Lipkin model with $\varepsilon = 1.0\,\text{MeV}$. Plot the ground-state energy as a function of the interaction strength $G$ for the BCS solution and the exact one.

**13.18.** Consider the BCS approach within the Lipkin model for two $j$ orbitals in the case of a general two-body interaction. Let

$$A \equiv \sum_J \hat{J}^2 \langle j^2\,;\,J|V|j^2\,;\,J\rangle , \quad B \equiv 2 \sum_{J=\text{even}} \hat{J}^2 \langle j^2\,;\,J|V|j^2\,;\,J\rangle . \qquad (13.128)$$

Calculate the BCS quantities $\Delta$, $\lambda$, $v_1$ and $v_2$ as functions of $A$ and $B$.

**13.19.** Take $j = \frac{7}{2}$ in Exercise 13.18 and calculate numerical values for the quantities $\Delta$, $\lambda$, $v_1$ and $v_2$ using the SDI with strength $A_1 = 1.0\,\text{MeV}$.

# 14

## Quasiparticle Mean Field: BCS and Beyond

### Prologue

In the previous two chapters we have laid the foundation for the BCS theory to describe open-shell nuclei. The properties of BCS solutions were compared with exact results from schematic solvable models. In this chapter we go into the details of numerical solution of the BCS equations. The implications of these solutions are discussed through applications to ds- and pf-shell nuclei.

Later in the chapter we introduce an improved version of BCS theory, namely the Lipkin–Nogami BCS (LNBCS) approach. We discuss the properties of the solutions of this theoretical approach and compare them with the results of plain BCS theory. Especially illuminating are the analyses performed within frameworks of exactly solvable models.

### 14.1 Numerical Solution of the BCS Equations

The BCS equations were derived in Sect. 13.3 and their solutions were devised for some schematic models in Sect. 13.5. We now set out to find a general solution. Specifically, the equations to be solved simultaneously are the BCS equations (13.56), (13.57) and (13.61), and the particle number condition (13.72). Collected together, the equations are

$$u_a = \theta^{(l_a)} \frac{1}{\sqrt{2}} \sqrt{1 + \frac{\eta_a}{E_a}} \quad \text{and}$$

$$v_a = \frac{1}{\sqrt{2}} \sqrt{1 - \frac{\eta_a}{E_a}} \quad \text{(occupation amplitudes)},$$

$$E_a = \sqrt{\eta_a^2 + \Delta_a^2} \quad \text{(quasiparticle energy)}, \tag{14.1}$$

$$2\widehat{j_a}\Delta_a = -\sum_b \frac{\widehat{j_b}\Delta_b}{\sqrt{\eta_b^2 + \Delta_b^2}} \langle a\,a\,;\,0|V|b\,b\,;\,0\rangle \quad \text{(gap equation)},$$

$$\bar{n} = \sum_a \widehat{j_a}^2 v_a^2 = \frac{1}{2}\sum_a \widehat{j_a}^2 \left(1 - \frac{\eta_a}{E_a}\right) \quad \text{(particle number)}.$$

These equations crystallize the information contained in the BCS framework. They must be solved numerically, which requires iterative methods. A suitable method is presented below.

### 14.1.1 Iterative Numerical Procedure

Equations (14.1) can be solved *iteratively*. We give the pairing gap and chemical potential initial, guessed values $\Delta = \Delta_0$ and $\lambda = \lambda_0$. From these values we calculate the other relevant quantities by using the BCS equations (13.56) and (13.57) plus those in (14.1) and the relations (13.45). The calculated quantities serve to define new values for $\Delta$ and $\lambda$ which, in turn, are used to generate new values for the relevant BCS quantities. This loop of computation continues until convergence is reached.

A flow chart of the computation is shown below. It contains the auxiliary quantities

$$S \equiv \frac{1}{2}\sum_a \widehat{j_a}^2 \left(1 - \frac{\varepsilon_a - \mu_a}{E_a}\right), \quad R \equiv \frac{1}{2}\sum_a \frac{\widehat{j_a}^2}{E_a} \tag{14.2}$$

used to express $\lambda$ as (Exercise 14.1)

$$\lambda = \frac{\bar{n} - S}{R}. \tag{14.3}$$

Convergence is judged in terms of a chosen small number $\epsilon$ for the difference between the computed particle number and the required number $\bar{n}$. The flow chart is as follows:

$$\boxed{\text{Set } \Delta = \Delta_0, \quad \lambda = \lambda_0} \tag{14.4}$$

$$\Downarrow$$

$$E_a^{(0)} = \sqrt{(\varepsilon_a - \lambda_0)^2 + \Delta_0^2}$$

$$u_a^{(0)} = \theta^{(l_a)} \frac{1}{\sqrt{2}} \sqrt{1 + \frac{\varepsilon_a - \lambda_0}{E_a^{(0)}}}, \quad v_a^{(0)} = \sqrt{1 - \left(u_a^{(0)}\right)^2} \tag{14.5}$$

$$\Downarrow$$

$$
\Delta_a^{(0)} = -\widehat{\jmath}_a^{-1} \sum_b \widehat{\jmath}_b u_b^{(0)} v_b^{(0)} \langle a\, a\,;\, 0 | V | b\, b\,;\, 0 \rangle
$$

$$
\mu_a^{(0)} = -\widehat{\jmath}_a^{-2} \sum_b \left( v_b^{(0)} \right)^2 \sum_J \widehat{J}^2 [\mathcal{N}_{ab}(J)]^{-2} \langle a\, b\,;\, J | V | a\, b\,;\, J \rangle \qquad (14.6)
$$

$$
\eta_a^{(0)} = \varepsilon_a - \lambda_0 - \mu_a^{(0)}
$$

$$\Downarrow$$

$$
E_a^{(1)} = \sqrt{\left( \eta_a^{(0)} \right)^2 + \left( \Delta_a^{(0)} \right)^2}
$$

$$
u_a^{(1)} = \theta^{(l_a)} \frac{1}{\sqrt{2}} \sqrt{1 + \frac{\eta_a^{(0)}}{E_a^{(1)}}} \,, \qquad v_a^{(1)} = \sqrt{1 - \left( u_a^{(1)} \right)^2}
$$

$$
n^{(1)} = \sum_a \widehat{\jmath}_a^{\,2} \left( v_a^{(1)} \right)^2 \qquad\qquad\qquad (14.7)
$$

$$
\lambda^{(1)} = \frac{\bar{n} - S}{R} \quad \text{where} \quad E_a = E_a^{(1)} \,, \quad \mu_a = \mu_a^{(0)}
$$

$$\Downarrow$$

$$
\delta n = \bar{n} - n^{(1)} \qquad\qquad (14.8)
$$

$$\Downarrow$$

If $|\delta n| \le \epsilon$ calculation finished

If $|\delta n| > \epsilon$ then $u_a^{(0)} \to u_a^{(1)}$ , $\quad v_a^{(0)} \to v_a^{(1)}$ , $\quad \lambda_0 \to \lambda_1 = \lambda^{(1)} + \delta\lambda$  (14.9)

Go to (14.6)

Note that in (14.7) we use the desired value $\bar{n}$ of the average particle number rather than the approximation $n^{(1)}$ in the evaluation of $\lambda^{(1)}$. This increases the stability of convergence.

The increment $\delta\lambda$ in (14.9) can be computed from the change $\delta n$ given in (14.8). For the link between $\delta\lambda$ and $\delta n$ we can write

$$
\delta\lambda = \kappa \delta n \,. \qquad (14.10)
$$

The proportionality factor $\kappa$ can be obtained by differentiation (Exercise 14.2):

$$
\kappa^{-1} = \frac{\partial n}{\partial \lambda} = \sum_a \frac{\partial n}{\partial \eta_a} \frac{\partial \eta_a}{\partial \lambda} = \tfrac{1}{2} \sum_a \widehat{\jmath}_a^{\,2} \frac{\Delta_a^2}{E_a^3} \,. \qquad (14.11)
$$

Alternatively $\kappa$ can be taken simply as an arbitrary energy of the order of 0.1 MeV.

The numerical examples of the following subsection show that for the particular case of the SDI all the self-energies $\mu_a$ are degenerate at $\mu_a \equiv \mu$. Then the pairing gaps $\Delta_a$, the quasiparticle energies $E_a$ and the occupation amplitudes $u_a$ and $v_a$ are independent of the self-energy $\mu$; only the chemical potential $\lambda$ depends on it. Therefore, for the SDI the physics of the problem is not affected by setting $\mu_a = 0$ at every iteration step. This improves the convergence of the procedure sketched in the flow chart (14.4)–(14.9). For a realistic interaction no such simplification occurs.

### 14.1.2 Application to Nuclei in the d-s and f-p-0g$_{9/2}$ Shells

To illustrate the numerical method we have carried out BCS calculations for nuclei in the d-s and f-p-0g$_{9/2}$ shells. The results are stated in Tables 14.1–14.3. The calculations cover the entire d-s shell and the upper two-thirds of the f-p-0g$_{9/2}$ shell; the first-third was omitted for lack of experimental data. The calculations were done with two-body matrix elements from the SDI with strength $A_1 = 1.0\,\mathrm{MeV}$. In the d-s shell we used the single-particle energies

$$\varepsilon_{0d_{5/2}} = 0 \,, \quad \varepsilon_{1s_{1/2}} = 0.87\,\mathrm{MeV} \,, \quad \varepsilon_{0d_{3/2}} = 5.08\,\mathrm{MeV} \qquad (14.12)$$

for both protons and neutrons. In the f-p shell we used the energies

$$\varepsilon_{0f_{7/2}} = 0 \,, \quad \varepsilon_{1p_{3/2}} = 4.80\,\mathrm{MeV} \,, \quad \varepsilon_{1p_{1/2}} = 6.82\,\mathrm{MeV} \,,$$
$$\varepsilon_{0f_{5/2}} = 8.40\,\mathrm{MeV} \,, \quad \varepsilon_{0g_{9/2}} = 8.80\,\mathrm{MeV} \qquad (14.13)$$

for both protons and neutrons. The energies (14.12) and (14.13) are somewhat different from those in Fig. 9.2; the inclusion of the 0g$_{9/2}$ orbital constitutes a qualitative difference.

### Quasiparticle Energies and the Effective Fermi Surface

Tables 14.1–14.3 include the computed occupation probabilities $v_a^2$ and quasiparticle energies $E_a$ for the orbitals $a$ of each valence space, and the chemical potential $\lambda$, the energy gap $\Delta$ and the self-energy $\mu$. Since the single-particle energies are the same for protons and neutrons the results are the same for both types of nucleon.

Note the absence of the orbital index $a$ on $\Delta$ and $\mu$. Like the pure pairing force, the SDI produces the same pairing gap and self-energy for all the orbitals. This is not the case for a general nucleon–nucleon interaction; for a realistic two-body force each level has its own pairing gap and self-energy. The last row of each table gives the chemical potential for the case where the self-energy is set equal to zero: $\lambda + \mu = \lambda(\mu = 0)$.

In the d-s valence space of Table 14.1 the $l$ quantum numbers are even $(l = 2, 0)$, so in both the CS and the BR phase conventions the amplitudes $u_a$

**Table 14.1.** BCS occupation probabilities $v_a^2$, quasiparticle energies $E_a$, chemical potential $\lambda$, pairing gap $\Delta$ and self-energy $\mu$ for different particle numbers $\bar{n}$ in the d-s shell

| Quantity | $\bar{n} = 2$ | $\bar{n} = 4$ | $\bar{n} = 6$ | $\bar{n} = 8$ | $\bar{n} = 10$ |
|---|---|---|---|---|---|
| $v_{0d_{5/2}}^2$ | 0.2661 | 0.5183 | 0.7504 | 0.9477 | 0.9780 |
| $v_{1s_{1/2}}^2$ | 0.1527 | 0.3457 | 0.5993 | 0.9133 | 0.9692 |
| $v_{0d_{3/2}}^2$ | 0.0245 | 0.0497 | 0.0748 | 0.1217 | 0.5485 |
| $E_{0d_{5/2}}$ (MeV) | 2.257 | 2.411 | 2.674 | 3.602 | 5.478 |
| $E_{1s_{1/2}}$ (MeV) | 2.772 | 2.533 | 2.362 | 2.849 | 4.654 |
| $E_{0d_{3/2}}$ (MeV) | 6.451 | 5.543 | 4.399 | 2.451 | 1.616 |
| $\lambda$ (MeV) | $-2.056$ | $-1.912$ | $-1.661$ | $-0.775$ | 0.237 |
| $\Delta$ (MeV) | 1.995 | 2.409 | 2.315 | 1.603 | 1.609 |
| $\mu$ (MeV)[a] | 1.000 | 2.000 | 3.000 | 4.000 | 5.000 |
| $\lambda(\mu = 0)$ (MeV) | $-1.056$ | 0.088 | 1.339 | 3.225 | 5.237 |

The last line gives the chemical potential calculated with zero self-energy. The calculation used the single-particle energies (14.12) and the SDI with $A_1 = 1.0\,\mathrm{MeV}$.
[a] The round numbers follow merely from $A_1 = 1.0\,\mathrm{MeV}$.

**Table 14.2.** BCS occupation probabilities $v_a^2$, quasiparticle energies $E_a$, chemical potential $\lambda$, pairing gap $\Delta$ and self-energy $\mu$ for different particle numbers $\bar{n}$ in the f-p-$0g_{9/2}$ shell

| Quantity | $\bar{n} = 10$ | $\bar{n} = 12$ | $\bar{n} = 14$ | $\bar{n} = 16$ | $\bar{n} = 18$ |
|---|---|---|---|---|---|
| $v_{0f_{7/2}}^2$ | 0.7117 | 0.7719 | 0.8163 | 0.8511 | 0.8800 |
| $v_{1p_{3/2}}^2$ | 0.3462 | 0.4499 | 0.5461 | 0.6312 | 0.7052 |
| $v_{1p_{1/2}}^2$ | 0.2258 | 0.3080 | 0.3962 | 0.4856 | 0.5729 |
| $v_{0f_{5/2}}^2$ | 0.1622 | 0.2239 | 0.2946 | 0.3724 | 0.4554 |
| $v_{0g_{9/2}}^2$ | 0.1497 | 0.2066 | 0.2725 | 0.3461 | 0.4261 |
| $E_{0f_{7/2}}$ (MeV) | 6.702 | 7.638 | 8.558 | 9.436 | 10.267 |
| $E_{1p_{3/2}}$ (MeV) | 6.380 | 6.442 | 6.656 | 6.961 | 7.318 |
| $E_{1p_{1/2}}$ (MeV) | 7.261 | 6.942 | 6.775 | 6.720 | 6.745 |
| $E_{0f_{5/2}}$ (MeV) | 8.234 | 7.688 | 7.269 | 6.948 | 6.700 |
| $E_{0g_{9/2}}$ (MeV) | 8.509 | 7.916 | 7.443 | 7.060 | 6.747 |
| $\lambda$ (MeV) | $-2.162$ | $-1.846$ | $-1.586$ | $-1.374$ | $-1.197$ |
| $|\Delta|$ (MeV) | 6.071 | 6.410 | 6.628 | 6.717 | 6.673 |
| $\mu$ (MeV) | 5.000 | 6.000 | 7.000 | 8.000 | 9.000 |
| $\lambda(\mu = 0)$ (MeV) | 2.838 | 4.154 | 5.414 | 6.626 | 7.803 |

The last line gives the chemical potential calculated with zero self-energy. The calculation used the single-particle energies (14.13) and the SDI with $A_1 = 1.0\,\mathrm{MeV}$.

**Table 14.3.** Continuation of Table 14.2 for particle numbers $\bar{n} = 20$–$28$

| Quantity | $\bar{n} = 20$ | $\bar{n} = 22$ | $\bar{n} = 24$ | $\bar{n} = 26$ | $\bar{n} = 28$ |
|---|---|---|---|---|---|
| $v^2_{0f_{7/2}}$ | 0.9049 | 0.9271 | 0.9473 | 0.9660 | 0.9835 |
| $v^2_{1p_{3/2}}$ | 0.7696 | 0.8261 | 0.8762 | 0.9214 | 0.9624 |
| $v^2_{1p_{1/2}}$ | 0.6561 | 0.7345 | 0.8078 | 0.8762 | 0.9402 |
| $v^2_{0f_{5/2}}$ | 0.5423 | 0.6319 | 0.7230 | 0.8152 | 0.9076 |
| $v^2_{0g_{9/2}}$ | 0.5116 | 0.6019 | 0.6963 | 0.7943 | 0.8957 |
| $E_{0f_{7/2}}$ (MeV) | 11.053 | 11.797 | 12.504 | 13.177 | 13.820 |
| $E_{1p_{3/2}}$ (MeV) | 7.699 | 8.092 | 8.485 | 8.876 | 9.260 |
| $E_{1p_{1/2}}$ (MeV) | 6.826 | 6.945 | 7.091 | 7.255 | 7.432 |
| $E_{0f_{5/2}}$ (MeV) | 6.508 | 6.359 | 6.244 | 6.155 | 6.087 |
| $E_{0g_{9/2}}$ (MeV) | 6.486 | 6.266 | 6.076 | 5.911 | 5.766 |
| $\lambda$ (MeV) | $-1.049$ | $-0.923$ | $-0.815$ | $-0.720$ | $-0.637$ |
| $|\Delta|$ (MeV) | 6.485 | 6.134 | 5.588 | 4.778 | 3.525 |
| $\mu$ (MeV) | 10.000 | 11.000 | 12.000 | 13.000 | 14.000 |
| $\lambda(\mu = 0)$ (MeV) | 8.951 | 10.077 | 11.185 | 12.280 | 13.363 |

are positive according to (13.56). Because of the relation (13.60) the pairing gap $\Delta$ is then also positive.

In the f-p-$g_{9/2}$ valence space of Tables 14.2 and 14.3 we have both odd and even $l$ values. For the former ($l = 3, 1$) the $u_a$ amplitudes are negative in the CS phase convention, while for the 'intruder' orbital $0g_{9/2}$ ($l = 4$) they are positive. Accordingly, $\Delta$ is *negative* for the f-p orbitals and *positive* for the $0g_{9/2}$ orbital with CS phases. With BR phases $\Delta$ is always positive. To avoid any phase confusion we record $|\Delta|$ in Tables 14.2 and 14.3.

Tables 14.1–14.3 display the obvious result that the occupation probabilities increase with increasing number of valence particles. At the same time the chemical potential $\lambda$ rises monotonically as more particles are added to the valence space. As discussed in Subsect. 13.5.4, the increase in the chemical potential reflects the Pauli principle: the fuller the valence space, the less ground-state binding energy is gained by adding a couple of paired nucleons.

The magnitude of the pairing gap $\Delta$ attains its maximum value when half of the valence space is occupied. This is the behaviour already encountered in the single-$j$-shell model and the Lipkin model in Sect. 13.5. The Pauli principle hinders the increase in stability of the BCS ground state beyond the middle of the shell.

For a clear illustration of the trends of the quasiparticle energies with increasing number $N_{\rm act}$ of valence particles we have plotted the quasiparticle energies of Tables 14.1–14.3 in Figs. 14.1 and 14.2. The results are the same for protons and neutrons due to the same single-particle energies used.

In the d-s shell the $0d_{3/2}$ quasiparticle energy drops with increasing number of valence particles while the $0d_{5/2}$ and $1s_{1/2}$ energies increase. This can be understood from the expression (13.57) for the quasiparticle energy. For the

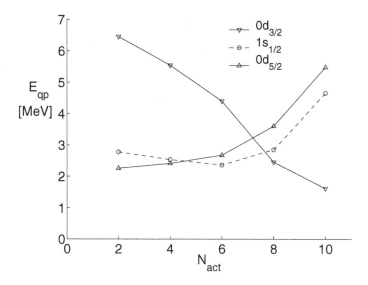

**Fig. 14.1.** Quasiparticle energies in the d-s valence space as functions of the number $N_{\mathrm{act}}$ of valence particles in the shell

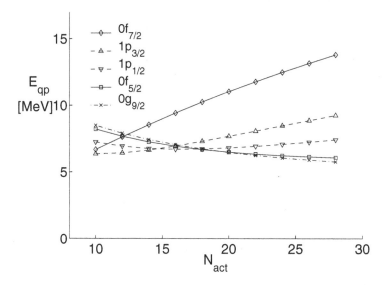

**Fig. 14.2.** Quasiparticle energies in the f-p-$0g_{9/2}$ valence space as functions of the number $N_{\mathrm{act}}$ of valence particles in the shell

SDI the pairing gap is common to all the orbitals, so the ordering of the quasiparticle energies $E_a$ is determined solely by the parameter $\eta_a$. Equation (13.45) shows that the smaller is $\eta_a$ the closer the single-particle energy is to the effective Fermi level $\lambda(\mu = 0)$. Table 14.1 shows that $\lambda(\mu = 0)$ is close to the $0d_{5/2}$ and $1s_{1/2}$ orbitals in the beginning of the d-s space and close to the $0d_{3/2}$ orbital at the end of it.

Quasiparticle energies behave similarly in the f-p-$0g_{9/2}$ valence space, as shown by Tables 14.2 and 14.3 and Fig. 14.2. This behaviour is a general feature of BCS theory, independent of the particular valence space and two-body interaction. In short, quasiparticle energies $E_a$ are governed by the differences $\varepsilon_a - \lambda(\mu_a = 0)$.

## Quasiparticle Spectra

A BCS calculation is always done for an even–even nucleus, which we call the *reference nucleus*. The calculation concerns only one kind of nucleon, with no regard to the other kind.[1] Let the reference nucleus have $Z$ protons and $N$ neutrons. The BCS calculation creates the ground state $|BCS\rangle$ of the reference nucleus and a set of quasiparticle creation operators $a_\alpha^\dagger$. Suppose that the calculation was done for protons, so we have proton quasiparticles. Then the results provide the quasiparticle energies $E_a$ for all nuclei, within the valence space of the reference nucleus, with $Z \pm 1$ protons and $N, N \pm 2, \ldots$ neutrons. The relative energies of the quasiparticles constitute the same low-energy spectrum for all such nuclei. The interpretation of neutron quasiparticles is analogous.

The relative energies of quasiparticles change with the number of particles in the valence space, as is clearly seen from Figs. 14.1 and 14.2. Comparison of quasiparticle spectra with experiment is shown in Figs. 14.3 and 14.4 for the d-s shell and in Figs. 14.5–14.7 for the f-p-$0g_{9/2}$ shell. Note that the calculated proton and neutron quasiparticle spectra are the same due to the assumed identical single-particle energies. However, the experimental proton and neutron single-particle energies differ due to Coulomb effects, increasingly with increasing $A$. Even the proton and neutron valence spaces can become different, which results in dissimilar quasiproton and quasineutron spectra.

In Figs. 14.3 and 14.4 we can follow the change of the ground state of the odd-$A$ system as a function of the number of valence particles in the d-s shell. For $N_{\text{act}} = 2$ the ground state is $0d_{5/2}$. This agrees with the experimental ground state of $^{19}_{8}O_{11}$, but there is no agreement for the other two states. For $N_{\text{act}} = 4$ the BCS quasiparticle spectrum qualitatively reproduces the experimental spectra of $^{25}_{12}Mg_{13}$ and $^{25}_{13}Al_{12}$. For $N_{\text{act}} = 6$ the ground state comes out as $1s_{1/2}$, in agreement with the spectra of $^{29}_{14}Si_{15}$ and $^{29}_{15}P_{14}$. The other two states come out wrong.

The failure of the BCS calculations to predict the higher states can stem from the interaction type and strength or from the chosen single-particle energies (the Hartree–Fock single-particle field depends on the proton and neutron

---

[1] However, in a case like the present where the single-particle energies and two-body interactions are the same for protons and neutrons, the same calculation applies to either kind.

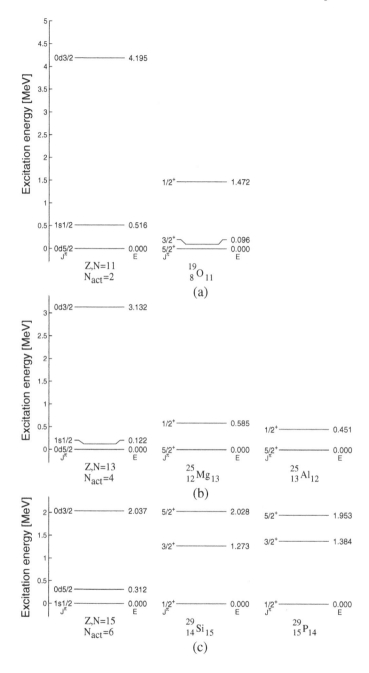

**Fig. 14.3.** Comparison of the calculated BCS quasiparticle spectrum for 2 (a), 4 (b) and 6 (c) active particles in the d-s shell with the corresponding experimental spectra of neutron-odd and proton-odd nuclei. The single-particle energies (14.12) and the SDI with $A_1 = 1.0\,\mathrm{MeV}$ were used in the calculations

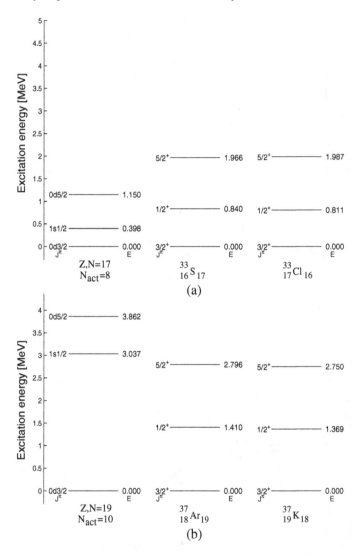

**Fig. 14.4.** Comparison of the calculated BCS quasiparticle spectrum for 8 (a) and 10 (b) active particles in the d-s shell with the correponding experimental spectra of neutron-odd and proton-odd nuclei. The single-particle energies (14.12) and the SDI with $A_1 = 1.0\,\text{MeV}$ were used in the calculations

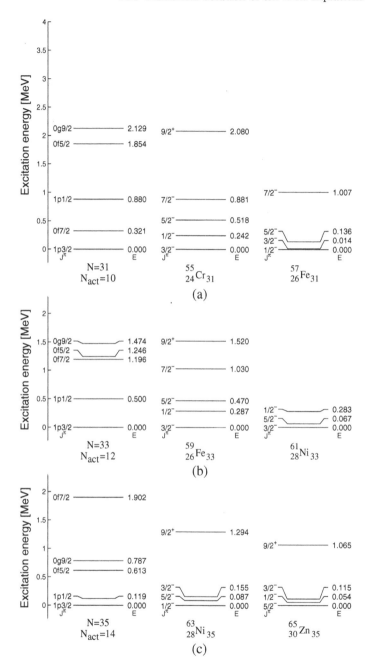

**Fig. 14.5.** Comparison of the calculated BCS quasiparticle spectrum for 10 (a), 12 (b) and 14 (c) active particles in the f-p-$0g_{9/2}$ shell with the corresponding experimental spectra of neutron-odd nuclei. The single-particle energies (14.13) and the SDI with $A_1 = 1.0$ MeV were used in the calculations

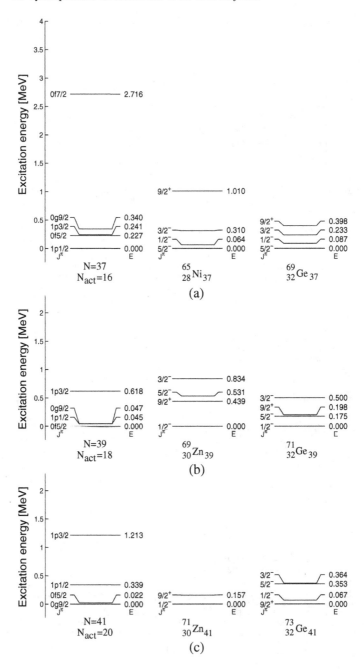

**Fig. 14.6.** Comparison of the calculated BCS quasiparticle spectrum for 16 (a), 18 (b) and 20 (c) active particles in the f-p-$0g_{9/2}$ shell with the corresponding experimental spectra of neutron-odd nuclei. The single-particle energies (14.13) and the SDI with $A_1 = 1.0$ MeV were used in the calculations

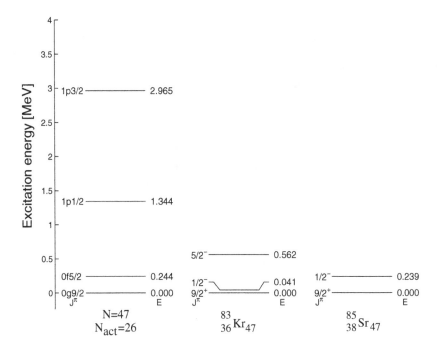

**Fig. 14.7.** Comparison of the calculated BCS quasiparticle spectrum for 26 active particles in the f-p-$0g_{9/2}$ shell with the correponding experimental spectra of neutron-odd nuclei. The single-particle energies (14.13) and the SDI with $A_1 = 1.0\,\mathrm{MeV}$ were used in the calculations

numbers). Yet another possible reason is that three-quasiparticle components may mix into the one-quasiparticle states.

Figure 14.4 shows that for $N_{\mathrm{act}} = 8$ the $0d_{3/2}$ quasiparticle has the lowest energy. The calculated spectrum agrees even semi-quantitatively with the experimental spectra of $^{33}_{16}S_{17}$ and $^{33}_{17}Cl_{16}$. Finally, the $N_{\mathrm{act}} = 10$ case is in qualitative agreement with the spectra of $^{37}_{18}Ar_{19}$ and $^{37}_{19}K_{18}$.

Figures 14.5–14.7 show the quasiparticle spectra of the f-p-$0g_{9/2}$ shell nuclei. The orbital $1p_{3/2}$ has the lowest quasiparticle energy for $N_{\mathrm{act}} = 10$–$14$. The nuclei in Fig. 14.5 have a $3/2^-$ state as the ground state or a low-lying excited state. Where known, the order of the $0f$ quasiparticle states is reversed from that predicted. The observed $9/2^+$ states are well reproduced by the $0g_{9/2}$ quasiparticle. A one-to-one correspondence can be assigned between all the experimental levels of $^{55}_{24}Cr_{31}$ and $^{59}_{26}Fe_{33}$ and the BCS levels.

For $N_{\mathrm{act}} = 16$ the $1p_{1/2}$ quasiparticle energy becomes the lowest one. The corresponding state is the first excited state in $^{65}_{28}Ni_{37}$ and $^{69}_{32}Ge_{37}$ at very low excitation. In the experimental spectra the $5/2^-$ state has become the ground state, so the two lowest BCS states have reversed their order relative to the experimental spectrum.

For the $N_{\text{act}} = 18$ case the lowest quasiparticle is $0f_{5/2}$, with $1p_{1/2}$ and $0g_{9/2}$ nearly degenerate with it. Small perturbations to the single-particle energies or the interaction could cause the computed spectrum to correspond to the experimental sequence of energies in $^{69}_{30}\text{Zn}_{39}$ and $^{71}_{32}\text{Ge}_{39}$.

Finally, for $N_{\text{act}} = 20\text{--}28$, the $0g_{9/2}$ quasiparticle lies the lowest. As shown in Fig. 14.6, in the $N_{\text{act}} = 20$ case this is borne out by $^{73}_{32}\text{Ge}_{41}$ and nearly so by $^{71}_{30}\text{Zn}_{41}$. Also the other levels of $^{73}\text{Ge}$ correspond rather well to the calculated quasiparticle energies. For $N_{\text{act}} = 26$ in Fig. 14.7, the BCS calculation correctly predicts the ground states of $^{83}_{36}\text{Kr}_{47}$ and $^{85}_{38}\text{Sr}_{47}$.

In conclusion of this section we recall that all our quasiparticle calculations used the same single-particle energies (14.12) and (14.13) and the same SDI strength $A_1 = 1.0\,\text{MeV}$. Fine-tuning the single-particle energies and the SDI strength when progressing along the mass scale would surely improve the correspondence between the quasiparticle spectra and the experimental ones. It is also a fact that the low-energy spectra of the nuclei discussed can be contaminated by contributions from three-quasiparticle and higher degrees of freedom.

## 14.2 Lipkin–Nogami BCS Theory

In Subsect. 13.5.2 we discovered that there exists a critical value $G_{\text{crit}}$ below which the interaction strength $G$ of the pure pairing force is not large enough to support a BCS solution. Then the pairing gap collapses and the sharp Fermi surface of the particle–hole vacuum is recovered. This happens when the interaction strength is small relative to the average spacing of the single-particle levels around the Fermi surface.

It was seen in connection with the Lipkin model in Subsect. 13.5.2 that the *exact* solutions do not display this critical behaviour. Instead, they behave smoothly as $G \to 0$, as shown in Fig. 13.2. It turns out that the critical behaviour of the BCS solution can be cured by resorting to an improved model Hamiltonian within the so-called LN BCS approach.

### 14.2.1 The Lipkin–Nogami Model Hamiltonian

The unwanted critical behaviour of the BCS solutions derives from the nucleon number fluctuations caused by the ground-state ansatz (13.1) of standard BCS theory. These fluctuations can be reduced by replacing the BCS auxiliary Hamiltonian (13.40)

$$\mathcal{H}_{\text{BCS}} = H - \lambda \hat{n} \tag{14.14}$$

by the *Lipkin–Nogami* auxiliary Hamiltonian

$$\boxed{\mathcal{H}_{\mathrm{LNBCS}} = H - \lambda_1 \hat{n} - \lambda_2 \hat{n}^2 .} \tag{14.15}$$

The Hamiltonian (14.15) was first suggested by Lipkin [74] and Nogami [75] in the early 1960s. The variation is now applied to minimize the ground-state energy in the trial form

$$\begin{aligned}
E_0^{\mathrm{LN}} &= \langle \mathrm{LNBCS}|H|\mathrm{LNBCS}\rangle - \lambda_1 \langle \mathrm{LNBCS}|\hat{n} - \bar{n}|\mathrm{LNBCS}\rangle \\
&\quad - \lambda_2 \langle \mathrm{LNBCS}|\hat{n}^2 - \bar{n}^2|\mathrm{LNBCS}\rangle \\
&= \langle \mathrm{LNBCS}|\mathcal{H}_{\mathrm{LNBCS}}|\mathrm{LNBCS}\rangle + \lambda_1 \bar{n} + \lambda_2 \bar{n}^2 ,
\end{aligned} \tag{14.16}$$

where $\bar{n}$ is the average particle number according to

$$\bar{n} = \langle \mathrm{LNBCS}|\hat{n}|\mathrm{LNBCS}\rangle . \tag{14.17}$$

The state $|\mathrm{LNBCS}\rangle$ is the vacuum of the Lipkin–Nogami quasiparticles that are assumed to have the Bogoliubov–Valatin form contained in (13.10). The $\lambda_2$ term serves to reduce the width of the nucleon number distribution in the quasiparticle vacuum. This makes the Lipkin–Nogami extension more accurate than plain BCS theory. Note that for $\lambda_2 = 0$ the energy $E_0^{\mathrm{LN}}$ above coincides with the energy $H_0$ in (13.66). For more information on the original work see [76, 77].

To proceed with the new Hamiltonian (14.15) we will express the square of the number operator $\hat{n}$ in terms of uncoupled quasiparticle operators. To that end we rewrite (13.38) in the uncoupled form

$$\hat{n} = \sum_a \hat{j}_a^2 v_a^2 + \sum_\alpha (u_a^2 - v_a^2) a_\alpha^\dagger a_\alpha - \sum_\alpha u_a v_a (a_\alpha^\dagger \tilde{a}_\alpha^\dagger + \tilde{a}_\alpha a_\alpha) . \tag{14.18}$$

From this we see that only the constant term of $\hat{n}$ contributes to (14.17), with the result

$$\boxed{\bar{n} = \sum_a \hat{j}_a^2 v_a^2 .} \tag{14.19}$$

This is the same as the BCS result (13.39).

The square of the number operator (14.18) consists of terms of the form

$$\hat{n}^2 = (\hat{n}^2)_{00} + (\hat{n}^2)_{11} + (\hat{n}^2)_{20} + (\hat{n}^2)_{40} + (\hat{n}^2)_{31} + (\hat{n}^2)_{22} , \tag{14.20}$$

where the subscripts indicate the numbers of creation and annihilation operators commuted to normal order. Evaluated, the terms are

$$(\hat{n}^2)_{00} = \left( \sum_a \hat{j}_a^2 v_a^2 \right)^2 + 2 \sum_a \hat{j}_a^2 u_a^2 v_a^2 , \tag{14.21}$$

$$(\hat{n}^2)_{11} = \sum_\alpha \left[ \left( 2 \sum_b \hat{j}_b^2 v_b^2 + u_a^2 - v_a^2 \right)(u_a^2 - v_a^2) - 4 u_a^2 v_a^2 \right] a_\alpha^\dagger a_\alpha , \tag{14.22}$$

$$(\hat{n}^2)_{20} = -2\sum_{\alpha}\left(\sum_{b}\hat{j}_b^2 v_b^2 + u_a^2 - v_a^2\right)u_a v_a(a_\alpha^\dagger\tilde{a}_\alpha^\dagger + \tilde{a}_\alpha a_\alpha)\,, \qquad (14.23)$$

$$(\hat{n}^2)_{40} = \sum_{\alpha\beta}u_a v_a u_b v_b(a_\alpha^\dagger\tilde{a}_\alpha^\dagger a_\beta^\dagger\tilde{a}_\beta^\dagger + \tilde{a}_\beta a_\beta\tilde{a}_\alpha a_\alpha)\,, \qquad (14.24)$$

$$(\hat{n}^2)_{31} = -2\sum_{\alpha\beta}(u_a^2 - v_a^2)u_b v_b(a_\beta^\dagger\tilde{a}_\beta^\dagger a_\alpha^\dagger a_\alpha + a_\alpha^\dagger a_\alpha\tilde{a}_\beta a_\beta)\,, \qquad (14.25)$$

$$(\hat{n}^2)_{22} = 2\sum_{\alpha\beta}u_a v_a u_b v_b a_\alpha^\dagger\tilde{a}_\alpha^\dagger\tilde{a}_\beta a_\beta - \sum_{\alpha\beta}(u_a^2 - v_a^2)(u_b^2 - v_b^2)a_\alpha^\dagger a_\beta^\dagger a_\alpha a_\beta\,.$$
$$\qquad (14.26)$$

We set out to calculate the vacuum expectation value of the auxiliary Hamiltonian (14.15), i.e. evaluate the expression

$$\mathcal{H}_0^{\mathrm{LN}} \equiv \langle\mathrm{LNBCS}|\mathcal{H}_{\mathrm{LNBCS}}|\mathrm{LNBCS}\rangle = \langle\mathrm{LNBCS}|H|\mathrm{LNBCS}\rangle$$
$$- \lambda_1\langle\mathrm{LNBCS}|\hat{n}|\mathrm{LNBCS}\rangle - \lambda_2\langle\mathrm{LNBCS}|\hat{n}^2|\mathrm{LNBCS}\rangle\,. \qquad (14.27)$$

With use of (13.43) this becomes

$$\mathcal{H}_0^{\mathrm{LN}} = \sum_a(\varepsilon_a - \lambda_1)\hat{j}_a^2 v_a^2 + \frac{1}{2}\sum_{abJ}v_a^2 v_b^2\hat{J}^2[\mathcal{N}_{ab}(J)]^{-2}\langle ab\,; J|V|ab\,; J\rangle$$
$$+ \frac{1}{2}\sum_{ab}\hat{j}_a\hat{j}_b u_a v_a u_b v_b\langle aa\,; 0|V|bb\,; 0\rangle$$
$$- \lambda_2\left(\sum_u\hat{j}_a^2 v_a^2\right)^2 - 2\lambda_2\sum_a\hat{j}_a^2 u_a^2 v_a^2\,. \qquad (14.28)$$

The BCS quantities $\Delta_a$, $\mu_a$ and $\eta_a$ are defined by Eq. (13.45). We can recognize them among the terms of (14.28). Note, however, that due to the role played by $\lambda_2$ in Lipkin–Nogami theory, the LNBCS parameter $\lambda_1$ is not the same as the BCS parameter $\lambda$. Thus, keeping in mind that the LNBCS quantities of BCS form relate to $\lambda_1(\neq \lambda)$, we write (14.28) as

$$\mathcal{H}_0^{\mathrm{LN}} = \sum_a(\eta_a + \mu_a)\hat{j}_a^2 v_a^2 - \frac{1}{2}\sum_a\hat{j}_a^2 v_a^2\mu_a - \frac{1}{2}\sum_a\hat{j}_a^2 u_a v_a\Delta_a$$
$$- \lambda_2\sum_{ab}\hat{j}_a^2\hat{j}_b^2 v_a^2 v_b^2 - 2\lambda_2\sum_a\hat{j}_a^2 u_a^2 v_a^2$$
$$= \sum_a\hat{j}_a^2\left[v_a^2\left(\eta_a + \tfrac{1}{2}\mu_a - \lambda_2\sum_b\hat{j}_b^2 v_b^2 - 2\lambda_2 u_a^2\right) - \tfrac{1}{2}u_a v_a\Delta_a\right]\,. \qquad (14.29)$$

Denoting

$$\xi_a \equiv \eta_a - \lambda_2\sum_b\hat{j}_b^2 v_b^2 - 2\lambda_2 u_a^2 \qquad (14.30)$$

we obtain

$$\mathcal{H}_0^{\mathrm{LN}} = \sum_a \widehat{j_a}^2 \left[ v_a^2 (\xi_a + \tfrac{1}{2}\mu_a) - \tfrac{1}{2} u_a v_a \Delta_a \right] , \qquad (14.31)$$

which has the form of the expression (13.46) for the BCS quantity $\mathcal{H}_0$.

For impending use, we define

$$\tilde{\varepsilon}_a \equiv \varepsilon_a + 4\lambda_2 v_a^2 , \quad \tilde{\lambda} \equiv \lambda_1 + 2\lambda_2 \left( 1 + \sum_b \widehat{j_b}^2 v_b^2 \right) , \quad \tilde{\eta}_a \equiv \tilde{\varepsilon}_a - \tilde{\lambda} - \mu_a . \quad (14.32)$$

These definitions lead to

$$\tilde{\eta}_a = \eta_a + 4\lambda_2 v_a^2 - 2\lambda_2 \left( 1 + \sum_b \widehat{j_b}^2 v_b^2 \right) , \qquad (14.33)$$

$$\xi_a = \tilde{\eta}_a - 2\lambda_2 v_a^2 + \lambda_2 \sum_b \widehat{j_b}^2 v_b^2 . \qquad (14.34)$$

Note that the LNBSC quantities with tildes are reduced to the corresponding BCS quantities, and $\xi_a$ to $\eta_a$, when $\lambda_2 \to 0$.

### 14.2.2 Derivation of the Lipkin–Nogami BCS Equations

As seen from (14.16) and (14.27), the ground-state energy $E_0^{\mathrm{LN}}$ and the auxiliary quantity $\mathcal{H}_0^{\mathrm{LN}}$ differ only by constant terms. Minimizing $\mathcal{H}_0^{\mathrm{LN}}$ is therefore equivalent to minimizing $E_0^{\mathrm{LN}}$. It follows that the variational problem in Lipkin–Nogami theory is analogous to (13.49), so we require

$$\frac{\partial}{\partial v_c} \mathcal{H}_0^{\mathrm{LN}} = 0 \quad \text{for all } c . \qquad (14.35)$$

We already have the BCS part of this, $\partial \mathcal{H}_0 / \partial v_c$, in (13.53). Differentiating the $\lambda_2$ terms as stated in (14.28), with use of (13.50), gives

$$\begin{aligned}
0 &= \frac{\partial}{\partial v_c} \mathcal{H}_0 - \lambda_2 \frac{\partial}{\partial v_c} \left[ \left( \sum_a \widehat{j_a}^2 v_a^2 \right)^2 + 2 \sum_a \widehat{j_a}^2 u_a^2 v_a^2 \right] \\
&= 2\widehat{j_c}^2 v_c \eta_c + \widehat{j_c}^2 \frac{v_c^2 - u_c^2}{u_c} \Delta_c - 4\lambda_2 \widehat{j_c}^2 v_c \left( \sum_a \widehat{j_a}^2 v_a^2 + u_c^2 - v_c^2 \right) \\
&= 2\widehat{j_c}^2 v_c \tilde{\eta}_c + \widehat{j_c}^2 \frac{v_c^2 - u_c^2}{u_c} \Delta_c ,
\end{aligned} \qquad (14.36)$$

where (14.33) was used in the last step. It follows that

$$(u_c^2 - v_c^2) \Delta_c = 2 u_c v_c \tilde{\eta}_c , \qquad (14.37)$$

which is the same as the condition (13.54) of BCS theory except that $\eta_c$ is replaced by $\tilde{\eta}_c$.

Because of the analogy of (14.37) to (13.54) we can immediately convert the BCS equations (13.56) into their Lipkin–Nogami counterparts

$$u_c = \frac{1}{\sqrt{2}} \theta^{(l_c)} \sqrt{1 + \frac{\tilde{\eta}_c}{\sqrt{\tilde{\eta}_c^2 + \Delta_c^2}}} \; , \qquad v_c = \frac{1}{\sqrt{2}} \sqrt{1 - \frac{\tilde{\eta}_c}{\sqrt{\tilde{\eta}_c^2 + \Delta_c^2}}} \; , \qquad (14.38)$$

together with the gap equation

$$2\widehat{j_a} \Delta_a = -\sum_b \frac{\widehat{j_b} \Delta_b}{\sqrt{\tilde{\eta}_b^2 + \Delta_b^2}} \langle a\, a\, ;\, 0|V|b\, b\, ;\, 0\rangle \qquad (14.39)$$

following from (13.61). The phase factor $\theta^{(l_c)}$ is given in (13.58). Equations (14.38) and (14.39) are the *LNBCS equations*.

The BCS quasiparticles emerged from the term

$$\mathcal{H}_{11} \equiv \sum_b \mathcal{H}_{11}(b) \big[a_b^\dagger \tilde{a}_b\big]_{00} \qquad (14.40)$$

of the auxiliary Hamiltonian $\mathcal{H}_{\mathrm{BCS}}$, with $\mathcal{H}_{11}(b)$ given by (13.47). To establish the concept of LNBCS quasiparticle, let us construct the corresponding term $\mathcal{H}_{11}^{\mathrm{LN}}$. This is done by including the $\lambda_2$ contribution (14.22) according to the scheme (14.15). Since we used in Chap. 13 the BCS Hamiltonian in the coupled form (13.32), we also need the $\hat{n}^2$ contribution in coupled form. Equation (14.22) yields

$$(\hat{n}^2)_{11} = \sum_b \widehat{j_b} \Big[\Big(2\sum_a \widehat{j_a}^2 v_a^2 + u_b^2 - v_b^2\Big)(u_b^2 - v_b^2) - 4u_b^2 v_b^2\Big]\big[a_b^\dagger \tilde{a}_b\big]_{00} \; . \qquad (14.41)$$

For each term $b$, we then have

$$\begin{aligned}
\mathcal{H}_{11}^{\mathrm{LN}}(b) &= \mathcal{H}_{11}(b) - \lambda_2(\hat{n}^2)_{11}(b) = \widehat{j_b}\big[(u_b^2 - v_b^2)\eta_b + 2u_b v_b \Delta_b\big] \\
&\quad - \lambda_2 \widehat{j_b}\Big[\Big(2\sum_a \widehat{j_a}^2 v_a^2 + u_b^2 - v_b^2\Big)(u_b^2 - v_b^2) - 4u_b^2 v_b^2\Big] \\
&= \widehat{j_b}\big[(u_b^2 - v_b^2)(\tilde{\eta} - 2\lambda_2 v_b^2 + \lambda_2) + 2u_b v_b \Delta_b + 4\lambda_2 u_b^2 v_b^2\big] \\
&= \widehat{j_b}\Big(\sqrt{\tilde{\eta}_b^2 + \Delta_b^2} + \lambda_2\Big) \equiv \widehat{j_b} E_b^{\mathrm{LN}} \; , \qquad (14.42)
\end{aligned}$$

where (14.33) was used for the second equality and (14.37) and (14.38) for the third. This establishes the Lipkin–Nogami quasiparticle energy as

$$E_b^{\mathrm{LN}} = \sqrt{\tilde{\eta}_b^2 + \Delta_b^2} + \lambda_2 \; . \qquad (14.43)$$

The expression (14.43) differs from the BCS expression (13.57) through the replacement of $\eta_b$ by $\tilde{\eta}_b$ and the addition of the term $\lambda_2$. Applications reveal that the LNBCS result is a significant improvement over the BCS result. An example of this is given by the single-$j$-shell model in Subsect. 14.3.1.

Similar to the construction of $\mathcal{H}_{11}^{LN}$, let us now construct $\mathcal{H}_{20}^{LN}$. The BCS contribution is given through (13.48). The $\hat{n}^2$ contribution (14.23) is in coupled form

$$(\hat{n}^2)_{20} = 2 \sum_b \hat{j}_b \Big( \sum_a \hat{j}_a^2 v_a^2 + u_b^2 - v_b^2 \Big) u_b v_b \Big( \big[a_b^\dagger a_b^\dagger\big]_{00} - \big[\tilde{a}_b \tilde{a}_b\big]_{00} \Big) . \tag{14.44}$$

Substitution gives

$$\begin{aligned}
\mathcal{H}_{20}^{LN}(b) &= \mathcal{H}_{20}(b) - \lambda_2 (\hat{n}^2)_{20}(b) = \hat{j}_b \big[ u_b v_b \eta_b - \tfrac{1}{2}(u_b^2 - v_b^2)\Delta_b \big] \\
&\quad - 2\lambda_2 \hat{j}_b \Big( \sum_a \hat{j}_a^2 v_a^2 + u_b^2 - v_b^2 \Big) u_b v_b \\
&= \hat{j}_b u_b v_b \Big[ \eta_b - \tilde{\eta}_b - 2\lambda_2 \Big( \sum_a \hat{j}_a^2 v_a^2 + u_b^2 - v_b^2 \Big) \Big] = 0 ,
\end{aligned} \tag{14.45}$$

where we used (14.37) and then (14.33). This repeats the BCS result (13.64): also in the LNBCS the $\mathcal{H}_{20}$ part of the auxiliary Hamiltonian vanishes. This concludes the derivation of the LNBCS mean field with its single-quasiparticle energies (14.43).

We still do not know the value of $\lambda_2$. To find it one has to delve into the Lipkin–Nogami residual interaction analogous to (13.33). We omit the long derivation and only state the result. With the abbreviation

$$V_{abcd}^{(J)} \equiv \langle a\,b;\, J|V|c\,d;\, J \rangle \tag{14.46}$$

it is

$$\lambda_2 = \frac{(\sum_a u_a^2 v_a^2)^2 \sum_J \hat{J}^2 [\mathcal{N}_{ab}(J)]^{-2} V_{abab}^{(J)} - \sum_{ab} \hat{j}_a \hat{j}_b u_a^3 v_a u_b v_b^3 V_{aabb}^{(0)}}{2(\sum_a \hat{j}_a^2 u_a^2 v_a^2)^2 - 4 \sum_a \hat{j}_a^2 u_a^4 v_a^4} . \tag{14.47}$$

A simplification of (14.47) is obtained for the pure pairing force by using the relation (12.11) for the monopole matrix element and setting all the other multipole matrix elements to zero. The final result is

$$\lambda_2 = \frac{G}{4} \frac{\zeta^{(l_a l_b)} \sum_a \hat{j}_a^2 u_a^3 v_a \sum_b \hat{j}_b^2 u_b v_b^3 - 2 \sum_a \hat{j}_a^2 u_a^4 v_a^4}{\big(\sum_a \hat{j}_a^2 u_a^2 v_a^2\big)^2 - 2 \sum_a \hat{j}_a^2 u_a^4 v_a^4} \geq \frac{G}{4} . \tag{14.48}$$

The phase factor $\zeta^{(l_a l_b)}$ is defined in (12.11). The relation on the right-hand side of (14.48) can be proved by means of the Schwartz inequality. Equation (14.48) is due to Nogami [75–77].

By way of summary we repeat the expression for the LNBCS ground-state energy, originally stated in (14.16),

$$E_0^{LN} = \mathcal{H}_0^{LN} + \lambda_1 \bar{n} + \lambda_2 \bar{n}^2 . \tag{14.49}$$

The corresponding BCS relation (13.66) is

$$H_0 = \mathcal{H}_0 + \lambda \bar{n} . \tag{14.50}$$

In the limit $\lambda_2 \to 0$ the Lipkin–Nogami result (14.49) reduces to the BCS result (14.50) with $\lambda_1 = \lambda$.

## 14.3 Lipkin–Nogami BCS Theory in Simple Models

In this section we discuss simple applications of the Lipkin–Nogami approach in the context of exactly solvable schematic models. These models include the single-$j$-shell model and the two-level Lipkin model.

### 14.3.1 Single $j$ Shell

Consider the case of a single $j$ shell with single-particle energy $\varepsilon_j = 0$. The BCS solution for this simple model was discussed in Subsect. 13.5.1. For the pure pairing force the LNBCS gap equation (14.39) yields

$$\sqrt{\tilde{\eta}^2 + \Delta^2} = \tfrac{1}{2}\Omega G , \tag{14.51}$$

where $\Omega = \tfrac{1}{2}\widehat{j}^2$. Equation (14.48) gives

$$\lambda_2 = \tfrac{1}{4}G . \tag{14.52}$$

From (14.43), (14.51) and (14.52) we obtain the quasiparticle energy

$$E^{\mathrm{LN}} = \tfrac{1}{2}\Omega G + \tfrac{1}{4}G = \tfrac{1}{2}\Omega G\left(1 + \frac{1}{2\Omega}\right) . \tag{14.53}$$

We observe that this expression includes a correction to the BCS quasiparticle energy (13.94). The result (14.53) can be compared with the result of the seniority model of Sect. 12.4. To enable direct comparison we write from (12.28) and (12.38) the energy difference between the seniority $v = 1$ state and the $v = 0$ ground state:

$$E_{v=1}(N) - E_{v=0}(N) = \tfrac{1}{2}\Omega G\left(1 + \frac{1}{2\Omega}\right) . \tag{14.54}$$

The LNBCS result (14.53) is in *exact* agreement with this. So we see that the Lipkin–Nogami correction to the BCS quasiparticle energy is needed to reproduce the result (14.54) of the seniority model.

Equation (14.19) gives

$$\bar{n} = 2\Omega v_j^2 , \tag{14.55}$$

so that

$$v_j^2 = \frac{\bar{n}}{2\Omega} , \quad u_j^2 = 1 - v_j^2 . \tag{14.56}$$

With the phase factor $\theta^{(l)}$ as in (14.38), the LNBCS occupation amplitudes $u_j, v_j$ resulting from (14.56) are the same as the BCS values (13.92). It then follows from (13.45) that the LNBCS pairing gap $\Delta$ and self-energy $\mu$ are those given by the BCS equations (13.96) and (13.97), respectively. By comparing (13.54) and (14.37) we see that $\tilde\eta$ is equal to the BCS quantity $\eta$ calculated with the parameter $\lambda(\neq \lambda_1)$. The quantity $\eta(\lambda)$ is given by (13.98), whence

$$\tilde\eta = \eta(\lambda) = \tfrac{1}{2}\Omega G\left(1 - \frac{\bar{n}}{\Omega}\right). \tag{14.57}$$

Equation (14.34) gives now

$$\xi = \tilde\eta - \tfrac{1}{4}G\frac{\bar{n}}{\Omega} + \tfrac{1}{4}G\bar{n} = \tfrac{1}{2}\Omega G\left[1 - \frac{\bar{n}}{2\Omega}\left(1 + \frac{1}{\Omega}\right)\right]. \tag{14.58}$$

Calculating $\mathcal{H}_0^{\mathrm{LN}}$ from (14.31) we obtain

$$\begin{aligned}
\mathcal{H}_0^{\mathrm{LN}} &= 2\Omega\left[\frac{\bar{n}}{2\Omega}(\xi + \tfrac{1}{2}\mu) - \tfrac{1}{2}u_j v_j \Delta\right] \\
&= \bar{n}\left\{\tfrac{1}{2}\Omega G\left[1 - \frac{\bar{n}}{2\Omega}\left(1 + \frac{1}{\Omega}\right)\right] + \tfrac{1}{2}\frac{\bar{n}G}{2\Omega}\right\} - \tfrac{1}{2}\left(1 - \frac{\bar{n}}{2\Omega}\right)\bar{n}\Omega G = 0.
\end{aligned} \tag{14.59}$$

Let us continue with the analysis and calculate further relevant quantities. From (14.32) we obtain

$$\tilde\varepsilon = 0 + 4 \times \tfrac{1}{4}G\frac{\bar{n}}{2\Omega} = \frac{G\bar{n}}{2\Omega} \tag{14.60}$$

and further, with use of (13.97),

$$\tilde\eta = \tilde\varepsilon - \tilde\lambda - \mu = \frac{G\bar{n}}{2\Omega} - \tilde\lambda - \frac{\bar{n}}{2\Omega}G = -\tilde\lambda. \tag{14.61}$$

Solving now for $\lambda_1$ from (14.32) yields

$$\lambda_1 = -\tilde\eta - \tfrac{1}{2}G(1 + \bar{n}) = -\tfrac{1}{2}G(\Omega + 1). \tag{14.62}$$

Comparison with (13.99) shows that $\lambda_1$ and $\lambda$ are indeed different.

With the quantities $\mathcal{H}_0^{\mathrm{LN}}$, $\lambda_1$ and $\lambda_2$ given by (14.59), (14.62) and (14.52), respectively, we are now in a position to find the ground-state energy (14.49). The result is

$$E_0^{\mathrm{LN}} = 0 - \tfrac{1}{2}G(\Omega + 1)\bar{n} + \tfrac{1}{4}G\bar{n}^2 = -\tfrac{1}{4}G\bar{n}(2\Omega - \bar{n} + 2). \tag{14.63}$$

Unlike the BCS result (13.100), this result agrees *exactly* with the expression (12.28) of the seniority model.

In this subsection we applied LNBCS theory to a single $j$ shell. The main results are that the computed LNBCS ground-state energy and quasiparticle energy agree with the exact results of the seniority model of Chap. 12. This is a significant improvement over plain BCS theory.

## 14.3.2 The Lipkin Model

In this subsection we apply the LNBCS formalism to the Lipkin model. The model has two shells $j_1 l_1 = j_2 l_2 = jl$ with single-particle energies $\varepsilon_1 = -\frac{1}{2}\varepsilon$ and $\varepsilon_2 = \frac{1}{2}\varepsilon$, and the total number of particles is $2\Omega = 2j + 1$. We use the pure pairing force as our two-body interaction, with matrix elements given by (12.11). Plain BCS theory was similarly applied in Subsect. 13.5.2.

The BCS expressions (13.104) and (13.105) are valid also for the LNBCS pairing gaps and self-energies, as one can see by tracing back their derivation. From the gap equation (14.39) we obtain

$$\frac{2}{\Omega G} = \frac{1}{\sqrt{\tilde{\eta}_1^2 + \Delta^2}} + \frac{1}{\sqrt{\tilde{\eta}_2^2 + \Delta^2}} . \tag{14.64}$$

Equation (14.19) for the average particle number gives

$$2\Omega = \bar{n} = 2\Omega(v_1^2 + v_2^2) . \tag{14.65}$$

We assume $l =$ even so the $u$ coefficients are positive also in the CS phase convention. As in the BCS case, it follows that

$$v_2 = u_1 , \quad v_1 = u_2 . \tag{14.66}$$

From the condition (14.37) we can derive the *general* result (Exercise 14.12)

$$\boxed{u_c v_c = \frac{\Delta_c}{2\sqrt{\tilde{\eta}_c^2 + \Delta_c^2}} .} \tag{14.67}$$

The relations (14.66) now lead to

$$\tilde{\eta}_1^2 = \tilde{\eta}_2^2 \tag{14.68}$$

and Eq. (14.38) imply

$$\tilde{\eta}_1 = -\tilde{\eta}_2 . \tag{14.69}$$

We note from (14.64) that

$$\sqrt{\tilde{\eta}_1^2 + \Delta^2} = \sqrt{\tilde{\eta}_2^2 + \Delta^2} = \Omega G . \tag{14.70}$$

Our aim is to derive equations for the solution of the occupation amplitudes and the $\lambda$ parameters. This involves the calculation of various auxiliary quantities appearing at the end of Subsect. 14.2.1. Equations (13.105) and (14.66) give for the self-energies the relations

$$\mu_1 = \frac{1}{2}(1 + v_1^2)G , \quad \mu_2 = \frac{1}{2}(1 + v_2^2)G , \quad \mu_1 + \mu_2 = \frac{3}{2}G . \tag{14.71}$$

The definition (14.32) of the shifted single-particle energy, again with (14.66), gives

$$\tilde{\varepsilon}_1 = -\tfrac{1}{2}\varepsilon + 4\lambda_2 v_1^2 , \quad \tilde{\varepsilon}_2 = \tfrac{1}{2}\varepsilon + 4\lambda_2 v_2^2 , \quad \tilde{\varepsilon}_1 + \tilde{\varepsilon}_2 = 4\lambda_2 . \tag{14.72}$$

To find the shifted Fermi level $\tilde{\lambda}$ we use the definition of $\tilde{\eta}$ in (14.32) and form $\tilde{\eta}_1 + \tilde{\eta}_2$. Equations (14.69), (14.71) and (14.72) then give

$$\tilde{\lambda} = 2\lambda_2 - \tfrac{3}{4}G . \tag{14.73}$$

On the other hand, we may write $\tilde{\lambda}$ as defined in (14.32) and apply (14.65), which results in

$$\tilde{\lambda} = \lambda_1 + 2\lambda_2(1 + \bar{n}) = \lambda_1 + 2\lambda_2(1 + 2\Omega) . \tag{14.74}$$

Equating (14.73) and (14.74) leads to

$$\lambda_1 = -\tfrac{3}{4}G - 4\Omega\lambda_2 . \tag{14.75}$$

Having found the relation (14.75) between $\lambda_1$ and $\lambda_2$ we proceed to find sufficient relations between $\lambda_2$ and the occupation amplitudes such that they can all be solved. Substituting (14.71)–(14.73) into the definition of $\tilde{\eta}_a$ in (14.32) gives

$$\begin{aligned}\tilde{\eta}_1 = -\tilde{\eta}_2 &= -\tfrac{1}{2}\varepsilon + 4\lambda_2 v_1^2 - (2\lambda_2 - \tfrac{3}{4}G) - \tfrac{1}{2}(1 + v_1^2)G \\ &= -\tfrac{1}{2}\varepsilon + \tfrac{1}{4}G - 2\lambda_2 + (4\lambda_2 - \tfrac{1}{2}G)v_1^2 . \end{aligned} \tag{14.76}$$

By forming $v_1^2$ from (14.38), using (14.70) and substituting the expression (14.76) for $\tilde{\eta}_1$ we get

$$v_1^2 = \frac{(\Omega - \tfrac{1}{4})G + \tfrac{1}{2}\varepsilon + 2\lambda_2}{(2\Omega - \tfrac{1}{2})G + 4\lambda_2} . \tag{14.77}$$

This is one equation between $v_1$ and $\lambda_2$. Another equation is provided by (14.48), in simplified form (Exercise 14.13)

$$\lambda_2 = \tfrac{1}{4}G \frac{\Omega - 2u_1^2 v_1^2}{(4\Omega - 2)u_1^2 v_1^2} . \tag{14.78}$$

These two equations, together with (13.9) and (14.66), yield the solution for $\lambda_2$ and all occupation amplitudes when $G$ is given.

As the first step to solve Eqs. (14.77) and (14.78) simultaneously, we define an auxiliary variable $z$ as

$$z \equiv (\Omega - \tfrac{1}{4})G + 2\lambda_2 . \tag{14.79}$$

In terms of $z$ the expression (14.77) for $v_1^2$ becomes

$$v_1^2 = \frac{z + \tfrac{1}{2}\varepsilon}{2z} . \tag{14.80}$$

We write the quantity $u_1^2 v_1^2$ appearing in (14.78) as $(1 - v_1^2)v_1^2$ and substitute from (14.80), with the result

$$u_1^2 v_1^2 = \frac{z^2 - \frac{1}{4}\varepsilon^2}{4z^2} \, . \tag{14.81}$$

We insert $\lambda_2$ from (14.79) into the left-hand side of (14.78) and $u_1^2 v_1^2$ from (14.81) into the right-hand side. After simplification this results in the cubic equation

$$z^3 - G(\Omega + \tfrac{1}{4})z^2 - \tfrac{1}{4}\varepsilon^2 z + \tfrac{1}{16}G\varepsilon^2 \frac{2\Omega(4\Omega - 3) - 1}{2\Omega - 1} = 0 \, . \tag{14.82}$$

To complete the solution of the Lipkin model within the Lipkin–Nogami theory we need to determine the roots of Eq. (14.82). Applying the inequality condition in (14.48)–(14.79) gives

$$z \geq (\Omega + \tfrac{1}{4})G \, . \tag{14.83}$$

A graphical inspection shows that (14.82) has only one positive, unique solution compatible with the condition (14.83). After obtaining this solution, $z = z_0$, we can solve for $\lambda_2$ from (14.79) to obtain

$$\lambda_2 = \tfrac{1}{2}[z_0 - (\Omega - \tfrac{1}{4})G] \, . \tag{14.84}$$

Inserting this $\lambda_2$ into (14.77) gives $v_1^2$.

The ground-state energy $E_0^{LN}$ is found from (14.49) with $\mathcal{H}_0^{LN}$ given by (14.31), $\lambda_2$ by (14.84) and $\lambda_1$ then by (14.75). However, before we can compute $\mathcal{H}_0^{LN}$ we must find and collect the input parameters appearing in (14.75). For the Lipkin model at hand we can express them in terms of $v_1^2$.

Using the results (14.67) and (14.70) yields

$$u_1 v_1 = u_2 v_2 = \frac{\Delta}{2\Omega G} \, , \tag{14.85}$$

which, together with (14.37) and (14.69), leads to

$$\tilde{\eta}_1 = \Omega G(1 - 2v_1^2) = -\tilde{\eta}_2 \, . \tag{14.86}$$

Equation (14.34) now gives

$$\xi_1 = \Omega G(1 - 2v_1^2) + 2\lambda_2(\Omega - v_1^2) \, , \tag{14.87}$$
$$\xi_2 = -\Omega G(1 - 2v_1^2) + 2\lambda_2(\Omega - 1 + v_1^2) \, . \tag{14.88}$$

Both self-energies (14.71) expressed in terms of $v_1^2$ are

$$\mu_1 = \tfrac{1}{2}G(1 + v_1^2) \, , \quad \mu_2 = \tfrac{1}{2}G(2 - v_1^2) \, . \tag{14.89}$$

From (14.85) we have $\Delta = 2\Omega G u_1 v_1$, so multiplication by $u_1 v_1$ gives

$$u_1 v_1 \Delta = u_2 v_2 \Delta = 2 \Omega G (1 - v_1^2) v_1^2 . \qquad (14.90)$$

Equations (14.87)–(14.90) give the input parameters for calculating $\mathcal{H}_0^{LN}$ from (14.31) in terms of $v_1^2$, which in turn is obtained numerically from the chain of Eqs. (14.82), (14.84) and (14.77). In the next subsection we discuss this process for the $j = \frac{7}{2}$ Lipkin model.

### 14.3.3 Example: The $j = \frac{7}{2}$ Case

We now give the results of applying the formalism of the previous subsection to the case of $j = \frac{7}{2}$. The occupation amplitudes $v_1$ and $v_2$ are plotted in Fig. 14.8 as functions of the quantity $y = \varepsilon/G$. This ratio is a measure of the relative magnitudes of the energy gap between the two $j$ orbitals and the pairing interaction strength. In the figure the LNBCS results are compared with the exact solution and with the BCS solution of Subsect. 13.5.2. Because the variable $y$ of the present figure is inversely proportional to the variable $x$ defined in (13.121), the present curves look different from those in Fig. 13.2.

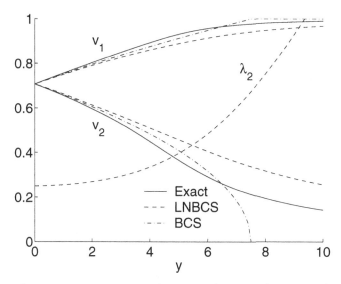

**Fig. 14.8.** Occupation amplitudes $v_1$ (lower level) and $v_2$ (upper level) of the two-level Lipkin model for $j = \frac{7}{2}$ ($\Omega = 4$) as functions of the variable $y = \varepsilon/G$. The exact, BCS and LNBCS solutions are shown. Also the LNBCS parameter $\lambda_2$ is displayed. All computations used the pure pairing force

Figure 14.8 shows that for $y$ near zero, i.e. $G \gg \varepsilon$, the pairing force is strong enough to scatter pairs from the lower level to the higher one so efficiently that both levels are nearly equally occupied. With increasing level separation $\varepsilon$ (or decreasing pairing strength $G$) the diminished scattering depletes the upper level. This happens initially in much the same way for all three solutions. Increasing the level separation further to the value $\varepsilon = 7.5G$ makes the BCS solution collapse to the trivial case $v_1 = 1, v_2 = 0$; this was discussed in Subsect. 13.5.3. Contrariwise, the LNBCS solution continues smoothly across this point and follows the trend of the exact solution.

The absence of the BCS collapse in the Lipkin–Nogami extension of the theory is a general one, not limited to the special case of the two-level Lipkin model. In Fig. 14.8 we also plot the LNBCS parameter $\lambda_2$. It is seen to be a parabolically increasing function of $y$.

Figure 14.9 shows the pairing gap $\Delta$, deduced from (14.90), as a function of $y$ for the LNBCS solution of the $j = \frac{7}{2}$ Lipkin model. The BCS solution (13.122) is shown for comparison. The BCS pairing gap is seen to vanish at $y = 7.5$, which corresponds to a small relative pairing strength. This does not happen in the LNBCS description, where the pairing gap diminishes smoothly with increasing $y$.

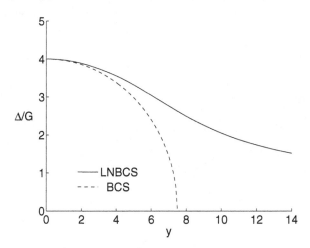

**Fig. 14.9.** Pairing gap $\Delta$ in the BCS and LNBCS approaches as a function of the variable $y = \varepsilon/G$ for the $j = \frac{7}{2}$ ($\Omega = 4$) Lipkin model with pure pairing force

In Fig. 14.10 the ground-state energy of the $j = \frac{7}{2}$ Lipkin model is plotted as a function of $y$ for the exact, BCS and LNBCS solutions. The BCS solution is the one of (13.124). The LNBCS solution is from (14.49), obtained through the numerical and algebraic scheme presented at the end of Subsect. 14.3.2. From the figure we can see that the LNBCS solution is practically identical to the exact one, whereas the BCS solution ceases to exist at $\varepsilon = 7.5G$.

The following generally valid characteristics of LNBCS theory are evident from the foregoing discussion of the special cases of the single-$j$ model and the Lipkin model.

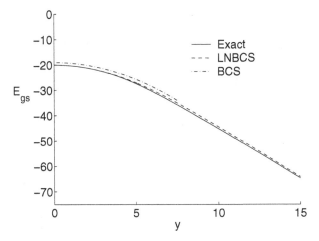

**Fig. 14.10.** Ground-state energy of the $j = \frac{7}{2}$ ($\Omega = 4$) Lipkin model calculated by using the pure pairing force for a range of values of the variable $y = \varepsilon/G$. The LNBCS result is compared with the exact and BCS solutions. Note that the BCS solution ceases to exist at $y = 7.5$

- The LNBCS description does not break down for small pairing strength (or, equivalently, a large energy gap between the single-particle levels) as does the BCS description.
- The quality of the computed occupation amplitudes is similar in the LNBCS and BCS approaches until the vicinity of the BCS breakdown.
- The LNBCS ground-state energies are somewhat higher than the exact ones and notably lower than the valid BCS ones.

## 14.4 The Two-Level Model for $j = j' = \frac{7}{2}$

Let us extend the discussion of the previous section to the case of the $j = j' = \frac{7}{2}$ two-level model. There is no simple BCS or LNBCS solution, except for the Lipkin case $\bar{n} = 2\Omega$. Therefore all the relevant quantities of the two approaches must be computed numerically.

Figure 14.11 shows the ground-state energy of the two-level model as a function of the number or particles occupying the valence space. The energy is plotted for two values of the level separation, $\varepsilon = 1.0\,\text{MeV}$ and $\varepsilon = 6.0\,\text{MeV}$, with $G = 1.0\,\text{MeV}$ ($y = 1.0$ and $y = 6.0$). The solid line represents both the LNBCS and exact solutions; there is no visible difference between them. Even the BCS solution is seen to be rather good. This demonstrates the striking accuracy of the LNBCS approach in producing ground-state energies.

Each curve in Fig. 14.11 has a minimum at midshell. For the BCS case this was discussed in Subsect. 13.5.4, and the behaviour was explained in terms

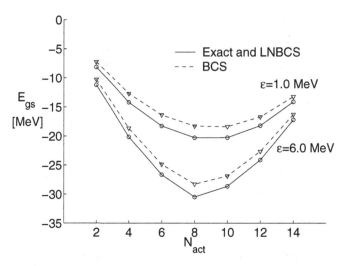

**Fig. 14.11.** Ground-state energy as a function of the number or active particles occupying the $j = j' = \frac{7}{2}$ levels of a two-level model. The interaction is the pure pairing force with strength $G = 1.0\,\text{MeV}$, and $\varepsilon$ is the energy separation between the two levels

of the Pauli principle. The chemical potential $\lambda$ is negative at the beginning of a shell, then increases and crosses zero immediately after midshell; see Table 13.1. According to (13.71), a negative $\lambda$ means that when a particle is added to the system, the ground-state energy acquires a negative contribution. After midshell the additional energy becomes positive, which increases the energy of the ground state. Thus the binding energy has a maximum at midshell.

To explain the behaviour of the LNBCS curves of Fig. 14.11, let us derive the chemical potential in LNBCS theory in analogy to the derivation of (13.71). From (14.49) we have, with $n$ as a free variable,

$$\mathcal{H}_0^{\text{LN}} = E_0^{\text{LN}} - \lambda_1 n - \lambda_2 n^2 \,, \tag{14.91}$$

and we require that $\partial \mathcal{H}_0^{\text{LN}}/\partial n = 0$ at the requested particle number $n = \bar{n}$. This leads to

$$0 = \left. \frac{\partial \mathcal{H}_0^{\text{LN}}}{\partial n} \right|_{n=\bar{n}} = \left. \frac{\partial E_0^{\text{LN}}}{\partial n} \right|_{n=\bar{n}} - \lambda_1 - 2\lambda_2 \bar{n} \,. \tag{14.92}$$

To have the normal meaning of a chemical potential, the LNBCS chemical potential has to be defined as

$$\lambda_{\text{LNBCS}} \equiv \left. \frac{\partial E_0^{\text{LN}}}{\partial n} \right|_{n=\bar{n}} = \lambda_1 + 2\lambda_2 \bar{n} \,. \tag{14.93}$$

Equation (14.32), together with (14.19), provides for an alternative expression in terms of the quantity $\tilde{\lambda}$, so that

$$\boxed{\lambda_{\mathrm{LNBCS}} = \lambda_1 + 2\lambda_2 \bar{n} = \tilde{\lambda} - 2\lambda_2 .} \tag{14.94}$$

The chemical potential (14.94) and its various terms are plotted in Fig. 14.12 as functions of the number of active particles in the $j = j' = \frac{7}{2}$ two-level model. The relative pairing strength is given by $y = \varepsilon/G = 6.0$. The figure shows that $\lambda_{\mathrm{LNBCS}}$ crosses zero at midshell. Thus $\lambda_{\mathrm{LNBCS}}$ behaves the same way as the BCS chemical potential $\lambda$ recorded in Table 13.1. At $\bar{n} = 8$, both $\lambda_{\mathrm{LNBCS}}$ and $\lambda$ equal $-0.5\,\mathrm{MeV}$; even at other particle numbers they are close to each other.

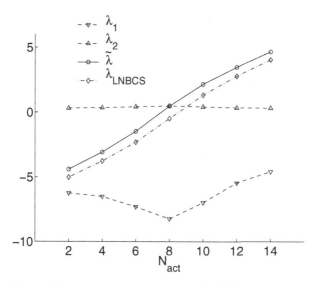

**Fig. 14.12.** Values of the various $\lambda$ parameters of LNBCS theory as functions of the number of active particles in the $j = j' = \frac{7}{2}$ two-level model. The relative pairing strength is given by $y = \varepsilon/G = 6.0$

## 14.5 Application of Lipkin–Nogami Theory to Realistic Calculations

To make a realistic LNBCS calculation we can use the BCS flow chart in Subsect. 14.1.1 with the following modifications.

At stage (14.6) compute $\lambda_2^{(0)}$ from (14.47) or (14.48) with the amplitudes $u_a^{(0)}$ and $v_a^{(0)}$. Replace the computation of $\eta_a^{(0)}$ by the computation of $\tilde{\eta}_a^{(0)}$ according to (14.32):

$$\tilde{\eta}_a^{(0)} = \tilde{\varepsilon}_a^{(0)} - \tilde{\lambda}_0^{(0)} - \mu_a^{(0)} . \tag{14.95}$$

Compute $\tilde{\lambda}_0^{(0)}$ from (14.32) as

$$\tilde{\lambda}_0^{(0)} = \tilde{\lambda}_0 + 2\lambda_2^{(0)}(1 + \bar{n}) , \tag{14.96}$$

where $\tilde{\lambda}_0$ is given as an initial value at stage (14.4). Calculate the shifted single-particle energy in (14.95) from (14.32) as

$$\tilde{\varepsilon}_a^{(0)} = \varepsilon_a + 4\lambda_2^{(0)}\left(v_a^{(0)}\right)^2 . \tag{14.97}$$

At stage (14.7) replace $\eta_a^{(0)}$ by $\tilde{\eta}_a^{(0)}$ and $\lambda^{(1)}$ by $\tilde{\lambda}^{(1)}$. To compute $\tilde{\lambda}^{(1)}$, replace $\varepsilon_a$ by $\tilde{\varepsilon}_a^{(0)}$ .

At stage (14.9) replace $\lambda_1$ by $\tilde{\lambda}_1$ and $\lambda^{(1)}$ by $\tilde{\lambda}^{(1)}$.

**Table 14.4.** LNBCS occupation probabilities $v_a^2$, quasiparticle energies $E_a^{\mathrm{LN}}$, auxiliary quantities $\lambda_1$, $\lambda_2$ and $\tilde{\lambda}$, chemical potential $\lambda_{\mathrm{LNBCS}}$, pairing gap $\Delta$ and self-energy $\mu$ for different particle numbers $\bar{n}$ in the d-s shell

| Quantity | $\bar{n} = 2$ | $\bar{n} = 4$ | $\bar{n} = 6$ | $\bar{n} = 8$ | $\bar{n} = 10$ |
|---|---|---|---|---|---|
| $v_{0d_{5/2}}^2$ | 0.2654 | 0.5149 | 0.7393 | 0.9066 | 0.9741 |
| $v_{1s_{1/2}}^2$ | 0.1541 | 0.3513 | 0.6033 | 0.8537 | 0.9634 |
| $v_{0d_{3/2}}^2$ | 0.0248 | 0.0520 | 0.0893 | 0.2133 | 0.5572 |
| $E_{0d_{5/2}}^{\mathrm{LN}}$ (MeV) | 2.302 | 2.489 | 2.846 | 3.799 | 5.412 |
| $E_{1s_{1/2}}^{\mathrm{LN}}$ (MeV) | 2.806 | 2.603 | 2.568 | 3.180 | 4.604 |
| $E_{0d_{3/2}}^{\mathrm{LN}}$ (MeV) | 6.461 | 5.518 | 4.306 | 2.782 | 1.861 |
| $\lambda_1$ (MeV) | −2.261 | −2.463 | −3.243 | −5.239 | −3.534 |
| $\lambda_2$ (MeV) | 0.041 | 0.067 | 0.139 | 0.285 | 0.192 |
| $\tilde{\lambda}$ (MeV) | −2.018 | −1.789 | −1.292 | −0.109 | 0.699 |
| $\lambda_{\mathrm{LNBCS}}$ (MeV) | −2.099 | −1.924 | −1.571 | −0.679 | 0.315 |
| $\Delta$ (MeV) | 1.997 | 2.421 | 2.377 | 2.046 | 1.658 |
| $\mu$ (MeV) | 1.000 | 2.000 | 3.000 | 4.000 | 5.000 |

The calculation used the single-particle energies (14.12) and the SDI with $A_1 = 1.0\,\mathrm{MeV}$.

We have applied the BCS and LNBCS computation procedures to the d-s shell using the single-particle energies (14.12) and the SDI with $A_1 = 1.0\,\mathrm{MeV}$. The results are summarized in Table 14.4 and Fig. 14.13. Comparison of the table with Table 14.1 reveals that all the LNBCS quantities are quite close to their BCS counterparts. The figure shows that the LNBCS ground-state energy follows the BCS curve but is somewhat lower, as one would expect.

## Epilogue

In the first part of this chapter we used the BCS formalism for numerical applications in the d-s and f-p-$0g_{9/2}$ shells. In the second part we extended the formalism to the LNBCS approach. The difference between BCS and LNBCS

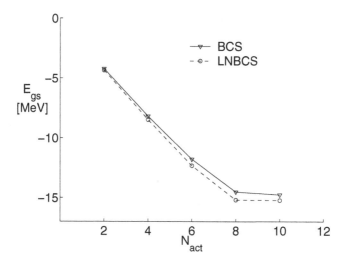

**Fig. 14.13.** Ground-state energy as a function of the number of active particles in the d-s shell for the BCS and LNBCS solutions. The calculation used the single-particle energies (14.12) and the SDI with $A_1 = 1.0$ MeV

results was found relatively small except that the more sophisticated Lipkin–Nogami scheme avoids the BCS collapse of the pairing gap at small interaction strength. Both schemes give rise to independent quasiparticles and an associated quasiparticle mean field. In the following chapter we discuss transition amplitudes for few-quasiparticle systems within the BCS mean-field scheme. In later chapters, BCS quasiparticles are allowed to interact via the residual Hamiltonian to produce quasiparticle configuration mixing.

## Exercises

**14.1.** Derive the result (14.3) by starting from the definition of $\eta_b$ in (13.45) and the relation (13.72).

**14.2.** Give a detailed derivation of (14.11).

**14.3.** Do a BCS calculation in the d-s shell by using the single-particle energies (14.12) and the SDI with strength $A_1 = 2.0$ MeV. Calculate the $u$ and $v$ amplitudes, the pairing gap $\Delta$ and the chemical potential $\lambda(\mu = 0)$. Plot these quantities as functions of the number of active particles. Compare with the values in Table 14.1 and comment.

**14.4.** Use the BCS approach with the SDI in the d-s shell. Vary the parameter $A_1$ and the single-particle energies so as to roughly reproduce the energies of the three lowest states in $^{37}$Ar and $^{37}$K.

**14.5.** Starting from the expression (14.18) for the number operator $\hat{n}$ derive the term (14.21) of $\hat{n}^2$.

**14.6.** Starting from the expression (14.18) for the number operator $\hat{n}$ derive the term (14.22) of $\hat{n}^2$.

**14.7.** Fill in the details of the derivation of (14.29).

**14.8.** Fill in the details of the derivation of (14.36).

**14.9.** Derive (14.48) by starting from (14.47) and the definition (12.11) of the pure pairing force.

**14.10.** Do a numerical application of the calculations of Subsect. 14.3.1 for $j = \frac{7}{2}$, $\varepsilon_{7/2} = 0$ and $G = 1.0\,\text{MeV}$.

**14.11.** Plot the LNBCS ground-state energy of Exercise 14.10 as a function of the mean particle number $\bar{n}$. Plot the corresponding BCS ground-state energy for comparison.

**14.12.** Derive the relation (14.67).

**14.13.** Derive the relation (14.78).

**14.14.** Find the occupation amplitudes, the pairing gap and the chemical potential by using
(a) the BCS approach,
(b) the LNBCS approach
for the $j = \frac{7}{2}$ Lipkin model. Use the pure pairing force with strength $G = 1.0\,\text{MeV}$ and an energy splitting $\varepsilon = 6.0\,\text{MeV}$.

**14.15.** Consider the $j = \frac{1}{2}$ Lipkin model with $\varepsilon = 1.0\,\text{MeV}$. Plot the ground-state energy as a function of the strength $G$ of the pure pairing force for the LNBCS solution and for the exact one.

# Transitions in the Quasiparticle Picture

## Prologue

In this chapter we deal with electromagnetic and beta-decay transitions in terms of independent quasiparticles. Transition amplitudes are derived for transitions between one-quasiparticle states and between two-quasiparticle states. Derivations and applications are made within the BCS framework, but the expressions for the amplitudes are valid also in the LNBCS description.

## 15.1 Quasiparticle Representation of a One-Body Transition Operator

Equation (4.22) gives the coupled representation of a one-body tensor operator $T_\lambda$ as

$$T_{\lambda\mu} = \hat{\lambda}^{-1} \sum_{ab} (a\|T_\lambda\|b) \left[ c_a^\dagger \tilde{c}_b \right]_{\lambda\mu} . \tag{15.1}$$

Using (13.12) we express this in terms of quasiparticle operators according to

$$
\begin{aligned}
\left[ c_a^\dagger \tilde{c}_b \right]_{\lambda\mu} &= \sum_{m_\alpha m_\beta} (j_a\, m_\alpha\, j_b\, m_\beta | \lambda\, \mu) c_\alpha^\dagger \tilde{c}_\beta \\
&= \sum_{m_\alpha m_\beta} (j_a\, m_\alpha\, j_b\, m_\beta | \lambda\, \mu) \left( u_a a_\alpha^\dagger - v_a \tilde{a}_\alpha \right) \left( u_b \tilde{a}_\beta + v_b a_\beta^\dagger \right) \\
&= u_a u_b \left[ a_a^\dagger \tilde{a}_b \right]_{\lambda\mu} + u_a v_b \left[ a_a^\dagger a_b^\dagger \right]_{\lambda\mu} - v_a u_b \left[ \tilde{a}_a \tilde{a}_b \right]_{\lambda\mu} - v_a v_b \left[ \tilde{a}_a a_b^\dagger \right]_{\lambda\mu} .
\end{aligned}
\tag{15.2}
$$

Commuting the last term of (15.2) into normal order we find

$$\left[ \tilde{a}_a a_b^\dagger \right]_{\lambda\mu} = -\delta_{ab}\delta_{\lambda 0}\delta_{\mu 0} \hat{j}_a - (-1)^{j_a + j_b + \lambda} \left[ a_b^\dagger \tilde{a}_a \right]_{\lambda\mu} , \tag{15.3}$$

so that (15.2) becomes

$$\boxed{\begin{aligned}
\left[c_a^\dagger \tilde{c}_b\right]_{\lambda\mu} &= u_a u_b \left[a_a^\dagger \tilde{a}_b\right]_{\lambda\mu} + (-1)^{j_a+j_b+\lambda} v_a v_b \left[a_b^\dagger \tilde{a}_a\right]_{\lambda\mu} \\
&\quad + u_a v_b \left[a_a^\dagger a_b^\dagger\right]_{\lambda\mu} - v_a u_b \left[\tilde{a}_a \tilde{a}_b\right]_{\lambda\mu} + \delta_{ab}\delta_{\lambda 0}\delta_{\mu 0}\widehat{j_a} v_a v_b \,.
\end{aligned}} \tag{15.4}$$

The last term does not contribute to gamma transitions; only electron conversion allows an E0 transition. Also, it does not contribute to beta decay because $\lambda = 0$ operators like the Fermi operator connect different nucleon labels, $a \neq b$.

## 15.2 Transition Densities for Few-Quasiparticle Systems

In the following we derive transition densities for one- and two-quasiparticle states. The expressions contain the BCS occupation amplitudes $u_a$ and $v_a$. The derivations are done by using the BCS vacuum $|\text{BCS}\rangle$ as reference, but the results are equally valid for LNBCS occupation amplitudes.

### 15.2.1 Transitions Between One-Quasiparticle States

The reduced one-body transition density $(\Psi_f \| \left[c_a^\dagger \tilde{c}_b\right]_\lambda \| \Psi_i)$ is defined in (4.25). We now take the initial state $|\Psi_i\rangle$ and final state $|\Psi_f\rangle$ to be one-quasiparticle states,

$$|\Psi_i\rangle = a_i^\dagger |\text{BCS}\rangle \,, \quad |\Psi_f\rangle = a_f^\dagger |\text{BCS}\rangle \,, \tag{15.5}$$

where the labels $i$ and $f$ carry the appropriate single-particle quantum numbers $nljm$. The Wigner–Eckart theorem (2.27) gives

$$\langle\Psi_f|\left[c_a^\dagger \tilde{c}_b\right]_{\lambda\mu}|\Psi_i\rangle = \widehat{j_f}^{-1}(j_i\, m_i\, \lambda\, \mu|j_f\, m_f)(\Psi_f\|\left[c_a^\dagger \tilde{c}_b\right]_\lambda\|\Psi_i) \,. \tag{15.6}$$

To calculate the matrix element (15.6) we insert (15.4) and (15.5) into it, which results in

$$\begin{aligned}
\langle\Psi_f|\left[c_a^\dagger \tilde{c}_b\right]_{\lambda\mu}|\Psi_i\rangle &= u_a u_b\langle\text{BCS}|a_f\left[a_a^\dagger \tilde{a}_b\right]_{\lambda\mu}a_i^\dagger|\text{BCS}\rangle \\
&\quad + (-1)^{j_a+j_b+\lambda} v_a v_b\langle\text{BCS}|a_f\left[a_b^\dagger \tilde{a}_a\right]_{\lambda\mu}a_i^\dagger|\text{BCS}\rangle \,. \tag{15.7}
\end{aligned}$$

With the recipe (13.17) for contractions we obtain

$$\begin{aligned}
\langle\text{BCS}|a_f&\left[a_a^\dagger \tilde{a}_b\right]_{\lambda\mu}a_i^\dagger|\text{BCS}\rangle \\
&= \sum_{m_\alpha m_\beta}(-1)^{j_b+m_\beta}(j_a\, m_\alpha\, j_b\, m_\beta|\lambda\, \mu)\langle\text{BCS}|a_f a_\alpha^\dagger a_{-\beta}a_i^\dagger|\text{BCS}\rangle \\
&= \delta_{af}\delta_{bi}(-1)^{j_i-m_i}(j_f\, m_f\, j_i\, -m_i|\lambda\, \mu) \,. \tag{15.8}
\end{aligned}$$

Combining Eqs. (15.6)–(15.8) and using the Clebsch–Gordan symmetry property (1.37) gives the reduced transition density

$$\boxed{(\Psi_f\|\left[c_a^\dagger \tilde{c}_b\right]_\lambda\|\Psi_i) = \widehat{\lambda}\left[\delta_{af}\delta_{bi}u_i u_f + (-1)^{j_i+j_f+\lambda}\delta_{ai}\delta_{bf}v_i v_f\right] \,.} \tag{15.9}$$

### 15.2.2 Transitions Between a Two-Quasiparticle State and the BCS Vacuum

We write a normalized two-quasiparticle state as

$$|a\,b;\,J\,M\rangle = \mathcal{N}_{ab}(J)\big[a_a^\dagger a_b^\dagger\big]_{JM}|\mathrm{BCS}\rangle = A_{ab}^\dagger(JM)|\mathrm{BCS}\rangle\,, \qquad (15.10)$$

where the definition (11.17) was used in the last step. We set out to calculate the reduced one-body transition density for a transition from this state to the BCS vacuum. The Wigner–Eckart theorem together with (1.42) gives

$$\langle\mathrm{BCS}|\big[c_c^\dagger \tilde{c}_d\big]_{\lambda\mu}|a\,b;\,J\,M\rangle = (-1)^{J+M}\widehat{J}^{-1}\delta_{\lambda J}\delta_{\mu,-M}(\mathrm{BCS}\|\big[c_c^\dagger \tilde{c}_d\big]_\lambda\|a\,b;\,J)\,. \qquad (15.11)$$

Only the fourth term of (15.4) contributes to this, and we obtain

$$\langle\mathrm{BCS}|\big[c_c^\dagger \tilde{c}_d\big]_{\lambda\mu}|a\,b;\,J\,M\rangle = -\mathcal{N}_{ab}(J)v_c u_d\langle\mathrm{BCS}|\big[\tilde{a}_c\tilde{a}_d\big]_{\lambda\mu}\big[a_a^\dagger a_b^\dagger\big]_{JM}|\mathrm{BCS}\rangle$$

$$= -\mathcal{N}_{ab}(J)v_c u_d \sum_{\substack{m_\gamma m_\delta \\ m_\alpha m_\beta}} (j_c\,m_\gamma\,j_d\,m_\delta|\lambda\,\mu)(j_a\,m_\alpha\,j_b\,m_\beta|J\,M)$$

$$\times (-1)^{j_c+m_\gamma+j_d+m_\delta}\langle\mathrm{BCS}|a_{-\gamma}a_{-\delta}a_\alpha^\dagger a_\beta^\dagger|\mathrm{BCS}\rangle$$

$$= \mathcal{N}_{ab}(J)\delta_{\lambda J}\delta_{-\mu,M}\big[\delta_{ca}\delta_{db}(-1)^{J+M}v_a u_b - \delta_{cb}\delta_{da}(-1)^{j_b+j_a-M}v_b u_a\big]\,. \qquad (15.12)$$

Comparison with (15.11) shows that[1]

$$\boxed{\begin{aligned}(\mathrm{BCS}\|\big[c_c^\dagger \tilde{c}_d\big]_\lambda\|a\,b;\,J) &= \delta_{\lambda J}\mathcal{N}_{ab}(J)\widehat{J} \\ &\quad \times \big[\delta_{ca}\delta_{db}v_a u_b - (-1)^{j_a+j_b+J}\delta_{cb}\delta_{da}v_b u_a\big]\,.\end{aligned}} \qquad (15.13)$$

A similar derivation leads to the reversed transition density

$$\boxed{\begin{aligned}(a\,b;\,J\|\big[c_c^\dagger \tilde{c}_d\big]_\lambda\|\mathrm{BCS}) &= \delta_{\lambda J}\mathcal{N}_{ab}(J)\widehat{J} \\ &\quad \times \big[\delta_{ca}\delta_{db}u_a v_b - (-1)^{j_a+j_b+J}\delta_{cb}\delta_{da}u_b v_a\big]\,.\end{aligned}} \qquad (15.14)$$

On the right-hand side this differs from (15.13) only through the exchange $u \leftrightarrow v$.

### 15.2.3 Transitions Between Two-Quasiparticle States

Consider transitions between two two-quasiparticle states of the form (15.10). Like the cases in the preceding sections, the operator (15.4) and the Wigner–Eckart theorem lead to a reduced transition density

---

[1] Although $\delta_{\lambda J}$ formally cancels, we include it in the reduced matrix element to indicate that only $\lambda = J$ is allowed, as was done in connection with (6.118) and elsewhere.

$$
\begin{aligned}
(a_f\, b_f\,;\, J_f &\| [c_a^\dagger \tilde{c}_b]_\lambda \| a_i\, b_i\,;\, J_i) \\
&= u_a u_b \mathcal{K}_{ab}^\lambda(fi) + (-1)^{j_a+j_b+\lambda} v_a v_b \mathcal{K}_{ba}^\lambda(fi)\,,
\end{aligned}
\tag{15.15}
$$

where it remains to evaluate

$$
\begin{aligned}
\mathcal{K}_{ab}^\lambda(fi) &\equiv \widehat{J_f}(J_i\, M_i\, \lambda\, \mu | J_f\, M_f)^{-1} \\
&\quad \times \langle \mathrm{BCS} | A_{a_f b_f}(J_f M_f) [a_a^\dagger \tilde{a}_b]_{\lambda\mu} A_{a_i b_i}^\dagger (J_i M_i) | \mathrm{BCS} \rangle\,.
\end{aligned}
\tag{15.16}
$$

The quantity $\mathcal{K}_{ab}^\lambda(fi)$ is analogous to the reduced one-body transition density (6.95). The only difference is that particles are replaced by quasiparticles, but particles and quasiparticles have the same contractions with respect to the appropriate vacuum. Therefore the final expression for $\mathcal{K}_{ab}^\lambda(fi)$ can be deduced from (6.22) and (6.99), and it reads

$$
\begin{aligned}
\mathcal{K}_{ab}^\lambda(fi) = \widehat{\lambda}\widehat{J_i}\widehat{J_f}\, \mathcal{N}_{a_i b_i}(J_i)\, \mathcal{N}_{a_f b_f}(J_f) \\
\times \Bigg[ \delta_{b_i b_f}\delta_{a a_f}\delta_{b a_i}(-1)^{j_{a_f}+j_{b_f}+J_i+\lambda} 
\begin{Bmatrix} J_f & J_i & \lambda \\ j_{a_i} & j_{a_f} & j_{b_f} \end{Bmatrix} \\
+ \delta_{a_i b_f}\delta_{a a_f}\delta_{b b_i}(-1)^{j_{a_f}+j_{b_i}+\lambda}
\begin{Bmatrix} J_f & J_i & \lambda \\ j_{b_i} & j_{a_f} & j_{b_f} \end{Bmatrix} \\
+ \delta_{a_i a_f}\delta_{a b_f}\delta_{b b_i}(-1)^{j_{a_i}+j_{b_i}+J_f+\lambda}
\begin{Bmatrix} J_f & J_i & \lambda \\ j_{b_i} & j_{b_f} & j_{a_f} \end{Bmatrix} \\
+ \delta_{b_i a_f}\delta_{a b_f}\delta_{b a_i}(-1)^{J_i+J_f+\lambda+1}
\begin{Bmatrix} J_f & J_i & \lambda \\ j_{a_i} & j_{b_f} & j_{a_f} \end{Bmatrix} \Bigg]\,.
\end{aligned}
\tag{15.17}
$$

The reduced transition densities derived in this section are used in the next section to produce expressions for electromagnetic and beta-decay transition amplitudes in the framework of independent quasiparticles.

## 15.3 Transitions in Odd-$A$ Nuclei

In this section we consider open shell-nuclei with an odd number of protons or neutrons. States of such nuclei can be described, to first approximation, as one-quasiparticle excitations of the BCS vacuum, recorded in (15.5). Number parity, discussed in Subsect. 13.4.2, means that the first corrections to one-quasiparticle states are admixtures of three-quasiparticle states. The higher the excitation energy, the more the mixing. Finally the three-quasiparticle components become dominant. Then even five-quasiparticle excitations begin to play a role. However, in some cases the lowest-lying states in an odd-$A$ nucleus are of three-quasiparticle character.

A transparent way to describe three-quasiparticle states is to couple an extra quasiparticle with the two-quasiparticle excitations of the even–even reference nucleus. This is called quasiparticle–phonon coupling. For an example see [65].

### 15.3.1 Transition Amplitudes

Here we describe the electromagnetic and beta decays of odd-$A$ nuclei in the one-quasiparticle approximation.

#### Electromagnetic Transitions

The initial and final one-quasiparticle states are those of (15.5). They are connected by the electromagnetic operator $\mathcal{M}_{\sigma\lambda}$ given in (6.5)–(6.7). Substituting the reduced one-body transition density (15.9) into the definition (4.25) of the transition amplitude then leads to the result

$$
\begin{aligned}
(\Psi_f\|\mathcal{M}_{\sigma\lambda}\|\Psi_i) &= \mathcal{D}_{if}^{(\lambda)}(f\|\mathcal{M}_{\sigma\lambda}\|i)\ , \\
\mathcal{D}_{if}^{(\lambda)} &\equiv \eta^{(\lambda)}(|u_i u_f| \mp v_i v_f)\ , \\
&\quad - \text{ for } \sigma = \text{E}\ ,\ + \text{ for } \sigma = \text{M}\ .
\end{aligned}
\tag{15.18}
$$

Here the different phases arise from the symmetry properties (6.27)–(6.30) for the single-particle matrix elements and from the phase factor (13.58) for the $u$ amplitudes. The prefactor is given by

$$
\eta^{(\lambda)} = \begin{cases} (-1)^{\lambda} & \text{CS phase convention}\ ,\ \sigma = \text{E}\ , \\ (-1)^{\lambda+1} & \text{CS phase convention}\ ,\ \sigma = \text{M}\ , \\ 1 & \text{BR phase convention}\ . \end{cases}
\tag{15.19}
$$

The quantity $\mathcal{D}_{if}^{(\lambda)}$ is expressed in an alternative form in (15.137).

Note in (15.18) the distinction between the single-particle matrix element $(f\|\mathcal{M}_{\sigma\lambda}\|i)$ and the transition matrix element $(\Psi_f\|\mathcal{M}_{\sigma\lambda}\|\Psi_i)$ involving many-nucleon states. The same distinction is apparent in the analogous particle relation (6.66). Only the quasiparticle picture is complicated through the occupation amplitudes contained in the factor $\mathcal{D}_{if}^{(\lambda)}$.

The first term of $\mathcal{D}_{if}^{(\lambda)}$ corresponds to a transition between two particle states relative to the Fermi surface of the even–even reference nucleus. Likewise the second term corresponds to a transition between two hole states. In the quasiparticle picture these two possibilities arise from the first two terms in the operator (15.4). These terms describe transitions in opposite directions between the two one-quasiparticle states.

#### Beta-decay Transitions

In beta decay the initial state is a neutron state and the final state a proton state ($\beta^-$ transition) or vice versa ($\beta^+$/EC transition). The corresponding initial and final states are

$$|\Psi_i\rangle = a^\dagger_{\nu_i}|\text{BCS}\rangle \; , \; a^\dagger_{\pi_i}|\text{BCS}\rangle \; , \tag{15.20}$$

$$|\Psi_f\rangle = a^\dagger_{\pi_f}|\text{BCS}\rangle \; , \; a^\dagger_{\nu_f}|\text{BCS}\rangle \; . \tag{15.21}$$

One-quasiparticle beta transitions are classified into particle-type transitions and hole-type transitions according to the scheme

$$
\begin{array}{|ll|}
\hline
n_i \xrightarrow{\beta^-} p_f \; , \quad p_i \xrightarrow{\beta^+/\text{EC}} n_f \quad \text{particle type} \; , \\
p_i \xrightarrow{\beta^-} n_f \; , \quad n_i \xrightarrow{\beta^+/\text{EC}} p_f \quad \text{hole type} \; . \\
\hline
\end{array}
\tag{15.22}
$$

The examples of the subsequent subsections clarify this classification.

Similar to the electromagnetic case, the particle and hole types of beta transition arise from the first two terms of the operator (15.4). For example, in a $\beta^-$ transition the first term describes a neutron quasiparticle transforming into a proton quasiparticle and the second term describes a proton quasiparticle transforming into a neutron quasiparticle. Contrary to the electromagnetic case, these two channels belong to different nuclear transitions.

Using the reduced transition density (15.9), the general formulas (7.18) and (7.19) and the symmetry relation (7.23) we obtain the Fermi and Gamow–Teller matrix elements

$$
\begin{array}{|ll|}
\hline
\mathcal{M}_\mathrm{F} = u_i u_f \hat{j}_i \delta_{fi} \; , \quad \mathcal{M}_\mathrm{GT} = \sqrt{3}\, u_i u_f \mathcal{M}_\mathrm{GT}(fi) \quad \text{particle type} \; , \\
\mathcal{M}_\mathrm{F} = -v_i v_f \hat{j}_i \delta_{fi} \; , \quad \mathcal{M}_\mathrm{GT} = \sqrt{3}\, v_i v_f \mathcal{M}_\mathrm{GT}(fi) \quad \text{hole type} \; . \\
\hline
\end{array}
\tag{15.23}
$$

These matrix elements are analogous to the corresponding ones in (7.46) and (7.47) written in the particle and hole pictures. The Gamow–Teller single-particle matrix elements are given by (7.21), and their values are listed for several single-particle orbitals in Table 7.3.

For $K$th-forbidden unique beta decay the reduced transition density (15.9) is substituted into (7.187). With use of the symmetry relations (7.190) and (7.191) the nuclear matrix elements become

$$
\begin{array}{|ll|}
\hline
\mathcal{M}_{K\mathrm{u}} = \sqrt{2K+3}\, u_i u_f \mathcal{M}^{(K\mathrm{u})}(fi) \qquad \text{particle type} \; , \\
\mathcal{M}_{K\mathrm{u}} = \theta^{(K)}\sqrt{2K+3}\, v_i v_f \mathcal{M}^{(K\mathrm{u})}(fi) \quad \text{hole type} \; , \\
\hline
\end{array}
\tag{15.24}
$$

where the phase factor is

$$
\theta^{(K)} = \begin{cases} (-1)^K & \text{Condon–Shortley phase convention} \; , \\ 1 & \text{Biedenharn–Rose phase convention} \; . \end{cases}
\tag{15.25}
$$

The single-particle matrix elements $\mathcal{M}^{(K\mathrm{u})}(fi)$ are given by (7.188) and (7.189), and tabulated in Tables 7.6, 7.8 and 7.9 for 1st-, 2nd- and 3rd-forbidden transitions.

As described in Chap. 13, one-quasiparticle states are computed by starting from the BCS (or LNBCS) description of an even–even reference nucleus. Now we have to choose the reference nucleus according to the beta-decay transition we want to calculate. We must also check whether the transition is of the particle type or hole type. The procedure to select the reference nucleus and the transition type is illustrated in the examples below.

### 15.3.2 Beta and Gamma Decays in the $A = 25$ Chain of Isobars

Figure 15.1 shows the electromagnetic and beta decays involving the lowest states of $^{25}_{13}\text{Al}_{12}$ and $^{25}_{12}\text{Mg}_{13}$. We take the even–even reference nucleus to be $^{24}_{12}\text{Mg}_{12}$, denoting this as

$$|\text{BCS}\rangle = |^{24}_{12}\text{Mg}_{12}\rangle , \qquad (15.26)$$

and do a BCS calculation in the d-s shell with $^{16}_{8}\text{O}_8$ as the core. There are thus four valence protons and four valence neutrons, $Z_{\text{act}} = 4 = N_{\text{act}}$. Such a calculation, using the SDI with strength $A_1 = 1.0\,\text{MeV}$, is recorded in Table 14.1, whence we take the occupation amplitudes needed:

$$u_{0d_{5/2}} = 0.6940 , \ v_{0d_{5/2}} = 0.7199 , \qquad (15.27)$$

$$u_{1s_{1/2}} = 0.8089 , \ v_{1s_{1/2}} = 0.5880 . \qquad (15.28)$$

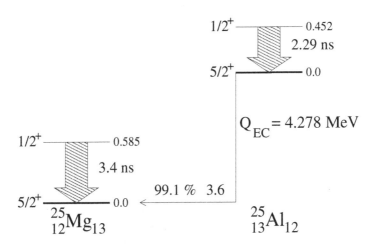

**Fig. 15.1.** Beta decay ($\beta^+$/EC) of the $5/2^+$ ground state of $^{25}\text{Al}$ to the $5/2^+$ ground state of $^{25}\text{Mg}$. The decay $Q$ value, branching and $\log ft$ value are given. Also given are the electromagnetic half-lives of the $1/2^+$ first excited states

The quasiparticle spectrum, together with the experimental spectra of $^{25}$Mg and $^{25}$Al, is shown in Fig. 14.3(b).

Both electromagnetic transitions in Fig. 15.1 have multipolarity $\sigma\lambda =$ E2. Their transition amplitudes are given by (15.18) and Table 6.4 as

$$
\begin{aligned}
(5/2^+\|\boldsymbol{Q}_2\|1/2^+) &= (u_{1s_{1/2}}u_{0d_{5/2}} - v_{1s_{1/2}}v_{0d_{5/2}})(0d_{5/2}\|\boldsymbol{Q}_2\|1s_{1/2}) \\
&= (0.8089 \times 0.6940 - 0.5880 \times 0.7199)(-2.185 e_{\mathrm{eff}}b^2) \\
&= -0.3017 e_{\mathrm{eff}}b^2 = -1.003 e_{\mathrm{eff}}\,\mathrm{fm}^2 \ ,
\end{aligned} \tag{15.29}
$$

where we have inserted an effective charge and the oscillator length $b = 1.823\,\mathrm{fm}$ given by (3.43) and (3.45). The $B(\mathrm{E2})$ value (6.4) becomes

$$
B(\mathrm{E2}\,;\,1/2_1^+ \to 5/2_1^+) = 0.503 e_{\mathrm{eff}}^2\,\mathrm{fm}^4 \ . \tag{15.30}
$$

Using the experimental half-lives we extract the experimental $B(\mathrm{E2})$ values following the recipes of Chap. 6. The result is

$$
B(\mathrm{E2}\,;\,{}^{25}\mathrm{Al})_{\mathrm{exp}} = 13.1\,e^2\mathrm{fm}^4 \ , \quad B(\mathrm{E2}\,;\,{}^{25}\mathrm{Mg})_{\mathrm{exp}} = 2.4\,e^2\mathrm{fm}^4 \ . \tag{15.31}
$$

From (6.26) we have $e_{\mathrm{eff}}^{\mathrm{p}} = (1+\chi)e$ and $e_{\mathrm{eff}}^{\mathrm{n}} = \chi e$, so we can fit the experimental values with the electric polarization constants

$$
\chi(^{25}\mathrm{Al}) = 4.1 \ , \quad \chi(^{25}\mathrm{Mg}) = 2.2 \ . \tag{15.32}
$$

These values are very large. However, the decay amplitudes (15.18) are sensitive to the difference $|u_i u_f| - v_i v_f$. Hence small improvements in the occupation amplitudes can lead to large improvements in the $B(\mathrm{E2})$ values.

Let us now turn to the beta decay. The angular momenta allow both Fermi and Gamow–Teller transitions. The decay proceeds in the $\beta^+/\mathrm{EC}$ mode from a proton-quasiparticle nucleus to a neutron-quasiparticle nucleus. The classification in (15.22) then means that this as a particle type of transition. Equations (15.23) and Table 7.3 give the matrix elements

$$
\mathcal{M}_{\mathrm{F}} = 0.6940^2 \times \sqrt{6} = 1.180 \ , \tag{15.33}
$$

$$
\mathcal{M}_{\mathrm{GT}} = \sqrt{3} \times 0.6940^2 \times \sqrt{\frac{14}{5}} = 1.396 \ , \tag{15.34}
$$

whence (7.14) and (7.15) give the reduced transition probabilities

$$
B_{\mathrm{F}} = 0.232 \ , \quad B_{\mathrm{GT}} = 0.508 \ . \tag{15.35}
$$

The $\log ft$ value is now obtained from (7.13) and (7.33) as

$$
\log ft = 3.92 \ . \tag{15.36}
$$

This is fairly close to the experimental value 3.6 quoted in Fig. 15.1.

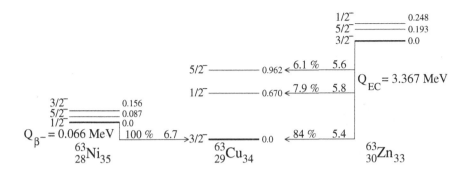

**Fig. 15.2.** Beta-minus decay of the $1/2^-$ ground state of $^{63}$Ni to the $3/2^-$ ground state of $^{63}$Cu, and $\beta^+$/EC decay of the $3/2^-$ ground state of $^{63}$Zn to the ground and excited states of $^{63}$Cu. The decay $Q$ values, branchings and log $ft$ values are given

### 15.3.3 Beta Decays in the $A = 63$ Chain of Isobars

We discuss next $\beta^-$ decay of $^{63}$Ni and $\beta^+$/EC decay of $^{63}$Zn to the ground and excited states of $^{63}$Cu. The relevant experimental data are shown in Fig. 15.2.

Consider first the $\beta^-$ decay of the $1/2^-$ ground state of $^{63}$Ni to the $3/2^-$ ground state of $^{63}$Cu. Since we have a $\beta^-$ decay from a neutron-odd to a proton-odd nucleus this is a particle-type transition according to (15.22). We choose $^{62}_{28}$Ni$_{34}$ as the even–even reference nucleus and denote

$$|\text{BCS}\rangle_1 = |^{62}_{28}\text{Ni}_{34}\rangle \ . \tag{15.37}$$

The BCS calculation is done in the f-p-$0g_{9/2}$ valence space for 8 active protons and 14 active neutrons.

The BCS calculation for $N = 35$, $N_{\text{act}} = 14$ was performed in Sect. 14.1, and the resulting quasiparticle spectrum is shown in Fig. 14.5 (c). For odd-proton $^{63}$Cu the situation simplifies since the reference nucleus $^{62}$Ni is semi-magic with $Z = 28$. Then the final proton states can be taken as pure particle states in the 1p-$0f_{5/2}$-$0g_{9/2}$ shell; the experimental levels qualitatively agree with the single-particle levels shown in Figs. 3.3 and 9.2(c).

From Table 14.2 we find the occupation amplitudes for $N_{\text{act}} = 14$. On the above interpretation the eight active BCS protons are locked in the $0f_{7/2}$ shell, and in the $^{63}$Cu ground state the odd proton is with equal probability in any of the four $1p_{3/2}$ substates. In the CS phase convention we then have

$$N_{\text{act}} = 14 : \quad u_{1p_{1/2}} = -0.7770 \ , \ v_{1p_{1/2}} = 0.6294 \ , \tag{15.38}$$

$$Z_{\text{act}} = 8 : \quad u_{1p_{3/2}} = -\sqrt{\frac{3}{4}} = -0.8660 \ , \ v_{1p_{3/2}} = \sqrt{\frac{1}{4}} = 0.5000 \ . \tag{15.39}$$

Equation (15.23) and Table 7.3 give the Gamow–Teller matrix element

$$\mathcal{M}_{GT} = \sqrt{3} \times (-0.7770) \times (-0.8660) \times \left(-\frac{4}{3}\right) = -1.554 \,, \qquad (15.40)$$

whence, by the equations of Chap. 7,

$$B_{GT} = 1.887 \,, \quad \log ft = 3.51 \,. \qquad (15.41)$$

The computed $\log ft$ is far too small against the experimental value 6.7. This can be traced back to admixtures of three-quasiparticle components in the low-energy states of $^{63}$Ni.

Next we address the $\beta^+/$EC decay of the $3/2^-$ ground state of $^{63}$Zn to the $3/2^-$ ground and excited states of $^{63}$Cu. Since the transitions are now from a neutron-odd nucleus to a proton-odd one, (15.22) indicates that these are hole-type transitions. We choose the even–even reference nucleus as

$$|BCS\rangle_2 = |^{64}_{30}Zn_{34}\rangle \,. \qquad (15.42)$$

Based on this reference we can write

$$|3/2^- \,; {}^{63}Cu\rangle = a^\dagger_{\pi 1p_{3/2}} |BCS\rangle_2 \,, \qquad (15.43)$$

$$|1/2^- \,; {}^{63}Cu\rangle = a^\dagger_{\pi 1p_{1/2}} |BCS\rangle_2 \,, \qquad (15.44)$$

$$|5/2^- \,; {}^{63}Cu\rangle = a^\dagger_{\pi 0f_{5/2}} |BCS\rangle_2 \,, \qquad (15.45)$$

$$|3/2^- \,; {}^{63}Zn\rangle = a^\dagger_{\nu 1p_{3/2}} |BCS\rangle_2 \,. \qquad (15.46)$$

Table 14.2 gives the occupation amplitudes (CS phases)

$$Z_{act} = 10: \quad u_{1p_{3/2}} = -0.8086 \,, \ v_{1p_{3/2}} = 0.5884 \,,$$

$$u_{1p_{1/2}} = -0.8799 \,, \ v_{1p_{1/2}} = 0.4752 \,,$$

$$u_{0f_{5/2}} = -0.9153 \,, \ v_{0f_{5/2}} = 0.4027 \,, \qquad (15.47)$$

$$N_{act} = 14: \quad u_{1p_{3/2}} = -0.6737 \,, \ v_{1p_{3/2}} = 0.7390 \,. \qquad (15.48)$$

The relevant quasiparticle spectra are shown in Fig. 14.5 in panel (a) for protons and in panel (c) for neutrons. From the expressions (15.23) we now obtain for the hole transitions

$$\mathcal{M}_F(3/2^- \rightarrow 3/2^-) = -0.7390 \times 0.5884 \times 2 = -0.870 \,, \qquad (15.49)$$

$$\mathcal{M}_{GT}(3/2^- \rightarrow 3/2^-) = \sqrt{3} \times 0.7390 \times 0.5884 \times \frac{2\sqrt{5}}{3} = 1.123 \,, \qquad (15.50)$$

$$\mathcal{M}_{GT}(3/2^- \rightarrow 1/2^-) = \sqrt{3} \times 0.7390 \times 0.4752 \times \frac{4}{3} = 0.811 \,, \qquad (15.51)$$

$$\mathcal{M}_{GT}(3/2^- \rightarrow 5/2^-) = 0 \,, \qquad (15.52)$$

The last zero is due to the vanishing single-particle matrix element, so three-quasiparticle admixtures are needed to explain the observed transition. For the remaining two transitions the reduced probabilities are

$$B_{\mathrm{F}}(3/2^- \to 3/2^-) = 0.189 \,, \quad B_{\mathrm{GT}}(3/2^- \to 3/2^-) = 0.493 \,, \quad (15.53)$$
$$B_{\mathrm{GT}}(3/2^- \to 1/2^-) = 0.257 \,. \quad (15.54)$$

These lead to the log $ft$ values

$$\log ft(3/2^- \to 3/2^-) = 3.95 \,, \quad (15.55)$$
$$\log ft(3/2^- \to 1/2^-) = 4.38 \,, \quad (15.56)$$

which are much smaller than the experimental values. Again, three-quasiparticle admixtures are needed for a better description.

## 15.4 Transitions Between a Two-Quasiparticle State and the BCS Vacuum

In this section we first address the electromagnetic decay of excited states of an even–even nucleus to its BCS ground state. These excited states are proton–proton and neutron–neutron two-quasiparticle states. Secondly we study beta decay involving two-quasiparticle states. They are proton–neutron states built on the BCS vacuum of the even–even reference nucleus. These states reside in odd–odd nuclei next to the even–even reference nucleus. Beta decay occurs then between an odd–odd nucleus and the adjacent even–even one.

### 15.4.1 Formalism for Transition Amplitudes

We set out to develop the necessary formalism for transitions involving two-quasiparticle states and the BCS vacuum. In our two-part study of such transitions we begin with electromagnetic transition amplitudes. The second part takes up beta-decay transitions in an analogous way.

#### Electromagnetic Transitions to the BCS Ground State

To begin with, we assume that the initial two-quasiparticle state has only one component. For an electromagnetic transition the initial state is then of the form (15.10). Using the transition density (15.13) and the symmetry properties (6.27)–(6.30) of the single-particle matrix elements, we obtain for the decay of a proton–proton or neutron–neutron two-quasiparticle state the amplitude

$$\boxed{(\mathrm{BCS}\|\mathcal{M}_{\sigma\lambda}\|a\,b;\,J) = \delta_{\lambda J}\mathcal{N}_{ab}(J)\theta^{(l_b)}(v_a|u_b| \pm v_b|u_a|)(a\|\mathcal{M}_{\sigma\lambda}\|b)\,, \\ + \text{ for } \sigma = \mathrm{E}\,, \; - \text{ for } \sigma = \mathrm{M}\,.}$$

$$(15.57)$$

The phase factor $\theta^{(l)}$ is defined in (15.25).

As an extension of (15.10) the wave function of a nuclear state consisting of two or more two-quasiparticle configurations is

$$|J^\pi M\rangle = \sum_{a\leq b} X_{ab}^{J^\pi} A_{ab}^\dagger(JM)|\text{BCS}\rangle . \qquad (15.58)$$

The indices $a$ and $b$ are both either proton or neutron indices. Equation (15.57) is immediately extended to

$$(\text{BCS}\|\mathcal{M}_{\sigma\lambda}\|J^\pi) = \delta_{\lambda J} \sum_{a\leq b} X_{ab}^{J^\pi} \mathcal{N}_{ab}(J)\theta^{(l_b)}(v_a|u_b| \pm v_b|u_a|)(a\|\mathcal{M}_{\sigma\lambda}\|b) .$$

$$(15.59)$$

The case $Z = N$ is an interesting special case of the state (15.58). Then the valence space and single-particle energies to be used are the same for protons and neutrons. In addition, within the SDI the same strength $A_1$ is to be used for protons and neutrons in the BCS calculations. In this particular case the two-quasiparticle states have good isospin and they can be written as

$$|J\,M\,\pm\rangle \equiv \frac{1}{\sqrt{2}} \sum_{a\leq b} \mathcal{N}_{ab}(J)\left([a_{\pi a}^\dagger a_{\pi b}^\dagger]_{JM} \pm [a_{\nu a}^\dagger a_{\nu b}^\dagger]_{JM}\right)|\text{BCS}\rangle . \qquad (15.60)$$

The $+$ sign corresponds to an isoscalar and the $-$ sign to an isovector combination of the two-quasiparticle components.

The special wave function (15.60) leads to a simplified electric decay amplitude. With effective charges inserted, (15.57) yields

$$(\text{BCS}\|Q_\lambda\|J\pm)$$
$$= \frac{1}{\sqrt{2}}\delta_{\lambda J}(e_{\text{eff}}^p \pm e_{\text{eff}}^n)b^\lambda \sum_{a\leq b} \mathcal{N}_{ab}(J)\theta^{(l_b)}(v_a|u_b| + v_b|u_a|)\overline{(a\|Q_\lambda\|b)} . \qquad (15.61)$$

Here the single-particle matrix element $\overline{(a\|Q_{\sigma\lambda}\|b)}$ is the one defined in (6.46). The $+$ sign between the effective charges corresponds to an isoscalar and the $-$ sign to an isovector transition.

## Beta Transitions to and from the BCS Ground State

A BCS (or LNBCS) calculation of the even–even reference nucleus produces the quasiparticles that can be used to describe the states of an adjacent odd–odd nucleus. These states are proton–neutron two-quasiparticle states

$$|pn;J\,M\rangle = [a_p^\dagger a_n^\dagger]_{JM}|\text{BCS}\rangle . \qquad (15.62)$$

Applying the transition density (15.13), with $a = p$ and $b = n$, to (7.18) and (7.19) we obtain the amplitudes for beta transitions from the state (15.62) to the BCS vacuum. For a non-vanishing contribution to $\beta^-$ decay the index $d$ in (15.13) must be a neutron index and $c$ a proton index. It follows that the first term of (15.13) carries $\beta^-$ decay. With the single-particle matrix elements (7.20) and (7.21) we obtain the $\beta^-$ transition amplitudes

$$
\boxed{
\begin{aligned}
\mathcal{M}_{\mathrm{F}}^{(-)}(pnJ \to \mathrm{BCS}) &= \delta_{J0}\delta_{pn}\hat{j}_n v_p u_n \,, \\
\mathcal{M}_{\mathrm{GT}}^{(-)}(pnJ \to \mathrm{BCS}) &= \delta_{J1}\sqrt{3}\, v_p u_n \mathcal{M}_{\mathrm{GT}}(pn) \,,
\end{aligned}
}
\tag{15.63}
$$

where $\delta_{pn}$ means that the quantum numbers of the proton and neutron single-particle states must be the same for non-vanishing.

For a non-vanishing contribution to $\beta^+/\mathrm{EC}$ decay the index $d$ in (15.13) must be a proton index and $c$ a neutron index. Accordingly the second term of (15.13) is now effective. With use of the symmetry relation (7.23) we then have the $\beta^+/\mathrm{EC}$ transition amplitudes

$$
\boxed{
\begin{aligned}
\mathcal{M}_{\mathrm{F}}^{(+)}(pnJ \to \mathrm{BCS}) &= \delta_{J0}\delta_{np}\hat{j}_p v_n u_p \,, \\
\mathcal{M}_{\mathrm{GT}}^{(+)}(pnJ \to \mathrm{BCS}) &= -\delta_{J1}\sqrt{3}\, v_n u_p \mathcal{M}_{\mathrm{GT}}(pn) \,.
\end{aligned}
}
\tag{15.64}
$$

The matrix elements for beta transitions $\mathrm{BCS} \to pnJ$ are derived similarly by using the transition density (15.14). The results are

$$
\boxed{
\begin{aligned}
\mathcal{M}_{\mathrm{F}}^{(-)}(\mathrm{BCS} \to pnJ) &= \delta_{J0}\delta_{pn}\hat{j}_n u_p v_n \,, \\
\mathcal{M}_{\mathrm{GT}}^{(-)}(\mathrm{BCS} \to pnJ) &= \delta_{J1}\sqrt{3}\, u_p v_n \mathcal{M}_{\mathrm{GT}}(pn)
\end{aligned}
}
\tag{15.65}
$$

and

$$
\boxed{
\begin{aligned}
\mathcal{M}_{\mathrm{F}}^{(+)}(\mathrm{BCS} \to pnJ) &= \delta_{J0}\delta_{np}\hat{j}_p u_n v_p \,, \\
\mathcal{M}_{\mathrm{GT}}^{(+)}(\mathrm{BCS} \to pnJ) &= -\delta_{J1}\sqrt{3}\, u_n v_p \mathcal{M}_{\mathrm{GT}}(pn) \,.
\end{aligned}
}
\tag{15.66}
$$

We can see from (15.63)–(15.66) that corresponding transitions $pnJ \to \mathrm{BCS}$ and $\mathrm{BCS} \to pnJ$ are related through the exchange $u \leftrightarrow v$, the property relating the transition densities (15.13) and (15.14). Furthermore, the equations are related according to

$$
\mathcal{M}_J^{\mp}(\mathrm{BCS} \to pnJ) = (-1)^J \mathcal{M}_J^{\pm}(pnJ \to \mathrm{BCS}) \,,
\tag{15.67}
$$

where $J = 0$ indicates a Fermi transition and $J = 1$ a Gamow–Teller transition. This result can also be obtained from the relation (2.32) by recognizing that the Fermi and Gamow–Teller operators are Hermitian tensors.

To illustrate the physical meaning of our results and to make connection to Subsect. 7.4.1 let us consider $\beta^-$ decay in the limit $|\mathrm{BCS}\rangle \to |\mathrm{HF}\rangle$. Figure 15.3(a) illustrates the $\beta^-$ transitions (15.65): the initial state is the particle–hole vacuum and the final state is a proton-particle–neutron-hole state. We see from the figure that $v_n = 1$ and $u_p = 1$. With these values our $\beta^-$ results (15.65) agree with (7.63) and (7.64).

Figure 15.3(b) represents the particle–hole limit of (15.63): the initial state is a proton-hole–neutron-particle state and the final state is the particle–hole vacuum. We see from the figure that $u_n = 1$ and $v_p = 1$. With these values the expressions (15.63) agree with (7.60) and (7.61); the apparent sign difference in the Gamow–Teller result is due to the different coupling order.

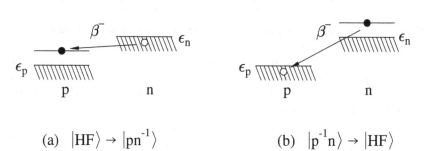

(a) $|\text{HF}\rangle \to |pn^{-1}\rangle$                    (b) $|p^{-1}n\rangle \to |\text{HF}\rangle$

**Fig. 15.3.** Qualitative view of $\beta^-$ decay in the limit $|\text{BCS}\rangle \to |\text{HF}\rangle$ to and from an odd–odd nucleus. The hatched blocks enclose the occupied states belonging to the particle–hole vacuum

For $K$th-forbidden unique beta decay the transition density (15.13) substituted into the definition (7.187), with use of the symmetry relations (7.190) and (7.191), gives the transition amplitudes

$$
\begin{aligned}
\mathcal{M}_{Ku}^{(-)}(pnJ \to \text{BCS}) &= \delta_{J,K+1} \hat{J} \, v_p u_n \mathcal{M}^{(Ku)}(pn) \,, \\
\mathcal{M}_{Ku}^{(+)}(pnJ \to \text{BCS}) &= -\delta_{J,K+1} \hat{J} \, v_n u_p \theta^{(K)} \mathcal{M}^{(Ku)}(pn) \,,
\end{aligned}
\tag{15.68}
$$

where $\theta^{(K)}$ is the phase factor (15.25). Again the transition densities for $\text{BCS} \to pnJ$ are obtained from (15.68) by the exchange $u \leftrightarrow v$. The two decay directions are also related through

$$
\mathcal{M}_{Ku}^{(\mp)}(\text{BCS} \to pnJ) = -\theta^{(K)} \mathcal{M}_{Ku}^{(\pm)}(pnJ \to \text{BCS}) \,.
\tag{15.69}
$$

It is often necessary to deal with wave functions that are linear combinations of proton–neutron two-quasiparticle components. Such wave functions have the form

$$
|J^\pi M\rangle = \sum_{pn} X_{pn}^{J^\pi} \left[ a_p^\dagger a_n^\dagger \right]_{JM} |\text{BCS}\rangle \,.
\tag{15.70}
$$

When computing beta-decay rates with these wave functions, use Eqs. (15.63)–(15.69) by combining them with amplitudes $X_{pn}$. This is analogous to the electromagnetic case with the linear combinations (15.58).

## 15.4.2 Beta and Gamma Decays in the $A = 30$ Chain of Isobars

Let us consider electromagnetic and $\beta^+/\mathrm{EC}$ decays of the $A = 30$ nuclei depicted in Fig. 15.4. This case involves two BCS vacua, namely

$$|\mathrm{BCS}\rangle_1 = |^{30}_{14}\mathrm{Si}_{16}\rangle \,, \quad Z_{\mathrm{act}} = 6 \,, \ N_{\mathrm{act}} = 8 \,, \tag{15.71}$$

$$|\mathrm{BCS}\rangle_2 = |^{30}_{16}\mathrm{S}_{14}\rangle \,, \quad Z_{\mathrm{act}} = 8 \,, \ N_{\mathrm{act}} = 6 \,. \tag{15.72}$$

The quasiparticle spectra in Fig. 14.3(c) for $Z_{\mathrm{act}}, N_{\mathrm{act}} = 6$ and Fig. 14.4(a) for $Z_{\mathrm{act}}, N_{\mathrm{act}} = 8$ provide guidance for constructing the two-quasiparticle states needed. From the one or two lowest levels in the figures we write the following ansatz for two-quasiparticle wave functions with reference to $|\mathrm{BCS}\rangle_1$:

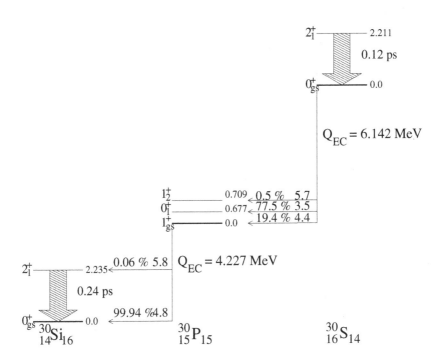

**Fig. 15.4.** Beta decay of the $0^+$ ground state of $^{30}\mathrm{S}$ to the $0^+$ and $1^+$ states in $^{30}\mathrm{P}$. Also shown is the beta decay of the $1^+$ ground state of $^{30}\mathrm{P}$ to the $0^+$ and $2^+$ states in $^{30}\mathrm{Si}$. The decays proceed via the $\beta^+/\mathrm{EC}$ mode. The experimental $Q$ values, excitation energies, decay branchings, $\log ft$ values and E2 gamma-decay half-lives are shown

$$|^{30}\text{Si}\,;\,0_{\text{gs}}^{+}\rangle = |\text{BCS}\rangle_1\,, \tag{15.73}$$

$$|^{30}\text{Si}\,;\,2_1^{+}\rangle = \frac{1}{\sqrt{2}}\Big([a_{\pi 1 s_{1/2}}^{\dagger}a_{\pi 0 d_{5/2}}^{\dagger}]_2 + \frac{1}{\sqrt{2}}[a_{\nu 0 d_{3/2}}^{\dagger}a_{\nu 0 d_{3/2}}^{\dagger}]_2\Big)|\text{BCS}\rangle_1\,, \tag{15.74}$$

$$|^{30}\text{P}\,;\,1_{\text{gs}}^{+}\rangle = \alpha_1\,[a_{\pi 1 s_{1/2}}^{\dagger}a_{\nu 0 d_{3/2}}^{\dagger}]_1|\text{BCS}\rangle_1$$
$$+ \frac{\beta_1}{\sqrt{2}}\Big([a_{\pi 0 d_{5/2}}^{\dagger}a_{\nu 0 d_{3/2}}^{\dagger}]_1 + [a_{\pi 1 s_{1/2}}^{\dagger}a_{\nu 1 s_{1/2}}^{\dagger}]_1\Big)|\text{BCS}\rangle_1\,. \tag{15.75}$$

The amplitudes $\alpha_1$ and $\beta_1$, with normalization $\alpha_1^2 + \beta_1^2 = 1$, are to be adjusted to the experimental data.

With reference to the vacuum $|\text{BCS}\rangle_2$ we write similarly the wave functions

$$|^{30}\text{S}\,;\,0_{\text{gs}}^{+}\rangle = |\text{BCS}\rangle_2\,, \tag{15.76}$$

$$|^{30}\text{S}\,;\,2_1^{+}\rangle = \frac{1}{\sqrt{2}}\Big(\frac{1}{\sqrt{2}}[a_{\pi 0 d_{3/2}}^{\dagger}a_{\pi 0 d_{3/2}}^{\dagger}]_2 + [a_{\nu 1 s_{1/2}}^{\dagger}a_{\nu 0 d_{5/2}}^{\dagger}]_2\Big)|\text{BCS}\rangle_2\,, \tag{15.77}$$

$$|^{30}\text{P}\,;\,0_1^{+}\rangle = [a_{\pi 1 s_{1/2}}^{\dagger}a_{\nu 1 s_{1/2}}^{\dagger}]_0|\text{BCS}\rangle_2\,, \tag{15.78}$$

$$|^{30}\text{P}\,;\,1_{\text{gs}}^{+}\rangle = \alpha_2\,[a_{\pi 0 d_{3/2}}^{\dagger}a_{\nu 1 s_{1/2}}^{\dagger}]_1|\text{BCS}\rangle_2$$
$$+ \frac{\beta_2}{\sqrt{2}}\Big([a_{\pi 0 d_{3/2}}^{\dagger}a_{\nu 0 d_{5/2}}^{\dagger}]_1 + [a_{\pi 1 s_{1/2}}^{\dagger}a_{\nu 1 s_{1/2}}^{\dagger}]_1\Big)|\text{BCS}\rangle_2\,. \tag{15.79}$$

We write the wave function of the second $1^{+}$ state in $^{30}\text{P}$ to be orthogonal to the $1^{+}$ ground state. With reference to $|\text{BCS}\rangle_2$ the wave function is thus

$$|^{30}\text{P}\,;\,1_2^{+}\rangle = \beta_2\,[a_{\pi 0 d_{3/2}}^{\dagger}a_{\nu 1 s_{1/2}}^{\dagger}]_1|\text{BCS}\rangle_2$$
$$- \frac{\alpha_2}{\sqrt{2}}\Big([a_{\pi 0 d_{3/2}}^{\dagger}a_{\nu 0 d_{5/2}}^{\dagger}]_1 + [a_{\pi 1 s_{1/2}}^{\dagger}a_{\nu 1 s_{1/2}}^{\dagger}]_1\Big)|\text{BCS}\rangle_2\,. \tag{15.80}$$

### Beta Decay

Let us first consider the $\beta^{+}/\text{EC}$ decays. For the decay of $^{30}\text{P}$ to $^{30}\text{Si}$ the wave functions are (15.73) and (15.75). The transition is of Gamow–Teller type (15.64). We thus obtain

$$\mathcal{M}_{\text{GT}}^{(+)}(1_{\text{gs}}^{+} \to 0_{\text{gs}}^{+}) = -\sqrt{3}\Big\{\alpha_1 v_{\nu 0 d_{3/2}} u_{\pi 1 s_{1/2}}\mathcal{M}_{\text{GT}}(s_{1/2}d_{3/2})$$
$$+ \frac{\beta_1}{\sqrt{2}}\big[v_{\nu 0 d_{3/2}} u_{\pi 0 d_{5/2}}\mathcal{M}_{\text{GT}}(d_{5/2}d_{3/2}) + v_{\nu 1 s_{1/2}} u_{\pi 1 s_{1/2}}\mathcal{M}_{\text{GT}}(s_{1/2}s_{1/2})\big]\Big\}\,. \tag{15.81}$$

With the $u$ and $v$ amplitudes from Table 14.1 and the Gamow–Teller single-particle matrix elements from Table 7.3 this becomes

$$\mathcal{M}_{\text{GT}}^{(+)}(1_{\text{gs}}^{+} \to 0_{\text{gs}}^{+}) = -\sqrt{3}\Big\{\alpha_1 \times 0.3489 \times 0.6330 \times 0$$
$$+ \frac{\beta_1}{\sqrt{2}}\Big[0.3489 \times 0.4996 \times \Big(-\frac{4}{\sqrt{5}}\Big) + 0.9557 \times 0.6330 \times \sqrt{2}\Big]\Big\} = -0.666\beta_1\,. \tag{15.82}$$

With the usual relations from Subsect. 7.2.1 this gives

$$B_{\mathrm{GT}} = 0.231\beta_1^2 \ , \quad \log ft = 4.43 - 2\log\beta_1 \ . \tag{15.83}$$

To reproduce the experimental value $\log ft = 4.8$, we thus need to take

$$\beta_1 = 0.65 \ , \quad \alpha_1 = \sqrt{1 - \beta_1^2} = 0.76 \ . \tag{15.84}$$

Next consider the $\beta^+/\mathrm{EC}$ transitions from the ground state of $^{30}\mathrm{S}$ to the states of $^{30}\mathrm{P}$. For the Fermi transition from the initial state (15.76) to the $0_1^+$ final state (15.78) the first equation (15.66) gives

$$\mathcal{M}_{\mathrm{F}}^{(+)}(0_{\mathrm{gs}}^+ \to 0_1^+) = \sqrt{2} \times 0.6330 \times 0.9557 = 0.856 \ . \tag{15.85}$$

This leads to

$$\log ft(0_{\mathrm{gs}}^+ \to 0_1^+) = 3.92 \ . \tag{15.86}$$

Compared with the experimental value 3.5 the computed value indicates a transition that is somewhat too slow. Improved agreement can be expected if more components are included in the wave function (15.78).

For the decays to the $1^+$ states the wave functions are (15.76), (15.79) and (15.80). With the second equation (15.66) we calculate

$$\mathcal{M}_{\mathrm{GT}}^{(+)}(0_{\mathrm{gs}}^+ \to 1_{\mathrm{gs}}^+) = -\sqrt{3}\Big\{\alpha_2 \times 0 + \frac{\beta_2}{\sqrt{2}}\big[u_{\nu 0d_{5/2}} v_{\pi 0d_{3/2}} \mathcal{M}_{\mathrm{GT}}(d_{3/2}d_{5/2})$$
$$+ u_{\nu 1s_{1/2}} v_{\pi 1s_{1/2}} \mathcal{M}_{\mathrm{GT}}(s_{1/2}s_{1/2})\big]\Big\}$$
$$= -\beta_2 \sqrt{\frac{3}{2}}\Big(0.4996 \times 0.3489 \times \frac{4}{\sqrt{5}} + 0.6330 \times 0.9557 \times \sqrt{2}\Big) = -1.430\beta_2 \ . \tag{15.87}$$

From the structure of the wave functions (15.79) and (15.80) we see that the matrix element for the transition to the $1_2^+$ state is obtained from (15.87) by the replacement $\beta_2 \to -\alpha_2$, so that

$$\mathcal{M}_{\mathrm{GT}}^{(+)}(0_{\mathrm{gs}}^+ \to 1_2^+) = 1.430\alpha_2 \ . \tag{15.88}$$

The states (15.75) and (15.79) are the same except that protons and neutrons are exchanged. We can therefore take $\alpha_2 = \alpha_1 = 0.76$, $\beta_2 = \beta_1 = 0.65$ as a first guess to predict the $\log ft$ values for the $0^+ \to 1^+$ transitions. The outcome is

$$\log ft(0_{\mathrm{gs}}^+ \to 1_{\mathrm{gs}}^+) = 3.66 \ , \quad \log ft(0_{\mathrm{gs}}^+ \to 1_2^+) = 3.52 \ . \tag{15.89}$$

These values are rather far from their experimental counterparts 4.4 and 5.7. We return to this matter when discussing the mixing of two-quasiparticle configurations within the quasiparticle versions of the TDA and RPA.

**Electromagnetic Transitions**

We calculate the E2 transition in $^{30}$Si with the wave functions (15.73) and (15.74). Equation (15.59), together with Table 14.1 for the occupation amplitudes and Table 6.4 for the single-particle matrix elements, gives the decay amplitude

$$(\text{BCS}\|Q_2\|^{30}\text{Si}\,;\,2_1^+)$$

$$= \frac{1}{\sqrt{2}}\Big[(v_{\pi 1s_{1/2}}u_{\pi 0d_{5/2}} + v_{\pi 0d_{5/2}}u_{\pi 1s_{1/2}})(\pi 1s_{1/2}\|Q_2\|\pi 0d_{5/2})$$

$$+ \frac{1}{\sqrt{2}} \times 2v_{\nu 0d_{3/2}}u_{\nu 0d_{3/2}}(\nu 0d_{3/2}\|Q_2\|\nu 0d_{3/2})\Big]$$

$$= \frac{1}{\sqrt{2}}\Big[(0.7741 \times 0.4996 + 0.8663 \times 0.6330)(-2.185e^{\text{p}}_{\text{eff}}b^2)$$

$$+ \sqrt{2} \times 0.3489 \times 0.9372(-1.975e^{\text{n}}_{\text{eff}}b^2)\Big]$$

$$= -(1.445e^{\text{p}}_{\text{eff}} + 0.6458e^{\text{n}}_{\text{eff}})b^2 = -(5.031e^{\text{p}}_{\text{eff}} + 2.249e^{\text{n}}_{\text{eff})\,\text{fm}^2\,, \qquad (15.90)$$

where the value $b = 1.866\,\text{fm}$ for the oscillator parameter was used.

By taking $e^{\text{p}}_{\text{eff}} = (1+\chi)e$ and $e^{\text{n}}_{\text{eff}} = \chi e$ in (15.90) we obtain the reduced transition probability

$$B(\text{E2}\,;\,^{30}\text{Si}) = \frac{1}{5}(5.031 + 7.280\chi)^2\,e^2\text{fm}^4\,. \qquad (15.91)$$

The measured half-life and excitation energy of the $2_1^+$ state give the experimental value $B(\text{E2}\,;\,^{30}\text{Si}) = 42\,e^2\text{fm}^4$. This is fitted by

$$\chi\left(^{30}\text{Si}\right) = 1.3\,. \qquad (15.92)$$

This appreciable value of $\chi$ points to collectivity of the $2_1^+$ state. Such collectivity can be achieved theoretically through a quasiparticle-TDA or -RPA description.

For the E2 decay of the $2_1^+$ state of $^{30}$S we obtain

$$(\text{BCS}\|Q_2\|^{30}\text{S}\,;\,2_1^+)$$

$$= \frac{1}{\sqrt{2}}\Big[\frac{1}{\sqrt{2}} \times 2v_{\pi 0d_{3/2}}u_{\pi 0d_{3/2}}(\pi 0d_{3/2}\|Q_2\|\pi 0d_{3/2})$$

$$+ (v_{\nu 1s_{1/2}}u_{\nu 0d_{5/2}} + v_{\nu 0d_{5/2}}u_{\nu 1s_{1/2}})(\nu 1s_{1/2}\|Q_2\|\nu 0d_{5/2})\Big]$$

$$= \frac{1}{\sqrt{2}}\Big[\sqrt{2} \times 0.3489 \times 0.9372(-1.975e^{\text{p}}_{\text{eff}}b^2)$$

$$+ (0.7741 \times 0.4996 + 0.8663 \times 0.6330)(-2.185e^{\text{n}}_{\text{eff}}b^2)\Big]$$

$$= -(0.6458e^{\text{p}}_{\text{eff}} + 1.445e^{\text{n}}_{\text{eff}})b^2 = -(2.249e^{\text{p}}_{\text{eff}} + 5.031e^{\text{n}}_{\text{eff}})\,\text{fm}^2\,. \qquad (15.93)$$

This is the same as (15.90) with protons and neutrons exchanged, as suggested by the proton–neutron symmetry between the states (15.74) and (15.77).

Equation (15.93) leads to

$$B(E2\,;{}^{30}S) = \frac{1}{5}(2.249 + 7.280\chi)^2\,e^2\mathrm{fm}^4\,. \qquad (15.94)$$

The experimental $B(E2)$ value extracted from the data of Fig. 15.4 is $89\,e^2\mathrm{fm}^4$, whence

$$\chi({}^{30}S) = 2.6\,. \qquad (15.95)$$

This large value of the electric polarization constant clearly indicates that a collective wave function is needed to reproduce the experimental decay rate.

## 15.5 Transitions Between Two-Quasiparticle States

In this section we discuss electromagnetic and beta-decay transitions involving two-quasiparticle states of the proton–proton, neutron–neutron and proton–neutron types.

### 15.5.1 Electromagnetic Transitions

#### Even–Even Nuclei

Our starting points for an even–even nucleus are the initial and final states written in terms of two-quasiparticle states of like nucleons. From (15.10) they are

$$|a_i\,b_i\,;\,J_i\,M_i\rangle = A^\dagger_{a_i b_i}(J_i M_i)|\mathrm{BCS}\rangle\,, \qquad (15.96)$$

$$|a_f\,b_f\,;\,J_f\,M_f\rangle = A^\dagger_{a_f b_f}(J_f M_f)|\mathrm{BCS}\rangle\,. \qquad (15.97)$$

With these wave functions and the transition density (15.15) in (6.22), we have the decay amplitude

$$(a_f\,b_f\,;\,J_f\|\mathcal{M}_{\sigma\lambda}\|a_i\,b_i\,;\,J_i)$$

$$= \widehat{\lambda}^{-1}\sum_{ab}(a\|\mathcal{M}_{\sigma\lambda}\|b)\big[u_a u_b \mathcal{K}^\lambda_{ab}(fi) + v_a v_b(-1)^{j_a+j_b+\lambda}\mathcal{K}^\lambda_{ba}(fi)\big]$$

$$= \widehat{\lambda}^{-1}\sum_{ab}(a\|\mathcal{M}_{\sigma\lambda}\|b)\mathcal{D}^{(\lambda)}_{ab}\mathcal{K}^\lambda_{ab}(fi)\,, \qquad (15.98)$$

where $\mathcal{D}^{(\lambda)}_{ab}$ is defined in (15.18). Substituting the expression (15.17) for $\mathcal{K}^\lambda_{ab}(fi)$ we obtain the final form

$$
\begin{aligned}
(a_f\, b_f\,;\, J_f \| &\boldsymbol{\mathcal{M}}_{\sigma\lambda} \| a_i\, b_i\,;\, J_i) = \widehat{J_i}\widehat{J_f}\mathcal{N}_{a_i b_i}(J_i)\mathcal{N}_{a_f b_f}(J_f) \\
\times \Big[ &\delta_{b_i b_f}(-1)^{j_{a_f}+j_{b_f}+J_i+\lambda}
\left\{\begin{matrix} J_f & J_i & \lambda \\ j_{a_i} & j_{a_f} & j_{b_f} \end{matrix}\right\}
\mathcal{D}^{(\lambda)}_{a_i a_f}(a_f\|\boldsymbol{\mathcal{M}}_{\sigma\lambda}\|a_i) \\
&+ \delta_{a_i b_f}(-1)^{j_{a_f}+j_{b_i}+\lambda}
\left\{\begin{matrix} J_f & J_i & \lambda \\ j_{b_i} & j_{a_f} & j_{b_f} \end{matrix}\right\}
\mathcal{D}^{(\lambda)}_{b_i a_f}(a_f\|\boldsymbol{\mathcal{M}}_{\sigma\lambda}\|b_i) \\
&+ \delta_{a_i a_f}(-1)^{j_{a_i}+j_{b_i}+J_f+\lambda}
\left\{\begin{matrix} J_f & J_i & \lambda \\ j_{b_i} & j_{b_f} & j_{a_f} \end{matrix}\right\}
\mathcal{D}^{(\lambda)}_{b_i b_f}(b_f\|\boldsymbol{\mathcal{M}}_{\sigma\lambda}\|b_i) \\
&+ \delta_{b_i a_f}(-1)^{J_i+J_f+\lambda+1}
\left\{\begin{matrix} J_f & J_i & \lambda \\ j_{a_i} & j_{b_f} & j_{a_f} \end{matrix}\right\}
\mathcal{D}^{(\lambda)}_{a_i b_f}(b_f\|\boldsymbol{\mathcal{M}}_{\sigma\lambda}\|a_i) \Big], \\
a_i b_i = n_i n_i' \ &\text{or} \ p_i p_i'\,, \qquad a_f b_f = n_f n_f' \ \text{or} \ p_f p_f'\,.
\end{aligned}
\tag{15.99}
$$

A simple special case of (15.99) is decay to a $J_f^\pi = 0^+$ excited state. Then we have $a_f = b_f$, $\lambda = J_i$. The result is (Exercise 15.31)

$$
\begin{aligned}
(a_f\, a_f\,;\, 0^+\|&\boldsymbol{\mathcal{M}}_{\sigma\lambda}\|a_i\, b_i\,;\, J_i) = \sqrt{2}\,\delta_{\lambda J_i}\widehat{j_{a_f}}^{\,-1}\mathcal{N}_{a_i b_i}(J_i) \\
\times [\,\delta_{a_i a_f}&\mathcal{D}^{(\lambda)}_{b_i a_f}(a_f\|\boldsymbol{\mathcal{M}}_{\sigma\lambda}\|b_i) - \delta_{b_i a_f}(-1)^{j_{a_i}+j_{a_f}+J_i}\mathcal{D}^{(\lambda)}_{a_i a_f}(a_f\|\boldsymbol{\mathcal{M}}_{\sigma\lambda}\|a_i)]\,.
\end{aligned}
\tag{15.100}
$$

We note that Eqs. (15.99) and (15.100) differ from (6.99) and (6.103), respectively, only through the $\mathcal{D}$ factors. In the Hartree–Fock limit, quasiparticles become particles when $u \to 1$, $v \to 0$. From (15.18) and (15.19) we then see that $\mathcal{D} \to \pm 1$. The results thus become identical except for an unimportant overall phase factor. In the BR convention the phase factor is always 1.

## Odd–Odd Nuclei

For an odd–odd nucleus the initial and final states are written as proton–neutron two-quasiparticle states. This is clear by the number-parity argument of Subsect. 13.4.2. The states are

$$
|p_i\, n_i\,;\, J_i\, M_i\rangle = [a_{p_i}^\dagger a_{n_i}^\dagger]_{J_i M_i}|\text{BCS}\rangle\,,
\tag{15.101}
$$

$$
|p_f\, n_f\,;\, J_f\, M_f\rangle = [a_{p_f}^\dagger a_{n_f}^\dagger]_{J_f M_f}|\text{BCS}\rangle\,.
\tag{15.102}
$$

The decay amplitude for these states can be written directly from (15.99) by setting $a_i = p_i$, $b_i = n_i$, $a_f = p_f$, $b_f = n_f$. Because $\delta_{pn} = 0$ only the first and third terms contribute, and we have

$$
\begin{aligned}
(p_f\, n_f\,;\, J_f\|&\boldsymbol{\mathcal{M}}_{\sigma\lambda}\|p_i\, n_i\,;\, J_i) \\
= \widehat{J_i}\widehat{J_f}\Big[&\delta_{n_i n_f}(-1)^{j_{p_f}+j_{n_f}+J_i+\lambda}
\left\{\begin{matrix} J_f & J_i & \lambda \\ j_{p_i} & j_{p_f} & j_{n_f} \end{matrix}\right\}
\mathcal{D}^{(\lambda)}_{p_i p_f}(p_f\|\boldsymbol{\mathcal{M}}_{\sigma\lambda}\|p_i) \\
&+ \delta_{p_i p_f}(-1)^{j_{p_i}+j_{n_i}+J_f+\lambda}
\left\{\begin{matrix} J_f & J_i & \lambda \\ j_{n_i} & j_{n_f} & j_{p_f} \end{matrix}\right\}
\mathcal{D}^{(\lambda)}_{n_i n_f}(n_f\|\boldsymbol{\mathcal{M}}_{\sigma\lambda}\|n_i) \Big]\,,
\end{aligned}
\tag{15.103}
$$

As in the case of (15.100), by setting $J_f = 0$ in (15.103) we obtain the simple expression

$$(p_f\, n_f\,;\, 0\|\mathcal{M}_{\sigma\lambda}\|p_i\, n_i\,;\, J_i) = \delta_{\lambda J_i}\delta_{j_{p_f}j_{n_f}}(-1)^{j_{p_i}+j_{p_f}+1}\widehat{j}_{p_f}^{\,-1}$$
$$\times\, [\delta_{n_i n_f}(-1)^{J_i}\mathcal{D}_{p_i p_f}^{(\lambda)}\,(p_f\|\mathcal{M}_{\sigma\lambda}\|p_i) + \delta_{p_i p_f}\mathcal{D}_{n_i n_f}^{(\lambda)}\,(n_f\|\mathcal{M}_{\sigma\lambda}\|n_i)]\,. \quad (15.104)$$

In the Hartree–Fock limit with $\mathcal{D} \to \pm 1$, the expressions (15.103) and (15.104), respectively, become (6.102) and (6.104) to within sign.

### 15.5.2 Beta-Decay Transitions

Beta-decay transitions proceed between an even–even nucleus and an adjacent odd–odd one. The states of the even–even nucleus are two-quasiparticle states of like nucleons; the states of the odd–odd nucleus are proton–neutron two-quasiparticle states. As in the case of odd-$A$ nuclei discussed in Subsect. 15.3, the transitions can be classified into two categories, particle type and hole type. Stated like (15.22), the classification is

Particle type

$$n_i n_i' \xrightarrow{\beta^-} p_f n_f\,, \quad p_i n_i \xrightarrow{\beta^-} p_f p_f'\,, \quad p_i p_i' \xrightarrow{\beta^+/\text{EC}} p_f n_f\,, \quad p_i n_i \xrightarrow{\beta^+/\text{EC}} n_f n_f'\,,$$

Hole type

$$p_i p_i' \xrightarrow{\beta^-} p_f n_f\,, \quad p_i n_i \xrightarrow{\beta^-} n_f n_f'\,, \quad n_i n_i' \xrightarrow{\beta^+/\text{EC}} p_f n_f\,, \quad p_i n_i \xrightarrow{\beta^+/\text{EC}} p_f p_f'\,.$$

$$(15.105)$$

### Allowed Beta Decay

Equations (7.105)–(7.108) give the transition amplitudes for allowed beta decay between two-particle states. The transition amplitudes for quasiparticles are straightforward generalizations of these equations. For particle-type transitions the single-particle matrix elements are multiplied by $u_i u_f$. The corresponding factor for hole-type transitions is $(-1)^{L+1}v_i v_f$, where the phase factor is from (7.117). To carry these occupation factors we define

$$\mathcal{B}_L^{(-)}(if) \equiv u_i u_f \qquad\qquad \text{particle type}\,, \qquad\qquad (15.106)$$
$$\mathcal{B}_L^{(+)}(if) \equiv (-1)^{L+1}v_i v_f \quad \text{hole type}\,. \qquad\qquad (15.107)$$

The Fermi and Gamow–Teller single-particle matrix elements from Chap. 7 are

$$\mathcal{M}_0(pn) = \mathcal{M}_\text{F}(pn) = \widehat{j}_p\delta_{pn}\,, \qquad\qquad (15.108)$$
$$\mathcal{M}_1(pn) = \mathcal{M}_\text{GT}(pn)\,. \qquad\qquad (15.109)$$

Inserting the factors (15.106) and (15.107) into Eqs. (7.105)–(7.108) we obtain the transition amplitudes for allowed beta decay between two-quasiparticle states. With the notation $\mathcal{M}^{(\mp)}$ for particle-type $\beta^-$ decay or hole-type $\beta^+/$EC decay, according to (15.105), we then have

$$
\mathcal{M}_L^{(\mp)}(n_i\, n_i'\,;\, J_i \to p_f\, n_f\,;\, J_f) = \widehat{L}\widehat{J_i}\widehat{J_f}\mathcal{N}_{n_i n_i'}(J_i)
$$
$$
\times \left[ \delta_{n_i' n_f}(-1)^{j_{p_f}+j_{n_f}+J_i+L} \left\{ \begin{array}{ccc} J_i & J_f & L \\ j_{p_f} & j_{n_i} & j_{n_f} \end{array} \right\} \mathcal{B}_L^{(\mp)}(p_f n_i)\mathcal{M}_L(p_f n_i) \right.
$$
$$
\left. + \delta_{n_i n_f}(-1)^{j_{p_f}+j_{n_i'}+L} \left\{ \begin{array}{ccc} J_i & J_f & L \\ j_{p_f} & j_{n_i'} & j_{n_f} \end{array} \right\} \mathcal{B}_L^{(\mp)}(p_f n_i')\mathcal{M}_L(p_f n_i') \right],
$$

$$(15.110)$$

$$
\mathcal{M}_L^{(\mp)}(p_i\, n_i\,;\, J_i \to p_f\, p_f'\,;\, J_f) = \widehat{L}\widehat{J_i}\widehat{J_f}\mathcal{N}_{p_f p_f'}(J_f)
$$
$$
\times \left[ \delta_{p_i p_f'}(-1)^{j_{p_f}+j_{n_i}+L} \left\{ \begin{array}{ccc} J_i & J_f & L \\ j_{p_f} & j_{n_i} & j_{p_f'} \end{array} \right\} \mathcal{B}_L^{(\mp)}(p_f n_i)\mathcal{M}_L(p_f n_i) \right.
$$
$$
\left. + \delta_{p_i p_f}(-1)^{j_{p_f}+j_{n_i}+J_f+L} \left\{ \begin{array}{ccc} J_i & J_f & L \\ j_{p_f'} & j_{n_i} & j_{p_f} \end{array} \right\} \mathcal{B}_L^{(\mp)}(p_f' n_i)\mathcal{M}_L(p_f' n_i) \right],
$$

$$(15.111)$$

$$
\mathcal{M}_L^{(\pm)}(p_i\, p_i'\,;\, J_i \to p_f\, n_f\,;\, J_f) = \widehat{L}\widehat{J_i}\widehat{J_f}\mathcal{N}_{p_i p_i'}(J_i)
$$
$$
\times \left[ \delta_{p_i p_f}(-1)^{j_{n_f}+j_{p_f}+J_f+L} \left\{ \begin{array}{ccc} J_i & J_f & L \\ j_{n_f} & j_{p_i'} & j_{p_f} \end{array} \right\} \mathcal{B}_L^{(\mp)}(p_i' n_f)\mathcal{M}_L(p_i' n_f) \right.
$$
$$
\left. + \delta_{p_i' p_f}(-1)^{j_{p_i}+j_{n_f}+J_i+J_f+L} \left\{ \begin{array}{ccc} J_i & J_f & L \\ j_{n_f} & j_{p_i} & j_{p_f} \end{array} \right\} \mathcal{B}_L^{(\mp)}(p_i n_f)\mathcal{M}_L(p_i n_f) \right],
$$

$$(15.112)$$

$$
\mathcal{M}_L^{(\pm)}(p_i\, n_i\,;\, J_i \to n_f\, n_f'\,;\, J_f) = \widehat{L}\widehat{J_i}\widehat{J_f}\mathcal{N}_{n_f n_f'}(J_f)
$$
$$
\times \left[ \delta_{n_i n_f'}(-1)^{j_{p_i}+j_{n_i}+J_i+L} \left\{ \begin{array}{ccc} J_i & J_f & L \\ j_{n_f} & j_{p_i} & j_{n_i} \end{array} \right\} \mathcal{B}_L^{(\mp)}(p_i n_f)\mathcal{M}_L(p_i n_f) \right.
$$
$$
\left. + \delta_{n_i n_f}(-1)^{j_{p_i}+j_{n_f'}+J_i+J_f+L} \left\{ \begin{array}{ccc} J_i & J_f & L \\ j_{n_f'} & j_{p_i} & j_{n_f} \end{array} \right\} \mathcal{B}_L^{(\mp)}(p_i n_f')\mathcal{M}_L(p_i n_f') \right].
$$

$$(15.113)$$

Equations (15.110)–(15.113) simplify in the case of a zero-coupled quasiparticle pair, i.e. when $J_f = 0$ or $J_i = 0$. The resulting expressions can be written down directly by inserting the occupation factors (15.106) and (15.107) into Eqs. (7.109)–(7.116):

$$\mathcal{M}_L^{(\mp)}(n_i\, n_i'\,;\, J_i \to p_f\, n_f\,;\, J_f = 0) = \delta_{J_i L}\delta_{j_{p_f} j_{n_f}}\widehat{J_i}\widehat{j_{p_f}}^{-1}\mathcal{N}_{n_i n_i'}(J_i)$$
$$\times \left[\delta_{n_i' n_f}(-1)^{j_{n_i}+j_{n_f}+J_i+1}\mathcal{B}_L^{(\mp)}(p_f n_i)\mathcal{M}_L(p_f n_i)\right.$$
$$\left. + \delta_{n_i n_f}\mathcal{B}_L^{(\mp)}(p_f n_i')\mathcal{M}_L(p_f n_i')\right], \qquad (15.114)$$

$$\mathcal{M}_L^{(\mp)}(n_i\, n_i'\,;\, J_i = 0 \to p_f\, n_f\,;\, J_f) = \delta_{J_f L}\delta_{j_{n_i} j_{n_i'}}\widehat{J_f}\widehat{j_{n_i}}^{-1}\mathcal{N}_{n_i n_i'}(0)$$
$$\times \left[\delta_{n_i' n_f}\mathcal{B}_L^{(\mp)}(p_f n_i)\mathcal{M}_L(p_f n_i) + \delta_{n_i n_f}\mathcal{B}_L^{(\mp)}(p_f n_i')\mathcal{M}_L(p_f n_i')\right], \quad (15.115)$$

$$\mathcal{M}_L^{(\mp)}(p_i\, n_i\,;\, J_i \to p_f\, p_f'\,;\, J_f = 0) = \delta_{J_i L}\delta_{j_{p_f} j_{p_f'}}\widehat{J_i}\widehat{j_{p_f}}^{-1}\mathcal{N}_{p_f p_f'}(0)$$
$$\times \left[\delta_{p_i p_f'}\mathcal{B}_L^{(\mp)}(p_f n_i)\mathcal{M}_L(p_f n_i) + \delta_{p_i p_f}\mathcal{B}_L^{(\mp)}(p_f' n_i)\mathcal{M}_L(p_f' n_i)\right], \quad (15.116)$$

$$\mathcal{M}_L^{(\mp)}(p_i\, n_i\,;\, J_i = 0 \to p_f\, p_f'\,;\, J_f) = \delta_{J_f L}\delta_{j_{n_i} j_{p_i}}\widehat{J_f}\widehat{j_{n_i}}^{-1}\mathcal{N}_{p_f p_f'}(J_f)$$
$$\times \left[\delta_{p_i p_f'}\mathcal{B}_L^{(\mp)}(p_f n_i)\mathcal{M}_L(p_f n_i)\right.$$
$$\left. + \delta_{p_i p_f}(-1)^{j_{p_f}+j_{p_f'}+J_f+1}\mathcal{B}_L^{(\mp)}(p_f' n_i)\mathcal{M}_L(p_f' n_i)\right], \quad (15.117)$$

$$\mathcal{M}_L^{(\pm)}(p_i\, p_i'\,;\, J_i \to p_f\, n_f\,;\, J_f = 0) = \delta_{J_i L}\delta_{j_{n_f} j_{p_f}}\widehat{J_i}\widehat{j_{n_f}}^{-1}\mathcal{N}_{p_i p_i'}(J_i)$$
$$\times \left[\delta_{p_i p_f}(-1)^{j_{p_f}+j_{p_i'}+1}\mathcal{B}_L^{(\mp)}(p_i' n_f)\mathcal{M}_L(p_i' n_f)\right.$$
$$\left. + \delta_{p_i' p_f}(-1)^{J_i}\mathcal{B}_L^{(\mp)}(p_i n_f)\mathcal{M}_L(p_i n_f)\right], \qquad (15.118)$$

$$\mathcal{M}_L^{(\pm)}(p_i\, p_i'\,;\, J_i = 0 \to p_f\, n_f\,;\, J_f) = \delta_{J_f L}\delta_{j_{p_i} j_{p_i'}}(-1)^{J_f}\widehat{J_f}\widehat{j_{p_i}}^{-1}\mathcal{N}_{p_i p_i'}(0)$$
$$\times \left[\delta_{p_i p_f}\mathcal{B}_L^{(\mp)}(p_i' n_f)\mathcal{M}_L(p_i' n_f) + \delta_{p_i' p_f}\mathcal{B}_L^{(\mp)}(p_i n_f)\mathcal{M}_L(p_i n_f)\right], \quad (15.119)$$

$$\mathcal{M}_L^{(\pm)}(p_i\, n_i\,;\, J_i \to n_f\, n_f'\,;\, J_f = 0) = \delta_{J_i L}\delta_{j_{n_f} j_{n_f'}}(-1)^{J_i}\widehat{J_i}\widehat{j_{n_f}}^{-1}\mathcal{N}_{n_f n_f'}(0)$$
$$\times \left[\delta_{n_i n_f'}\mathcal{B}_L^{(\mp)}(p_i n_f)\mathcal{M}_L(p_i n_f) + \delta_{n_i n_f}\mathcal{B}_L^{(\mp)}(p_i n_f')\mathcal{M}_L(p_i n_f')\right], \quad (15.120)$$

$$\mathcal{M}_L^{(\pm)}(p_i\, n_i\,;\, J_i = 0 \to n_f\, n_f'\,;\, J_f) = \delta_{J_f L}\delta_{j_{p_i} j_{n_i}}\widehat{J_f}\widehat{j_{p_i}}^{-1}\mathcal{N}_{n_f n_f'}(J_f)$$
$$\times \left[\delta_{n_i n_f'}(-1)^{j_{n_i}+j_{n_f}+1}\mathcal{B}_L^{(\mp)}(p_i n_f)\mathcal{M}_L(p_i n_f)\right.$$
$$\left. + \delta_{n_i n_f}(-1)^{J_f}\mathcal{B}_L^{(\mp)}(p_i n_f')\mathcal{M}_L(p_i n_f')\right]. \qquad (15.121)$$

## $K$th-forbidden Unique Beta Decay

The formulas (15.110)–(15.121) can be carried over to the case of $K$th-forbidden unique beta-decay transitions by using the conversion recipe (7.210) for $\beta^-$ decay and (7.213) for $\beta^+/$EC decay.

### 15.5.3 Beta Decay of $^{30}$P

Let us compute the log $ft$ value of the $\beta^+/$EC decay of the $1^+$ ground state of $^{30}_{15}$P$_{15}$ to the first excited $2^+$ state in $^{30}_{14}$Si$_{16}$; see Fig. 15.4. For the initial and final states we assume (15.79) and (15.77) respectively. The transition is of Gamow–Teller type. Inspection of the states shows that there are three particle-type contributions to the transition. Their decay amplitudes are given by (15.113). However, two of these vanish because of vanishing single-particle matrix elements:

$$M^{(+)}_{GT}\left(\pi 1s_{1/2}\,\nu 0d_{3/2}\,;\,1 \rightarrow (\nu 0d_{3/2})^2\,;\,2\right) = 0\,, \tag{15.122}$$

$$M^{(+)}_{GT}\left(\pi 1s_{1/2}\,\nu 1s_{1/2}\,;\,1 \rightarrow (\nu 0d_{3/2})^2\,;\,2\right) = 0\,. \tag{15.123}$$

The two terms of the non-vanishing decay amplitude contribute equally. For the $\mathcal{B}$ factors (15.106) and (15.107) we need the occupation amplitudes. They are given ($Z_{\text{act}} = 6$, $N_{\text{act}} = 8$) by Table 14.1, while Table 7.3 gives the single-particle matrix elements. Substitution yields

$$M^{(+)}_{GT}\left(\pi 0d_{5/2}\,\nu 0d_{3/2}\,;\,1 \rightarrow (\nu 0d_{3/2})^2\,;\,2\right) = 2 \times \sqrt{3} \times \sqrt{3} \times \sqrt{5} \times \frac{1}{\sqrt{2}}$$

$$\times (-1)^{\frac{5}{2}+\frac{3}{2}+1+1}\left\{\begin{matrix} 1 & 2 & 1 \\ \frac{3}{2} & \frac{5}{2} & \frac{3}{2} \end{matrix}\right\} u_{\pi 0d_{5/2}} u_{\nu 0d_{3/2}} M_{GT}(d_{5/2}\,d_{3/2})$$

$$= 3\sqrt{10}\left(-\frac{1}{10\sqrt{6}}\right)0.4996 \times 0.9372 \left(-\frac{4}{\sqrt{5}}\right) = 0.3244\,. \tag{15.124}$$

In addition there are three hole-type contributions, to be calculated from (15.111). Two of them vanish because of vanishing single-particle matrix elements and Kronecker deltas in (15.111):

$$M^{(+)}_{GT}(\pi 1s_{1/2}\,\nu 1s_{1/2}\,;\,1 \rightarrow \pi 1s_{1/2}\,\pi 0d_{5/2}\,;\,2) = 0\,, \tag{15.125}$$

$$M^{(+)}_{GT}(\pi 0d_{5/2}\,\nu 0d_{3/2}\,;\,1 \rightarrow \pi 1s_{1/2}\,\pi 0d_{5/2}\,;\,2) = 0\,. \tag{15.126}$$

The third one is

$$M^{(+)}_{GT}(\pi 1s_{1/2}\,\nu 0d_{3/2}\,;\,1 \rightarrow \pi 1s_{1/2}\,\pi 0d_{5/2}\,;\,2) = \sqrt{3} \times \sqrt{3} \times \sqrt{5}$$

$$\times \left[0 + (-1)^{\frac{1}{2}+\frac{3}{2}+2+1}\left\{\begin{matrix} 1 & 2 & 1 \\ \frac{5}{2} & \frac{3}{2} & \frac{1}{2} \end{matrix}\right\}(-1)^{1+1} v_{\pi 0d_{5/2}} v_{\nu 0d_{3/2}} M_{GT}(d_{5/2}\,d_{3/2})\right]$$

$$= -3\sqrt{5}\left(-\frac{1}{2\sqrt{5}}\right)0.8663 \times 0.3489 \left(-\frac{4}{\sqrt{5}}\right) = -0.8110\,. \tag{15.127}$$

With the values $\alpha_1 = 0.76$ and $\beta_1 = 0.65$ of the wave-function amplitudes from (15.84) the complete Gamow–Teller matrix element becomes

$$\mathcal{M}_{\mathrm{GT}}^{(+)}\left({}^{30}\mathrm{P}(1_{\mathrm{gs}}^+) \to {}^{30}\mathrm{Si}(2_1^+)\right) = \frac{\beta_1}{\sqrt{2}} \times \frac{1}{\sqrt{2}} \times 0.3244 + \alpha_1 \frac{1}{\sqrt{2}}(-0.8110)$$

$$= -0.33 . \tag{15.128}$$

The usual relations of Subsect. 7.2.1 give

$$\log ft(1_{\mathrm{gs}}^+ \to 2_1^+) = 5.0 . \tag{15.129}$$

The experimental value in Fig. 15.3 is 5.8. Thus the decay of the ${}^{30}\mathrm{P}$ ground state to the ground state and first excited $2^+$ state in ${}^{30}\mathrm{Si}$ are seen to be fairly consistently described by the simple wave functions (15.73)–(15.75).

### 15.5.4 Magnetic Dipole Decay in ${}^{30}$P

We continue our discussion of the $A = 30$ chain of nuclei. Consider the M1 decay of the $0^+$ first excited state of ${}^{30}\mathrm{P}$ to the $1^+$ ground state, as shown in Fig. 15.5. We use the wave functions (15.78) and (15.79). Equation (15.104) gives the reduced matrix element except that it is stated in reverse order, $J_i \to 0$. However, as seen from (2.32) or (11.290), this only introduces a phase factor that disappears from the reduced transition probability (6.4).

The states (15.78) and (15.79) give three contributions to the transition. Because of vanishing single-particle matrix elements, two of them are zero:

$$(\pi 1s_{1/2}\, \nu 1s_{1/2} ;\, 0\|\boldsymbol{M}_1\|\pi 0d_{3/2}\, \nu 1s_{1/2} ;\, 1) = 0 , \tag{15.130}$$

$$(\pi 1s_{1/2}\, \nu 1s_{1/2} ;\, 0\|\boldsymbol{M}_1\|\pi 0d_{3/2}\, \nu 0d_{5/2} ;\, 1) = 0 . \tag{15.131}$$

The non-zero contribution is

$$(\pi 1s_{1/2}\, \nu 1s_{1/2} ;\, 0\|\boldsymbol{M}_1\|\pi 1s_{1/2}\, \nu 1s_{1/2} ;\, 1) = (-1)^{\frac{1}{2}+\frac{1}{2}+1}\frac{1}{\sqrt{2}}$$

$$\times \left[ (-1)^1 \mathcal{D}_{\pi 1s_{1/2}\, \pi 1s_{1/2}}^{(1)}(\pi 1s_{1/2}\|\boldsymbol{M}_1\|\pi 1s_{1/2}) \right.$$

$$\left. + \mathcal{D}_{\nu 1s_{1/2}\, \nu 1s_{1/2}}^{(1)}(\nu 1s_{1/2}\|\boldsymbol{M}_1\|\nu 1s_{1/2}) \right] . \tag{15.132}$$

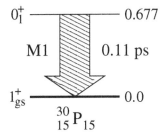

**Fig. 15.5.** Decay of the first excited state to the ground state in ${}^{30}$P. The excitation energy in Mega-electronvolts and the half-life are indicated

For the magnetic transition at hand the $\mathcal{D}$ factors are given by (15.18) and (15.19) as

$$\mathcal{D}^{(1)}_{\pi 1s_{1/2}\,\pi 1s_{1/2}} = u^2_{\pi 1s_{1/2}} + v^2_{\pi 1s_{1/2}} = 1 \,, \tag{15.133}$$

$$\mathcal{D}^{(1)}_{\nu 1s_{1/2}\,\nu 1s_{1/2}} = u^2_{\nu 1s_{1/2}} + v^2_{\nu 1s_{1/2}} = 1 \,. \tag{15.134}$$

Tables 6.6 and 6.7 give the single-particle matrix elements. With the bare $g$ factors (6.25), the reduced matrix element (15.132) becomes

$$(\pi 1s_{1/2}\,\nu 1s_{1/2}\,;\,0\|\boldsymbol{M}_1\|\pi 1s_{1/2}\,\nu 1s_{1/2}\,;\,1)$$

$$= \frac{1}{\sqrt{2}}[-(\pi 1s_{1/2}\|\boldsymbol{M}_1\|\pi 1s_{1/2}) + (\nu 1s_{1/2}\|\boldsymbol{M}_1\|\nu 1s_{1/2})]$$

$$= \frac{1}{\sqrt{2}}[-(1\times 0) - (5.586\times 0.598) + 0\times 0 + (-3.826)\times 0.598]\,\mu_N/c$$

$$= -3.980\,\mu_N/c\,. \tag{15.135}$$

Including the factor $\beta_2/\sqrt{2}$ from the state (15.79), we obtain the reduced transition probability

$$B(\mathrm{M1}\,;\,0^+_1 \to 1^+_{\mathrm{gs}}) = \left(\frac{\beta_2}{\sqrt{2}}\right)^2 (-3.980\,\mu_N/c)^2 = 7.92\beta_2^2\,(\mu_N/c)^2\,. \tag{15.136}$$

From Table 6.9 and the data in Fig. 15.5 we calculate an experimental $B(\mathrm{M1})$ value of $1.14\,(\mu_N/c)^2$. Equating it with the theoretical value (15.136) we find $\beta_2 = 0.38$. This is not very far from the value 0.65 determined from the beta-decay fit in Subsect. 15.4.2.

## Epilogue

We have discussed electromagnetic and beta-decay transitions with the BCS vacuum, one-quasiparticle states and two-quasiparticle states as initial and final states in even–even, odd–odd and odd-$A$ nuclei. This paves the way to treating decays of more complicated nuclear states by methods involving two-quasiparticle configuration mixing. In the rest of the book we discuss these methods extensively. The first one to be discussed, in the next chapter, is the quasiparticle-TDA.

## Exercises

**15.1.** Prove the relation (15.3).

**15.2.** Derive the relation (15.14).

**15.3.** Derive the relation (15.15).

**15.4.** Carry out the details leading to the result (15.18).

**15.5.** Show that the factor $\mathcal{D}_{if}^{(\lambda)}$ defined in (15.18) can be written as

$$\mathcal{D}_{if}^{(\lambda)} = u_i u_f - \zeta^{(\lambda)} v_i v_f \,, \tag{15.137}$$

where the phase factor $\zeta^{(\lambda)}$ was defined in (11.223).

**15.6.** Derive the beta-decay matrix elements (15.23).

**15.7.** Verify the values of the reduced E2 transition probabilities in (15.31).

**15.8.** Compute the $\log ft$ values for the $\beta^-$ decay of the ground state of $^{25}$Na to the ground state and first excited $3/2^+$ state in $^{25}$Mg. Take the occupation amplitudes from Table 14.1. Compare with experimental data and comment.

**15.9.** Calculate the decay half-life of the first excited $3/2^+$ state in $^{25}$Na. Take the occupation amplitudes from Table 14.1. Compare with experimental data and comment. Discuss also the decay of the first excited state of $^{25}$Mg.

**15.10.** Compute the $\log ft$ values for the $\beta^-$ decay of the $1/2^+$ ground state of $^{27}$Mg to the first $1/2^+$ and $3/2^+$ states in $^{27}$Al. Take the occupation amplitudes from Table 14.1. Compare with experimental data and comment.

**15.11.** Compute the $\log ft$ values for the $\beta^+$/EC decay of the $5/2^+$ ground state of $^{27}$Si to the $5/2^+$ ground state and $3/2^+$ excited state of $^{27}$Al. Take the occupation amplitudes from Table 14.1. Compare with experimental data and try to identify which one of the excited $3/2^+$ states in $^{27}$Al most closely corresponds to a one-quasiparticle state.

**15.12.** Compute the electromagnetic decay half-lives of the first excited states of $^{27}$Mg, $^{27}$Al and $^{27}$Si. Take the occupation amplitudes from Table 14.1. Compare with experimental data and comment.

**15.13.** Compute the $\log ft$ value for the $\beta^+$/EC decay of the ground state of $^{31}$S to the ground state of $^{31}$P. Take the occupation amplitudes from Table 14.1. Compare with experimental data and comment.

**15.14.** Compute the $\log ft$ values for the $\beta^+$/EC decay of the $3/2^+$ ground state of $^{33}$Cl to the $3/2^+$ ground state and $5/2^+$ excited state of $^{33}$S. Take the occupation amplitudes from Table 14.1. Compare with experimental data and try to identify which one of the excited $5/2^+$ states in $^{33}$S most closely corresponds to a one-quasiparticle state.

**15.15.** Compute the $\log ft$ values for the $\beta^-$ decay of the $3/2^+$ ground state of $^{35}$S and the $\beta^+$/EC decay of the $3/2^+$ ground state of $^{35}$Ar to the $3/2^+$ ground state of $^{35}$Cl. Take the occupation amplitudes from Table 14.1. Compare with experimental data and comment.

**15.16.** Compute the $\log ft$ values for the transitions from ground state to ground state in the decay chain $^{61}$Zn $\rightarrow$ $^{61}$Cu $\rightarrow$ $^{61}$Ni. Assume the relevant states to be one-quasiparticle states and take the occupation amplitudes from Table 14.2. Compare with experimental data and comment.

**15.17.** Derive in detail the transition amplitude (15.57).

**15.18.** Derive the transition matrix elements (15.63) and (15.64).

**15.19.** Derive the relation (15.67) by means of the Hermitian tensor property (2.32) and its extension (11.290).

**15.20.** Compute the $\log ft$ value of the 2nd-forbidden unique $\beta^+$/EC decay of the $3^+$ ground state of $^{22}$Na to the $0^+$ ground state of $^{22}$Ne. Use Fig. 14.3 as a guide to construct approximate wave functions. Take the occupation amplitudes from Table 14.1. Compare with experimental data and comment.

**15.21.** Discuss the beta and gamma decays involving the nuclei $^{24}$Na, $^{24}$Mg and $^{24}$Al.
(a) Use the quasiparticle spectra of Subsect. 14.1.2 to construct approximate wave functions of the $0^+_{\mathrm{gs}}$, $2^+_1$ and $4^+_1$ states of $^{24}$Mg.
(b) Similarly construct wave functions for the $1^+_1$ states of $^{24}$Na and $^{24}$Al.

**15.22.** This is continuation of Exercise 15.21. By using the occupation amplitudes of Table 14.1 determine the structure of the $1^+_1$ state in $^{24}$Al by comparing the computed $\log ft$ value with the experimental one for the transition $^{24}$Al$(1^+_1) \rightarrow$ $^{24}$Mg$(0^+_{\mathrm{gs}})$.

**15.23.** This is continuation of Exercise 15.22. By using the occupation amplitudes of Table 14.1 compute the half-life of the $2^+_1$ state of $^{24}$Mg. Determine the electric polarization constant $\chi$ by comparison with the experimental half-life.

**15.24.** This is continuation of Exercise 15.23. Calculate the $\log ft$ value for the $\beta^+$/EC decay of the $1^+_1$ state in $^{24}$Al to the $2^+_1$ state in $^{24}$Mg. Compare with experimental data and comment.

**15.25.** By using the quasiparticle spectra of Subsect. 14.1.2 and the occupation amplitudes of Table 14.1 compute the half-life of the $1^+_1$ state of $^{26}$Al. Assume a reasonable structure for the wave functions. Compare with experimental data and comment.

**15.26.** Compute the $\log ft$ values for the $\beta^+$/EC decay of the $0^+$ ground state of $^{34}$Ar to the $0^+$ ground state and first two excited $1^+$ states in $^{34}$Cl. Take the occupation amplitudes from Table 14.1. Form the final states by inspecting the single-quasiparticle spectra in Fig. 14.4. Compare with experimental data and comment.

**15.27.** Compute the $\log ft$ values for the $\beta^-$ decay of the $1^+$ ground state of $^{34}$P and the $\beta^+$/EC decay of the $0^+$ ground state of $^{34}$Cl to the $0^+$ ground state of $^{34}$S. Take the occupation amplitudes from Table 14.1. Compare with experimental data and comment.

**15.28.** Compute the electromagnetic decay half-lives of the first excited states in $^{34}$S and $^{34}$Ar by using a common value for the electric polarization constant. Take the occupation amplitudes from Table 14.1. Compare with experimental data and comment.

**15.29.** Verify the experimental $B(E2)$ value used to obtain (15.92) and (15.95).

**15.30.** Verify the second equality in (15.98).

**15.31.** Derive the relation (15.100) from (15.99).

**15.32.** Derive the relation (15.104) from (15.103).

**15.33.** Derive the relations (15.114) and (15.115) from (15.110).

**15.34.** Compute the $\log ft$ value for the $\beta^-$ decay of the $1^+$ ground state of $^{34}$P to the $2_1^+$ excited state in $^{34}$S. Use the quasiparticle spectra of Subsect. 14.1.2 to build reasonable wave functions for the initial and final states. Use the occupation amplitudes of Table 14.1.

# Mixing of Two-Quasiparticle Configurations

## Prologue

In this chapter we discuss configuration mixing of two-quasiparticle states. It is caused by the residual interaction remaining beyond the quasiparticle mean field defined in Chap. 13. We derive the equations of motion by the EOM method developed in Sect. 11.1. To accomplish this we need to express the residual Hamiltonian in terms of quasiparticles.

Applied to like-quasiparticle (proton–proton, neutron–neutron) pairs, the EOM leads to the quasiparticle TDA (QTDA) for an even–even nucleus. Applied to proton–neutron quasiparticle pairs, the EOM leads to the proton–neutron QTDA (pnQTDA) for odd–odd nuclei. The QTDA and pnQTDA are the subjects of this chapter and the next.

## 16.1 Quasiparticle Representation of the Residual Interaction

In this section we finish the development of the quasiparticle representation of the nuclear Hamiltonian. This consists in expressing the residual interaction in (13.65) in terms of the quasiparticle creation and annihilation operators (13.10). The residual interaction was given in compact form in (13.33),

$$ V_{\mathrm{RES}} = \tfrac{1}{4} \sum_{\alpha\beta\gamma\delta} \bar{v}_{\alpha\beta\gamma\delta} \mathcal{N}\big[ c_\alpha^\dagger c_\beta^\dagger c_\delta c_\gamma \big]_{\mathrm{BCS}} \,, \qquad (16.1) $$

where $\mathcal{N}[\ldots]_{\mathrm{BCS}}$ denotes normal ordering with respect to the BCS vacuum $|\mathrm{BCS}\rangle$.

Substitution of (13.12) and (13.13) into (16.1) gives

$$
\begin{aligned}
V_{\mathrm{RES}} = \tfrac{1}{4}\sum_{\alpha\beta\gamma\delta} \bar{v}_{\alpha\beta\gamma\delta}\mathcal{N}\big[(u_a a_\alpha^\dagger - v_a \tilde{a}_\alpha)(u_b a_\beta^\dagger - v_b \tilde{a}_\beta) \\
\times (u_d a_\delta - v_d \tilde{a}_\delta^\dagger)(u_c a_\gamma - v_c \tilde{a}_\gamma^\dagger)\big]_{\mathrm{BCS}} \\
= \tfrac{1}{4}\sum_{\alpha\beta\gamma\delta} \bar{v}_{\alpha\beta\gamma\delta}\mathcal{N}\big[(u_a u_b a_\alpha^\dagger a_\beta^\dagger - 2 u_a v_b a_\alpha^\dagger \tilde{a}_\beta + v_a v_b \tilde{a}_\alpha \tilde{a}_\beta) \\
\times (v_c v_d \tilde{a}_\delta^\dagger \tilde{a}_\gamma^\dagger - 2 u_c v_d \tilde{a}_\delta^\dagger a_\gamma + u_c u_d a_\delta a_\gamma)\big]_{\mathrm{BCS}}\,, \quad (16.2)
\end{aligned}
$$

where terms were combined by changing summation indices and using (4.29). We next expand (16.2) completely. The result is

$$
\begin{aligned}
V_{\mathrm{RES}} = \tfrac{1}{4}\sum_{\alpha\beta\gamma\delta} \bar{v}_{\alpha\beta\gamma\delta}\big[ & u_a u_b v_c v_d\, a_\alpha^\dagger a_\beta^\dagger \tilde{a}_\delta^\dagger \tilde{a}_\gamma^\dagger - 2 u_a u_b u_c v_d\, a_\alpha^\dagger a_\beta^\dagger \tilde{a}_\delta^\dagger a_\gamma \\
& + u_a u_b u_c u_d\, a_\alpha^\dagger a_\beta^\dagger a_\delta a_\gamma - 2 u_a v_b v_c v_d\, a_\alpha^\dagger \tilde{a}_\delta^\dagger \tilde{a}_\gamma^\dagger \tilde{a}_\beta - 4 u_a v_b u_c v_d\, a_\alpha^\dagger \tilde{a}_\delta^\dagger \tilde{a}_\beta a_\gamma \\
& - 2 u_a v_b u_c u_d\, a_\alpha^\dagger \tilde{a}_\beta a_\delta a_\gamma + v_a v_b v_c v_d\, \tilde{a}_\delta^\dagger \tilde{a}_\gamma^\dagger \tilde{a}_\alpha \tilde{a}_\beta - 2 v_a v_b u_c v_d\, \tilde{a}_\delta^\dagger \tilde{a}_\alpha \tilde{a}_\beta a_\gamma \\
& + v_a v_b u_c u_d\, \tilde{a}_\alpha \tilde{a}_\beta a_\delta a_\gamma \big]\,. \quad (16.3)
\end{aligned}
$$

By changes of summation indices, use of the symmetry relations (4.29) and (13.127) with real $\bar{v}_{\alpha\beta\gamma\delta}$, and recognition of Hermitian conjugates we find

$$
\begin{aligned}
V_{\mathrm{RES}} = \sum_{\alpha\beta\gamma\delta} \bar{v}_{\alpha\beta\gamma\delta}\big[ & \tfrac{1}{4} u_a u_b v_c v_d (a_\alpha^\dagger a_\beta^\dagger \tilde{a}_\delta^\dagger \tilde{a}_\gamma^\dagger + \mathrm{H.c.}) \\
& - \tfrac{1}{2}(u_a u_b u_c v_d - v_a v_b v_c u_d)(a_\alpha^\dagger a_\beta^\dagger \tilde{a}_\delta^\dagger a_\gamma + \mathrm{H.c.}) \\
& + \tfrac{1}{4}(u_a u_b u_c u_d + v_a v_b v_c v_d) a_\alpha^\dagger a_\beta^\dagger a_\delta a_\gamma - u_a v_b u_c v_d\, a_\alpha^\dagger \tilde{a}_\delta^\dagger \tilde{a}_\beta a_\gamma\big] \\
\equiv H_{40} + H_{31} + H_{22}\,, & \quad (16.4)
\end{aligned}
$$

where the three terms stand for the three lines of the equation.

We want to express (16.4) in angular-momentum-coupled form. To do so we use the relation (8.17) for the two-body matrix element $\bar{v}_{\alpha\beta\gamma\delta}$ and express all creation and annihilation operators in proper tensor form. For $H_{40}$ we calculate by standard angular momentum techniques

$$
\begin{aligned}
& \sum_{\substack{m_\alpha m_\beta \\ m_\gamma m_\delta}} \bar{v}_{\alpha\beta\gamma\delta}(-1)^{j_d + m_\delta + j_c + m_\gamma} a_\alpha^\dagger a_\beta^\dagger a_{-\delta}^\dagger a_{-\gamma}^\dagger \\
& = \sum_{JM} [\mathcal{N}_{ab}(J)\mathcal{N}_{cd}(J)]^{-1}\langle a\,b;\, J|V|c\,d;\, J\rangle \sum_{\substack{m_\alpha m_\beta \\ m_\gamma m_\delta}} (-1)^{j_d + m_\delta + j_c + m_\gamma} \\
& \quad \times (j_a\, m_\alpha\, j_b\, m_\beta|J\,M)(j_c\, m_\gamma\, j_d\, m_\delta|J\,M) a_\alpha^\dagger a_\beta^\dagger a_{-\delta}^\dagger a_{-\gamma}^\dagger \\
& = \sum_{J}(-1)^{J+1}[\mathcal{N}_{ab}(J)\mathcal{N}_{cd}(J)]^{-1}\langle a\,b;\, J|V|c\,d;\, J\rangle \big[a_a^\dagger a_b^\dagger\big]_J \cdot \big[a_c^\dagger a_d^\dagger\big]_J\,, \quad (16.5)
\end{aligned}
$$

where we use the dot product notation (2.51).

For $H_{31}$ we calculate similarly

$$\sum_{\substack{m_\alpha m_\beta \\ m_\gamma m_\delta}} \bar{v}_{\alpha\beta\gamma\delta}(-1)^{j_d+m_\delta+j_c-m_\gamma} a_\alpha^\dagger a_\beta^\dagger a_{-\delta}^\dagger \tilde{a}_{-\gamma}$$

$$= \sum_J (-1)^J [\mathcal{N}_{ab}(J)\mathcal{N}_{cd}(J)]^{-1} \langle a\,b\,;\,J|V|d\,c\,;\,J \rangle \left[ a_a^\dagger a_b^\dagger \right]_J \cdot \left[ a_d^\dagger \tilde{a}_c \right]_J, \quad (16.6)$$

where we used the symmetry property (8.30).

The calculation for the first term of $H_{22}$ is essentially the same as (16.5), whence

$$\sum_{\substack{m_\alpha m_\beta \\ m_\gamma m_\delta}} \bar{v}_{\alpha\beta\gamma\delta}(-1)^{j_d-m_\delta+j_c-m_\gamma} a_\alpha^\dagger a_\beta^\dagger \tilde{a}_{-\delta} \tilde{a}_{-\gamma}$$

$$= \sum_J (-1)^{J+1} [\mathcal{N}_{ab}(J)\mathcal{N}_{cd}(J)]^{-1} \langle a\,b\,;\,J|V|c\,d\,;\,J \rangle \left[ a_a^\dagger a_b^\dagger \right]_J \cdot \left[ \tilde{a}_c \tilde{a}_d \right]_J. \quad (16.7)$$

As above, the calculation for the second term of $H_{22}$ begins with

$$\sum_{\substack{m_\alpha m_\beta \\ m_\gamma m_\delta}} \bar{v}_{\alpha\beta\gamma\delta}(-1)^{j_d+m_\delta+j_c-m_\gamma} a_\alpha^\dagger a_{-\delta}^\dagger \tilde{a}_\beta \tilde{a}_{-\gamma}$$

$$= \sum_{JM} [\mathcal{N}_{ab}(J)\mathcal{N}_{cd}(J)]^{-1} \langle a\,b\,;\,J|V|c\,d\,;\,J \rangle \sum_{\substack{m_\alpha m_\beta \\ m_\gamma m_\delta}} (-1)^{j_d+m_\delta+j_c-m_\gamma}$$

$$\times (j_a\, m_\alpha\, j_b\, m_\beta|J\,M)(j_c\, m_\gamma\, j_d\, m_\delta|J\,M) a_\alpha^\dagger a_{-\delta}^\dagger \tilde{a}_\beta \tilde{a}_{-\gamma}. \quad (16.8)$$

The recoupling needed to bring the operator part of this expression into the form in (16.7) is accomplished through $6j$ symbols. From Eq. (1.58) that defines the $6j$ symbol one can derive the identity (Exercise 16.3)

$$\sum_M (j_a\, m_\alpha\, j_b\, m_\beta|J\,M)(j_c\, m_\gamma\, j_d\, m_\delta|J\,M) = \hat{J}^2 \sum_{J'M'} (-1)^{J'+j_c+j_d+m_\alpha-m_\gamma}$$

$$\times (j_a\, m_\alpha\, j_d\, -m_\delta|J'\, -M')(j_c\, -m_\gamma\, j_b\, m_\beta|J'M') \begin{Bmatrix} j_a & j_d & J' \\ j_c & j_b & J \end{Bmatrix}. \quad (16.9)$$

Substituting this into (16.8) results in

$$\sum_{\substack{m_\alpha m_\beta \\ m_\gamma m_\delta}} \bar{v}_{\alpha\beta\gamma\delta}(-1)^{j_d+m_\delta+j_c-m_\gamma} a_\alpha^\dagger a_{-\delta}^\dagger \tilde{a}_\beta \tilde{a}_{-\gamma} = \sum_J [\mathcal{N}_{ab}(J)\mathcal{N}_{cd}(J)]^{-1}$$

$$\times (-1)^{j_b+j_c} \hat{J}^2 \langle a\,b\,;\,J|V|c\,d\,;\,J \rangle \sum_{J'} \begin{Bmatrix} j_a & j_d & J' \\ j_c & j_b & J \end{Bmatrix} \left[ a_a^\dagger a_d^\dagger \right]_{J'} \cdot \left[ \tilde{a}_b \tilde{a}_c \right]_{J'}. \quad (16.10)$$

To obtain final expressions for $H_{40}$, $H_{31}$ and $H_{22}$ it remains to attach the $u, v$ factors and numerical constants in (16.4). With substitution from (16.5) $H_{40}$ becomes

$$H_{40} = \frac{1}{4} \sum_{\substack{abcd \\ J}} (-1)^{J+1} [\mathcal{N}_{ab}(J)\mathcal{N}_{cd}(J)]^{-1} u_a u_b v_c v_d$$

$$\times \langle a\,b;\,J|V|c\,d;\,J\rangle \left( \left[ a_a^\dagger a_b^\dagger \right]_J \cdot \left[ a_c^\dagger a_d^\dagger \right]_J + \text{H.c.} \right). \qquad (16.11)$$

To construct $H_{31}$ we substitute from (16.6). To have the same order $abcd$ in the labels of the operator as in (16.11) we exchange the summation indices $c$ and $d$. The result is

$$H_{31} = \frac{1}{2} \sum_{\substack{abcd \\ J}} (-1)^{J+1} [\mathcal{N}_{ab}(J)\mathcal{N}_{cd}(J)]^{-1} (u_a u_b v_c u_d - v_a v_b u_c v_d)$$

$$\times \langle a\,b;\,J|V|c\,d;\,J\rangle \left( \left[ a_a^\dagger a_b^\dagger \right]_J \cdot \left[ a_c^\dagger \tilde{a}_d \right]_J + \text{H.c.} \right). \qquad (16.12)$$

The term $H_{22}$ is more complicated than the other two. With substitution from (16.7) and (16.10) it becomes

$$H_{22} = \frac{1}{4} \sum_{\substack{abcd \\ J}} (-1)^{J+1} [\mathcal{N}_{ab}(J)\mathcal{N}_{cd}(J)]^{-1} (u_a u_b u_c u_d + v_a v_b v_c v_d)$$

$$\times \langle a\,b;\,J|V|c\,d;\,J\rangle \left[ a_a^\dagger a_b^\dagger \right]_J \cdot \left[ \tilde{a}_c \tilde{a}_d \right]_J$$

$$+ \sum_{\substack{abcd \\ JJ'}} (-1)^{j_b+j_c+1} [\mathcal{N}_{ab}(J)\mathcal{N}_{cd}(J)]^{-1} \hat{J}^2 u_a v_b u_c v_d$$

$$\times \langle a\,b;\,J|V|c\,d;\,J\rangle \left\{ \begin{matrix} j_a & j_d & J' \\ j_c & j_b & J \end{matrix} \right\} \left[ a_a^\dagger a_d^\dagger \right]_{J'} \cdot \left[ \tilde{a}_b \tilde{a}_c \right]_{J'}. \qquad (16.13)$$

We combine the two terms by changing summation indices in the second term and anticommuting the annihilation operators, with the result

$$H_{22} = \sum_{\substack{abcd \\ J}} (-1)^J \left[ -\frac{1}{4} [\mathcal{N}_{ab}(J)\mathcal{N}_{cd}(J)]^{-1} (u_a u_b u_c u_d + v_a v_b v_c v_d) \right.$$

$$\times \langle a\,b;\,J|V|c\,d;\,J\rangle$$

$$+ u_a v_b u_c v_d \sum_{J'} [\mathcal{N}_{ad}(J')\mathcal{N}_{cb}(J')]^{-1} \hat{J'}^2 \left\{ \begin{matrix} j_a & j_b & J \\ j_c & j_d & J' \end{matrix} \right\}$$

$$\left. \times \langle a\,d;\,J'|V|c\,b;\,J'\rangle \right] \left[ a_a^\dagger a_b^\dagger \right]_J \cdot \left[ \tilde{a}_c \tilde{a}_d \right]_J. \qquad (16.14)$$

Certain conventional abbreviations are used for lumped terms occurring in (16.11), (16.12) and (16.14). They are

$$V_{abcd}^{(40)}(J) \equiv -\frac{1}{2} [\mathcal{N}_{ab}(J)\mathcal{N}_{cd}(J)]^{-1} u_a u_b v_c v_d \langle a\,b;\,J|V|c\,d;\,J\rangle$$

$$\equiv u_a u_b v_c v_d G(abcdJ), \qquad (16.15)$$

$$V_{abcd}^{(31)}(J) \equiv -\frac{1}{2} [\mathcal{N}_{ab}(J)\mathcal{N}_{cd}(J)]^{-1} (u_a u_b v_c u_d - v_a v_b u_c v_d) \langle a\,b;\,J|V|c\,d;\,J\rangle$$

$$= (u_a u_b v_c u_d - v_a v_b u_c v_d) G(abcdJ) \,, \tag{16.16}$$

$$V_{abcd}^{(22)}(J) \equiv -\tfrac{1}{2}[\mathcal{N}_{ab}(J)\mathcal{N}_{cd}(J)]^{-1}(u_a u_b u_c u_d + v_a v_b v_c v_d)\langle a\, b\,;\, J|V|c\, d\,;\, J\rangle$$

$$+ 2u_a v_b u_c v_d \sum_{J'}[\mathcal{N}_{ad}(J')\mathcal{N}_{cb}(J')]^{-1}\hat{J}'^2 \begin{Bmatrix} j_a & j_b & J \\ j_c & j_d & J' \end{Bmatrix}$$

$$\times \langle a\, d\,;\, J'|V|c\, b\,;\, J'\rangle$$

$$\equiv (u_a u_b u_c u_d + v_a v_b v_c v_d) G(abcdJ) + 4u_a v_b u_c v_d F(abcdJ)\,. \tag{16.17}$$

With these abbreviations the residual interaction is stated compactly as

$$\boxed{\begin{aligned}
V_{\text{RES}} &= \tfrac{1}{4}\sum_{\alpha\beta\gamma\delta}\bar{v}_{\alpha\beta\gamma\delta}\mathcal{N}\big[c_\alpha^\dagger c_\beta^\dagger c_\delta c_\gamma\big]_{\text{BCS}} = H_{40} + H_{31} + H_{22}\,, \\
H_{40} &= \tfrac{1}{2}\sum_{\substack{abcd\\J}}(-1)^J V_{abcd}^{(40)}(J)\Big(\big[a_a^\dagger a_b^\dagger\big]_J \cdot \big[a_c^\dagger a_d^\dagger\big]_J + \text{H.c.}\Big)\,, \\
H_{31} &= \sum_{\substack{abcd\\J}}(-1)^J V_{abcd}^{(31)}(J)\Big(\big[a_a^\dagger a_b^\dagger\big]_J \cdot \big[a_c^\dagger \tilde{a}_d\big]_J + \text{H.c.}\Big)\,, \\
H_{22} &= \tfrac{1}{2}\sum_{\substack{abcd\\J}}(-1)^J V_{abcd}^{(22)}(J)\big[a_a^\dagger a_b^\dagger\big]_J \cdot \big[\tilde{a}_c \tilde{a}_d\big]_J\,.
\end{aligned}} \tag{16.18}$$

Except for the normalization factor and sign, the $J'$ sum in (16.14) is recognized to be identical with the right-hand side of the Pandya transformation (9.22). This leads us to define a *generalized particle–hole matrix element* by the *generalized Pandya transformation*

$$\boxed{\begin{aligned}
&\langle a\, b^{-1}\,;\, J|V_{\text{RES}}|c\, d^{-1}\,;\, J\rangle \\
&\equiv -\sum_{J'}[\mathcal{N}_{ad}(J')\mathcal{N}_{cb}(J')]^{-1}\hat{J}'^2 \begin{Bmatrix} j_a & j_b & J \\ j_c & j_d & J' \end{Bmatrix}\langle a\, d\,;\, J'|V|c\, b\,;\, J'\rangle\,.
\end{aligned}} \tag{16.19}$$

The Pandya transformation (9.22) and its generalization (16.19) differ only through the presence of the normalization factor in the latter. This factor takes into account that the same orbital may be both a 'particle orbital' and a 'hole orbital'. The orbitals of a particle–hole pair $ab^{-1}$ are not constrained to lie above (particles $a$) or below (holes $b^{-1}$) a Fermi surface. An open-shell nucleus has several neutrons and protons outside a closed major shell. Therefore we cannot define a Fermi surface and the related particles and holes.

The generalized particle–hole matrix element (16.19) reduces to the ordinary particle–hole matrix element of (9.22) for pure particle–hole excitations at a closed major shell. The normalization constants, defined in (5.21), are then $\mathcal{N}_{ad}(J') = 1$ and $\mathcal{N}_{cb}(J') = 1$.

Equations (16.15) and (16.17) defined the abbreviations $G$ and $F$, originally due to Baranger [31]. We now restate them explicitly as

$$G(abcdJ) = -\tfrac{1}{2}[\mathcal{N}_{ab}(J)\mathcal{N}_{cd}(J)]^{-1}\langle a\,b\,;\,J|V|c\,d\,;\,J\rangle$$

$$= -\tfrac{1}{2}\sqrt{1+\delta_{ab}(-1)^J}\sqrt{1+\delta_{cd}(-1)^J}\langle a\,b\,;\,J|V|c\,d\,;\,J\rangle\,, \quad (16.20)$$

$$F(abcdJ) = -\tfrac{1}{2}\langle a\,b^{-1}\,;\,J|V_{\mathrm{RES}}|c\,d^{-1}\,;\,J\rangle\,. \tag{16.21}$$

The explicit normalization factors in (16.20) have been written so that they take into account the vanishing of the two-body matrix element. Note that the indices $a$, $b$, $c$ and $d$ contain not only the orbital information but also the proton and neutron labels $\pi$ and $\nu$. For proton–neutron two-body matrix elements (16.20) thus gives

$$G(pnp'n'J) = -\tfrac{1}{2}\langle p\,n\,;\,J|V|p'\,n'\,;\,J\rangle\,. \tag{16.22}$$

This concludes the derivation of the quasiparticle transformation of the residual Hamiltonian. Next we derive the quasiparticle-TDA equation.

## 16.2 Derivation of the Quasiparticle-TDA Equation

In Sect. 11.1 we used the EOM method to derive the particle–hole RPA equation. In a similar manner we can use the EOM to derive the quasiparticle-TDA (QTDA) equation. In the QTDA formalism we start from the proton–proton and neutron–neutron two-quasiparticle configurations. By the number-parity principle of Subsect. 13.4.2 both types of two-quasiparticle excitation describe states of an even–even nucleus. Due to the number constraint (13.72) of the BCS, or (14.19) of the LNBCS, the dominant component in a QTDA wave function has the nucleon numbers of the even–even reference nucleus for which the BCS or LNBCS calculation was done. Accordingly the QTDA solutions are assumed to describe states of the even–even reference nucleus.

We now start the derivation of the QTDA equation. As stated in Table 11.1, the relevant operators are

$$\delta Q = A_{ab}(JM)\,, \quad Q_\omega^\dagger = \sum_{cd} X_{cd}^\omega A_{cd}^\dagger(JM)\,, \tag{16.23}$$

where $\omega = nJ^\pi M$ is the full set of quantum numbers. The quasiparticle pair creation and annihilation operators were given in (11.17) and (11.19) as

$$A_{ab}^\dagger(JM) = \mathcal{N}_{ab}(J)\big[a_a^\dagger a_b^\dagger\big]_{JM}\,, \quad A_{ab}(JM) = -\mathcal{N}_{ab}(J)(-1)^{J+M}\big[\tilde{a}_a \tilde{a}_b\big]_{J,-M}\,. \tag{16.24}$$

The parity of the quasiparticle pair created by (16.24) is

$$\pi = (-1)^{l_a+l_b}\,. \tag{16.25}$$

The commutator between two quasiparticle pair operators is (Exercise 16.5)

$$\left[A_{ab}(JM), A_{cd}^\dagger(J'M')\right] = \delta_{JJ'}\delta_{MM'}\mathcal{N}_{ab}^2(J)[\delta_{ac}\delta_{bd} - (-1)^{j_a+j_b+J}\delta_{ad}\delta_{bc}]$$
$$+ \text{ terms with } a^\dagger a \,. \tag{16.26}$$

The expectation value of the $a^\dagger a$ terms vanishes for the BCS vacuum $|\mathrm{BCS}\rangle$, whence

$$\langle\mathrm{BCS}|\left[A_{ab}(JM), A_{cd}^\dagger(J'M')\right]|\mathrm{BCS}\rangle$$
$$= \delta_{JJ'}\delta_{MM'}\mathcal{N}_{ab}^2(J)[\delta_{ac}\delta_{bd} - (-1)^{j_a+j_b+J}\delta_{ad}\delta_{bc}] \,. \tag{16.27}$$

The starting point of the derivation of the QTDA equation is the equation of motion (11.11). Written for the present case of Bose-like basic excitations and the BCS vacuum it is

$$\langle\mathrm{BCS}|[\delta Q, \mathcal{H}, Q_\omega^\dagger]|\mathrm{BCS}\rangle = E_\omega\langle\mathrm{BCS}|[\delta Q, Q_\omega^\dagger]|\mathrm{BCS}\rangle \,, \tag{16.28}$$

where $\mathcal{H}$ is the complete Hamiltonian in terms of quasiparticles. Inserting (16.23) into the right-hand side gives

$$\langle\mathrm{BCS}|[\delta Q, Q_\omega^\dagger]|\mathrm{BCS}\rangle = \sum_{cd} X_{cd}^\omega\langle\mathrm{BCS}|\left[A_{ab}(JM), A_{cd}^\dagger(JM)\right]|\mathrm{BCS}\rangle \,. \tag{16.29}$$

Two situations arise when applying (16.27) to (16.29) because the indices carry the proton and neutron labels.

- If $a = p$ (proton) and $b = n$ (neutron), then the sum is $\sum_{cd} = \sum_{p'n'}$ and $\delta_{ad} = \delta_{pn'} = 0$, $\delta_{bc} = \delta_{np'} = 0$. In this case we have

$$\langle\mathrm{BCS}|[\delta Q, Q_\omega^\dagger]|\mathrm{BCS}\rangle = X_{pn}^\omega \,. \tag{16.30}$$

- If $a$ and $b$ both are either proton or neutron labels, we must restrict the sum so as to avoid double counting. For $a \le b$ we then carry the sum for $c \le d$, i.e. $\sum_{c\le d}$. In this case (16.27) yields

$$\langle\mathrm{BCS}|[\delta Q, Q_\omega^\dagger]|\mathrm{BCS}\rangle = \mathcal{N}_{ab}^2(J)\left[X_{ab}^\omega + (-1)^J\delta_{ab}X_{aa}^\omega\right]$$
$$= \frac{1 + \delta_{ab}(-1)^J}{(1+\delta_{ab})^2}\left[X_{ab}^\omega + (-1)^J\delta_{ab}X_{aa}^\omega\right]$$
$$= \left[\frac{1 + \delta_{ab}(-1)^J}{1+\delta_{ab}}\right]^2 X_{ab}^\omega = X_{ab}^\omega \,, \tag{16.31}$$

where we have inserted the normalization constant from (5.21). The final step relies on the fact that the operator $A_{aa}(JM)$ vanishes for odd $J$.

The results (16.30) and (16.31) are remarkably simple. To derive the QTDA equation we use (16.31), i.e. we choose the two-quasiparticle basis with $a \le b$ and the associated restricted summation. The result (16.30) applies to the proton–neutron form of the QTDA, the pnQTDA, to be treated in the following chapter.

It remains to evaluate the left-hand side of (16.28). The Hamiltonian is that given by (13.65), i.e.

$$\mathcal{H} = \mathcal{H}_0 + \mathcal{H}_{11} + V_{\mathrm{RES}} , \quad \mathcal{H}_{11} = \sum_b \hat{j}_b E_b \left[ a_b^\dagger \tilde{a}_b \right]_{00} = \sum_\beta E_b a_\beta^\dagger a_\beta . \quad (16.32)$$

Here $\mathcal{H}_0$ is the energy of the quasiparticle mean field and $\mathcal{H}_{11}$ contains the single-quasiparticle energies. The residual interaction $V_{\mathrm{RES}}$ is given in terms of quasiparticles through Eqs. (16.15)–(16.18).

Let us consider the commutators of $A$ and $A^\dagger$ with the various terms of $\mathcal{H}$. All commutators with $\mathcal{H}_0$ vanish since this term is a c-number. To find the contribution of the term $\mathcal{H}_{11}$ we start with the commutator (Exercise 16.6)

$$\left[ A_{ab}(JM), \mathcal{H}_{11} \right] = (E_a + E_b) A_{ab}(JM) . \quad (16.33)$$

This gives us

$$\left[ \mathcal{H}_{11}, A_{cd}^\dagger(JM) \right] = \left[ A_{cd}(JM), \mathcal{H}_{11} \right]^\dagger = (E_c + E_d) A_{cd}^\dagger(JM) . \quad (16.34)$$

It follows from (16.33) and (16.34) that

$$\left[ A_{ab}(JM), \left[ \mathcal{H}_{11}, A_{cd}^\dagger(JM) \right] \right] = (E_c + E_d) \left[ A_{ab}(JM), A_{cd}^\dagger(JM) \right] , \quad (16.35)$$

$$\left[ \left[ A_{ab}(JM), \mathcal{H}_{11} \right], A_{cd}^\dagger(JM) \right] = (E_a + E_b) \left[ A_{ab}(JM), A_{cd}^\dagger(JM) \right] . \quad (16.36)$$

With $\delta Q$ and $Q_\omega^\dagger$ given by (16.23) and the definition (11.12) of the symmetrized double commutator, we can write the contribution of $\mathcal{H}_{11}$ to the left-hand side of (16.28). To carry out the restricted summation we assume $a \le b$ and substitute (16.27) for the expectation value. This gives

$$\langle \mathrm{BCS} | \left[ \delta Q, \mathcal{H}_{11}, Q_\omega^\dagger \right] | \mathrm{BCS} \rangle = \tfrac{1}{2} \sum_{c \le d} X_{cd}^\omega (E_c + E_d + E_a + E_b)$$

$$\times \mathcal{N}_{ab}^2(J) [\delta_{ac}\delta_{bd} - (-1)^{j_a + j_b + J} \delta_{ad}\delta_{bc}] = (E_a + E_b) X_{ab}^\omega , \quad (16.37)$$

where the second equality is obtained in the same way as in (16.31).

The remaining commutators are between the quasiparticle pair operators and the residual interaction. We insert into the equation of motion (16.28) the parts already calculated, namely the right-hand side given by (16.31) and the $\mathcal{H}_{11}$ term (16.37) of the left-hand side. Equation (16.28) then becomes

$$(E_a + E_b) X_{ab}^\omega + \sum_{c \le d} X_{cd}^\omega \langle \mathrm{BCS} | \left[ A_{ab}(JM), V_{\mathrm{RES}}, A_{cd}^\dagger(JM) \right] | \mathrm{BCS} \rangle = E_\omega X_{ab}^\omega .$$

$$(16.38)$$

The remaining task is to calculate the BCS expectation value present in (16.38). Equation (16.18) shows the structure of $V_{\mathrm{RES}}$. The BCS expectation value of a string of quasiparticle creation and annihilation operators can be non-zero only when there are the same number of both kinds. Commutation with the pair operators $A$ and $A^\dagger$ cannot change the difference in number of the two kinds. Therefore only the term $H_{22}$ can give a finite contribution.

To calculate the $H_{22}$ term given in (16.18) we express it in terms of the quasiparticle pair operators (16.24) according to

$$H_{22} = -\tfrac{1}{2} \sum_{\substack{abcd \\ JM}} [\mathcal{N}_{ab}(J)\mathcal{N}_{cd}(J)]^{-1} V^{(22)}_{abcd}(J) A^\dagger_{ab}(JM) A_{cd}(JM) \ . \tag{16.39}$$

We can identify terms that give zero without going through a detailed calculation. The double commutator consists of terms of the type $[A, A^\dagger A, A^\dagger]$. Expanding and dropping the commutators $[A^\dagger, A^\dagger] = 0 = [A, A]$, we find

$$\begin{aligned}
[A, A^\dagger A, A^\dagger] &= \tfrac{1}{2}\{AA^\dagger[A, A^\dagger] + [A, A^\dagger]AA^\dagger\} \\
&= \tfrac{1}{2}\{AA^\dagger(\delta + a^\dagger a) + (\delta + a^\dagger a)AA^\dagger\}
\end{aligned} \tag{16.40}$$

in a schematic notation for the commutators (16.26). The terms with $a^\dagger a$ vanish in the BCS expectation value, so effectively we are left with the $\delta$ terms only. In the expectation value we can also replace the remaining $AA^\dagger$ terms with $[A, A^\dagger]$, and inspection shows that the two terms of (16.40) contribute equally.

Spelt out in detail from (16.26), the BCS expectation value of (16.40) is

$$\begin{aligned}
\langle \mathrm{BCS}| &\big[A_{ab}(JM), A^\dagger_{a'b'}(J'M')\big]\big[A_{c'd'}(J'M'), A^\dagger_{cd}(JM)\big]|\mathrm{BCS}\rangle \\
&= \delta_{JJ'}\delta_{MM'}\mathcal{N}^2_{ab}(J)[\delta_{aa'}\delta_{bb'} - (-1)^{j_a+j_b+J}\delta_{ab'}\delta_{ba'}] \\
&\quad \times \mathcal{N}^2_{c'd'}(J)[\delta_{c'c}\delta_{d'd} - (-1)^{j_{c'}+j_{d'}+J}\delta_{c'd}\delta_{d'c}] \ .
\end{aligned} \tag{16.41}$$

With $H_{22}$ in place of $V_{\mathrm{RES}}$ in (16.38), we insert (16.41) into the equation and calculate

$$\begin{aligned}
\sum_{c\le d} & X^\omega_{cd}\langle \mathrm{BCS}|\big[A_{ab}(JM), H_{22}, A^\dagger_{cd}(JM)\big]|\mathrm{BCS}\rangle \\
&= -\tfrac{1}{2}\sum_{\substack{c\le d \\ a'b'c'd'}} X^\omega_{cd}[\mathcal{N}_{a'b'}(J)\mathcal{N}_{c'd'}(J)]^{-1} V^{(22)}_{a'b'c'd'}(J) \\
&\qquad \times \mathcal{N}^2_{ab}(J)[\delta_{aa'}\delta_{bb'} - (-1)^{j_a+j_b+J}\delta_{ab'}\delta_{ba'}] \\
&\qquad \times \mathcal{N}^2_{c'd'}(J)[\delta_{c'c}\delta_{d'd} - (-1)^{j_{c'}+j_{d'}+J}\delta_{c'd}\delta_{d'c}] \\
&= -\tfrac{1}{2}\sum_{c\le d} X^\omega_{cd}\mathcal{N}_{ab}(J)\mathcal{N}_{cd}(J)\big[V^{(22)}_{abcd}(J) - (-1)^{j_c+j_d+J} V^{(22)}_{abdc}(J) \\
&\qquad - (-1)^{j_a+j_b+J} V^{(22)}_{bacd}(J) + (-1)^{j_a+j_b+j_c+j_d} V^{(22)}_{badc}(J)\big] \ .
\end{aligned} \tag{16.42}$$

The quantities $V^{(22)}_{abcd}$ can be expressed more explicitly through Eqs. (16.17) and (16.19):

$$V_{abcd}^{(22)}(J) = -\tfrac{1}{2}[\mathcal{N}_{ab}(J)\mathcal{N}_{cd}(J)]^{-1}(u_a u_b u_c u_d + v_a v_b v_c v_d)\langle a\,b\,;\,J|V|c\,d\,;\,J\rangle$$

$$-\,2u_a v_b u_c v_d\langle a\,b^{-1}\,;\,J|V_{\text{RES}}|c\,d^{-1}\,;\,J\rangle\,. \tag{16.43}$$

Furthermore, inspection of (16.19) shows that

$$\langle b\,a^{-1}\,;\,J|V_{\text{RES}}|d\,c^{-1}\,;\,J\rangle = (-1)^{j_a+j_b+j_c+j_d}\langle a\,b^{-1}\,;\,J|V_{\text{RES}}|c\,d^{-1}\,;\,J\rangle\,. \tag{16.44}$$

This relation allows us to combine terms expressed according to (16.43), and (16.42) becomes

$$\sum_{c\leq d} X_{cd}^{\omega}\langle\text{BCS}|\big[A_{ab}(JM),H_{22},A_{cd}^{\dagger}(JM)\big]|\text{BCS}\rangle$$

$$= \sum_{c\leq d} X_{cd}^{\omega}\Big\{(u_a u_b u_c u_d + v_a v_b v_c v_d)\langle a\,b\,;\,J|V|c\,d\,;\,J\rangle$$

$$+\,\mathcal{N}_{ab}(J)\mathcal{N}_{cd}(J)\big[(u_a v_b u_c v_d + v_a u_b v_c u_d)\langle a\,b^{-1}\,;\,J|V_{\text{RES}}|c\,d^{-1}\,;\,J\rangle$$

$$-\,(-1)^{j_c+j_d+J}(u_a v_b v_c u_d + v_a u_b u_c v_d)\langle a\,b^{-1}\,;\,J|V_{\text{RES}}|d\,c^{-1}\,;\,J\rangle\big]\Big\}\,. \tag{16.45}$$

This concludes the evaluation of the interaction term in (16.38). To be able to express the final result in a compact form we define

$$\boxed{\begin{aligned}A_{ab,cd} &\equiv (E_a + E_b)\delta_{ac}\delta_{bd} + (u_a u_b u_c u_d + v_a v_b v_c v_d)\langle a\,b\,;\,J|V|c\,d\,;\,J\rangle\\ &+\,\mathcal{N}_{ab}(J)\mathcal{N}_{cd}(J)\big[(u_a v_b u_c v_d + v_a u_b v_c u_d)\langle a\,b^{-1}\,;\,J|V_{\text{RES}}|c\,d^{-1}\,;\,J\rangle\\ &-\,(-1)^{j_c+j_d+J}(u_a v_b v_c u_d + v_a u_b u_c v_d)\langle a\,b^{-1}\,;\,J|V_{\text{RES}}|d\,c^{-1}\,;\,J\rangle\big]\,.\end{aligned}}$$
$$\tag{16.46}$$

The quantities $A_{ab,cd}$ are the elements of the QTDA matrix A. From the equation of motion (16.38) we have thus derived the *QTDA equation*

$$\boxed{\sum_{c\leq d} A_{ab,cd}X_{cd}^{\omega} = E_{\omega}X_{ab}^{\omega}\,.} \tag{16.47}$$

Written as a matrix equation this is

$$\mathsf{A}X^{\omega} = E_{\omega}X^{\omega}\,. \tag{16.48}$$

For the QTDA the indices $ab$ and $cd$ describe either proton–proton ($ab = pp'$) or neutron–neutron ($ab = nn'$) quasiparticle pairs. The pair indices play the roles of the row and column indices of a matrix. Since the matrix elements $A_{ab,cd}$ are independent of the $M$ quantum number, the amplitudes $X_{ab}^{\omega}$ must also be independent of $M$.

Since the quasiparticle indices carry also the nucleon kind, the formalism for proton–neutron excitations can be obtained as a modification of the above. The two-body matrix elements $\langle p\,p'\,;\,J'|V|n'\,n\,;\,J'\rangle$ are zero because of charge conservation. It follows from the Pandya relation (16.19) that also $\langle p\,n^{-1}\,;\,J|V_{\mathrm{RES}}|n'\,p'^{-1}\,;\,J\rangle = 0$. Equation (16.47) is then modified to read

$$\sum_{p'n'} A_{pn,p'n'} X_{p'n'}^{\omega} = E_{\omega} X_{pn}^{\omega} \,. \tag{16.49}$$

This is known as the *pnQTDA equation*. From (16.46) we read the pnQTDA matrix as

$$
\begin{aligned}
A_{pn,p'n'}(J) &= (E_p + E_n)\delta_{pp'}\delta_{nn'} \\
&+ (u_p u_n u_{p'} u_{n'} + v_p v_n v_{p'} v_{n'})\langle p\,n\,;\,J|V|p'\,n'\,;\,J\rangle \\
&+ (u_p v_n u_{p'} v_{n'} + v_p u_n v_{p'} u_{n'})\langle p\,n^{-1}\,;\,J|V_{\mathrm{RES}}|p'\,n'^{-1}\,;\,J\rangle \,. 
\end{aligned} \tag{16.50}
$$

The Pandya transformation (16.19) is now

$$\langle p\,n^{-1}\,;\,J|V_{\mathrm{RES}}|p'\,n'^{-1}\,;\,J\rangle = -\sum_{J'} \hat{J}'^2 \left\{ \begin{matrix} j_p & j_n & J \\ j_{p'} & j_{n'} & J' \end{matrix} \right\} \langle p\,n'\,;\,J'|V|p'\,n\,;\,J'\rangle \,. \tag{16.51}$$

The pnQTDA formalism describes states of odd–odd nuclei and is further developed in Chap. 17.

Anticipating a closer examination of the general properties of QTDA solutions we note some basic properties of the QTDA matrix A. From parity conservation of the nuclear force it follows that

$$l_a + l_b + l_c + l_d = \text{even} \,. \tag{16.52}$$

Taking into account the phase conventions in (8.29), (13.56) and (13.58) we can use (16.52) to prove that the QTDA and pnQTDA matrices satisfy the relation

$$A_{ab,cd}^{(\mathrm{BR})}(J) = (-1)^{\frac{1}{2}(l_c+l_d-l_a-l_b)} A_{ab,cd}^{(\mathrm{CS})}(J) \,, \tag{16.53}$$

where BR denotes the Biedenharn–Rose phase convention and CS the Condon–Shortley one.

We can write the expressions (16.46) and (16.50) also by using the quantities $G$ and $F$ given in (16.20) and (16.21). The elements of the QTDA matrix then become[1]

$$
\begin{aligned}
A_{ab,cd}(J) &= (E_a + E_b)\delta_{ac}\delta_{bd} \\
&- 2\mathcal{N}_{ab}(J)\mathcal{N}_{cd}(J)[(u_a u_b u_c u_d + v_a v_b v_c v_d)G(abcdJ) \\
&+ (u_a v_b u_c v_d + v_a u_b v_c u_d)F(abcdJ) \\
&- (-1)^{j_c+j_d+J}(u_a v_b v_c u_d + v_a u_b u_c v_d)F(abdcJ)] \,. 
\end{aligned} \tag{16.54}
$$

---

[1] Baranger's [31] quasiparticle pair operators do not contain the normalization factor $\mathcal{N}$, so his expression has a somewhat different appearance.

## 16.3 General Properties of QTDA Solutions

As mentioned in the beginning of Sect. 16.2, the number-parity condition of Subsect. 13.4.2 dictates that the proton–proton and neutron–neutron two-quasiparticle excitations of the QTDA describe states of even–even nuclei. In particular we assume that the QTDA gives a reasonable description of the even–even reference nucleus for which the BCS (or LNBCS) calculation was done. This can be assumed since the calculation was constrained to have the average particle numbers peak at the nucleon numbers of the reference nucleus.

With reference to Subsect. 11.1.1, we note that the QTDA and pnQTDA equations (16.47) and (16.49) result from a variational principle. This is because the equation of motion (16.28) was written with the exact vacuum $|\text{QTDA}\rangle = |\text{BCS}\rangle$ substituted for $|\Psi_0\rangle$ in (11.11). Likewise the particle–hole TDA equation (9.36) is of variational origin because it can be derived from (11.11) with $|\Psi_0\rangle = |\text{HF}\rangle = |\text{TDA}\rangle$.

The QTDA wave function is given by (16.23), with the restricted summation denoted, as

$$\boxed{|\omega\rangle = Q_\omega^\dagger|\text{BCS}\rangle = \sum_{a\leq b} X_{ab}^\omega A_{ab}^\dagger(JM)|\text{BCS}\rangle\,.} \tag{16.55}$$

This excitation is often called a phonon, a term used occasionally earlier in the book. In the present context the term refers to the collective properties of the QTDA, in particular to its lowest collective solution. We discuss next the orthogonality and completeness properties of the state (16.55).

### 16.3.1 Orthogonality

Two different solutions (16.55), $|\omega\rangle$ and $|\omega'\rangle$, of the QTDA equation must be orthogonal. Additionally we wish to have the states normalized. By requiring orthonormality and using (16.27) we obtain

$$\begin{aligned}
\langle\omega|\omega'\rangle &= \delta_{\omega\omega'} = \delta_{nn'}\delta_{JJ'}\delta_{MM'}\delta_{\pi\pi'} \\
&= \sum_{\substack{a\leq b \\ a'\leq b'}} X_{ab}^{\omega*}X_{a'b'}^{\omega'}\langle\text{BCS}|A_{ab}(JM)A_{a'b'}^\dagger(J'M')|\text{BCS}\rangle \\
&= \sum_{\substack{a\leq b \\ a'\leq b'}} X_{ab}^{\omega*}X_{a'b'}^{\omega'}\delta_{JJ'}\delta_{MM'}\mathcal{N}_{ab}^2(J)[\delta_{aa'}\delta_{bb'} - (-1)^{j_a+j_b+J}\delta_{ab'}\delta_{ba'}] \\
&= \delta_{JJ'}\delta_{MM'}\delta_{\pi\pi'}\sum_{a\leq b} X_{ab}^{\omega*}X_{ab}^{\omega'}\,,
\end{aligned} \tag{16.56}$$

where the final step was taken from (16.31). The orthogonality with respect to the quantum numbers $J^\pi M$ results from the construction of the basis states. The remaining orthonormality condition is

$$\boxed{\sum_{a \leq b} X_{ab}^{nJ^{\pi}*} X_{ab}^{n'J^{\pi}} = \delta_{nn'} \quad \text{(QTDA orthonormality)}.}$$  (16.57)

### 16.3.2 Completeness

Throughout this subsection we assume $a \leq b$ and $c \leq d$ and consider the values $J = J'$, $M = M'$ as fixed. The commutator (16.27) now becomes

$$\langle \text{BCS}| \big[ A_{ab}(JM), A_{cd}^{\dagger}(JM) \big] |\text{BCS} \rangle = \mathcal{N}_{ab}^2(J) \big[ \delta_{ac}\delta_{bd} + (-1)^J \delta_{ab}\delta_{cd} \big]$$
$$= \mathcal{N}_{ab}^2(J) \delta_{ac}\delta_{bd} \big[ 1 + (-1)^J \delta_{ab} \big] = \delta_{ac}\delta_{bd} .$$  (16.58)

By using this result and the completeness condition

$$\sum_n |n\, J^{\pi} M \rangle \langle n\, J^{\pi} M| = 1$$  (16.59)

we obtain

$$\delta_{ac}\delta_{bd} = \langle \text{BCS}| \big[ A_{ab}(JM), A_{cd}^{\dagger}(JM) \big] |\text{BCS} \rangle$$
$$= \langle \text{BCS}| A_{ab}(JM) A_{cd}^{\dagger}(JM) |\text{BCS} \rangle$$
$$= \sum_n \langle \text{BCS}| A_{ab}(JM) |n\, J^{\pi} M \rangle \langle n\, J^{\pi} M| A_{cd}^{\dagger}(JM) |\text{BCS} \rangle$$
$$= \sum_n \sum_{\substack{a' \leq b' \\ c' \leq d'}} X_{a'b'}^{nJ^{\pi}} X_{c'd'}^{nJ^{\pi}*} \langle \text{BCS}| A_{ab}(JM) A_{a'b'}^{\dagger}(JM) |\text{BCS} \rangle$$
$$\times \langle \text{BCS}| A_{c'd'}(JM) A_{cd}^{\dagger}(JM) |\text{BCS} \rangle$$
$$= \sum_n \sum_{\substack{a' \leq b' \\ c' \leq d'}} X_{a'b'}^{nJ^{\pi}} X_{c'd'}^{nJ^{\pi}*} \delta_{aa'}\delta_{bb'}\delta_{cc'}\delta_{dd'} = \sum_n X_{ab}^{nJ^{\pi}} X_{cd}^{nJ^{\pi}*} .$$  (16.60)

This concludes the derivation of the completeness relation

$$\boxed{\sum_n X_{ab}^{nJ^{\pi}} X_{cd}^{nJ^{\pi}*} = \delta_{ac}\delta_{bd} \quad \text{(QTDA completeness)}.}$$  (16.61)

## 16.4 Excitation Spectra of Open-Shell Even–Even Nuclei

The aim of the QTDA is to describe the lowest excited states of open-shell even–even nuclei by two-quasiparticle configurations and their mixing. This approach involves a matrix structure like that stated in (9.83) for the particle–hole TDA.

## 16.4.1 Explicit Form of the QTDA Matrix

Following the construction of the TDA matrix in Subsect. 9.3.1 we can write for the QTDA matrix $\mathsf{A}$ the schematic expression

$$
\begin{aligned}
&\mathsf{A}_{\text{QTDA}} \\
&= \begin{pmatrix} \mathsf{E}(pp-pp) + \mathsf{V}_{\text{QTDA}}(pp-pp) & \mathsf{V}_{\text{QTDA}}(pp-nn) \\ \mathsf{V}_{\text{QTDA}}(nn-pp) & \mathsf{E}(nn-nn) + \mathsf{V}_{\text{QTDA}}(nn-nn) \end{pmatrix} .
\end{aligned}
$$

$$(16.62)$$

Details of the matrix can be inferred from (16.46). On general principles the matrix is Hermitian, and because our matrix elements are always real, it is *symmetric*. This facilitates considerably the numerical work. Let us examine the block decomposition of (16.62) in more detail.

As deduced from (16.46), $\mathsf{E}(pp-pp)$ and $\mathsf{E}(nn-nn)$ are diagonal matrices containing the proton and neutron two-quasiparticle energies, i.e.

$$E(p_1 p_2 - p_3 p_4) = (E_{p_1} + E_{p_2})\delta_{p_1 p_3}\delta_{p_2 p_4} , \tag{16.63}$$

$$E(n_1 n_2 - n_3 n_4) = (E_{n_1} + E_{n_2})\delta_{n_1 n_3}\delta_{n_2 n_4} . \tag{16.64}$$

Reduced to two-body matrix elements in isospin notation, the proton–proton interaction terms of (16.46) contain

$V_{\text{QTDA}}(p_1 p_2 - p_3 p_4) :$

$\langle a_1 a_2 ; J\, T = 1|V|a_3 a_4 ; J\, T = 1 \rangle ,$

$\langle p_1 p_2^{-1}; J|V_{\text{RES}}|p_3 p_4^{-1}; J\rangle \rightarrow \langle a_1 a_4 ; J'\, T = 1|V|a_3 a_2 ; J'\, T = 1 \rangle ,$

$\langle p_1 p_2^{-1}; J|V_{\text{RES}}|p_4 p_3^{-1}; J\rangle \rightarrow \langle a_1 a_3 ; J'\, T = 1|V|a_4 a_2 ; J'\, T = 1 \rangle , \quad (16.65)$

where the arrows indicate breakdown by the Pandya transformation (16.19) followed by application of (8.25). The same scheme with $p_i \rightarrow n_i$ applies to the neutron–neutron block.

As pointed out before (16.49), the $pp - nn$ and $nn - pp$ blocks have only particle–hole matrix elements. The reduction similar to (16.65) is

$V_{\text{QTDA}}(p_1 p_2 - n_3 n_4) :$

$\langle p_1 p_2^{-1}; J|V_{\text{RES}}|n_3 n_4^{-1}; J\rangle \rightarrow \langle a_1 a_4 ; J'\, T' = 1,0|V|a_3 a_2 ; J'\, T' = 1,0 \rangle ,$

$\langle p_1 p_2^{-1}; J|V_{\text{RES}}|n_4 n_3^{-1}; J\rangle \rightarrow \langle a_1 a_3 ; J'\, T' = 1,0|V|a_4 a_2 ; J'\, T' = 1,0 \rangle ,$

$$(16.66)$$

where the arrows now indicate a non-trivial two-stage reduction, first by the Pandya transformation (16.19) and then by (8.27). Equation (8.27) shows that the same scheme with $p_i \leftrightarrow n_i$ applies to the $nn - pp$ block.

Numerical values of the particle–hole matrix elements (16.19) are needed in applications. As before, we use the SDI with its strength parameters $A_T$ for our numerical examples. The particle–hole matrix elements in (16.65) and (16.66) can be conveniently tabulated in terms of the auxiliary quantities

$$\mathcal{M}_{abcd}^{(1)}(J) \equiv -\sum_{J'} \hat{J}'^2 [\mathcal{N}_{ad}(J')\mathcal{N}_{cb}(J')]^{-1} \begin{Bmatrix} j_a & j_b & J \\ j_c & j_d & J' \end{Bmatrix}$$

$$\times \langle a\,d\,;\, J'\,1|V_{\mathrm{SDI}}|c\,b\,;\, J'\,1\rangle_{A_1=1}\,, \quad (16.67)$$

$$\mathcal{M}_{abcd}^{(2)}(JT) \equiv \tfrac{1}{2}(-1)^T \sum_{J'} \hat{J}'^2 \sqrt{[1-(-1)^{J'+T}\delta_{ad}][1-(-1)^{J'+T}\delta_{cb}]}$$

$$\times \begin{Bmatrix} j_a & j_b & J \\ j_c & j_d & J' \end{Bmatrix} \langle a\,d\,;\, J'\,T|V_{\mathrm{SDI}}|c\,b\,;\, J'\,T\rangle_{A_T=1}\,. \quad (16.68)$$

The particle–hole matrix elements are given by $\mathcal{M}_{phph}^{(1)}(J)$ and $\mathcal{M}_{phph}^{(2)}(JT)$ as

$$\langle p_1\,p_2^{-1}\,;\, J|V_{\mathrm{RES}}|p_3\,p_4^{-1}\,;\, J\rangle = \langle n_1\,n_2^{-1}\,;\, J|V_{\mathrm{RES}}|n_3\,n_4^{-1}\,;\, J\rangle$$

$$= A_1 \mathcal{M}_{a_1 a_2 a_3 a_4}^{(1)}(J)\,, \quad (16.69)$$

$$\langle p_1\,p_2^{-1}\,;\, J|V_{\mathrm{RES}}|n_3\,n_4^{-1}\,;\, J\rangle = \langle n_1\,n_2^{-1}\,;\, J|V_{\mathrm{RES}}|p_3\,p_4^{-1}\,;\, J\rangle$$

$$= A_1 \mathcal{M}_{a_1 a_2 a_3 a_4}^{(2)}(J1) + A_0 \mathcal{M}_{a_1 a_2 a_3 a_4}^{(2)}(J0)\,. \quad (16.70)$$

Taking the SDI two-body matrix elements from Table 8.2 we can calculate and tabulate the auxiliary matrix elements $\mathcal{M}_{abcd}^{(1)}(J)$ and $\mathcal{M}_{abcd}^{(2)}(JT)$ for the 0d-1s valence space. They are listed in Tables 16.1–16.3.

Calculations of the QTDA matrix $\mathbf{A}$ are simplified by the fact that the complete matrix is always symmetric. In the case that the proton and neutron two-quasiparticle bases are the same also the non-diagonal blocks are symmetric. If, furthermore, the single-particle energies and interactions are the same for protons and neutrons, the diagonal blocks are the same. These symmetries are present in our applications.

In the following we apply the formalism to a simple example that displays the essential features of even more realistic cases.

### 16.4.2 Excitation Energies of 2$^+$ States in $^{24}$Mg

Consider the $2^+$ states of the nucleus $^{24}_{12}\mathrm{Mg}_{12}$ in the $0\mathrm{d}_{5/2}$-$1\mathrm{s}_{1/2}$ valence space. The two-quasiparticle basis states are

$$\{|\pi_1\rangle,\, |\pi_2\rangle,\, |\nu_1\rangle,\, |\nu_2\rangle\} = \{|(\pi 0\mathrm{d}_{5/2})^2\,;\, 2^+\rangle,\, |\pi 0\mathrm{d}_{5/2}\,\pi 1\mathrm{s}_{1/2}\,;\, 2^+\rangle,$$

$$|(\nu 0\mathrm{d}_{5/2})^2\,;\, 2^+\rangle,\, |\nu 0\mathrm{d}_{5/2}\,\nu 1\mathrm{s}_{1/2}\,;\, 2^+\rangle\}\,. \quad (16.71)$$

We take the same single-particle energies and SDI parameters for protons and neutrons, namely

$$\varepsilon_{0\mathrm{d}_{5/2}} = 0\,,\ \varepsilon_{0\mathrm{s}_{1/2}} = 0.87\,\mathrm{MeV}\,,\quad A_0 = A_1 = 1.0\,\mathrm{MeV}\,. \quad (16.72)$$

**Table 16.1.** Quantities $\mathcal{M}_{abcd}^{(1)}(J)$ for the 0d-1s shells in the CS phase convention

| abcd | J | $\mathcal{M}^{(1)}$ | J | $\mathcal{M}^{(1)}$ | J | $\mathcal{M}^{(1)}$ | J | $\mathcal{M}^{(1)}$ | J | $\mathcal{M}^{(1)}$ |
|------|---|------|---|------|---|------|---|------|---|------|
| 1111 | 0 | −3.0000 | 1 | 1.6286 | 2 | −0.6857 | 3 | 0.9143 | 4 | −0.2857 |
| 1111 | 5 | 1.4286 | | | | | | | | |
| 1112 | 2 | −0.6414 | 3 | 0.9389 | | | | | | |
| 1113 | 1 | −1.2828 | 2 | −0.3429 | 3 | −0.4199 | 4 | −0.4041 | | |
| 1121 | 2 | −0.6414 | 3 | 0.9389 | | | | | | |
| 1122 | 0 | −1.7320 | 1 | 1.1832 | | | | | | |
| 1123 | 1 | 0.4781 | 2 | −0.5237 | | | | | | |
| 1131 | 1 | 1.2828 | 2 | 0.3429 | 3 | 0.4199 | 4 | 0.4041 | | |
| 1132 | 1 | −0.4781 | 2 | 0.5237 | | | | | | |
| 1133 | 0 | −2.4495 | 1 | −1.1759 | 2 | −0.5237 | 3 | −0.3429 | | |
| 1212 | 2 | −0.2000 | 3 | 1.0000 | | | | | | |
| 1213 | 2 | 0.2138 | 3 | −0.5111 | | | | | | |
| 1221 | 2 | −1.0000 | 3 | 1.0000 | | | | | | |
| 1223 | 2 | 0.0000 | | | | | | | | |
| 1231 | 2 | 0.8552 | 3 | 0.5111 | | | | | | |
| 1232 | 2 | 0.9798 | | | | | | | | |
| 1233 | 2 | −0.4899 | 3 | −0.1565 | | | | | | |
| 1313 | 1 | 1.8000 | 2 | 0.5429 | 3 | 0.3714 | 4 | 0.1429 | | |
| 1321 | 2 | −0.8552 | 3 | −0.5111 | | | | | | |
| 1322 | 1 | −1.2649 | | | | | | | | |
| 1323 | 1 | 0.4472 | 2 | 0.3928 | | | | | | |
| 1331 | 1 | −1.8000 | 2 | 0.8857 | 3 | −0.3714 | 4 | 1.2857 | | |
| 1332 | 1 | −0.4472 | 2 | 0.9165 | | | | | | |
| 1333 | 1 | 0.4000 | 2 | −0.2619 | 3 | −0.2799 | | | | |
| 2121 | 2 | −0.2000 | 3 | 1.0000 | | | | | | |
| 2123 | 2 | −0.9798 | | | | | | | | |
| 2131 | 2 | −0.2138 | 3 | 0.5111 | | | | | | |
| 2132 | 2 | 0.0000 | | | | | | | | |
| 2133 | 2 | −0.4899 | 3 | −0.1565 | | | | | | |
| 2222 | 0 | −1.0000 | 1 | 1.0000 | | | | | | |
| 2223 | 1 | 0.0000 | | | | | | | | |
| 2231 | 1 | 1.2649 | | | | | | | | |
| 2232 | 1 | 0.0000 | | | | | | | | |
| 2233 | 0 | −1.4142 | 1 | −0.6325 | | | | | | |
| 2323 | 1 | 1.0000 | 2 | 0.2000 | | | | | | |
| 2331 | 1 | −0.4472 | 2 | 0.9165 | | | | | | |
| 2332 | 1 | −1.0000 | 2 | 1.0000 | | | | | | |
| 2333 | 1 | −0.8944 | 2 | −0.4000 | | | | | | |
| 3131 | 1 | 1.8000 | 2 | 0.5429 | 3 | 0.3714 | 4 | 0.1429 | | |
| 3132 | 1 | 0.4472 | 2 | 0.3928 | | | | | | |
| 3133 | 1 | −0.4000 | 2 | 0.2619 | 3 | 0.2799 | | | | |
| 3232 | 1 | 1.0000 | 2 | 0.2000 | | | | | | |
| 3233 | 1 | 0.8944 | 2 | 0.4000 | | | | | | |
| 3333 | 0 | −2.0000 | 1 | 1.2000 | 2 | −0.4000 | 3 | 1.2000 | | |

The states are numbered $1 = 0d_{5/2}$, $2 = 1s_{1/2}$ and $3 = 0d_{3/2}$. The first column gives the state labels, and the following columns give $J$ and $\mathcal{M}^{(1)}$.

**Table 16.2.** Quantities $\mathcal{M}^{(2)}_{abcd}(JT)$ for the 0d-1s shells in the CS phase convention

| abcd | JT | $\mathcal{M}^{(2)}$ | JT | $\mathcal{M}^{(2)}$ | JT | $\mathcal{M}^{(2)}$ | JT | $\mathcal{M}^{(2)}$ | JT | $\mathcal{M}^{(2)}$ |
|------|----|------|----|------|----|------|----|------|----|------|
| 1111 | 00 | −4.5000 | 01 | −1.5000 | 10 | −0.8143 | 11 | 0.8143 | 20 | −1.0286 |
| 1111 | 21 | −0.3429 | 30 | −0.4571 | 31 | 0.4571 | 40 | −0.4286 | 41 | −0.1429 |
| 1111 | 50 | −0.7143 | 51 | 0.7143 | | | | | | |
| 1112 | 20 | −0.9621 | 21 | −0.3207 | 30 | −0.4695 | 31 | 0.4695 | | |
| 1113 | 10 | 0.6414 | 11 | −0.6414 | 20 | −0.5143 | 21 | −0.1714 | 30 | 0.2100 |
| 1113 | 31 | −0.2100 | 40 | −0.6061 | 41 | −0.2020 | | | | |
| 1121 | 20 | −0.9621 | 21 | −0.3207 | 30 | −0.4695 | 31 | 0.4695 | | |
| 1122 | 00 | −2.5981 | 01 | −0.8660 | 10 | −0.5916 | 11 | 0.5916 | | |
| 1123 | 10 | −0.2390 | 11 | 0.2390 | 20 | −0.7856 | 21 | −0.2619 | | |
| 1131 | 10 | −0.6414 | 11 | 0.6414 | 20 | 0.5143 | 21 | 0.1714 | 30 | −0.2100 |
| 1131 | 31 | 0.2100 | 40 | 0.6061 | 41 | 0.2020 | | | | |
| 1132 | 10 | 0.2390 | 11 | −0.2390 | 20 | 0.7856 | 21 | 0.2619 | | |
| 1133 | 00 | −3.6742 | 01 | −1.2247 | 10 | 0.5880 | 11 | −0.5880 | 20 | −0.7856 |
| 1133 | 21 | −0.2619 | 30 | 0.1714 | 31 | −0.1714 | | | | |
| 1212 | 20 | −1.1000 | 21 | −0.1000 | 30 | −0.5000 | 31 | 0.5000 | | |
| 1213 | 20 | −0.7483 | 21 | 0.1069 | 30 | 0.2556 | 31 | −0.2556 | | |
| 1221 | 20 | −0.7000 | 21 | −0.5000 | 30 | −0.5000 | 31 | 0.5000 | | |
| 1223 | 20 | −0.9798 | 21 | 0.0000 | | | | | | |
| 1231 | 20 | 0.2138 | 21 | 0.4276 | 30 | −0.2556 | 31 | 0.2556 | | |
| 1232 | 20 | 0.4899 | 21 | 0.4899 | | | | | | |
| 1233 | 20 | −0.7348 | 21 | −0.2449 | 30 | 0.0782 | 31 | −0.0782 | | |
| 1313 | 10 | −0.9000 | 11 | 0.9000 | 20 | −0.6143 | 21 | 0.2714 | 30 | −0.1857 |
| 1313 | 31 | 0.1857 | 40 | −1.2143 | 41 | 0.0714 | | | | |
| 1321 | 20 | −0.2138 | 21 | −0.4276 | 30 | 0.2556 | 31 | −0.2556 | | |
| 1322 | 10 | 0.6325 | 11 | −0.6325 | | | | | | |
| 1323 | 10 | −0.2236 | 11 | 0.2236 | 20 | −0.7201 | 21 | 0.1964 | | |
| 1331 | 10 | 0.9000 | 11 | −0.9000 | 20 | −0.1000 | 21 | 0.4429 | 30 | 0.1857 |
| 1331 | 31 | −0.1857 | 40 | 0.5000 | 41 | 0.6429 | | | | |
| 1332 | 10 | 0.2236 | 11 | −0.2236 | 20 | 0.0655 | 21 | 0.4583 | | |
| 1333 | 10 | −0.2000 | 11 | 0.2000 | 20 | −0.3928 | 21 | −0.1309 | 30 | 0.1400 |
| 1333 | 31 | −0.1400 | | | | | | | | |
| 2121 | 20 | −1.1000 | 21 | −0.1000 | 30 | −0.5000 | 31 | 0.5000 | | |
| 2123 | 20 | −0.4899 | 21 | −0.4899 | | | | | | |
| 2131 | 20 | 0.7483 | 21 | −0.1069 | 30 | −0.2556 | 31 | 0.2556 | | |
| 2132 | 20 | 0.9798 | 21 | 0.0000 | | | | | | |
| 2133 | 20 | −0.7348 | 21 | −0.2449 | 30 | 0.0782 | 31 | −0.0782 | | |
| 2222 | 00 | −1.5000 | 01 | −0.5000 | 10 | −0.5000 | 11 | 0.5000 | | |
| 2223 | 10 | 0.0000 | 11 | 0.0000 | | | | | | |
| 2231 | 10 | −0.6325 | 11 | 0.6325 | | | | | | |
| 2232 | 10 | 0.0000 | 11 | 0.0000 | | | | | | |
| 2233 | 00 | −2.1213 | 01 | −0.7071 | 10 | 0.3162 | 11 | −0.3162 | | |
| 2323 | 10 | −0.5000 | 11 | 0.5000 | 20 | −0.9000 | 21 | 0.1000 | | |

The states are numbered $1 = 0d_{5/2}$, $2 = 1s_{1/2}$ and $3 = 0d_{3/2}$. The first column gives the state labels, and the following columns give the $JT$ combinations and $\mathcal{M}^{(2)}$.

**Table 16.3.** Continuation of Table 16.2

| abcd | JT | $\mathcal{M}^{(2)}$ | JT | $\mathcal{M}^{(2)}$ | JT | $\mathcal{M}^{(2)}$ | JT | $\mathcal{M}^{(2)}$ | JT | $\mathcal{M}^{(2)}$ |
|------|----|------|----|------|----|------|----|------|----|------|
| 2331 | 10 | 0.2236 | 11 | −0.2236 | 20 | 0.0655 | 21 | 0.4583 | | |
| 2332 | 10 | 0.5000 | 11 | −0.5000 | 20 | 0.3000 | 21 | 0.5000 | | |
| 2333 | 10 | 0.4472 | 11 | −0.4472 | 20 | −0.6000 | 21 | −0.2000 | | |
| 3131 | 10 | −0.9000 | 11 | 0.9000 | 20 | −0.6143 | 21 | 0.2714 | 30 | −0.1857 |
| 3131 | 31 | 0.1857 | 40 | −1.2143 | 41 | 0.0714 | | | | |
| 3132 | 10 | −0.2236 | 11 | 0.2236 | 20 | −0.7201 | 21 | 0.1964 | | |
| 3133 | 10 | 0.2000 | 11 | −0.2000 | 20 | 0.3928 | 21 | 0.1309 | 30 | −0.1400 |
| 3133 | 31 | 0.1400 | | | | | | | | |
| 3232 | 10 | −0.5000 | 11 | 0.5000 | 20 | −0.9000 | 21 | 0.1000 | | |
| 3233 | 10 | −0.4472 | 11 | 0.4472 | 20 | 0.6000 | 21 | 0.2000 | | |
| 3333 | 00 | −3.0000 | 01 | −1.0000 | 10 | −0.6000 | 11 | 0.6000 | 20 | −0.6000 |
| 3333 | 21 | −0.2000 | 30 | −0.6000 | 31 | 0.6000 | | | | |

The first task is to solve the BCS equations for our two-level valence space. Using the procedure described in Subsect. 14.1.1 we obtain the BCS occupation amplitudes and quasiparticle energies given in Table 16.4. The same values apply for protons and neutrons.

**Table 16.4.** BCS occupation amplitudes $u$ and $v$ and quasiparticle energies $E$ calculated for $^{24}_{12}\text{Mg}_{12}$ with single-particle energies $\varepsilon_{0d_{5/2}} = 0$ and $\varepsilon_{1s_{1/2}} = 0.87\,\text{MeV}$ and SDI interaction strength $A_1 = 1.0\,\text{MeV}$

| Orbital | $u$ | $v$ | $E$ (MeV) |
|---------|-----|-----|-----------|
| $0d_{5/2}$ | 0.6685 | 0.7437 | 1.977 |
| $1s_{1/2}$ | 0.8119 | 0.5838 | 2.073 |

Next we turn to forming the QTDA matrix A. With the four basis states (16.71) its dimension is 4-by-4, but because of the symmetries there are only six different matrix elements: $A_{\pi_1\pi_1}$, $A_{\pi_1\pi_2}$, $A_{\pi_2\pi_2}$, $A_{\pi_1\nu_1}$, $A_{\pi_1\nu_2}$, $A_{\pi_2\nu_2}$. These are given by Eqs. (16.46), (16.69) and (16.70) with numerical data from Tables 8.2, 16.1, 16.2 and 16.4:

$$
\begin{aligned}
A_{\pi_1\pi_1} &= A\big((\pi 0d_{5/2})^2, (\pi 0d_{5/2})^2\big) \\
&= 2 \times 1.977\,\text{MeV} + (0.6685^4 + 0.7437^4)(-0.6857 A_1) \\
&\quad + \frac{1}{2}\big[2 \times 0.6685^2 \times 0.7437^2(-0.6857 A_1) \\
&\qquad - (-1)^{\frac{5}{2}+\frac{5}{2}+2} \times 2 \times 0.6685^2 \times 0.7437^2(-0.6857 A_1)\big] \\
&= 3.954\,\text{MeV} - 0.686 A_1 ,
\end{aligned}
\tag{16.73}
$$

$$A_{\pi_1\pi_2} = A\big((\pi 0\mathrm{d}_{5/2})^2, \pi 0\mathrm{d}_{5/2}\pi 1\mathrm{s}_{1/2}\big)$$

$$= (0.6685^3 \times 0.8119 + 0.7437^3 \times 0.5838)(-0.9071 A_1)$$

$$+ \frac{1}{\sqrt{2}}\big[0.6685^2 \times 0.7437 \times 0.5838 + 0.7437^2 \times 0.6685 \times 0.8119\big)$$

$$\times (-0.6414 A_1) - (-1)^{\frac{5}{2}+\frac{1}{2}+2} \times (0.6685 \times 0.7437^2 \times 0.8119$$

$$+ 0.7437 \times 0.6685^2 \times 0.5838)(-0.6414 A_1)\big]$$

$$= -0.886 A_1 , \tag{16.74}$$

$$A_{\pi_2\pi_2} = A\big(\pi 0\mathrm{d}_{5/2}\pi 1\mathrm{s}_{1/2}, \pi 0\mathrm{d}_{5/2}\pi 1\mathrm{s}_{1/2}\big) = 1.977\,\mathrm{MeV} + 2.073\,\mathrm{MeV}$$

$$+ (0.6685^2 \times 0.8119^2 + 0.7437^2 \times 0.5838^2)(-1.2000 A_1)$$

$$+ (0.6685^2 \times 0.5838^2 + 0.7437^2 \times 0.8119^2)(-0.2000 A_1)$$

$$- (-1)^{\frac{5}{2}+\frac{1}{2}+2} \times 2 \times 0.6685 \times 0.5838 \times 0.7437 \times 0.8119$$

$$\times (-1.0000 A_1) = 4.050\,\mathrm{MeV} - 1.154 A_1 , \tag{16.75}$$

$$A_{\pi_1\nu_1} = A\big((\pi 0\mathrm{d}_{5/2})^2, (\nu 0\mathrm{d}_{5/2})^2\big)$$

$$= \frac{1}{2}\big[2 \times 0.6685^2 \times 0.7437^2(-0.3429 A_1 - 1.0286 A_0)$$

$$- (-1)^{\frac{5}{2}+\frac{5}{2}+2} \times 2 \times 0.6685^2 \times 0.7437^2(-0.3429 A_1 - 1.0286 A_0)\big]$$

$$= -0.508 A_0 - 0.170 A_1 , \tag{16.76}$$

$$A_{\pi_1\nu_2} = A\big((\pi 0\mathrm{d}_{5/2})^2, \nu 0\mathrm{d}_{5/2}\nu 1\mathrm{s}_{1/2}\big)$$

$$= \frac{1}{\sqrt{2}}\big[(0.6685^2 \times 0.7437 \times 0.5838 + 0.7437^2 \times 0.6685 \times 0.8119)$$

$$\times (-0.3207 A_1 - 0.9621 A_0)$$

$$- (-1)^{\frac{5}{2}+\frac{1}{2}+2} \times (0.6685 \times 0.7437^2 \times 0.8119$$

$$+ 0.7437 \times 0.6685^2 \times 0.5838)(-0.3207 A_1 - 0.9621 A_0)\big]$$

$$= -0.672 A_0 - 0.224 A_1 , \tag{16.77}$$

$$A_{\pi_2\nu_2} = A\big(\pi 0\mathrm{d}_{5/2}\pi 1\mathrm{s}_{1/2}, \nu 0\mathrm{d}_{5/2}\nu 1\mathrm{s}_{1/2}\big)$$

$$= \big[(0.6685^2 \times 0.5838^2 + 0.7437^2 \times 0.8119^2)(-0.1000 A_1 - 1.1000 A_0)$$

$$- (-1)^{\frac{5}{2}+\frac{1}{2}+2} \times 2 \times 0.6685 \times 0.5838 \times 0.7437 \times 0.8119$$

$$\times (-0.5000 A_1 - 0.7000 A_0)\big] = -0.898 A_0 - 0.287 A_1 . \tag{16.78}$$

We can now form the matrix $\mathsf{A}_{\mathrm{QTDA}}$ from the elements (16.73)–(16.78) in the pattern (16.62). The complete matrix is symmetric; in this case the proton–neutron block is also symmetric and the diagonal blocks are identical. The complete matrix is

$A_{QTDA}(2^+)$

$$
= \begin{pmatrix}
3.954 - 0.686A_1 & -0.886A_1 & -0.508A_0 - 0.170A_1 & -0.672A_0 - 0.224A_1 \\
-0.886A_1 & 4.050 - 1.154A_1 & -0.672A_0 - 0.224A_1 & -0.898A_0 - 0.287A_1 \\
\dots & \dots & 3.954 - 0.686A_1 & -0.886A_1 \\
\dots & \dots & -0.886A_1 & 4.050 - 1.154A_1
\end{pmatrix},
$$
(16.79)

where the quasiparticle energies are in Mega-electronvolts and the dots represent the neutron–proton block that is identical with the proton–neutron block.

Diagonalizing the matrix (16.79) with SDI parameters $A_0 = A_1 = 1.0\,\mathrm{MeV}$ gives the eigenvalues

$$
E(2_1^+) = 0.313\,\mathrm{MeV}\,, \quad E(2_2^+) = 3.945\,\mathrm{MeV}\,,
$$
$$
E(2_3^+) = 3.986\,\mathrm{MeV}\,, \quad E(2_4^+) = 4.082\,\mathrm{MeV}\,.
$$
(16.80)

Experimentally, the first excited state of $^{24}$Mg is at $1.369\,\mathrm{MeV}$, so the default values $A_0 = A_1 = 1.0\,\mathrm{MeV}$ do not produce realistic results. A better choice, made also in the BCS calculation, is $A_0 = 0.7\,\mathrm{MeV}$, $A_1 = 1.4\,\mathrm{MeV}$. The results are shown in Fig. 16.1 for our chosen valence space $0d_{5/2}$-$1s_{1/2}$ and also for the complete 0d-1s valence space.

Although the description of the low-lying states shown is quite good, the $0^+$ states pose a problem. The lowest $0^+$ state of two-quasiparticle nature falls below the QTDA ground state for our SDI parameters. Its energy is $E(0_1^+) = -3.389\,\mathrm{MeV}$ in the $0d_{5/2}$-$1s_{1/2}$ valence space and $E(0_1^+) = -3.488\,\mathrm{MeV}$ in the complete 0d-1s valence space. This effect is due to a monopole residual interaction that is too strong in the $T = 1$ channel. Since monopole correlations of this type are largely taken into account already at the BCS level, a smaller $A_1$ is appropriate. Reducing $A_1$ close to zero for states with $J^\pi = 0^+$ improves the description of the $0^+$ spectrum.

### 16.4.3 Pairing Strength Parameters from Empirical Pairing Gaps

The problem with $0^+$ states discussed above suggests the use of different $T = 1$ interactions in the BCS calculation and in the subsequent QTDA matrix diagonalization. Let us therefore develop a scheme for determining the SDI parameters $A_1$ for protons and neutrons within the BCS calculation. To distinguish these parameters from those of a later QTDA calculation, we call them *pairing parameters* $A_{\mathrm{pair}}^{(p)}$ and $A_{\mathrm{pair}}^{(n)}$. We proceed to demonstrate how they can be determined by fitting the lowest proton and neutron quasiparticle energies to the empirical pairing gaps extracted from the proton and neutron separation energies.

The proton and neutron *separation energies* $S_{\mathrm{p}}(A, Z)$ and $S_{\mathrm{n}}(A, Z)$ are defined in terms of atomic masses $m(A, Z)$ as

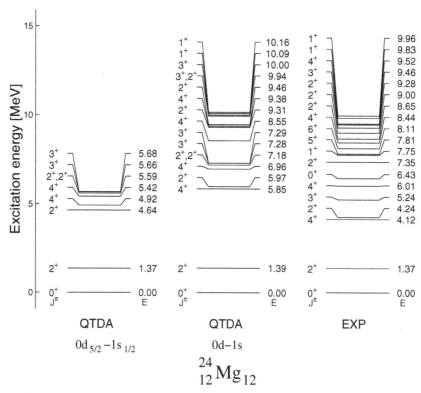

**Fig. 16.1.** QTDA spectrum of $^{24}$Mg in the valence spaces shown and the experimental spectrum. The single-particle energies are from (14.12). The SDI was used with parameters $A_0 = 0.7\,\text{MeV}$ and $A_1 = 1.4\,\text{MeV}$

$$S_\text{p}(A, Z) \equiv m(A - 1, Z - 1)c^2 + m(^1\text{H})c^2 - m(A, Z)c^2 \,, \qquad (16.81)$$
$$S_\text{n}(A, Z) \equiv m(A - 1, Z)c^2 + M_\text{n}c^2 - m(A, Z)c^2 \,, \qquad (16.82)$$

where $M_\text{n}$ is the mass of the neutron and $m(^1\text{H})$ is the mass of the hydrogen atom. From the relation (12.2) between the atomic mass $m(A, Z)$ and the nuclear mass $M(A, Z)$ we have

$$m(A, Z) = M(A, Z) + Zm_\text{e} - B_\text{e}(Z)/c^2 \,, \qquad (16.83)$$
$$m(^1\text{H}) = M_\text{p} + m_\text{e} - B_\text{e}(1)/c^2 \,, \qquad (16.84)$$

where $M_\text{p}$ is the proton mass. The atomic mass takes into account the mass of the $Z$ electrons and their total binding energy $B_\text{e}(Z)$. Substituting (16.83) and (16.84) for the atomic masses in (16.81) and (16.82) gives the separation energies in terms of nuclear masses as[2]

---

[2] In $S_\text{p}$ the difference $B_\text{e}(Z) - B_\text{e}(Z - 1)$ is negligible to a very good approximation.

$$S_p(A, Z) = M(A - 1, Z - 1)c^2 + M_p c^2 - M(A, Z)c^2 \,, \tag{16.85}$$
$$S_n(A, Z) = M(A - 1, Z)c^2 + M_n c^2 - M(A, Z)c^2 \,. \tag{16.86}$$

Equation (12.1) gives the pairing gap $\Delta$ as

$$\Delta = (2M_A - M_{A-1} - M_{A+1})c^2 \,, \tag{16.87}$$

where the change in $A$ is due to changing the number of only one kind of nucleon. By means of (16.85) and (16.86) the pairing gaps can be expressed in terms of the separation energies, with the result

$$\Delta_p(A, Z) = \tfrac{1}{2}[S_p(A + 1, Z + 1) - S_p(A, Z)] \,, \tag{16.88}$$
$$\Delta_n(A, Z) = \tfrac{1}{2}[S_n(A + 1, Z) - S_n(A, Z)] \,. \tag{16.89}$$

These equations can be replaced by more accurate interpolation formulas such as the three-point formulas [78, 79]

$$\boxed{\begin{aligned}
\Delta_p(A, Z) &= \tfrac{1}{4}(-1)^{Z+1}[S_p(A + 1, Z + 1) - 2S_p(A, Z) + S_p(A - 1, Z - 1)] \,, \\
\Delta_n(A, Z) &= \tfrac{1}{4}(-1)^{A-Z+1}[S_n(A + 1, Z) - 2S_n(A, Z) + S_n(A - 1, Z)] \,.
\end{aligned}}$$
$$\tag{16.90}$$

In contrast to (16.90) we have the very simple formula (12.3), derived from the liquid-drop model and empirical parameters,

$$\Delta \approx 12A^{-1/2} \,\mathrm{MeV} \tag{16.91}$$

for protons and neutrons alike. We also note that it is not safe to use (16.90) to determine a pairing gap in the immediate vicinity of a major shell closure because the separation energies there behave irregularly. In the case of zero or two particles or holes in a major shell it is better to use the simple smooth formula (16.91).

In Subsect. 13.4.1 we discussed the relation between the pairing gap and the quasiparticle energies. In particular, from (13.69) we have

$$E_{qp} \geq \Delta_{smallest} \,. \tag{16.92}$$

We interpret $\Delta_{smallest}$, separately for protons and neutrons, as the experimental pairing gap determined from e.g. (16.90). The lowest calculated quasiparticle energies are then to be fitted according to

$$E_{qp}^p(\text{lowest}) = \Delta_p \,, \quad E_{qp}^n(\text{lowest}) = \Delta_n \,. \tag{16.93}$$

The pairing strength parameters $A_{pair}^{(p)}$ and $A_{pair}^{(n)}$ are varied in the BCS calculations until the conditions (16.93) are satisfied with desired accuracy. This procedure is illustrated by the following examples.

It is well to note that the lowest two-quasiparticle states used to determine $A_{pair}^{(p,n)}$ according to (16.93) occur at $2\Delta_{p,n}$. This energy is much higher than the lowest excitation energies subsequently produced by the QTDA, as will be seen from the examples.

## 16.4.4 Excitation Spectrum of $^{24}_{12}\mathrm{Mg}_{12}$

Our first example is $^{24}\mathrm{Mg}$. By using (16.90) with the separation energies taken from the atomic mass evaluations of [78, 79] yields the pairing gaps

$$\Delta_{\mathrm{p}}(^{24}\mathrm{Mg}) = -\tfrac{1}{4}[S_{\mathrm{p}}(^{25}\mathrm{Al}) - 2S_{\mathrm{p}}(^{24}\mathrm{Mg}) + S_{\mathrm{p}}(^{23}\mathrm{Na})]$$
$$= -\tfrac{1}{4}[2271 - 2 \times 11691 + 8794]\,\mathrm{keV} = 3079\,\mathrm{keV}\,, \tag{16.94}$$
$$\Delta_{\mathrm{n}}(^{24}\mathrm{Mg}) = -\tfrac{1}{4}[S_{\mathrm{n}}(^{25}\mathrm{Mg}) - 2S_{\mathrm{n}}(^{24}\mathrm{Mg}) + S_{\mathrm{n}}(^{23}\mathrm{Mg})]$$
$$= -\tfrac{1}{4}[7331 - 2 \times 16531 + 13148]\,\mathrm{keV} = 3146\,\mathrm{keV}\,. \tag{16.95}$$

We now make a set of BCS calculations for $^{24}_{12}\mathrm{Mg}_{12}$, varying the parameters $A_{\mathrm{pair}}^{(\mathrm{p})}$ and $A_{\mathrm{pair}}^{(\mathrm{n})}$ until the lowest proton and neutron quasiparticle energies agree with the empirical pairing gaps according to Eq. (16.93). The parameters found are recorded in Table 16.5.

The BCS stage is followed by QTDA calculations. The SDI parameters $A_0$ and $A_1$ are adjusted with the aim of reproducing the lowest levels. In particular, an approximate fit is made to the first excited $0^+$ and $2^+$ levels. Having recognized the special nature of the $0^+$ excitations at the end of Subsect. 16.4.2, we make separate fits for $A_1(J^\pi \neq 0^+)$ and $A_1(J^\pi = 0^+)$. The interaction parameters are listed in Table 16.5, while the single-particle energies are from (14.12). The energy spectra are shown in Fig. 16.2.

**Table 16.5.** SDI pairing parameters $A_{\mathrm{pair}}^{(\mathrm{p})}$ and $A_{\mathrm{pair}}^{(\mathrm{n})}$ and strengths $A_0$ and $A_1$ for the QTDA calculation of the excitation spectrum of $^{24}\mathrm{Mg}$ in two different valence spaces

| Valence space | $A_{\mathrm{pair}}^{(\mathrm{p})}$ | $A_{\mathrm{pair}}^{(\mathrm{n})}$ | $A_0$ | $A_1(J^\pi \neq 0^+)$ | $A_1(J^\pi = 0^+)$ |
|---|---|---|---|---|---|
| $0d_{5/2}\text{-}1s_{1/2}$ | 1.55 | 1.58 | 0.94 | 1.53 | 0.00 |
| $0d\text{-}1s$ | 1.23 | 1.26 | 0.27 | 1.43 | 0.00 |

All energies are in Mega-electronvolts.

Comparison of Figs. 16.1 and 16.2 shows that the separate treatment of the pairing and the QTDA degrees of freedom significantly improves the theoretical description of the excitation spectrum of $^{24}\mathrm{Mg}$. It is expected that this would be the case for all open-shell even–even nuclei.

The SDI parameters are seen to be valence-space dependent, with the larger values associated with the smaller valence space. This demonstrates the general feature that a bare two-body interaction has to be *renormalized* more in a small valence space than in a large one to produce a suitable *effective interaction*. The present renormalization amounts to a crude overall scaling of the two-body matrix elements. In a more systematic approach each individual two-body matrix element of the bare interaction is renormalized separately.

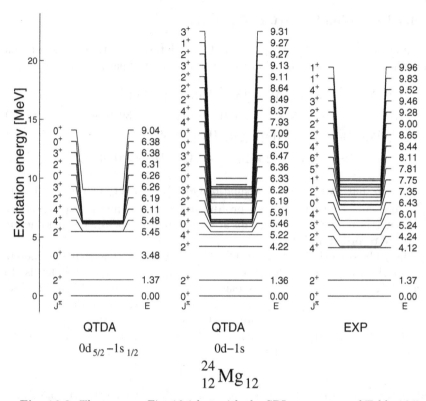

**Fig. 16.2.** The same as Fig. 16.1 but with the SDI parameters of Table 16.5

For detailed information on renormalization and effective interactions see e.g. [16].

### 16.4.5 Excitation Spectra of the Mirror Nuclei $^{30}_{14}\mathrm{Si}_{16}$ and $^{30}_{16}\mathrm{S}_{14}$

We discuss next the excitation spectra of the mirror nuclei $^{30}_{14}\mathrm{Si}_{16}$ and $^{30}_{16}\mathrm{S}_{14}$. These nuclei were treated in connection with beta decay in Subsect. 15.4.2. Only rudimentary, qualitative configuration mixing was included there. We now proceed to a quantitative QTDA calculation of the excitation spectra.

Because of the charge independence of the nuclear force, we assume the excitation spectra of a pair of mirror nuclei to be the same. This property of the nuclear force was discussed at the end of Subsect. 5.5.1, in the context of isospin symmetry of excitation spectra. As seen in the experimental spectra of Chap. 5, the assumption of isospin symmetry is supported by the data. We can thus describe a pair of mirror nuclei by one QTDA calculation.

We can build the spectra of both $^{30}\mathrm{Si}$ and $^{30}\mathrm{S}$ by taking $^{30}\mathrm{Si}$ as the reference nucleus. Equations (16.90) and data from [78, 79] give the pairing gaps

$$\Delta_p(^{30}_{14}\text{Si}_{16}) = 2308\,\text{keV}\,, \quad \Delta_n(^{30}_{14}\text{Si}_{16}) = 1539\,\text{keV}\,, \tag{16.96}$$

$$\Delta_p(^{30}_{16}\text{S}_{14}) = 1442\,\text{keV}\,, \quad \Delta_n(^{30}_{16}\text{S}_{14}) = 2387\,\text{keV}\,. \tag{16.97}$$

Thus for six active particles (protons in silicon, neutrons in sulphur) the pairing gap is about 2.3 MeV. For eight active particles (neutrons in silicon, protons in sulphur) the pairing gap is about 1.5 MeV. The BCS calculations reproduce these values with the pairing parameters given in Table 16.6.

**Table 16.6.** SDI pairing parameters $A_{\text{pair}}^{(p)}$ and $A_{\text{pair}}^{(n)}$ and strengths $A_0$ and $A_1$ for the QTDA calculation of the excitation spectrum of $^{30}$Si in two different valence spaces

| Valence space | $A_{\text{pair}}^{(p)}$ | $A_{\text{pair}}^{(n)}$ | $A_0$ | $A_1(J^\pi \neq 0^+)$ | $A_1(J^\pi = 0^+)$ |
|---|---|---|---|---|---|
| $0d_{5/2}\text{-}1s_{1/2}$ | 1.30 | 1.30 | not active | 1.58 | 0.00 |
| $0d\text{-}1s$ | 1.00 | 1.00 | 0.25 | 1.10 | 0.00 |

All energies are in Mega-electronvolts.

The results of the subsequent QTDA calculations are shown in Fig. 16.3. There the calculated spectra are compared with the experimental spectrum of $^{30}$Si. As in the case of $^{24}$Mg, the strength parameters $A_0$ and $A_1$ were fixed by fitting the first $0^+$ and $2^+$ excitation energies.

The smaller valence space $0d_{5/2}\text{-}1s_{1/2}$ becomes a special case when the number of active protons or neutrons is eight. Eight particles fill the valence space completely, and only nucleons of the other kind are active in the calculations. Then only the parameter $A_1$ of the SDI is operative. In this situation the number of QTDA states is drastically reduced, as is visible in Fig. 16.3.

The correspondence between the theoretical and experimental spectra can be further improved by using a more realistic two-body interaction or making changes to the single-particle energies. Better single-particle energies can come from a mean-field calculation of the Hartree–Fock type or from the use of a Woods–Saxon potential. These approaches were discussed in Chap. 3. The single-particle energies can also be adjusted by hand, e.g. by comparing the lowest quasiparticle energies of the BCS calculation with the low-energy spectra of the neighbouring odd-mass nuclei.

From Table 16.6 we can see that the interaction strengths decrease when going from the restricted basis to the complete 0d-1s basis. This behaviour was recognized and discussed at the end of the previous subsection.

### 16.4.6 Excitation Spectrum of $^{66}$Zn

Let us finally consider QTDA calculations in the 0f-1p and 0f-1p-0g$_{9/2}$ valence spaces, with $^{66}_{30}\text{Zn}_{36}$ as a representative case. The single-particle energies are taken from (14.13). Above the $N = Z = 20$ core we have 10 active protons

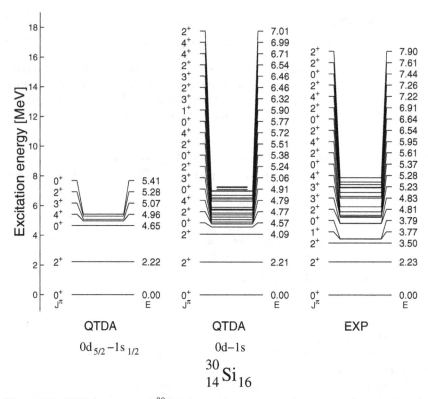

**Fig. 16.3.** QTDA spectra of $^{30}$Si in the valence spaces shown and the experimental spectrum. The single-particle energies are from (14.12). The SDI was used with the parameters of Table 16.6

and 16 active neutrons. The corresponding quasiparticle spectra are presented and compared with data in Figs. 14.5(a) and 14.6(a) for the SDI strength parameter $A_1 = 1.0\,\text{MeV}$. Relative energies of single-quasiparticle states are not much affected by changes in $A_1$, so we can expect a similar correspondence between computed quasiparticle spectra and experiment also for other values of $A_1$.

The empirical pairing gaps, found as in the previous examples, are

$$\Delta_\text{p}(^{66}_{30}\text{Zn}_{36}) = 1283\,\text{keV}\,, \quad \Delta_\text{n}(^{66}_{30}\text{Zn}_{36}) = 1772\,\text{keV}\,. \tag{16.98}$$

Adjusting the pairing parameters to these values according to (16.93) results in the parameter values recorded in Table 16.7.

The SDI strength parameters $A_0$ and $A_1$ of the QTDA calculation are determined the same way as in the previous examples. The values obtained are stated in Table 16.7, and the resulting energy spectra are depicted in Fig. 16.4 for the two valence spaces.

**Table 16.7.** SDI pairing parameters $A_{\text{pair}}^{(\text{p})}$ and $A_{\text{pair}}^{(\text{n})}$ and strengths $A_0$ and $A_1$ for the QTDA calculation of the excitation spectrum of $^{66}$Zn in two different valence spaces

| Valence space | $A_{\text{pair}}^{(\text{p})}$ | $A_{\text{pair}}^{(\text{n})}$ | $A_0$ | $A_1(J^\pi \neq 0^+)$ | $A_1(J^\pi = 0^+)$ |
|---|---|---|---|---|---|
| 0f-1p | 0.57 | 0.64 | 0.30 | 0.70 | 0.00 |
| 0f-1p-0g$_{9/2}$ | 0.43 | 0.37 | 0.25 | 0.60 | 0.00 |

All energies are in Mega-electronvolts.

The spectrum from the small valence space in Fig. 16.4 has no negative-parity states. This is because without orbitals of both parities available the condition (16.25) allows only positive-parity two-quasiparticle states. In the large valence space, negative-parity two-quasiparticle states can be generated by combining the 0g$_{9/2}$ orbital with the orbitals of the 0f-1p shell. Figure 16.4 also shows that the density of states increases notably through the inclusion

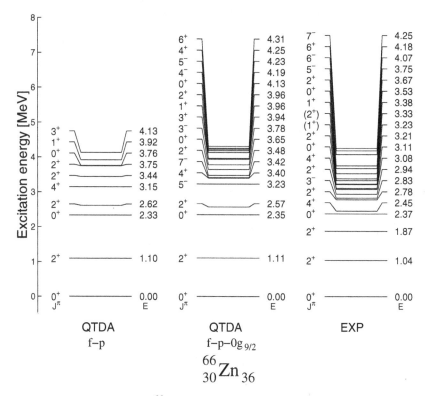

**Fig. 16.4.** QTDA spectra of $^{66}$Zn in the valence spaces shown and the experimental spectrum. The single-particle energies are from (14.13). The SDI was used with the parameters of Table 16.7

of the $0g_{9/2}$ orbital, although the energies of corresponding states do not shift very much. To further illustrate the effect of enlarging the valence space, Table 16.8 shows how the matrix dimensions grow when going from the truncated 1d-0s shell to the full 0f-1p-$0g_{9/2}$ shell.

**Table 16.8.** Dimension of the QTDA matrix for states of various $J^\pi$ in the 0d-1s and 0f-1p-$0g_{9/2}$ valence spaces and in their truncations

| Valence space | Dimension of QTDA matrix | | | | | | | | | |
|---|---|---|---|---|---|---|---|---|---|---|
| | $0^+$ | $2^+$ | $3^+$ | $4^+$ | $6^+$ | $3^-$ | $4^-$ | $5^-$ | $6^-$ | $7^-$ |
| $1s_{1/2}$-$0d_{5/2}$ | 4 | 4 | 2 | 2 | | | | | | |
| 1s-0d | 6 | 10 | 4 | 4 | | | | | | |
| 1p-0f | 8 | 16 | 10 | 12 | 4 | | | | | |
| 1p-0f-$0g_{9/2}$ | 10 | 18 | 10 | 14 | 6 | 6 | 8 | 8 | 6 | 4 |

As discussed in Subsect. 16.4.5, the correspondence between experimental and theoretical spectra can be improved, e.g. by fine tuning the single-particle energies to better reproduce the low-energy spectra of the neighbouring even–odd and odd–even nuclei at the BCS stage of the calculation. At the QTDA stage one can try multipole-dependent strengths $A_1(J^\pi)$, as we have indeed done when using a separate $A_1(0^+)$. More generally and beyond the SDI, one can introduce a multipole–multipole effective interaction in the spirit of (9.44), where each multipole component was multliplied with a phenomenological strength constant $\chi_J$.

To end this section, we comment on nuclear deformation. Many nuclei in the middle region of a major shell have a non-spherical, deformed shape. Spectra produced by theories built on the assumption of spherical shape are not readily comparable with experimental spectra of deformed nuclei. Identification of corresponding theoretical and experimental states is not straightforward.

In the following sections we test our wave functions by computing electromagnetic decay rates in open-shell even–even nuclei.

## 16.5 Electromagnetic Transitions to the Ground State

In this section we consider electromagnetic transitions to the ground state. Here we use the CS phase convention throughout. The corresponding expressions for BR phasing can be read directly from Sect. 15.4.

### 16.5.1 Decay Amplitude

The wave function of a QTDA excited state is a linear combination of two-quasiparticle components according to (16.55). The reduced matrix element for an electromagnetic transition from such a state to the ground state $|\text{BCS}\rangle = |\text{QTDA}\rangle$ was given in (15.59). With the CS phase factor from (15.25) this decay amplitude is

$$(\text{BCS}\|\mathcal{M}_{\sigma\lambda}\|\omega) = \delta_{\lambda J}\sum_{a\leq b} X_{ab}^{\omega}\mathcal{N}_{ab}(J)(-1)^{l_b}(v_a|u_b| \pm v_b|u_a|)(a\|\mathcal{M}_{\sigma\lambda}\|b) \,,$$

(16.99)

where the upper sign is for E$\lambda$ and the lower sign for M$\lambda$ transitions. Let us discuss a few examples of application of (16.99).

### 16.5.2 E2 Decay of the Lowest 2$^+$ State in $^{24}$Mg

Consider the E2 decay of the lowest 2$^+$ state in $^{24}$Mg. We follow the evolution of the decay rate as a function of the size of the valence space of the QTDA calculation. This will give information about the convergence of the involved nuclear wave function.

### The 0d$_{5/2}$-1s$_{1/2}$ Valence Space

We start with the 1s$_{1/2}$-0d$_{5/2}$ valence space. The BCS problem was solved in Subsect. 16.4.4, and the pairing strengths were listed in Table 16.5. The occupation amplitudes are given in Table 16.9.

**Table 16.9.** BCS occupation amplitudes $u$ and $v$ and quasiparticle energies $E$ calculated for protons and neutrons in $^{24}_{12}$Mg$_{12}$

| Orbital | $u_p$ | $v_p$ | $E^{\text{p}}$ (MeV) | $u_n$ | $v_n$ | $E^{\text{n}}$ (MeV) |
|---------|-------|-------|----------------------|-------|-------|----------------------|
| 0d$_{5/2}$ | 0.6821 | 0.7312 | 3.0848 | 0.6826 | 0.7308 | 3.1451 |
| 1s$_{1/2}$ | 0.7773 | 0.6291 | 3.1464 | 0.7760 | 0.6307 | 3.2055 |

The SDI pairing parameters of Table 16.5 and the single-particle energies $\varepsilon_{0d_{5/2}} = 0.0$ and $\varepsilon_{1s_{1/2}} = 0.87$ MeV were used in the calculation.

Proceeding as in Subsect. 16.4.2, we build the QTDA matrix for the 2$^+$ states. With the SDI strengths $A_0 = 0.94$ MeV and $A_1 = 1.53$ MeV from Table 16.5, the QTDA matrix becomes

$$\mathsf{A}_{\text{QTDA}}(2^+) = \begin{pmatrix} 5.120 & -1.374 & -0.742 & -0.982 \\ -1.374 & 4.425 & -0.982 & -1.296 \\ \cdots & \cdots & 5.241 & -1.375 \\ \cdots & \cdots & -1.375 & 4.543 \end{pmatrix} \text{MeV} \,.$$

(16.100)

Diagonalization yields the wave function for the lowest 2$^+$ state as

$$|^{24}\text{Mg}; 2_1^+\rangle_{\text{QTDA}} = X_1|(\pi 0\text{d}_{5/2})^2; 2^+\rangle + X_2|\pi 0\text{d}_{5/2}\,\pi 1\text{s}_{1/2}; 2^+\rangle$$
$$+ X_3|(\nu 0\text{d}_{5/2})^2; 2^+\rangle + X_4|\nu 0\text{d}_{5/2}\,\nu 1\text{s}_{1/2}; 2^+\rangle,$$
$$\tag{16.101}$$

where the amplitudes are given by

$$X_1 = 0.437\,, \quad X_2 = 0.568\,, \quad X_3 = 0.425\,, \quad X_4 = 0.552\,. \tag{16.102}$$

By use of (16.99), (16.102) and Tables 6.4 and 16.9 we obtain the transition amplitude

$$(\text{BCS}\|\boldsymbol{Q}_2\|2_1^+) = X_1 \times \frac{1}{\sqrt{2}} \times 2 \times 0.7312 \times 0.6821(-2.585e_{\text{eff}}^{\text{p}}b^2)$$
$$+ X_2 \times (0.7312 \times 0.7773 + 0.6291 \times 0.6821)(-2.185e_{\text{eff}}^{\text{p}}b^2)$$
$$+ X_3 \times \frac{1}{\sqrt{2}} \times 2 \times 0.7308 \times 0.6826(-2.585e_{\text{eff}}^{\text{n}}b^2)$$
$$+ X_4 \times (0.7308 \times 0.7760 + 0.6307 \times 0.6826)(-2.185e_{\text{eff}}^{\text{n}}b^2)$$
$$= -(1.8233X_1 + 2.1795X_2)e_{\text{eff}}^{\text{p}}b^2 - (1.8236X_3 + 2.1798X_4)e_{\text{eff}}^{\text{n}}b^2$$
$$= -(2.035e_{\text{eff}}^{\text{p}} + 1.978e_{\text{eff}}^{\text{n}})b^2\,. \tag{16.103}$$

Equations (3.43) and (3.45) give the oscillator length $b = 1.813\,\text{fm}$, and we express the effective charges in the usual form $e_{\text{eff}}^{\text{p}} = (1 + \chi)e$, $e_{\text{eff}}^{\text{n}} = \chi e$. The transition amplitude (16.103) now becomes

$$(\text{BCS}\|\boldsymbol{Q}_2\|2_1^+) = -(6.69 + 13.19\chi)e\,\text{fm}^2\,. \tag{16.104}$$

This gives the reduced transition probability

$$B(\text{E2}; 2_1^+ \to 0_{\text{gs}}^+) = \frac{1}{5}(6.69 + 13.19\chi)^2\,e^2\text{fm}^4\,. \tag{16.105}$$

The experimental half-life of the $2_1^+$ state is $1.44\,\text{ps}$ and its excitation energy is $1.3686\,\text{MeV}$. From Table 6.8 we find

$$B(\text{E2}; 2_1^+ \to 0_{\text{gs}}^+)_{\text{exp}} = 82.0\,e^2\text{fm}^4\,. \tag{16.106}$$

This value is reproduced by choosing

$$\chi = 1.03 \tag{16.107}$$

in (16.105). This large electric polarization constant indicates that a larger valence space is needed for a reasonable theoretical description.

**The 0d-1s Valence Space**

We now repeat the previous calculation in the complete 0d-1s valence space. Table 16.5 gives the pairing parameters and QTDA strength parameters also for this case. The proton basis states for $J^\pi = 2^+$ are

$$\{|\pi_1\rangle, |\pi_2\rangle, |\pi_3\rangle, |\pi_4\rangle, |\pi_5\rangle\}$$
$$= \{|(\pi 0d_{5/2})^2; 2^+\rangle, |\pi 0d_{5/2}\,\pi 1s_{1/2}; 2^+\rangle, |\pi 0d_{5/2}\,\pi 0d_{3/2}; 2^+\rangle,$$
$$|\pi 1s_{1/2}\,\pi 0d_{3/2}; 2^+\rangle, |(\pi 0d_{3/2})^2; 2^+\rangle\}, \qquad (16.108)$$

and the neutron basis states are labelled similarly. The QTDA matrix is constructed the same way as in Subsect. 16.4.2. Diagonalization then gives the lowest $2^+$ state as

$$|^{24}\mathrm{Mg}; 2_1^+\rangle_{\mathrm{QTDA}} = 0.409|\pi_1\rangle + 0.522|\pi_2\rangle + 0.147|\pi_3\rangle + 0.232|\pi_4\rangle$$
$$+ 0.110|\pi_5\rangle + 0.385|\nu_1\rangle + 0.493|\nu_2\rangle + 0.141|\nu_3\rangle$$
$$+ 0.223|\nu_4\rangle + 0.106|\nu_5\rangle. \qquad (16.109)$$

The BCS calculation has produced the $u$ and $v$ occupation amplitudes that modify and extend those in Table 16.9. With these amplitudes we proceed as in (16.103) and obtain

$$(\mathrm{BCS}\|Q_2\|2_1^+) = -(2.451 e_{\mathrm{eff}}^{\mathrm{p}} + 2.325 e_{\mathrm{eff}}^{\mathrm{n}})b^2, \qquad (16.110)$$

whence

$$B(\mathrm{E2};\, 2_1^+ \to 0_{\mathrm{gs}}^+) = \frac{1}{5}(8.06 + 15.70\chi)^2\, e^2\mathrm{fm}^4. \qquad (16.111)$$

Fitting the experimental value (16.106) gives the electric polarization constant

$$\chi = 0.78. \qquad (16.112)$$

This is notably smaller than 1.03 in (16.107) but still large. Smaller, more realistic values of $\chi$ can be obtained when QRPA wave functions are introduced in Chap. 18.

### 16.5.3 Collective States and Electric Transitions

Let us examine properties of E2 transitions from $2^+$ states to the ground state in even–even nuclei. We start by continuing with the examples of the previous subsection. The results are from the same calculations. However, instead of using the fitted polarization constants (16.107) and (16.112), we here choose $\chi = 0.3$, leading to the effective charges $e_{\mathrm{eff}}^{\mathrm{p}} = 1.3e$ and $e_{\mathrm{eff}}^{\mathrm{n}} = 0.3e$. Such effective charges are generally considered reasonable.

All values of $B(\mathrm{E2};\, 2_n^+ \to 0_{\mathrm{gs}}^+)$ for $^{24}_{12}\mathrm{Mg}_{12}$ are displayed in Fig. 16.5, with panel (a) for the $0d_{5/2}$-$1s_{1/2}$ valence space and panel (b) for the complete $0d$-$1s$ valence space. Figure 16.5 shows not only the $B(\mathrm{E2})$ values resulting from the multi-component QTDA wave functions but also the values for pure BCS states of two quasiparticles. They are labelled by $n = 1$–$4$ according to (16.101) for the small space and by $n = 1$–$10$ according to (16.109) for the large space. From the figure we see that the BCS strength is radically

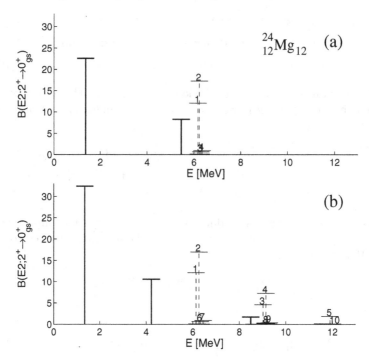

**Fig. 16.5.** $B(E2)$ values for $^{24}$Mg in units of $e^2\text{fm}^4$. The solid bars give the QTDA results. The dashed bars give two-quasiparticle BCS values at the unperturbed energies; the numbering is explained in the text. The valence space for panel (a) is $0d_{5/2}$-$1s_{1/2}$ and that for panel (b) is $0d$-$1s$. The model parameters are from Table 16.5. The electric polarization constant is $\chi = 0.3$

redistributed by the QTDA procedure. The lowest transition collects most of the strength, leaving only a small portion for the high-energy transitions.

From (16.102) and (16.103) we see that the $X$ amplitudes of the lowest $2^+$ state, as well as the BCS amplitudes $u$ and $v$ of the valence space, are all positive while the single-particle matrix elements are all negative. Such coherence is a general feature of the lowest $2^+$ state but does not occur for the higher states. The coherence produces a large $B(E2)$ value, which signifies collective behaviour of the nucleons.

The value $B(E2\,;\,2_1^+ \to 0_{gs}^+) = 32.4\,e^2\text{fm}^4$ in Fig. 16.5(b), obtained with the reasonable polarization constant $\chi = 0.3$, is still only 40 % of the experimental value (16.106). This deficiency is remedied in Chap. 18 by the use of QRPA wave functions. They contain ground-state correlations that lead to enhancement of collective transitions. Such enhancement was observed when comparing TDA and RPA results in Chap. 11.

The nucleus $^{30}_{14}$Si$_{16}$ provides our next example. This nucleus was discussed in Subsect. 16.4.5, with its energy spectrum in Fig. 16.3. Like Fig. 16.5, Fig. 16.6 presents the E2 transitions from the $2^+$ states to the ground state.

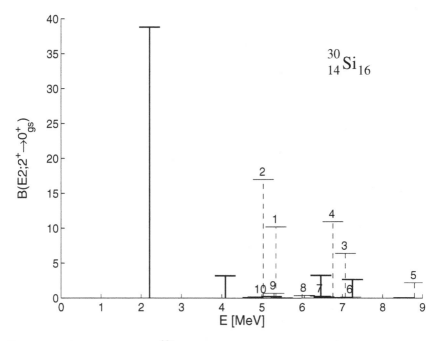

**Fig. 16.6.** E2 transitions in $^{30}$Si presented as in Fig. 16.5. The valence space is 0d-1s, the model parameters are from Table 16.6, and the polarization constant is $\chi = 0.3$

The calculations were performed with the model parameters of Table 16.6. As in the case of $^{24}$Mg, the polarization constant was taken to be $\chi = 0.3$.

We compare the computed and experimental values of $B(\text{E2}\,;\,2^+_1 \to 0^+_{\text{gs}})$. The experimental half-life 0.24 ps and energy 2.2355 MeV of the $2^+_1$ state give

$$B(\text{E2}\,;\,2^+_1 \to 0^+_{\text{gs}})_{\text{exp}} = 42\,e^2\text{fm}^4\,. \tag{16.113}$$

Read off Fig. 16.6, the computed value is $B(\text{E2}) \approx 39\,e^2\text{fm}^4$, so we have good agreement between theory and experiment. However, with the QRPA we can expect good agreement with a smaller value of $\chi$. As in $^{24}$Mg, the QTDA redistributes the E2 strength from the pure two-quasiparticle case in favour of the lowest-lying $2^+$ state.

Our final example is $^{66}_{30}\text{Zn}_{36}$, whose energy spectrum was discussed in Subsect. 16.4.6. Figure 16.7 shows the values of $B(\text{E2}\,;\,2^+_n \to 0^+_{\text{gs}})$ for the valence spaces 0f-1p (a) and 0f-1p-0g$_{9/2}$ (b). The parameters are from Table 16.7, and the chosen polarization constant is $\chi = 0.5$. The two-quasiparticle components for protons are numbered as

$$\{|\pi_1\rangle\,,\,|\pi_2\rangle\,,\,|\pi_3\rangle\,,\,|\pi_4\rangle\,,\,|\pi_5\rangle\,,\,|\pi_6\rangle\,,\,|\pi_7\rangle\,,\,|\pi_8\rangle\,,\,|\pi_9\rangle\}$$
$$= \{|(\pi 0\text{f}_{7/2})^2\,;\,2^+\rangle\,,\,|\pi 0\text{f}_{7/2}\,\pi 0\text{p}_{3/2}\,;\,2^+\rangle\,,\,|\pi 0\text{f}_{7/2}\,\pi 0\text{f}_{5/2}\,;\,2^+\rangle\,,$$

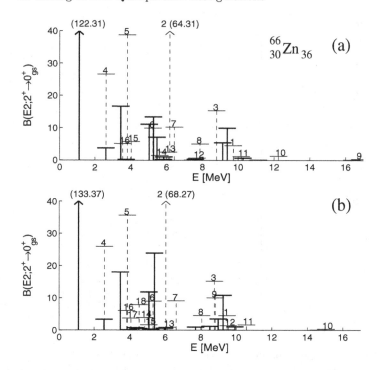

**Fig. 16.7.** E2 transitions in $^{66}$Zn presented as in Fig. 16.5. The valence space for panel (a) is 0f-1p and that for panel (b) is 0f-1p-0g$_{9/2}$. The model parameters are from Table 16.7, and the polarization constant is $\chi = 0.5$. The numbering of the two-quasiparticle states is explained in the text

$$\{ |\pi 1p_{3/2}\, \pi 1p_{3/2}\,; 2^+\rangle,\ |\pi 1p_{3/2}\, \pi 1p_{1/2}\,; 2^+\rangle,\ |\pi 1p_{3/2}\, \pi 0f_{5/2}\,; 2^+\rangle,$$

$$|\pi 1p_{1/2}\, \pi 0f_{5/2}\,; 2^+\rangle,\ |\pi 0f_{5/2}\, \pi 0f_{5/2}\,; 2^+\rangle,\ |\pi 0g_{9/2}\, \pi 0g_{9/2}\,; 2^+\rangle \}\ . \tag{16.114}$$

The corresponding neutron states are numbered 10–18. As in the previous examples, the QTDA brings a strong concentration of E2 strength on the lowest-lying $2^+$ state. That strength is greater in the large valence space than in the small, although the relative difference is less than it is in the case of $^{24}$Mg.

From the experimental excitation energy $E(2_1^+) = 1.0394\,\mathrm{MeV}$ and decay half-life 1.6 ps we find

$$B(\mathrm{E2}\,;\, 2_1^+ \to 0_{\mathrm{gs}}^+)_{\mathrm{exp}} = 290\, e^2\mathrm{fm}^4\ . \tag{16.115}$$

In spite of the large polarization constant used, the computed values are much smaller. As noted before, the situation is remedied by the QRPA approach.

## 16.6 QTDA Sum Rule for Electromagnetic Transitions

In this section we discuss the NEWSR for the QTDA. This sum rule is analogous to the one in Subsect. 9.4.2 for the particle–hole TDA. First we derive the formalism and then present some applications.

### 16.6.1 Formalism

Consider the sum over all $n$ of the quantities $|(n\,\lambda\|\mathcal{M}_{\sigma\lambda}\|\mathrm{BCS})|^2$ within the QTDA. Using the result (15.57) we obtain

$$\sum_n |(n\,\lambda\|\mathcal{M}_{\sigma\lambda}\|\mathrm{BCS})|^2 = \sum_n |(\mathrm{BCS}\|\mathcal{M}_{\sigma\lambda}\|n\,\lambda)|^2$$

$$= \sum_n \left| \sum_{a\leq b} X_{ab}^{n\lambda} \mathcal{N}_{ab}(\lambda)\theta^{(l_b)}(v_a|u_b| \pm v_b|u_a|)(a\|\mathcal{M}_{\sigma\lambda}\|b) \right|^2$$

$$= \sum_n \sum_{\substack{a\leq b \\ a'\leq b'}} X_{ab}^{n\lambda} X_{a'b'}^{n\lambda*} \mathcal{N}_{ab}(\lambda)\mathcal{N}_{a'b'}(\lambda)\theta^{(l_b)}\theta^{(l_{b'})}(v_a|u_b| \pm v_b|u_a|)$$

$$\times (v_{a'}|u_{b'}| \pm v_{b'}|u_{a'}|)(a\|\mathcal{M}_{\sigma\lambda}\|b)(a'\|\mathcal{M}_{\sigma\lambda}\|b')^* , \quad (16.116)$$

where the upper signs are for E$\lambda$ and the lower signs for M$\lambda$ transitions. Substituting here the completeness relation (16.61) gives

$$\boxed{\sum_n |(n\,\lambda\|\mathcal{M}_{\sigma\lambda}\|\mathrm{BCS})|^2 = \sum_{a\leq b} |\mathcal{N}_{ab}(\lambda)(v_a|u_b| \pm v_b|u_a|)(a\|\mathcal{M}_{\sigma\lambda}\|b)|^2 .}$$

$$(16.117)$$

The relation (16.117) is the NEWSR of the QTDA. Except for the quasiparticle properties, it is like the NEWSR (9.103) of the particle–hole TDA. The left-hand side depends on the structure of the QTDA states, whereas the right-hand side depends only on the single-particle properties of the quasiparticles and the single-particle matrix elements. The NEWSR can be used to check a QTDA calculation of electromagnetic transition rates.

The squared amplitude on the left-hand side of (16.117) is called the transition strength to the $|n\,\lambda\rangle$ state. It is the same as the reduced transition probability $B(\mathrm{E}\lambda)$ to that state. Because of the definition (6.4) it is $2\lambda + 1$ times greater than the $B(\mathrm{E}\lambda)$ value in the opposite direction, which is the actual direction of decay.

For electric transitions the sum rule (16.117) can be expressed in a form displaying the effective charges, in analogy to (9.104). With the abbreviation

$$d_{ab}^{(\lambda)} \equiv \mathcal{N}_{ab}(\lambda)(v_a|u_b| + v_b|u_a|)(a\|\mathbf{Q}_\lambda\|b) , \quad (16.118)$$

Eq. (16.117) gives

$$\boxed{\sum_k \left| (k\,\lambda \| Q_\lambda \| \mathrm{BCS}) \right|^2 = (e_{\mathrm{eff}}^{\mathrm{p}}/e)^2 \sum_{p \le p'} |d_{pp'}^{(\lambda)}|^2 + (e_{\mathrm{eff}}^{\mathrm{n}}/e)^2 \sum_{n \le n'} |d_{nn'}^{(\lambda)}|^2 \, .}$$

(16.119)

Below we apply this form of the sum rule to our previous examples of E2 transitions.

### 16.6.2 Examples of the NEWSR in the 0d-1s and 0f-1p-0g$_{9/2}$ Shells

We test the NEWSR on our previous examples in the 0d-1s and 0f-1p-0g$_{9/2}$ shells. The E2 strength distribution is given for $^{24}$Mg in Tables 16.10 and 16.11, and for $^{66}$Zn in Table 16.12. The first column labelled 'Strength' in the tables lists individual terms of the right-hand side of (16.119) while the second column so labelled lists $k$ contributions from the left-hand side.

**Table 16.10.** Two-quasiparticle configurations, their energies and their E2 strengths for $^{24}$Mg

| Configuration | $E_{2\mathrm{qp}}$ (MeV) | Strength $(e^2\mathrm{fm}^4)$ | $k$ | $E_k$ (MeV) | Strength $(e^2\mathrm{fm}^4)$ |
|---|---|---|---|---|---|
| $\pi 0d_{5/2}\pi 0d_{5/2}$ | 6.170 | 60.39 | 1 | 1.374 | 112.88 |
| $\pi 0d_{5/2}\pi 1s_{1/2}$ | 6.231 | 86.25 | 2 | 5.453 | 41.40 |
| $\nu 0d_{5/2}\nu 0d_{5/2}$ | 6.290 | 3.22 | 3 | 6.191 | 0.09 |
| $\nu 0d_{5/2}\nu 1s_{1/2}$ | 6.351 | 4.59 | 4 | 6.311 | 0.08 |
| Sum | | 154.5 | | | 154.5 |

Column four labels the QTDA states, column five gives their energies and column six the E2 decay strengths to them. The calculation was done in the 0d$_{5/2}$-1s$_{1/2}$ valence space with details reported in Subsect. 16.5.3.

We can conclude from the tables that

- the NEWSR is satisfied,
- the dominant contribution to QTDA strength comes from the lowest-lying 2$^+$ state,
- the neutron two-quasiparticle contributions are much smaller than the proton ones because of the small effective charge of the neutrons.

Tables 16.10, 16.11 and 16.12, respectively, essentially repeat the information in Figs. 16.5(a), 16.5(b) and 16.7(b). Only the strengths in the tables are to be divided by 5 to get the $B(\mathrm{E2})$ values in the figures. Also it is not quantitatively clear from the figures that the NEWSR is satisfied.

**Table 16.11.** Two-quasiparticle configurations, their energies and their E2 strengths for $^{24}$Mg

| Configuration | $E_{2\mathrm{qp}}$ (MeV) | Strength $(e^2\mathrm{fm}^4)$ | $k$ | $E_k$ (MeV) | Strength $(e^2\mathrm{fm}^4)$ |
|---|---|---|---|---|---|
| $\pi 0d_{5/2}\pi 0d_{5/2}$ | 6.143 | 60.68 | 1 | 1.357 | 162.08 |
| $\pi 0d_{5/2}\pi 1s_{1/2}$ | 6.267 | 84.95 | 2 | 4.221 | 53.07 |
| $\pi 0d_{5/2}\pi 0d_{3/2}$ | 9.018 | 22.95 | 3 | 6.186 | 0.13 |
| $\pi 1s_{1/2}\pi 0d_{3/2}$ | 9.142 | 36.41 | 4 | 6.362 | 0.03 |
| $\pi 0d_{3/2}\pi 0d_{3/2}$ | 11.893 | 9.44 | 5 | 8.493 | 8.66 |
| $\nu 0d_{5/2}\nu 0d_{5/2}$ | 6.318 | 3.23 | 6 | 8.640 | 0.04 |
| $\nu 0d_{5/2}\nu 1s_{1/2}$ | 6.443 | 4.52 | 7 | 9.112 | 1.31 |
| $\nu 0d_{5/2}\nu 0d_{3/2}$ | 9.164 | 1.23 | 8 | 9.266 | 0.27 |
| $\nu 1s_{1/2}\nu 0d_{3/2}$ | 9.289 | 1.96 | 9 | 11.577 | 0.33 |
| $\nu 0d_{3/2}\nu 0d_{3/2}$ | 12.010 | 0.52 | 10 | 11.697 | 0.00 |
| Sum | | 225.9 | | | 225.9 |

Column four labels the QTDA states, column five gives their energies and column six the E2 decay strengths to them. The calculation was done in the 0d-1s valence space with details reported in Subsect. 16.5.3.

## 16.7 Transitions Between QTDA Excited States

In this section we consider electric and magnetic transitions between two excited QTDA states of the form (16.55). The transition amplitudes are analogous to those of the particle–hole TDA in Subsect. 9.4.7.

### 16.7.1 Transition Amplitudes

The reduced matrix element for an electromagnetic transition between two two-quasiparticle states was derived in Subsect. 15.5.1. The result is given in (15.99) in its general form; in (15.100) it is specified to final angular momentum $J_f = 0$.

From (16.55), the initial and final QTDA wave functions are

$$|\omega_i\rangle = Q_{\omega_i}^\dagger|\mathrm{BCS}\rangle = \sum_{a_i \leq b_i} X_{a_i b_i}^{\omega_i} A_{a_i b_i}^\dagger (J_i M_i)|\mathrm{BCS}\rangle , \qquad (16.120)$$

$$|\omega_f\rangle = Q_{\omega_f}^\dagger|\mathrm{BCS}\rangle = \sum_{a_f \leq b_f} X_{a_f b_f}^{\omega_f} A_{a_f b_f}^\dagger (J_f M_f)|\mathrm{BCS}\rangle . \qquad (16.121)$$

The transition amplitude between these states is

$$\boxed{(\omega_f\|\mathcal{M}_{\sigma\lambda}\|\omega_i)_{\mathrm{QTDA}} = \sum_{\substack{a_i \leq b_i \\ a_f \leq b_f}} X_{a_f b_f}^{\omega_f *} X_{a_i b_i}^{\omega_i} (a_f b_f ; J_f\|\mathcal{M}_{\sigma\lambda}\|a_i b_i ; J_i) ,}$$

$$(16.122)$$

where the two-quasiparticle matrix element (15.99) appears on the right-hand side. We next apply (16.122) to an example.

**Table 16.12.** Two-quasiparticle configurations, their energies and their E2 strengths for $^{66}$Zn

| Configuration | $E_{2qp}$ (MeV) | Strength $(e^2\text{fm}^4)$ | $k$ | $E_k$ (MeV) | Strength $(e^2\text{fm}^4)$ |
|---|---|---|---|---|---|
| $\pi 0f_{7/2}\pi 0f_{7/2}$ | 9.495 | 22.99 | 1 | 1.106 | 666.83 |
| $\pi 0f_{7/2}\pi 1p_{3/2}$ | 6.042 | 341.51 | 2 | 2.568 | 16.71 |
| $\pi 0f_{7/2}\pi 0f_{5/2}$ | 8.781 | 76.53 | 3 | 3.479 | 89.95 |
| $\pi 1p_{3/2}\pi 1p_{3/2}$ | 2.589 | 129.96 | 4 | 3.962 | 0.73 |
| $\pi 1p_{3/2}\pi 1p_{1/2}$ | 3.877 | 178.76 | 5 | 4.328 | 3.81 |
| $\pi 1p_{3/2}\pi 0f_{5/2}$ | 5.328 | 45.47 | 6 | 4.529 | 2.73 |
| $\pi 1p_{1/2}\pi 0f_{5/2}$ | 6.616 | 45.95 | 7 | 5.098 | 59.17 |
| $\pi 0f_{5/2}\pi 0f_{5/2}$ | 8.066 | 22.93 | 8 | 5.191 | 1.71 |
| $\pi 0g_{9/2}\pi 0g_{9/2}$ | 8.829 | 50.55 | 9 | 5.404 | 119.46 |
| $\nu 0f_{7/2}\nu 0f_{7/2}$ | 15.055 | 1.88 | 10 | 5.942 | 0.04 |
| $\nu 0f_{7/2}\nu 1p_{3/2}$ | 10.590 | 8.68 | 11 | 5.969 | 3.04 |
| $\nu 0f_{7/2}\nu 0f_{5/2}$ | 9.567 | 7.27 | 12 | 7.715 | 4.77 |
| $\nu 1p_{3/2}\nu 1p_{3/2}$ | 6.126 | 4.77 | 13 | 8.525 | 6.25 |
| $\nu 1p_{3/2}\nu 1p_{1/2}$ | 4.869 | 19.55 | 14 | 8.996 | 17.51 |
| $\nu 1p_{3/2}\nu 0f_{5/2}$ | 5.103 | 8.45 | 15 | 9.275 | 54.11 |
| $\nu 1p_{1/2}\nu 0f_{5/2}$ | 3.846 | 30.32 | 16 | 9.444 | 0.50 |
| $\nu 0f_{5/2}\nu 0f_{5/2}$ | 4.080 | 18.44 | 17 | 9.944 | 5.40 |
| $\nu 0g_{9/2}\nu 0g_{9/2}$ | 4.552 | 39.10 | 18 | 14.644 | 0.09 |
| Sum | | 1053 | | | 1053 |

Column four labels the QTDA states, column five gives their energies and column six gives the E2 decay strengths to them. The calculation was done in the $0f-1p-0g_{9/2}$ valence space with details reported in Subsection 16.5.3.

## 16.7.2 Example: The $0_1^+ \rightarrow 2_1^+$ Transition in $^{24}$Mg

Consider an E2 transition between the first excited $0^+$ state and the first $2^+$ state in $^{24}$Mg. The experimental data for this transition are shown in Fig. 16.8. We assume QTDA wave functions calculated in the $0d_{5/2}$-$1s_{1/2}$ valence space with the parameters of Table 16.5. The resulting level scheme is shown on the left in Fig. 16.2. The initial and final wave functions are

$$|^{24}\text{Mg}; 0_1^+\rangle = 0.629|\pi_1\rangle_0 + 0.340|\pi_2\rangle_0 + 0.615|\nu_1\rangle_0 + 0.334|\nu_2\rangle_0 , \quad (16.123)$$

$$|^{24}\text{Mg}; 2_1^+\rangle = 0.437|\pi_1\rangle_2 + 0.568|\pi_2\rangle_2 + 0.425|\nu_1\rangle_2 + 0.552|\nu_2\rangle_2 . \quad (16.124)$$

The $0^+$ basis states for protons are

$$|\pi_1\rangle_0 = |(\pi 0d_{5/2})^2; 0^+\rangle , \quad |\pi_2\rangle_0 = |(\pi 1s_{1/2})^2; 0^+\rangle \quad (16.125)$$

and similarly for neutrons. The proton $2^+$ basis states are

$$|\pi_1\rangle_2 = |(\pi 0d_{5/2})^2; 2^+\rangle , \quad |\pi_2\rangle_2 = |\pi 0d_{5/2}\,\pi 1s_{1/2}; 2^+\rangle \quad (16.126)$$

and similarly for neutrons. The $2^+$ state (16.124) is the one given by (16.101) and (16.102).

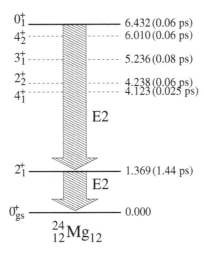

$0_1^+$ ——————— 6.432 (0.06 ps)
$4_2^+$ - - - - - - - - - - - - 6.010 (0.06 ps)

$3_1^+$ - - - - - - - - - - - - 5.236 (0.08 ps)

$2_2^+$ - - - - - - - - - - - - 4.238 (0.06 ps)
$4_1^+$ - - - - - - - - - - - - 4.123 (0.025 ps)

E2

$2_1^+$ ——————— 1.369 (1.44 ps)

E2

$0_{gs}^+$ ——————— 0.000

$_{12}^{24}\text{Mg}_{12}$

**Fig. 16.8.** Experimental low-energy spectrum of $^{24}$Mg. Measured half-lives are given. The main decay cascade of the first excited $0^+$ state is shown with arrows

Equation (15.100) gives the transition amplitude between two-quasiparticle states when the final state has $J = 0$. In the present case it is the initial state that has $J = 0$. We use the CS phase convention, and then the transition amplitude is unchanged because of the symmetry property (2.32). Applied to the $2^+ \to 0^+$ transition, (15.100) gives

$$((a_f)^2 \,;\, 0^+ \| \boldsymbol{Q}_2 \| a_i \, b_i \,;\, 2^+) = \sqrt{2} \, \widehat{j_{a_f}}^{-1} \mathcal{N}_{a_i b_i}(2)$$
$$\times \left[ \delta_{a_i a_f} (u_{b_i} u_{a_f} - v_{b_i} v_{a_f})(a_f \| \boldsymbol{Q}_2 \| b_i) \right.$$
$$\left. - \delta_{b_i a_f} (u_{a_i} u_{a_f} - v_{a_i} v_{a_f})(-1)^{j_{a_i} + j_{a_f}} (a_f \| \boldsymbol{Q}_2 \| a_i) \right] . \quad (16.127)$$

The BCS occupation amplitudes are listed in Table 16.9 and the single-particle matrix elements in Table 6.4. The oscillator length is $b = 1.813\,\text{fm}$.

It is useful to produce a table of the relevant quantities to sort out the contributions to the sum (16.122). This is done in Table 16.13. The table lists the contributing initial and final configurations in columns one and two and their QTDA amplitudes in columns three and four. Furthermore, we divide the two-quasiparticle transition matrix element (16.127) into two parts according to

$$((a_f)^2 \,;\, 0^+ \| \boldsymbol{Q}_2 \| a_i \, b_i \,;\, 2^+) = \sqrt{2} \, \widehat{j_{a_f}}^{-1} \mathcal{N}_{a_i b_i}(2)(M_1 - M_2) . \quad (16.128)$$

Here $M_1$ and $M_2$ are the two terms in the square brackets in (16.127), and they are tabulated in columns five and six of Table 16.13. The last column lists the total individual contributions to the sum (16.122).

The QTDA transition amplitude (16.122) is the sum of the entries in the last column of Table 16.13, so that

**Table 16.13.** Contributions to the sum (16.122) for the transition $0_1^+ \rightarrow 2_1^+$ in $^{24}\mathrm{Mg}$

| $a_i b_i (2^+)$ | $(a_f)^2(0^+)$ | $X_{a_i b_i}^{\omega_i}$ | $X_{a_f b_f}^{\omega_f}$ | $M_1\,(\mathrm{fm}^2)$ | $M_2\,(\mathrm{fm}^2)$ | Total $(\mathrm{fm}^2)$ |
|---|---|---|---|---|---|---|
| $\pi 0d_{5/2}\pi 0d_{5/2}$ | $(\pi 0d_{5/2})^2$ | 0.437 | 0.629 | $0.590e_{\mathrm{eff}}^{\mathrm{p}}$ | $-0.590e_{\mathrm{eff}}^{\mathrm{p}}$ | $0.132e_{\mathrm{eff}}^{\mathrm{p}}$ |
| $\pi 0d_{5/2}\pi 0d_{5/2}$ | $(\pi 1s_{1/2})^2$ | 0.437 | 0.340 | 0 | 0 | 0 |
| $\pi 0d_{5/2}\pi 1s_{1/2}$ | $(\pi 0d_{5/2})^2$ | 0.568 | 0.629 | $-0.504e_{\mathrm{eff}}^{\mathrm{p}}$ | 0 | $-0.104e_{\mathrm{eff}}^{\mathrm{p}}$ |
| $\pi 0d_{5/2}\pi 1s_{1/2}$ | $(\pi 1s_{1/2})^2$ | 0.568 | 0.340 | 0 | $0.504e_{\mathrm{eff}}^{\mathrm{p}}$ | $-0.097e_{\mathrm{eff}}^{\mathrm{p}}$ |
| $\nu 0d_{5/2}\nu 0d_{5/2}$ | $(\nu 0d_{5/2})^2$ | 0.425 | 0.615 | $0.579e_{\mathrm{eff}}^{\mathrm{n}}$ | $-0.579e_{\mathrm{eff}}^{\mathrm{n}}$ | $0.124e_{\mathrm{eff}}^{\mathrm{n}}$ |
| $\nu 0d_{5/2}\nu 0d_{5/2}$ | $(\nu 1s_{1/2})^2$ | 0.425 | 0.334 | 0 | 0 | 0 |
| $\nu 0d_{5/2}\nu 1s_{1/2}$ | $(\nu 0d_{5/2})^2$ | 0.552 | 0.615 | $-0.494e_{\mathrm{eff}}^{\mathrm{n}}$ | 0 | $-0.097e_{\mathrm{eff}}^{\mathrm{n}}$ |
| $\nu 0d_{5/2}\nu 1s_{1/2}$ | $(\nu 1s_{1/2})^2$ | 0.552 | 0.334 | 0 | $0.494e_{\mathrm{eff}}^{\mathrm{n}}$ | $-0.091e_{\mathrm{eff}}^{\mathrm{n}}$ |

The initial and final configurations and their QTDA amplitudes are listed in the first four columns. The two terms in (16.128) are given in columns five and six. The total contribution of the pair of configurations is given in the last column.

$$
\begin{aligned}
(0_1^+\|\boldsymbol{Q}_2\|2_1^+) &= (0.132 - 0.104 - 0.097)e_{\mathrm{eff}}^{\mathrm{p}}\,\mathrm{fm}^2 \\
&\quad + (0.124 - 0.097 - 0.091)e_{\mathrm{eff}}^{\mathrm{n}}\,\mathrm{fm}^2 \\
&= -0.069e_{\mathrm{eff}}^{\mathrm{p}}\,\mathrm{fm}^2 - 0.064e_{\mathrm{eff}}^{\mathrm{n}}\,\mathrm{fm}^2 \\
&= -(0.069 + 0.133\chi)e\,\mathrm{fm}^2 \ .
\end{aligned} \tag{16.129}
$$

In subsect. 16.5.2 we found $\chi = 1.03$ by fitting $B(\mathrm{E2}\,; 2_1^+ \rightarrow 0_{\mathrm{gs}}^+)$. This $\chi$ gives

$$
(0_1^+\|\boldsymbol{Q}_2\|2_1^+) = -0.206e\,\mathrm{fm}^2 \ , \tag{16.130}
$$

whence

$$
B(\mathrm{E2}\,; 0_1^+ \rightarrow 2_1^+) = 0.042\,e^2\mathrm{fm}^4 \ . \tag{16.131}
$$

As shown in Fig. 16.8, the experimental half-life of the $0_1^+$ state is 0.06 ps. The main decay branch is the direct transition to the $2_1^+$ state, so that this transition essentially determines the value of the half-life. With the experimental transition energy of 5.063 MeV, Table 6.8 gives

$$
B(\mathrm{E2}\,; 0_1^+ \rightarrow 2_1^+)_{\mathrm{exp}} = 3\,e^2\mathrm{fm}^4 \ . \tag{16.132}
$$

We see that the computed transition is two orders of magnitude too slow. The discrepancy can come from several sources. The two-quasiparticle terms (16.127) are sensitive to the BCS occupation amplitudes. The computed transition rates are sensitive to the choices of model space, single-particle energies and pairing strengths. Small variations in them can cause large variations in the decay rates.

## Epilogue

This chapter constitutes the first level of handling quasiparticle configuration mixing. The formalism, the quasiparticle TDA, is analogous to the particle–hole TDA and describes excitations in even–even nuclei. Similarities between these two formalisms carry over into their results, in particular as regards collective states.

In the following chapter we extend our investigations to proton–neutron two-quasiparticle excitations and their mixing. This enables the description of states and transitions in odd–odd nuclei.

## Exercises

**16.1.** Derive (16.4) from (16.3).

**16.2.** Give a detailed derivation of (16.5).

**16.3.** Derive the identity (16.9).

**16.4.** Complete the details leading to the expression (16.10).

**16.5.** Derive the commutator relation (16.26).

**16.6.** Derive the commutator result (16.33).

**16.7.** Prove the condition (16.52).

**16.8.** By diagonalizing the matrix (16.79) verify the values of the eigenenergies (16.80).

**16.9.** Derive the expressions (16.85) and (16.86) for the proton and neutron separation energies.

**16.10.** Derive the expressions (16.88) and (16.89) for the pairing gaps $\Delta_{\mathrm{p}}(A, Z)$ and $\Delta_{\mathrm{n}}(A, Z)$.

**16.11.** Verify the values of the pairing gaps (16.96) and (16.97).

**16.12.** Verify the values of the pairing gaps (16.98).

**16.13.** Derive the elements of the matrix (16.100).

**16.14.** By diagonalizing the matrix (16.100) verify the wave function given by (16.101) and (16.102).

**16.15.** Derive the $B(\mathrm{E}2)$ values (16.106), (16.113) and (16.115) from the experimental data provided.

**16.16.** Consider $2^+$ excitations in $^{20}$Ne. You may use the BCS results of Table 14.1 computed for the complete 0d-1s valence space.

(a) Form the QTDA matrix for the $2^+$ states in the subset basis (dictated by the lowest quasiparticle energies) constructed from the $0\mathrm{d}_{5/2}$ and $1\mathrm{s}_{1/2}$ proton and neutron orbitals.

(b) Diagonalize the QTDA matrix to find the eigenenergies and eigenstates. Use the SDI with parameters $A_0 = A_1 = 1.0$ MeV. Compare the eigenenergies with experiment and comment.

**16.17.** Continue Exercise 16.16 and compute the reduced transition probability $B(E2\,;\,2_1^+ \to 0_{gs}^+)$ for $^{20}$Ne. By comparing with the experimental decay half-life determine the electric polarization constant $\chi$.

**16.18.** Consider $2^+$ excitations in $^{26}$Mg. You may use the BCS results of Table 14.1 computed for the complete 0d-1s valence space.

(a) Form the QTDA matrix for the $2^+$ states in the subset basis (dictated by the lowest quasiparticle energies) constructed from the $0d_{5/2}$ and $1s_{1/2}$ proton orbitals and the $1s_{1/2}$ neutron orbital.
(b) Diagonalize the QTDA matrix to find the eigenenergies and eigenstates. Use the SDI with parameters $A_0 = A_1 = 1.0$ MeV. Compare the eigenenergies with experiment and comment.

**16.19.** Continuation of Exercise 16.18.

(a) Compute the reduced transition probability $B(E2\,;\,2_1^+ \to 0_{gs}^+)$ for $^{26}$Mg. Determine the electric polarization constant $\chi$ by using available experimental data.
(b) Compute the $B(E2\,;\,2_n^+ \to 0_{gs}^+)$ for the remaining values of $n$. Compute the summed E2 strength. Check that the QTDA sum rule is satisfied.

**16.20.** Continuation of Exercise 16.18.

(a) Compute the QTDA energy for the first $3^+$ state in $^{26}$Mg by using the SDI with parameters $A_0 = A_1 = 1.0$ MeV.
(b) Compute the decay half-life of the $3_1^+$ state by assuming that it decays to the first $2^+$ state by an M1 transition. Use the bare gyromagnetic ratios and the experimental gamma energy. Compare with experimental data and comment.

**16.21.** Consider $2^+$ excitations in $^{30}$Si. You may use the BCS results of Table 14.1 computed for the complete 0d-1s valence space.

(a) Form the QTDA matrix for the $2^+$ states in the subset basis (dictated by the lowest quasiparticle energies) constructed from the $0d_{5/2}$ and $1s_{1/2}$ proton orbitals and the $1s_{1/2}$ and $0d_{3/2}$ neutron orbitals.
(b) Diagonalize the QTDA matrix to find the eigenenergies and eigenstates. Use the SDI with parameters $A_0 = A_1 = 1.0$ MeV. Compare the eigenenergies with experiment and with the complete d-s calculation of Fig. 16.3, and comment.

**16.22.** Continuation of Exercise 16.21.

(a) Compute the wave function of the first $2^+$ state in $^{30}$Si and compare with the rudimentary first guess (15.74).

(b) Compute the reduced transition probability $B(E2\,;\, 2_1^+ \to 0_{\mathrm{gs}}^+)$ for $^{30}$Si. By comparing with the experimental decay half-life determine the electric polarization constant $\chi$.

**16.23.** Consider $2^+$ excitations in $^{34}$S. You may use the BCS results of Table 14.1 computed for the complete 0d-1s valence space.

(a) Form the QTDA matrix for the $2^+$ states in the subset basis (dictated by the lowest quasiparticle energies) constructed from the $1s_{1/2}$ and $0d_{3/2}$ proton orbitals and the $0d_{3/2}$ neutron orbital.
(b) Diagonalize the QTDA matrix to find the eigenenergies and eigenstates. Use the SDI with parameters $A_0 = A_1 = 1.0\,\mathrm{MeV}$. Compare the eigenenergies with experiment and comment.

**16.24.** Continuation of Exercise 16.23.

(a) Compute the reduced transition probability $B(E2\,;\, 2_1^+ \to 0_{\mathrm{gs}}^+)$ for $^{34}$S. Determine the electric polarization constant $\chi$ by using available experimental data.
(b) Compute the $B(E2\,;\, 2_n^+ \to 0_{\mathrm{gs}}^+)$ for the remaining values of $n$. Compute the summed E2 strength. Check that the QTDA sum rule is satisfied.

**16.25.** Continuation of Exercise 16.24.
Compute the reduced transition probability $B(E2\,;\, 2_2^+ \to 2_1^+)$ for $^{34}$S by using the polarization constant found in Exercise 16.24. Determine the decay half-life of the $2_2^+$ state by taking into account also its decay to the ground state. Use experimental gamma energies. Compare with experiment and comment.

**16.26.** Continuation of Exercise 16.23.

(a) Compute the excitation energy of the first $0^+$ state in $^{34}$S. Use the subset basis of Exercise 16.23 and the SDI with parameters $A_0 = 1.0\,\mathrm{MeV}$ and $A_1 = 0.0\,\mathrm{MeV}$.
(b) Determine the decay half-life of the $0_1^+$ state by assuming that it decays mainly to the $2_1^+$ state. Use the electric polarization constant of Exercise 16.24 and the experimental gamma energy. Compare with experiment and comment.

**16.27.** Consider $^{38}_{18}\mathrm{Ar}_{20}$ as a d-s-shell two-hole nucleus. Apply the quasiparticle formalism in the 0d-1s-0f$_{7/2}$ valence space. Use the single-particle energies $\varepsilon_{0d_{5/2}} = 0$, $\varepsilon_{1s_{1/2}} = 1.5\,\mathrm{MeV}$, $\varepsilon_{0d_{3/2}} = 4.0\,\mathrm{MeV}$ and $\varepsilon_{0f_{7/2}} = 7.0\,\mathrm{MeV}$. Do a BCS calculation for protons and neutrons and determine the SDI pairing parameters by comparing with the liquid-drop pairing-gap formula (16.91). Note that it is not safe to use separation energies to determine the empirical pairing gap at a major-shell closure.

**16.28.** Continuation of Exercise 16.27.

(a) Form the QTDA matrix for the $2^+$ states in $^{38}$Ar in the subset basis constructed from the $1s_{1/2}$ and $0d_{3/2}$ proton orbitals and the $0d_{3/2}$ and $0f_{7/2}$ neutron orbitals. Use the SDI with parameters $A_0 = A_1 = 1.0$ MeV.
(b) Diagonalize the QTDA matrix to find the eigenenergies and eigenstates. Compare the eigenenergies with experiment and with the two-hole calculation of Fig. 8.7, and comment.

**16.29.** Continuation of Exercise 16.28.

(a) Compute the reduced transition probability $B(E2\,;\, 2^+_1 \to 0^+_{gs})$ for $^{38}$Ar. Determine the electric polarization constant $\chi$ by using available experimental data.
(b) Compute the $B(E2\,;\, 2^+_n \to 0^+_{gs})$ for the remaining values of $n$. Compute the summed E2 strength. Check that the QTDA sum rule is satisfied.

**16.30.** Continuation of Exercise 16.27.

(a) Form the QTDA matrix for the $3^-$ states in the subset basis constructed from the $0d_{3/2}$ and $0f_{7/2}$ proton and neutron orbitals. Use the SDI with parameters $A_0 = A_1 = 1.0$ MeV.
(b) Diagonalize the QTDA matrix to find the eigenenergies and eigenstates. Compare the eigenenergies with experiment and comment.

**16.31.** Continuation of Exercise 16.30.

(a) Compute the reduced transition probabilities $B(E3\,;\, 3^-_n \to 0^+_{gs})$ for $^{38}$Ar. Use the electric polarization constant found in Exercise 16.29.
(b) Check that the QTDA sum rule is satisfied.

**16.32.** Continuation of Exercise 16.31.
Compute the decay half-life of the $3^-_1$ state in $^{38}$Ar by considering its decay branchings to the $0^+_{gs}$ and $2^+_1$ states. Use experimental gamma energies and the electric polarization constant determined in Exercise 16.29. Compare with experiment and comment.

**16.33.** Verify the numbers in Table 16.10.

**16.34.** Calculate the wave functions (16.123) and (16.124).

**16.35.** Verify the numbers in Table 16.13.

# Two-Quasiparticle Mixing in Odd–Odd Nuclei

## Prologue

In Chap. 16 the residual Hamiltonian was used to mix proton–proton and neutron–neutron two-quasiparticle configurations. The resulting wave functions described states in even–even open-shell nuclei. In this chapter we develop a corresponding formalism, the proton–neutron QTDA, for mixing proton–neutron two-quasiparticle configurations. This mixing produces wave functions that describe states in odd–odd open-shell nuclei. The quasiparticles are obtained from a BCS calculation for an even–even reference nucleus next to the odd–odd nucleus of interest.

Of the decay transitions we take up the particularly interesting case of charge-changing Gamow–Teller transitions to highly excited giant resonance states. The Gamow–Teller giant resonance (GTGR) region accounts for the lion's share of a sum rule for transitions from the even–even reference nucleus to the $1^+$ states of the two neighbouring odd–odd isobars.

## 17.1 The Proton–Neutron QTDA

In Chap. 16 the EOM method was used to derive both the QTDA and the pnQTDA, proton–neutron QTDA. By the number-parity principle of Subsect. 13.4.2 the QTDA describes excited states in the even–even reference nucleus for which we perform the BCS or LNBCS calculation. The associated two-quasiparticle configurations are of the proton–proton and neutron–neutron types.

In the pnQTDA the basic excitations are of the proton–neutron type and thus describe states of odd–odd nuclei, again following the number-parity principle. As discussed in the beginning of Sect. 16.3, the QTDA and the pnQTDA come from a variational principle because the *exact* ground state was used in (11.11).

We assume that the pnQTDA wave functions give a reasonable description of the two odd–odd isobars adjacent to the even–even reference nucleus. This assumption is supported by the structure of the BCS (or LNBCS) ground state. The number constraint condition (13.72) or (14.19) makes the particle numbers peak at the nucleon numbers of the reference nucleus.

### 17.1.1 Equation of Motion

The pnQTDA equations (16.49)–(16.51) were derived as a by-product of the QTDA equations. The eigenvalue equation is

$$\sum_{p'n'} A_{pn,p'n'} X_{p'n'}^{\omega} = E_{\omega} X_{pn}^{\omega} \,, \tag{17.1}$$

where $\omega = kJ^{\pi}$ contains the eigenvalue index $k$ and the spin–parity $J^{\pi}$. The matrix $\mathsf{A}$ is given by

$$\begin{aligned}
A_{pn,p'n'}(J) =\ & (E_p + E_n)\delta_{pp'}\delta_{nn'} \\
& + (u_p u_n u_{p'} u_{n'} + v_p v_n v_{p'} v_{n'})\langle p\,n\,;\,J|V|p'\,n'\,;\,J\rangle \\
& + (u_p v_n u_{p'} v_{n'} + v_p u_n v_{p'} u_{n'})\langle p\,n^{-1}\,;\,J|V_{\mathrm{RES}}|p'\,n'^{-1}\,;\,J\rangle \,.
\end{aligned} \tag{17.2}$$

In the Baranger notation (16.20) and (16.21), this is

$$\begin{aligned}
A_{pn,p'n'}(J) =\ & (E_p + E_n)\delta_{pp'}\delta_{nn'} - 2(u_p u_n u_{p'} u_{n'} + v_p v_n v_{p'} v_{n'})G(pnp'n'J) \\
& - 2(u_p v_n u_{p'} v_{n'} + v_p u_n v_{p'} u_{n'})F(pnp'n'J) \,.
\end{aligned} \tag{17.3}$$

The matrix is Hermitian, so with real elements it is symmetric. Numerical examples are given in Sect. 17.2.

The proton–neutron two-body matrix elements for (17.2) are decomposed into their isospin components according to (8.26). For the SDI these components are given by Tables 8.1–8.4. The particle–hole matrix elements for (17.2) are reduced to two-body matrix elements by the Pandya transformation (16.51),

$$\langle p\,n^{-1}\,;\,J|V_{\mathrm{RES}}|p'\,n'^{-1}\,;\,J\rangle = -\sum_{J'}\hat{J}'^{2}\left\{\begin{matrix} j_p & j_n & J \\ j_{p'} & j_{n'} & J' \end{matrix}\right\}\langle p\,n'\,;\,J'|V|p'\,n\,;\,J'\rangle \,. \tag{17.4}$$

Again (8.26) serves to decompose the two-body matrix elements into their isospin components. For the SDI the result is then

$$\langle p_1\,n_2^{-1}\,;\,J|V_{\mathrm{RES}}|p_3\,n_4^{-1}\,;\,J\rangle = A_1\mathcal{M}^{(2)}_{a_1 a_2 a_3 a_4}(J1) - A_0\mathcal{M}^{(2)}_{a_1 a_2 a_3 a_4}(J0) \tag{17.5}$$

with $\mathcal{M}^{(2)}_{a_1 a_2 a_3 a_4}(JT)$ defined in (16.68). The auxiliary matrix elements are listed in Tables 16.2 and 16.3 for the 0d-1s valence space. Note that the different structure of the particle–hole matrix elements (16.70) and (17.5) results in a sign difference on the right-hand side.

To end this subsection we consider the phase conventions. In analogy to (16.53) the pnQTDA matrix obeys the relation

$$A^{(\mathrm{BR})}_{pn,p'n'}(J) = (-1)^{\frac{1}{2}(l_{p'}+l_{n'}-l_p-l_n)} A^{(\mathrm{CS})}_{pn,p'n'}(J) \,, \qquad (17.6)$$

where CS denotes the Condon–Shortley and BR the Biedenharn–Rose phase convention. The phase differences show up in the solutions of (17.1) as different phases for the $X$ amplitudes. These phases, in turn, carry into the expressions for decay amplitudes, as shown in Sects. 17.3 and 17.4.

### 17.1.2 Properties of Solutions

The solutions of the pnQTDA equation (17.1) have orthonormality and completeness properties similar to those of the QTDA, derived in Sect. 16.3. The pnQTDA wave function is

$$|\omega\rangle = Q^\dagger_\omega|\mathrm{BCS}\rangle = \sum_{pn} X^\omega_{pn} A^\dagger_{pn}(JM)|\mathrm{BCS}\rangle \,, \quad A^\dagger_{pn}(JM) = \left[a^\dagger_p a^\dagger_n\right]_{JM} .$$

$$(17.7)$$

The derivations of the orthonormality and completeness conditions can be traced back to Sect. 16.3, with the simplifying difference that now there is no summation restriction. The pnQTDA orthonormality relation thus becomes

$$\sum_{pn} X^{kJ^\pi *}_{pn} X^{k'J^\pi}_{pn} = \delta_{kk'} \quad (\mathrm{pnQTDA\ orthonormality}) \,, \qquad (17.8)$$

which is like (16.57) except for the summation difference. For the completeness relation of the pnQTDA we have

$$\sum_k X^{kJ^\pi}_{pn} X^{kJ^\pi *}_{p'n'} = \delta_{pp'}\delta_{nn'} \quad (\mathrm{pnQTDA\ completeness}) \,, \qquad (17.9)$$

which is exactly like (16.61).

## 17.2 Excitation Spectra of Open–Shell Odd–Odd Nuclei

By the number-parity principle of Subsect. 13.4.2, states of odd–odd nuclei can be described by proton–neutron two-quasiparticle configurations and their mixing. This means that the pnQTDA is applicable to open-shell odd–odd nuclei. The matrix structure is that of the pnTDA, given in (10.7). In detail, the pnQTDA matrix (17.2) differs from the pnTDA matrix (10.7) in three ways: the particle–hole energies are replaced by two-quasiparticle energies; not only the particle–hole matrix elements but also the two-particle matrix elements are present; and the BCS occupation amplitudes enter.

Below we apply the pnQTDA formalism to examples within the 0d-1s shell.

## 17.2.1  $1^+$ States in the Mirror Nuclei $^{24}$Na and $^{24}$Al

Consider the $1^+$ states of the mirror nuclei $^{24}_{11}$Na$_{13}$ and $^{24}_{13}$Al$_{11}$ in the simple $0d_{5/2}$-$1s_{1/2}$ valence space. As explained in Subsect. 16.4.5, we can assume that mirror nuclei have the same structure. Thus we propose to describe the energy spectra of $^{24}$Na and $^{24}$Al by a single pnQTDA calculation. We choose $^{24}$Mg as the reference nucleus.

In the $0d_{5/2}$-$1s_{1/2}$ valence space the proton–neutron two-quasiparticle basis states are

$$\{|1\rangle, |2\rangle\} = \{|\pi 0d_{5/2}\,\nu 0d_{5/2}\,; 1^+\rangle, |\pi 1s_{1/2}\,\nu 1s_{1/2}\,; 1^+\rangle\}. \qquad (17.10)$$

We take the BCS results for $^{24}$Mg from Table 16.5; the table caption explains the input into the calculation. In particular the table gives the $u,v$ occupation amplitudes and the quasiparticle energies needed in (17.2).

The pnQTDA matrix elements can now be computed from (17.2). Substituting from the sources listed below the equation yields

$$
\begin{aligned}
A_{11}(1^+) = {}& 3.0848\,\text{MeV} + 3.1451\,\text{MeV} \\
& + (0.6821^2 \times 0.6826^2 + 0.7312^2 \times 0.7308^2)(-1.6286A_0) \\
& + (0.6821^2 \times 0.7308^2 + 0.7312^2 \times 0.6826^2)(0.8143A_1 + 0.8143A_0) \\
= {}& 6.2299\,\text{MeV} - 0.4129A_0 + 0.4052A_1\,,
\end{aligned}
\qquad (17.11)
$$

$$
\begin{aligned}
A_{12}(1^+) = {}& (0.6821 \times 0.6826 \times 0.7773 \times 0.7760 \\
& + 0.7312 \times 0.7308 \times 0.6291 \times 0.6307)(-1.1832A_0) \\
& + (0.6821 \times 0.7308 \times 0.7773 \times 0.6307 + 0.7312 \times 0.6826 \times 0.6291 \times 0.7760) \\
& \times (0.5916A_1 + 0.5916A_0) = -0.2944A_0 + 0.2887A_1\,,
\end{aligned}
\qquad (17.12)
$$

$$
\begin{aligned}
A_{22}(1^+) = {}& 3.1464\,\text{MeV} + 3.2055\,\text{MeV} \\
& + (0.7773^2 \times 0.7760^2 + 0.6291^2 \times 0.6307^2)(-1.000A_0) \\
& + (0.7773^2 \times 0.6307^2 + 0.6291^2 \times 0.7760^2)(0.5000A_1 + 0.5000A_0) \\
= {}& 6.3519\,\text{MeV} - 0.2819A_0 + 0.2393A_1\,.
\end{aligned}
\qquad (17.13)
$$

With the default choice $A_0 = A_1 = 1.0\,\mathrm{MeV}$ the pnQTDA matrix given by (17.11)–(17.13) becomes

$$\mathbf{A}_{\mathrm{pnQTDA}}(1^+) = \begin{pmatrix} 6.2222 & -0.0057 \\ -0.0057 & 6.3093 \end{pmatrix} \mathrm{MeV} \, . \tag{17.14}$$

Diagonalization, which can be done by hand according to Subsect. 8.3.2, gives the eigenvalues

$$E(2_1^+) = 6.222\,\mathrm{MeV} \, , \quad E(2_2^+) = 6.310\,\mathrm{MeV} \, , \tag{17.15}$$

and eigenvectors

$$|1_1^+\rangle = 0.998|1\rangle + 0.065|2\rangle \, , \tag{17.16}$$

$$|1_2^+\rangle = -0.065|1\rangle + 0.998|2\rangle \, . \tag{17.17}$$

As pointed out in the beginning of this subsection, the computed $1^+$ states (17.16) and (17.17) describe the $1^+$ spectra of both $^{24}$Na and $^{24}$Al. The energy difference between the computed $1^+$ states is $0.088\,\mathrm{MeV}$, which is far short of the measured value of $0.874\,\mathrm{MeV}$ in $^{24}$Na. For $^{24}$Al the experimental energy difference is not known. Below we study the energy spectrum of $^{24}$Na more extensively.

## 17.2.2 Energy Spectra in the d-s and f-p-0g$_{9/2}$ Shells

We now continue the investigation of pnQTDA spectra started in the previous subsection. Figure 17.1 shows two computed excitation spectra of $^{24}$Na. The left panel displays all energy levels produced by the $0d_{5/2}$-$1s_{1/2}$ model space, calculated for all $J$ with the SDI parameters of Subsect. 17.2.1. The middle panel shows the spectrum from a full 0d-1s calculation, where we used the SDI pairing parameters of Table 16.5 and $A_0 = A_1 = 1.0\,\mathrm{MeV}$ for all $J$. The right panel shows all known experimental levels up to half an MeV.

Figure 17.1 shows that the density of states in the calculated spectra is much too high. This is partly due to our choice $A_0 = A_1$. Breaking this equality diminishes the level density but still does not reproduce the observed sequence of states. The level density and sequence are influenced by the single-particle energies and the characteristics of the two-body interaction. The spacing increases with the model space because the higher multipoles of an interaction operate more efficiently in a larger space.

The computed spectrum of $^{24}$Na can be interpreted to be a linear combination of the spectra of $^{24}$Na and $^{24}$Al. This stems from the basic feature of losing good nucleon number in the two-quasiparticle configurations built on a BCS (or LNBCS) vacuum. The nucleon number uncertainties, inherent in the BCS (LNBCS) vacuum, show up as particle number fluctuations. These fluctuations were addressed in Subsect. 13.4.2, and they are amplified by forming

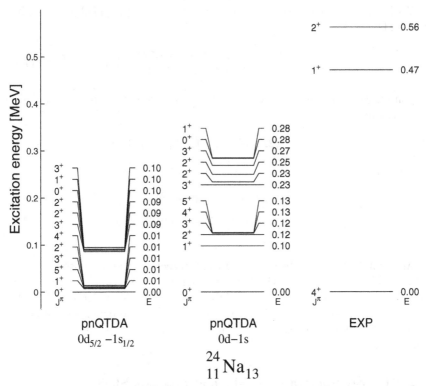

**Fig. 17.1.** Proton–neutron QTDA spectra of $^{24}$Na in two different valence spaces, together with the experimental spectrum shown up to the $2_1^+$ state. The single-particle energies are from (14.12). The SDI was used with the pairing parameters of Table 16.5 and the pnQTDA parameters $A_0 = A_1 = 1.0\,\mathrm{MeV}$ for all multipoles

proton–neutron two-quasiparticle excitations on top of a BCS (LNBCS) vacuum. Then the calculated energy spectrum mixes states of all nuclei that are within one particle or one hole of the even–even reference nucleus. This in fact partly explains the excessive density of low-lying states in the computed spectra of $^{24}$Na.

As our next example we discuss the excitation spectrum of the self-conjugate $(N = Z)$ nucleus $^{30}_{15}\mathrm{P}_{15}$. The wave functions of its three lowest states were discussed in Subsect. 15.4.2 in a very schematic fashion. Also the beta and M1 decays of $^{30}$P were studied within a rudimentary configuration mixing scheme in Subsects. 15.5.3 and 15.5.4.

The computed and experimental energy spectra of $^{30}$P are shown in Fig. 17.2. The complete 0d–1s valence space was used in the computations. Now the reference nucleus is $^{30}$Si, and its BCS ground state was computed by using the pairing parameters of Table 16.6. The pnQTDA calculation was done by using the SDI parameters $A_0 = A_1 = 1.0\,\mathrm{MeV}$ for all multipoles.

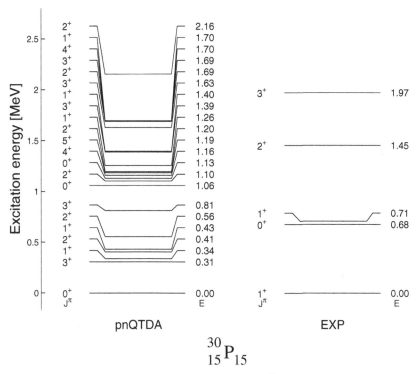

**Fig. 17.2.** Proton–neutron QTDA spectrum of $^{30}$P in the 0d-1s valence space, together with the experimental spectrum shown up to the $3_1^+$ state. The single-particle energies are from (14.12). The SDI was used with the pairing parameters of Table 16.6 and the pnQTDA parameters $A_0 = A_1 = 1.0$ MeV for all multipoles

As in the case of $^{24}$Na, the computed spectrum of $^{30}$P is much compressed relative to the experimental one. As discussed in the context of $^{24}$Na, this phenomenon can be explained by the particular choice $A_0 = A_1$ of the SDI parameters and particle number fluctuations.

Consider next the nucleus $^{66}_{29}$Zn$_{37}$ as a representative case of nuclei in the 0f-1p-0g$_{9/2}$ shells. Figure 17.3 presents a comparison between the experimental spectrum of $^{66}$Cu and two pnQTDA calculations. The calculations were done in the 0f-1p (left) and 0f-1p-0g$_{9/2}$ (middle) valence spaces by using the single-particle energies of (14.13) and the pairing parameters of Table 16.7. The SDI parameters $A_0 = 1.0$ MeV and $A_1 = 0.8$ MeV were used in the pnQTDA calculation.

In the pnQTDA calculations the order of the $0_1^+$ state and the $2_1^+$ state is sensitive to the relative magnitudes of the $A_0$ and $A_1$ strengths of the SDI. In the present calculations the parameters were chosen such that the $1_1^+$ state became the ground state. The correspondence between the experimental

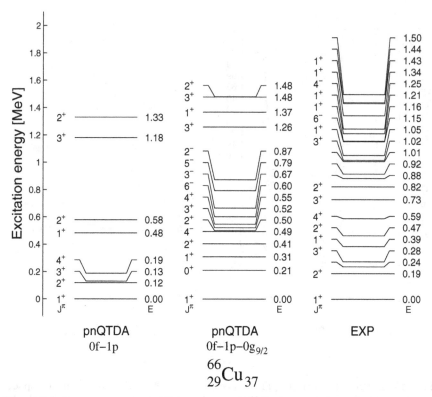

**Fig. 17.3.** Proton–neutron QTDA spectrum of $^{66}$Cu in two different valence spaces, together with the experimental spectrum. The complete spectra are shown up to an energy of 1.5 MeV. The single-particle energies are from (14.13). The SDI was used with the pairing parameters of Table 16.7 and the pnQTDA parameters $A_0 = 1.0$ MeV and $A_1 = 0.8$ MeV for all multipoles

spectrum and the calculated spectra in Fig. 17.3 is appreciably better than in our previous examples.

The spectrum computed in the extended basis, which contains the intruder state $0g_{9/2}$, is in fair qualitative agreement with the experimental spectrum. The intruder allows the generation of negative-parity states. These states are seen in the experimental spectrum, but half an MeV higher than the calculated ones. This is due to the chosen, slightly too low single-particle energy of the $0g_{9/2}$ state.

### 17.2.3 Average Particle Number in the pnQTDA

Particle-number fluctuations are associated with the wave functions computed in the pnQTDA. These fluctuations mix states of different nuclei in the pn-QTDA wave functions. The most serious mixing occurs among nuclei that are immediate neighbours of the even–even reference nucleus for which the BCS (LNBCS) calculation is carried out. The excessive density of states in light nuclei was ascribed to this mixing.

Particle number fluctuations always contaminate computed quasiparticle wave functions. It is therefore of interest to study quantitatively the average particle number of a pnQTDA wave function. At the BCS (LNBCS) level the average proton and neutron numbers were constrained to equal the correct nucleon numbers of the even–even reference nucleus. We expect that the average nucleon numbers of pnQTDA wave functions closely coincide with those of the reference. This is so because particle number fluctuations can be expected to mix odd–odd nuclei adjacent to the reference in a democratic fashion, which roughly preserves the average proton and neutron numbers of the BCS (LNBCS).

### Formalism

To access the average nucleon numbers we start from the proton and neutron number operators $\hat{n}_{\rm p}$ and $\hat{n}_{\rm n}$ given in (4.16). With the pnQTDA wave function (17.7) the effective, or average, number of protons in the pnQTDA state is

$$
\begin{aligned}
Z_{\rm eff} &\equiv \langle \omega | \hat{n}_{\rm p} | \omega \rangle \\
&= \sum_{\substack{pn \\ p'n'}} X^{\omega *}_{pn} X^{\omega}_{p'n'} \sum_{\substack{m_\pi m_\nu \\ m_{\pi'} m_{\nu'}}} (j_p \, m_\pi \, j_n \, m_\nu | J \, M)(j_{p'} \, m_{\pi'} \, j_{n'} \, m_{\nu'} | J \, M)\langle \hat{n}_{\rm p} \rangle \,,
\end{aligned}
$$

(17.18)

where

$$
\langle \hat{n}_{\rm p} \rangle \equiv \langle {\rm BCS} | a_\nu a_\pi \hat{n}_{\rm p} a^\dagger_{\pi'} a^\dagger_{\nu'} | {\rm BCS} \rangle \,.
$$

(17.19)

We proceed to evaluate this expression.

Equation (13.38) gives the number operator in terms of quasiparticles, whence

$$
\langle \hat{n}_{\rm p} \rangle = \delta_{\pi\pi'} \delta_{\nu\nu'} Z_{\rm act} + \delta_{\nu\nu'} \sum_{\pi''} (u^2_{p''} - v^2_{p''}) \langle {\rm BCS} | a_\pi a^\dagger_{\pi''} a_{\pi''} a^\dagger_{\pi'} | {\rm BCS} \rangle \,,
$$

(17.20)

where the number $Z_{\rm act}$ of valence protons is the average number of active protons in the BCS vacuum according to (13.37) and (13.39). To the second term we apply Wick's theorem of Subsect. 4.3.3 and find

$$
\langle \hat{n}_{\rm p} \rangle = \delta_{\pi\pi'} \delta_{\nu\nu'} (Z_{\rm act} + u^2_p - v^2_p) \,.
$$

(17.21)

An analogous relation applies to the neutrons. Substituting (17.21) into (17.18) and using the orthonormality relation (17.8), we obtain

$$
\begin{aligned}
Z_{\rm eff} &= Z_{\rm act} + \sum_{pn} (u^2_p - v^2_p)|X^\omega_{pn}|^2 \,, \\
N_{\rm eff} &= N_{\rm act} + \sum_{pn} (u^2_n - v^2_n)|X^\omega_{pn}|^2 \,.
\end{aligned}
$$

(17.22)

From Eqs. (17.22) we see that the effective proton and neutron numbers depend on the state $\omega$ and are not equal to the particle numbers of the odd–odd nucleus. However, the basic relation (13.9) implies that $-1 \leq u_a^2 - v_a^2 \leq 1$, which together with (17.8) leads to the conclusion

$$|Z_{\text{eff}} - Z_{\text{act}}| \leq 1 , \quad |N_{\text{eff}} - N_{\text{act}}| \leq 1 . \tag{17.23}$$

Thus the average particle numbers in the states of an odd–odd nucleus stay within unity of those of the even–even reference nucleus.

### Examples

Table 17.1 gives effective particle numbers calculated from (17.22) with the reference nucleus $^{24}_{12}\text{Mg}_{12}$. The pnQTDA wave functions were computed in the 0d-1s valence space. The excitation spectrum is shown in Fig. 17.1, with the parameters stated in the figure caption. The table shows that the average nucleon numbers of the pnQTDA wave functions are very close to the average nucleon numbers of the reference nucleus and thus relatively far from the nucleon numbers of the odd–odd nuclei described.

**Table 17.1.** Effective proton and neutron numbers (17.22) of the lowest pnQTDA states for the reference nucleus $^{24}\text{Mg}$

| $J^\pi$ | $E_{\text{ex}}$ (MeV) | $Z_{\text{eff}}$ | $N_{\text{eff}}$ |
|---|---|---|---|
| $0^+$ | 0.000 | 4.07 | 4.07 |
| $1^+$ | 0.098 | 4.02 | 4.03 |
| $2^+$ | 0.122 | 4.01 | 4.01 |
| $3^+$ | 0.125 | 4.00 | 4.01 |
| $4^+$ | 0.126 | 4.00 | 4.01 |
| $5^+$ | 0.126 | 4.00 | 4.01 |
| $3^+$ | 0.228 | 4.15 | 4.13 |
| $2^+$ | 0.234 | 4.16 | 4.12 |
| $2^+$ | 0.250 | 4.12 | 4.16 |
| $3^+$ | 0.269 | 4.13 | 4.15 |
| $0^+$ | 0.284 | 4.23 | 4.23 |
| $1^+$ | 0.285 | 4.26 | 4.26 |

The numbers of active particles in the reference BCS vacuum are $Z_{\text{act}} = 4$, $N_{\text{act}} = 4$. The energies $E_{\text{ex}}$ are the excitation energies.

Another example of the effective nucleon numbers of the pnQTDA wave functions is given in Table 17.2. Here the reference nucleus is $^{66}_{30}\text{Zn}_{36}$. The computations were done in the 0f-1p-0g$_{9/2}$ valence space. The resulting spectrum is displayed in Fig. 17.3, where also the relevant parameters are quoted. In this case the deviations from the particle numbers $Z_{\text{act}} = 10$ and $N_{\text{act}} = 16$ of the reference BCS vacuum are greater than in the case of $^{24}\text{Mg}$. The state dependence of the average nucleon numbers is clearly visible in the table.

**Table 17.2.** Effective proton and neutron numbers (17.22) of the lowest pnQTDA states for the reference nucleus $^{66}$Zn

| $J^\pi$ | $E_{ex}$ (MeV) | $Z_{eff}$ | $N_{eff}$ |
|---------|----------------|-----------|-----------|
| $1^+$ | 0.000 | 10.37 | 16.48 |
| $0^+$ | 0.211 | 10.80 | 16.31 |
| $1^+$ | 0.310 | 10.18 | 15.76 |
| $2^+$ | 0.405 | 10.19 | 15.98 |
| $4^-$ | 0.495 | 10.35 | 16.64 |
| $2^+$ | 0.501 | 10.20 | 16.25 |
| $3^+$ | 0.522 | 10.24 | 16.52 |
| $4^+$ | 0.548 | 10.17 | 16.53 |
| $6^-$ | 0.602 | 10.18 | 16.65 |
| $3^-$ | 0.667 | 10.20 | 16.65 |
| $5^-$ | 0.794 | 10.24 | 16.65 |
| $2^-$ | 0.874 | 10.95 | 16.59 |
| $3^+$ | 1.260 | 10.80 | 16.49 |
| $1^+$ | 1.369 | 10.51 | 15.85 |
| $3^+$ | 1.479 | 10.07 | 15.24 |
| $2^+$ | 1.481 | 10.69 | 16.26 |

The numbers of active particles in the reference BCS vacuum are $Z_{act} = 10$, $N_{act} = 16$. The energies $E_{ex}$ are the excitation energies.

## 17.3 Electromagnetic Transitions in the pnQTDA

In this section we discuss electromagnetic transitions in an odd–odd open-shell nucleus. The wave functions are obtained by performing a pnQTDA calculation in the adjacent even–even reference nucleus. The initial and final wave functions, of the form (17.7), are

$$|\omega_i\rangle = \sum_{p_i n_i} X_{p_i n_i}^{\omega_i} \left[ a_{p_i}^\dagger a_{n_i}^\dagger \right]_{J_i M_i} |BCS\rangle \,, \tag{17.24}$$

$$|\omega_f\rangle = \sum_{p_f n_f} X_{p_f n_f}^{\omega_f} \left[ a_{p_f}^\dagger a_{n_f}^\dagger \right]_{J_f M_f} |BCS\rangle \,. \tag{17.25}$$

Using these wave functions we can immediately write the decay amplitude as

$$\boxed{(\omega_f \| \mathcal{M}_{\sigma\lambda} \| \omega_i) = \sum_{\substack{p_i n_i \\ p_f n_f}} X_{p_f n_f}^{\omega_f *} X_{p_i n_i}^{\omega_i} (p_f\, n_f \,;\, J_f \| \mathcal{M}_{\sigma\lambda} \| p_i\, n_i \,;\, J_i) \,,} \tag{17.26}$$

where the reduced matrix element $(p_f\, n_f \,;\, J_f \| \mathcal{M}_{\sigma\lambda} \| p_i\, n_i \,;\, J_i)$ is the proton–neutron two-quasiparticle transition matrix element given in (15.103) for the general case and in (15.104) for $J_f = 0$.

We next discuss an example that clarifies the use of (17.26) in practical applications.

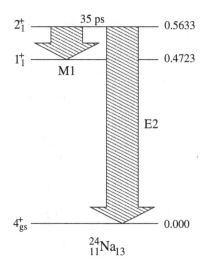

**Fig. 17.4.** Experimental low-energy spectrum of $^{24}$Na. The measured half-life of the $2_1^+$ state and the decay branchings are indicated

### 17.3.1 Decay of the $2_1^+$ State in $^{24}$Na

Consider the decay of the $2_1^+$ state of $^{24}$Na. The decay channels are shown in Fig. 17.4. We choose the valence space $0d_{5/2}$-$1s_{1/2}$ and use the SDI with the parameters stated in the caption to Fig. 17.1, which displays the computed spectrum. We use the CS phase convention throughout the calculations.

For our calculation we need the structure of the nuclear wave functions involved. Equations (17.10) and (17.16) give the $1_1^+$ state as

$$|^{24}\text{Na}; 1_1^+\rangle = 0.998|1\rangle_1 + 0.065|2\rangle_1 \qquad (17.27)$$

with

$$|1\rangle_1 = |\pi 0d_{5/2}\,\nu 0d_{5/2}; 1^+\rangle\,, \quad |2\rangle_1 = |\pi 1s_{1/2}\,\nu 1s_{1/2}; 1^+\rangle\,. \qquad (17.28)$$

The wave functions of the other relevant states, $4_{\text{gs}}^+$ and $2_1^+$, we obtain by forming the corresponding pnQTDA matrices and diagonalizing them. For the $2^+$ states we have the proton–neutron basis

$$\{|1\rangle_2\,, |2\rangle_2\,, |3\rangle_2\}$$
$$= \{|\pi 0d_{5/2}\,\nu 0d_{5/2}; 2^+\rangle\,, |\pi 0d_{5/2}\,\nu 1s_{1/2}; 2^+\rangle\,, |\pi 1s_{1/2}\,\nu 0d_{5/2}; 2^+\rangle\}\,. \qquad (17.29)$$

By proceeding as in Subsect. 17.2.1 we find the pnQTDA matrix

$$A_{\text{pnQTDA}}(2^+) = \begin{pmatrix} 6.227 & 0.003 & 0.003 \\ 0.003 & 6.305 & -0.001 \\ 0.003 & -0.001 & 6.306 \end{pmatrix} \text{ MeV}\,. \qquad (17.30)$$

Diagonalization gives the lowest eigenvalue

$$E(2_1^+) = 6.226\,\text{MeV} \tag{17.31}$$

with eigenvector

$$|^{24}\text{Na}\,;\, 2_1^+\rangle = 0.998|1\rangle_2 - 0.040|2\rangle_2 - 0.039|3\rangle_2 \,. \tag{17.32}$$

In the $0\text{d}_{5/2}$-$1\text{s}_{1/2}$ valence space we have only one $4^+$ state, so that

$$|^{24}\text{Na}\,;\, 4_{\text{gs}}^+\rangle = |\pi 0\text{d}_{5/2}\,\nu 0\text{d}_{5/2}\,;\, 4^+\rangle \,. \tag{17.33}$$

Let us first discuss the M1 decay $2_1^+ \to 1_1^+$. We write the two-quasiparticle matrix element (15.103) in (17.26) as

$$(p_f\,n_f\,;\, 1\|\boldsymbol{M}_1\|p_i\,n_i\,;\, 2) = M^{(1)}(p_f n_f p_i n_i) + M^{(2)}(p_f n_f p_i n_i) \,, \tag{17.34}$$

where we have abbreviated

$$M^{(1)}(p_f n_f p_i n_i) \equiv \delta_{n_i n_f}(-1)^{j_{p_f}+j_{n_f}+1}\sqrt{15}\left\{\begin{matrix} 1 & 2 & 1 \\ j_{p_i} & j_{p_f} & j_{n_f} \end{matrix}\right\}$$
$$\times (u_{p_i}u_{p_f} + v_{p_i}v_{p_f})(p_f\|\boldsymbol{M}_1\|p_i) \,, \tag{17.35}$$

$$M^{(2)}(p_f n_f p_i n_i) \equiv \delta_{p_i p_f}(-1)^{j_{p_i}+j_{n_i}}\sqrt{15}\left\{\begin{matrix} 1 & 2 & 1 \\ j_{n_i} & j_{n_f} & j_{p_f} \end{matrix}\right\}$$
$$\times (u_{n_i}u_{n_f} + v_{n_i}v_{n_f})(n_f\|\boldsymbol{M}_1\|n_i) \tag{17.36}$$

with the $\mathcal{D}$ factors inserted from (15.18).

The M1 single-particle matrix elements (6.47) are given by Tables 6.6 and 6.7. Table 16.9 gives the BCS occupation amplitudes. Inspection of the terms (17.35) and (17.36) with the input of the present example reveals that, because of the Kronecker deltas and $6j$ symbols, the only non-zero terms are obtained for $p_i n_i = p_f n_f = 0\text{d}_{5/2}0\text{d}_{5/2}$. Evaluated, these terms are

$$M^{(1)}(0\text{d}_{5/2}0\text{d}_{5/2}0\text{d}_{5/2}0\text{d}_{5/2})$$
$$= \sqrt{15}\left(-\frac{4}{15}\sqrt{\frac{2}{7}}\right) \times 1 \times (2.832 \times 1 + 0.708 g_{\text{p}})\,\mu_{\text{N}}/c = -3.747\,\mu_{\text{N}}/c \,, \tag{17.37}$$

$$M^{(2)}(0\text{d}_{5/2}0\text{d}_{5/2}0\text{d}_{5/2}0\text{d}_{5/2})$$
$$= -\sqrt{15}\left(-\frac{4}{15}\sqrt{\frac{2}{7}}\right) \times 1 \times (0 + 0.708 g_{\text{n}})\,\mu_{\text{N}}/c = -1.495\,\mu_{\text{N}}/c \,, \tag{17.38}$$

where we inserted the gyromagnetic factors (6.8). With the amplitudes from (17.27) and (17.32), Eq. (17.26) now yields

$$(1_1^+\|\boldsymbol{M}_1\|2_1^+) = 0.998 \times 0.998 \times (-3.747 - 1.495)\,\mu_{\text{N}}/c = -5.221\,\mu_{\text{N}}/c \,. \tag{17.39}$$

This gives the reduced transition probability

$$B(\text{M1}\,;\, 2_1^+ \rightarrow 1_1^+) = \frac{1}{5}(-5.221\,\mu_N/c)^2 = 5.452\,(\mu_N/c)^2\,. \tag{17.40}$$

With the experimental gamma energy from Fig. 17.4, Table 6.9 gives the decay probability

$$T(\text{M1}\,;\, {}^{24}\text{Na}) = 1.779 \times 10^{13} \times 0.0910^3 \times 5.452\,1/\text{s} = 7.31 \times 10^{10}\,1/\text{s}\,. \tag{17.41}$$

For the E2 transition $2_1^+ \rightarrow 4_{\text{gs}}^+$ we obtain from (15.103)

$$(p_f\,n_f\,;\, 4\|\boldsymbol{Q}_2\|p_i\,n_i\,;\, 2) = Q^{(1)}(p_f n_f p_i n_i) + Q^{(2)}(p_f n_f p_i n_i)\,, \tag{17.42}$$

where

$$Q^{(1)}(p_f n_f p_i n_i) \equiv \delta_{n_i n_f}(-1)^{j_{p_f}+j_{n_f}}\sqrt{45}\left\{\begin{matrix} 4 & 2 & 2 \\ j_{p_i} & j_{p_f} & j_{n_f} \end{matrix}\right\}$$
$$\times\,(u_{p_i}u_{p_f} - v_{p_i}v_{p_f})(p_f\|\boldsymbol{Q}_2\|p_i)\,, \tag{17.43}$$

$$Q^{(2)}(p_f n_f p_i n_i) \equiv \delta_{p_i p_f}(-1)^{j_{p_i}+j_{n_i}}\sqrt{45}\left\{\begin{matrix} 4 & 2 & 2 \\ j_{n_i} & j_{n_f} & j_{p_f} \end{matrix}\right\}$$
$$\times\,(u_{n_i}u_{n_f} - v_{n_i}v_{n_f})(n_f\|\boldsymbol{Q}_2\|n_i)\,. \tag{17.44}$$

With the single-particle matrix elements of Table 6.4 and the BCS occupation amplitudes of Table 16.9 we produce the values of $Q^{(1)}$ and $Q^{(2)}$ given in Table 17.3. Using these numbers and (17.26) we have the decay amplitude

$$\begin{aligned}(4_1^+\|\boldsymbol{Q}_2\|2_1^+) &= 1 \times 0.998 \times (0.0815e_{\text{eff}}^{\text{p}} + 0.0800e_{\text{eff}}^{\text{n}})b^2 \\ &\quad + 1 \times (-0.040) \times (-0.1063e_{\text{eff}}^{\text{n}}b^2) \\ &\quad + 1 \times (-0.039) \times (-0.1085e_{\text{eff}}^{\text{p}}b^2) \\ &= 0.281e_{\text{eff}}^{\text{p}}\,\text{fm}^2 + 0.276e_{\text{eff}}^{\text{n}}\,\text{fm}^2\,, \end{aligned} \tag{17.45}$$

where the $A = 24$ oscillator length $b = 1.813\,\text{fm}$ was inserted.

**Table 17.3.** Contributions to the sum (17.26) for the $2_1^+ \rightarrow 4_{\text{gs}}^+$ transition in $^{24}\text{Na}$

| $p_i n_i(2^+)$ | $p_f n_f(4^+)$ | $X_{p_i n_i}^{\omega_i}$ | $X_{p_f n_f}^{\omega_f}$ | $Q^{(1)}\,(e_{\text{eff}}^{\text{p}}b^2)$ | $Q^{(2)}\,(e_{\text{eff}}^{\text{n}}b^2)$ |
|---|---|---|---|---|---|
| $0d_{5/2}0d_{5/2}$ | $0d_{5/2}0d_{5/2}$ | 0.998 | 1 | 0.0815 | 0.0800 |
| $0d_{5/2}1s_{1/2}$ | $0d_{5/2}0d_{5/2}$ | -0.040 | 1 | 0 | -0.1063 |
| $1s_{1/2}0d_{5/2}$ | $0d_{5/2}0d_{5/2}$ | -0.039 | 1 | -0.1085 | 0 |

The initial and final quasiparticle configurations and the corresponding pnQTDA amplitudes are listed in the first four columns. The quantities (17.43) and (17.44) are given in columns five and six. CS phases are assumed.

We can now take the electric polarization constant $\chi = 1.03$ from (16.107). It was obtained from a QTDA calculation for the even–even reference nucleus $^{24}$Mg in our presently adopted valence space. Substituted into (17.45) this gives

$$(4_1^+ \| \boldsymbol{Q}_2 \| 2_1^+) = 0.855 \, e \, \text{fm}^2 \, , \tag{17.46}$$

leading to the reduced transition probability

$$B(\text{E2} \, ; \, 2_1^+ \to 4_{\text{gs}}^+) = 0.146 \, e^2 \text{fm}^4 \, . \tag{17.47}$$

This, in turn, leads to transition probability by the use of Table 6.8 and the experimental gamma energy. The result is

$$T(\text{E2} \, ; \, ^{24}\text{Na}) = 1.223 \times 10^9 \times 0.5633^5 \times 0.146 \, 1/\text{s} = 1.0 \times 10^7 \, 1/\text{s} \, . \tag{17.48}$$

The transition probability (17.48) is four orders of magnitude smaller than the M1 transition probability (17.41), so that the M1 transition in fact determines the decay half-life of the $2_1^+$ state. The half-life is then

$$t_{1/2}(2_1^+) = \frac{\ln 2}{T(\text{M1})} = 9.5 \, \text{ps} \, . \tag{17.49}$$

This is roughly one fourth of the experimental half-life of 35 ps, shown in Fig. 17.4. It is possible to improve on the calculation by increasing the size of the model space and performing a more careful analysis of the single-particle energies and the interaction parameters.

## 17.4 Beta-Decay Transitions in the pnQTDA

In this section we discuss beta-decay transitions between an open-shell odd–odd nucleus and the neighbouring even–even reference nucleus. This reference nucleus is used to generate, in a pnQTDA calculation, the levels of the odd–odd nucleus in question. First we discuss transitions involving the BCS ground state of the reference nucleus.

### 17.4.1 Transitions to and from an Even–Even Ground State

As pointed out at the end of Subsect. 15.4.1, the decay amplitude for a beta transition from a pnQTDA excited state $|\omega\rangle$ to the BCS ground state can be written immediately as

$$\boxed{(\text{BCS} \| \beta_{\text{F/GT}}^{\mp} \| \omega) = \sum_{pn} X_{pn}^{\omega} \mathcal{M}_{\text{F/GT}}^{(\mp)}(pnJ \to \text{BCS}) \, .} \tag{17.50}$$

The appropriate Fermi ($\mathcal{M}_{\text{F}}^{(\mp)}$) and Gamow–Teller ($\mathcal{M}_{\text{GT}}^{(\mp)}$) matrix elements for $\beta^-$ and $\beta^+$ decay are given by (15.63) and (15.64). When the initial state is the even–even ground state we have similarly

$$\boxed{(\omega\|\beta^{\mp}_{\text{F/GT}}\|\text{BCS}) = \sum_{pn} X^{\omega*}_{pn} \mathcal{M}^{(\mp)}_{\text{F/GT}}(\text{BCS} \to pnJ)} \qquad (17.51)$$

with the appropriate transition matrix elements given in (15.65) and (15.66).

The corresponding formulas for $K$th-forbidden unique transitions can be obtained from (17.50) and (17.51) by making the replacements

$$\mathcal{M}^{(\mp)}_{\text{F/GT}}(pnJ \to \text{BCS}) \to \mathcal{M}^{(\mp)}_{Ku}(pnJ \to \text{BCS}), \qquad (17.52)$$

$$\mathcal{M}^{(\mp)}_{\text{F/GT}}(\text{BCS} \to pnJ) \to \mathcal{M}^{(\mp)}_{Ku}(\text{BCS} \to pnJ), \qquad (17.53)$$

where the replacing matrix elements are given by (15.68) and (15.69).

Next we discuss an example which clarifies the use of the formalism.

### 17.4.2 Gamow–Teller Beta Decay of $^{30}$S

Consider the $\beta^+$ decay of the $0^+$ ground state of $^{30}_{16}\text{S}_{14}$ to the first two $1^+$ states in $^{30}_{15}\text{P}_{15}$. This decay pattern was presented in Fig. 15.4 and discussed in the example of Subsect. 15.4.2. There we made a rudimentary, ad hoc ansatz for two-quasiparticle mixing in the $1^+$ wave functions. In Exercise 17.19 these guessed wave functions are compared with wave functions obtained by diagonalizing the pnQTDA matrix in a sub-basis of the 0d-1s shell. In this subsection we obtain the wave functions from a pnQTDA calculation in the 0d-1s valence space using $^{30}$S as the reference nucleus.

The scheme is to adjust the SDI pairing parameters to reproduce the empirical pairing gaps (16.97). However, we need not do this because we have already determined the pairing parameters for $^{30}$Si, which is the mirror nucleus of $^{30}$S, and we can assume the pairing gaps of mirror nuclei to be the same. Our pairing parameters are then those in Table 16.6, i.e. $A^{(p)}_{\text{pair}} = A^{(n)}_{\text{pair}} = 1.0 \,\text{MeV}$. The resulting BCS occupation amplitudes are listed in Table 14.1, and again in Table 17.4 for convenience. If we make the usual choice $A_0 = A_1 = 1.0 \,\text{MeV}$ for the pnQTDA calculation we obtain the spectrum of Fig. 17.2 for $^{30}$P.

The two $1^+$ states in $^{30}$P that we need are written as

$$|^{30}\text{P} ; 1^+_1\rangle = \sum_{k=1}^{7} X^{1^+_1}_k |k\rangle_1, \quad |^{30}\text{P} ; 1^+_2\rangle = \sum_{k=1}^{7} X^{1^+_2}_k |k\rangle_1, \qquad (17.54)$$

where the configurations for the basis states $|k\rangle_1$ are listed in Table 17.4. After forming the pnQTDA matrix and diagonalizing it we obtain the states (17.54) as the first two eigenvectors. The $X$ amplitudes are listed in Table 17.5. It also gives the beta transition matrix elements in (17.51), computed by using (15.66) and the occupation amplitudes and Gamow–Teller single-particle matrix elements of Table 17.4. The fourth and sixth columns list the products of the beta transition matrix element and the $X$ amplitude of the $1^+_1$ and $1^+_2$ states, respectively. Each product is a contribution to the sum (17.51), given in the last row of the table.

**Table 17.4.** Proton–neutron quasiparticle configurations, labelled with $k$, used to build the $1^+$ states of $^{30}$P in the 0d-1s valence space

| $k$ | Configuration | $E_{pn}$ (MeV) | $u_p$ | $v_p$ | $u_n$ | $v_n$ | $\mathcal{M}_{GT}(pn)$ |
|---|---|---|---|---|---|---|---|
| 1 | $\pi 0d_{5/2}\nu 0d_{5/2}$ | 6.275 | 0.2285 | 0.9735 | 0.4996 | 0.8662 | $\sqrt{\frac{14}{5}}$ |
| 2 | $\pi 0d_{5/2}\nu 0d_{3/2}$ | 8.001 | 0.2285 | 0.9735 | 0.9619 | 0.2735 | $-\frac{4}{\sqrt{5}}$ |
| 3 | $\pi 1s_{1/2}\nu 1s_{1/2}$ | 5.210 | 0.2943 | 0.9557 | 0.6330 | 0.7741 | $\sqrt{2}$ |
| 4 | $\pi 1s_{1/2}\nu 0d_{3/2}$ | 7.248 | 0.2943 | 0.9557 | 0.9619 | 0.2735 | 0 |
| 5 | $\pi 0d_{3/2}\nu 0d_{5/2}$ | 5.125 | 0.9372 | 0.3488 | 0.4996 | 0.8662 | $\frac{4}{\sqrt{5}}$ |
| 6 | $\pi 0d_{3/2}\nu 1s_{1/2}$ | 4.813 | 0.9372 | 0.3488 | 0.6330 | 0.7741 | 0 |
| 7 | $\pi 0d_{3/2}\nu 0d_{3/2}$ | 6.850 | 0.9372 | 0.3488 | 0.9619 | 0.2735 | $-\frac{2}{\sqrt{5}}$ |

The reference nucleus is $^{30}$S. Two-quasiparticle energies $E_{pn} \equiv E_p + E_n$, BCS occupation amplitudes and Gamow–Teller single-particle matrix elements from Table 7.3 are also given.

**Table 17.5.** Contributions to the sum (17.51) for the $\beta^+$/EC decay of $^{30}$S to the two lowest $1^+$ states of $^{30}$P

| $k$ | $\mathcal{M}_{GT}^{(+)}(k)$ | $X_k^{1_1^+}$ | Contribution | $X_k^{1_2^+}$ | Contribution |
|---|---|---|---|---|---|
| 1 | $-1.4096$ | 0.470 | $-0.6625$ | $-0.039$ | 0.0550 |
| 2 | 2.9014 | 0.113 | 0.3279 | $-0.015$ | $-0.0435$ |
| 3 | $-1.4818$ | 0.863 | $-1.2788$ | $-0.126$ | 0.1867 |
| 4 | 0 | $-0.028$ | 0 | $-0.007$ | 0 |
| 5 | $-0.5399$ | 0.080 | $-0.0432$ | 0.134 | $-0.0723$ |
| 6 | 0 | $-0.120$ | 0 | $-0.956$ | 0 |
| 7 | 0.5198 | $-0.012$ | $-0.0062$ | $-0.226$ | $-0.1175$ |
| Sum | | | $-1.6628$ | | 0.0084 |

The last row gives the sum of these contributions. The label $k$ is that of Table 17.4. The second column lists the Gamow–Teller matrix elements, and the third and fifth columns list the amplitudes of the $1_1^+$ and $1_2^+$ wave functions in (17.54). CS phases are assumed

With the sums in Table 17.5, Eqs. (7.14) and (7.15) give

$$B(GT\,;\,0_{gs}^+ \to 1_1^+) = 1.25^2(-1.6628)^2 = 4.32\,, \tag{17.55}$$

$$B(GT\,;\,0_{gs}^+ \to 1_2^+) = 1.25^2 \times 0.0084^2 = 1.10 \times 10^{-4}\,. \tag{17.56}$$

Equations (7.16) and (7.33) then give

$$\log ft(1_1^+) = 3.15\,, \quad \log ft(1_2^+) = 7.75\,. \tag{17.57}$$

Figure 15.4 gives the experimental $\log ft$ values 4.4 and 5.7 for decay to the $1_1^+$ and $1_2^+$ states, respectively. Thus we predict the first decay to be too fast and second too slow. However, the qualitative trend is correctly reproduced, contrary to the results (15.89) of the rudimentary approach, where the wave functions (15.79) and (15.80) were merely an educated guess. In fact, comparison of the wave functions (15.79) and (15.80) with those in Table 17.5 reveals major differences.

We continue our analysis by studying the $\beta^+/\text{EC}$ transition

$$^{30}\text{P}(1^+_{\text{gs}}) \xrightarrow{\beta^+/\text{EC}} {}^{30}\text{Si}(0^+_{\text{gs}}) , \quad \log ft_{\text{exp}} = 4.8 , \qquad (17.58)$$

shown in Fig. 15.4. The $1^+_{\text{gs}}$ wave function of $^{30}\text{P}$ is now computed by starting from $^{30}\text{Si}$ as the reference nucleus. Since $^{30}\text{Si}$ is the mirror of $^{30}\text{P}$ we proceed by exchanging the proton and neutron labels of the configurations and occupation amplitudes in Table 17.4. The transition amplitude is now given by (17.50).

The Gamow–Teller matrix elements are computed from (15.64) and the last column of Table 17.4. The $1^+_{\text{gs}}$ wave function is obtained from the previously computed $1^+_1$ wave function of $^{30}\text{P}$. Because the coupling order changes when the proton and neutron labels are exchanged the new basis states differ from the previous ones by a factor of $(-1)^{j_p+j_n}$; see (5.25). The resulting wave function is presented in Table 17.6, where we also list the Gamow–Teller matrix elements.

The sum (17.50) stated in the last row of Table 17.6 gives the reduced beta transition probability

$$B(\text{GT}; 1^+_{\text{gs}} \rightarrow 0^+_{\text{gs}}) = \frac{1.25^2}{3} \times 1.6628^2 = 1.440 , \qquad (17.59)$$

which leads to

$$\log ft(1^+_{\text{gs}} \rightarrow 0^+_{\text{gs}}) = 3.63 . \qquad (17.60)$$

**Table 17.6.** Contributions to the sum (17.50) for the $\beta^+/\text{EC}$ decay of $^{30}\text{P}$ to the ground state of $^{30}\text{Si}$

| $k$ | Configuration | $\mathcal{M}^{(+)}_{\text{GT}}(k)$ | $X^{1^+_{\text{gs}}}_k$ | Contribution |
|---|---|---|---|---|
| 1 | $\pi 0d_{5/2} \nu 0d_{5/2}$ | −1.4096 | −0.470 | 0.6625 |
| 2 | $\pi 0d_{5/2} \nu 0d_{3/2}$ | 0.5399 | 0.080 | 0.0432 |
| 3 | $\pi 1s_{1/2} \nu 1s_{1/2}$ | −1.4818 | −0.863 | 1.2788 |
| 4 | $\pi 1s_{1/2} \nu 0d_{3/2}$ | 0 | −0.120 | 0 |
| 5 | $\pi 0d_{3/2} \nu 0d_{5/2}$ | −2.9014 | 0.113 | −0.3279 |
| 6 | $\pi 0d_{3/2} \nu 1s_{1/2}$ | 0 | −0.028 | 0 |
| 7 | $\pi 0d_{3/2} \nu 0d_{3/2}$ | 0.5198 | 0.012 | 0.0062 |
| Sum | | | | 1.6628 |

The last row gives the sum of the contributions. The configurations, labelled with $k$ as in Table 17.4, the Gamow–Teller matrix elements and the wave function amplitudes are also given. CS phases are assumed.

The experimental value from Fig. 15.4 is 4.8, which means that the theoretical transition rate is too high.

We conclude the present examples by noting that

$$\left|(^{30}\mathrm{Si}\,;\,0_{\mathrm{gs}}^{+}\|\beta_{\mathrm{GT}}^{+}\|^{30}\mathrm{P}\,;\,1_{\mathrm{gs}}^{+})\right| = \left|(^{30}\mathrm{P}\,;\,1_{1}^{+}\|\beta_{\mathrm{GT}}^{+}\|^{30}\mathrm{S}\,;\,0_{\mathrm{gs}}^{+})\right| = 1.663\ . \quad (17.61)$$

The equality is due to the symmetry under exchange of protons and neutrons. The symmetry in turn comes from the use of the same single-particle energies and interaction parameters for protons and neutrons. Such symmetries cease to exist for heavier nuclei since their proton and neutron valence spaces are different.

### 17.4.3 The Ikeda Sum Rule and the pnQTDA

Let us continue to examine beta-decay transitions where the even–even ground state is either the initial or final state. Consider Gamow–Teller $\beta^-$ and $\beta^+$ transitions from an initial state $|0_i^+\rangle$. These transitions reach two *complete sets* of final states, one for the beta-minus and one for the beta-plus transitions. The states of these sets have angular momentum $J_f = 1$. We denote them according to

$$|f_{\mp}\rangle \equiv |n_{\mp}\,1^+\,M_f\rangle\ , \qquad \mp \text{ for } \beta^{\mp} \text{ transitions}\ . \quad (17.62)$$

Here $n_{\mp}$ enumerates the $1^+$ states and $M_f$ is the $z$ projection of the angular momentum for each set of states.

The initial state is the ground state of an even–even nucleus, and the final states belong to the two adjacent odd–odd isobars. The charge-changing transitions thus go from an even–even $0^+$ ground state to the $1^+$ states of the two neighbouring odd–odd nuclei. These transitions are characterized by the Gamow–Teller $\beta^-$ and $\beta^+$ *total transition strengths*

$$\boxed{S_{\mp} \equiv \sum_{\substack{M_f \mu \\ n_{\mp}}} \left|\langle f_{\mp}|\beta_{\mathrm{GT}}^{\mp}(\mu)|0_i^+\rangle\right|^2 \quad (\text{total } \beta^{\mp} \text{ strength})\ ,} \quad (17.63)$$

where $\beta_{\mathrm{GT}}^{\mp}(\mu)$ is the spherical component $\mu$ of the Gamow–Teller operator as defined in (7.69). The sum over $n_-$ or $n_+$ runs over the respective complete set. By means of the Wigner–Eckart theorem (2.27) Eq. (17.63) can be put into the alternative form

$$S_{\mp} = \sum_{n_{\mp}} \left|(f_{\mp}\|\beta_{\mathrm{GT}}^{\mp}\|0_i^+)\right|^2\ , \quad (17.64)$$

where $M_f$ is no longer included in the quantum numbers of $f_{\mp}$.

Next we derive an important property of the total strengths $S_{\mp}$, the so-called Ikeda sum rule.

**Derivation of the Ikeda Sum Rule**

Starting from (17.63) we write

$$S_- = \sum_{\substack{M_f\mu \\ n_-}} \langle 0_i^+|(\beta_{\mathrm{GT}}^-(\mu))^\dagger|f_-\rangle\langle f_-|\beta_{\mathrm{GT}}^-(\mu)|0_i^+\rangle$$

$$= \sum_\mu \langle 0_i^+|(\beta_{\mathrm{GT}}^-(\mu))^\dagger \beta_{\mathrm{GT}}^-(\mu)|0_i^+\rangle , \tag{17.65}$$

where we used the completeness of the states $|f_-\rangle$,

$$\sum_{M_f n_-} |f_-\rangle\langle f_-| = \sum_{M_f n_-} |n_- 1^+ M_f\rangle\langle n_- 1^+ M_f| = 1 . \tag{17.66}$$

In coordinate representation the Gamow–Teller operator is

$$\beta_{\mathrm{GT}}^-(\mu) = \sum_{k=1}^A \sigma_\mu(k) t_-(k) , \tag{17.67}$$

where the isospin lowering operator $t_- = t_1 - t_2$ changes a neutron into a proton according to (5.75). The Hermitian conjugate of this is

$$(\beta_{\mathrm{GT}}^-(\mu))^\dagger = \sum_{k=1}^A (\sigma_\mu(k))^\dagger (t_-(k))^\dagger = (-1)^\mu \sum_{k=1}^A \sigma_{-\mu}(k) t_+(k)$$

$$= (-1)^\mu \beta_{\mathrm{GT}}^+(-\mu) , \tag{17.68}$$

where we used (2.15) and the fact that the Cartesian components of $\boldsymbol{\sigma}$ are Hermitian. Note also that the spin and isospin operators commute. The expression (17.65) for $S_-$ now becomes

$$S_- = \sum_\mu (-1)^\mu \langle 0_i^+|\beta_{\mathrm{GT}}^+(-\mu)\beta_{\mathrm{GT}}^-(\mu)|0_i^+\rangle . \tag{17.69}$$

Exchanging the plus and minus indices gives the corresponding expression for $S_+$.

We wish to evaluate the difference $S_- - S_+$. With the change $\mu \to -\mu$ in the summation index of $S_+$ the difference becomes

$$S_- - S_+ = \sum_\mu (-1)^\mu \langle 0_i^+|\beta_{\mathrm{GT}}^+(-\mu)\beta_{\mathrm{GT}}^-(\mu) - \beta_{\mathrm{GT}}^-(\mu)\beta_{\mathrm{GT}}^+(-\mu)|0_i^+\rangle$$

$$= \sum_\mu (-1)^\mu \langle 0_i^+|[\beta_{\mathrm{GT}}^+(-\mu), \beta_{\mathrm{GT}}^-(\mu)]|0_i^+\rangle$$

$$= \sum_\mu (-1)^\mu \sum_{k,k'=1}^A \langle 0_i^+|[\sigma_{-\mu}(k)t_+(k), \sigma_\mu(k')t_-(k')]|0_i^+\rangle . \tag{17.70}$$

We use the identity (11.22) and the spin and isospin relations (2.47), (2.49) and (5.74) to evaluate the commutator. The result is

$$
\begin{aligned}
&\left[\sigma_{-\mu}(k)t_+(k), \sigma_\mu(k')t_-(k')\right] \\
&= 0 + \sigma_{-\mu}(k)\sigma_\mu(k')[t_+(k), t_-(k')] + [\sigma_{-\mu}(k), \sigma_\mu(k')]t_-(k')t_+(k) + 0 \\
&= \delta_{kk'}\Big\{\big[(-1)^\mu 1_2 - 2\sqrt{2}(1 -\mu\, 1\, \mu|1\,0)\sigma_0(k)\big] 2t_3(k) \\
&\quad - 2\sqrt{2}(1 -\mu\, 1\, \mu|1\,0)\sigma_0(k)t_-(k)t_+(k)\Big\} \,.
\end{aligned}
\tag{17.71}
$$

With the last two terms combined and $k' = k$, the sum over $\mu$ in (17.70) gives

$$
\sum_\mu \left\{2t_3(k) - 2\sqrt{2}(-1)^\mu(1 -\mu\, 1\, \mu|1\,0)\sigma_0(k)[2t_3(k) + t_-(k)t_+(k)]\right\}
$$

$$
= 6t_3(k) \,. \tag{17.72}
$$

Clebsch–Gordan orthogonality (1.26) causes the second term to vanish since $(-1)^\mu = -\sqrt{3}(1 -\mu\, 1\, \mu|0\,0)$ according to (1.34).

We are now in a position to proceed to the final result. Equation (17.70) and the isospin relation (5.81) yield

$$
S_- - S_+ = 6\sum_{k=1}^{A}\langle 0_i^+|t_3(k)|0_i^+\rangle = 6\langle 0_i^+|T_3|0_i^+\rangle = 3(N - Z) \,. \tag{17.73}
$$

The *Ikeda sum rule* [80] is thus

$$
\boxed{S_- - S_+ = 3(N - Z) \,,} \tag{17.74}
$$

where $N$ and $Z$ are the neutron and proton numbers of the initial $0^+$ state, as indicated by (17.73).

Our derivation of the Ikeda sum rule shows that the rule is independent of the structure of the initial $0^+$ state. It is only assumed that the state is normalized, as indicated by the last step in (17.73). The sum rule is also independent of the structure of the final $1^+$ states; the only assumption was that they form two complete sets. This means that the Ikeda sum rule is model independent. Any two complete sets of model wave functions should satisfy it. In this way the Ikeda sum rule serves as an independent test of any calculated complete sets of $1^+$ wave functions connected by Gamow–Teller transitions to a $0^+$ state.

## The Ikeda Sum Rule for the pnQTDA

Because of the model independence of the Ikeda sum rule we can use it to test pnQTDA wave functions. We start from the BCS ground state of the even–even reference nucleus and apply the beta-plus and beta-minus Gamow–Teller operators to it. This gives a set of $1^+$ states described by the pnQTDA wave functions (17.7). These are states in the two odd–odd isobars next to the reference nucleus. Note that a single pnQTDA calculation produces the $1^+$ states of both odd–odd nuclei in question. This is possible because wave functions computed with BCS (LNBCS) quasiparticles do not have a good nucleon number.

We take the states (17.7) as our (single) set of pnQTDA states. Their quantum numbers are $\omega = n_\omega, 1^+$. Expressed in terms of reduced matrix elements, $S_-$ is given by (17.64) as

$$S_- = \sum_{n_\omega} \left| (\omega \| \beta_{\mathrm{GT}}^- \| \mathrm{BCS}) \right|^2 . \tag{17.75}$$

The reduced matrix element is now the transition amplitude (17.51), whence

$$
\begin{aligned}
S_- &= \sum_{n_\omega} \left| \sum_{pn} X_{pn}^\omega \mathcal{M}_{\mathrm{GT}}^{(-)}(\mathrm{BCS} \to pnJ) \right|^2 \\
&= \sum_{n_\omega} \sum_{\substack{pn \\ p'n'}} X_{pn}^\omega X_{p'n'}^{\omega*} \mathcal{M}_{\mathrm{GT}}^{(-)}(\mathrm{BCS} \to pnJ) \left[ \mathcal{M}_{\mathrm{GT}}^{(-)}(\mathrm{BCS} \to p'n'J) \right]^* . 
\end{aligned} \tag{17.76}
$$

Use of the completeness relation (17.9) simplifies this to

$$S_- = \sum_{pn} \left| \mathcal{M}_{\mathrm{GT}}^{(-)}(\mathrm{BCS} \to pnJ) \right|^2 . \tag{17.77}$$

We substitute for the transition matrix element from (15.65), which results in

$$S_- = 3 \sum_{pn} u_p^2 v_n^2 [\mathcal{M}_{\mathrm{GT}}(pn)]^2 . \tag{17.78}$$

The strength $S_+$ is completely analogous. The minus indices are replaced by plus indices and Eq. (15.66) is used instead of (15.65). The result is

$$S_+ = \sum_{n_\omega} \left| (\omega \| \beta_{\mathrm{GT}}^+ \| \mathrm{BCS}) \right|^2 = 3 \sum_{pn} u_n^2 v_p^2 [\mathcal{M}_{\mathrm{GT}}(pn)]^2 . \tag{17.79}$$

As in the case of the Ikeda sum rule, we now form the difference

$$S_- - S_+ = 3 \sum_{pn} (u_p^2 v_n^2 - u_n^2 v_p^2)[\mathcal{M}_{\mathrm{GT}}(pn)]^2 = 3 \sum_{pn} (v_n^2 - v_p^2)[\mathcal{M}_{\mathrm{GT}}(pn)]^2 . \tag{17.80}$$

Substituting the expression (7.21) for the Gamow–Teller single-particle matrix element gives

$$S_- - S_+ = 6 \sum_{pn} (v_n^2 - v_p^2)\delta_{n_p n_n} \delta_{l_p l_n} \hat{j}_p^2 \hat{j}_n^2 \left\{ \begin{matrix} \frac{1}{2} & \frac{1}{2} & 1 \\ j_n & j_p & l_p \end{matrix} \right\}^2 . \tag{17.81}$$

Remembering that $l_p = l_n$ we can separate the sums and write (17.81) as

$$S_- - S_+ = 6 \sum_n v_n^2 \hat{j}_n^2 \sum_{j_p} \hat{j}_p^2 \left\{ \begin{array}{ccc} \frac{1}{2} & 1 & \frac{1}{2} \\ j_n & l_n & j_p \end{array} \right\}^2$$

$$- 6 \sum_p v_p^2 \hat{j}_p^2 \sum_{j_n} \hat{j}_n^2 \left\{ \begin{array}{ccc} \frac{1}{2} & 1 & \frac{1}{2} \\ j_p & l_p & j_n \end{array} \right\}^2 . \tag{17.82}$$

By the $6j$ unitarity relation (1.66) the sums over $j_p$ and $j_n$ are both equal to $\frac{1}{2}$. Hence the final result is

$$S_- - S_+ = 3 \sum_n \hat{j}_n^2 v_n^2 - 3 \sum_p \hat{j}_p^2 v_p^2 = 3 \left( N_{\text{act}} - Z_{\text{act}} \right) , \tag{17.83}$$

where the BCS number constraint (13.72) was used in the last step.

The result (17.83) shows that indeed the wave functions from a single pnQTDA calculation satisfy the model-independent Ikeda sum rule, but with the difference that the true nucleon numbers are replaced by the active numbers chosen for the reference nucleus. For the pnQTDA the Ikeda sum rule thus reads

$$\boxed{S_- - S_+ = 3(N_{\text{act}} - Z_{\text{act}}) \quad \text{(pnQTDA)} ,} \tag{17.84}$$

where the two total strengths are given by (17.75) and (17.79).

Note that the $j_p$ and $j_n$ sums in (17.82) run over the two *spin–orbit partners* $j = l \pm \frac{1}{2}$. If one of the partners is missing from the valence space the sum rule is not satisfied. This is demonstrated by an example in the following subsection.

In what follows we study the implications of the Ikeda sum rule for actual pnQTDA calculations.

### 17.4.4 Examples of the Ikeda Sum Rule

We consider the Ikeda sum rule for a few pnQTDA calculations in the d-s and 0f-1p-0g$_{9/2}$ valence spaces. Our first example is the reference nucleus $^{24}$Mg in the d-s shell. In the BCS calculation we use the single-particle energies (14.12) and the pairing parameters of Table 16.5. For the pnQTDA we use the SDI interaction parameters $A_0 = A_1 = 1.0\,\text{MeV}$. The resulting pnQTDA energies and transition strengths are listed in Table 17.7. The transition strengths are those in the Ikeda sum rule (17.84). The summed strengths are stated in the last line of the table.

Table 17.7 gives the strengths $S_-$ and $S_+$. For the reference nucleus $^{24}_{12}\text{Mg}_{12}$ we have $Z_{\text{act}} = 4$, $N_{\text{act}} = 4$. The left side and right side of the Ikeda sum rule (17.84) are now

$$S_- - S_+ = 8.424 - 8.424 = 0.000 ,$$
$$3(N_{\text{act}} - Z_{\text{act}}) = 3(4 - 4) = 0 . \tag{17.85}$$

**Table 17.7.** Ikeda sum rule for a pnQTDA calculation with the reference nucleus $^{24}$Mg

| $n$ | $E(1_n^+)$ (MeV) | $\left|(1_n^+\|\beta_{GT}^-\|BCS)\right|^2$ | $\left|(1_n^+\|\beta_{GT}^+\|BCS)\right|^2$ |
|---|---|---|---|
| 1 | 6.202 | 2.9826 | 2.9960 |
| 2 | 6.389 | 0.9636 | 0.9526 |
| 3 | 8.389 | 0.7442 | 0.7533 |
| 4 | 8.700 | 2.5529 | 2.6202 |
| 5 | 9.317 | 0.1644 | 0.1419 |
| 6 | 9.548 | 1.0111 | 0.9557 |
| 7 | 11.292 | 0.0056 | 0.0047 |
| Sum |  | 8.424 | 8.424 |

The second column lists the pnQTDA energies. Columns three and four give the $\beta^-$ and $\beta^+$ Gamow–Teller strengths and the last row gives their sums.

Thus the sum rule is indeed satisfied. The table also shows that the $\beta^-$ and $\beta^+$ strength distributions are nearly the same.

The reference nucleus of our next example is $^{30}$Si. Its $\beta^-$ and $\beta^+$ strength distributions are shown in Fig. 17.5. Again the valence space is 0d-1s with the single-particle energies (14.12), the pairing parameters are from Table 16.6, and the pnQTDA parameters are $A_0 = A_1 = 1.0$ MeV.

In this case the $\beta^-$ and $\beta^+$ strength distributions are not as symmetric as in the previous case. Now the total $\beta^-$ strength is larger than the total $\beta^+$ strength as required by the Ikeda sum rule. With the reference nucleus $^{30}_{14}$Si$_{16}$ we have $Z_{\mathrm{act}} = 6$, $N_{\mathrm{act}} = 8$. The two sides of the sum rule give

$$S_- - S_+ = 13.163 - 7.163 = 6.000 \ ,$$
$$3(N_{\mathrm{act}} - Z_{\mathrm{act}}) = 3(8 - 6) = 6 \ , \tag{17.86}$$

so the rule is satisfied. Figure 17.5 shows that a large part of the sum rule is exhausted by a $\beta^-$ transition to a state slightly below 10 MeV of excitation in $^{30}$P. On the $\beta^+$ side there is a smaller major peak at around 6 MeV of excitation in $^{30}$Al.

Our final example concerns the 0f-1p-0g$_{9/2}$ shell. The reference nucleus is $^{66}$Zn. We make two calculations, one in the 0f-1p valence space and the other in the 0f-1p-0g$_{9/2}$ valence space. The single-particle energies are from (14.13) and the pairing parameters from Table 16.6. In both pnTDA calculations the SDI parameters are $A_0 = 1.0$ MeV and $A_1 = 0.8$ MeV.

The $\beta^-$ and $\beta^+$ strength distributions from our two calculations are displayed in Fig. 17.6. The dashed bars represent the 0f-1p calculation and the solid bars the 0f-1p-0g$_{9/2}$ calculation. For both valence spaces a considerable part of the sum rule is exhausted by a single beta-minus transition, as was the case in the $^{30}$Si example. The strong $1^+$ state resides at around 14 MeV of excitation in $^{66}$Ga. On the beta-plus side we have for both valence spaces one strong peak at around 8 MeV of excitation in $^{66}$Cu.

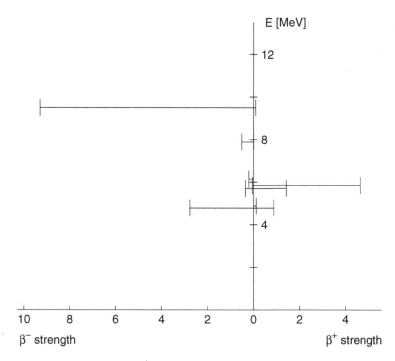

**Fig. 17.5.** Calculated $\beta^-$ and $\beta^+$ strengths for Gamow–Teller transitions from the BCS ground state of $^{30}$Si to the $1^+$ pnQTDA states in $^{30}$P and $^{30}$Al

One can see from Fig. 17.6 that the total $\beta^-$ strength far exceeds the total $\beta^+$ strength. This is consistent with the Ikeda sum rule (17.84). With the reference nucleus $^{66}_{30}\text{Zn}_{36}$ we have $Z_{\text{act}} = 10$, $N_{\text{act}} = 16$. For the 0f-1p calculation the two sides of the sum rule give

$$S_- - S_+ = 26.269 - 8.269 = 18.000 \,,$$
$$3(N_{\text{act}} - Z_{\text{act}}) = 3(16 - 10) = 18 \,. \tag{17.87}$$

For the 0f-1p-0g$_{9/2}$ calculation we have similarly

$$S_- - S_+ = 27.047 - 11.797 = 15.250 \,,$$
$$3(N_{\text{act}} - Z_{\text{act}}) = 3(16 - 10) = 18 \,. \tag{17.88}$$

From (17.87) and (17.88) we see that while the Ikeda sum rule is satisfied for the 0f-1p valence space it is *not* satisfied for the 0f-1p-g$_{9/2}$ space. This is because the 1p-0f-0g$_{9/2}$ valence space does not contain the spin–orbit partner 0g$_{7/2}$ of the intruder orbital 0g$_{9/2}$. This failure of the sum rule was discussed at the end of Subsect. 17.4.3.

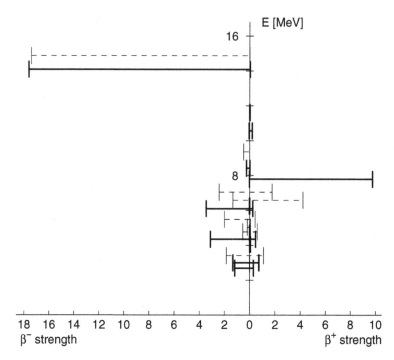

**Fig. 17.6.** Calculated $\beta^-$ and $\beta^+$ strengths for Gamow–Teller transitions from the BCS ground state of $^{66}$Zn to the $1^+$ pnQTDA states in $^{66}$Ga and $^{66}$Cu. The dashed bars represent the 0f-1p calculation and the solid bars the 0f-1p-0g$_{9/2}$ calculation

### 17.4.5 Gamow–Teller Giant Resonance

Figures 17.5 and 17.6 contain one peak at $E \gtrsim 10$ MeV that towers above the rest. This peak appears on the $\beta^-$ side and represents a high-lying excited $1^+$ state in the pnQTDA spectrum (in these cases the highest). Because the state is strongly connected by the Gamow–Teller operator to the ground state of the even–even reference nucleus it bears the name *Gamow–Teller giant resonance*, GTGR for short.

In both cases the GTGR wave function is dominated by one component, namely $|\pi 0d_{3/2}\nu 0d_{5/2}; 1^+\rangle$ for $^{30}$Si and $|\pi 0f_{5/2}\nu 0f_{7/2}; 1^+\rangle$ for $^{66}$Zn. These configurations consist of spin–orbit partners. The Gamow–Teller operator connects this type of configuration strongly to the BCS ground state of the reference nucleus. This is because the magnitude of the BCS factor $u_p v_n$ in the $\beta^-$ amplitude (15.65) is nearly unity, due to a nearly empty, high-lying proton orbital and a nearly full, low-lying neutron orbital. The configuration is nearly of particle–hole character and has a high two-quasiparticle energy, which leads to the high excitation energy of the GTGR.

Figures 17.5 and 17.6 display one rather strong peak also on the $\beta^+$ side. This peak stems from the reverse combination of the spin–orbit partners, namely the component $|\pi 0d_{5/2}\nu 0d_{3/2}; 1^+\rangle$ for $^{30}$Si and $|\pi 0f_{5/2}\nu 0f_{7/2}; 1^+\rangle$ for $^{66}$Zn. The occupation factor $u_n v_p$, from (15.66), is enhanced by the same mechanism as for the GTGR, but not as strongly. This is because the proton side is less occupied than the neutron side ($v_p < v_n$) and the neutron side is less empty than the proton side ($|u_n| < |u_p|$), so that $|u_n v_p| < |u_p v_n|$.

The strength difference between the GTGR and the reverse spin–orbit state grows with increasing proton–neutron asymmetry. This happens when we go to heavier nuclei or further away from the valley of beta stability towards neutron-rich nuclei. The reverse state may lose most of its strength through mixing with other low-lying $1^+$ excitations. This mixing can redistribute the spin–orbit strength in such a way that no clearly visible peak is detected on the $\beta^+$ side. The residual interaction can redistribute some of the spin–orbit configuration strength also in the vicinity of the GTGR. In spite of this the GTGR remains visible with some acquired finite width.

There are empirical formulas for the energy of the GTGR, and we present one of them [81, 82]. Denote the energy of the GTGR $(Z+1, N-1)$ relative to the ground state of the reference nucleus $(Z, N)$ by

$$\Delta E_{\mathrm{GT}} \equiv E(1^+_{\mathrm{GTGR}}) - E(0^+_{\mathrm{gs}}) . \tag{17.89}$$

The empirical formula for $\Delta E_{\mathrm{GT}}$ is composed of two terms. One of them is the Coulomb energy, which we denote by $\Delta E_{\mathrm{C}}$. To define it, recall the isospin multiplets (triplets) in Figs. 5.4–5.11. For example, Fig. 5.4 shows the even–even nucleus $^{6}_{2}$He$_4$ and the adjacent odd–odd nucleus $^{6}_{3}$Li$_3$. The $0^+$ ground state of $^6$He, of isospin $T = 1$ and $M_T = +1$, and the $0^+$ state of $^6$Li, of isospin $T = 1$ and $M_T = 0$, are an example of a pair of *isobaric analogue states* (IAS). They are shown level in the figure, but this is because the Coulomb energy has been subtracted. In reality, the energy difference between the $0^+$ states of $^6$Li and $^6$He is just $\Delta E_{\mathrm{C}}$.

The second term entering the empirical expression for (17.89) is the energy difference, within the odd–odd $(Z+1, N-1)$ nucleus, between the GTGR and the $0^+$ state that is the IAS of the ground state of the reference nucleus. This difference is $E(1^+_{\mathrm{GTGR}}) - E(0^+_{\mathrm{IAS}}) \equiv \Delta E_{Z+1,N-1}$. Empirical systematics of $\Delta E_{\mathrm{C}}$ and $\Delta E_{Z+1,N-1}$ have produced the formula

$$\begin{aligned}
\Delta E_{\mathrm{GT}} &= \Delta E_{\mathrm{C}} + \Delta E_{Z+1,N-1} \\
&= [1.444(Z + \tfrac{1}{2})A^{-1/3} - 30.0(N - Z - 2)A^{-1} + 5.57]\,\mathrm{MeV} . \tag{17.90}
\end{aligned}$$

Since the spin–orbit partner configuration driving the GTGR is largely a particle–hole configuration, the particle–hole part of the pnQTDA matrix element (17.2) is dominant. In some studies that part has therefore been multiplied by a phenomenological constant $g_{\mathrm{ph}}$, called the *particle–hole interaction strength*. The value of $g_{\mathrm{ph}}$ can be fixed by fitting the empirical location (17.90) of the GTGR. For further information see Subsect. 19.7.4 and [51].

### 17.4.6 Beta-Decay Transitions Between a QTDA and a pnQTDA State

Consider $\beta^-$ and $\beta^+$ decay transitions between a pnQTDA excitation and a QTDA excitation. As discussed in Chap. 16, the QTDA describes excited states of an even–even reference nucleus. Now we take the QTDA states (16.55) to be final states,

$$|\omega_f\rangle = \sum_{a_f \leq b_f} X^{\omega_f}_{a_f b_f} A^\dagger_{a_f b_f}(J_f M_f)|\text{BCS}\rangle . \tag{17.91}$$

As an initial state we have a pnQTDA excitation (17.7),

$$|\omega_i\rangle = \sum_{p_i n_i} X^{\omega_i}_{p_i n_i} A^\dagger_{p_i n_i}(J_i M_i)|\text{BCS}\rangle . \tag{17.92}$$

For allowed beta decay, the contributions from the wave functions (17.91) and (17.92) are the two-quasiparticle matrix elements given by (15.111) and (15.113). Thus we can write directly the decay amplitude

$$
\begin{aligned}
(\omega_f\|\beta^\mp_{\text{F/GT}}\|\omega_i) = &\sum_{\substack{p_i n_i \\ p_f \leq p'_f}} X^{\omega_f *}_{p_f p'_f} X^{\omega_i}_{p_i n_i} \mathcal{M}^{(\mp)}_{\text{F/GT}}(p_i\, n_i\,;\, J_i \to p_f\, p'_f\,;\, J_f) \\
&+ \sum_{\substack{p_i n_i \\ n_f \leq n'_f}} X^{\omega_f *}_{n_f n'_f} X^{\omega_i}_{p_i n_i} \mathcal{M}^{(\mp)}_{\text{F/GT}}(p_i\, n_i\,;\, J_i \to n_f\, n'_f\,;\, J_f) .
\end{aligned}
$$

$$\tag{17.93}$$

For $K$th-forbidden unique beta decay the transition matrix elements in (17.93) are replaced as explained at the end of Subsect. 15.5.2.

### 17.4.7 Gamow–Teller Beta Decay of $^{30}$P

Let us apply (17.93) to the $\beta^+$/EC decay of the $1^+$ ground state of $^{30}$P to the $2^+_1$ state of $^{30}$Si; see Fig. 15.4. The reference nucleus is $^{30}$Si. We take the complete 0d-1s shell as the valence space and use the parameters from Subsect. 17.2.2. The calculated pnQTDA spectrum of $^{30}$P is shown in Fig. 17.2. The same parameters were used in the QTDA calculation in Subsect. 16.4.5, with the resulting excitation spectrum of $^{30}$Si shown in Fig. 16.3.

The pnQTDA wave function for the state $|1^+_1\rangle \equiv |1^+_{\text{gs}}\rangle$ is given in (17.54), with the $X$ amplitudes listed in Table 17.6. The structure of the $2^+_1$ QTDA wave function is

$$|^{30}\text{Si}\,;\, 2^+_1\rangle = \sum_{k=1}^{10} X^{2^+_1}_k |k\rangle_2 . \tag{17.94}$$

The two-quasiparticle configurations $k$ of the basis and the $X$ amplitudes are given in Table 17.8, together with the BCS occupation amplitudes needed to compute the $1^+_{\text{gs}} \to 2^+_1$ decay amplitude.

**Table 17.8.** Structure of the $2_1^+$ state of $^{30}$Si

| $k$ | Configuration | $X_k^{2_1^+}$ | $u_1$ | $v_1$ | $u_2$ | $v_2$ |
|---|---|---|---|---|---|---|
| 1 | $\pi 0d_{5/2}\pi 0d_{5/2}$ | 0.462 | 0.4996 | 0.8662 | 0.4996 | 0.8662 |
| 2 | $\pi 0d_{5/2}\pi 1s_{1/2}$ | 0.679 | 0.4996 | 0.8662 | 0.6330 | 0.7741 |
| 3 | $\pi 0d_{5/2}\pi 0d_{3/2}$ | 0.135 | 0.4996 | 0.8662 | 0.9619 | 0.2735 |
| 4 | $\pi 1s_{1/2}\pi 0d_{3/2}$ | 0.261 | 0.6330 | 0.7741 | 0.9619 | 0.2735 |
| 5 | $\pi 0d_{3/2}\pi 0d_{3/2}$ | 0.115 | 0.9619 | 0.2735 | 0.9619 | 0.2735 |
| 6 | $\nu 0d_{5/2}\nu 0d_{5/2}$ | 0.161 | 0.2285 | 0.9735 | 0.2285 | 0.9735 |
| 7 | $\nu 0d_{5/2}\nu 1s_{1/2}$ | 0.260 | 0.2285 | 0.9735 | 0.2943 | 0.9557 |
| 8 | $\nu 0d_{5/2}\nu 0d_{3/2}$ | 0.110 | 0.2285 | 0.9735 | 0.9372 | 0.3488 |
| 9 | $\nu 1s_{1/2}\nu 0d_{3/2}$ | 0.272 | 0.2943 | 0.9557 | 0.9372 | 0.3488 |
| 10 | $\nu 0d_{3/2}\nu 0d_{3/2}$ | 0.214 | 0.9372 | 0.3488 | 0.9372 | 0.3488 |

The second column lists the configurations and the third column the QTDA amplitudes. The last four columns list the BCS occupation amplitudes of the first (index 1) and second (index 2) orbitals in column two

In the present case the $\beta^+$ transition matrix elements in (17.93) are

$$\mathcal{M}_{\mathrm{GT}}^{(+)}(p_i\,n_i\,;\,1 \rightarrow p_f\,p'_f\,;\,2) = 3\sqrt{5}(-1)^{j_{p_f}+j_{n_i}+1}\mathcal{N}_{p_f p'_f}(2)v_{n_i}\left\{\begin{array}{ccc} 2 & 1 & 1 \\ j_{n_i} & j_{p_f} & j_{p'_f} \end{array}\right\}$$

$$\times\,[\delta_{p_i p'_f}v_{p_f}\mathcal{M}_{\mathrm{GT}}(p_f n_i) + \delta_{p_i p_f}v_{p'_f}\mathcal{M}_{\mathrm{GT}}(p'_f n_i)] \qquad (17.95)$$

and

$$\mathcal{M}_{\mathrm{GT}}^{(+)}(p_i\,n_i\,;\,1 \rightarrow n_f\,n'_f\,;\,2) = 3\sqrt{5}(-1)^{j_{p_i}+j_{n'_f}}\mathcal{N}_{n_f n'_f}(2)u_{p_i}\left\{\begin{array}{ccc} 2 & 1 & 1 \\ j_{p_i} & j_{n_f} & j_{n'_f} \end{array}\right\}$$

$$\times\,[\delta_{n_i n'_f}u_{n_f}\mathcal{M}_{\mathrm{GT}}(p_i n_f) + \delta_{n_i n_f}u_{n'_f}\mathcal{M}_{\mathrm{GT}}(p_i n'_f)]\,. \qquad (17.96)$$

The non-zero matrix elements (17.95) and (17.96) are listed in Table 17.9. Most of the matrix elements turn out to be zero because of the Kronecker deltas and the selection rules on the Gamow–Teller single-particle matrix elements. From the table we can see that the dominating contribution comes from the pnQTDA component $|\pi 0d_{5/2}\nu 0d_{5/2}\,;\,1^+\rangle$ and the QTDA component $|(\pi 0d_{5/2})^2\,;\,2^+\rangle$. These components are based solely on the $0d_{5/2}$ orbital. The remaining contributions together change the final sum by only 6 %.

The same transition amplitude was calculated in the example of Subsect. 15.5.3. There the structures (15.74) and (15.75) of the wave functions were based on an educated guess. It is remarkable that these wave functions do not contain the $0d_{5/2}$ orbital, which now turned out to be responsible for practically all of the transition amplitude.

We can now use the sum of the contributions, given in the last line of Table 17.9, to write down the value of the final transition amplitude:

$$(2_1^+\|\beta_{\mathrm{GT}}^+\|1_1^+) = 0.437\,. \qquad (17.97)$$

**Table 17.9.** Non-zero contributions to the sum (17.93) for the $\beta^+$/EC transition $^{30}\text{P}(1^+_{\text{gs}}) \rightarrow {}^{30}\text{Si}(2^+_1)$

| $k$(pnQTDA) | $k'$(QTDA) | $\mathcal{M}^{(+)}_{\text{GT}}(k \rightarrow k')$ | $X_k^{1^+_{\text{gs}}} X_{k'}^{2^+_1}$ | Contribution |
|:---:|:---:|:---:|:---:|:---:|
| 1 | 1 | −1.9082 | −0.217 | 0.414 |
| 1 | 3 | 0.3195 | −0.063 | −0.020 |
| 1 | 6 | 0.2584 | −0.076 | −0.020 |
| 1 | 8 | −0.5619 | −0.052 | 0.029 |
| 2 | 1 | −0.6395 | 0.037 | −0.024 |
| 2 | 3 | 0.1071 | 0.011 | 0.001 |
| 2 | 8 | 0.1282 | 0.009 | 0.001 |
| 2 | 10 | 0.3244 | 0.017 | 0.006 |
| 4 | 2 | −0.8107 | −0.081 | 0.066 |
| 4 | 4 | 0.0522 | −0.031 | −0.002 |
| 4 | 9 | 0.5102 | −0.033 | −0.017 |
| 5 | 3 | 0.9466 | 0.015 | 0.014 |
| 5 | 5 | 0.1845 | 0.013 | 0.002 |
| 5 | 6 | −0.4653 | 0.018 | −0.008 |
| 5 | 8 | 1.0119 | 0.012 | 0.012 |
| 6 | 4 | 2.0262 | −0.007 | −0.014 |
| 6 | 7 | −0.5898 | −0.007 | 0.004 |
| 6 | 9 | 0.4938 | −0.008 | −0.004 |
| 7 | 3 | 0.6783 | 0.002 | 0.001 |
| 7 | 5 | 0.1322 | 0.001 | 0.000 |
| 7 | 8 | −0.4935 | 0.001 | −0.000 |
| 7 | 10 | −1.2491 | 0.003 | −0.004 |
| Sum | | | | 0.437 |

The first two columns list the pnQTDA and QTDA configurations indexed according to Tables 17.6 and 17.8. Column three gives the matrix elements (17.95) and (17.96). Column four lists the products of the pnQTDA and QTDA wave-function amplitudes from Tables 17.6 and 17.8. The last column gives the contributions to the sum (17.93). The sum is stated in the bottom row.

This gives

$$B_{\text{GT}} = \frac{1.25^2}{3} \times 0.437^2 = 0.0995 , \quad \log ft = 4.79 . \tag{17.98}$$

Comparison with the experimental value $\log ft = 5.8$ from Fig. 15.4 shows that we predict too fast a transition.

From Table 17.9 it is evident that there are few strong transition matrix elements $\mathcal{M}^{(+)}_{\text{GT}}(k \rightarrow k')$. The largest occurs for $k = 6, k' = 4$ and the next largest for $k = 1 = k'$; the latter also provides the lion's share to the final decay amplitude. The $k = 6, k' = 4$ contribution is negative while the dominant contribution is positive. This indicates that relatively small changes in the $X$ amplitudes can result in a notable change in the final outcome.

# Epilogue

In this chapter we have developed a formalism which is able to describe states of open-shell odd–odd nuclei by starting from the BCS ground state of the neighbouring even–even reference nucleus. In the remaining two chapters of the book we increase the level of sophistication in our attempt to describe states of even–even and odd–odd open-shell nuclei. The framework to accomplish this is the QRPA, which builds on a correlated ground state consisting of the BCS vacuum and its many-quasiparticle excitations.

# Exercises

**17.1.** Derive the orthogonality relation (17.8).

**17.2.** Derive the completeness relation (17.9).

**17.3.** Complete the details of the derivation of Eqs. (17.22).

**17.4.** Verify the numbers in Table 17.1.

**17.5.** Verify the numbers in the matrix (17.30).

**17.6.** Verify the wave function (17.32).

**17.7.** Show that there is only one combination of the indices $p_i n_i$ and $p_f n_f$ that produces a non-zero M1 transition matrix element in (17.34)–(17.36).

**17.8.** Verify the numbers in the second column of Table 17.5.

**17.9.** Verify the numbers in Table 17.6.

**17.10.** Derive (17.64) from (17.63).

**17.11.** Verify the numbers in the third column of Table 17.7.

**17.12.** Verify the numbers in the first two rows of Table 17.9.

**17.13.** Continuation of Exercise 16.16.
Use the reference nucleus $^{20}$Ne to discuss the $2^+$ states of $^{20}$Na. Use the SDI with parameters $A_0 = A_1 = 1.0$ MeV.

(a) Form the pnQTDA matrix for the $2^+$ states.
(b) Diagonalize the matrix and find the wave function of the lowest $2^+$ state of $^{20}$Na.

**17.14.** Continuation of Exercise 17.13.
Calculate the $\log ft$ value for the $\beta^+$/EC decay of the $2^+$ ground state of $^{20}$Na to the first $2^+$ state in $^{20}$Ne. The wave function of that state was calculated in Exercise 16.16. Compare the result with experimental data and comment.

**17.15.** Continuation of Exercise 16.18.
Calculate the $\log ft$ values for the second-forbidden unique $\beta^+/$EC decays of the $5^+$ ground state of $^{26}$Al to the two lowest $2^+$ states in $^{26}$Mg. Assume a one-component structure for the $5^+$ state. Compute also the decay half-life of the $5^+$ state. Compare with experimental data and comment.

**17.16.** Continuation of Exercise 17.15.
Calculate the $\log ft$ value for to the $\beta^-$ decay of the $3^+$ ground state of $^{26}$Na to the first $2^+$ state in $^{26}$Mg. Compare with experimental data and comment.

**17.17.** Calculate the $\log ft$ values for the $\beta^-$ and $\beta^+/$EC decays of the $3^+$ ground states of $^{28}$Al and $^{28}$P to the first $2^+$ state in $^{28}$Si. For the reference nucleus $^{28}$Si use the BCS results of Table 14.1. Diagonalize the QTDA and pnQTDA matrices in the $0d_{5/2}$-$1s_{1/2}$ valence space to find the wave functions. Use the SDI with parameters $A_0 = A_1 = 1.0\,$MeV. Compare with experimental data and comment.

**17.18.** Consider $1^+$ states in $^{30}$P. Start from the reference nucleus $^{30}$Si and use the BCS results of Table 14.1, computed in the full d-s valence space. Write down the pnQTDA matrix in the sub-basis $1s_{1/2}$-$0d_{3/2}$ for protons and $0d_{5/2}$-$1s_{1/2}$ for neutrons. Use the SDI with parameters $A_0 = A_1 = 1.0\,$MeV.

**17.19.** Continuation of Exercise 17.18.
Diagonalize the pnQTDA matrix to find the eigenenergies and eigenstates. Compare the eigenenergies with experimental data and with the results of the complete d-s shell calculation of Fig. 17.2. Compare the eigenstates with the guessed wave functions of Subsect. 15.4.2.

**17.20.** Continuation of Exercise 17.19.
Calculate the $\log ft$ values for the $\beta^+/$EC decay of $^{30}$S to the $1^+$ states in $^{30}$P. Compare with experimental data and comment.

**17.21.** Continuation of Exercise 17.20.
Evaluate the difference $S_- - S_+$ of the total strengths for Gamow–Teller transitions from the ground state of $^{30}$Si to the $1^+$ states of $^{30}$P and $^{30}$Al. Compare with the value of the Ikeda sum rule. Explain your observation.

**17.22.** Consider $1^+$ states in $^{34}$P and $^{34}$Cl. Start from the reference nucleus $^{34}$S and use the BCS results of Table 14.1, computed in the full d-s valence space.

(a) Form the pnQTDA matrix for the $1^+$ states of $^{34}$P and $^{34}$Cl in the sub-basis (dictated by the lowest quasiparticle energies) $0d_{3/2}$ for neutrons and $1s_{1/2}$-$0d_{3/2}$ for protons.
(b) Diagonalize the pnQTDA matrix to find the eigenenergies and eigenstates. Use the SDI with parameters $A_0 = A_1 = 1.0\,$MeV. Compare with experimental data and comment.

**17.23.** Continuation of Exercise 17.22.
Calculate the $\log ft$ values for the beta decays
(a) $^{34}\mathrm{P}(1^+_{\mathrm{gs}}) \rightarrow {}^{34}\mathrm{S}(0^+_{\mathrm{gs}})$,
(b) $^{34}\mathrm{Cl}(0^+_{\mathrm{gs}}) \rightarrow {}^{34}\mathrm{S}(0^+_{\mathrm{gs}})$.
Compare with experimental data.

**17.24.** Study the validity of the Ikeda sum rule for the Gamow–Teller transitions from the ground state of $^{34}\mathrm{S}$ to the $1^+$ states in $^{34}\mathrm{P}$ and $^{34}\mathrm{Cl}$. Perform a pnQTDA calculation set up in Exercise 17.22.

**17.25.** Continuation of Exercises 17.22 and 17.23.
Calculate the $\log ft$ values for the beta decays
(a) $^{34}\mathrm{P}(1^+_{\mathrm{gs}}) \rightarrow {}^{34}\mathrm{S}(2^+_1)$,
(b) $^{34}\mathrm{Cl}(3^+_1) \rightarrow {}^{34}\mathrm{S}(2^+_1)$.
The wave function of the $2^+_1$ state has been calculated in Exercise 16.23.
Compare with experimental data.

**17.26.** Continuation of Exercise 17.22.
Calculate the partial half-lives of the following electromagnetic transitions in $^{34}\mathrm{Cl}$:
(a) $1^+_1 \rightarrow 3^+_1$,
(b) $1^+_1 \rightarrow 0^+_{\mathrm{gs}}$.
(c) Calculate also the total half-life of the $1^+_1$ state.
Use experimental gamma energies. Compare with experimental data and comment.

**17.27.** Continuation of Exercise 16.27.
Consider the $1^+$ states of $^{38}\mathrm{K}$ by using $^{38}\mathrm{Ar}$ as the reference nucleus.

(a) Form the pnQTDA matrix in the sub-basis $1\mathrm{s}_{1/2}$-$0\mathrm{d}_{3/2}$ for both protons and neutrons.
(b) Diagonalize the pnQTDA matrix to find the eigenenergies and eigenstates. Use the SDI with parameters $A_0 = A_1 = 1.0\,\mathrm{MeV}$. Compare with the experimental spectrum and the result of the two-hole calculation of Fig. 8.8. Comment on the similarities and differences.

**17.28.** Continuation of Exercise 17.27.
Calculate the $\log ft$ values for the $\beta^+/\mathrm{EC}$ decays $^{38}\mathrm{Ca}(0^+_{\mathrm{gs}}) \rightarrow {}^{38}\mathrm{K}(1^+_m)$, $m = 1, 2, 3, 4$. Note that by isospin symmetry

$$\mathcal{M}^{(+)}_{\mathrm{GT}}(^{38}\mathrm{Ca} \rightarrow {}^{38}\mathrm{K}) = \mathcal{M}^{(-)}_{\mathrm{GT}}(^{38}\mathrm{Ar} \rightarrow {}^{38}\mathrm{K}) . \qquad (17.99)$$

Compare with experimental data by plotting the theoretical and experimental $1^+$ spectra and indicating the $\beta^+/\mathrm{EC}$ feeding. Comment on the results.

**17.29.** Continuation of Exercise 17.28.
Evaluate the difference $S_- - S_+$ of the total strengths for transitions from the ground state of $^{38}\mathrm{Ar}$ to the $1^+$ states of $^{38}\mathrm{K}$ and $^{38}\mathrm{Cl}$. Apply the Ikeda sum rule and comment.

# Two-Quasiparticle Mixing by the QRPA

## Prologue

In the previous two chapters we introduced two-quasiparticle configuration mixing. The method was based on the QTDA. In this chapter we extend the formalism to the QRPA. We derive the QRPA equations by the equations-of-motion method. Due to approximations in the derivation the resulting equations do not satisfy a variational principle. The properties of QRPA solutions are similar to those of the particle–hole RPA of Chap. 11.

The most significant improvement of the QRPA over the QTDA is the replacement of the BCS vacuum by a correlated vacuum as the ground state of an even–even nucleus. The ground-state correlations promote collectivity of electromagnetic decays, which was the case also in the particle–hole RPA. In both theories, electromagnetic transitions obey the EWSR.

## 18.1 The QRPA Equations

As was the case for the RPA in Sect. 11.2, the QRPA equations can be derived in several ways. We choose the EOM method since it has been systematically applied earlier in this book.

We derived the QTDA equation (16.47) by the EOM. The derivation was based on the use of the BCS vacuum as the exact vacuum of the QTDA. Thus the QTDA equations emerge from a variational principle. Below we derive also the QRPA equations by using the BCS vacuum as the operative vacuum during the derivation. This vacuum, however, is not the exact vacuum of the QRPA. The resulting QRPA equations are therefore approximate and do not satisfy a variational principle. This is analogous to the particle–hole RPA of Chap. 11 where the operative vacuum $|\Psi_0\rangle = |\mathrm{HF}\rangle$ was not the exact ground state.

### 18.1.1 Derivation of the QRPA Equations by the EOM

The starting point in the EOM is the definition of a suitable excitation operator according to (11.1). Table 11.1 contains a compilation of excitation operators and related basic elements. It gives the QRPA excitation operator as

$$Q^\dagger_\omega = \sum_{a \leq b} \left[ X^\omega_{ab} A^\dagger_{ab}(JM) - Y^\omega_{ab} \tilde{A}_{ab}(JM) \right] , \qquad (18.1)$$

where $\omega = nJ^\pi M$ and the summation is restricted to avoid double counting. The two-quasiparticle operators were defined in (11.17) and (11.19),

$$A^\dagger_{ab}(JM) = \mathcal{N}_{ab}(J) \left[ a^\dagger_a a^\dagger_b \right]_{JM} , \qquad (18.2)$$

$$\tilde{A}_{ab}(JM) = (-1)^{J+M} A_{ab}(J-M) = -\mathcal{N}_{ab}(J) \left[ \tilde{a}_a \tilde{a}_b \right]_{JM} . \qquad (18.3)$$

From (18.1) we see immediately that $Q_\omega|\text{BCS}\rangle \neq 0$, so the simple BCS vacuum is not the vacuum of the QRPA. Instead, many-quasiparticle components are expected to appear in the QRPA vacuum, in analogy to (11.91). That is why we call it a correlated vacuum.

Hermitian conjugation of the basic excitation (18.1) gives

$$Q_\omega = \sum_{a \leq b} \left[ X^{\omega*}_{ab} A_{ab}(JM) - Y^{\omega*}_{ab} \tilde{A}^\dagger_{ab}(JM) \right] . \qquad (18.4)$$

From Table 11.1 we obtain the basic variations as

$$\delta Q = A_{ab}(JM) , \quad \delta Q = \tilde{A}^\dagger_{ab}(JM) , \qquad (18.5)$$

and they are Bose-like entities. With $|\text{BCS}\rangle$ as the approximate vacuum $|\Psi_0\rangle$ the equation of motion (11.11) becomes

$$\langle \text{BCS}|[\delta Q, \mathcal{H}, Q^\dagger_\omega]|\text{BCS}\rangle = E_\omega \langle \text{BCS}|[\delta Q, Q^\dagger_\omega]|\text{BCS}\rangle , \qquad (18.6)$$

where $\mathcal{H}$ is the complete Hamiltonian expressed in terms of quasiparticle operators in (16.32).

To evaluate the right-hand side of (18.6) we need the BCS expectation values of the commutators of the quasiparticle pair operators $A$ and $A^\dagger$. Equations (16.27) and (16.58), together with (18.3), give

$$\langle \text{BCS}|[A_{ab}(JM), A^\dagger_{cd}(J'M')]|\text{BCS}\rangle = \delta_{ac}\delta_{bd}\delta_{JJ'}\delta_{MM'} , \qquad (18.7)$$

$$\langle \text{BCS}|[\tilde{A}_{ab}(JM), \tilde{A}^\dagger_{cd}(J'M')]|\text{BCS}\rangle = \delta_{ac}\delta_{bd}\delta_{JJ'}\delta_{MM'} . \qquad (18.8)$$

Here and in the sequel we assume the restrictions

$$a \leq b , \quad c \leq d . \qquad (18.9)$$

Equations (18.7) and (18.8) are analogous to the RPA equations (11.44) and (11.45) respectively.

The following steps of the derivation are obtained directly from the results of Subsect. 11.2.1 by analogy. Here the BCS vacuum $|\text{BCS}\rangle$ plays the same role for the quasiparticle pair operators $A^\dagger, A$ as the particle–hole vacuum $|\text{HF}\rangle$ played for the particle–hole operators $\mathcal{A}^\dagger, \mathcal{A}$. We define analogously to (11.48)

$$
\begin{aligned}
A_{ab,cd}(J) &\equiv \langle \text{BCS}|\,[A_{ab}(JM), \mathcal{H}, A^\dagger_{cd}(JM)]\,|\text{BCS}\rangle \,, \\
B_{ab,cd}(J) &\equiv -\langle \text{BCS}|\,[A_{ab}(JM), \mathcal{H}, \tilde{A}_{cd}(JM)]\,|\text{BCS}\rangle \,.
\end{aligned}
\tag{18.10}
$$

In analogy with (11.54) and (11.55) we then obtain

$$
\sum_{c \le d} A_{ab,cd} X^\omega_{cd} + \sum_{c \le d} B_{ab,cd} Y^\omega_{cd} = E_\omega X^\omega_{ab} \,,
\tag{18.11}
$$

$$
-\sum_{c \le d} \left(\mathsf{B}^\dagger\right)_{ab,cd} X^\omega_{cd} - \sum_{c \le d} \left(\mathsf{A}^{\mathrm{T}}\right)_{ab,cd} Y^\omega_{cd} = E_\omega Y^\omega_{ab} \,.
\tag{18.12}
$$

These are the *QRPA equations*. Combined into a matrix equation they read

$$
\begin{pmatrix} \mathsf{A} & \mathsf{B} \\ -\mathsf{B}^* & -\mathsf{A}^* \end{pmatrix} \begin{pmatrix} X^\omega \\ Y^\omega \end{pmatrix} = E_\omega \begin{pmatrix} X^\omega \\ Y^\omega \end{pmatrix} \,.
\tag{18.13}
$$

Since the matrix elements (18.10) are independent of $M$, the $X$ and $Y$ amplitudes must also be independent of $M$.

Equation (18.13) is formally identical with the RPA equation (11.60). Both constitute a non-Hermitian eigenvalue problem. As can be seen from (16.38), the matrix $\mathsf{A}$ defined in (18.10) is nothing but the QTDA matrix (16.46). The relations (11.58) apply here too and show that the matrix $\mathsf{A}$ is Hermitian and the matrix $\mathsf{B}$ is symmetric. In analogy to (11.63) the correlation matrix $\mathsf{B}$ can be written as

$$
B_{ab,cd}(J) = \langle \text{BCS}| A_{ab}(JM) \tilde{A}_{cd}(JM) \mathcal{H} | \text{BCS}\rangle \,.
\tag{18.14}
$$

### 18.1.2 Explicit Form of the Correlation Matrix

In this subsection we derive an explicit expression for the elements (18.14) of the correlation matrix $\mathsf{B}$. From (18.14) we see that only the $H_{40}$ term contributes to $\mathsf{B}$. We expand the angular momentum couplings in $A_{ab}$ and $\tilde{A}_{cd}$ and use the uncoupled expression for $H_{40}$ in (16.4). Substitution into (18.14), with the summation index changes $\gamma' \to -\gamma'$ and $\delta' \to -\delta'$, yields

$$B_{ab,cd}(J)$$

$$= \tfrac{1}{4}(-1)^{J+M}\mathcal{N}_{ab}(J)\mathcal{N}_{cd}(J) \sum_{\substack{m_\alpha m_\beta \\ m_\gamma m_\delta}} (j_a\, m_\alpha\, j_b\, m_\beta | J\, M)(j_c\, m_\gamma\, j_d\, m_\delta | J\, -M)$$

$$\times \sum_{\alpha'\beta'\gamma'\delta'} u_{a'} u_{b'} v_{c'} v_{d'} (-1)^{j_{c'}-m_{\gamma'}+j_{d'}-m_{\delta'}} \bar{v}_{\alpha'\beta',-\gamma',-\delta'}$$

$$\times \langle \mathrm{BCS}|a_\beta a_\alpha a_\delta a_\gamma a_{\alpha'}^\dagger a_{\beta'}^\dagger a_{\delta'}^\dagger a_{\gamma'}^\dagger |\mathrm{BCS}\rangle \equiv \sum_{i=1}^{24} B_i \ . \tag{18.15}$$

The 24 terms $B_i$ come from all possible contractions in the BCS vacuum expectation value:

$$\langle \mathrm{BCS}|a_\beta a_\alpha a_\delta a_\gamma a_{\alpha'}^\dagger a_{\beta'}^\dagger a_{\delta'}^\dagger a_{\gamma'}^\dagger |\mathrm{BCS}\rangle$$

$$= -\delta_{\alpha\alpha'}\delta_{\beta\beta'}\delta_{\gamma\gamma'}\delta_{\delta\delta'} + \delta_{\alpha\alpha'}\delta_{\beta\beta'}\delta_{\gamma\delta'}\delta_{\delta\gamma'} + \delta_{\alpha\alpha'}\delta_{\beta\delta'}\delta_{\delta\beta'}\delta_{\gamma\gamma'}$$

$$- \delta_{\alpha\alpha'}\delta_{\beta\delta'}\delta_{\delta\gamma'}\delta_{\gamma\beta'} - \delta_{\alpha\alpha'}\delta_{\beta\gamma'}\delta_{\delta\beta'}\delta_{\gamma\delta'} + \delta_{\alpha\alpha'}\delta_{\beta\gamma'}\delta_{\delta\delta'}\delta_{\gamma\beta'}$$

$$- \delta_{\alpha\beta'}\delta_{\beta\alpha'}\delta_{\delta\gamma'}\delta_{\gamma\delta'} + \delta_{\alpha\beta'}\delta_{\beta\alpha'}\delta_{\delta\delta'}\delta_{\gamma\gamma'} - \delta_{\alpha\beta'}\delta_{\beta\delta'}\delta_{\delta\alpha'}\delta_{\gamma\gamma'}$$

$$+ \delta_{\alpha\beta'}\delta_{\beta\delta'}\delta_{\delta\gamma'}\delta_{\gamma\alpha'} - \delta_{\alpha\beta'}\delta_{\beta\gamma'}\delta_{\delta\delta'}\delta_{\gamma\alpha'} + \delta_{\alpha\beta'}\delta_{\beta\gamma'}\delta_{\delta\alpha'}\delta_{\gamma\delta'}$$

$$- \delta_{\alpha\delta'}\delta_{\beta\alpha'}\delta_{\delta\beta'}\delta_{\gamma\gamma'} + \delta_{\alpha\delta'}\delta_{\beta\alpha'}\delta_{\delta\gamma'}\delta_{\gamma\beta'} + \delta_{\alpha\delta'}\delta_{\beta\beta'}\delta_{\delta\alpha'}\delta_{\gamma\gamma'}$$

$$- \delta_{\alpha\delta'}\delta_{\beta\beta'}\delta_{\delta\gamma'}\delta_{\gamma\alpha'} + \delta_{\alpha\delta'}\delta_{\beta\gamma'}\delta_{\delta\beta'}\delta_{\gamma\alpha'} - \delta_{\alpha\delta'}\delta_{\beta\gamma'}\delta_{\delta\alpha'}\delta_{\gamma\beta'}$$

$$- \delta_{\alpha\gamma'}\delta_{\beta\alpha'}\delta_{\delta\delta'}\delta_{\gamma\beta'} + \delta_{\alpha\gamma'}\delta_{\beta\alpha'}\delta_{\delta\beta'}\delta_{\gamma\delta'} - \delta_{\alpha\gamma'}\delta_{\beta\beta'}\delta_{\delta\alpha'}\delta_{\gamma\delta'}$$

$$+ \delta_{\alpha\gamma'}\delta_{\beta\beta'}\delta_{\delta\delta'}\delta_{\gamma\alpha'} - \delta_{\alpha\gamma'}\delta_{\beta\delta'}\delta_{\delta\beta'}\delta_{\gamma\alpha'} + \delta_{\alpha\gamma'}\delta_{\beta\delta'}\delta_{\delta\alpha'}\delta_{\gamma\beta'} \ , \tag{18.16}$$

where the sequence of terms corresponds to the numbering in (18.15).

Because of the symmetry relations (4.29), the contributions from (18.16) to (18.15) form six groups of four equal terms according to

$$B_1 = B_2 = B_7 = B_8 \ , \qquad B_3 = B_5 = B_9 = B_{12} \ ,$$
$$B_4 = B_6 = B_{10} = B_{11} \ , \qquad B_{13} = B_{15} = B_{20} = B_{21} \ ,$$
$$B_{14} = B_{16} = B_{19} = B_{22} \ , \qquad B_{17} = B_{18} = B_{23} = B_{24} \ . \tag{18.17}$$

Thus it remains to calculate

$$B_{ab,cd} = 4(B_1 + B_3 + B_4 + B_{13} + B_{14} + B_{17}) \ . \tag{18.18}$$

The six different terms are processed below.

For the term $B_1$ of (18.15) we obtain

$$B_1 = \tfrac{1}{4}(-1)^{J+M+1}\mathcal{N}_{ab}(J)\mathcal{N}_{cd}(J)u_a u_b v_c v_d$$

$$\times \sum_{\substack{m_\alpha m_\beta \\ m_\gamma m_\delta}} (j_a\, m_\alpha\, j_b\, m_\beta | J\, M)(j_c\, m_\gamma\, j_d\, m_\delta | J\, -M)(-1)^{j_c-m_\gamma+j_d-m_\delta}\bar{v}_{\alpha\beta,-\gamma,-\delta} \ .$$

$$\tag{18.19}$$

Inverting the two-body matrix element to coupled form by (8.17) gives

$$B_1 = \tfrac{1}{4}(-1)^{J+M+1}\mathcal{N}_{ab}(J)\mathcal{N}_{cd}(J)u_a u_b v_c v_d$$

$$\times \sum_{\substack{m_\alpha m_\beta \\ m_\gamma m_\delta}} (j_a\, m_\alpha\, j_b\, m_\beta | J\, M)(j_c\, m_\gamma\, j_d\, m_\delta | J\, -M)(-1)^{j_c - m_\gamma + j_d - m_\delta}$$

$$\times \sum_{J'M'} [\mathcal{N}_{ab}(J')\mathcal{N}_{cd}(J')]^{-1}(j_a\, m_\alpha\, j_b\, m_\beta | J'\, M')(j_c\, -m_\gamma\, j_d\, -m_\delta | J'\, M')$$

$$\times \langle a\, b;\, J'|V|c\, d;\, J' \rangle \,. \tag{18.20}$$

Basic properties of Clebsch–Gordan coefficients allow the sums to be performed, with the result

$$B_1 = -\tfrac{1}{4} u_a u_b v_c v_d \langle a\, b;\, J|V|c\, d;\, J \rangle \,. \tag{18.21}$$

As in (18.20) the term $B_3$ becomes

$$B_3 = \tfrac{1}{4}(-1)^{J+M}\mathcal{N}_{ab}(J)\mathcal{N}_{cd}(J)u_a u_d v_b v_c$$

$$\times \sum_{\substack{m_\alpha m_\beta \\ m_\gamma m_\delta}} (j_a\, m_\alpha\, j_b\, m_\beta | J\, M)(j_c\, m_\gamma\, j_d\, m_\delta | J\, -M)(-1)^{j_c - m_\gamma + j_b - m_\beta}$$

$$\times \sum_{J'M'} [\mathcal{N}_{ad}(J')\mathcal{N}_{cb}(J')]^{-1}(j_a\, m_\alpha\, j_d\, m_\delta | J'\, M')(j_c\, -m_\gamma\, j_b\, -m_\beta | J'\, M')$$

$$\times \langle a\, d;\, J'|V|c\, b;\, J' \rangle \,. \tag{18.22}$$

Now the sums cannot be evaluated immediately. However, the four Clebsch–Gordan coefficients with the phase factors sum into a $6j$ symbol. The result for $B_3$ is (Exercise 18.3)

$$B_3 = -\tfrac{1}{4}\mathcal{N}_{ab}(J)\mathcal{N}_{cd}(J)u_a u_d v_b v_c \sum_{J'} [\mathcal{N}_{ad}(J')\mathcal{N}_{cb}(J')]^{-1}$$

$$\times \widehat{J'}^2 \begin{Bmatrix} j_a & j_b & J \\ j_c & j_d & J' \end{Bmatrix} \langle a\, d;\, J'|V|c\, b;\, J' \rangle$$

$$= \tfrac{1}{4}\mathcal{N}_{ab}(J)\mathcal{N}_{cd}(J)u_a u_d v_b v_c \langle a\, b^{-1};\, J|V_{\mathrm{RES}}|c\, d^{-1};\, J \rangle \,, \tag{18.23}$$

where the generalized particle–hole matrix element (16.19) was identified in the last step.

Treating $B_4$ in the same way as $B_3$ leads to the result

$$B_4 = (-1)^{j_c + j_d + J + 1} B_3(c \leftrightarrow d)$$

$$= \tfrac{1}{4}(-1)^{j_c + j_d + J + 1}\mathcal{N}_{ab}(J)\mathcal{N}_{cd}(J)u_a u_c v_b v_d \langle a\, b^{-1};\, J|V_{\mathrm{RES}}|d\, c^{-1};\, J \rangle \,. \tag{18.24}$$

Substitution of the Kronecker deltas into (18.15) shows that $B_{13}$ is related to $B_4$ through the symmetry relations (4.29) and (13.127). The term $B_{13}$ can then be deduced from $B_4$ as

$$B_{13} = \tfrac{1}{4}(-1)^{j_c+j_d+J+1} \mathcal{N}_{ab}(J)\mathcal{N}_{cd}(J)u_b u_d v_a v_c \langle a\, b^{-1}\,;\, J|V_{\mathrm{RES}}|d\, c^{-1}\,;\, J\rangle \,.$$
$$(18.25)$$

The term $B_{14}$ is similarly obtained from $B_3$,

$$B_{14} = \tfrac{1}{4}\mathcal{N}_{ab}(J)\mathcal{N}_{cd}(J)u_b u_c v_a v_d \langle a\, b^{-1}\,;\, J|V_{\mathrm{RES}}|c\, d^{-1}\,;\, J\rangle \,. \qquad (18.26)$$

Finally, $B_{17}$ is found similarly from the expression (18.21) for $B_1$, with the result

$$B_{17} = -\tfrac{1}{4}u_c u_d v_a v_b \langle a\, b\,;\, J|V|c\, d\,;\, J\rangle \,. \qquad (18.27)$$

With the six different terms calculated, (18.18) gives for the correlation matrix

$$\boxed{\begin{aligned}
B_{ab,cd}(J) = {}&-(u_a u_b v_c v_d + v_a v_b u_c u_d)\langle a\, b\,;\, J|V|c\, d\,;\, J\rangle \\
&+ \mathcal{N}_{ab}(J)\mathcal{N}_{cd}(J)\big[(u_a v_b v_c u_d + v_a u_b u_c v_d)\langle a\, b^{-1}\,;\, J|V_{\mathrm{RES}}|c\, d^{-1}\,;\, J\rangle \\
&- (-1)^{j_c+j_d+J}(u_a v_b u_c v_d + v_a u_b v_c u_d)\langle a\, b^{-1}\,;\, J|V_{\mathrm{RES}}|d\, c^{-1}\,;\, J\rangle\big]\,.
\end{aligned}}$$
$$(18.28)$$

Written in terms of the Baranger matrix elements (16.20) and (16.21) the B matrix is

$$\begin{aligned}
B_{ab,cd}(J) = 2\mathcal{N}_{ab}(J)\mathcal{N}_{cd}(J)\big[&(u_a u_b v_c v_d + v_a v_b u_c u_d)G(abcdJ) \\
&- (u_a v_b v_c u_d + v_a u_b u_c v_d)F(abcdJ) \\
&+ (-1)^{j_c+j_d+J}(u_a v_b u_c v_d + v_a u_b v_c u_d)F(abdcJ)\big]\,. \qquad (18.29)
\end{aligned}$$

This expression differs from the original one [31] through the normalization factors $\mathcal{N}$; see the footnote at the end of Sect. 16.2.

To end this section we note the difference of the CS and BR phase conventions as regards the correlation matrix. The relation is the same as for A in (16.53), namely

$$B_{ab,cd}^{(\mathrm{BR})}(J) = (-1)^{\frac{1}{2}(l_c+l_d-l_a-l_b)} B_{ab,cd}^{(\mathrm{CS})}(J)\,. \qquad (18.30)$$

## 18.2 General Properties of QRPA Solutions

In this section we list the main properties of solutions of the QRPA equation (18.13). These properties are one-to-one with those of the RPA solutions as discussed in Sect. 11.3. The QRPA solutions possess a richer structure than the QTDA solutions discussed in Sect. 16.3. This is due to the complicated structure of the correlated ground state of the QRPA.

### 18.2.1 QRPA Energies and Wave Functions

As an extension of (11.1), a QRPA excitation is written as

$$|\omega\rangle = Q_\omega^\dagger |\text{QRPA}\rangle \qquad (18.31)$$

with the phonon operator $Q_\omega^\dagger$ given in (18.1). The ket vector $|\text{QRPA}\rangle$ designates the correlated ground state of the QRPA, the QRPA vacuum. Its explicit form can be determined from the condition

$$Q_\omega |\text{QRPA}\rangle = 0 \quad \text{for all } \omega , \qquad (18.32)$$

where $Q_\omega$ is given by (18.4).

We consider next the orthogonality and completeness of the QRPA solutions (18.31).

### Normalization and Orthogonality

The orthogonality of two QRPA states (18.31) produces a constraint on the $X$ and $Y$ amplitudes of the QRPA phonon operator (18.1). We also require normalization and write

$$\delta_{\omega\omega'} = \langle\omega|\omega'\rangle = \langle\text{QRPA}|Q_\omega Q_{\omega'}^\dagger|\text{QRPA}\rangle = \langle\text{QRPA}|[Q_\omega, Q_{\omega'}^\dagger]|\text{QRPA}\rangle$$

$$= \sum_{\substack{a\leq b \\ c\leq d}} \Big\{ X_{ab}^{\omega*} X_{cd}^{\omega'} \langle\text{QRPA}|[A_{ab}(JM), A_{cd}^\dagger(J'M')]|\text{QRPA}\rangle$$

$$+ Y_{ab}^{\omega*} Y_{cd}^{\omega'} \langle\text{QRPA}|[\tilde{A}_{ab}^\dagger(JM), \tilde{A}_{cd}(J'M')]|\text{QRPA}\rangle \Big\} . \quad (18.33)$$

We can now use the quasiboson approximation (QBA), which was introduced in Chap. 11 for the RPA. In the present application we replace the correlated QRPA vacuum by the simple BCS vacuum when taking vacuum expectation values of commutators. By use of (18.7) and (18.8) the vacuum expectation values in (18.33) then become

$$\langle\text{QRPA}|[A_{ab}(JM), A_{cd}^\dagger(J'M')]|\text{QRPA}\rangle$$

$$\stackrel{\text{QBA}}{\approx} \langle\text{BCS}|[A_{ab}(JM), A_{cd}^\dagger(J'M')]|\text{BCS}\rangle$$

$$= \delta_{ac}\delta_{bd}\delta_{JJ'}\delta_{MM'} \quad (a \leq b , \ c \leq d) , \qquad (18.34)$$

$$\langle\text{QRPA}|[\tilde{A}_{ab}^\dagger(JM), \tilde{A}_{cd}(J'M')]|\text{QRPA}\rangle$$

$$\stackrel{\text{QBA}}{\approx} \langle\text{BCS}|[\tilde{A}_{ab}^\dagger(JM), \tilde{A}_{cd}(J'M')]|\text{BCS}\rangle$$

$$= -\delta_{ac}\delta_{bd}\delta_{JJ'}\delta_{MM'} \quad (a \leq b , \ c \leq d) . \quad (18.35)$$

In this approximation (18.33) becomes

$$\delta_{\omega\omega'} = \delta_{nn'}\delta_{JJ'}\delta_{MM'}\delta_{\pi\pi'} = \delta_{JJ'}\delta_{MM'}\delta_{\pi\pi'} \sum_{a\leq b} \left( X_{ab}^{\omega*} X_{ab}^{\omega'} - Y_{ab}^{\omega*} Y_{ab}^{\omega'} \right) , \quad (18.36)$$

where the $\delta_{\pi\pi'}$ on the right-hand side is implied by the factor $\delta_{ac}\delta_{bd}$ in (18.34) and (18.35).

Equation (18.36) gives the orthonormality relation of the QRPA as

$$\sum_{a\leq b}\left(X_{ab}^{nJ^\pi *}X_{ab}^{n'J^\pi} - Y_{ab}^{nJ^\pi *}Y_{ab}^{n'J^\pi}\right) = \delta_{nn'} \quad \text{(QRPA orthonormality)}.$$

(18.37)

A special case of this is the normalization condition

$$\sum_{a\leq b}\left(|X_{ab}^\omega|^2 - |Y_{ab}^\omega|^2\right) = 1.$$

(18.38)

Except for the QRPA summation restriction, Eqs. (18.37) and (18.38) are the same as the respective RPA equations (11.82) and (11.83).

**Completeness**

Equation (11.84) states the two completeness relations of the RPA. Their proof was omitted, with reference to a similar proof in this subsection.

We now set out to derive the QRPA completeness relations. The QBA result (18.34) gives

$$\delta_{ac}\delta_{bd} = \langle\text{QRPA}|[A_{ab}(JM), A_{cd}^\dagger(JM)]|\text{QRPA}\rangle$$
$$= \langle\text{QRPA}|A_{ab}(JM)A_{cd}^\dagger(JM)|\text{QRPA}\rangle$$
$$- \langle\text{QRPA}|A_{cd}^\dagger(JM)A_{ab}(JM)|\text{QRPA}\rangle$$

(18.39)

with the usual restrictions $a\leq b$, $c\leq d$. For the QRPA states

$$|\omega\rangle = |n\,J^\pi\,M\rangle = Q_\omega^\dagger|\text{QRPA}\rangle,$$

(18.40)

$$|\tilde\omega\rangle \equiv (-1)^{J+M}|n\,J^\pi\,-M\rangle \equiv \tilde Q_\omega^\dagger|\text{QRPA}\rangle$$

(18.41)

with fixed values of $J^\pi$ and $M$ the completeness relations are

$$\sum_n|\omega\rangle\langle\omega| = \sum_n|n\,J^\pi\,M\rangle\langle n\,J^\pi\,M| = 1,$$

(18.42)

$$\sum_n|\tilde\omega\rangle\langle\tilde\omega| = \sum_n|n\,J^\pi\,-M\rangle\langle n\,J^\pi\,-M| = 1.$$

(18.43)

Inserting (18.42) and (18.43) into (18.39) gives

$$\delta_{ac}\delta_{bd}$$

$$= \sum_n \langle \text{QRPA}|A_{ab}(JM)Q_\omega^\dagger|\text{QRPA}\rangle\langle\text{QRPA}|Q_\omega A_{cd}^\dagger(JM)|\text{QRPA}\rangle$$

$$- \sum_n \langle \text{QRPA}|A_{cd}^\dagger(JM)\tilde{Q}_\omega^\dagger|\text{QRPA}\rangle\langle\text{QRPA}|\tilde{Q}_\omega A_{ab}(JM)|\text{QRPA}\rangle$$

$$= \sum_n \langle \text{QRPA}|\big[A_{ab}(JM),Q_\omega^\dagger\big]|\text{QRPA}\rangle\langle\text{QRPA}|\big[Q_\omega,A_{cd}^\dagger(JM)\big]|\text{QRPA}\rangle$$

$$- \sum_n \langle \text{QRPA}|\big[A_{cd}^\dagger(JM),\tilde{Q}_\omega^\dagger\big]|\text{QRPA}\rangle\langle\text{QRPA}|\big[\tilde{Q}_\omega,A_{ab}(JM)\big]|\text{QRPA}\rangle .$$

$$(18.44)$$

Noting that the coefficients $X$ and $Y$ are independent of $M$, we insert the expansions of the QRPA phonon operators, apply again the QBA to the commutators and use (18.7). This yields

$$\delta_{ac}\delta_{bd} \overset{\text{QBA}}{\approx} \sum_n \sum_{\substack{a'\le b' \\ c'\le d'}} X_{a'b'}^\omega X_{c'd'}^{\omega*} \langle\text{BCS}|\big[A_{ab}(JM),A_{a'b'}^\dagger(JM)\big]|\text{BCS}\rangle$$

$$\times \langle\text{BCS}|\big[A_{c'd'}(JM),A_{cd}^\dagger(JM)\big]|\text{BCS}\rangle$$

$$- \sum_n \sum_{\substack{c'\le d' \\ a'\le b'}} (-Y_{c'd'}^\omega)(-Y_{a'b'}^{\omega*})\langle\text{BCS}|\big[A_{cd}^\dagger(JM),A_{c'd'}(JM)\big]|\text{BCS}\rangle$$

$$\times \langle\text{BCS}|\big[A_{a'b'}^\dagger(JM),A_{ab}(JM)\big]|\text{BCS}\rangle$$

$$= \sum_n X_{ab}^\omega X_{cd}^{\omega*} - \sum_n Y_{cd}^\omega Y_{ab}^{\omega*} . \qquad (18.45)$$

This concludes the derivation of *completeness relation I* of the QRPA:

$$\boxed{\sum_n \big(X_{ab}^{nJ^\pi} X_{cd}^{nJ^\pi*} - Y_{ab}^{nJ^\pi*} Y_{cd}^{nJ^\pi}\big) = \delta_{ac}\delta_{bd} , \quad a \le b , \; c \le d .} \qquad (18.46)$$

To derive the second completeness relation of the QRPA we start from the expression for the vacuum expectation value of the commuting operators $A$ and $\tilde{A}$. Insertion of the completeness relations (18.42) and (18.43) then gives

$$0 = \langle\text{QRPA}|\big[A_{ab}(JM),\tilde{A}_{cd}(JM)\big]|\text{QRPA}\rangle$$

$$= \sum_n \langle\text{QRPA}|A_{ab}(JM)Q_\omega^\dagger|\text{QRPA}\rangle\langle\text{QRPA}|Q_\omega\tilde{A}_{cd}(JM)|\text{QRPA}\rangle$$

$$- \sum_n \langle\text{QRPA}|\tilde{A}_{cd}(JM)\tilde{Q}_\omega^\dagger|\text{QRPA}\rangle\langle\text{QRPA}|\tilde{Q}_\omega A_{ab}(JM)|\text{QRPA}\rangle .$$

$$(18.47)$$

As explained above, we expand the QRPA phonon operators and use the quasiboson approximation. The result is

$$0 \overset{\text{QBA}}{\approx} \sum_n \sum_{\substack{a' \leq b' \\ c' \leq d'}} X^\omega_{a'b'}(-Y^{\omega*}_{c'd'}) \langle \text{BCS}| \left[ A_{ab}(JM), A^\dagger_{a'b'}(JM) \right] |\text{BCS}\rangle$$

$$\times \langle \text{BCS}| \left[ \tilde{A}^\dagger_{c'd'}(JM), \tilde{A}_{cd}(JM) \right] |\text{BCS}\rangle$$

$$- \sum_n \sum_{\substack{c' \leq d' \\ a' \leq b'}} X^\omega_{c'd'}(-Y^{\omega*}_{a'b'}) \langle \text{BCS}| \left[ \tilde{A}_{cd}(JM), \tilde{A}^\dagger_{c'd'}(JM) \right] |\text{BCS}\rangle$$

$$\times \langle \text{BCS}| \left[ A^\dagger_{a'b'}(JM), A_{ab}(JM) \right] |\text{BCS}\rangle$$

$$= \sum_n X^\omega_{ab} Y^{\omega*}_{cd} - \sum_n X^\omega_{cd} Y^{\omega*}_{ab} . \tag{18.48}$$

This is *completeness relation II* of the QRPA,

$$\boxed{\sum_n \left( X^{nJ^\pi}_{ab} Y^{nJ^\pi *}_{cd} - Y^{nJ^\pi *}_{ab} X^{nJ^\pi}_{cd} \right) = 0 , \quad a \leq b , \ c \leq d .} \tag{18.49}$$

The completeness relations (11.84) stated for the RPA have now become derived together with the QRPA completeness relations (18.46) and (18.49). The correspondence follows from the fact that the QRPA equation (18.13) is formally identical with the RPA equation (11.60). The correspondence implies that the conclusions on positive- and negative-energy states at the end of Subsect. 11.3.1 apply to the QRPA as well. Thus the QRPA solutions are doubled and only the positive-energy solutions are physical. Hence the eigenvalue index $n$ in (18.46) and (18.49) runs only over the *positive-energy solutions*.

The completeness conditions (18.46) and (18.49) can be combined into a matrix equation:

$$\sum_{\substack{n \\ E_n > 0}} \left[ \begin{pmatrix} X^\omega \\ Y^\omega \end{pmatrix} \left( X^{\omega\dagger}, -Y^{\omega\dagger} \right) - \begin{pmatrix} Y^{\omega*} \\ X^{\omega*} \end{pmatrix} \left( Y^{\omega T}, -X^{\omega T} \right) \right] = \begin{pmatrix} 1 & 0 \\ 0 & 1 \end{pmatrix} . \tag{18.50}$$

This form of the completeness relations is suitable for formal derivation of many results on QRPA solutions.

### Positive- and Negative-energy Solutions

As asserted above, the positive- and negative-energy solutions of the QRPA behave exactly as those of the RPA. The discussion at the end of Subsect. 11.3.1, including Eqs. (11.86)–(11.88), is directly applicable to the QRPA.

Just as the RPA represents a refinement of the TDA, so is the QRPA a refinement of the QTDA. Equation (11.88) describes this condition,

$$|Y^\omega_{ab}| \ll 1 \quad \text{for all } \omega, ab , \tag{18.51}$$

also for the QRPA solutions. Only the positive-energy solutions conform to (18.51). Its physical meaning is that the ground-state correlations remain small, so that we can write schematically

$$|\text{QRPA}\rangle = |\text{BCS}\rangle + \text{ small corrections} . \qquad (18.52)$$

The small corrections are four-quasiparticle, eight-quasiparticle, etc., components as demonstrated in the first part of Subsect. 18.2.2.

### Breaking Point of the QRPA

For a sufficiently strong two-body interaction the QRPA formalism breaks down. At this *breaking point* the lowest root of the QRPA equation (18.13) becomes imaginary. This circumstance implies the need of a different type of mean field, namely a deformed one, as the starting point of the calculation.

The dynamics of the breaking point can be tangibly seen in the elementary case of one active two-quasiparticle excitation. Then the QRPA equation becomes a two-by-two matrix equation like (11.136). The positive, physical root in (11.138) becomes imaginary as soon as $b^2$ exceeds $a^2$, i.e. at the critical strength of the two-body interaction. The numbers in (11.133) and (11.135) demonstrate that this happens when the two-body interaction becomes large compared with the single-particle energies.

### 18.2.2 The QRPA Ground State and Transition Densities

### The QRPA Ground State

The QRPA ground state (vacuum) is defined by the annihilation condition (18.32). This condition is analogous to the RPA condition (11.89). The construction of the QRPA ground state proceeds in analogy with Subsect. 11.3.2. Thus we write the Thouless theorem as

$$|\text{QRPA}\rangle = \mathcal{N}_0 e^S |\text{BCS}\rangle , \qquad (18.53)$$

where $\mathcal{N}_0$ is a normalization constant and

$$S = \tfrac{1}{2} \sum_{JM} \sum_{\substack{a \leq b \\ c \leq d}} C_{abcd}(J) A_{ab}^\dagger(JM) \tilde{A}_{cd}^\dagger(JM) , \quad C_{cdab}(J) = C_{abcd}(J) . \qquad (18.54)$$

This operator $S$ differs from (11.92) through the presence of quasiparticle pair operators $A^\dagger$ instead of particle–hole operators $\mathcal{A}^\dagger$ and through the summation restrictions. The coefficients $C_{abcd}(J)$ are to be determined so that (18.32) is satisfied.

The result is analogous to (11.101) and reads

$$\sum_{a \leq b} X_{ab}^{\omega*} C_{abcd}(J) = Y_{cd}^{\omega*} , \quad \text{for all } \omega, c \leq d . \qquad (18.55)$$

Solving this set of linear equations yields the $C$ coefficients. From (18.53) and (18.54) we see that the first term in the expansion of the exponent is the BCS vacuum and the succeeding terms introduce contributions with $4, 8, 12, \dots$ quasiparticles.

**One-quasiparticle Densities**

The quantity

$$\langle QRPA|a^\dagger a|QRPA\rangle \tag{18.56}$$

is known as the *one-quasiparticle density*. It is a useful concept for discussing many-body aspects of the QRPA vacuum. The corresponding one-particle densities were defined in (11.102) to study correlations in the RPA vacuum.

The evaluation of (18.56) proceeds in analogy to the derivation of the one-particle density (11.108). The starting point is the commutation relation

$$\sum_{m\alpha'} \left[a_{\alpha'}^\dagger a_{\alpha'}, A_{ab}^\dagger(JM)\right] = \left(\delta_{a'a} + \delta_{a'b}\right) A_{ab}^\dagger(JM), \tag{18.57}$$

derived same as (11.103). The analogy between (11.103) and (18.57) is not complete because the $\delta_{a'b}$ term appears only here. However, this second term can be taken into account without a detailed calculation following Subsect. 11.3.3. The Kronecker delta $\delta_{a'a}$ causes the replacement $a \to a'$ in the intermediate equations in the RPA derivation. To accomplish the QRPA derivation we add to the RPA-like term a second term where the replacement is $b \to a'$, due to $\delta_{a'b}$.

Another change from the RPA form of the intermediate equations is the QRPA summation restriction $a \le b$ for operators like $A_{ab}^\dagger(JM)$. Tracing the changes through the equations of Subsect. 11.3.3 we see that they carry through to the final result (11.108) for particle operators. For the one-quasiparticle density we can then write

$$\widehat{j_{a'}}^2 \langle QRPA|a_{\alpha'}^\dagger a_{\alpha'}|QRPA\rangle = \sum_{\substack{b(\ge a')Jn \\ E_n>0}} \widehat{J}^2 \left|Y_{a'b}^{nJ^\pi}\right|^2 + \sum_{\substack{a(\le a')Jn \\ E_n>0}} \widehat{J}^2 \left|Y_{aa'}^{nJ^\pi}\right|^2 . \tag{18.58}$$

The index $a'$ is shown in parenthesis under the sums to make clear that it is a fixed index and not summed over. We write (18.58) in the final form

$$\boxed{\langle QRPA|a_\alpha^\dagger a_\alpha|QRPA\rangle = \widehat{j_a}^{-2} \sum_{\substack{nJ \\ E_n>0}} \widehat{J}^2 \left( \sum_{b(\ge a)} \left|Y_{ab}^{nJ^\pi}\right|^2 + \sum_{b(\le a)} \left|Y_{ba}^{nJ^\pi}\right|^2 \right).}$$
$$\tag{18.59}$$

For the QRPA ground state to consist mainly of the quasiparticle vacuum $|BCS\rangle$, the right-hand side of (18.59) must be small. This requires that

$$\left|Y_{ab}^{nJ^\pi}\right| \ll 1 \quad \text{for all } nJ, ab . \tag{18.60}$$

If this condition holds, the quasiboson approximation is also good.

The one-quasiparticle densities (18.59) open up a way to improve the QRPA description towards *higher-QRPA* frameworks. The procedure was briefly discussed with regard to higher-RPA frameworks at the end of Subsect. 11.3.3. The point of departure is to retain the exact QRPA ground state in the equation of motion (11.11). As a result the matrices (18.10) are replaced with the exact expressions

$$A_{ab,cd} \equiv \langle \mathrm{QRPA}| \big[ A_{ab}, \mathcal{H}, A_{cd}^\dagger \big] |\mathrm{QRPA}\rangle \,, \qquad (18.61)$$

$$B_{ab,cd} \equiv -\langle \mathrm{QRPA}| \big[ A_{ab}, \mathcal{H}, \tilde{A}_{cd} \big] |\mathrm{QRPA}\rangle \,. \qquad (18.62)$$

Use of (18.61) and (18.62) leads to a self-consistent iterative solution of the higher-QRPA equations, which are of the general form (11.114). In practice, approximations are necessary. The degree of approximation in handling Eqs. (18.61) and (18.62) can be governed by the degree of approximation adopted for the evaluation of the one-quasiparticle densities (18.59). As was discussed in the RPA context, a higher-QRPA framework accounts more accurately for the Pauli principle than does the standard one.

## 18.3 QRPA Description of Open-Shell Even–Even Nuclei

The QRPA is designed for the description of excited states of spherical even–even nuclei. Based on BCS quasiparticles, it is suited for applications to open-shell nuclei whenever the BCS produces meaningful results. In practice we apply the QRPA in the same way as the QTDA.

### 18.3.1 Structure of the Correlation Matrix

Equation (16.62) states the structure of the QTDA matrix A. The same matrix goes into the QRPA equation (18.13). Additionally we need to include the correlation matrix B, given by (18.28), in the matrix equation (18.13). The general form of the B matrix is

$$\boxed{\mathsf{B} = \begin{pmatrix} \mathsf{V}_{\mathrm{QRPA}}(pp-pp) & \mathsf{V}_{\mathrm{QRPA}}(pp-nn) \\ \mathsf{V}_{\mathrm{QRPA}}(nn-pp) & \mathsf{V}_{\mathrm{QRPA}}(nn-nn) \end{pmatrix}}\,. \qquad (18.63)$$

The isospin structure of the blocks is shown schematically in (16.65) and (16.66), and the particle–hole matrix elements are given specifically by (16.69) and (16.70). The auxiliary particle–hole matrix elements used are defined in (16.67) and (16.68) and tabulated for the 0d-1s shell in Tables 16.1–16.3.

As pointed out before, the matrix B is symmetric. If the proton and neutron two-quasiparticle bases are the same, also the non-diagonal blocks are symmetric. In addition, when the single-particle energies and interactions are the same for protons and neutrons the diagonal blocks are the same. All these symmetries are present in the applications below.

We illustrate the computation of the elements of the B matrix and the QRPA eigenvalue problem by extending the QTDA example of Subsect. 16.4.2 to the QRPA.

## 18.3.2 Excitation Energies of $2^+$ States in $^{24}$Mg

Let us calculate the $2^+$ energy spectrum of $^{24}_{12}$Mg$_{12}$ in the $0d_{5/2}$-$1s_{1/2}$ valence space in the QRPA. The two-quasiparticle basis states are given in (16.71) and the single-particle energies in (16.72). The input into the BCS calculation and the resultant occupation amplitudes and quasiparticle energies are stated in Table 16.4.

Equations (16.73)–(16.78) show the calculation of the A matrix. We now calculate the correlation matrix B. Because of the symmetries, like the A matrix, the B matrix has only six different elements. They are constructed by use of (18.28), (16.69) and (16.70) with numerical data from Tables 8.2, 16.1, 16.2 and 16.4.

To illustrate the method we form in detail the three different elements of the $pp - pp$ block:

$$\begin{aligned}
B_{\pi_1 \pi_1} &= B\left((\pi 0d_{5/2})^2, (\pi 0d_{5/2})^2\right) \\
&= -2 \times 0.6685^2 \times 0.7437^2(-0.6857A_1) \\
&\quad + \frac{1}{2}\Big[2 \times 0.6685^2 \times 0.7437^2(-0.6857A_1) \\
&\qquad - (-1)^{\frac{5}{2}+\frac{5}{2}+2} \times 2 \times 0.6685^2 \times 0.7437^2(-0.6857A_1)\Big] \\
&= 0.0000A_1 ,
\end{aligned} \tag{18.64}$$

$$\begin{aligned}
B_{\pi_1 \pi_2} &= B\left((\pi 0d_{5/2})^2, \pi 0d_{5/2}\pi 1s_{1/2}\right) \\
&= -(0.6685^2 \times 0.7437 \times 0.5838 \\
&\quad + 0.7437^2 \times 0.6685 \times 0.8119)(-0.9071A_1) \\
&\quad + \frac{1}{\sqrt{2}}\Big[(0.6685 \times 0.7437^2 \times 0.8119 \\
&\qquad + 0.7437 \times 0.6685^2 \times 0.5838)(-0.6414A_1) \\
&\qquad - (-1)^{\frac{5}{2}+\frac{1}{2}+2} \times (0.6685^2 \times 0.7437 \times 0.5838 \\
&\qquad\qquad + 0.7437^2 \times 0.6685 \times 0.8119)(-0.6414A_1)\Big] \\
&= 0.0000A_1 ,
\end{aligned} \tag{18.65}$$

$$\begin{aligned}
B_{\pi_2 \pi_2} &= B\left(\pi 0d_{5/2}\pi 1s_{1/2}, \pi 0d_{5/2}\pi 1s_{1/2}\right) \\
&= -2 \times 0.6685 \times 0.8119 \times 0.7437 \times 0.5838(-1.2000A_1) \\
&\quad + 2 \times 0.6685 \times 0.5838 \times 0.7437 \times 0.8119(-0.2000A_1) \\
&\quad - (-1)^{\frac{5}{2}+\frac{1}{2}+2} \times (0.6685^2 \times 0.5838^2 \\
&\qquad\qquad + 0.7437^2 \times 0.8119^2)(-1.0000A_1) \\
&= -0.0456A_1 .
\end{aligned} \tag{18.66}$$

The $nn$–$nn$ block is the same as the $pp$–$pp$ block due to the same single-particle energies and interaction parameters for protons and neutrons. The $pp$–$nn$ block is handled same as the calculation of (16.76)–(16.78). We thus obtain for the complete B matrix

$$B_{QRPA}(2^+)$$
$$= \begin{pmatrix} 0 & 0 & -0.508A_0 - 0.169A_1 & -0.672A_0 - 0.224A_1 \\ 0 & -0.046A_1 & -0.672A_0 - 0.224A_1 & -0.880A_0 - 0.306A_1 \\ \cdots & \cdots & 0 & 0 \\ \cdots & \cdots & 0 & -0.046A_1 \end{pmatrix} , \quad (18.67)$$

where exact zeros are placed for (18.64) and (18.65) and the dots represent the $nn$–$pp$ block which is identical to the $pp$–$nn$ block.

We now diagonalize the supermatrix of Eq. (18.13) by the Ullah–Rowe procedure developed for the RPA in Subsect. 11.5.2. This means that we insert the matrices (16.79) and (18.67) into (18.13) and transform the equation into a real symmetric eigenvalue problem. There is a choice between proceeding from one of the matrices $M_\mp = A \mp B$. The resultant eigenvalue equation (11.161) or (11.162), respectively, is then solved for the eigenstates $R_-^\omega$ or $R_+^\omega$. After this we obtain the X and Y amplitudes by using (11.165) or (11.167). The procedure was applied in detail to two examples in Subsects. 11.5.3 and 11.5.4.

We choose for the SDI parameters the values $A_0 = A_1 = 0.6\,\text{MeV}$ so as to roughly reproduce the experimental location of the first $2^+$ state in $^{24}$Mg. Going through the steps of the Ullah–Rowe method we find the QRPA eigenvalues

$$E_n = (1.388, 3.852, 3.988, 4.012)\,\text{MeV} . \tag{18.68}$$

To write down the eigenvectors we denote by $|\pi^{-1}\rangle$ and $|\nu^{-1}\rangle$ the two-quasiparticle annihilations corresponding to the two-quasiparticle basis states $|\pi\rangle$ and $|\nu\rangle$ of (16.71). The X amplitudes multiply the two-quasiparticle basis states and the Y amplitudes the two-quasiparticle annihilation part in the QRPA wave function (18.1). The eigenvectors are

$$|^{24}\text{Mg}; 2_1^+\rangle = 0.468|\pi_1\rangle + 0.164|\pi_1^{-1}\rangle + 0.595|\pi_2\rangle + 0.215|\pi_2^{-1}\rangle$$
$$+ 0.468|\nu_1\rangle + 0.164|\nu_1^{-1}\rangle + 0.595|\nu_2\rangle + 0.215|\nu_2^{-1}\rangle , \tag{18.69}$$

$$|^{24}\text{Mg}; 2_2^+\rangle = 0.621|\pi_1\rangle - 0.057|\pi_1^{-1}\rangle + 0.351|\pi_2\rangle - 0.072|\pi_2^{-1}\rangle$$
$$- 0.621|\nu_1\rangle + 0.057|\nu_1^{-1}\rangle - 0.351|\nu_2\rangle + 0.072|\nu_2^{-1}\rangle , \tag{18.70}$$

$$|^{24}\text{Mg}; 2_3^+\rangle = 0.554|\pi_1\rangle - 0.002|\pi_1^{-1}\rangle - 0.439|\pi_2\rangle - 0.004|\pi_2^{-1}\rangle$$
$$+ 0.554|\nu_1\rangle - 0.002|\nu_1^{-1}\rangle - 0.439|\nu_2\rangle - 0.004|\nu_2^{-1}\rangle , \tag{18.71}$$

$$|^{24}\text{Mg}; 2_4^+\rangle = -0.345|\pi_1\rangle - 0.024|\pi_1^{-1}\rangle + 0.619|\pi_2\rangle - 0.029|\pi_2^{-1}\rangle$$
$$+ 0.345|\nu_1\rangle + 0.024|\nu_1^{-1}\rangle - 0.619|\nu_2\rangle + 0.029|\nu_2^{-1}\rangle . \tag{18.72}$$

Next we continue with further examples of calculations in the 0d-1s shell.

### 18.3.3 Further Examples in the 0d-1s Shell

As discussed in Subsect. 16.4.3, we can use the empirical pairing gaps $\Delta_\mathrm{p}$ and $\Delta_\mathrm{n}$, given by (16.90), to determine the proton and neutron pairing strengths according to (16.93). In the subsequent QRPA calculation, the values of the SDI parameters $A_0$ and $A_1$ can be fixed by available spectroscopic data.

Figures 18.1 and 18.2 show the spectra of $^{24}$Mg and $^{30}$Si as examples of QRPA computations in the full 0d-1s shell. These nuclei were treated at the QTDA level in Sect. 16.4, with the energy spectra shown in Figs. 16.2 and 16.3. The SDI pairing parameters in the present calculations are those listed in Tables 16.5 and 16.6. The SDI parameters for the QRPA calculations are listed in Table 18.1. They differ somewhat from the corresponding QTDA parameters in Tables 16.5 and 16.6.

Comparison of the QRPA spectra of Figs. 18.1 and 18.2 with the corresponding QTDA spectra in Figs. 16.2 and 16.3 shows a remarkable similarity. Since the parameters used in the two approximations differed little, we can

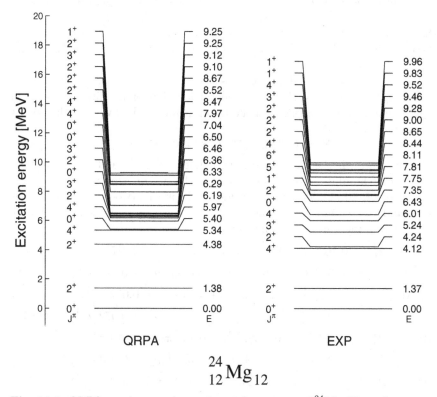

**Fig. 18.1.** QRPA spectrum and experimental spectrum of $^{24}$Mg. The valence space is 0d-1s with the single-particle energies (14.12). The SDI pairing parameters are given in Table 16.5 and the SDI QRPA parameters in Table 18.1

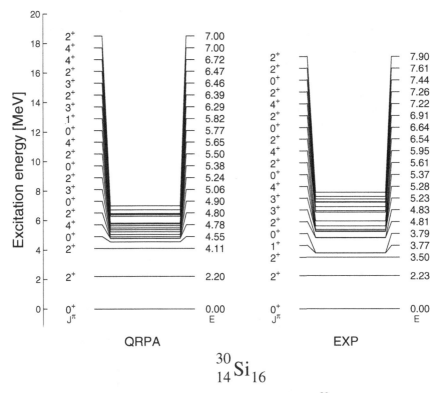

**Fig. 18.2.** QRPA spectrum and experimental spectrum of $^{30}$Si. The valence space is 0d-1s with the single-particle energies (14.12). The SDI pairing parameters are given in Table 16.6 and the SDI QRPA parameters in Table 18.1

**Table 18.1.** SDI parameters $A_0$ and $A_1$ used in the QRPA calculations of Figs. 18.1 and 18.2

| Nucleus | Valence space | $A_0$ | $A_1(J \neq 0)$ | $A_1(J = 0)$ |
|---------|---------------|-------|-----------------|--------------|
| $^{24}$Mg | 0d-1s | 0.27 | 1.25 | 0.00 |
| $^{30}$Si | 0d-1s | 0.25 | 1.00 | 0.00 |

All energies are in Mega-electron volts.

conclude that the ground-state correlations of the QRPA do not play a major role in these cases.

### 18.3.4 Spurious Contributions to $1^-$ States

One particular problem with nuclear structure calculations is the appearance of unphysical states or unphysical components of states. These unwanted *spurious* effects can invalidate computed wave functions or at least tend to contaminate them and thus reduce their physical rigour. We have already met this problem at the end of Subsect. 6.4.2 in connection with the electric dipole operator.

The problem arises from the fact that in calculations the nuclear centre, the geometrical origin of the nucleon coordinates, is kept fixed in space. When the nucleons move, the centre of mass of the nucleus moves about this origin. This is unphysical because in actual fact the centre of mass remains stationary and the nucleon motion occurs about it.

The momentum operator for the spurious centre-of-mass motion is

$$\boldsymbol{P} = -i\hbar\nabla_{\boldsymbol{R}} \,, \tag{18.73}$$

where

$$\boldsymbol{R} = \frac{1}{A}\sum_{i=1}^{A}\boldsymbol{r}_i \tag{18.74}$$

is the centre-of-mass coordinate and the $\boldsymbol{r}_i$ are the coordinates of the nucleons. Eigenstates of $\boldsymbol{P}$ mix with the physical states in typical nuclear structure calculations. This is to say that the simple translational motion, including zero motion, of the centre of mass of a nucleus does not separate from the relative motion of the nucleons, described by the relative coordinates $\boldsymbol{r}_{ij} = \boldsymbol{r}_i - \boldsymbol{r}_j$.

Since the momentum operator (18.73) is a vector, of spin–parity $J^\pi = 1^-$, spurious components tend to contaminate the computed $1^-$ states. As an example, we follow in Table 18.2 the evolution of the energies of the first two $1^-$ states in the TDA and RPA calculations for the particle–hole nucleus $^{16}$O. The results of the corresponding calculations for the $2^-$ and $3^-$ states were shown in Figs. 9.3 and 11.3.

From Table 18.2 we see that the TDA energy of the first $1^-$ state shifts considerably down when the size of the valence space is slightly increased. The situation is even more drastic with the RPA, for which the energy of the first $1^-$ state becomes imaginary and therefore unphysical. The energy of the second $1^-$ state is much less affected by the size of the valence space, so the state is expected to be less affected by spuriosities. A similar pattern emerges in the quasiparticle theories, QTDA and QRPA.

The sensitivity of the energy of the first $1^-$ state to the size of the model space is an indication of spuriosity in the wave functions. Projection methods

**Table 18.2.** TDA and RPA energies of the first two $1^-$ states of $^{16}$O in two different particle–hole valence spaces for the SDI parameter values $A_0 = A_1 = 1.0\,\text{MeV}$

| Valence space | $(0d_{5/2}\text{-}1s)\text{-}(0p)^{-1}$ | | $(0d\text{-}1s)\text{-}(0p)^{-1}$ | |
|---|---|---|---|---|
| Model | TDA | RPA | TDA | RPA |
| $E(1_1^-)\,(\text{MeV})$ | 10.186 | 7.816 | 7.795 | imaginary |
| $E(1_2^-)\,(\text{MeV})$ | 13.218 | 13.200 | 13.192 | 13.148 |

can be used to restore the translational invariance and to obtain nuclear states free of spuriosities. For further discussion of centre-of-mass spuriosities see [16].

## 18.4 Electromagnetic Transitions in the QRPA Framework

In this section we discuss electromagnetic transitions in the QRPA framework. As in the case of the QTDA in Sect. 16.5, we divide the discussion into two parts. They deal with electromagnetic decays to the QRPA ground state and with decay transitions between two QRPA states. It is to be expected that the $Y$ amplitudes of the QRPA phonon can make important contributions beyond the QTDA level of approximation.

### 18.4.1 Transitions to the QRPA Ground State

Consider a transition from a QRPA excitation (18.31) to the QRPA ground state $|\text{QRPA}\rangle$. We write the transition amplitude according to (4.25) and apply the Wigner–Eckart theorem (2.27) to the transition density to return to a non-reduced matrix element. The result is

$$(\text{QRPA}\|\mathcal{M}_{\sigma\lambda}\|\omega) = \widehat{\lambda}^{-1}\sum_{ab}(a\|\mathcal{M}_{\sigma\lambda}\|b)(\text{QRPA}\|\left[c_a^\dagger \tilde{c}_b\right]_\lambda\|\omega)$$

$$= (J\,M\,\lambda\,\mu|0\,0)^{-1}\widehat{\lambda}^{-1}\sum_{ab}(a\|\mathcal{M}_{\sigma\lambda}\|b)\langle\text{QRPA}|\left[c_a^\dagger \tilde{c}_b\right]_{\lambda\mu}Q_\omega^\dagger|\text{QRPA}\rangle\,. \quad (18.75)$$

To reduce the particle rank[1] of the operator in the last matrix element we introduce the commutator form

$$\langle\text{QRPA}|[c_a^\dagger \tilde{c}_b]_{\lambda\mu}Q_\omega^\dagger|\text{QRPA}\rangle = \langle\text{QRPA}|\left[[c_a^\dagger \tilde{c}_b]_{\lambda\mu}, Q_\omega^\dagger\right]|\text{QRPA}\rangle\,. \quad (18.76)$$

To evaluate (18.76) we substitute from (15.4) and (18.1) for the operators. This gives

$$\langle\text{QRPA}|\left[[c_a^\dagger \tilde{c}_b]_{\lambda\mu}, Q_\omega^\dagger\right]|\text{QRPA}\rangle$$

$$= \sum_{c\leq d}\Big(-v_a u_b X_{cd}^\omega\langle\text{QRPA}|\left[[\tilde{a}_a \tilde{a}_b]_{\lambda\mu}, A_{cd}^\dagger(JM)\right]|\text{QRPA}\rangle$$

$$- u_a v_b Y_{cd}^\omega\langle\text{QRPA}|\left[[a_a^\dagger a_b^\dagger]_{\lambda\mu}, \tilde{A}_{cd}(JM)\right]|\text{QRPA}\rangle\Big)$$

$$= \mathcal{N}_{ab}^{-1}(\lambda)\sum_{c\leq d}\Big(v_a u_b X_{cd}^\omega\langle\text{QRPA}|\left[\tilde{A}_{ab}(\lambda\mu), A_{cd}^\dagger(JM)\right]|\text{QRPA}\rangle$$

$$- u_a v_b Y_{cd}^\omega\langle\text{QRPA}|\left[A_{ab}^\dagger(\lambda\mu), \tilde{A}_{cd}(JM)\right]|\text{QRPA}\rangle\Big)\,, \quad (18.77)$$

---

[1] See Subsect. 11.1.1.

where (18.2) and (18.3) were used in the last step. We now make the quasi-boson approximation and apply (16.27), finding

$$\langle \text{QRPA}|[[c_a^\dagger \tilde{c}_b]_{\lambda\mu}, Q_\omega^\dagger]|\text{QRPA}\rangle \overset{\text{QBA}}{\approx} (-1)^{J+M}\delta_{\lambda J}\delta_{\mu,-M}\mathcal{N}_{ab}(J)$$
$$\times \sum_{c\le d}(v_a u_b X_{cd}^\omega + u_a v_b Y_{cd}^\omega)[\delta_{ac}\delta_{bd} - (-1)^{j_a+j_b+J}\delta_{ad}\delta_{bc}] . \quad (18.78)$$

Having found (18.78) we now substitute it into (18.75) through (18.76), which yields

$$(\text{QRPA}\|\mathcal{M}_{\sigma\lambda}\|\omega) = (-1)^{J+M}\delta_{\lambda J}\delta_{\mu,-M}(J\,M\,\lambda\,\mu|0\,0)^{-1}\hat{\lambda}^{-1}$$
$$\times \sum_{c\le d}\mathcal{N}_{cd}(J)\big[(c\|\mathcal{M}_{\sigma\lambda}\|d)(v_c u_d X_{cd}^\omega + u_c v_d Y_{cd}^\omega)$$
$$- (-1)^{j_c+j_d+J}(d\|\mathcal{M}_{\sigma\lambda}\|c)(v_d u_c X_{cd}^\omega + u_d v_c Y_{cd}^\omega)\big] . \quad (18.79)$$

From the symmetry properties (6.27)–(6.30) of the single-particle matrix elements we have

$$(-1)^{j_c+j_d+\lambda+1}(d\|\mathcal{M}_{\sigma\lambda}\|c) = \zeta^{(\lambda)}(c\|\mathcal{M}_{\sigma\lambda}\|d) , \quad (18.80)$$

where the phase factor $\zeta^{(\lambda)}$ was defined in (11.223) as

$$\zeta^{(\lambda)} = \begin{cases} (-1)^\lambda & \text{CS phase convention ;} \\ \pm 1 & \text{BR phase convention,} \\ & + \text{ for } \sigma = \text{E} , - \text{ for } \sigma = \text{M} . \end{cases} \quad (18.81)$$

This allows us to combine the terms in (18.79), and with use of (1.34) it becomes

$$(\text{QRPA}\|\mathcal{M}_{\sigma\lambda}\|\omega)$$
$$= \delta_{\lambda J}\sum_{c\le d}\mathcal{N}_{cd}(J)(c\|\mathcal{M}_{\sigma\lambda}\|d)(v_c u_d + \zeta^{(\lambda)}u_c v_d)(X_{cd}^\omega + \zeta^{(\lambda)}Y_{cd}^\omega) . \quad (18.82)$$

Equations (13.56) and (13.58) state the phases for the BCS occupation amplitudes in terms of the phase factor

$$\theta^{(l_c)} = \begin{cases} (-1)^{l_c} & \text{CS phase convention ,} \\ 1 & \text{BR phase convention .} \end{cases} \quad (18.83)$$

It follows that the occupation factor in (18.82) can be written as

$$v_c u_d + \zeta^{(\lambda)}u_c v_d = \theta^{(l_d)}(v_c|u_d| \pm v_d|u_c|) , \quad (18.84)$$

with the + sign for $\sigma = \text{E}$ and the minus sign for $\sigma = \text{M}$. With a change of the index names the final result for the decay amplitude is then

$$(\mathrm{QRPA}\|\mathcal{M}_{\sigma\lambda}\|\omega) = \delta_{\lambda J} \sum_{a \leq b} \mathcal{N}_{ab}(J)\theta^{(l_b)}(v_a|u_b| \pm v_b|u_a|)(a\|\mathcal{M}_{\sigma\lambda}|b)$$
$$\times (X_{ab}^{\omega} + \zeta^{(\lambda)}Y_{ab}^{\omega}),$$
$$+ \text{ for } \sigma = \mathrm{E}, \; - \text{ for } \sigma = \mathrm{M}.$$

(18.85)

With CS phases and $Y = 0$ the transition amplitude (18.85) is reduced to the corresponding QTDA result (16.99). The $Y$ term can have a large effect on the decay amplitude and half-life. Let us now study examples of the use of (18.85) in actual computations.

### 18.4.2 E2 Decays in the 0d-1s and 0f-1p-0g$_{9/2}$ Shells

### The $2_1^+$ State in $^{24}$Mg

We continue the example of Subsect. 18.3.2 and study the E2 decay of the $2_1^+$ state in $^{24}$Mg using the $0d_{5/2}$-$1s_{1/2}$ valence space. All parameters are the same as in Subsect. 18.3.2, and they reproduce the experimental energy of the $2_1^+$ state. The $X$ and $Y$ amplitudes of the $2_1^+$ state are given in (18.69) and recapitulated in Table 18.3. We also perform a QTDA calculation for comparison. In the QTDA the experimental energy of the $2_1^+$ state is reproduced by the SDI parameters $A_0 = A_1 = 0.71$ MeV.[2] The $X$ amplitudes produced by the QTDA calculation are listed in Table 18.3.

**Table 18.3.** Two-quasiparticle configuration $ab$ in the $0d_{5/2}$-$1s_{1/2}$ valence space and the quantity $Q_{ab}$ defined in (18.86) for the $2_1^+$ state of $^{24}$Mg

| | | QTDA | | QRPA | | |
|---|---|---|---|---|---|---|
| $ab$ | $Q_{ab}$ (fm$^2$) | $X_{ab}^{2_1^+}$ | $C_{ab}$ ($e_{\text{eff}}$fm$^2$) | $X_{ab}^{2_1^+}$ | $Y_{ab}^{2_1^+}$ | $C_{ab}$ ($e_{\text{eff}}$fm$^2$) |
| $\pi 0d_{5/2}\pi 0d_{5/2}$ | $-5.97e_{\text{eff}}^{\text{p}}$ | 0.439 | $-2.62$ | 0.468 | 0.164 | $-3.77$ |
| $\pi 0d_{5/2}\pi 1s_{1/2}$ | $-7.17e_{\text{eff}}^{\text{p}}$ | 0.554 | $-3.97$ | 0.595 | 0.215 | $-5.81$ |
| $\nu 0d_{5/2}\nu 0d_{5/2}$ | $-5.97e_{\text{eff}}^{\text{n}}$ | 0.439 | $-2.62$ | 0.468 | 0.164 | $-3.77$ |
| $\nu 0d_{5/2}\nu 1s_{1/2}$ | $-7.17e_{\text{eff}}^{\text{n}}$ | 0.554 | $-3.97$ | 0.595 | 0.215 | $-5.81$ |

The QTDA and QRPA amplitudes and the contributions $C_{ab}$ to the sum (18.87) are given in the following columns

To tabulate the contributions to (18.85) we define the abbreviation

$$Q_{ab} \equiv \mathcal{N}_{ab}(2)(v_a u_b + v_b u_a)(a\|\mathcal{M}_{\sigma\lambda}|b)\frac{e_{\text{eff}}}{e}.$$

(18.86)

---

[2] This calculation differs from those of Subsects. 16.4.2 and 16.5.2 because of different SDI parameters in the QTDA calculations.

This quantity is listed in Table 18.3 for the $^{24}$Mg oscillator length $b = 1.813$ fm. With effective charges inserted, the decay amplitude (18.85) now reads

$$(\text{QRPA}\|Q_2\|2_1^+) = \sum_{a \leq b} Q_{ab}(X_{ab}^{2_1^+} + Y_{ab}^{2_1^+}) \equiv \sum_{a \leq b} C_{ab} . \qquad (18.87)$$

In the present application, Eqs. (18.86) and (18.87) are independent of phase convention. The QTDA and QRPA contributions $C_{ab}$ of each configuration are stated in Table 18.3.

From Table 18.3 we can read for the summed QTDA and QRPA decay amplitudes

$$(\text{BCS}\|Q_2\|2_1^+)_{\text{QTDA}} = -6.59(e_{\text{eff}}^{\text{p}} + e_{\text{eff}}^{\text{n}}) \, \text{fm}^2 , \qquad (18.88)$$

$$(\text{QRPA}\|Q_2\|2_1^+)_{\text{QRPA}} = -9.58(e_{\text{eff}}^{\text{p}} + e_{\text{eff}}^{\text{n}}) \, \text{fm}^2 . \qquad (18.89)$$

By using the effective charges $e_{\text{eff}}^{\text{p}} = 1.3e$ and $e_{\text{eff}}^{\text{n}} = 0.3e$ we obtain the reduced transition probabilities

$$B(\text{E2}; 2_1^+ \to 0_{\text{gs}}^+)_{\text{QTDA}} = 22.2 \, e^2 \text{fm}^4 , \qquad (18.90)$$

$$B(\text{E2}; 2_1^+ \to 0_{\text{gs}}^+)_{\text{QRPA}} = 47.0 \, e^2 \text{fm}^4 . \qquad (18.91)$$

The results (18.90) and (18.91) demonstrate the enhancement of the decay probability when going from the QTDA to the QRPA. The enhancement is due to QRPA ground-state correlations, which show up in the presence of the $Y$ amplitudes and in the increase of the $X$ amplitudes in Table 18.3.

The transition amplitudes and reduced probabilities for all of the $2^+$ states are collected in Table 18.4. The E2 strength is seen to be strongly concentrated in the lowest state, indicating its collective (vibrational) character.

**Table 18.4.** QRPA energies and E2 decay amplitudes and reduced transition probabilities for $2^+$ states in $^{24}$Mg

| $n$ | $E_n$ (MeV) | $(\text{QRPA}\|Q_2\|2_n^+) \, (\text{fm}^2)$ | $B(\text{E2}; 2_n^+ \to 0_{\text{gs}}^+) \, (e^2\text{fm}^4)$ |
|---|---|---|---|
| 1 | 1.388 | $-9.58e_+$ | 47.0 |
| 2 | 3.852 | $-5.36e_-$ | 5.75 |
| 3 | 3.988 | $-0.136e_+$ | 0.01 |
| 4 | 4.012 | $-2.00e_-$ | 0.80 |

The calculation was done in the $0d_{5/2}$-$1s_{1/2}$ valence space with details given in the text. Note abbreviations $e_\pm \equiv e_{\text{eff}}^{\text{p}} \pm e_{\text{eff}}^{\text{n}}$

## Other Examples

We now continue with calculations using the full 0d-1s oscillator shell as valence space. As examples we take the nuclei $^{24}$Mg and $^{30}$Si and calculate the wave functions of their first $2^+$ states. We use the pairing parameters of Tables 16.5 and 16.6 and the QRPA parameters of Table 18.1. With effective charges $e_{\text{eff}}^{\text{p}} = 1.3e$ and $e_{\text{eff}}^{\text{n}} = 0.3e$ the resulting $B(\text{E2})$ values are

$$B(\text{E2}\,;\, 2_1^+ \to 0_{\text{gs}}^+)_{\text{QRPA}} = 78.8\, e^2\text{fm}^4 \quad \text{for } {}^{24}\text{Mg}\,, \qquad (18.92)$$

$$B(\text{E2}\,;\, 2_1^+ \to 0_{\text{gs}}^+)_{\text{QRPA}} = 50.0\, e^2\text{fm}^4 \quad \text{for } {}^{30}\text{Si}\,. \qquad (18.93)$$

These are very close to the experimental values quoted in (16.106) and (16.113).

The $B(\text{E2})$ values (18.92) and (18.93) are appreciably larger than the corresponding QTDA values displayed in Fig. 16.5(b) for $^{24}$Mg and in Fig. 16.6 for $^{30}$Si. So even when the QTDA and QRPA energies of the $2_1^+$ state are the same (adjusted to the experimental energy), their decay probabilities deviate considerably. The QRPA generates more collectivity than does the QTDA, with coherent contributions coming from both the $X$ and the $Y$ amplitudes. The ground-state correlations thus play an important role as generators of low-energy collectivity.

Our final example $^{66}$Zn illustrates E2 transitions in the 1p-0f-0g$_{9/2}$ major shell. Table 18.5 lists the calculated QRPA values of $B(\text{E2}\,;\, 2_n^+ \to 0_{\text{gs}}^+)$ for all $2^+$ states arising from the 1p-0f-0g$_{9/2}$ valence space. The calculation was done with the BCS pairing parameters of Table 16.7. The SDI parameters for the QRPA were chosen to reproduce the experimental energy of the $2_1^+$ state, with the result $A_0 = 0.25\,\text{MeV}$ and $A_1 = 0.53\,\text{MeV}$. The effective charges were selected as $e_{\text{eff}}^{\text{p}} = 1.5e$ and $e_{\text{eff}}^{\text{n}} = 0.5e$.

From Table 18.5 we see that the QRPA calculation reproduces the experimental decay probability (16.115) from the $2_1^+$ state. The same is not true of the corresponding QTDA calculation, as seen from Fig. 16.7(b). This is further evidence that the QRPA produces strong collectivity for low-lying states in spherical open-shell nuclei.

### 18.4.3 Energy-Weighted Sum Rule of the QRPA

The EWSR of the QRPA is exactly analogous to the EWSR of the particle–hole RPA, discussed in Subsect. 11.6.4 and stated in general form by (11.237). The EWSR of the RPA was put into compact matrix form (11.254), and we strive for a similar expression here.

Starting from the decay amplitude (18.85) and following Subsect. 11.6.4, we write

$$(n\,\lambda^\pi \|\mathcal{M}_{\sigma\lambda}\|\text{QRPA}) = \zeta^{(\lambda)} \sum_{ab} (X_{ab}^{\omega*} p_{ab}^\lambda + Y_{ab}^{\omega*} q_{ab}^\lambda)$$

$$= \zeta^{(\lambda)} \left(\mathsf{X}^{\omega\dagger},\, -\mathsf{Y}^{\omega\dagger}\right) \begin{pmatrix} \mathsf{p}^\lambda \\ -\mathsf{q}^\lambda \end{pmatrix}. \qquad (18.94)$$

This is identical in form with (11.240). As an extension of (11.239) the definitions are now

**Table 18.5.** Energies of $2^+$ states and reduced E2 transition probabilities for $^{66}$Zn

| $n$ | $E_n$ (MeV) | $B(\text{E2}\,;\,2_n^+ \rightarrow 0_{\text{gs}}^+)\,(e^2\text{fm}^4)$ |
|---|---|---|
| 1 | 1.070 | 284 |
| 2 | 2.522 | 0.751 |
| 3 | 3.473 | 19.3 |
| 4 | 3.957 | 0.073 |
| 5 | 4.328 | 0.813 |
| 6 | 4.543 | 1.44 |
| 7 | 5.083 | 17.5 |
| 8 | 5.177 | 0.024 |
| 9 | 5.386 | 21.8 |
| 10 | 5.950 | 0.009 |
| 11 | 5.976 | 0.881 |
| 12 | 7.729 | 0.586 |
| 13 | 8.532 | 1.03 |
| 14 | 8.997 | 3.54 |
| 15 | 9.170 | 16.2 |
| 16 | 9.505 | 0.253 |
| 17 | 9.885 | 1.12 |
| 18 | 14.667 | 0.007 |

The calculation was done in the 1p-0f-0g$_{9/2}$ valence space with details reported in the text

$$p_{ab}^\lambda \equiv \zeta^{(\lambda)} \mathcal{N}_{ab}(\lambda) \theta^{(l_b)} (v_a|u_b| \pm v_b|u_a|)(a\|\mathcal{M}_{\sigma\lambda}\|b)\,, \quad q_{ab}^\lambda \equiv \zeta^{(\lambda)} p_{ab}^\lambda\,;$$
$$+ \text{ for } \sigma = \text{E}\,, \; - \text{ for } \sigma = \text{M}\,. \tag{18.95}$$

The phase factors $\zeta^{(\lambda)}$ and $\theta^{(l_b)}$ are given by (18.81) and (18.83) respectively. By analogy with the RPA result (11.254) we can write the EWSR of the QRPA directly as

$$\boxed{\sum_n E_\omega \left|(n\,\lambda^\pi\|\mathcal{M}_{\sigma\lambda}\|\text{QRPA})\right|^2 = p^{\lambda\,\text{T}}\left[\text{A} - \zeta^{(\lambda)}\text{B}\right]p^\lambda\,.} \tag{18.96}$$

The elements of the column vector $p^\lambda$ are given by (18.95), and A is the QTDA matrix (16.46) and B the correlation matrix (18.28).

### 18.4.4 Electric Quadrupole Sum Rule in $^{24}$Mg

To see how the EWSR (18.96) works in practice, we continue with the first example of Subsect. 18.4.2. So we consider E2 decay of the $2^+$ states in $^{24}$Mg using the 0d$_{5/2}$-1s$_{1/2}$ valence space with the single-particle energies of (16.72). As before, we use the SDI with the BCS pairing parameters $A_{\text{pair}}^{(\text{p})} = A_{\text{pair}}^{(\text{n})} = 1.0$ MeV and the QRPA parameters $A_0 = A_1 = 0.6$ MeV. We use the CS phase convention.

We set out to evaluate the right-hand side of (18.96). Comparing (18.86) and (18.95), the latter with effective charges inserted, shows that

$$p_{ab}^2 = Q_{ab} \ . \tag{18.97}$$

The A and B matrices are given in (16.79) and (18.67). Their difference is

A − B

$$
= \begin{pmatrix} 3.542 & -0.532 & -0.407 & -0.538 \\ -0.532 & 3.358 & -0.538 & -0.711 \\ -0.407 & -0.538 & 3.542 & -0.532 \\ -0.538 & -0.711 & -0.532 & 3.358 \end{pmatrix} - \begin{pmatrix} 0 & 0 & -0.406 & -0.538 \\ 0 & -0.028 & -0.538 & -0.712 \\ -0.406 & -0.538 & 0 & 0 \\ -0.538 & -0.712 & 0 & -0.028 \end{pmatrix}
$$

$$
= \begin{pmatrix} 3.542 & -0.532 & -0.001 & 0.000 \\ -0.532 & 3.386 & 0.000 & 0.001 \\ -0.001 & 0.000 & 3.542 & -0.532 \\ 0.000 & 0.001 & -0.532 & 3.386 \end{pmatrix} , \tag{18.98}
$$

with all elements given in Mega-electronvolts. With the values of $Q_{ab} = p_{ab}^2$ from Table 18.3, the right-hand side of (18.96) becomes

$$
\mathbf{p}^{2\mathrm{T}}(\mathsf{A} - \mathsf{B})\mathbf{p}^2 = \left(-5.97e_{\mathrm{eff}}^{\mathrm{p}}, \ -7.17e_{\mathrm{eff}}^{\mathrm{p}}, \ -5.97e_{\mathrm{eff}}^{\mathrm{n}}, \ -7.17e_{\mathrm{eff}}^{\mathrm{n}}\right)
$$

$$
\times \begin{pmatrix} 3.542 & -0.532 & -0.001 & 0.000 \\ -0.532 & 3.386 & 0.000 & 0.001 \\ -0.001 & 0.000 & 3.542 & -0.532 \\ 0.000 & 0.001 & -0.532 & 3.386 \end{pmatrix} \begin{pmatrix} -5.97e_{\mathrm{eff}}^{\mathrm{p}} \\ -7.17e_{\mathrm{eff}}^{\mathrm{p}} \\ -5.97e_{\mathrm{eff}}^{\mathrm{n}} \\ -7.17e_{\mathrm{eff}}^{\mathrm{n}} \end{pmatrix} \mathrm{MeV\,fm}^4
$$

$$
= 255[(e_{\mathrm{eff}}^{\mathrm{p}})^2 + (e_{\mathrm{eff}}^{\mathrm{n}})^2] \, \mathrm{MeV\,fm}^4 \ . \tag{18.99}
$$

Next we evaluate the left-hand side of (18.96). With the energies and E2 matrix elements of Table 18.4 we obtain

$$
\sum_{n=1}^4 E_n \left| (2_n^+ \| Q_2 \| \mathrm{QRPA}) \right|^2
$$

$$
= 1.388 \, \mathrm{MeV} \times (-9.58e_+ \, \mathrm{fm}^2)^2 + 3.852 \, \mathrm{MeV} \times (-5.36e_- \, \mathrm{fm}^2)^2
$$

$$
+ 3.988 \, \mathrm{MeV} \times (-0.136e_+ \, \mathrm{fm}^2)^2 + 4.012 \, \mathrm{MeV} \times (-2.00e_- \, \mathrm{fm}^2)^2
$$

$$
= 254[(e_{\mathrm{eff}}^{\mathrm{p}})^2 + (e_{\mathrm{eff}}^{\mathrm{n}})^2] \, \mathrm{MeV\,fm}^4 + 1.5 e_{\mathrm{eff}}^{\mathrm{p}} e_{\mathrm{eff}}^{\mathrm{n}} \, \mathrm{MeV\,fm}^4 \ . \tag{18.100}
$$

This agrees with (18.99) to within one-half per cent. Rounding errors are responsible for the small discrepancy. In particular we note that the off-diagonal blocks of the matrix (18.98) would be zero in an accurate calculation. Also, the small $e_{\mathrm{eff}}^{\mathrm{p}} e_{\mathrm{eff}}^{\mathrm{n}}$ term in (18.100) would vanish in an accurate calculation. We conclude that within rounding errors the EWSR is indeed satisfied.

## 18.4.5 Electromagnetic Transitions Between Two QRPA Excitations

Equation (16.122) is a straightforward QTDA recipe for electromagnetic transitions from one QTDA excitation to another. To derive a corresponding expression for the QRPA we have to resort to the commutator technique of the RPA in order to include the $Y$ contributions to the transition amplitude. We first establish an approximate boson commutation rule for two QRPA phonons through the quasiboson approximation:

$$[Q_\omega, Q_{\omega'}^\dagger] \overset{\text{QBA}}{\approx} \langle \text{BCS}|[Q_\omega, Q_{\omega'}^\dagger]|\text{BCS}\rangle = \delta_{\omega\omega'} \,. \tag{18.101}$$

The second step follows from substituting (18.1) and (18.4) into the commutator, noting the expectation values (18.7) and (18.8) and applying the orthonormality relation (18.37). The approximate commutation result (18.101) is analogous to the RPA result (11.275).

By analogy with the derivation in Subsect. 11.6.9 we can write the electromagnetic transition amplitude for transitions between two QRPA excitations as

$$\langle n_f J_f M_f|\mathcal{M}_{\sigma\lambda\mu}|n_i J_i M_i\rangle \overset{\text{QBA}}{\approx} \langle \text{BCS}|[Q_{\omega_f}, \mathcal{M}_{\sigma\lambda\mu}, Q_{\omega_i}^\dagger]|\text{BCS}\rangle \,, \tag{18.102}$$

where

$$Q_{\omega_i}^\dagger = \sum_{a_i \leq b_i} \left[ X_{a_i b_i}^{\omega_i} A_{a_i b_i}^\dagger (J_i M_i) - Y_{a_i b_i}^{\omega_i} \tilde{A}_{a_i b_i}(J_i M_i) \right] \,, \tag{18.103}$$

$$Q_{\omega_f}^\dagger = \sum_{a_f \leq b_f} \left[ X_{a_f b_f}^{\omega_f} A_{a_f b_f}^\dagger (J_f M_f) - Y_{a_f b_f}^{\omega_f} \tilde{A}_{a_f b_f}(J_f M_f) \right] \,. \tag{18.104}$$

To proceed with the calculation we need to insert for the transition operator (15.1) its quasiparticle representation with the transition density (15.4). The resulting expression is simplified by the fact that

$$\langle \text{BCS}|[Q_{\omega_f}, [a_a^\dagger a_b^\dagger]_{\lambda\mu}, Q_{\omega_i}^\dagger]|\text{BCS}\rangle = 0 \,, \tag{18.105}$$

$$\langle \text{BCS}|[Q_{\omega_f}, [\tilde{a}_a \tilde{a}_b]_{\lambda\mu}, Q_{\omega_i}^\dagger]|\text{BCS}\rangle = 0 \,. \tag{18.106}$$

This leaves the two $a^\dagger a$ terms of (15.4) as the only non-zero contributions to the double commutator in (18.102), whence

$$\langle n_f J_f M_f|\mathcal{M}_{\sigma\lambda\mu}|n_i J_i M_i\rangle$$
$$= \hat{\lambda}^{-1} \sum_{ab} (a\|\mathcal{M}_{\sigma\lambda}\|b)\{u_a u_b \langle \text{BCS}|[Q_{\omega_f}, [a_a^\dagger \tilde{a}_b]_{\lambda\mu}, Q_{\omega_i}^\dagger]|\text{BCS}\rangle$$
$$+ (-1)^{j_a + j_b + \lambda} v_a v_b \langle \text{BCS}|[Q_{\omega_f}, [a_b^\dagger \tilde{a}_a]_{\lambda\mu}, Q_{\omega_i}^\dagger]|\text{BCS}\rangle\} \,. \tag{18.107}$$

By means of (18.80) and (15.137) this becomes

$$\langle n_f \, J_f \, M_f | \mathcal{M}_{\sigma\lambda\mu} | n_i \, J_i \, M_i \rangle$$

$$= \widehat{\lambda}^{-1} \sum_{ab} (a\|\boldsymbol{\mathcal{M}}_{\sigma\lambda}\|b) \mathcal{D}_{ab}^{(\lambda)} \langle \mathrm{BCS}| [Q_{\omega_f}, [a_a^\dagger \tilde{a}_b]_{\lambda\mu}, Q_{\omega_i}^\dagger] | \mathrm{BCS} \rangle \, , \quad (18.108)$$

where the factor $\mathcal{D}_{ab}^{(\lambda)}$ was defined in (15.18).

Expanded, the double commutator in (18.108) is like (11.280) with quasi-particle operators replacing the particle operators. Its BCS expectation value is like (11.281) with the BCS vacuum replacing the particle–hole vacuum and the quasiparticle pair operators $A$, $\tilde{A}$, $A^\dagger$ and $\tilde{A}^\dagger$ replacing the respective particle–hole operators $\mathcal{A}$, $\tilde{\mathcal{A}}$, $\mathcal{A}^\dagger$ and $\tilde{\mathcal{A}}^\dagger$. To proceed with the expectation value we need instead of (11.282) the commutator (Exercise 18.27)

$$\left[ A_{ab}(JM), A_{cd}^\dagger(J'M') \right] = \delta_{JJ'} \delta_{MM'} \mathcal{N}_{ab}^2(J) [\delta_{ac}\delta_{bd} - (-1)^{j_a+j_b+J} \delta_{ad}\delta_{bc}]$$

$$+ \mathcal{N}_{ab}(J) \mathcal{N}_{cd}(J') \sum_{\substack{m_\alpha m_\beta \\ m_\gamma m_\delta}} (j_a \, m_\alpha \, j_b \, m_\beta | J \, M)(j_c \, m_\gamma \, j_d \, m_\delta | J' \, M')$$

$$\times [\delta_{\alpha\delta} a_\gamma^\dagger a_\beta + \delta_{\beta\gamma} a_\delta^\dagger a_\alpha - \delta_{\alpha\gamma} a_\delta^\dagger a_\beta - \delta_{\beta\delta} a_\gamma^\dagger a_\alpha] \, .$$

$$(18.109)$$

Similar to the particle–hole case, this commutation relation implies that only the first two terms of the analogue of (11.281) are non-zero. Thus we have

$$\langle \mathrm{BCS}| [Q_{\omega_f}, [a_a^\dagger \tilde{a}_b]_{\lambda\mu}, Q_{\omega_i}^\dagger] | \mathrm{BCS} \rangle$$

$$= \sum_{\substack{a_i \leq b_i \\ a_f \leq b_f}} \left\{ X_{a_f b_f}^{\omega_f *} X_{a_i b_i}^{\omega_i} \langle \mathrm{BCS}| A_{a_f b_f}(J_f M_f) [a_a^\dagger \tilde{a}_b]_{\lambda\mu} A_{a_i b_i}^\dagger(J_i M_i) | \mathrm{BCS} \rangle \right.$$

$$\left. + Y_{a_f b_f}^{\omega_f *} Y_{a_i b_i}^{\omega_i} \langle \mathrm{BCS}| \tilde{A}_{a_i b_i}(J_i M_i) [a_a^\dagger \tilde{a}_b]_{\lambda\mu} \tilde{A}_{a_f b_f}^\dagger(J_f M_f) | \mathrm{BCS} \rangle \right\} \, .$$

$$(18.110)$$

The matrix elements on the right-hand side of (18.110) can be expressed in terms of the quantities $\mathcal{K}_{ab}^\lambda(fi)$ defined in (15.16). Doing this and using Clebsch–Gordan identities we find

$$\langle \mathrm{BCS}| [Q_{\omega_f}, [a_a^\dagger \tilde{a}_b]_{\lambda\mu}, Q_{\omega_i}^\dagger] | \mathrm{BCS} \rangle$$

$$= \sum_{\substack{a_i \leq b_i \\ a_f \leq b_f}} \left[ X_{a_f b_f}^{\omega_f *} X_{a_i b_i}^{\omega_i} \widehat{J_f}^{-1} (J_i \, M_i \, \lambda \, \mu | J_f \, M_f) \mathcal{K}_{ab}^\lambda(fi) \right.$$

$$\left. + Y_{a_f b_f}^{\omega_f *} Y_{a_i b_i}^{\omega_i} (-1)^{J_f+M_f+J_i+M_i} \widehat{J_i}^{-1} (J_f \, -M_f \, \lambda \, \mu | J_i \, -M_i) \mathcal{K}_{ab}^\lambda(if) \right]$$

$$= \widehat{J_f}^{-1} (J_i \, M_i \, \lambda \, \mu | J_f \, M_f) \sum_{\substack{a_i \leq b_i \\ a_f \leq b_f}} \left[ X_{a_f b_f}^{\omega_f *} X_{a_i b_i}^{\omega_i} \mathcal{K}_{ab}^\lambda(fi) + Y_{a_f b_f}^{\omega_f *} Y_{a_i b_i}^{\omega_i} \mathcal{K}_{ab}^\lambda(if) \right] \, .$$

$$(18.111)$$

We substitute this into (18.108) and apply the Wigner–Eckart theorem, which gives

$$(n_f J_f \| \mathcal{M}_{\sigma\lambda} \| n_i J_i)$$
$$= \hat{\lambda}^{-1} \sum_{\substack{ab \\ a_i \leq b_i \\ a_f \leq b_f}} (a \| \mathcal{M}_{\sigma\lambda} \| b) \mathcal{D}_{ab}^{(\lambda)} \left[ X_{a_f b_f}^{\omega_f *} X_{a_i b_i}^{\omega_i} \mathcal{K}_{ab}^{\lambda}(fi) + Y_{a_f b_f}^{\omega_f *} Y_{a_i b_i}^{\omega_i} \mathcal{K}_{ab}^{\lambda}(if) \right] .$$

$$(18.112)$$

With (15.98) this becomes

$$(n_f J_f \| \mathcal{M}_{\sigma\lambda} \| n_i J_i) = \sum_{\substack{a_i \leq b_i \\ a_f \leq b_f}} \left[ X_{a_f b_f}^{\omega_f *} X_{a_i b_i}^{\omega_i} (a_f b_f \,;\, J_f \| \mathcal{M}_{\sigma\lambda} \| a_i b_i \,;\, J_i) \right.$$

$$\left. + Y_{a_f b_f}^{\omega_f *} Y_{a_i b_i}^{\omega_i} (a_i b_i \,;\, J_i \| \mathcal{M}_{\sigma\lambda} \| a_f b_f \,;\, J_f) \right] .$$

$$(18.113)$$

We transpose the second reduced matrix element on the right-hand side the same way as in passing from (11.285) to (11.286). Equation (11.290) provides the necessary symmetry relation, and the two-quasiparticle matrix elements (15.99) are real. We then write the final result as

$$\boxed{\begin{aligned} (\omega_f \| \mathcal{M}_{\sigma\lambda} \| \omega_i)_{\mathrm{QRPA}} = \sum_{\substack{a_i \leq b_i \\ a_f \leq b_f}} \left[ X_{a_f b_f}^{\omega_f *} X_{a_i b_i}^{\omega_i} + (-1)^{\Delta J + \lambda} \zeta^{(\lambda)} Y_{a_f b_f}^{\omega_f *} Y_{a_i b_i}^{\omega_i} \right] \\ \times (a_f b_f \,;\, J_f \| \mathcal{M}_{\sigma\lambda} \| a_i b_i \,;\, J_i) . \end{aligned}}$$

$$(18.114)$$

The convention-dependent phase factor is given through (11.223). Our result has the same form as the RPA result (11.286), and in the limit of vanishing $Y$ terms it coincides with the QTDA expression (16.122).

Let us next put (18.114) into practice and present an example of its use.

## 18.4.6 Electric Quadrupole Transition $4_1^+ \rightarrow 2_1^+$ in $^{24}$Mg

We continue the example of Subsect. 18.4.2 and calculate the E2 transition probability for the $4_1^+ \rightarrow 2_1^+$ decay in $^{24}$Mg. As before, the valence space is $0d_{5/2}\text{-}1s_{1/2}$, and the parameters are unchanged. For comparison we continue also the QTDA calculation from Subsect. 18.4.2. The CS phase convention is used throughout. The QTDA and QRPA amplitudes for the $2_1^+$ state are listed in Table 18.3.

The Hamiltonian matrix for the $4^+$ states is formed in the basis

$$\{|\pi_1\rangle, |\nu_1\rangle\} = \{|(\pi 0d_{5/2})^2 \,;\, 4^+\rangle, |(\nu 0d_{5/2})^2 \,;\, 4^+\rangle\} , \qquad (18.115)$$

and its diagonalization yields the wave functions

$$|^{24}\mathrm{Mg}\,;\, 4_1^+\rangle_{\mathrm{QTDA}} = \frac{1}{\sqrt{2}} \left( |\pi_1\rangle + |\nu_1\rangle \right) , \qquad (18.116)$$

$$|^{24}\text{Mg}; 4_1^+\rangle_{\text{QRPA}} = 0.707|\pi_1\rangle + 0.017|\pi_1^{-1}\rangle + 0.707|\nu_1\rangle + 0.017|\nu_1^{-1}\rangle . \quad (18.117)$$

As before, the notation $|\pi_1^{-1}\rangle$ stands for two-quasiparticle annihilation from the QRPA ground state. This part of the wave function is multiplied by the $Y$ amplitude and corresponds to the two-proton configuration $\pi_1$ of (18.115).

Equation (16.122) gives the QTDA decay amplitude

$$(2_1^+\|\boldsymbol{Q}_2\|4_1^+)_{\text{QTDA}} = \big[X_1^f(\text{QTDA})X_1^i(\text{QTDA})Q_{11}^2$$
$$+ X_2^f(\text{QTDA})X_1^i(\text{QTDA})Q_{21}^2\big](e_{\text{eff}}^{\text{p}} + e_{\text{eff}}^{\text{n}}) , \quad (18.118)$$

and (18.114) gives the QRPA decay amplitude

$$(2_1^+\|\boldsymbol{Q}_2\|4_1^+)_{\text{QRPA}} = \big\{[X_1^f(\text{QRPA})X_1^i(\text{QRPA}) + Y_1^f(\text{QRPA})Y_1^i(\text{QRPA})]Q_{11}^2$$
$$+ [X_2^f(\text{QRPA})X_1^i(\text{QRPA}) + Y_2^f(\text{QRPA})Y_1^i(\text{QRPA})]Q_{21}^2\big\}(e_{\text{eff}}^{\text{p}} + e_{\text{eff}}^{\text{n}}) .$$
$$(18.119)$$

Table 18.3 and Eqs. (18.116) and (18.117) give the QTDA numerical values

$$X_1^f(\text{QTDA}) = 0.439 , \quad X_2^f(\text{QTDA}) = 0.554 , \quad X_1^i(\text{QTDA}) = 0.707$$
$$(18.120)$$

and the QRPA numerical values

$$X_1^f(\text{QRPA}) = 0.468 , \quad X_2^f(\text{QRPA}) = 0.595 , \quad X_1^i(\text{QRPA}) = 0.707 ,$$
$$Y_1^f(\text{QRPA}) = 0.164 , \quad Y_2^f(\text{QRPA}) = 0.215 , \quad Y_1^i(\text{QRPA}) = 0.017 .$$
$$(18.121)$$

The two-quasiparticle E2 matrix elements in (18.118) and (18.119) are abbreviated as

$$eQ_{11}^2 \equiv (0\text{d}_{5/2}\,0\text{d}_{5/2}; 2^+\|\boldsymbol{Q}_2\|0\text{d}_{5/2}\,0\text{d}_{5/2}; 4^+) , \quad (18.122)$$
$$eQ_{21}^2 \equiv (0\text{d}_{5/2}\,1\text{s}_{1/2}; 2^+\|\boldsymbol{Q}_2\|0\text{d}_{5/2}\,0\text{d}_{5/2}; 4^+) . \quad (18.123)$$

These are the same for protons and neutrons because we have the same pairing parameters for them. The effective charges account for protons and neutrons in (18.118) and (18.119).

The matrix elements (18.122) and (18.123) are evaluated by means of (15.99). The resulting expressions are

$$eQ_{11}^2 = -6\sqrt{5}\left\{\begin{matrix} 2 & 4 & 2 \\ \frac{5}{2} & \frac{5}{2} & \frac{5}{2} \end{matrix}\right\} (u_{0\text{d}_{5/2}}^2 - v_{0\text{d}_{5/2}}^2)(0\text{d}_{5/2}\|\boldsymbol{Q}_2\|0\text{d}_{5/2}) , \quad (18.124)$$

$$eQ_{21}^2 = -3\sqrt{10}\left\{\begin{matrix} 2 & 4 & 2 \\ \frac{5}{2} & \frac{1}{2} & \frac{5}{2} \end{matrix}\right\} (u_{0\text{d}_{5/2}}u_{1\text{s}_{1/2}} - v_{0\text{d}_{5/2}}v_{1\text{s}_{1/2}})(1\text{s}_{1/2}\|\boldsymbol{Q}_2\|0\text{d}_{5/2}) .$$
$$(18.125)$$

Table 6.4 gives the single-particle matrix elements and Table 16.4 gives the occupation amplitudes. Inserting them and the $6j$ symbols gives

$$Q_{11}^2 = -6\sqrt{5}\left(-\frac{3}{14\sqrt{10}}\right)(0.6685^2 - 0.7437^2)(-2.585b^2) = 0.8204\,\mathrm{fm}^2\ ,$$

$$\text{(18.126)}$$

$$Q_{21}^2 = -3\sqrt{10}\left(-\frac{1}{3\sqrt{10}}\right)(0.6685 \times 0.8119 - 0.7437 \times 0.5838)(-2.185b^2)$$

$$= -0.7798\,\mathrm{fm}^2\ , \tag{18.127}$$

where we used $b = 1.813\,\mathrm{fm}$.

Substituting the amplitudes (18.120) and the matrix elements (18.126) and (18.127) into (18.118) we find the QTDA transition amplitude

$$(2_1^+\|\boldsymbol{Q}_2\|4_1^+)_{\mathrm{QTDA}}$$
$$= [0.439 \times 0.707 \times 0.8204\,\mathrm{fm}^2 + 0.554 \times 0.707(-0.7798\,\mathrm{fm}^2)](e_{\mathrm{eff}}^{\mathrm{p}} + e_{\mathrm{eff}}^{\mathrm{n}})$$
$$= -0.0508(e_{\mathrm{eff}}^{\mathrm{p}} + e_{\mathrm{eff}}^{\mathrm{n}})\,\mathrm{fm}^2\ . \tag{18.128}$$

Likewise we substitute (18.121), (18.126) and (18.127) into (18.119) to find the QRPA transition amplitude

$$(2_1^+\|\boldsymbol{Q}_2\|4_1^+)_{\mathrm{QRPA}} = [(0.468 \times 0.707 + 0.164 \times 0.017) \times 0.8204\,\mathrm{fm}^2$$
$$+ (0.595 \times 0.707 + 0.215 \times 0.017)(-0.7798\,\mathrm{fm}^2)]$$
$$\times (e_{\mathrm{eff}}^{\mathrm{p}} + e_{\mathrm{eff}}^{\mathrm{n}}) = -0.0571(e_{\mathrm{eff}}^{\mathrm{p}} + e_{\mathrm{eff}}^{\mathrm{n}})\,\mathrm{fm}^2\ . \tag{18.129}$$

With the effective charges $e_{\mathrm{eff}}^{\mathrm{p}} = 1.3e$ and $e_{\mathrm{eff}}^{\mathrm{n}} = 0.3e$ these give

$$(2_1^+\|\boldsymbol{Q}_2\|4_1^+)_{\mathrm{QTDA}} = -0.081\,e\,\mathrm{fm}^2\ , \tag{18.130}$$

$$(2_1^+\|\boldsymbol{Q}_2\|4_1^+)_{\mathrm{QRPA}} = -0.091\,e\,\mathrm{fm}^2\ . \tag{18.131}$$

These lead to the $B(\mathrm{E2})$ values

$$B(\mathrm{E2}\,;\,4_1^+ \to 2_1^+)_{\mathrm{QTDA}} = 7.3 \times 10^{-4}e^2\mathrm{fm}^4\ , \tag{18.132}$$

$$B(\mathrm{E2}\,;\,4_1^+ \to 2_1^+)_{\mathrm{QTDA}} = 9.2 \times 10^{-4}e^2\mathrm{fm}^4\ . \tag{18.133}$$

With the experimental decay energy of 2.754 MeV we find from Table 6.8 the theoretical half-lives

$$t_{1/2}(\mathrm{E2})_{\mathrm{QTDA}} = 5\,\mathrm{ns}\ , \tag{18.134}$$

$$t_{1/2}(\mathrm{E2})_{\mathrm{QRPA}} = 4\,\mathrm{ns}\ . \tag{18.135}$$

The theoretical E2 decay half-lives (18.134) and (18.135) are close to each other but totally different from the experimental figure of 25 fs. One possible reason for the large deviation is the sign difference between $Q_{11}^2$ and $Q_{21}^2$ in (18.126) and (18.127). The relative sign is governed by the BCS occupation amplitudes. With equal signs the theoretical result would be about 10 ps. The result can be improved by a suitable choice of the SDI interaction strengths and by enlarging the valence space. Other possible explanations for the large measured $B(\mathrm{E2})$ value are surveyed in the following subsection.

### 18.4.7 Collective Vibrations and Rotations

In the previous subsection we saw that the experimental $B(\text{E2})$ value for the transition $4_1^+ \rightarrow 2_1^+$ in $^{24}\text{Mg}$ was several orders of magnitude larger than our computed values. One possibility to account for this discrepancy is to consider the $4_1^+$ state to be a collective *vibrational two-phonon state*. The $4_1^+$ state could also be a member of a *ground-state rotational band*. In both cases the transition would be strongly enhanced.

In the two-phonon case the enhancement is due to the strong collectivity of the one-phonon $2_1^+$ vibrational state, which is the building block of the $4_1^+$ state. In the rotational case the enhancement comes from collective rotation, which puts the $4_1^+ \rightarrow 2_1^+$ transition on an equal footing with the strongly collective $2_1^+ \rightarrow 0_{\text{gs}}^+$ transition. Rotational states are well established by experiments. Their description requires that the nucleus has a *deformed*, non-spherical shape. The most common deformed shape is spheroidal, prolate or oblate. Softer deformations occur for *gamma-unstable* nuclei, where the instability refers to departure from axial symmetry.

The rotational degree of freedom can couple to vibrational degrees of freedom. The best-known example are the *beta-vibrational* and *gamma-vibrational* rotation bands. Interaction between collective rotation and collective vibration can be treated in macroscopic collective models. Microscopically such complex interplay can be treated by coupling a deformed QTDA or QRPA phonon to a rotating nuclear core.

For more information on collective rotations and their coupling to vibrational degrees of freedom see, e.g. [16, 83]. The case of collective two-phonon vibrations is discussed in some detail in Sect. 18.5.

## 18.5 Collective Vibrational Two-Phonon States

Let us consider the first $4^+$ state as a vibrational two-phonon state in a spherical nucleus. A normalized *two-phonon state* has the form

$$|\omega\,\omega';\,JM\rangle = \frac{1}{\sqrt{1+\delta_{\omega\omega'}}}\left[Q_\omega^\dagger Q_{\omega'}^\dagger\right]_{JM}|\text{vac}\rangle . \qquad (18.136)$$

The phonon operators $Q_\omega^\dagger$ can be creation operators for collective QTDA or QRPA vibrational excitations, usually of the *quadrupole* ($2_1^+$ state) or *octupole* ($3_1^-$ state) type. The associated QTDA or QRPA vacuum is here denoted generally as $|\text{vac}\rangle$, with the defining property

$$Q_\omega|\text{vac}\rangle = 0 \quad \text{for all } \omega . \qquad (18.137)$$

Collective phonons of type $\omega$ are assumed to be identical bosons. In the case $\omega = \omega'$ the angular momentum $J$ can take only even values due to the exchange symmetry of two identical bosons, and the parity is positive. Combinations of two quadrupole phonons or two octupole phonons are well known experimentally. Also the combination of a quadrupole and an octupole phonon is known, with $J^\pi = 1^-, 2^-, 3^-, 4^-, 5^-$. For theoretical and experimental aspects of collective vibrational states see [84, 85].

An ideal state of two identical phonons is a degenerate multiplet of the allowed $J$ states at *twice* the energy of the one-phonon state. Two quadrupole phonons thus form a degenerate triplet of $J^\pi = 0^+, 2^+, 4^+$ states, and two octupole phonons form a degenerate quartet of $J^\pi = 0^+, 2^+, 4^+, 6^+$ states.

Experimentally, nearly degenerate two-phonon multiplets are common. Multiplets of three and even four quadrupole phonons have been suggested. Some cadmium isotopes are good examples of such multiphonon states [85].

Apart from belonging to a group of nearly degenerate states with proper quantum numbers, an $n$-phonon state is characterized by its strong electric decay to the $(n-1)$-phonon states. Consider a state of two phonons of multipolarity $\lambda$ and its E$\lambda$ decay to the state of one $\lambda$ phonon. It turns out that the $B(E\lambda)$ value of this transition is twice the $B(E\lambda)$ value of the ground-state decay of the one-phonon state. We proceed to derive this result on the basis of the commutation rule

$$\boxed{[Q_\omega, Q_{\omega'}^\dagger] = \delta_{\omega\omega'} \quad \text{for ideal bosons .}} \tag{18.138}$$

The matrix element for E$\lambda$ decay of the two-$\lambda$-phonon state to the one-$\lambda$-phonon state is

$$\langle \lambda\, M_f | \mathcal{M}_{\mathrm{E}\lambda\mu} | \lambda^2\,;\, J\,M_i \rangle = \frac{1}{\sqrt{2}} \langle \mathrm{vac} | Q_{\lambda M_f} \mathcal{M}_{\mathrm{E}\lambda\mu} [Q_\lambda^\dagger Q_\lambda^\dagger]_{JM_i} | \mathrm{vac} \rangle . \tag{18.139}$$

We expand the tensor product and make use of (18.137) to write

$$\langle \lambda\, M_f | \mathcal{M}_{\mathrm{E}\lambda\mu} | \lambda^2\,;\, J\,M_i \rangle$$
$$= \frac{1}{\sqrt{2}} \sum_{mm'} (\lambda\, m\, \lambda\, m' | J\, M_i) \langle \mathrm{vac} | [Q_{\lambda M_f}, \mathcal{M}_{\mathrm{E}\lambda\mu} Q_{\lambda m}^\dagger Q_{\lambda m'}^\dagger] | \mathrm{vac} \rangle . \tag{18.140}$$

Expanding the commutator and using (18.138) we obtain

$$[Q_{\lambda M_f}, \mathcal{M}_{\mathrm{E}\lambda\mu} Q_{\lambda m}^\dagger Q_{\lambda m'}^\dagger] = \mathcal{M}_{\mathrm{E}\lambda\mu} Q_{\lambda m}^\dagger [Q_{\lambda M_f}, Q_{\lambda m'}^\dagger]$$
$$+ \mathcal{M}_{\mathrm{E}\lambda\mu} [Q_{\lambda M_f}, Q_{\lambda m}^\dagger] Q_{\lambda m'}^\dagger + [Q_{\lambda M_f}, \mathcal{M}_{\mathrm{E}\lambda\mu}] Q_{\lambda m}^\dagger Q_{\lambda m'}^\dagger$$
$$= \mathcal{M}_{\mathrm{E}\lambda\mu} Q_{\lambda m}^\dagger \delta_{M_f m'} + \mathcal{M}_{\mathrm{E}\lambda\mu} \delta_{M_f m} Q_{\lambda m'}^\dagger + [Q_{\lambda M_f}, \mathcal{M}_{\mathrm{E}\lambda\mu}] Q_{\lambda m}^\dagger Q_{\lambda m'}^\dagger .$$
$$\tag{18.141}$$

When this is substituted into (18.140) the last term vanishes since it does not contain an equal number of quasiparticle annihilation and creation operators. The remaining terms give

$$\langle \lambda\, M_f | \mathcal{M}_{\mathrm{E}\lambda\mu} | \lambda^2\,;\, J\,M_i \rangle$$
$$= \frac{1}{\sqrt{2}} \sum_m [(\lambda\, m\, \lambda\, M_f | J\, M_i) + (\lambda\, M_f\, \lambda\, m | J\, M_i)] \langle \mathrm{vac} | \mathcal{M}_{\mathrm{E}\lambda\mu} | \lambda m \rangle , \tag{18.142}$$

where the summation index was renamed in the second term.

The matrix element on the right-hand side of (18.142) vanishes except for $m = -\mu$, and the two Clebsch–Gordan coefficients combine. We then have

$$\langle \lambda\, M_f | \mathcal{M}_{\mathrm{E}\lambda\mu} | \lambda^2 \,;\, J\, M_i \rangle$$
$$= \frac{1 + (-1)^{2\lambda - J}}{\sqrt{2}} (\lambda\, -\mu\, \lambda\, M_f | J\, M_i)\langle \mathrm{vac} | \mathcal{M}_{\mathrm{E}\lambda\mu} | \lambda\, -\mu \rangle \,. \quad (18.143)$$

This result in fact proves that $J$ is even. Application of the Wigner–Eckart theorem to both sides yields

$$\widehat{\lambda}^{-1}(J\, M_i\, \lambda\, \mu | \lambda\, M_f)(\lambda \| \boldsymbol{\mathcal{M}}_{\mathrm{E}\lambda} \| \lambda^2 \,;\, J)$$
$$= \sqrt{2}(\lambda\, -\mu\, \lambda\, M_f | J\, M_i)(\lambda\, -\mu\, \lambda\, \mu | 0\, 0)(\mathrm{vac} \| \boldsymbol{\mathcal{M}}_{\mathrm{E}\lambda} \| \lambda) \,. \quad (18.144)$$

With the Clebsch–Gordan relations (1.34) and (1.37) this becomes

$$\widehat{J}^{-1}(\lambda \| \boldsymbol{\mathcal{M}}_{\mathrm{E}\lambda} \| \lambda^2 \,;\, J) = \sqrt{2}\,\widehat{\lambda}^{-1}(\mathrm{vac} \| \boldsymbol{\mathcal{M}}_{\mathrm{E}\lambda} \| \lambda) \,. \quad (18.145)$$

Squaring both sides and identifying the vacuum as the $0^+$ ground state we have the final result

$$\boxed{B(\mathrm{E}\lambda \,;\, \lambda^2 J \to \lambda) = 2B(\mathrm{E}\lambda \,;\, \lambda \to 0^+_{\mathrm{gs}}) \,.} \quad (18.146)$$

The result (18.146) relies on the assumption (18.138) of ideal bosonic phonons. Note that the ratio is independent of the multipolarity $\lambda$ and the angular momentum $J$ of the two-phonon state. The rule has been found quite well obeyed in experiments for the $\lambda = 2$ case [85].

Another decay characteristic of the ideal two-phonon state is that

$$B(\mathrm{E}\lambda \,;\, \lambda^2 J \to 0^+_{\mathrm{gs}}) = 0 \,. \quad (18.147)$$

Also this characteristic has been experimentally verified as a valid approximation.

More information on collective vibrations and their coupling to rotational degrees of freedom can be found, e.g. in [16,83,84]. A microscopic approach to the interplay between two-phonon and one-phonon states is presented in [66].

## Epilogue

In this chapter we have extended the QTDA level of two-quasiparticle mixing to the more sophisticated QRPA level. The QRPA vacuum builds on many-quasiparticle correlations, which play an important role in producing states of great collectivity in open-shell even–even nuclei. Next, in the last chapter of this book, we will discuss the proton–neutron variant of the QRPA. There the ground-state correlations play an important role for the beta decay of odd–odd spherical nuclei.

## Exercises

**18.1.** Make a detailed derivation of Eqs. (18.11) and (18.12).

**18.2.** Verify the relations (18.17).

**18.3.** Make a detailed derivation of (18.23).

**18.4.** Show that the matrix equation (18.50) contains the completeness conditions (18.46) and (18.49).

**18.5.** Verify the commutation relation (18.57).

**18.6.** Produce the eigenenergies (18.68) by diagonalizing the supermatrix consisting of the QTDA matrix (16.79) and the correlation matrix (18.67).

**18.7.** Continue Exercise 18.6 and produce the eigenvectors (18.69)–(18.72).

**18.8.** Produce the spectrum of $^{24}$Mg by using the valence space and SDI parameters of Subsect. 18.3.2. Compare with the calculated and experimental spectra of Fig. 18.1 and comment.

**18.9.** Produce the spectrum of $^{30}$Si by using the $0d_{5/2}$-$1s_{1/2}$ valence space and the pairing and $A_1$ parameters of Table 16.6. Take the single-particle energies from (14.12). Compare with the calculated and experimental spectra of Fig. 18.2 and comment.

**18.10.** Verify the numbers in Table 18.3.

**18.11.** Verify the numbers in Table 18.4.

**18.12.** Continuation of Exercise 16.16.

(a) Form the correlation matrix for the $2^+$ states of $^{20}$Ne in the $0d_{5/2}$-$1s_{1/2}$ valence space.
(b) Form the QRPA matrix and diagonalize it to find the eigenenergies and eigenstates. Use the SDI with parameters $A_0 = A_1 = 1.0\,\text{MeV}$. Compare with the QTDA results of Exercise 16.16 and experimental data, and comment.

**18.13.** Continue Exercise 18.12 and compute the reduced transition probability $B(E2\,; 2_1^+ \rightarrow 0_{\text{gs}}^+)$ for $^{20}$Ne. By comparing with the experimental decay half-life determine the electric polarization constant $\chi$. Compare with the QTDA result of Exercise 16.17 and comment.

**18.14.** Continuation of Exercise 16.18.

(a) Form the correlation matrix for the $2^+$ states of $^{26}$Mg in the basis constructed from the $0d_{5/2}$ and $1s_{1/2}$ proton orbitals and the $1s_{1/2}$ neutron orbital.

(b) Form the QRPA matrix and diagonalize it to find the eigenenergies and eigenstates. Use the SDI with parameters $A_0 = A_1 = 1.0\,\mathrm{MeV}$. Compare with the QTDA results of Exercise 16.18 and experiment, and comment.

**18.15.** Continue Exercise 18.14 and compute the reduced transition probability $B(E2\,;\, 2_1^+ \to 0_{\mathrm{gs}}^+)$ for $^{26}\mathrm{Mg}$. By comparing with the experimental decay half-life determine the electric polarization constant $\chi$. Compare with the QTDA result of Exercise 16.19 and comment.

**18.16.** Continue Exercise 18.15 and compute $B(E2\,;\, 2_n^+ \to 0_{\mathrm{gs}}^+)$ for the remaining values of $n$. Compute the summed E2 strength. Check that the QRPA sum rule is satisfied.

**18.17.** Continuation of Exercise 16.21.

(a) Form the correlation matrix for the $2^+$ states of $^{30}\mathrm{Si}$ in the basis constructed from the $0\mathrm{d}_{5/2}$ and $1\mathrm{s}_{1/2}$ proton orbitals and the $1\mathrm{s}_{1/2}$ and $0\mathrm{d}_{3/2}$ neutron orbitals.
(b) Form the QRPA matrix and diagonalize it to find the eigenenergies and eigenstates. Use the SDI with parameters $A_0 = A_1 = 1.0\,\mathrm{MeV}$. Compare with the QTDA results of Exercise 16.21 and experiment, and comment.

**18.18.** Continue Exercise 18.17 and compute the reduced transition probability $B(E2\,;\, 2_1^+ \to 0_{\mathrm{gs}}^+)$ for $^{30}\mathrm{Si}$. By comparing with the experimental decay half-life determine the electric polarization constant $\chi$. Compare with the QTDA result of Exercise 16.22 and comment.

**18.19.** Continuation of Exercise 16.23.

(a) Form the correlation matrix for the $2^+$ states of $^{34}\mathrm{S}$ on the basis constructed from the $1\mathrm{s}_{1/2}$ and $0\mathrm{d}_{3/2}$ proton orbitals and the $0\mathrm{d}_{3/2}$ neutron orbital.
(b) Form the QRPA matrix and diagonalize it to find the eigenenergies and eigenstates. Use the SDI with parameters $A_0 = A_1 = 1.0\,\mathrm{MeV}$. Compare with the QTDA results of Exercise 16.23 and experiment, and comment.

**18.20.** Continue Exercise 18.19 and compute the reduced transition probability $B(E2\,;\, 2_1^+ \to 0_{\mathrm{gs}}^+)$ for $^{34}\mathrm{S}$. By comparing with the experimental decay half-life determine the electric polarization constant $\chi$. Compare with the QTDA result of Exercise 16.24 and comment.

**18.21.** Continue Exercise 18.20 and compute $B(E2\,;\, 2_n^+ \to 0_{\mathrm{gs}}^+)$ for the remaining values of $n$. Compute the summed E2 strength. Check that the QRPA sum rule is satisfied.

**18.22.** Continuation of Exercise 16.28.

(a) Form the correlation matrix for the $2^+$ states of $^{38}\mathrm{Ar}$ in the basis constructed from the $1\mathrm{s}_{1/2}$ and $0\mathrm{d}_{3/2}$ proton orbitals and the $0\mathrm{d}_{3/2}$ and $0\mathrm{f}_{7/2}$ neutron orbitals. Use the SDI with parameters $A_0 = A_1 = 1.0\,\mathrm{MeV}$.

(b) Form the QRPA matrix and diagonalize it to find the eigenenergies and eigenstates. Compare with the QTDA results of Exercise 16.28. Compare also with the experimental energies and two-hole calculation in Fig. 8.7 and comment.

**18.23.** Continue Exercise 18.22 and compute the reduced transition probability $B(E2\,;\,2_1^+ \to 0_{\mathrm{gs}}^+)$ for $^{38}\mathrm{Ar}$. By comparing with the experimental decay half-life determine the electric polarization constant $\chi$. Compare with the QTDA result of Exercise 16.29 and comment.

**18.24.** Continue Exercise 18.23 and compute $B(E2\,;\,2_n^+ \to 0_{\mathrm{gs}}^+)$ for the remaining values of $n$. Compute the summed E2 strength. Check that the QRPA sum rule is satisfied.

**18.25.** Continuation of Exercise 16.30.

(a) Form the QRPA matrix for the $3^-$ states in $^{38}\mathrm{Ar}$ in the basis constructed from the $0\mathrm{d}_{3/2}$ and $0\mathrm{f}_{7/2}$ proton and neutron orbitals. Use the SDI with parameters $A_0 = A_1 = 1.0\,\mathrm{MeV}$.
(b) Diagonalize the QRPA matrix to find the eigenenergies and eigenstates. Compare the eigenenergies with the QTDA results of Exercise 16.30 and experiment, and comment.

**18.26.** Continuation of Exercise 18.25.

(a) Calculate the reduced transition probabilities $B(E3\,;\,3_n^- \to 0_{\mathrm{gs}}^+)$ for the $3^-$ states of Exercise 18.25 by using the effective charges obtained in Exercise 18.23.
(b) Check that the QRPA sum rule is satisfied.

**18.27.** Derive the commutation relation (18.109).

**18.28.** Form the QRPA supermatrix for the $4^+$ states of the example in Subsect. 18.4.6.

**18.29.** Continue Exercise 18.28 and diagonalize the QRPA matrix to find the eigenenergies and eigenstates. Check with the wave function (18.117).

**18.30.** Continuation of Exercise 16.20.

(a) Compute the QRPA energy for the first $3^+$ state in $^{26}\mathrm{Mg}$ by using the SDI with parameters $A_0 = A_1 = 1.0\,\mathrm{MeV}$.
(b) Compute the decay half-life of the $3_1^+$ state by assuming that it decays to the first $2^+$ state by an M1 transition. Use the bare gyromagnetic ratios and the experimental gamma energy. Compare with experimental data and the QTDA result of Exercise 16.20, and comment.

**18.31.** Continuation of Exercise 18.21.

Compute the reduced transition probability $B(\text{E2}\,;\ 2_2^+ \to 2_1^+)$ for $^{34}$S by using the effective charges determined in Exercise 18.20. Determine the decay half-life of the $2_2^+$ state by taking into account also its decay to the ground state. Use experimental gamma energies. Compare with the QTDA result of Exercise 16.25 and experiment, and comment.

**18.32.** Continuation of Exercise 16.26.

(a) Compute the excitation energy of the first excited $0^+$ state $(0_1^+)$ in $^{34}$S. Use the basis of Exercise 18.19 and the SDI with parameters $A_0 = 1.0\,\text{MeV}$ and $A_1 = 0.0\,\text{MeV}$.

(b) Determine the decay half-life of the $0_1^+$ state by assuming that it decays mainly to the $2_1^+$ state. Use the electric polarization constant of Exercise 18.20 and the experimental gamma energy. Compare with the QTDA results of Exercise 16.26 and experiment, and comment.

**18.33.** Try to improve on the result (18.135) by changing the SDI pairing parameters and the parameters of the QRPA. You can also try to use the full 0d-1s valence space. Comment on the results.

**18.34.** Continuation of Exercise 18.26.

Compute the decay half-life of the $3_1^-$ state in $^{38}$Ar by considering its decay branchings to the $0_{\text{gs}}^+$ and $2_1^+$ states. Use experimental gamma energies and the electric polarization constant determined in Exercise 18.23. Compare with the QTDA result of Exercise 16.32 and experiment, and comment.

**18.35.** Derive the result (18.147).

# Proton–Neutron QRPA

## Prologue

We discussed proton–neutron two-quasiparticle excitations and their electromagnetic and beta decays in Chap. 15. In Chap. 17 we introduced the pnQTDA, the simplest configuration mixing scheme of these excitations. The vacuum of the pnQTDA is the BCS vacuum, and the solutions of the pnQTDA satisfy a variational principle.[1]

Our next task is to derive the equations of motion of the pnQRPA. This model is more sophisticated than the pnQTDA, but its solutions do not satisfy a variational principle. The vacuum of the pnQRPA contains correlations that affect quantitative and even qualitative features of electromagnetic and beta decays. The pnQRPA satisfies the Ikeda sum rule for Gamow–Teller transitions, as did the pnQTDA.

## 19.1 The pnQRPA Equation and its Basic Properties

In this section we introduce the pnQRPA equation and discuss properties of its solutions. These properties bear a close resemblance to those of the RPA and QRPA, discussed in detail in Chaps. 11 and 18.

### 19.1.1 The pnQRPA Equation

Just as the RPA and QRPA equations (11.60) and (18.13), the pnQRPA equation also can be derived by using the EOM method of Sect. 11.1. The basic excitation is

$$|\omega\rangle = Q_\omega^\dagger |\mathrm{pnQRPA}\rangle \,, \tag{19.1}$$

---

[1] For a discussion of variational principles in the present context, see the beginning of Sect. 18.1.

where the pnQRPA phonon creation operator is

$$Q_\omega^\dagger = \sum_{pn} \left[ X_{pn}^\omega A_{pn}^\dagger(JM) - Y_{pn}^\omega \tilde{A}_{pn}(JM) \right] \tag{19.2}$$

and $|\text{pnQRPA}\rangle$ denotes the pnQRPA vacuum.

The creation operator (19.2) can be viewed as a special case of (18.1). Then the derivation of the QRPA equation in Subsect. 18.1.1 serves also as a derivation of the pnQRPA equation. The matrix A for the pnQRPA is given by pnQTDA result (17.2). From (18.28) we write, by changing the $ab$ indices to $pn$ indices, the elements of matrix B as

$$\begin{aligned} B_{pn,p'n'}(J) = &-(u_p u_n v_{p'} v_{n'} + v_p v_n u_{p'} u_{n'})\langle pn\,;\, J|V|p'\,n'\,;\,J\rangle \\ &+ (u_p v_n v_{p'} u_{n'} + v_p u_n u_{p'} v_{n'})\langle pn^{-1}\,;\, J|V_{\text{RES}}|p'\,n'^{-1}\,;\,J\rangle . \end{aligned} \tag{19.3}$$

Because of charge conservation, the last term of (18.28) is zero for proton–neutron two-quasiparticle excitations. Equation (17.4) gives the particle–hole matrix element $\langle pn^{-1}\,;\, J|V_{\text{RES}}|p'\,n'^{-1}\,;\,J\rangle$. The relation between CS and BR phases is contained in (17.6) and (18.30).

The pnQRPA equations are given by (18.11) and (18.12), with the change of the restricted $ab$ sums to unrestricted $pn$ sums, as

$$\sum_{p'n'} A_{pn,p'n'} X_{p'n'}^\omega + \sum_{p'n'} B_{pn,p'n'} Y_{p'n'}^\omega = E_\omega X_{pn}^\omega , \tag{19.4}$$

$$-\sum_{p'n'} (B^\dagger)_{pn,p'n'} X_{p'n'}^\omega - \sum_{p'n'} (A^T)_{pn,p'n'} Y_{p'n'}^\omega = E_\omega Y_{pn}^\omega . \tag{19.5}$$

These are the *pnQRPA equations*. In matrix form they are identical with (18.13), i.e.

$$\begin{pmatrix} A & B \\ -B^* & -A^* \end{pmatrix} \begin{pmatrix} X^\omega \\ Y^\omega \end{pmatrix} = E_\omega \begin{pmatrix} X^\omega \\ Y^\omega \end{pmatrix} . \tag{19.6}$$

As in the case of the QRPA, A is Hermitian and B is symmetric. The eigenvalue problem is non-Hermitian.

Following the QRPA scheme and invoking analogies, we list below general properties of solutions of (19.6).

### 19.1.2 Basic Properties of the Solutions of the pnQRPA Equation

The structure of the pnQRPA vacuum can be determined from the annihilation condition

$$Q_\omega |\text{pnQRPA}\rangle = 0 \quad \text{for all } \omega , \tag{19.7}$$

where

$$Q_\omega = \sum_{pn} \left[ X_{pn}^{\omega*} A_{pn}(JM) - Y_{pn}^{\omega*} \tilde{A}_{pn}^\dagger(JM) \right] . \tag{19.8}$$

## Orthonormality

The orthonormality condition for pnQRPA solutions can be adopted from the QRPA result (18.37), so that it becomes

$$\sum_{pn} \left( X_{pn}^{kJ^\pi *} X_{pn}^{k'J^\pi} - Y_{pn}^{kJ^\pi *} Y_{pn}^{k'J^\pi} \right) = \delta_{kk'} \quad \text{(pnQRPA orthonormality)} .$$

$$(19.9)$$

A special case is the normalization condition

$$\sum_{pn} \left( |X_{pn}^\omega|^2 - |Y_{pn}^\omega|^2 \right) = 1 \quad \text{(pnQRPA normalization)} . \tag{19.10}$$

## Completeness

Completeness relations for the pnQRPA follow from Subsect. 18.2.1 by analogy. Thus (18.46) implies

$$\sum_{k} \left( X_{pn}^{kJ^\pi} X_{p'n'}^{kJ^\pi *} - Y_{pn}^{kJ^\pi *} Y_{p'n'}^{kJ^\pi} \right) = \delta_{pp'} \delta_{nn'} , \tag{19.11}$$

which is *completeness relation I* of the pnQRPA. Similarly, from Eq. (18.49) we have

$$\sum_{k} \left( X_{pn}^{kJ^\pi} Y_{p'n'}^{kJ^\pi *} - Y_{pn}^{kJ^\pi *} X_{p'n'}^{kJ^\pi} \right) = 0 , \tag{19.12}$$

which is *completeness relation II* of the pnQRPA.

Equations (19.11) and (19.12) combine into the matrix equation

$$\sum_{\substack{n \\ E_n > 0}} \left[ \begin{pmatrix} \mathsf{X}^\omega \\ \mathsf{Y}^\omega \end{pmatrix} \left( \mathsf{X}^{\omega\dagger}, -\mathsf{Y}^{\omega\dagger} \right) - \begin{pmatrix} \mathsf{Y}^{\omega *} \\ \mathsf{X}^{\omega *} \end{pmatrix} \left( \mathsf{Y}^{\omega\mathrm{T}}, -\mathsf{X}^{\omega\mathrm{T}} \right) \right] = \begin{pmatrix} 1 & 0 \\ 0 & 1 \end{pmatrix} . \tag{19.13}$$

This is identical with (18.50).

## Positive- and Negative-energy Solutions

Like the RPA and QRPA equations, the pnQRPA equation has positive- and negative-energy solutions. The discussion in Subsect. 18.2.1 applies essentially unchanged.

The condition (18.51) now takes the form

$$|Y_{pn}^\omega| \ll 1 \quad \text{for all } \omega, pn . \tag{19.14}$$

On this condition the pnQRPA solutions are close to the corresponding pnQTDA solutions. Then it is true that

$$|\text{pnQRPA}\rangle = |\text{BCS}\rangle + \text{ small corrections} . \tag{19.15}$$

The corrections are two-proton-quasiparticle–two-neutron-quasiparticle, four-proton-quasiparticle–four-neutron-quasiparticle, etc., components.

## The pnQRPA Ground State

In analogy to (18.53) and (18.54) we have for the pnQRPA ground state

$$|\text{pnQRPA}\rangle = \mathcal{N}_0 e^S |\text{BCS}\rangle \,, \tag{19.16}$$

where $\mathcal{N}_0$ is a normalization factor and

$$S = \tfrac{1}{2} \sum_{JM} \sum_{\substack{pn \\ p'n'}} C_{pnp'n'}(J) A^\dagger_{pn}(JM) \tilde{A}^\dagger_{p'n'}(JM) \,, \quad C_{p'n'pn}(J) = C_{pnp'n'}(J) \,.$$

$$\tag{19.17}$$

By analogy with (18.55) we write for the pnQRPA

$$\sum_{pn} X^{\omega *}_{pn} C_{pnp'n'}(J) = Y^{\omega *}_{p'n'} \quad \text{for all } p'n', \omega \,. \tag{19.18}$$

Solving this set of linear equations yields the $C$ coefficients.

From (19.16) and (19.17) we see that on expanding the exponential the first term of the pnQRPA vacuum is the BCS vacuum and the rest are $k$-proton-quasiparticle–$k$-neutron-quasiparticle terms for $k = 2, 4, 6, \ldots$.

## Breaking Point of the pnQRPA

As in the QRPA case, for a sufficiently strong two-body interaction the pn-QRPA formalism breaks down. At this *breaking point* the first root of the pnQRPA equations becomes imaginary. The interpretation is that the spherical vacuum $|\text{pnQRPA}\rangle$ is no longer appropriate and a deformed mean field has to be introduced. This matter is outside the scope of the present book and will not be elaborated.

The simple example of a single two-quasiparticle excitation discussed in the QRPA context applies here as well.

## One-quasiparticle Densities

The one-quasiparticle density of the QRPA was derived in Subsect. 18.2.2. From that derivation and its result (18.59) we deduce

$$\langle \text{pnQRPA}|a^\dagger_\pi a_\pi|\text{pnQRPA}\rangle = \hat{j}_p^{-2} \sum_{\substack{kJ \\ E_k > 0}} \hat{J}^2 \sum_n |Y^{kJ^\pi}_{pn}|^2 \,, \tag{19.19}$$

$$\langle \text{pnQRPA}|a^\dagger_\nu a_\nu|\text{pnQRPA}\rangle = \hat{j}_n^{-2} \sum_{\substack{kJ \\ E_k > 0}} \hat{J}^2 \sum_p |Y^{kJ^\pi}_{pn}|^2 \,. \tag{19.20}$$

This indicates that the condition (19.14) should be satisfied for (19.15) to be valid.

The one-quasiparticle densities (19.19) and (19.20) open up a way to improve on the pnQRPA description towards *higher-pnQRPA* frameworks. Similar frameworks were discussed briefly at the ends of Subsects. 11.3.3 and 18.2.2. In the present case the EOM can be used with the exact ground state to give the exact A and B matrices

$$A_{pn,p'n'} \equiv \langle \text{pnQRPA} | [A_{pn}, \mathcal{H}, A_{p'n'}^\dagger] | \text{pnQRPA} \rangle \,, \tag{19.21}$$

$$B_{pn,p'n'} \equiv -\langle \text{pnQRPA} | [A_{pn}, \mathcal{H}, \tilde{A}_{p'n'}] | \text{pnQRPA} \rangle \,. \tag{19.22}$$

This leads to a self-consistent, iterative solution of the higher-pnQRPA equations. The pair of Eqs. (19.21) and (19.22) is analogous to the RPA pair (11.112) and (11.113) and the QRPA pair (18.61) and (18.62).

The matrix elements (19.21) and (19.22) are evaluated in some approximation. The level of approximation is governed by the approximation adopted for the one-quasiparticle densities. In lowest order the self-consistent problem reads

$$\begin{pmatrix} \bar{A} & \bar{B} \\ -\bar{B}^* & -\bar{A}^* \end{pmatrix} \begin{pmatrix} \bar{X}^\omega \\ \bar{Y}^\omega \end{pmatrix} = E_\omega \begin{pmatrix} \bar{X}^\omega \\ \bar{Y}^\omega \end{pmatrix} \,. \tag{19.23}$$

The barred matrices $\bar{A}$, $\bar{B}$, $\bar{X}$ and $\bar{Y}$ are the usual matrices A, B, X and Y multiplied by combinations of the one-quasiparticle densities (19.19) and (19.20) in a way described in [86], where a *renormalized pnQRPA* (pnRQRPA) theory is introduced.

In all higher-QRPA theories the quasiboson approximation is abandoned and the bi-fermion commutator (18.109) is at least partly restored to better obey the Pauli principle.

# 19.2 Description of Open-Shell Odd–Odd Nuclei by the pnQRPA

The pnQTDA is the simplest scheme for describing odd–odd nuclei in terms of mixed proton–neutron two-quasiparticle excitations. This approach was discussed in Chap. 17. The extension to the pnQRPA requires evaluation of the correlation matrix (19.3). To illustrate how pnQRPA calculations are carried out in practice we take the following example.

### 19.2.1 Low-Lying $1^+$ States in $^{24}$Na and $^{24}$Al

In the example of Subsect. 17.2.1 the $1^+$ states of the mirror nuclei $^{24}_{11}\text{Na}_{13}$ and $^{24}_{13}\text{Al}_{11}$ were studied within the pnQTDA. We now continue the example into the pnQRPA within the same $0d_{5/2}$-$1s_{1/2}$ valence space and with the same energy parameters. Thus the even–even reference nucleus is $^{24}_{12}\text{Mg}_{12}$ with the pairing parameters $A_{\text{pair}}^{(\text{p})} = 1.55\,\text{MeV}$ and $A_{\text{pair}}^{(\text{n})} = 1.58\,\text{MeV}$ stated in Table 16.5. The resulting $u$ and $v$ amplitudes of the BCS are given in Table 16.9. Equation (17.10) defines the proton–neutron two-quasiparticle basis. The matrix A is given in (17.14) for the SDI parameters $A_0 = A_1 = 1.0\,\text{MeV}$.

We construct the B matrix (19.3) with the two-body interaction matrix elements from Table 8.2 and the particle–hole matrix elements from Eq. (17.5) and Table 16.2. The $u$ and $v$ amplitudes we take from Table 16.9. The elements of the B matrix are then

$$
\begin{aligned}
B_{11}(1^+) &= -2 \times 0.6821 \times 0.6826 \times 0.7312 \times 0.7308(-1.6286 A_0) \\
&\quad + 2 \times 0.6821 \times 0.7308 \times 0.7312 \times 0.6826(0.8143 A_1 + 0.8143 A_0) \\
&= 1.2156 A_0 + 0.4052 A_1 ,
\end{aligned}
\tag{19.24}
$$

$$
\begin{aligned}
B_{12}(1^+) &= -(0.6821 \times 0.6826 \times 0.6291 \times 0.6307 \\
&\qquad + 0.7312 \times 0.7308 \times 0.7773 \times 0.7760)(-1.1832 A_0) \\
&\quad + (0.6821 \times 0.7308 \times 0.6291 \times 0.7760 \\
&\qquad + 0.7312 \times 0.6826 \times 0.7773 \times 0.6307)(0.5916 A_1 + 0.5916 A_0) \\
&= 0.8887 A_0 + 0.2887 A_1 ,
\end{aligned}
\tag{19.25}
$$

$$
\begin{aligned}
B_{22}(1^+) &= -2 \times 0.7773 \times 0.7760 \times 0.6291 \times 0.6307(-1.0000 A_0) \\
&\quad + 2 \times 0.7773 \times 0.6307 \times 0.6291 \times 0.7760(0.5000 A_1 + 0.5000 A_0) \\
&= 0.7180 A_0 + 0.2393 A_1 .
\end{aligned}
\tag{19.26}
$$

With $A_0 = A_1 = 1.0\,\text{MeV}$ the correlation matrix becomes

$$
\mathsf{B}_{\text{pnQRPA}}(1^+) = \begin{pmatrix} 1.6208 & 1.1774 \\ 1.1774 & 0.9573 \end{pmatrix} \text{MeV} .
\tag{19.27}
$$

This matrix and the A matrix (17.14) constitute the pnQRPA matrix in (19.6).

The pnQRPA eigenenergies and eigenvectors are obtained from (19.6) by using the Ullah–Rowe diagonalization method presented in Subsect. 11.5.2. Diagonalization of (19.27) thus gives the eigenenergies

$$
E(1_1^+) = 5.718\,\text{MeV} , \quad E(1_2^+) = 6.286\,\text{MeV}
\tag{19.28}
$$

and the eigenvectors

$$
|1_1^+\rangle = 0.857|1\rangle - 0.171|1^{-1}\rangle + 0.558|2\rangle - 0.128|2^{-1}\rangle ,
\tag{19.29}
$$
$$
|1_2^+\rangle = 0.543|1\rangle + 0.009|1^{-1}\rangle - 0.840|2\rangle + 0.013|2^{-1}\rangle ,
\tag{19.30}
$$

where we use the notation $|1^{-1}\rangle$ and $|2^{-1}\rangle$, introduced in Subsect. 18.3.2, for the $Y$ terms.

Comparing the pnQRPA energies and wave functions (19.28)–(19.30) with the pnQTDA results (17.15)–(17.17) we see that the first pnQRPA state lies lower and is more mixed than its pnQTDA counterpart. This means that the pnQRPA state is more collective than the pnQTDA state. The mixing is carried mainly by the $X$ amplitudes, but additional collectivity arises from the ground-state correlations witnessed by the sizable $Y$ amplitudes.

Figure 19.1 shows two computed pnQRPA excitation spectra of $^{24}$Na and the complete experimental spectrum up to 1.5 MeV. The model space for the left-hand spectrum is $0d_{5/2}$-$1s_{1/2}$; the $1^+$ eigenvalues (19.28) are part of this spectrum. The middle spectrum is from a full $0d$-$1s$ calculation with $A_0 = 0.8$ MeV and $A_1 = 0.5$ MeV. In both calculations the SDI parameters are the same for all angular momenta $J$ and the pairing parameters are from Table 16.5.

Comparison of the theoretical spectra of Fig. 19.1 with those of Fig. 17.1 shows that the strong compression effect present in the pnQTDA spectra has now largely disappeared. However, the order of states remains badly incorrect. Even the two theoretical spectra are very different, which points to the need of a larger valence space. The theoretical description can also be improved by

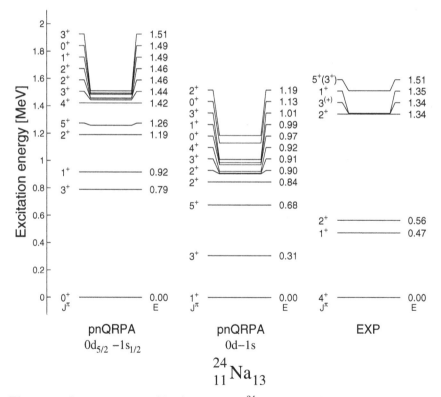

**Fig. 19.1.** Proton–neutron QRPA spectra of $^{24}$Na in the valence spaces shown and the complete experimental spectrum up to 1.5 MeV. The single-particle energies are from (14.12). The SDI was used with the parameter values stated in the text

adjusting the single-particle energies and by choosing the $A_0$ and $A_1$ parameters independently for each $J$.

## 19.2.2 Other Examples

We take up two more examples of the pnQRPA, one in the complete 0d-1s shell and the other one in the 0f-1p-0g$_{9/2}$ major shell.

Figure 19.2 shows the excitation spectrum of $^{30}_{15}$P$_{15}$ computed in the 0d-1s shell, together with the experimental spectrum. With the even–even reference nucleus $^{30}_{14}$Si$_{16}$, the SDI pairing parameters are from Table 16.6. The pnQRPA calculation was done with the SDI parameters $A_0 = 1.0$ MeV and $A_1 = 0.8$ MeV for all $J$. The computed spectrum is still compressed, but not as much as the pnQTDA spectrum in Fig. 17.2. Unlike the pnQTDA, the pnQRPA quite well reproduces the sequence of the lowest states. The theoretical description can be improved, e.g. by choosing the $A_0$ and $A_1$ parameters independently for each $J$.

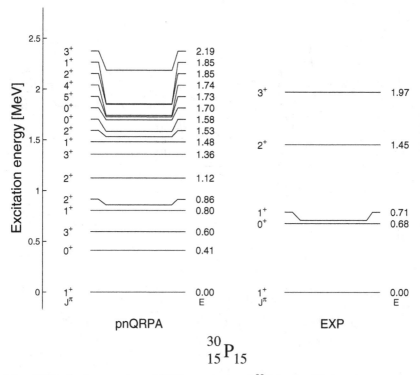

**Fig. 19.2.** Proton–neutron QRPA spectrum of $^{30}$P in the 0d-1s valence space and the complete experimental spectrum up to the $3^+_1$ level. The single-particle energies are from (14.12). The SDI was used with the parameters stated in the text

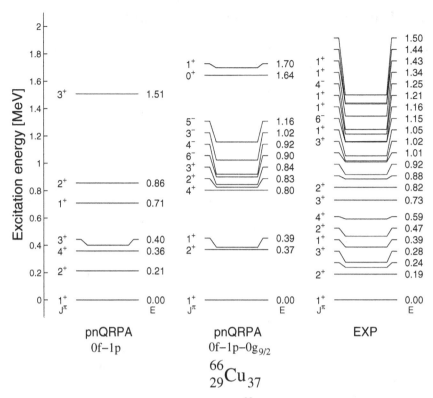

**Fig. 19.3.** Proton–neutron QRPA spectra of $^{66}$Cu in the valence spaces shown and the experimental spectrum. The single-particle energies are from (14.13). The SDI was used with the parameters stated in the text. All theoretical and experimental energy levels are shown up to 1.5 MeV of excitation

Figure 19.3 shows two computed pnQRPA spectra of $^{66}_{29}$Cu$_{37}$ and the experimental spectrum. The even–even reference nucleus is $^{66}_{30}$Zn$_{36}$. The calculations were done in the 0f-1p and the 0f-1p-0g$_{9/2}$ valence spaces with the single-particle energies (14.13) and the SDI pairing parameters of Table 16.7. For the pnQRPA calculations the SDI parameters were $A_0 = 0.6$ MeV and $A_1 = 0.4$ MeV for all $J^\pi$.

For both the pnQTDA and the pnQRPA the sequence of the first $0^+$, $1^+$ and $2^+$ states is sensitive to the relative magnitude of the $A_0$ and $A_1$ strengths of the SDI. The chosen parameters give a reasonable description of the experimental spectrum. The 0g$_{9/2}$intruder state is needed to generate negative-parity states, some of which are seen in the experimental spectrum. Contrary to the pnQTDA spectrum of Fig. 17.3, the pnQRPA spectrum has the negative-parity states roughly at the correct energy relative to the positive-parity states.

## 19.3 Average Particle Number in the pnQRPA

In Subsect. 17.2.3 we studied the effects of particle number fluctuation on pnQTDA wave functions. The wave functions were found to contain mainly admixtures of states of odd–odd nuclei next to the even–even reference nucleus. The average numbers of nucleons depend on the state and are near the nucleon numbers of the reference nucleus. In this section we extend the study to pnQRPA wave functions.

The average number of protons in a pnQRPA state (19.1) is

$$Z_{\text{eff}} = \langle \omega | \hat{n}_{\text{p}} | \omega \rangle = \langle \text{pnQRPA} | Q_\omega \hat{n}_{\text{p}} Q_\omega^\dagger | \text{pnQRPA} \rangle . \tag{19.31}$$

In analogy with (11.276) the matrix element can be written as

$$\langle \text{pnQRPA} | Q_\omega \hat{n}_{\text{p}} Q_\omega^\dagger | \text{pnQRPA} \rangle = \langle \text{pnQRPA} | [Q_\omega, \hat{n}_{\text{p}}, Q_\omega^\dagger] | \text{pnQRPA} \rangle$$
$$+ \tfrac{1}{2} \langle \text{pnQRPA} | Q_\omega Q_\omega^\dagger \hat{n}_{\text{p}} + \hat{n}_{\text{p}} Q_\omega Q_\omega^\dagger | \text{pnQRPA} \rangle . \tag{19.32}$$

In the second term we make the quasiboson approximation by using the commutation relation (11.275), whereupon the term becomes

$$\langle \text{pnQRPA} | \hat{n}_{\text{p}} | \text{pnQRPA} \rangle = Z_{\text{act}} . \tag{19.33}$$

This is the number of valence protons in the underlying BCS vacuum according to (13.37) and (13.39); we assume it to be the same in the BCS and pnQRPA vacua. The double commutator in (19.32) guarantees that the $Y$ contributions are properly taken into account in the final result.

The remaining task is to calculate the double-commutator term in (19.32). Expanded according to (11.12), it can be expressed as

$$\langle \text{pnQRPA} | [Q_\omega, \hat{n}_{\text{p}}, Q_\omega^\dagger] | \text{pnQRPA} \rangle$$
$$= \tfrac{1}{2} \langle \text{pnQRPA} | \left( [Q_\omega, [\hat{n}_{\text{p}}, Q_\omega^\dagger]] + \text{H.c.} \right) | \text{pnQRPA} \rangle$$
$$= \langle \text{pnQRPA} | [Q_\omega, [\hat{n}_{\text{p}}, Q_\omega^\dagger]] | \text{pnQRPA} \rangle . \tag{19.34}$$

When we insert the number operator $\hat{n}_{\text{p}}$ in its quasiparticle representation (13.38) we see that the only term contributing is

$$\sum_p \widehat{j_p} (u_p^2 - v_p^2) [a_p^\dagger \tilde{a}_p]_{00} = \sum_\pi (u_p^2 - v_p^2) a_\pi^\dagger a_\pi . \tag{19.35}$$

To proceed with (19.34) when $Q_\omega^\dagger$ is expanded in terms of the $X$ and $Y$ amplitudes, we need the commutation relations (Exercise 19.6)

$$\sum_\pi (u_p^2 - v_p^2) [a_\pi^\dagger a_\pi, A_{p'n'}^\dagger(JM)] = (u_{p'}^2 - v_{p'}^2) A_{p'n'}^\dagger(JM) , \tag{19.36}$$

$$\sum_\pi (u_p^2 - v_p^2) [a_\pi^\dagger a_\pi, \tilde{A}_{p'n'}(JM)] = -(u_{p'}^2 - v_{p'}^2) \tilde{A}_{p'n'}(JM) . \tag{19.37}$$

When both $Q_\omega$ and $Q_\omega^\dagger$ are expanded in (19.34) the remaining commutators are evaluated in the quasiboson approximation:

$$\langle \text{pnQRPA}|\left[A_{pn}(JM), A_{p'n'}^\dagger(JM)\right]|\text{pnQRPA}\rangle$$

$$\overset{\text{QBA}}{\approx} \langle \text{BCS}|\left[A_{pn}(JM), A_{p'n'}^\dagger(JM)\right]|\text{BCS}\rangle = \delta_{pp'}\delta_{nn'} \ . \qquad (19.38)$$

Through (19.32) and (19.33) this leads to the final result

$$\boxed{Z_{\text{eff}} = Z_{\text{act}} + \sum_{pn}(u_p^2 - v_p^2)\left(|X_{pn}^\omega|^2 + |Y_{pn}^\omega|^2\right) \ .} \qquad (19.39)$$

The analogous result for the effective neutron number is

$$\boxed{N_{\text{eff}} = N_{\text{act}} + \sum_{pn}(u_n^2 - v_n^2)\left(|X_{pn}^\omega|^2 + |Y_{pn}^\omega|^2\right) \ .} \qquad (19.40)$$

Equations (19.39) and (19.40) deviate from their pnQTDA counterparts in (17.22). The difference lies in the $|Y|^2$ contributions arising from the pnQRPA ground-state correlations. Increase in these contributions causes an increase in the $|X|^2$ contributions to maintain the normalization (19.10). This, in turn, can cause $Z_{\text{eff}}$ and $N_{\text{eff}}$ to deviate considerably from $Z_{\text{act}}$ and $N_{\text{act}}$, particularly for the large $Y$ magnitudes occurring near the breaking point of the pnQRPA. Contrariwise, the average particle numbers of pnQTDA excitations are constrained by (17.23).

## 19.4 Electromagnetic Transitions in the pnQRPA

In this section we want to extend our discussion of the electromagnetic observables of odd–odd nuclei from the pnQTDA formalism of Sect. 17.3 to the pnQRPA framework. It is reasonable to expect that the ground-state correlations, represented by the $Y$ amplitudes, play an important role in gamma decay rates.

### 19.4.1 Transition Amplitudes

The initial and final states for electromagnetic transitions in odd–odd nuclei are

$$|\omega_i\rangle = \sum_{p_i n_i}\left[X_{p_i n_i}^{\omega_i} A_{p_i n_i}^\dagger(J_i M_i) - Y_{p_i n_i}^{\omega_i} \tilde{A}_{p_i n_i}(J_i M_i)\right]|\text{pnQRPA}\rangle \ , \qquad (19.41)$$

$$|\omega_f\rangle = \sum_{p_f n_f}\left[X_{p_f n_f}^{\omega_f} A_{p_f n_f}^\dagger(J_f M_f) - Y_{p_f n_f}^{\omega_f} \tilde{A}_{p_f n_f}(J_f M_f)\right]|\text{pnQRPA}\rangle \ , \qquad (19.42)$$

where the vacuum |pnQRPA⟩ is based on the BCS vacuum computed for the even–even reference nucleus. The states (19.41) and (19.42) are similar to QRPA states created by the operators (18.103) and (18.104). This analogy allows us to write the transition amplitude on the pattern of the QRPA result (18.114). Thus we have

$$
(\omega_f \| \boldsymbol{M}_{\sigma\lambda} \| \omega_i)_{\mathrm{pnQRPA}} = \sum_{\substack{p_i n_i \\ p_f n_f}} \left[ X_{p_f n_f}^{\omega_f *} X_{p_i n_i}^{\omega_i} + (-1)^{\Delta J + \lambda} \zeta^{(\lambda)} Y_{p_f n_f}^{\omega_f *} Y_{p_i n_i}^{\omega_i} \right]
$$
$$
\times (p_f \, n_f \,;\, J_f \| \boldsymbol{M}_{\sigma\lambda} \| p_i \, n_i \,;\, J_i) \,,
$$
(19.43)

where $(p_f \, n_f \,;\, J_f \| \boldsymbol{M}_{\sigma\lambda} \| p_i \, n_i \,;\, J_i)$ is the two-quasiparticle transition matrix element according to (15.103). The phase factor $\zeta^{(\lambda)}$, given in (18.81), takes into account the CS or BR phase convention. The following example illustrates the use of the formalism.

### 19.4.2 Decay of the $2_1^+$ State in $^{24}$Na

Let us continue the example of Subsect. 17.3.1 on the decay of the $2_1^+$ state of $^{24}$Na to the $1_1^+$ and $4_{\mathrm{gs}}^+$ states, depicted in Fig. 17.4. In Subsect. 17.3.1 the pnQTDA wave functions were calculated in the $0d_{5/2}\text{-}1s_{1/2}$ valence space. We now proceed to calculate the pnQRPA wave functions in the same valence space.

We use the same parameters as in the pnQTDA calculation of Subsect. 17.3.1 and in the calculation of the $1^+$ states in Subsect. 19.2.1. As a result we obtain the pnQRPA wave functions

$$
|^{24}\mathrm{Na}\,;\, 1_1^+\rangle = 0.857|1\rangle_1 - 0.171|1^{-1}\rangle_1 + 0.558|2\rangle_1 - 0.128|2^{-1}\rangle_1 \,, \quad (19.44)
$$
$$
|^{24}\mathrm{Na}\,;\, 2_1^+\rangle = 0.719|1\rangle_2 - 0.093|1^{-1}\rangle_2 + 0.504|2\rangle_2 - 0.085|2^{-1}\rangle_2
$$
$$
+ 0.502|3\rangle_2 - 0.085|3^{-1}\rangle_2 \,, \quad (19.45)
$$
$$
|^{24}\mathrm{Na}\,;\, 4_{\mathrm{gs}}^+\rangle = 1.000|1\rangle_4 - 0.023|1^{-1}\rangle_4 \,, \quad (19.46)
$$

where the $1_1^+$ state repeats (19.29).

We first calculate the M1 transition $2_1^+ \to 1_1^+$. The $X$ and $Y$ amplitudes come from the wave functions (19.44) and (19.45). The two-quasiparticle matrix element was computed in Subsect. 17.3.1 and is given by (17.34), (17.37) and (17.38). Substitution into (19.43) then yields

$$
(1_1^+ \| \boldsymbol{M}_1 \| 2_1^+) = [0.857 \times 0.719 - (-0.171)(-0.093)](-3.747 - 1.495)\,\mu_{\mathrm{N}}/c
$$
$$
= -3.147\,\mu_{\mathrm{N}}/c \,. \quad (19.47)
$$

This gives the reduced transition probability

$$
B(\mathrm{M1}\,;\, 2_1^+ \to 1_1^+) = 1.981\,(\mu_{\mathrm{N}}/c)^2 \,. \quad (19.48)
$$

Table 6.9 with the experimental gamma energy 0.0910 MeV from Fig. 17.4 gives the decay probability

$$T(\text{M1}\,;\ ^{24}\text{Na}) = 2.656 \times 10^{10}\ 1/\text{s}\ . \tag{19.49}$$

Next we calculate the transition $2_1^+ \rightarrow 4_{\text{gs}}^+$. The run of the calculation is presented in Table 19.1. The $X$ and $Y$ amplitudes are from (19.45) and (19.46). The two-quasiparticle matrix elements

$$(p_f\, n_f\,;\ 4\|\boldsymbol{Q}_2\|p_i\, n_i\,;\ 2) \equiv Q_{if} \tag{19.50}$$

were essentially calculated in Subsect. 17.3.1. Table 17.3 gives the proton and neutron contributions $Q^{(1)}$ and $Q^{(2)}$, which are to be added according to (17.42). We insert the oscillator length $b = 1.813\,\text{fm}$ and the effective charges $e_{\text{eff}}^{\text{p}} = 1.3e$ and $e_{\text{eff}}^{\text{n}} = 0.3e$, suggested by the result (18.92).

**Table 19.1.** Contributions to the sum (19.43) for the $2_1^+ \rightarrow 4_{\text{gs}}^+$ transition in $^{24}$Na

| $p_i n_i(2^+)$ | $p_f n_f(4^+)$ | $X_{p_i n_i}^{\omega_i}$ | $Y_{p_i n_i}^{\omega_i}$ | $X_{p_f n_f}^{\omega_f}$ | $Y_{p_f n_f}^{\omega_f}$ | $Q_{if}\,(e\,\text{fm}^2)$ | $C_{if}\,(e\,\text{fm}^2)$ |
|---|---|---|---|---|---|---|---|
| $0d_{5/2}0d_{5/2}$ | $0d_{5/2}0d_{5/2}$ | 0.719 | $-0.093$ | 1.000 | $-0.023$ | 0.4271 | 0.308 |
| $0d_{5/2}1s_{1/2}$ | $0d_{5/2}0d_{5/2}$ | 0.504 | $-0.085$ | 1.000 | $-0.023$ | $-0.1048$ | $-0.053$ |
| $1s_{1/2}0d_{5/2}$ | $0d_{5/2}0d_{5/2}$ | 0.502 | $-0.085$ | 1.000 | $-0.023$ | $-0.4636$ | $-0.234$ |

The initial and final configurations and their $X$ and $Y$ amplitudes are listed in the first six columns. The two-quasiparticle transition matrix elements (15.103) are listed in column seven and the contributions in column eight.

The sum of the contributions in Table 19.1 is $0.021\, e\,\text{fm}^2$, whence

$$B(\text{E2}\,;\ 2_1^+ \rightarrow 4_{\text{gs}}^+) = \frac{1}{5}(0.021\, e\,\text{fm}^2)^2 = 8.8 \times 10^{-5}\, e^2\text{fm}^4\ . \tag{19.51}$$

With the gamma energy 0.5633 MeV from Fig. 17.4, Table 6.8 now gives the decay probability

$$T(\text{E2}\,;\ ^{24}\text{Na}) = 6.1 \times 10^3\ 1/\text{s}\ . \tag{19.52}$$

This is negligibly small in comparison with (19.49). Note that it is even three orders of magnitude smaller than the pnQTDA value (17.48) that was neglected in determining the $2_1^+$ half-life (17.49). The reason for this great difference between the pnQRPA and the pnQTDA is the fragmentation of states discussed below in the context of the M1 transition. With the E2 contribution negligible, (19.49) gives the decay half-life

$$t_{1/2}(2_1^+) = \frac{\ln 2}{T(\text{M1})} = 26\,\text{ps}\ , \tag{19.53}$$

which is remarkably close to the experimental value 35 ps in Fig. 17.4.

On comparison with the pnQTDA result (17.49) the pnQRPA result (19.53) is seen to be better by a factor of three. This is due to the improved description of the M1 transition $2_1^+ \rightarrow 1_1^+$. The pnQRPA wave functions (19.44) and (19.45) are strongly fragmented, while their pnQTDA counterparts (17.27) and (17.32) are 99.8 % pure $\pi 0d_{5/2}\nu 0d_{5/2}$ two-quasiparticle states. The fragmentation diminishes the leading strength.

## 19.5 Beta-Decay Transitions in the pnQRPA Framework

In this section we discuss beta-decay transitions between states of an open-shell odd–odd nucleus and states of the neighbouring even–even reference nucleus. In a pnQRPA calculation the reference nucleus serves to generate the levels of the odd–odd nucleus in question. First we discuss transitions involving the ground state of the reference nucleus.

### 19.5.1 Transitions Involving the Even–Even Ground State

Consider transitions taking the initial state (19.1) of an odd–odd nucleus to the ground state $|\text{pnQRPA}\rangle$ of the even–even reference nucleus. The decay amplitude for $\beta^-$ and $\beta^+$ transitions is

$$\langle \text{pnQRPA}|\beta_{LM}^{\mp}|\omega\rangle = \langle \text{pnQRPA}|\beta_{LM}^{\mp}Q_\omega^\dagger|\text{pnQRPA}\rangle$$
$$= \langle \text{pnQRPA}|[\beta_{LM}^{\mp}, Q_\omega^\dagger]|\text{pnQRPA}\rangle . \qquad (19.54)$$

As is usual in RPA work, we have introduced the commutator to access the $Y$ terms of the pnQRPA phonon.

The transition density (15.4) gives the quasiparticle representation of a $\beta^-$ operator as

$$\beta_{LM}^- = \sum_{pn} \mathcal{M}_L(pn)\big\{ u_p v_n A_{pn}^\dagger(LM) + v_p u_n \tilde{A}_{pn}(LM)$$
$$+ u_p u_n [a_p^\dagger \tilde{a}_n]_{LM} + (-1)^{j_p+j_n+L} v_p v_n [a_n^\dagger \tilde{a}_p]_{LM} \big\} . \qquad (19.55)$$

For allowed beta decay, (7.73) and (7.74) give the single-particle matrix elements $\mathcal{M}_0(pn) = \mathcal{M}_{\text{F}}(pn)$ and $\mathcal{M}_1(pn) = \mathcal{M}_{\text{GT}}(pn)$. For $K$th-forbidden unique beta decay the single-particle matrix element is $\mathcal{M}^{(K\text{u})}(pn)$ as given by (7.188).

The operators of allowed beta decay, $\mathbf{1}$ for Fermi and $\boldsymbol{\sigma}$ for Gamow–Teller type, are Hermitian tensor operators as defined in (2.31). The operator valid for both types was written specifically for beta-minus decay in (7.69),

$$\beta_{LM}^- = \hat{L}^{-1}\sum_{pn}(p\|\beta_L\|n)[c_p^\dagger \tilde{c}_n]_{LM} = \sum_{pn} \mathcal{M}_L(pn)[c_p^\dagger \tilde{c}_n]_{LM} . \qquad (19.56)$$

By taking the Hermitian conjugate of this and using the property (2.32) one can show that (Exercise 19.7)

$$\left(\beta^-_{LM}\right)^\dagger = (-1)^M \beta^+_{L,-M} \,, \tag{19.57}$$

where

$$\beta^+_{LM} = \widehat{L}^{-1} \sum_{np} (n\|\beta_L\|p)\left[c^\dagger_n \tilde{c}_p\right]_{LM} = \sum_{np} \mathcal{M}_L(np)\left[c^\dagger_n \tilde{c}_p\right]_{LM} \,. \tag{19.58}$$

The tensor operator $\beta^+_L$ appeared in (7.77). Combining (19.57) with the converse relation we have

$$\boxed{\left(\beta^\mp_{LM}\right)^\dagger = (-1)^M \beta^\pm_{L,-M} \,.} \tag{19.59}$$

The relations for allowed beta decay are the same for the CS and BR phase conventions. This is because the operators $\mathbf{1}$ and $\boldsymbol{\sigma}$ have no spatial dependence. For forbidden unique decay the symmetry relation corresponding to (19.59) is (7.212), and it has a convention-dependent phase factor due to spatial dependence.

We now apply the quasiboson approximation to (19.54) and find

$$\langle\mathrm{pnQRPA}|\beta^-_{LM}|\omega\rangle \overset{\mathrm{QBA}}{\approx} \langle\mathrm{BCS}|\left[\beta^-_{LM}, Q^\dagger_\omega\right]|\mathrm{BCS}\rangle$$

$$= \sum_{\substack{pn \\ p'n'}} \mathcal{M}_L(p'n')\Big\{ v_{p'}u_{n'}X^\omega_{pn}\langle\mathrm{BCS}|\left[\tilde{A}_{p'n'}(LM), A^\dagger_{pn}(JM_\omega)\right]|\mathrm{BCS}\rangle$$

$$- u_{p'}v_{n'}Y^\omega_{pn}\langle\mathrm{BCS}|\left[A^\dagger_{p'n'}(LM), \tilde{A}_{pn}(JM_\omega)\right]|\mathrm{BCS}\rangle\Big\} \,, \tag{19.60}$$

where we use the notation $M_\omega$ for the $z$ projection of $J$ to distinguish it from the $z$ projection $M$ of $L$. The matrix elements give Kronecker deltas so that

$$\langle\mathrm{pnQRPA}|\beta^-_{LM}|\omega\rangle = \delta_{LJ}\delta_{M,-M_\omega}(-1)^{L+M}\sum_{pn}\mathcal{M}_L(pn)(v_p u_n X^\omega_{pn}+u_p v_n Y^\omega_{pn}) \,. \tag{19.61}$$

By the Wigner–Eckart theorem this gives the transition amplitude

$$\boxed{(\mathrm{pnQRPA}\|\beta^-_L\|\omega) = \delta_{LJ}\widehat{L}\sum_{pn}\mathcal{M}_L(pn)(v_p u_n X^\omega_{pn}+u_p v_n Y^\omega_{pn}) \,.} \tag{19.62}$$

For forbidden unique decay the single-particle matrix elements $\mathcal{M}_L(pn)$ are replaced according to (7.210).

To find the transition amplitude for $\beta^+$ decay, we use (19.59), which gives

$$\langle\mathrm{pnQRPA}|\beta^+_{LM}|\omega\rangle = (-1)^M\langle\mathrm{pnQRPA}|\left(\beta^-_{L,-M}\right)^\dagger|\omega\rangle$$

$$= (-1)^M\langle\omega|\beta^-_{L,-M}|\mathrm{pnQRPA}\rangle^* = (-1)^M\langle\omega|\beta^-_{L,-M}|\mathrm{pnQRPA}\rangle \,, \tag{19.63}$$

where we have made the usual assumption that matrix elements are real. Analogous to (19.60) this becomes

$$\langle \text{pnQRPA}|\beta_{LM}^+|\omega\rangle \stackrel{\text{QBA}}{\approx} (-1)^M \langle \text{BCS}|[Q_\omega, \beta_{L,-M}^-]|\text{BCS}\rangle$$

$$= \sum_{\substack{pn \\ p'n'}} \mathcal{M}_L(p'n')\{u_{p'}v_{n'}X_{pn}^\omega \langle \text{BCS}|[A_{pn}(JM_\omega), A_{p'n'}^\dagger(L,-M)]|\text{BCS}\rangle$$

$$- v_{p'}u_{n'}Y_{pn}^\omega \langle \text{BCS}|[\tilde{A}_{pn}^\dagger(JM_\omega), \tilde{A}_{p'n'}(L,-M)]|\text{BCS}\rangle\}$$

$$= \delta_{LJ}\delta_{M,-M_\omega}(-1)^M \sum_{pn} \mathcal{M}_L(pn)(u_p v_n X_{pn}^\omega + v_p u_n Y_{pn}^\omega) . \tag{19.64}$$

The Wigner–Eckart theorem yields the reduced matrix element

$$\boxed{(\text{pnQRPA}\|\beta_L^+\|\omega) = \delta_{LJ}(-1)^L\hat{L}\sum_{pn} \mathcal{M}_L(pn)(u_p v_n X_{pn}^\omega + v_p u_n Y_{pn}^\omega) .}$$
$$\tag{19.65}$$

For forbidden unique decay the single-particle matrix element in the formula is replaced according to (7.213).

For transitions in the opposite direction we use the relation (19.59). For $\beta^-$ transitions it gives

$$\langle \omega|\beta_{LM}^-|\text{pnQRPA}\rangle = (-1)^M \langle \omega|(\beta_{L,-M}^+)^\dagger|\text{pnQRPA}\rangle$$

$$= (-1)^M \langle \text{pnQRPA}|\beta_{L,-M}^+|\omega\rangle$$

$$= \delta_{LJ}\delta_{MM_\omega} \sum_{pn} \mathcal{M}_L(pn)(u_p v_n X_{pn}^\omega + v_p u_n Y_{pn}^\omega) , \tag{19.66}$$

where the last step came from (19.64). The reduced matrix element becomes

$$\boxed{(\omega\|\beta_L^-\|\text{pnQRPA}) = \delta_{LJ}\hat{L}\sum_{pn} \mathcal{M}_L(pn)(u_p v_n X_{pn}^\omega + v_p u_n Y_{pn}^\omega) .} \tag{19.67}$$

The remaining transition amplitude is obtained similarly by use of (19.61), with the result

$$\boxed{(\omega\|\beta_L^+\|\text{pnQRPA}) = \delta_{LJ}(-1)^L\hat{L}\sum_{pn} \mathcal{M}_L(pn)(v_p u_n X_{pn}^\omega + u_p v_n Y_{pn}^\omega) .}$$
$$\tag{19.68}$$

We next discuss an application of the theory presented in this subsection.

## 19.5.2 Gamow–Teller Decay of the $1_1^+$ Isomer in $^{24}$Al

Consider the Gamow–Teller $\beta^+$ decay of the isomeric $1^+$ first excited state in $^{24}$Al to the ground state of its even–even neighbour nucleus $^{24}$Mg. The

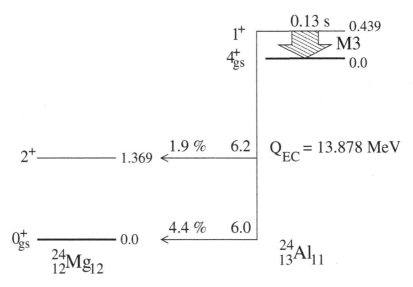

**Fig. 19.4.** Gamow–Teller $\beta^+$ decay of the first $1^+$ state of $^{24}$Al to the ground and excited states of $^{24}$Mg. The $1^+$ state decays also to the $4^+$ ground state of $^{24}$Al by an M3 gamma transition. The measured total half-life of this state is indicated. The experimental energies, $Q$ value, branchings and $\log ft$ values are given

experimental data are shown in Fig. 19.4. The $1_1^+$ state gamma decays by an M3 transition to the $4^+$ ground state of $^{24}$Al. Due to its long gamma-decay half-life it is an *isomeric state* and so its beta decay can be detected.

We follow the example of Subsect. 19.4.2 with the same valence space and parameters. Equation (19.65) is the appropriate formula for the transition under study. The transition is of Gamow–Teller type, and so the formula gives

$$(\text{pnQRPA}\|\beta_{\text{GT}}^+\|1^+) = -\sqrt{3}\sum_{pn}\mathcal{M}_{\text{GT}}(pn)\left(u_p v_n X_{pn}^{1^+} + v_p u_n Y_{pn}^{1^+}\right). \quad (19.69)$$

We take the single-particle matrix elements from Table 7.3 and the BCS occupation amplitudes from Table 16.9. The $X$ and $Y$ amplitudes are given by (19.44). We thus have

$$(\text{pnQRPA}\|\beta_{\text{GT}}^+\|1_1^+)$$

$$= -\sqrt{3}\Bigg\{\sqrt{\frac{14}{5}}[0.6821 \times 0.7308 \times 0.857 + 0.7312 \times 0.6826(-0.171)]$$

$$+ \sqrt{2}[0.7773 \times 0.6307 \times 0.558 + 0.6291 \times 0.7760(-0.128)]\Bigg\}$$

$$= -1.51 . \quad (19.70)$$

By the standard relations of Sect. 7.2 this gives

$$\log ft_+ = 3.71 \ . \tag{19.71}$$

This is far too small compared with the experimental value of 6.0.

We can extend our calculation to the complete 0d-1s valence space and use the interaction parameters quoted in Subsect. 19.2.1. We then obtain

$$(\text{pnQRPA}\|\beta_{\text{GT}}^+\|1_1^+)_{\text{0d-1s}} = -1.69 \ . \tag{19.72}$$

This leads to $\log_+ ft = 3.62$, which is even worse than (19.71). It appears that some other effects, like higher-order configurations, come into play and cause the decay rate to decrease drastically.

In the $1_1^+$ wave function (19.44) the $X$ and $Y$ amplitudes are of opposite sign. This is a general feature of $1_1^+$ states in odd–odd nuclei. It provides for the $Y$ amplitudes to reduce the magnitude of the Gamow–Teller matrix element, as seen in (19.70). With increasing two-body interaction strength the $Y$ magnitudes grow, which can bring the calculated Gamow–Teller strength down to the observed level. This behaviour, accompanied by a lowering of the $1_1^+$ energy, is considered collective.

However, the collective $1^+$ mechanism can produce $Y$ amplitudes comparable in magnitude to the $X$ amplitudes; this case is near in (19.44). Then the condition (19.14) is not satisfied, which invalidates the use of the pnQRPA. The final stage is the collapse of the pnQRPA as discussed in Subsect. 19.1.2.

## 19.6 The Ikeda Sum Rule for the pnQRPA

In this section we discuss beta-decay strength distributions computed by the pnQRPA. These distributions satisfy the Ikeda sum rule for Gamow–Teller transitions, as did the strength distributions computed by the pnQTDA in Subsect. 17.4.3.

### 19.6.1 Derivation of the Sum Rule

We defined in (17.63) the total $\beta^-$ and $\beta^+$ transition strengths $S_-$ and $S_+$ for Gamow–Teller transitions. We then showed that they satisfy the model-independent Ikeda sum rule (17.74),

$$S_- - S_+ = 3(N - Z) \ . \tag{19.73}$$

Furthermore, we derived the sum rule (17.84) as a special case applicable to the pnQTDA. Let us now derive a pnQRPA version of the sum rule.

The difference of the $\beta^-$ and $\beta^+$ total transition strengths (17.64) is

$$S_- - S_+ = \sum_{n_\omega} \left|(\omega\|\beta_{\text{GT}}^-\|\text{pnQRPA})\right|^2 - \sum_{n_\omega} \left|(\omega\|\beta_{\text{GT}}^+\|\text{pnQRPA})\right|^2 , \tag{19.74}$$

where $\omega = n_\omega, 1^+$. Substituting (19.68) and (19.67) into (19.74) gives

$$S_- - S_+ = 3 \sum_{n_\omega} \sum_{\substack{pn \\ p'n'}} \mathcal{M}_{GT}(pn) \mathcal{M}_{GT}(p'n')$$

$$\times \left[ (u_p v_n X_{pn}^\omega + v_p u_n Y_{pn}^\omega)(u_{p'} v_{n'} X_{p'n'}^{\omega*} + v_{p'} u_{n'} Y_{p'n'}^{\omega*}) \right.$$

$$\left. - (v_p u_n X_{pn}^{\omega*} + u_p v_n Y_{pn}^{\omega*})(v_{p'} u_{n'} X_{p'n'}^\omega + u_{p'} v_{n'} Y_{p'n'}^\omega) \right]$$

$$= \sum_{n_\omega} \sum_{\substack{pn \\ p'n'}} \mathcal{M}_{GT}(pn) \mathcal{M}_{GT}(p'n') \left[ u_p v_n u_{p'} v_{n'} (X_{pn}^\omega X_{p'n'}^{\omega*} - Y_{pn}^{\omega*} Y_{p'n'}^\omega) \right.$$

$$- v_p u_n v_{p'} u_{n'} (X_{pn}^{\omega*} X_{p'n'}^\omega - Y_{pn}^\omega Y_{p'n'}^{\omega*}) + u_p v_n v_{p'} u_{n'} (X_{pn}^\omega Y_{p'n'}^{\omega*} - Y_{pn}^{\omega*} X_{p'n'}^\omega)$$

$$\left. - v_p u_n u_{p'} v_{n'} (X_{pn}^{\omega*} Y_{p'n'}^\omega - Y_{pn}^\omega X_{p'n'}^{\omega*}) \right] . \tag{19.75}$$

The completeness relations (19.11) and (19.12) simplify this to

$$S_- - S_+ = 3 \sum_{pn} (u_p^2 v_n^2 - v_p^2 u_n^2)[\mathcal{M}_{GT}(pn)]^2 = 3 \sum_{pn} (v_n^2 - v_p^2)[\mathcal{M}_{GT}(pn)]^2 . \tag{19.76}$$

Since the right-hand sides of (19.76) and (17.80) coincide, the right-hand side of the pnQTDA final result (17.83) can be substituted into (19.76). The Ikeda sum rule for the pnQRPA thus becomes

$$\boxed{S_- - S_+ = 3(N_{act} - Z_{act}) \quad (\text{pnQRPA}) .} \tag{19.77}$$

The equation looks the same as (17.84), but the left-hand sides are calculated differently, in the pnQRPA and pnQTDA respectively.

### 19.6.2 Examples of the Sum Rule

We now present a few examples to test the validity of the Ikeda sum rule (19.77). The pnQRPA calculations were done in the 0d-1s and 1p-0f-0g$_{9/2}$ valence spaces.

Table 19.2 shows the results of a pnQRPA calculation for the reference nucleus $^{24}_{12}$Mg$_{12}$. The valence space is the complete 0d-1s major shell. The single-particle energies are from (14.12) and the SDI pairing parameters from Table 16.5. The SDI interaction parameters used in the pnQRPA calculation are $A_0 = 0.8$ MeV and $A_1 = 0.5$ MeV. The summed strengths from the table and the numbers of active nucleons give

$$S_- - S_+ = 5.393 - 5.393 = 0.000 ,$$

$$3(N_{act} - Z_{act}) = 3(4 - 4) = 0 , \tag{19.78}$$

so indeed the sum rule is satisfied by the pnQRPA calculation.

**Table 19.2.** Ikeda sum rule for a pnQRPA calculation with the reference nucleus $^{24}$Mg

| $n$ | $E(1_k^+)$ (MeV) | $\left\|(1_k^+\|\beta_{\mathrm{GT}}^-\|\mathrm{pnQRPA})\right\|^2$ | $\left\|(1_k^+\|\beta_{\mathrm{GT}}^+\|\mathrm{pnQRPA})\right\|^2$ |
|---|---|---|---|
| 1 | 5.358 | 2.8663 | 2.8764 |
| 2 | 6.344 | 0.0673 | 0.0641 |
| 3 | 8.187 | 0.9309 | 0.9565 |
| 4 | 8.578 | 0.3954 | 0.4163 |
| 5 | 9.273 | 0.2727 | 0.2526 |
| 6 | 9.392 | 0.8569 | 0.8243 |
| 7 | 11.372 | 0.0034 | 0.0026 |
| Sum | | 5.393 | 5.393 |

The second column lists the pnQRPA energies. Columns three and four give the $\beta^-$ and $\beta^+$ Gamow–Teller strengths and the last row gives their sums.

The reference nucleus of our second example is $^{30}_{14}$Si$_{16}$. Figure 19.5 shows its $\beta^-$ and $\beta^+$ strength distributions from a pnQRPA calculation. The valence space and single-particle energies are the same as in the $^{24}$Mg example. The SDI pairing parameters are from Table 16.6, and the pnQRPA parameters are $A_0 = 1.0\,\mathrm{MeV}$ and $A_1 = 0.8\,\mathrm{MeV}$. The two sides of the sum rule give

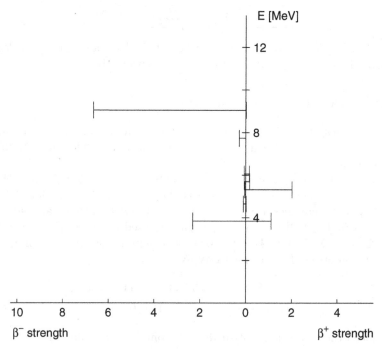

**Fig. 19.5.** Calculated $\beta^-$ and $\beta^+$ strengths for Gamow–Teller transitions from the pnQRPA ground state of $^{30}$Si to the $1^+$ pnQRPA states in $^{30}$P and $^{30}$Al

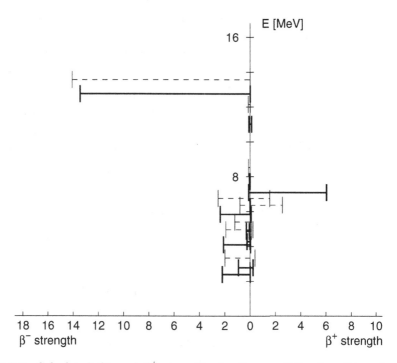

**Fig. 19.6.** Calculated $\beta^-$ and $\beta^+$ strengths for Gamow–Teller transitions from the pnQRPA ground state of $^{66}$Zn to the $1^+$ pnQRPA states in $^{66}$Ga and $^{66}$Cu. The dashed bars represent the 0f-1p calculation and the solid bars the 0f-1p-0g$_{9/2}$ calculation

$$S_- - S_+ = 9.510 - 3.510 = 6.000 \; ,$$
$$3(N_{\text{act}} - Z_{\text{act}}) = 3(8 - 6) = 6 \; , \tag{19.79}$$

so the sum rule is satisfied.

As our final example, Fig. 19.6 shows the $\beta^-$ and $\beta^+$ strength distributions for the reference nucleus $^{66}$Zn in the valence spaces 0f-1p and 0f-1p-0g$_{9/2}$. The single-particle energies are from (14.13) and the SDI pairing parameters from Table 16.7. The pnQRPA parameters are $A_0 = 0.6\,\text{MeV}$ and $A_1 = 0.4\,\text{MeV}$. For the 0f-1p calculation the two sides of the sum rule give

$$S_- - S_+ = 22.893 - 4.893 = 18.000 \; ,$$
$$3(N_{\text{act}} - Z_{\text{act}}) = 3(16 - 10) = 18 \; . \tag{19.80}$$

For the 0f-1p-0g$_{9/2}$ calculation we have

$$S_- - S_+ = 21.800 - 6.550 = 15.250 \; ,$$
$$3(N_{\text{act}} - Z_{\text{act}}) = 3(16 - 10) = 18 \; . \tag{19.81}$$

The total strengths in (19.81) do not satisfy the Ikeda sum rule. This is because the spin–orbit partner $0g_{7/2}$ of the orbital $0g_{9/2}$ is missing from the larger valence space. A general point was made of this at the end of Subsect. 17.4.3.

Figure 19.6 displays for both calculations a rather strong peak on the $\beta^+$ side at 6–7 MeV of excitation. For the larger valence space this peak is twice as strong as for the smaller one. The additional strength in the larger space comes from the $0g_{9/2}$ orbital. On the $\beta^-$ side there is a strong peak around 13 MeV for both valence spaces. This peak is similar to the one at 9 MeV in Fig. 19.5. The pnQTDA displayed similar behaviour in Figs. 17.5 and 17.6. In Subsect. 17.4.5 the strong $\beta^-$ peak was identified with the GTGR.

Comparison of the pnQTDA sum rule tests (17.85)–(17.88) with their pnQRPA counterparts (19.78)–(19.81) reveals that the total $\beta^-$ and $\beta^+$ strengths are typically 30% smaller in the pnQRPA. The reduction is caused by the $Y$ amplitudes acting out of phase with the $X$ amplitudes, as discussed at the end of Subsect. 19.5.2.

## 19.7 Beta-Decay Transitions Between a QRPA and a pnQRPA State

In this section we consider $\beta^-$ and $\beta^+$ decay of a pnQRPA state $|\omega_i\rangle$ in an odd–odd nucleus to a QRPA state $|\omega_f\rangle$ in a neighbouring even–even nucleus. The even–even nucleus serves as the reference nucleus for the QRPA and pnQRPA calculations.

### 19.7.1 Derivation of the Transition Amplitude

We take the beta-decaying state to be a pnQRPA state (19.41) of an odd–odd nucleus, i.e.

$$
\begin{aligned}
|\omega_i\rangle &= Q^\dagger_{\omega_i}|\text{pnQRPA}\rangle \\
&= \sum_{p_i n_i} \left[ X^{\omega_i}_{p_i n_i} A^\dagger_{p_i n_i}(J_i M_i) - Y^{\omega_i}_{p_i n_i} \tilde{A}_{p_i n_i}(J_i M_i) \right] |\text{pnQRPA}\rangle .
\end{aligned}
\tag{19.82}
$$

The final state in the even–even reference nucleus is given by (18.1) and (18.31) as

$$
\begin{aligned}
|\omega_f\rangle &= Q^\dagger_{\omega_f}|\text{QRPA}\rangle \\
&= \sum_{a_f \leq b_f} \left[ X^{\omega_f}_{a_f b_f} A^\dagger_{a_f b_f}(J_f M_f) - Y^{\omega_f}_{a_f b_f} \tilde{A}_{a_f b_f}(J_f M_f) \right] |\text{QRPA}\rangle .
\end{aligned}
\tag{19.83}
$$

To proceed in the usual RPA way, we express the matrix element in terms of a symmetrized double commutator. Analogously to (19.32) we have

$$\langle \mathrm{pnQRPA}|Q_{\omega_f}\beta^{\mp}_{LM}Q^{\dagger}_{\omega_i}|\mathrm{pnQRPA}\rangle = \langle \mathrm{pnQRPA}|[Q_{\omega_f},\beta^{\mp}_{LM},Q^{\dagger}_{\omega_i}]|\mathrm{pnQRPA}\rangle$$
$$+ \tfrac{1}{2}\langle \mathrm{pnQRPA}|Q_{\omega_f}Q^{\dagger}_{\omega_i}\beta^{\mp}_{LM} + \beta^{\mp}_{LM}Q_{\omega_f}Q^{\dagger}_{\omega_i}|\mathrm{pnQRPA}\rangle . \quad (19.84)$$

We make the quasiboson approximation, in the pattern of (18.101),

$$[Q_{\omega_i},Q^{\dagger}_{\omega_f}] \overset{\mathrm{QBA}}{\approx} \langle \mathrm{BCS}|[Q_{\omega_i},Q^{\dagger}_{\omega_f}]|\mathrm{BCS}\rangle = 0 . \quad (19.85)$$

The unequivocal zero results from the orthogonality of the pnQRPA initial state and the QRPA final state. Substitution of (19.85) into the second term of (19.84) makes the term vanish. Moreover, we make the quasiboson approximation in the first term by replacing the pnQRPA vacuum with the BCS vacuum. The matrix element in the quasiboson approximation is thus

$$\langle n_{\omega_f}\, J_f\, M_f|\beta^{\mp}_{LM}|n_{\omega_i}\, J_i\, M_i\rangle = \langle \mathrm{BCS}|[Q_{\omega_f},\beta^{\mp}_{LM},Q^{\dagger}_{\omega_i}]|\mathrm{BCS}\rangle . \quad (19.86)$$

To process (19.86) further we can insert the phonon operators of (19.82) and (19.83) to yield

$$\langle n_{\omega_f}\, J_f\, M_f|\beta^{\mp}_{LM}|n_{\omega_i}\, J_i\, M_i\rangle$$
$$= \sum_{\substack{p_i n_i \\ a_f \le b_f}} \Big\{ X^{\omega_f *}_{a_f b_f} X^{\omega_i}_{p_i n_i} \langle \mathrm{BCS}|[A_{a_f b_f},\beta^{\mp}_{LM},A^{\dagger}_{p_i n_i}]|\mathrm{BCS}\rangle$$
$$- X^{\omega_f *}_{a_f b_f} Y^{\omega_i}_{p_i n_i} \langle \mathrm{BCS}|[A_{a_f b_f},\beta^{\mp}_{LM},\tilde{A}_{p_i n_i}]|\mathrm{BCS}\rangle$$
$$- Y^{\omega_f *}_{a_f b_f} X^{\omega_i}_{p_i n_i} \langle \mathrm{BCS}|[\tilde{A}^{\dagger}_{a_f b_f},\beta^{\mp}_{LM},A^{\dagger}_{p_i n_i}]|\mathrm{BCS}\rangle$$
$$+ Y^{\omega_f *}_{a_f b_f} Y^{\omega_i}_{p_i n_i} \langle \mathrm{BCS}|[\tilde{A}^{\dagger}_{a_f b_f},\beta^{\mp}_{LM},\tilde{A}_{p_i n_i}]|\mathrm{BCS}\rangle \Big\} .$$
$$(19.87)$$

The commutator in the second (third) term contains an excess of quasiparticle annihilation (creation) operators, and so its expectation value vanishes. The first and fourth terms can be handled in analogy to the particle–hole case in Subsect. 11.6.9. A more direct alternative is to expand the double commutator and use (19.55). Most of the terms give zero either through $a|\mathrm{BCS}\rangle = 0$ or $\langle \mathrm{BCS}|a^{\dagger} = 0$, or through excess creation or annihilation operators. The result is (Exercise 19.8)

$$\langle \mathrm{BCS}|[A_{a_f b_f},\beta^{\mp}_{LM},A^{\dagger}_{p_i n_i}]|\mathrm{BCS}\rangle = \langle \mathrm{BCS}|A_{a_f b_f}\beta^{\mp}_{LM}A^{\dagger}_{p_i n_i}|\mathrm{BCS}\rangle , \quad (19.88)$$
$$\langle \mathrm{BCS}|[\tilde{A}^{\dagger}_{a_f b_f},\beta^{\mp}_{LM},\tilde{A}_{p_i n_i}]|\mathrm{BCS}\rangle = \langle \mathrm{BCS}|\tilde{A}_{p_i n_i}\beta^{\mp}_{LM}\tilde{A}^{\dagger}_{a_f b_f}|\mathrm{BCS}\rangle . \quad (19.89)$$

It is expedient to convert (19.89) into a form that describes a transition in the same direction as (19.88). Since the matrix element (19.89) is a real number we can take its complex conjugate to yield

$$\langle \mathrm{BCS}|\tilde{A}_{p_i n_i}(J_i M_i)\beta^{\mp}_{LM}\tilde{A}^{\dagger}_{a_f b_f}(J_f M_f)|\mathrm{BCS}\rangle$$
$$= \langle \mathrm{BCS}|\tilde{A}_{a_f b_f}(J_f M_f)\big(\beta^{\mp}_{LM}\big)^{\dagger}\tilde{A}^{\dagger}_{p_i n_i}(J_i M_i)|\mathrm{BCS}\rangle$$
$$= (-1)^{J_f+M_f+M+J_i+M_i}\langle \mathrm{BCS}|A_{a_f b_f}(J_f,-M_f)\beta^{\pm}_{L,-M}A^{\dagger}_{p_i n_i}(J_i,-M_i)|\mathrm{BCS}\rangle ,$$
$$(19.90)$$

where we used (19.59). Substituting (19.88) and (19.90) back into (19.87) and applying the Wigner–Eckart theorem we find

$$
(n_{\omega_f} J_f \| \beta_L^{\mp} \| n_{\omega_i} J_i) = \sum_{\substack{p_i n_i \\ a_f \leq b_f}} \left[ X_{a_f b_f}^{\omega_f *} X_{p_i n_i}^{\omega_i} (a_f\, b_f\, ;\, J_f \| \beta_L^{\mp} \| p_i\, n_i\, ;\, J_i) \right.
$$
$$
\left. + (-1)^L Y_{a_f b_f}^{\omega_f *} Y_{p_i n_i}^{\omega_i} (a_f\, b_f\, ;\, J_f \| \beta_L^{\pm} \| p_i\, n_i\, ;\, J_i) \right] .
$$

$$(19.91)$$

The two-quasiparticle matrix elements in (19.91) are given for final states of two quasiprotons and two quasineutrons by (15.111) and (15.113) respectively. With them specified, the final result becomes

$$
(\omega_f \| \beta_L^{\mp} \| \omega_i) = \sum_{\substack{p_i n_i \\ p_f \leq p'_f}} \left[ X_{p_f p'_f}^{\omega_f *} X_{p_i n_i}^{\omega_i} \mathcal{M}_L^{(\mp)}(p_i\, n_i\, ;\, J_i \to p_f\, p'_f\, ;\, J_f) \right.
$$
$$
\left. + (-1)^L Y_{p_f p'_f}^{\omega_f *} Y_{p_i n_i}^{\omega_i} \mathcal{M}_L^{(\pm)}(p_i\, n_i\, ;\, J_i \to p_f\, p'_f\, ;\, J_f) \right]
$$
$$
+ \sum_{\substack{p_i n_i \\ n_f \leq n'_f}} \left[ X_{n_f n'_f}^{\omega_f *} X_{p_i n_i}^{\omega_i} \mathcal{M}_L^{(\mp)}(p_i\, n_i\, ;\, J_i \to n_f\, n'_f\, ;\, J_f) \right.
$$
$$
\left. + (-1)^L Y_{n_f n'_f}^{\omega_f *} Y_{p_i n_i}^{\omega_i} \mathcal{M}_L^{(\pm)}(p_i\, n_i\, ;\, J_i \to n_f\, n'_f\, ;\, J_f) \right] ,
$$

$$(19.92)$$

This relation coincides with the pnQTDA relation (17.93) in the limit $Y \to 0$.

For $K$th-forbidden unique beta decay the Fermi ($L = 0$) and Gamow–Teller ($L = 1$) transition matrix elements in (19.92) are to be replaced as summarized at the end of Subsect. 15.5.2.

## 19.7.2 The Gamow–Teller Decay $^{24}$Al($1_1^+$) → $^{24}$Mg($2_1^+$)

We now apply the formula (19.92) to the $\beta^+$ decay of the isomeric first excited state of $^{24}$Al. This is continuation of the example of Subsect. 19.5.2, which dealt with the decay to the $0^+$ ground state. The experimental situation is shown in Fig. 19.4.

We want to compute the transition amplitude for the decay of the isomeric $1_1^+$ state to the $2_1^+$ state in $^{24}$Mg. The valence space and BCS parameters are the same as in Subsect. 19.5.2. Equation (19.44) gives the wave function $|^{24}\mathrm{Na}\,;\, 1_1^+\rangle$, calculated with the pnQRPA parameters $A_0 = A_1 = 1.0\,\mathrm{MeV}$. The mirror nucleus $^{24}$Al has the same wave function, i.e.

$$
|^{24}\mathrm{Al}\,;\, 1_1^+\rangle = 0.857|1\rangle_1 - 0.171|1^{-1}\rangle_1 + 0.558|2\rangle_1 - 0.128|2^{-1}\rangle_1 \quad (19.93)
$$

in the basis

$$
|1\rangle_1 = |\pi 0\mathrm{d}_{5/2}\, \nu 0\mathrm{d}_{5/2}\,;\, 1^+\rangle , \quad |2\rangle_1 = |\pi 1\mathrm{s}_{1/2}\, \nu 1\mathrm{s}_{1/2}\,;\, 1^+\rangle . \quad (19.94)
$$

For the QRPA calculation of the $2_1^+$ state in $^{24}$Mg we fit the $2_1^+$ and $2_2^+$ energies, which gives the SDI parameters $A_0 = 0.30$ MeV and $A_1 = 1.71$ MeV. With these parameters the QRPA wave function becomes

$$|^{24}\text{Mg}\,;\,2_1^+\rangle = 0.476|\pi_1\rangle + 0.178|\pi_1^{-1}\rangle + 0.620|\pi_2\rangle + 0.236|\pi_2^{-1}\rangle$$
$$+ 0.457|\nu_1\rangle + 0.179|\nu_1^{-1}\rangle + 0.596|\nu_2\rangle + 0.236|\nu_2^{-1}\rangle \,. \quad (19.95)$$

The basis states were given in (16.71) as

$$\{|\pi_1\rangle,\,|\pi_2\rangle,\,|\nu_1\rangle,\,|\nu_2\rangle\} = \{|(\pi 0\text{d}_{5/2})^2\,;\,2^+\rangle,\,|\pi 0\text{d}_{5/2}\,\pi 1\text{s}_{1/2}\,;\,2^+\rangle,$$
$$|(\nu 0\text{d}_{5/2})^2\,;\,2^+\rangle,\,|\nu 0\text{d}_{5/2}\,\nu 1\text{s}_{1/2}\,;\,2^+\rangle\} \,. \quad (19.96)$$

The only non-zero contributions to the two-quasiparticle matrix elements (15.111) and (15.113) come from the configurations where all orbitals are $0\text{d}_{5/2}$. Evaluated, the two-quasiparticle matrix elements are

$$\mathcal{M}_{\text{GT}}^{(\mp)}\big(\pi 0\text{d}_{5/2}\,\nu 0\text{d}_{5/2}\,;\,1^+ \to (\pi 0\text{d}_{5/2})^2\,;\,2^+\big)$$
$$= -\mathcal{M}_{\text{GT}}^{(\pm)}\big(\pi 0\text{d}_{5/2}\,\nu 0\text{d}_{5/2}\,;\,1^+ \to (\nu 0\text{d}_{5/2})^2\,;\,2^+\big)$$
$$= -\frac{8\sqrt{2}}{5}\mathcal{B}_1^{(\mp)}(\pi 0\text{d}_{5/2}\,\nu 0\text{d}_{5/2}) \quad (19.97)$$

with $\mathcal{B}_1^{(\mp)}$ given by (15.106) and (15.107).

Substituting the $X$ and $Y$ amplitudes from (19.93) and (19.95) and the matrix elements (19.97) with BCS amplitudes from Table 16.9 into the formula (19.92), we obtain

$$(2_1^+\|\beta_{\text{GT}}^+\|1_1^+) = 0.476 \times 0.857\Big(-\frac{8\sqrt{2}}{5} \times 0.7312 \times 0.7308\Big)$$
$$- 0.178(-0.171)\Big(-\frac{8\sqrt{2}}{5} \times 0.6821 \times 0.6826\Big)$$
$$+ 0.457 \times 0.857\Big(\frac{8\sqrt{2}}{5} \times 0.6821 \times 0.6826\Big)$$
$$- 0.179(-0.171)\Big(\frac{8\sqrt{2}}{5} \times 0.7312 \times 0.7308\Big)$$
$$= -0.076\,, \quad (19.98)$$

whence

$$\log ft_+ = 6.3 \,. \quad (19.99)$$

This $\log ft$ value is in perfect agreement with the experimental value of 6.2. No large modifications are expected when the $0\text{d}_{3/2}$ orbital is added to the valence space since the contribution of this orbital to the $1_1^+$ and $2_1^+$ wave functions is rather small.

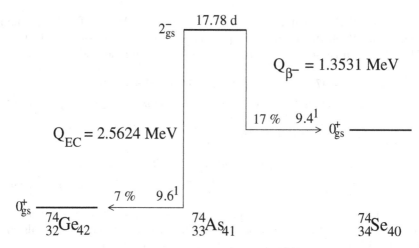

**Fig. 19.7.** First-forbidden unique $\beta^-$ and $\beta^+$ decay of $^{74}$As to the ground states of $^{74}$Se and $^{74}$Ge. The experimental $Q$ values, decay branchings and log $ft$ values are given. Use of the phase-space factor $f_{1u}$ in the log $ft$ value, as given in (7.165), is indicated by the superscript '1'

### 19.7.3 First-Forbidden Unique Beta Decay in the 0f-1p-0g$_{9/2}$ Shell

In this subsection we consider first-forbidden unique beta decay in the 0f-1p-0g$_{9/2}$ valence space. The transitions are of the type $2^- \rightarrow 0^+_{gs}$, and we use the pnQRPA to calculate the $2^-$ wave functions. A representative case is depicted in Fig. 19.7.

The decay amplitudes for first-forbidden unique transitions are obtained from (19.62) and (19.65) with the appropriate replacements (7.210) and (7.213) for the single-particle matrix elements. Using (7.167) and (7.168) with CS phasing we obtain

$$M^{(-)}_{1u} = \sqrt{5} \times 2.990 \times 10^{-3} \times b\,[\mathrm{fm}] \sum_{pn} m^{(1u)}(pn)(v_p u_n X^{2^-_1}_{pn} + u_p v_n Y^{2^-_1}_{pn})\,,$$

$$(19.100)$$

$$M^{(+)}_{1u} = \sqrt{5} \times 2.990 \times 10^{-3} \times b\,[\mathrm{fm}] \sum_{pn} m^{(1u)}(pn)(u_p v_n X^{2^-_1}_{pn} + v_p u_n Y^{2^-_1}_{pn})\,,$$

$$(19.101)$$

where the basic single-particle matrix elements $m^{(1u)}(pn)$ are given in (7.169) and Table 7.6.

For the $\beta^+$ decay in Fig. 19.7 the even–even reference nucleus is $^{74}$Ge and for the $\beta^-$ decay it is $^{74}$Se. We construct the $2^-_{gs}$ wave function of $^{74}$As by application of the pnQRPA to each reference nucleus. Thus the wave function is generated by two pnQRPA calculations, one starting from $^{74}$Ge and the other from $^{74}$Se. The computed $2^-_{gs}$ wave functions are not exactly the same

**Table 19.3.** Reference nuclei for pnQRPA calculations in the 0f-1p-0g$_{9/2}$ major shell

| Nucleus | $Z_{\text{act}}$ | $N_{\text{act}}$ | $\Delta_{\text{p}}$ (MeV) | $\Delta_{\text{n}}$ (MeV) | $A_{\text{pair}}^{(\text{p})}$ (MeV) | $A_{\text{pair}}^{(\text{n})}$ (MeV) | $b$ (fm) |
|---|---|---|---|---|---|---|---|
| $^{72}$Ge | 12 | 20 | 1.483 | 1.825 | 0.39 | 0.35 | 2.099 |
| $^{74}$Ge | 12 | 22 | 1.563 | 1.778 | 0.40 | 0.35 | 2.107 |
| $^{74}$Se | 14 | 20 | 1.800 | 1.931 | 0.41 | 0.37 | 2.107 |
| $^{76}$Ge | 12 | 24 | 1.509 | 1.570 | 0.39 | 0.33 | 2.115 |
| $^{76}$Se | 14 | 22 | 1.711 | 1.716 | 0.40 | 0.34 | 2.115 |
| $^{78}$Se | 14 | 24 | 1.618 | 1.654 | 0.39 | 0.34 | 2.123 |
| $^{82}$Kr | 16 | 26 | 0.977 | 1.648 | 0.23 | 0.34 | 2.138 |
| $^{84}$Kr | 16 | 28 | 1.424 | 1.615 | 0.30 | 0.35 | 2.145 |
| $^{84}$Sr | 18 | 26 | 1.886 | 1.614 | 0.36 | 0.34 | 2.145 |
| $^{86}$Sr | 18 | 28 | 1.619 | 1.505 | 0.32 | 0.33 | 2.152 |

Columns two and three give the numbers of active nucleons. The pairing gaps for protons and neutrons are given in columns four and five, followed by the SDI pairing parameters. The oscillator length $b$ is given in the last column.

but close to each other. In this respect the pnQRPA is different in philosophy from the nuclear shell model, where only one unique $2_{\text{gs}}^{-}$ wave function should be used.

Table 19.3 lists the reference nuclei for our pnQRPA calculations, where the single-particle energies were taken from (14.13). The proton and neutron pairing gaps in the table were extracted from the separation energies according to (16.90). The SDI pairing parameters $A_{\text{pair}}^{(\text{p})}$ and $A_{\text{pair}}^{(\text{n})}$ were fitted to the pairing gaps. The oscillator length $b$ is the Blomqvist–Molinari value defined by (3.43) and (3.45).

In Tables 19.4 and 19.5 we tabulate the occupation amplitudes $u$ and $v$ that are obtained from BCS calculations for the reference nuclei. The wave functions of the $2_1^{-}$ states in the odd–odd nuclei are found by diagonalizing the pnQRPA matrix. Table 19.6 lists the $X$ and $Y$ amplitudes of these wave functions. We have used the same SDI parameter values $A_0 = A_1 = 0.4\,\text{MeV}$ for all reference nuclei to better follow general trends in the calculated results.

We summarize the final computed results in Table 19.7. There the first two columns indicate the decay modes with the mother and daughter nuclei, and the third column gives the reference nuclei used in the pnQRPA calculations. The fourth column gives computed values of the transition matrix elements, i.e. the decay amplitudes for beta decays as given by (19.100) for the $\beta^{-}$ decays and by (19.101) for the $\beta^{+}$ decays. The resulting $\log ft$ values, as extracted from (7.165), have been given in the next column. We quote the experimental $\log ft$ values in the last column.

Table 19.7 shows that the computed $\beta^{+}/\text{EC}$ decays are always much too fast. The experimental $\log ft$ value exceeds the theoretical one by at least one unit. This behaviour is due to one strongly dominating two-quasiparticle

**Table 19.4.** BCS occupation amplitudes for proton single-particle orbitals contributing to $2^-$ states

| Orbital | $\pi 0f_{7/2}$ | | $\pi 0f_{5/2}$ | | $\pi 0g_{9/2}$ | |
|---|---|---|---|---|---|---|
| Nucleus | $u_p$ | $v_p$ | $u_p$ | $v_p$ | $u_p$ | $v_p$ |
| $^{72}$Ge | −0.1048 | 0.9945 | −0.9782 | 0.2078 | 0.9829 | 0.1840 |
| $^{74}$Ge | −0.1130 | 0.9936 | −0.9755 | 0.2199 | 0.9808 | 0.1952 |
| $^{74}$Se | −0.1311 | 0.9914 | −0.9288 | 0.3707 | 0.9455 | 0.3255 |
| $^{76}$Ge | −0.1048 | 0.9945 | −0.9782 | 0.2078 | 0.9829 | 0.1840 |
| $^{76}$Se | −0.1248 | 0.9922 | −0.9312 | 0.3645 | 0.9479 | 0.3186 |
| $^{78}$Se | −0.1186 | 0.9929 | −0.9337 | 0.3580 | 0.9503 | 0.3114 |
| $^{82}$Kr | −0.0527 | 0.9986 | −0.8934 | 0.4492 | 0.9422 | 0.3350 |
| $^{84}$Kr | −0.0821 | 0.9966 | −0.8856 | 0.4644 | 0.9249 | 0.3801 |
| $^{84}$Sr | −0.1135 | 0.9935 | −0.8000 | 0.6000 | 0.8530 | 0.5219 |
| $^{86}$Sr | −0.0965 | 0.9953 | −0.8024 | 0.5967 | 0.8623 | 0.5063 |

The calculations were done with the SDI pairing parameters in Table 19.3. CS phases are assumed.

**Table 19.5.** BCS occupation amplitudes for neutron single-particle orbitals contributing to $2^-$ states

| Orbital | $\nu 0f_{7/2}$ | | $\nu 0f_{5/2}$ | | $\nu 0g_{9/2}$ | |
|---|---|---|---|---|---|---|
| Nucleus | $u_n$ | $v_n$ | $u_n$ | $v_n$ | $u_n$ | $v_n$ |
| $^{72}$Ge | −0.1070 | 0.9943 | −0.7188 | 0.6952 | 0.7893 | 0.6140 |
| $^{74}$Ge | −0.1006 | 0.9949 | −0.6316 | 0.7753 | 0.7131 | 0.7011 |
| $^{74}$Se | −0.1145 | 0.9934 | −0.7188 | 0.6952 | 0.7849 | 0.6196 |
| $^{76}$Ge | −0.0849 | 0.9964 | −0.5350 | 0.8448 | 0.6272 | 0.7788 |
| $^{76}$Se | −0.0973 | 0.9952 | −0.6309 | 0.7758 | 0.7152 | 0.6989 |
| $^{78}$Se | −0.0877 | 0.9961 | −0.5362 | 0.8441 | 0.6253 | 0.7804 |
| $^{82}$Kr | −0.0738 | 0.9973 | −0.4296 | 0.9030 | 0.5150 | 0.8572 |
| $^{84}$Kr | −0.0550 | 0.9985 | −0.2992 | 0.9542 | 0.3659 | 0.9306 |
| $^{84}$Sr | −0.0738 | 0.9973 | −0.4296 | 0.9030 | 0.5150 | 0.8572 |
| $^{86}$Sr | −0.0519 | 0.9986 | −0.2971 | 0.9548 | 0.3684 | 0.9297 |

The calculations were done with the SDI pairing parameters in Table 19.3. CS phases are assumed.

contribution, namely $\pi 0f_{5/2}\nu 0g_{9/2}$ that produces a large $X$ amplitude; see the fourth column of Table 19.6. The rest of the contributions, be they $X$ or $Y$ terms, are not able to cancel the dominating term.

The prediction of too fast $\beta^+$ decay persists into larger valence spaces and more realistic interactions. The general feature is that the products $v_p u_n$ in front of the $Y$ amplitudes in (19.101) are not large enough to compensate for the $u_p v_n$ factors in front of the $X$ amplitudes. It seems that multi-quasiparticle

**Table 19.6.** pnQRPA amplitudes $X$ and $Y$ for $2_1^-$ states

| Config. | $\pi 0f_{7/2}\, \nu 0g_{9/2}$ | | $\pi 0f_{5/2}\, \nu 0g_{9/2}$ | | $\pi 0g_{9/2}\, \nu 0f_{7/2}$ | | $\pi 0g_{9/2}\, \nu 0f_{5/2}$ | |
|---|---|---|---|---|---|---|---|---|
| Nucleus | $X$ | $Y$ | $X$ | $Y$ | $X$ | $Y$ | $X$ | $Y$ |
| $^{72}$Ge | 0.015 | 0.055 | 0.951 | 0.052 | −0.041 | 0.012 | −0.318 | −0.047 |
| $^{74}$Ge | 0.018 | 0.044 | 1.003 | 0.042 | −0.035 | 0.009 | 0.020 | −0.037 |
| $^{74}$Se | 0.017 | 0.053 | 0.923 | 0.098 | −0.031 | 0.023 | −0.409 | −0.091 |
| $^{76}$Ge | 0.015 | 0.031 | 0.962 | 0.028 | −0.027 | 0.005 | 0.275 | −0.024 |
| $^{76}$Se | 0.022 | 0.047 | 1.003 | 0.085 | −0.030 | 0.017 | −0.105 | −0.076 |
| $^{78}$Se | 0.020 | 0.034 | 0.984 | 0.059 | −0.024 | 0.011 | 0.194 | −0.050 |
| $^{82}$Kr | 0.023 | 0.028 | 0.969 | 0.086 | −0.019 | 0.010 | 0.270 | −0.064 |
| $^{84}$Kr | 0.022 | 0.017 | 0.921 | 0.044 | −0.015 | 0.006 | 0.393 | −0.034 |
| $^{84}$Sr | 0.037 | 0.035 | 1.015 | 0.111 | −0.011 | 0.018 | 0.037 | −0.094 |
| $^{86}$Sr | 0.034 | 0.022 | 0.975 | 0.073 | −0.011 | 0.010 | 0.239 | −0.058 |

The proton–neutron two-quasiparticle configurations are given in the first row.

**Table 19.7.** First-forbidden unique beta-decay transitions in the 0f-1p-0g$_{9/2}$ valence space

| Transition | Mode | Reference | $M_{1u}^{(\pm)}$ | $\log f_{1u}t$ | $\log f_{1u}t(\mathrm{exp})$ |
|---|---|---|---|---|---|
| $^{72}$As$(2_1^-) \rightarrow\, ^{72}$Ge$(0_{gs}^+)$ | $\beta^+/$EC | $^{72}$Ge | −0.0457 | 8.05 | 9.8 |
| $^{74}$As$(2_1^-) \rightarrow\, ^{74}$Ge$(0_{gs}^+)$ | $\beta^+/$EC | $^{74}$Ge | −0.0394 | 8.18 | 9.6 |
| $^{74}$As$(2_1^-) \rightarrow\, ^{74}$Se$(0_{gs}^+)$ | $\beta^-$ | $^{74}$Se | 0.0152 | 9.01 | 9.4 |
| $^{76}$As$(2_1^-) \rightarrow\, ^{76}$Ge$(0_{gs}^+)$ | $\beta^+/$EC | $^{76}$Ge | −0.0300 | 8.42 | |
| $^{76}$As$(2_1^-) \rightarrow\, ^{76}$Se$(0_{gs}^+)$ | $\beta^-$ | $^{76}$Se | 0.0107 | 9.31 | 9.7 |
| $^{78}$As$(2_1^-) \rightarrow\, ^{78}$Se$(0_{gs}^+)$ | $\beta^-$ | $^{78}$Se | 0.0067 | 9.73 | 9.7 |
| $^{82}$Br$(2_1^-) \rightarrow\, ^{82}$Kr$(0_{gs}^+)$ | $\beta^-$ | $^{82}$Kr | 0.0042 | 10.1 | 8.9 |
| $^{84}$Br$(2_1^-) \rightarrow\, ^{84}$Kr$(0_{gs}^+)$ | $\beta^-$ | $^{84}$Kr | 0.0030 | 10.42 | 9.5 |
| $^{84}$Rb$(2_1^-) \rightarrow\, ^{84}$Kr$(0_{gs}^+)$ | $\beta^+/$EC | $^{84}$Kr | −0.0253 | 8.57 | 9.6 |
| $^{84}$Rb$(2_1^-) \rightarrow\, ^{84}$Sr$(0_{gs}^+)$ | $\beta^-$ | $^{84}$Sr | 0.0102 | 9.36 | 9.5 |
| $^{86}$Rb$(2_1^-) \rightarrow\, ^{86}$Sr$(0_{gs}^+)$ | $\beta^-$ | $^{86}$Sr | 0.0050 | 9.97 | 9.4 |

The third column gives the reference nucleus of the pnQRPA. The computed transition matrix element and $\log ft$ value are given in columns four and five. The last column lists the experimental $\log ft$ values

states are needed to reproduce the data. Deformation effects offer another possibility to that end.

For $\beta^-$ decay the computed $\log ft$ values range within two units of experiment. Unlike the $\beta^+$ transitions they are sensitive to small changes in the $2_1^-$ wave function and hence to the interaction parameters. The $uv$ products in the $\beta^-$ decay amplitude (19.100) make it possible for the $Y$ terms to largely cancel the leading $X$ terms. In this case the ground-state correlations play an essential role for obtaining the correct transition matrix element.

In conclusion we make the general remark that strong cancellation effects can show up in pnQRPA calculations of Gamow–Teller $\beta^-$ decay. Such cancellations occur when the $1_1^+$ state of an odd–odd nucleus decays to the ground state of the even–even daughter nucleus. While the $\beta^-$ branch is sensitive to ground-state correlations, i.e. to the interaction strength, the $\beta^+$ branch is essentially independent of them. As discussed in the following subsection, the sensitivity of the $\beta^-$ branch can be exploited to constrain parameters of model Hamiltonians.

### 19.7.4 Empirical Particle–Hole and Particle–Particle Forces

As noted at the end of Subsect. 17.4.5, the particle–hole parts of the pnQTDA matrix A govern the energy of the $1^+$ GTGR. For the pnQRPA the same happens with the A and B matrices, i.e. their particle–hole parts determine the location of the GTGR. In fact, in a large number of calculations (see e.g. [51]) this part has been multiplied by an overall scaling factor $g_{\mathrm{ph}}$ whose value is fixed by fitting the experimental energy of the GTGR. This energy can be obtained from the measured energies of the GTGR states as described in Subsect. 17.4.5 and summarized in the empirical formula (17.90).

Also the particle–particle parts of the A and B matrices can be multiplied by a common scaling factor $g_{\mathrm{pp}}$. Increasing this parameter increases the ground-state correlations and thus the magnitudes of the $Y$ amplitudes. This leads to cancellations in the Gamow–Teller $\beta^-$ decay amplitude, as discussed in the previous subsection. The calculated $\log ft$ value is sensitive to $g_{\mathrm{pp}}$, which allows fitting the parameter to experimental data [51].

Above we have described how to create separate phenomenological particle–hole and particle–particle forces. In doing this we violate the Pandya relation (17.4) whenever we choose $g_{\mathrm{pp}} \neq g_{\mathrm{ph}}$.

## Epilogue

This chapter has dealt with the pnQRPA and its applications. We found that ground-state correlations play an important role in beta-decay transitions.

This chapter ends the book. During our long journey through various formalisms and derivations we have learned the basic properties of nuclear structure and nuclear transitions. Although not necessary for applications, the derivations help grasp the essence and limitations of nuclear structure models. They also serve as training for theoretical work.

Nuclear states owe their structure to a mean field and configuration mixing. We have treated this mixing by all basic microscopic approaches, within the realms of particles, holes and quasiparticles. Applications have been worked through and discussed in detail. To be tractable, the applications were limited to light and medium-heavy nuclei. Extension of these applications to heavy spherical nuclei is straightforward but tedious.

# Exercises

**19.1.** Derive (19.3) from the general expression (18.28).

**19.2.** Produce the eigenenergies (19.28) by diagonalizing the supermatrix consisting of the pnQTDA matrix (17.14) and the correlation matrix (19.27).

**19.3.** Produce the eigenvectors (19.29) and (19.30) by using the matrices of Exercise 19.2.

**19.4.** Reproduce the pnQRPA spectrum of $^{24}$Na on the left in Fig. 19.1 by using the $0d_{5/2}$-$1s_{1/2}$ valence space and the SDI parameters of Subsect. 19.2.1. Take the single-particle energies from (14.12).

**19.5.** Produce the pnQRPA spectrum of $^{30}$P by using the $0d_{5/2}$-$1s_{1/2}$ valence space and the SDI parameters of Table 16.6. Take the single-particle energies from (14.12). Compare with the calculated and experimental spectra of Fig. 19.2 and comment.

**19.6.** Derive in detail the commutator relations (19.36) and (19.37).

**19.7.** Derive the relation (19.57).

**19.8.** Carry out the derivation of (19.88) and (19.89).

**19.9.** Verify the result (19.91).

**19.10.** Continuation of Exercises 16.16 and 17.13.
Apply the pnQRPA to $^{20}$Na with $^{20}$Ne as the reference nucleus. Use the SDI with parameters $A_0 = A_1 = 1.0$ MeV.

(a) Form the correlation matrix for the $2^+$ states.
(b) Diagonalize the pnQRPA matrix by taking the pnQTDA matrix from Exercise 17.13. Find the wave function of the lowest $2^+$ state of $^{20}$Na.

**19.11.** Continuation of Exercise 19.10.
Calculate the $\log ft$ value for the $\beta^+$/EC decay of the $2^+$ ground state of $^{20}$Na to the first $2^+$ state in $^{20}$Ne. The QRPA wave function of the latter state was calculated in Exercise 18.12. Compare your result with the result of Exercise 17.14 and with experimental data, and comment.

**19.12.** Continuation of Exercises 16.18 and 17.15.
Calculate the $\log ft$ values for the second-forbidden unique $\beta^+$/EC decay of the $5^+$ ground state of $^{26}$Al to the two lowest $2^+$ states in $^{26}$Mg. The QRPA wave functions of these states were computed in Exercise 18.14. Compute a pnQRPA wave function for the $5^+$ state. Use the SDI with parameters $A_0 = A_1 = 1.0$ MeV. Compute also the decay half-life of the $5^+$ state. Compare with the results of Exercise 17.15 and with experimental data, and comment.

**19.13.** Continuation of Exercise 19.12.
Calculate the $\log ft$ value for the $\beta^-$ decay of the $3^+$ ground state of $^{26}$Na to the first $2^+$ state in $^{26}$Mg. Compare with the result of Exercise 17.16 and with experimental data, and comment.

**19.14.** Continuation of Exercise 17.17.
Calculate the $\log ft$ values for the $\beta^-$ and $\beta^+$/EC decays of the $3^+$ ground states of $^{28}$Al and $^{28}$P to the first $2^+$ state in $^{28}$Si. For the reference nucleus $^{28}$Si use the BCS results of Table 14.1. Diagonalize the QRPA and pnQRPA matrices in the $0d_{5/2}$-$1s_{1/2}$ valence space to find the wave functions. Use the SDI with parameters $A_0 = A_1 = 1.0$ MeV. Compare with the results of Exercise 17.17 and with experimental data, and comment.

**19.15.** Continuation of Exercise 17.18.
Consider $1^+$ states in $^{30}$P. Start from the reference nucleus $^{30}$Si and use the BCS results of Table 14.1, computed in the full d-s valence space. Form the pnQRPA matrix in the sub-basis of Exercise 17.18. Use the SDI with parameters $A_0 = A_1 = 1.0$ MeV.

**19.16.** Continuation of Exercise 19.15.
Diagonalize the pnQRPA matrix to find the eigenenergies and eigenstates. Compare with experimental data, the results of Exercise 17.19 and the complete d-s shell calculation of Fig. 19.2. Comment on the results.

**19.17.** Continuation of Exercise 19.16.
Calculate the $\log ft$ values for the $\beta^+$/EC decay of $^{30}$S to the $1^+$ states in $^{30}$P. Compare with the results of Exercise 17.20 and with experimental data, and comment.

**19.18.** Continuation of Exercise 19.17.
Evaluate the difference $S_- - S_+$ of the total strengths for Gamow–Teller transitions from the ground state of $^{30}$Si to the $1^+$ states of $^{30}$P and $^{30}$Al. Compare with the Ikeda sum rule. Explain your observation.

**19.19.** Continuation of Exercise 17.22.
Consider the $1^+$ states in $^{34}$P and $^{34}$Cl. Start from the reference nucleus $^{34}$S and use the SDI parameters and pnQTDA matrices of Exercise 17.22. Form the pnQRPA matrix for the $1^+$ states of $^{34}$P and $^{34}$Cl in the sub-basis of Exercise 17.22.

**19.20.** Continuation of Exercise 19.19.
Diagonalize the pnQRPA matrix to find the eigenenergies and eigenstates. Use the SDI with parameters $A_0 = A_1 = 1.0$ MeV. Compare with the results of Exercise 17.22 and with experimental data, and comment.

**19.21.** Continuation of Exercise 19.19.
Compute the pnQRPA eigenenergies and eigenstates of the $0^+$ and $3^+$ states in $^{34}$P and $^{34}$Cl by using the sub-basis of Exercise 17.22 and the SDI with parameters $A_0 = A_1 = 1.0$ MeV.

**19.22.** Continuation of Exercises 19.20 and 19.21.

Calculate the $\log ft$ values for the beta decays

(a) $^{34}\mathrm{P}(1^+_{\mathrm{gs}}) \to {}^{34}\mathrm{S}(0^+_{\mathrm{gs}})$,

(b) $^{34}\mathrm{Cl}(0^+_{\mathrm{gs}}) \to {}^{34}\mathrm{S}(0^+_{\mathrm{gs}})$.

Compare with the results of Exercise 17.23 and with experimental data.

**19.23.** Continuation of Exercise 19.20.

Study the validity of the Ikeda sum rule for the Gamow–Teller transitions from the ground state of $^{34}\mathrm{S}$ to the $1^+$ states in $^{34}\mathrm{P}$ and $^{34}\mathrm{Cl}$.

**19.24.** Continuation of Exercises 19.20 and 19.21.

Calculate the $\log ft$ values for the beta decays

(a) $^{34}\mathrm{P}(1^+_{\mathrm{gs}}) \to {}^{34}\mathrm{S}(2^+_1)$,

(b) $^{34}\mathrm{Cl}(3^+_1) \to {}^{34}\mathrm{S}(2^+_1)$.

The wave function of the $2^+_1$ state was calculated in Exercise 18.19. Compare with the results of Exercise 17.25 and with experimental data.

**19.25.** Continuation of Exercises 19.20 and 19.21.

Calculate the partial half-lives of the following electromagnetic transitions in $^{34}\mathrm{Cl}$:

(a) $1^+_1 \to 3^+_1$,

(b) $1^+_1 \to 0^+_{\mathrm{gs}}$.

(c) Calculate also the total half-life of the $1^+_1$ state. Use experimental gamma energies. Compare with the results of Exercise 17.26 and with experimental data, and comment.

**19.26.** Continuation of Exercises 16.27 and 17.27.

Consider the $1^+$ states of $^{38}\mathrm{K}$ by using $^{38}\mathrm{Ar}$ as the reference nucleus.

(a) Form the pnQRPA matrix by using the sub-basis of Exercise 17.27.

(b) Diagonalize the pnQRPA matrix to find the eigenenergies and eigenstates. Use the SDI with parameters $A_0 = A_1 = 1.0\,\mathrm{MeV}$. Compare with the theoretical and experimental spectra in Fig. 8.8 and with the results of Exercise 17.27. Comment on the similarities and differences.

**19.27.** Continuation of Exercise 19.26.

Calculate the $\log ft$ values for the $\beta^+/\mathrm{EC}$ decays $^{38}\mathrm{Ca}(0^+_{\mathrm{gs}}) \to {}^{38}\mathrm{K}(1^+_m)$, $m = 1, 2, 3, 4$. Note the isospin symmetry (17.99). Compare with the results of Exercise 17.28 and with experimental data by plotting the theoretical and experimental $1^+$ spectra and indicating the $\beta^+/\mathrm{EC}$ feeding. Comment on the results.

**19.28.** Continuation of Exercise 19.27.

Evaluate the difference $S_- - S_+$ of the total strengths for transitions from the ground state of $^{38}\mathrm{Ar}$ to the $1^+$ states of $^{38}\mathrm{K}$ and $^{38}\mathrm{Cl}$. Apply the Ikeda sum rule and comment.

**19.29.** Verify the relations (19.100) and (19.101).

# References

1. M.E. Rose: *Elementary Theory of Angular Momentum* (Wiley New York 1957)
2. A.R. Edmonds: *Angular Momentum in Quantum Mechanics* (Princeton Universtiy Press, Princeton 1960)
3. A. Messiah: *Quantum Mechanics*, vol 1 (North-Holland Amsterdam 1961)
4. A. Messiah: *Quantum Mechanics*, vol 2 (North-Holland Amsterdam 1964)
5. D.M. Brink, G.R. Satchler: *Angular Momentum* (Clarendon Oxford 1968)
6. E. Merzbacher: *Quantum Mechanics* (Wiley New York 1970)
7. D.A. Varshalovich, A.N. Moskalev, V.K. Khersonskii: *Quantum Theory of Angular Momentum* (World Scientific Singapore 1988)
8. A.M. Lane: *Nuclear Theory* (Benjamin New York 1964)
9. A. deShalit, H. Feshbach: *Theoretical Nuclear Physics* (Wiley New York 1974)
10. A. de-Shalit, I. Talmi: *Nuclear Shell Theory* (Academic New York 1963)
11. G.E. Brown: *Unified Theory of Nuclear Models* (North-Holland Amsterdam 1964)
12. A. Bohr, B.R. Mottelson: *Nuclear Structure*, vol 1 (Benjamin New York 1969)
13. J.M. Eisenberg, W. Greiner: *Nuclear Models* (North-Holland Amsterdam 1970)
14. D.J. Rowe: *Nuclear Collective Motion* (Methuen London 1970)
15. J.M. Blatt, V.F. Weisskopf: *Theoretical Nuclear Physics* (Springer New York 1979)
16. P. Ring, P. Schuck: *The Nuclear Many-Body Problem* (Springer New York 1980)
17. K.L.G. Heyde: *The Nuclear Shell Model* (Springer Berlin 1990)
18. I. Talmi: *Simple Models of Complex Nuclei* (Harwood Switzerland 1993)
19. R.D. Lawson: *Theory of the Nuclear Shell Model* (Clarendon Oxford 1980)
20. M. Rotenberg, R. Bivins, N. Metropolis, J.K. Wooten Jr: *The 3j and 6j Symbols* (M.I.T. Technology Press Cambridge MA 1959)
21. B.L. Cohen: *Concepts of Nuclear Physics* (TATA McGraw-Hill New Delhi 1971)
22. A. Lindner: *Drehimpulse in der Quantenmechanik* (Taubner Stuttgart 1984)
23. W.T. Sharp et al.: *Tables of Coefficients for Angular Distribution Analysis*, AECL No 97 (Atomic Energy of Canada Ltd 1954)
24. E.P. Wigner: *Gruppentheorie* (Vieweg Braunschweig 1931)
25. C. Eckart: Rev. Mod. Phys. **2**, 305 (1930)
26. G. Racah: Phys. Rev. **62**, 438 (1942)
27. W. Pauli: Z. Phys. **43**, 601 (1927)

630    References

28. P.J. Brussaard, P.W.M Glaudemans: *Shell-Model Applications in Nuclear Spectroscopy* (North-Holland Amsterdam 1977)
29. W. Magnus, F. Oberhettinger, R.P. Soni: *Formulas and Theorems for the Special Functions of Mathematical Physics* (Springer Berlin 1966)
30. J. Blomqvist, A. Molinari: Nucl. Phys. A **106**, 545 (1968)
31. M. Baranger: Phys. Rev. **120**, 957 (1960)
32. E.U. Condon, G.H. Shortley: *The Theory of Atomic Spectra* (Cambridge University Press, Cambridge, 1935)
33. L.C. Biedenharn, M.E. Rose: Rev. Mod. Phys. **25**, 729 (1953)
34. G.C. Wick: Phys. Rev. **80**, 268 (1950)
35. D.R. Hartree: Proc. Camb. Phil. Soc. **24**, 89 (1928)
36. V.A. Fock: Z. Phys. **61**, 126 (1930)
37. R.B. Firestone, V.S. Shirley, S.Y.F. Chu, C.M. Baglin, J. Zipkin: *Table of Isotopes* (Wiley-Interscience New York 1996)
38. See website http://ie.lbl.gov/ensdf/
39. S.A. Moszkowski: Theory of Multipole Radiation. In: *Alpha-, Beta- and Gamma-Ray Spectroscopy*, ed by K. Siegbahn (North-Holland Amsterdam 1965) pp 863–886
40. S.G. Nilsson: Kgl. Danske Videnskab. Selskab, Mat. Fys. Medd. **29**, 16 (1955)
41. V.F. Weisskopf: Phys. Rev. **83**, 1073 (1951)
42. See website http://www.nndc.bnl.gov/nndc/stone_moments/moments.html
43. S. Weinberg: Phys. Rev. Lett. **32**, 438 (1974)
44. A. Salam: *Proc. 8th Nobel Symposium*, ed by N. Svartholm (Stockholm 1968)
45. S.L. Glashow, J. Iliopoulos, L. Maiani: Phys. Rev. D **2**, 1285 (1970)
46. Super-Kamiokande Collaboration, S. Fukuda et al.: Phys. Rev. Lett. **86**, 5651 (2001)
47. SNO Collaboration, Q.R. Ahmad et al.: Phys. Rev. Lett. **89**, 011302 (2002)
48. KamLAND Collaboration, K. Eguchi et al.: Phys. Rev. Lett. **90**, 021802 (2003)
49. M. Apollonio et al.: Phys. Lett. B **466**, 415 (1999)
50. B. Pontecorvo: Zh. Eksp. Teor. Fiz. **33**, 549 (1957)
51. J. Suhonen, O. Civitarese: Phys. Rep. **300**, 123 (1998)
52. J.D. Vergados: Phys. Rep. **361**, 1 (2002)
53. F. Halzen, A.D. Martin: *Quarks and Leptons: An Introductory Course in Modern Particle Physics* (Wiley New York 1984)
54. H.F. Schopper: *Weak Interactions and Nuclear Beta Decay* (North-Holland Amsterdam 1966)
55. H. Behrens, W. Bühring: *Electron Radial Wave Functions and Nuclear Beta Decay* (Clarendon Oxford 1982)
56. J.C. Hardy, I.S. Towner, V.T. Koslowsky, E. Hagberg, H. Schmeing: Nucl. Phys. A **509**, 429 (1990)
57. E. Fermi: Z. Phys. **88**, 161 (1934)
58. G. Gamow, E. Teller: Phys. Rev. **49**, 895 (1936)
59. N.B. Gove, M.J. Martin: Nucl. Data Tables **10**, 205 (1971)
60. H. Primakoff, S.P. Rosen: Rep. Prog. Phys. **22**, 121 (1959)
61. C.S. Wu, S.A. Moszkowski: *Beta Decay* (Interscience New York 1966)
62. M.T. Mustonen, M. Aunola, J. Suhonen: Phys. Rev. C **73**, 054301 (2006)
63. A. Balysh et al.: Phys. Rev. Lett. **77**, 5186 (1996)
64. V. Brudanin et al.: Phys. Lett. B **495**, 63 (2000)
65. J. Toivanen, J. Suhonen: Phys. Rev. C **57**, 1237 (1998)

66. D.S. Delion, J. Suhonen: Phys. Rev. C **67**, 034301 (2003)
67. N. Ullah, D.J. Rowe: Nucl. Phys. A **163**, 257 (1971)
68. H.J. Lipkin, N. Meshkov, A.J. Glick: Nucl. Phys. **62**, 188, 199, 211 (1965)
69. J. Bardeen, L.N. Cooper, J.R. Schrieffer: Phys. Rev. **108**, 1175 (1957)
70. A. Bohr, B.R. Mottelson, D. Pines: Phys. Rev. **110**, 936 (1958)
71. S.T. Belyaev: Mat. Fys. Medd. Dan. Vid. Selsk. **31**, no 11, 1 (1959)
72. N.N. Bogoliubov: Sov. Phys. JETP **7**, 41 (1958)
73. J.G. Valatin: Phys. Rev. **122**, 1012 (1961)
74. H.J. Lipkin: Ann. Phys. (N.Y.) **9**, 272 (1960)
75. Y. Nogami: Phys. Rev. **134**, B313 (1964)
76. Y. Nogami, I.J. Zucker: Nucl. Phys. **60**, 203 (1964)
77. Y. Nogami: Phys. Lett. **15**, 335 (1965)
78. G. Audi, A.H. Wapstra: Nucl. Phys. A **565**, 1 (1993)
79. G. Audi, A.H. Wapstra: Nucl. Phys. A **595**, 409 (1995)
80. K. Ikeda: Prog. Theor. Phys. **31**, 434 (1964)
81. J.D. Anderson, C. Wong, J.W. McLure: Phys. Rev. B **138**, 615 (1965)
82. D.J. Horen et al.: Phys. Lett. B **99**, 383 (1981)
83. A. Bohr, B.R. Mottelson: *Nuclear Structure*, vol 2 (Benjamin Reading 1975)
84. R.F. Casten: *Nuclear Structure from a Simple Perspective* (Oxford University Press New York 1990)
85. R.F. Casten, N.V. Zamfir: Phys. Rep. **264**, 81 (1996)
86. J. Toivanen, J. Suhonen: Phys. Rev. Lett. **75**, 410 (1995)

# Index

$D$ function   25, 26
$LS$ coupling scheme   21
$Q$ value   161, 171
$\log ft$ value (*see* $ft$ value in beta decay, electron capture)   166
$ft$ value (*see* beta decay, electron capture)   166
$g$ factor   119, 123
$jj$ coupling scheme   21
$l$-forbidden allowed beta decay (*see also* beta decay)   172
$m$ table   11, 382

Allowed beta transitions (*see also* beta decay)
  and pnQRPA   610–621
  and pnQTDA   539–554
  and pnTDA   298–303
  Fermi and Gamow–Teller matrix elements   167
  Ikeda sum rule   543–549, 614–618
  in odd-$A$ nuclei   454, 456, 458–461
  in one-particle and one-hole nuclei   173–175
  in particle–hole nuclei   176–180, 182
  in two-particle and two-hole nuclei   182–189, 238–240
  involving two-quasiparticle states   462–467, 471–475
  phase-space factors   168–170
  selection rules   165
Angular momentum
  coupling   6, 7, 11, 15
  definition   4

orbital   55
  raising and lowering operators   4, 5
  total   55
Annihilation operator for
  BCS quasiparticles   395
  particles   64
Anticommutation of
  particle operators   65
  Pauli spin matrices   31, 32
  quasiparticle operators   396
Anticommutator relations   311, 312
Antiparticle   160
Antisymmetry of
  many-nucleon states   57
  two-nucleon interaction matrix elements   69
Associated Laguerre polynomial   49, 50
Atomic mass   373, 500, 501
Average particle number   400, 401, 406, 419, 420, 433, 438, 440, 532–535, 606, 607
Axial-vector coupling constant   166

Baranger's notation for
  isospin representation   107
  QTDA matrix   491
  single-particle orbitals   55, 394
  two-body matrix elements   485, 486
Bardeen–Cooper–Schrieffer theory (*see* BCS)   393
Barn   120
Barrier
  centrifugal   51

# Theoretical and Mathematical Physics

**From Nucleons to Nucleus**
Concepts of Microscopic Nuclear Theory
By J. Suhonen

**Concepts and Results in Chaotic Dynamics:**
**A Short Course**
By P. Collet and J.-P. Eckmann

**The Theory of Quark and Gluon Interactions**
4th Edition
By F. J. Ynduráin

**From Microphysics to Macrophysics**
Methods and Applications of Statistical Physics
Volume I
By R. Balian

**From Microphysics to Macrophysics**
Methods and Applications of Statistical Physics
Volume II
By R. Balian

---

**Quantum Non-linear Sigma Models**
From Quantum Field Theory
to Supersymmetry, Conformal Field Theory,
Black Holes and Strings
By S. V. Ketov

**Perturbative Quantum Electrodynamics and
Axiomatic Field Theory**
By O. Steinmann

**The Nuclear Many-Body Problem**
By P. Ring and P. Schuck

**Magnetism and Superconductivity**
By L.-P. Lévy

**Information Theory and Quantum Physics**
Physical Foundations for Understanding the
Conscious Process
By H. S. Green

**Quantum Field Theory in Strongly
Correlated Electronic Systems**
By N. Nagaosa

**Quantum Field Theory in Condensed Matter
Physics**
By N. Nagaosa

**Conformal Invariance and Critical
Phenomena**
By M. Henkel

**Statistical Mechanics of Lattice Systems**
Volume 1: Closed-Form and Exact Solutions
2nd Edition
By D. A. Lavis and G. M. Bell

**Statistical Mechanics of Lattice Systems**
Volume 2: Exact, Series
and Renormalization Group Methods
By D. A. Lavis and G. M. Bell

**Fields, Symmetries, and Quarks**
2nd Edition
By U. Mosel

**Renormalization** An Introduction
By M. Salmhofer

**Multi-Hamiltonian Theory of Dynamical
Systems**
By M. Błaszak

**Quantum Groups and Their Representations**
By A. Klimyk and K. Schmüdgen

**Quantum** The Quantum Theory of Particles,
Fields, and Cosmology
By E. Elbaz

**Effective Lagrangians for the Standard Model**
By A. Dobado, A. Gómez-Nicola,
A. L. Maroto and J. R. Peláez

**Scattering Theory of Classical
and Quantum $N$-Particle Systems**
By. J. Derezinski and C. Gérard

**Quantum Relativity** A Synthesis
of the Ideas of Einstein and Heisenberg
By D. R. Finkelstein

**The Mechanics and Thermodynamics
of Continuous Media**
By M. Šilhavý

**Local Quantum Physics** Fields, Particles,
Algebras
2nd Edition
By R. Haag

**Relativistic Quantum Mechanics
and Introduction to Field Theory**
By F. J. Ynduráin

**Supersymmetric Methods in Quantum
and Statistical Physics**
By G. Junker

**Path Integral Approach
to Quantum Physics** An Introduction
2nd printing
By G. Roepstorff

**Finite Quantum Electrodynamics**
The Causal Approach
2nd edition
By G. Scharf

**From Electrostatics to Optics**
A Concise Electrodynamics Course
By G. Scharf

**Geometry of the Standard Model
of Elementary Particles**
By A. Derdzinski

**Quantum Mechanics II**
By A. Galindo and P. Pascual

**Generalized Coherent States**
and Their Applications
By A. Perelomov

**The Elements of Mechanics**
By G. Gallavotti

**Essential Relativity** Special, General,
and Cosmological Revised
2nd edition
By W. Rindler